THIRD EDITION

The Changing Earth

Exploring Geology and Evolution

CDROM

James S. Monroe
Reed Wicander
Central Michigan University

BROOKS/COLE

THOMSON LEARNING™

Australia • Canada • Mexico • Singapore • Spain • United Kingdom • United States

BROOKS/COLE

THOMSON LEARNING

Sponsoring Editors: *Keith Dodson, Nina Horne*
Project Development Editor: *Marie Carigma-Sambilay*
Marketing Team: *Tom Ziolkowski, Christine Davis*
Editorial Assistant: *Faith Riley*
Project Editor: *Keith Faivre*
Production Service: *Nancy Shammas, New Leaf Publishing Services*
Manuscript Editor: *Patterson Lamb*
Permissions Editor: *Lillian Campobasso*
Interior Design: *Carolyn Deacy Design*

Cover Design: *Denise Davidson*
Cover Photo: *James Kay Photography*
Illustrations: *Precision Graphics*
Photo Researcher: *Kathleen Olson, Kofoto*
Indexer: *Kay Banning*
Print Buyer: *Jessica Reed*
Typesetting: *Carlisle Communications, Inc.*
Cover Printing: *Phoenix Color Corporation*
Printing and Binding: *The Courier Company, Inc., Kendallville*

For more information about this or any other Brooks/Cole product, contact:
BROOKS/COLE
511 Forest Lodge Road
Pacific Grove, CA 93950 USA
www.brookscole.com
1-800-423-0563 (Thomson Learning Academic Resource Center)

For permission to use material from this text, contact us by
www.thomsonrights.com
fax: 1-800-730-2215
phone: 1-800-730-2214

Printed in the United States of America

10   9   8   7   6   5   4   3   2   1

Library of Congress Cataloging-in-Publication Data

Monroe, James S. (James Stewart), [date]
    The changing earth : exploring geology and evolution / James S. Monroe, Reed Wicander.—3rd ed.
        p. cm.
    Includes bibliographical references and index.
    ISBN 0-534-37550-2 (alk. paper)
    1. Geology. I. Wicander, Reed. II. Title.
QE28 .M686 2002
550—dc21   2001025372

About the Authors

JAMES S. MONROE

JAMES S. MONROE is professor emeritus of geology at Central Michigan University where he taught physical geology, historical geology, prehistoric life, and stratigraphy and sedimentology since 1975. He has co-authored several textbooks with Reed Wicander and has interests in Cenozoic geology and geologic education.

Photo courtesy of Melanie Wicander

REED WICANDER

REED WICANDER is a geology professor at Central Michigan University where he teaches physical geology, historical geology, prehistoric life, and invertebrate paleontology. He has co-authored several geology textbooks with James S. Monroe. His main research interests involve various aspects of Paleozoic palynology, specifically the study of acritarchs, on which he has published many papers. He is a past president of the American Association of Stratigraphic Palynologists and currently a councillor of the International Federation of Palynological Societies.

Brief Contents

Contents

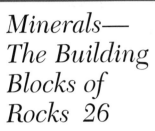

v

CHAPTER 8

The Seafloor 210

CHAPTER 9

Plate Tectonics: A Unifying Theory 234

CHAPTER 10

Deformation, Mountain Building, and the Continents 266

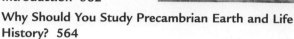

CHAPTER 22

Paleozoic Life History 630

CHAPTER 23

Mesozoic Earth and Life History 662

Preface

Earth is a dynamic planet that has changed continuously during its 4.6 billion years of existence. The size, shape, and geographic distribution of the continents and ocean basins have changed through time, as have the atmosphere and biota. We have become increasingly aware of how fragile our planet is and, more importantly, how interdependent all of its various systems are. We have learned that we cannot continually pollute our environment and that our natural resources are limited and, in most cases, nonrenewable. Furthermore, we are coming to realize how central geology is to our everyday lives. For these and other reasons, geology is one of the most important college or university courses a student can take.

The Changing Earth: Exploring Geology and Evolution is designed for an introductory course in geology that can serve both majors and nonmajors in geology and the Earth sciences. One of the problems with any introductory science course is that students are overwhelmed by the amount of material that must be learned. Furthermore, most of the material does not seem to be linked by any unifying theme and does not always appear to be relevant to their lives.

The goals of this book are to provide students with a basic understanding of geology and its processes and, more importantly, with an understanding of how geology relates to the human experience; that is, how geology affects not only individuals, but society in general. It is also our intention to provide students with an overview of the geologic and biologic history of Earth, not as a set of encyclopedic facts to memorize, but rather as a continuum of interrelated events that reflect the underlying geologic and biologic principles and processes that have shaped our planet and life upon it. With these goals in mind, we introduce the major themes of the book in the first chapter to provide students with an overview of the subject and enable them to see how Earth's various systems are interrelated. We also discuss the economic and environmental aspects of geology throughout the book rather than treating these topics in separate chapters. In this way students can see, through relevant and interesting examples, how geology impacts our lives.

New Features in the Third Edition

The third edition has undergone considerable rewriting and updating to produce a book that is easier to read, with a high level of current information, many new photographs, figures, Prologues, and Perspectives. Drawing on the comments and suggestions of reviewers, we have incorporated many new features into this edition.

New material in this edition of *The Changing Earth* includes an expanded section on Earth systems and an added emphasis on the systems approach throughout the book. There is updated information in every chapter, particularly such recent events as the earthquakes in Turkey (Chapter 7) and the flooding and landslides in Venezuela (Chapter 11). The chapters on surface processes (Chapters 11–16) are still largely descriptive, but many of the sections of these chapters have been rewritten to emphasize the systems approach in discussing the dynamic nature of these processes.

Updated information on mineral and energy resources and environmental issues has been added to many chapters, as well as new Prologues and Perspectives, such as precious metals (Chapter 2), oceanic resources (Chapter 8), and dam failures and their resulting floods (Chapter 12).

Other important changes include a number of new Prologues, such as granitic rocks and their occurrence at various national parks (Chapter 3), the eruption of Mt. Vesuvius (Chapter 4), the 1999 earthquakes in Turkey (Chapter 7), and flooding and landslides in Venezuela (Chapter 11). New Perspectives also appear, such as columnar jointing (Chapter 4), dams, reservoirs, and hydroelectric power (Chapter 12), arsenic in groundwater (Chapter 13), and geologic time and climate change (Chapter 17).

Many photographs in the second edition have been replaced, including most of the chapter-opening photographs. In addition, a number of photographs within chapters have been enlarged to enhance their visual impact.

We feel that the rewriting and updating done in the text and the addition of new photographs greatly improves the third edition by making it easier to read and comprehend, as well as a more effective teaching tool. Additionally, improvements have been made in the ancillary package that accompanies the book.

Text Organization

Plate tectonic theory is the unifying theme of geology and this book. This theory has revolutionized geology because it provides a global perspective of Earth and allows geologists to treat many seemingly unrelated geologic phenomena as part of a total planetary system. Because plate tectonic theory is so important, it is introduced in Chapter 1, and is integrated into the discussion in most of the subsequent chapters. It is treated in depth in Chapter 9.

Another theme of this book is that Earth is a complex, dynamic planet that has changed continually since its origin some 4.6 billion years ago. We can better understand this complexity by using a systems approach in the study of Earth. As mentioned earlier, we have expanded the section on Earth Systems in Chapter 1 and emphasized this approach throughout the book.

We have organized *The Changing Earth: Exploring Geology and Evolution,* 3rd edition, into several informal categories. Chapter 1 is an introduction to geology and Earth systems, its relevance to the human experience, plate tectonic theory, the rock cycle, organic evolution, geologic time and uniformitarianism, and the origin of the solar system and Earth. Chapters 2–6 examine Earth's materials (minerals and igneous, sedimentary, and metamorphic rocks) and the geologic processes associated with them, including the role of plate tectonics in their origin and distribution. Chapters 7–10 deal with the related topics of earthquakes and Earth's interior, the seafloor, plate tectonics, and deformation and mountain building. Chapters 11–16 cover Earth's surface processes. Chapter 17 discusses geologic time, introduces several dating methods, and explains how geologists correlate rocks. Chapter 18 explores fossils and evolution, and Chapter 19 discusses the origin of the universe and solar system and Earth's place in the evolution of these larger systems. Chapters 20 through 24 constitute our chronological treatment of the geologic and biologic history of Earth. These chapters are arranged so that the geologic history is followed by a discussion of the biologic history during that time interval. We think that this format allows easier integration of life history with geologic history.

Of particular assistance to students are the end-of-chapter summary tables found in chapters 21–23. These tables are designed to give an overall perspective of the geologic and biologic events that occurred during a particular time interval and to show how the events are interrelated. The emphasis in these tables is on the geologic evolution of North America. Global tectonic events and sea-level changes are also incorporated into these tables to provide global insights.

We have found that presenting the material in the order we have discussed above works well for most students. We know, however, that many instructors prefer an entirely different order of topics depending on the emphasis in their course. We have therefore written this book so that instructors can present the chapters in any order that suits the needs of their course.

Chapter Organization

All chapters have the same organizational format. Each opens with a photograph that relates to the chapter material, a detailed outline that engages the student by having many of the headings as questions, a Chapter Objectives outline, followed by a Prologue that is intended to stimulate interest in the chapter by discussing some aspect of the material. Following the Introduction, each chapter has a section titled Why Should You Study . . . ? This section is written to show students the relevance of the chapter material and how it fits into the larger geologic perspective.

The text is written in a clear informal style, making it easy for students to comprehend. Numerous color diagrams and photographs complement the text, providing a visual representation of the concepts and information presented. In addition icons are provided throughout the text identifying topics that are covered in the *Earth Systems Today* CD-ROM or on the World Wide Web.

Each chapter contains at least one Perspective that presents a brief discussion of an interesting aspect of geology or geological research. What Would You Do? boxes are a new feature in each chapter. These boxes are designed to encourage critical thinking by students as they attempt to solve a hypothetical problem or issue on the local, national, and global level. Each chapter has at least two What Would You Do? boxes.

The topics of environmental and economic geology are discussed throughout the text. Integrating economic and environmental geology with the chapter material helps students see the importance and relevance of geology to their lives. Mineral and energy resources are discussed in the final sections of a number of chapters to provide interesting, relevant information in the context of the chapter topics. In addition, each of the chapters on geologic history in the second half of the text contains a final section on mineral resources characteristic of that time period. These sections provide applied economic material of interest to students.

The end-of-chapter materials begin with a concise review of important concepts and ideas in the Chapter Summary. The Important Terms, which are printed in boldface type in the chapter text, are listed at the end of each chapter for easy review, and a full glossary of important terms appears at the end of the text. The Review Questions are another important feature of this book; they include multiple-choice questions with answers as well as short essay questions and thought-provoking and quantitative questions. Many new multiple-choice, short essay, and quantitative questions have been added in each chapter for this edition. Each chapter concludes with World Wide Web Activities keyed to World Wide Web sites students can visit (through links from the book's website) to get additional information about the topics covered in that chapter, and a CD-ROM exploration activity. This edition of *The Changing Earth* is accompanied by an *Earth Systems Today* CD-ROM created to promote interactive learning of difficult-to-demonstrate geological concepts. The end-of-chapter CD-ROM activities are meant to encourage students to explore the various modules and thus facilitate their understanding of such concepts. For students interested in pursuing a particular topic, an up-to-date list of Additional Readings for each chapter is available at **http://www.brookscole.com/geo**

Ancillary Materials

For Instructors

We are pleased to offer a full suite of text and multimedia products to accompany the third edition of *The Changing Earth*.

The Brooks/Cole Earth Science Resource Center on the World Wide Web
http://www.brookscole.com/geo
An award-winning site that makes an encyclopedia's worth of online resources easy to find.

The Changing Earth, 3E, Book-Specific Web Site
http://www.brookscole.com/geo
Through Brooks/Cole's main Earth Science Resource Center site, you can also access resources specific to the book, including hyper-contents (where links for each chapter expand the book's coverage), critical thinking questions, and self-quizzes for students.

Multimedia Manager for *The Changing Earth:* A Microsoft® PowerPoint® Tool
The Multimedia Manager is the one instructor's tool that will allow you to build years' worth of multimedia presentations. Through a friendly interface, users are provided a bank of all the images from the text as well as prepared PowerPoint lecture outlines for each chapter in order to assemble, edit, publish, and present custom lectures. Hybrid CD-ROM ISBN 0-534-38624-5

CNN's Physical Geology Today Videos
Brooks/Cole has partnered with CNN to bring you videotapes each with 45 minutes worth of recent news coverage of major topics in physical geology: volcanoes, earthquakes, natural resources, and more. Produced by Turner Learning, Inc.

Volume One, ISBN 0-534-53783-9
Volume Two, ISBN 0-534-54780-X
Volume Three, ISBN 0-534-38173-1

Instructor's Manual with Test Bank
A full array of teaching tips and resources: chapter outlines/overviews, learning objectives, lecture suggestions, important terms, and a list of key resources. This manual also contains approximately 2000 test questions. ISBN 0-534-37552-9

ExamView
A CD-ROM that allows you to create, customize, and deliver tests and study guides in minutes with an easy-to-use assessment and tutorial system. Hybrid CD ISBN 0-534-37555-3

InfoTrac College Edition
Updated daily, this massive on-line library of authoritative sources gives you access to full-text articles from over 600 periodicals, going back as far as four years.

Transparencies
193 transparencies selected by the authors, providing clear and effective illustrations of important figures and maps from the text. ISBN 0-534-37556-1

Slide Set
ISBN 0-534-37554-5
277 slides featuring key photos from the text.

For Students

Study Guide
A valuable tool to help you excel in this course! Filled with sample test questions, learning objectives, useful analogies, key terms, drawings and figures, and vocabulary reviews to guide your study. ISBN 0-534-37553-7

Current Perspectives in Geology, 2000 Edition
Michael McKinney, Kathleen McHugh, and Susan Meadows (University of Tennessee, Knoxville)
This book of 42 current readings is designed to supplement any geology textbook and is ideal for instructors who include a writing component in their course. The articles are culled from a number of popular science magazines (such as *American Scientist, National Wildlife, Discover, Science, New Scientist,* and *Nature*). Available for sale to students or bundled at a discount with any Brooks/Cole geology text. ISBN 0-534-37213-9

Essential Study Skills for Science Students
Daniel Chiras (University of Colorado–Denver)
Designed to accompany any introductory science text. It offers tips on improving your memory, learning more quickly, getting the most out of lectures, preparing for tests,

producing first-rate term papers, and improving critical thinking skills. For just one dollar extra, it can be bundled with every student copy of the text.
ISBN 0-534-37595-2

Earth Systems Today™: The Brooks/Cole Earth Sciences CD-ROM for Students

Predict a volcanic eruption. Locate the epicenter of an earthquake. Or explore other key topics in geology through some 50 interactive exercises on CD-ROM. In each case, students can manipulate variables and data and view the results of their selections. Over 35 minutes of full motion video clips, plus animations, help illustrate difficult concepts. This CD-ROM offers an easy-to-use, graphically oriented interface, a searchable glossary, and user-tracking for professors to monitor student progress.
ISBN 0-534-37733-5

InfoTrac College Edition Student Guide for Earth Science

This 24-page booklet provides a detailed user guide for students, illustrating how to use the InfoTrac College Edition database. Special features include log-in help, a complete search tips sheet filled with helpful hints, and a topic list of suggested keyword search terms for earth science.
ISBN 0-534-24742-3

Earth Online

Michael Ritter (University of Wisconsin–Stevens Point)
An inexpensive, hands-on Internet guide written for the novice. It provides a tool for students to get "up and running" on the Internet with homework exercises, lab exercises, Web searches, and more. To keep the book as useful as possible, the author maintains an Earth Online home page with exercises, tips, new links, and constant updates of the exercises and reference sites. Access it through the Brooks/Cole Earth Science Resource Center at http://www.brookscole.com/geo.
ISBN 0-534-51707-2

Acknowledgments

As the authors, we are, of course, responsible for the organization, style, and accuracy of the text, and any mistakes, omissions, or errors are our responsibility. The finished product is the culmination of many years of work during which we received numerous comments and advice from many geologists who reviewed parts of the text: Deborah Caskey, El Paso Community College; Richard Diecchio, George Mason University; Robert Ewing, Portland Community College; David J. Fitzgerald, St. Mary's University; Dann M. Halverson, University of Southwestern Louisiana; Ray Kenny, New Mexico Highlands University; Glenn B. Stracher, East Georgia College; Monte D. Wilson, Boise State University.

We wish to express our sincere appreciation to the reviewers who reviewed the second edition and made many helpful and useful comments that led to the improvements seen in this third edition: Kenneth Beem, Montgomery College; Patricia J. Bush, Dèlgado Community College; Paul J. Bybee, Utah Valley State College; William C. Cornell, University of Texas at El Paso; Kathleen Devaney, El Paso Community College; Guy Worthey, St. Ambrose University.

We also wish to thank Kathy Benison, Richard V. Dietrich (Professor Emeritus), David J. Matty, Jane M. Matty, Wayne E. Moore (Professor Emeritus), and Sven Morgan of the Geology Department, and Bruce M. C. Pape of the Geography Department of Central Michigan University, as well as Eric Johnson (Hartwick College, New York), and Stephen D. Stahl (SUNY, College at Oneonta, New York) for providing us with photographs and answering our questions concerning various topics. We also thank Pam Iacco of the Geology Department, whose general efficiency was invaluable during the preparation of this book. Anna Dutton and Kelly Wallington, also of the Geology Department, helped with some of the clerical tasks associated with this book. We are also grateful for the generosity of the various agencies and individuals from many countries who provided photographs.

Special thanks must go to Marie Carigma-Sambilay, project development editor at Brooks/Cole Publishing Company, who saw this edition through to completion. We are equally indebted to our acquisition editors, Nina Horne and Keith Dodson, and our project editor, Keith Faivre, whose attention to detail and consistency is greatly appreciated. We thank Nancy Shammas, production manager, for all her help and Patterson Lamb for her copyediting skills. Because geology is largely a visual science, we extend special thanks to Carlyn Iverson, who rendered the reflective art and to the artists at Precision Graphics, who were responsible for much of the rest of the art program. We also thank the artists at Magellan Geographix, who rendered many of the maps. They all did an excellent job, and we enjoyed working with them.

As always, our families were very patient and encouraging when much of our spare time and energy were devoted to this book. We again thank them for their continued support and understanding.

James S. Monroe
Reed Wicander

Developing Critical Thinking and Study Skills

Introduction

College is a demanding and important time, a time when your values will be challenged, and you will try out new ideas and philosophies. You will make personal and career decisions that will affect your entire life. One of the most important lessons you can learn in college is how to balance your time among work, study, and recreation. If you develop good time management and study skills early in your college career, you will find that your college years will be successful and rewarding.

This section offers some suggestions to help you maximize your study time and develop critical thinking and study skills that will benefit you, not only in college, but throughout your life. While mastering the content of a course is obviously important, learning how to study and to think critically is, in many ways, far more important. Like most things in life, learning to think critically and study efficiently will initially require additional time and effort, but once mastered, these skills will save you time in the long run.

You may already be familiar with many of the suggestions and may find that others do not directly apply to you. Nevertheless, if you take the time to read this section and apply the appropriate suggestions to your own situation, we are confident that you will become a better and more efficient student, find your classes more rewarding, have more time for yourself, and get better grades. We have found that the better students are usually also the busiest. Because these students are busy with work or extracurricular activities, they have had to learn to study efficiently and manage their time effectively.

One of the keys to success in college is avoiding procrastination. While procrastination provides temporary satisfaction because you have avoided doing something you did not want to do, in the long run it leads to stress. While a small amount of stress can be beneficial, waiting until the last minute usually leads to mistakes and a subpar performance. By setting clear, specific goals and working toward them on a regular basis, you can greatly reduce the temptation to procrastinate. It is better to work efficiently for short periods of time than to put in long, unproductive hours on a task, which is usually what happens when you procrastinate.

Another key to success in college is staying physically fit. It is easy to fall into the habit of eating junk food and never exercising. To be mentally alert, you must be physically fit. Try to develop a program of regular exercise. You will find that you have more energy, feel better, and study more efficiently.

General Study Skills

Most courses, and geology in particular, build upon previous material, so it is extremely important to keep up with the coursework and set aside regular time for study in each of your courses. Try to follow these hints, and you will find you do better in school and have more time for yourself:

- Develop the habit of studying on a daily basis.
- Set aside a specific time each day to study. Some people are day people, and others are night people. Determine when you are most alert and use that time for study.
- Have an area dedicated for study. It should include a well-lighted space with a desk and the study materials you need, such as a dictionary, thesaurus, paper, pens and pencils, and a computer if you have one.
- Study for short periods and take frequent breaks, usually after an hour of study. Get up and move around and do something completely different. This will help you stay alert, and you'll return to your studies with renewed vigor.
- Try to review each subject every day or at least the day of the class. Develop the habit of reviewing lecture material from a class the same day.
- Become familiar with the vocabulary of the course. Look up any unfamiliar words in the glossary of your textbook or in a dictionary. Learning the language of the discipline will help you learn the material.

Getting the Most from Your Notes

If you are to get the most out of a course and do well on exams, you must learn to take good notes. Taking good notes does not mean you should try to write down every word your professor says. Part of being a good note taker is knowing what is important and what you can safely leave out.

Early in the semester, try to determine whether the lecture will follow the textbook or be predominantly new material. If much of the material is covered in the textbook, your notes do not have to be as extensive or detailed as when the material is new. In any case, the following suggestions should make you a better note taker and enable you to derive the maximum amount of information from a lecture:

- Regardless of whether the lecture discusses the same material as the textbook or supplements the reading assignment, read or scan the chapter the lecture will cover *before* class. This way you will be somewhat familiar with the concepts and can listen critically to what is being said rather than trying to write down everything. Later a few key words or phrases will jog your memory about what was said.

- Before each lecture, briefly review your notes from the previous lecture. Doing this will refresh your memory and provide a context for the new material.

- Develop your own style of note taking. Do not try to write down every word. These are notes you're taking, not a transcript. Learn to abbreviate and develop your own set of abbreviations and symbols for common words and phrases: for example, w/o (without), w (with), = (equals), ∧ (above or increases), ∨ (below or decreases), < (less than), > (greater than), & (and), u (you).

- Geology lends itself to many abbreviations that can increase your note-taking capability: for example, pt (plate tectonics), ig (igneous), meta (metamorphic), sed (sedimentary), rx (rock or rocks), ss (sandstone), my (million years), and gts (geologic time scale).

- Rewrite your notes soon after the lecture. Rewriting your notes helps reinforce what you heard and gives you an opportunity to determine whether you understand the material.

- By learning the vocabulary of the discipline before the lecture, you can cut down on the amount you have to write—you won't have to write down a definition if you already know the word.

- Learn the mannerisms of the professor. If he or she says something is important or repeats a point, be sure to write it down and highlight it in some way. Students have told me (RW) that when I stated something twice during a lecture, they knew it was important and probably would appear on a test. (They were usually right!)

- Check any unclear points in your notes with a classmate or look them up in your textbook. Pay particular attention to the professor's examples, which usually elucidate and clarify an important point and are easier to remember than an abstract concept.

- Go to class regularly, and sit near the front of the class if possible. It is easier to hear and see what is written on the board or projected onto the screen, and there are fewer distractions.

- If the professor allows it, tape record the lecture, but don't use the recording as a substitute for notes. Listen carefully to the lecture and write down the important points; then fill in any gaps when you replay the tape.

- If your school allows it, and they are available, buy class lecture notes. These are usually taken by a graduate student who is familiar with the material; typically they are quite comprehensive. Again use these notes to supplement your own.

- Ask questions. If you don't understand something, ask the professor. Many students are reluctant to do this, especially in a large lecture hall, but if you don't understand a point, other people are probably confused as well. If you can't ask questions during a lecture, talk to the professor after the lecture or during office hours.

Getting the Most out of What You Read

The old adage that "you get out of something what you put into it" is true when it comes to reading textbooks. By carefully reading your text and following these suggestions, you can greatly increase your understanding of the subject:

- Look over the chapter outline to see what the material is about and how it flows from topic to topic. If you have time, skim through the chapter before you start to read in depth.

- Pay particular attention to the tables, charts, and figures. They contain a wealth of information in abbreviated form and illustrate important concepts and ideas. Geology, in particular, is a visual science, and the figures and photographs will help you visualize what is being discussed in the text and provide actual examples of features such as faults or unconformities.

- As you read your textbook, highlight or underline key concepts or sentences, but make sure you don't highlight everything. Make notes in the margins. If you don't understand a term or concept, look it up in the glossary.

- Read the chapter summary carefully. Be sure you understand all of the key terms, especially those in boldface or italic type. Because geology builds on previous material, it is imperative that you understand the terminology.

- Go over the end-of-chapter questions. Write out your answers as if you were taking a test. Only when you see your answer in writing will you know if you really understood the material.
- Access the latest geologic information on the internet. The end-of-chapter World Wide Web Activities will enhance your understanding of the chapter concepts and the way geologic information is disseminated today. Knowing how to search the Internet is an essential skill.

Developing Critical Thinking Skills

Few things in life are black and white, and it is important to be able to examine an issue from all sides and come to a logical conclusion. One of the most important things you will learn in college is to think critically and not accept everything you read and hear at face value. Thinking critically is particularly important in learning new material and relating it to what you already know. Although you can't know everything, you can learn to question effectively and arrive at conclusions consistent with the facts. Thus, these suggestions for critical thinking can help you in all your courses:

- Whenever you encounter new facts, ideas, or concepts, be sure you understand and can define all of the terms used in the discussion.
- Determine how the facts or information was derived. If the facts were derived from experiments, were the experiments well executed and free of bias? Can they be repeated? The controversy over cold fusion is an excellent example. Two scientists claimed to have produced cold fusion reactions using simple experimental laboratory apparatus, yet other scientists have never been able to achieve the same reaction by repeating the experiments.
- Do not accept any statement at face value. What is the source of the information? How reliable is the source?
- Consider whether the conclusions follow from the facts. If the facts do not appear to support the conclusions, ask questions and try to determine why they don't. Is the argument logical or is it somehow flawed?
- Be open to new ideas. After all, the underlying principles of plate tectonic theory were known early in this century yet were not accepted until the 1970s despite overwhelming evidence.
- Look at the big picture to determine how various elements are related. For example, how will constructing a dam across a river that flows to the sea affect the stream's profile? What will be the consequences to the beaches that will be deprived of sediment from the river? One of the most important lessons you can learn from your geology course is how interrelated the various Earth systems are. When you alter one feature, you affect numerous other features as well.

Improving Your Memory

Why do you remember some things and not others? The reason is that the brain stores information in different ways and forms, making it easy to remember some things and difficult to remember others. Because college requires that you learn a vast amount of information, any suggestions that can help you retain more material will help you in your studies:

- Pay attention to what you read or hear. Focus on the task at hand, and avoid daydreaming. Repetition of any sort will help you remember material. Review the previous lecture before going to class, or look over the last chapter before beginning the next. Ask yourself questions as you read.
- Use mnemonic devices to help you learn unfamiliar material. For example, the order of the Paleozoic periods (Cambrian, Ordovician, Silurian, Devonian, Mississippian, Pennsylvanian, and Permian) of the geologic time scale can be remembered by the phrase, Campbell's Onion Soup Does Make Peter Pale, or the order of the Cenozoic epochs (Paleocene, Eocene, Oligocene, Miocene, Pliocene, and Pleistocene) can be remembered by the phrase, Put Eggs On My Plate Please. Using rhymes can also be helpful.
- Look up the roots of important terms. If you understand where a word comes from, its meaning will be easier to remember. For example, *pyroclastic* comes from *pyro* meaning "fire" and *clastic* meaning "broken pieces." Hence a pyroclastic rock is one formed by volcanism and composed of pieces of other rocks. We have provided the roots of many important terms throughout this text to help you remember their definitions.
- Outline the material you are studying. This practice will help you see how the various components are interrelated. Learning a body of related material is much easier than learning unconnected and discrete facts. Looking for relationships is particularly helpful in geology because so many things are interrelated. For example, plate tectonics explains how mountain building, volcanism, and earthquakes are all related. The rock cycle relates the three major groups of rocks to each other and to subsurface and surface processes (Chapter 1).

- Use deductive reasoning to tie concepts together. Remember that geology builds on what you learned previously. Use that material as your foundation and see how the new material relates to it.

- Draw a picture. If you can draw a picture and label its parts, you probably understand the material. Geology lends itself very well to this type of memory device because so much is visual. For example, instead of memorizing a long list of glacial terms, draw a picture of a glacier and label its parts and the type of topography it forms.

- Focus on what is important. You can't remember everything, so focus on the important points of the lecture or the chapter. Try to visualize the big picture and use the facts to fill in the details.

Preparing for Exams

For most students, tests are the critical part of a course. To do well on an exam, you must be prepared. These suggestions will help you focus on preparing for examinations:

- The most important advice is to study regularly rather than try to cram everything into one massive study session. Get plenty of rest the night before an exam, and stay physically fit to avoid becoming susceptible to minor illnesses that sap your strength and lessen your ability to concentrate on the subject at hand.

- Set up a schedule so that you cover small parts of the material on a regular basis. Learning some concrete examples will help you understand and remember the material.

- Review the chapter summaries. Construct an outline to make sure you understand how everything fits together. Drawing diagrams will help you remember key points. Make flash cards to help you remember terms and concepts.

- Form a study group, but make sure your group focuses on the task at hand, not on socializing. Quiz each other and compare notes to be sure you have covered all the material. We have found that students dramatically improved their grades after forming or joining a study group.

- Write out answers to all of the Review Questions. Before doing so, however, become thoroughly familiar with the subject matter by reviewing your lecture notes and reading the chapter. Otherwise, you will spend an inordinate amount of time looking up answers.

- If you have any questions, visit the professor or teaching assistant. If review sessions are offered, be sure to attend. If you are having problems with the material, ask for help as soon as you have difficulty. Don't wait until the end of the semester.

- If old exams are available, look at them to see what is emphasized and what type of questions are asked. Find out whether the exam will be all objective or all essay or a combination. If you have trouble with a particular type of question (such as multiple choice or essay), practice answering questions of that type—your study group or a classmate may be able to help.

Taking Exams

The most important thing to remember when taking an exam is not to panic. This, of course, is easier said than done. Almost everyone suffers from test anxiety to some degree. Usually, it passes as soon as the exam begins, but in some cases, it is so debilitating that the individual does not perform as well as he or she could. If you are one of those people, get help as soon as possible. Most colleges and universities have a program to help students overcome test anxiety or at least keep it in check. Don't be afraid to seek help if you suffer test anxiety. Your success in college depends to a large extent on how well you perform on exams, so by not seeking help, you are only hurting yourself. In addition, the following suggestions may be helpful:

- First of all, relax. Then look over the exam briefly to see its format and determine which questions are worth the most points. If it helps, quickly jot down any information you are afraid you might forget or particularly want to remember for a question.

- Answer the questions that you know the best first. Make sure, however, that you don't spend too much time on any one question or on one that is worth only a few points.

- If the exam is a combination of multiple choice and essay, answer the multiple-choice questions first. If you are not sure of an answer, go on to the next one. Sometimes the answer to one question can be found in another question. Furthermore, the multiple-choice questions may contain many of the facts needed to answer some of the essay questions.

- Read the question carefully and answer only what it asks. Save time by not repeating the question as your opening sentence to the answer. Get right to the point. Jot down a quick outline for longer essay questions to make sure you cover everything.

- If you don't understand a question, ask the examiner. Don't assume anything. After all, it is your grade that will suffer if you misinterpret the question.

- If you have time, review your exam to make sure you covered all the important points and answered all the questions.
- If you have followed our suggestions, by the time you finish the exam, you should feel confident that you did well and will have cause for celebration.

Concluding Comments

We hope that the suggestions we have offered will be of benefit to you not only in this course, but throughout your college career. While it is difficult to break old habits and change a familiar routine, we are confident that following these suggestions will make you a better student. Furthermore, many of the suggestions will help you work more efficiently, not only in college, but also throughout your career. Learning is a lifelong process that does not end when you graduate. The critical thinking skills that you learn now will be invaluable throughout your life, both in your career and as an informed citizen.

Understanding Earth: An Introduction

Apollo 17 *view of Earth. Almost the entire coastline of Africa is clearly shown in this view with Madagascar visible off its eastern coast. The Arabian Peninsula can be seen at the northeastern edge of Africa, while the Asian mainland is on the horizon toward the northeast. The present location of continents and ocean basins is the result of plate movement. The interaction of plates through time has affected the physical and biological history of Earth.*

PROLOGUE

OBJECTIVES

At the end of this chapter, you will have learned that

- Geology is the study of Earth.

- Earth is a complex, integrated system.

- Geology plays an important role in the human experience, affecting us as individuals and as members of society and nation-states.

- The solar system and planets evolved from a turbulent, rotating cloud of material surrounding the embryonic Sun.

- Theories are based on the scientific method.

- Plate tectonic theory revolutionized geology.

- The rock cycle illustrates the interrelationships between Earth's internal and external processes and shows how and why the three major rock groups are related to each other.

- The theory of organic evolution provides the conceptual framework for understanding the history of life.

- An appreciation of geologic time and the principle of uniformitarianism is central to understanding the evolution of Earth and its biota.

- Geology is an integral part of our lives.

If we could film the history of Earth from its beginning 4.6 billion years ago to the present, and show it as a feature-length movie, we would witness a planet undergoing remarkable change. We would see the geography of the planet continuously changing as continents repeatedly collided and broke apart. As a result of these collisions and breakups, mountain ranges would rise up, only to eventually be worn down, while ocean basins would open and close between the moving continents. Oceanic and atmospheric circulation patterns would shift in response to the movement of continents, and hence would affect global climate and weather patterns. At various times we would see massive ice sheets appearing, growing, and then melting away. At other times, the landscape would be dominated by vast interior deserts or extensive swamps.

As our imaginary camera focused closer to Earth's surface, it would capture a breathtaking panorama of life. We would witness the first living cells evolving from a primordial organic soup sometime between 4.6 and 3.5 billion years ago. About 2 billion years later, cells with a nucleus would make their appearance. It would not be until around 700 million years ago that we would see the first multicelled soft-bodied organisms evolving in the oceans, followed by small animals with skeletons, and then animals with backbones.

Meanwhile, the various continents would appear barren and seemingly devoid of life until about 450 million years ago. At that time, the landscape would change forever with the evolution of land plants. In a relatively short period of time, insects would make their appearance as well as primitive amphibians, and then reptiles. Soon the continents would be teeming with life. However, some 250 million years ago, near the close of the Paleozoic Era when the continents were coalescing into one supercontinent called Pangaea, a major catastrophe occurred, resulting in the greatest mass extinction event the world has ever seen. About 90% of all species in the oceans became extinct, while 30% of insect orders died out, and more than 65% of all amphibian and reptile families disappeared.

During the next period of Earth history, known as the Mesozoic Era, new organisms evolved, filling and expanding the niches left vacant by the Permian mass extinction event. The stars during this part of our movie are the dinosaurs who ruled the land, while flying reptiles circled overhead and marine reptiles dominated the seas. Mammals, as well as birds, made their debut during the Mesozoic, but were relegated to supporting roles. Also during this time, Pangaea broke apart into various continental landmasses and the Atlantic Ocean came into being. Just as at the end of the Paleozoic Era, another mass extinction event occurred at the end of the Mesozoic Era. This global catastrophe wiped out the dinosaurs, flying reptiles, and marine reptiles, as well as various marine invertebrate animals.

The last part of our epic movie features the rise and dominance of mammals and flowering plants, and in the grand finale, the evolution and appearance of humans. During this time, known as the Cenozoic Era, the continents continued moving toward their current locations while the landscape was being sculpted to its present-day topography. In the last second or so of the movie we would see multiple advances and retreats of glaciers in the Northern Hemisphere, and the migration and settlement of humans to all continents. As the movie ends and fades to the closing credits, the impact of human activity on the global ecosystem will be quite evident. It seems only fitting that the final image in the movie will be of Earth, a shimmering blue oasis in the black void of space.

In our analogy of Earth history as a feature-length film, three themes stand out. The first is that Earth's outermost part is composed of a series of moving plates whose interactions have affected the physical and biological history of the planet. The second is that Earth's biota has evolved, or changed, throughout its history. The third is that the physical and biological changes have taken place over long periods of time.

The first of these themes, *plate tectonics*, provides geologists with a unifying theory that explains Earth's internal workings and accounts for how many seemingly unrelated geologic features and events are connected. The second theme, the theory of *organic evolution*, illustrates how life has changed through time, based on the idea that all living organisms are the evolutionary descendants of life-forms in the past. The last theme, the concept of *geologic time*, shows how small, almost imperceptible changes over vast periods of time have resulted in significant changes in Earth.

These three interrelated themes—plate tectonics, organic evolution, and geologic time—are central to our understanding and appreciation of the workings and history of our planet. As you read this book, keep in mind that the different topics you are studying concern parts of dynamic interrelated systems and are not isolated pieces of information. For example, volcanic eruptions are the product of complex interactions involving Earth's interior and surface. The impact of these eruptions is felt not just locally, but also globally. This is because the tremendous amount of ash and gases thrown into the atmosphere by volcanic eruptions reduces the amount of solar radiation reaching Earth, thus contributing to climatic changes that affect the entire planet.

Introduction to Earth Systems

In this book, we focus on Earth as a complex, dynamic planet that has been changing continually since it was formed some 4.6 billion years ago. These changes and the present-day features we observe resulted from interactions between the various internal and external Earth subsystems and cycles. Earth is unique among the planets of our solar system in that it supports life and has oceans of water, a hospitable atmosphere, and a variety of climates. It supports life as we know it because of a combination of factors, such as its distance from the Sun, the evolution of its interior, crust, oceans, and atmosphere. In turn, life processes, over time, have influenced the evolution of Earth's atmosphere, oceans, and to some extent, its crust. These physical factors and the changes they have brought about have also affected the evolution of life.

If we view Earth as a whole, we can see innumerable interactions occurring between its various components. Furthermore, these components do not act in isolation but are interconnected—so that when one part changes, it affects the other parts.

We can better appreciate the complexity of Earth by thinking of it as a system. The system concept makes Earth easier to study because it divides the whole into smaller components that we can understand better without losing sight of how the components all fit together as a system. A **system** is defined as a combination of related parts that interact in an organized fashion (Figure 1.1). Information, materials, and energy entering the system from the outside are *inputs*, whereas information, materials, and energy that leave the system are *outputs*. An automobile is a good example of a system. Its various subsystems include the engine, transmission, steering, and brakes. These subsystems are interconnected in such a way that a change in

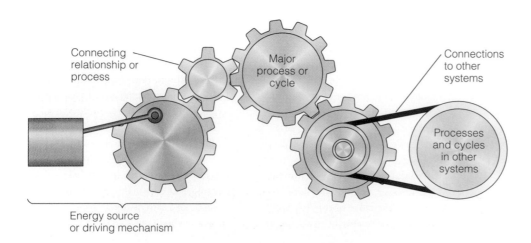

Connecting relationship or process

Major process or cycle

Connections to other systems

Processes and cycles in other systems

Energy source or driving mechanism

Figure 1.1 *A series of gears can be used to illustrate how some of Earth's systems and processes interact. Pistons and driving rods represent energy sources or driving mechanisms, large gears represent important processes or cycles, and small gears represent connecting processes or relationships. Pulleys are used to show relationships to other systems. Although gears are a useful way to represent systems diagrammatically, remember that real Earth systems are far more complex.*

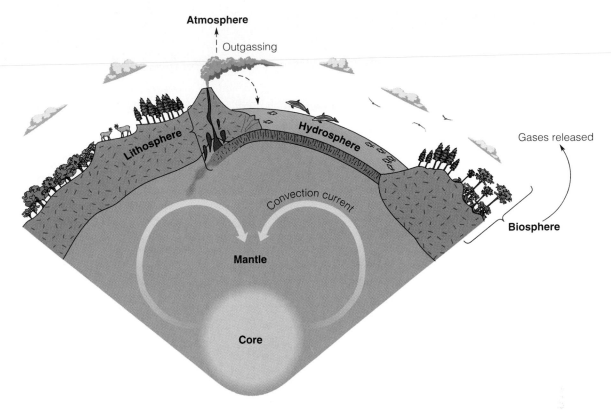

Figure 1.2 *The atmosphere, biosphere, hydrosphere, lithosphere, mantle, and core can all be thought of as subsystems of Earth. The interactions among these subsystems are what make Earth a dynamic planet, which has evolved and continues to change since its origin 4.6 billion years ago.*

any one of them affects the others. The main input into the automobile system is gasoline, and its outputs are movement, heat, and pollutants.

We can examine Earth in the same way we view an automobile—that is, as a system of interconnected components that interact and affect each other in many ways. The principal subsystems of Earth are the *atmosphere, hydrosphere, biosphere, lithosphere, mantle*, and *core* (Figure 1.2). The complex interactions between these subsystems result in a dynamically changing body that exchanges matter and energy and recycles them into different forms (Table 1.1). The rock cycle is an excellent example. It illustrates how the interaction between Earth's internal and external processes recycles Earth materials to form the three major rock groups (see The Rock Cycle p. 19).

We must also not forget that humans are part of the Earth system, and our presence alone affects this system to some extent. Accordingly, we must understand that actions we take can produce changes with wide-ranging effects that we might not be aware of (see Perspective 1.1). For this reason, we need to put paramount importance in understanding geology in particular, and science in general. If the human species is to survive, we must understand how the various Earth systems work and interact

with each other, and more importantly, how our actions affect the delicate balance between these systems.

What Is Geology?

What is geology and what do geologists do? **Geology,** from the Greek *geo* and *logos*, is defined as the study of Earth. It is generally divided into two broad areas—physical geology and historical geology. *Physical geology* studies Earth materials, such as minerals and rocks, as well as the processes operating within Earth and upon its surface. *Historical geology* examines the origin and evolution of Earth, its continents, oceans, atmosphere, and life.

The discipline of geology is so broad that it is subdivided into many fields or specialties. Table 1.2 shows many of the diverse fields of geology and their relationship to the sciences of astronomy, physics, chemistry, and biology.

Nearly every aspect of geology has some economic or environmental relevance. Many geologists are involved in exploration for mineral and energy resources, using their specialized knowledge to locate the natural resources on which our industrialized society is based. As the demand for these nonrenewable resources increases, geologists

The Aral Sea

The Aral Sea in the Central Asian desert region of the former Soviet Union (Figure 1) is a continuing environmental and human disaster. In 1960 it was the world's fourth largest lake. Since then its size has steadily decreased from 67,000 km^2 to about 34,000 km^2, and its volume has fallen from 1090 km^3 to 300 km^3, so that today it is only the sixth largest lake. Presently, it is divided into two separate basins, a small northern basin and a larger southern basin. At its present rate of reduction, the Aral Sea could disappear within the next 30 years.

What could have caused such a disaster? For thousands of years the Aral Sea was fed by two large rivers, the Amu Dar'ya and Syr Dar'ya. The headwaters of both rivers begin in the mountains more than 2000 km to the southeast, and flow northward through the Kyzyl Kum and Kara Kum deserts into the Aral Sea. Until the early part of the 20th century, a balance existed between the water supplied by the Amu Dar'ya and the Syr Dar'ya rivers and the rate of evaporation of the Aral Sea, which has no outlet. In 1918, however, it was decreed that waters from the two rivers supplying the Aral Sea would be diverted to irrigate millions of acres of cotton so that the Soviet Union could become self-sufficient in cotton production. As a result of this decision, the Aral Sea has become an ecological and environmental nightmare affecting some 35 million people.

To gain an appreciation of the magnitude of this disaster, consider that as late as the 1960s the Aral Sea was home to 24 native species of fish and supported a major fishing industry (Figure 2). The town of Muynak was the major fishing port and produced 3% of the Soviet Union's annual catch. The fishing industry provided about 60,000 jobs, with approximately 10,000 fishermen working out of Muynak. Today, Muynak is more than 20 km from the shoreline of the Aral Sea, and there are no native fish species left in the sea.

As the Aral Sea shrinks, vast areas of the former sea bottom are exposed. This new sediment, far from being rich and productive, contains large amounts of sodium chloride and sodium sulfate. The concentration of salts is so high that only one plant species grows in it. It is too salty for anything else.

As the winds blow across the near-barren land, salt and dust are picked up and carried throughout the Aral region. These salts cause great damage to the cotton crops and other vegetation of the Aral basin and also exact a heavy toll on humans. The dry, salty dust has caused respiratory and eye diseases to increase dramatically during the past 30 years, as well as the reported number of cases of throat cancer. In addition, the drinking water supply has become so polluted that many people suffer from intestinal disorders.

Figure 1 *Location and 1985 space shuttle view of the Aral Sea.*

As a result of the diversion of water from the Amu Dar'ya and the Syr Dar'ya rivers and the resulting reduction in the Aral Sea, desertification has become a major problem in the Aral basin. This has caused a change in the weather patterns of the

Figure 2 *Fishing boats lie abandoned in the dry seabed that was once part of the Aral Sea. As recently as the 1960s, the Aral Sea supported a large fishing industry, but today the water is too saline for any fish to survive. Consequently, frozen fish are shipped in from the Pacific for processing in plants that are now more than 20 km from the Aral Sea.*

region, so that it is now colder in the winter and hotter and drier in the summer.

Despite the damage done to the region by the dying of the Aral Sea, irrigation for cotton production continues. However, the soil in many of the fields has become increasingly salty due to lower stream and groundwater levels, requiring more irrigation to maintain the same production levels. This means that even more water is diverted from the Aral Sea, causing it to shrink still further.

Can the Aral Sea be saved? Only by a concerted and cooperative effort by the countries that comprise the Aral basin. Realizing this, the now independent states of Kazakhstan, Uzbekistan, Kyrgyzstan, Tadzhikistan, and Turkmenistan took the first steps in 1992 by signing a formal agreement on sharing the waters of the Amu Dar'ya and Syr Dar'ya. In addition, these states entered into discussions to create a council to oversee and coordinate the management of the basin's resources. Although full restoration of the Aral Sea is probably not possible, maintaining its current surface level and even raising the surface level of the small northern Aral Sea is feasible. If this could be accomplished, salinity levels would be reduced, making it possible to reintroduce native fish species and revive commercial fishing.

Table 1.1

Interactions Between Earth's Principal Subsystems

	Atmosphere	Hydrosphere	Biosphere	Lithosphere
Atmosphere	Interaction among various air masses	Surface currents driven by wind Evaporation	Gases for respiration Dispersal of spores, pollen, and seeds by wind	Weathering and erosion by wind Transport of water vapor for precipitation of rain and snow
Hydrosphere	Input of water vapor and stored solar heat	Hydrologic cycle	Water for life	Precipitation Weathering and erosion
Biosphere	Gases from respiration	Removal of disolved materials by organisms	Global ecosystems Food cycles	Modification of weathering and erosion processes Formation of soil
Lithosphere	Input of stored solar heat Landscapes affect air movements	Source of solid and dissolved materials	Source of mineral nutrients Modification of ecosystems by plate movements	Plate tectonics

Table 1.2

Specialties of Geology and Their Broad Relationship to the Other Sciences

Specialty	Area of Study	Related Science
Geochronology	Time and history of Earth	Astronomy
Planetary geology	Geology of the planets	
Paleontology	Fossils	Biology
Economic geology	Mineral and energy resources	
Environmental geology	Environment	
Geochemistry	Chemistry of Earth	Chemistry
Hydrogeology	Water resources	
Mineralogy	Minerals	
Petrology	Rocks	
Geophysics	Earth's interior	Physics
Structural geology	Rock deformation	
Seismology	Earthquakes	
Geomorphology	Landforms	
Oceanography	Oceans	
Paleogeography	Ancient geographic features and locations	
Stratigraphy/sedimentology	Layered rocks and sediments	

Figure 1.3 *Geologists use computers and other sophisticated technology in their search for petroleum and other natural resources.*

apply the basic principles of geology in increasingly sophisticated ways to help focus their attention on areas with a high potential for economic success (Figure 1.3).

While some geologists work on locating mineral and energy resources, an extremely important role, other geologists also use their expertise to help solve many environmental problems. Some geologists find groundwater for the ever-burgeoning needs of communities and industries, or monitor surface and underground water pollution and suggest ways to clean it up. Geological engineers help find safe locations for dams, waste disposal sites, and power plants, and design earthquake-resistant buildings.

Geologists also make short- and long-range predictions about earthquakes and volcanic eruptions and the potential destruction they may cause. In addition, they work with civil defense planners to help draw up contingency plans should such natural disasters occur.

Figure 1.4 Kindred Spirits *by Asher Brown Durand (1849) realistically depicts the layered rocks occurring along gorges in the Catskill Mountains of New York State. Asher Brown Durand was one of numerous artists of the 19th-century Hudson River School, who were known for their realistic landscapes.*

As this brief survey illustrates, geologists pursue a wide variety of careers and roles. As the world's population increases and makes greater demands on Earth's limited resources, we will depend even more on geologists and their expertise.

How Does Geology Relate to the Human Experience?

Many people are surprised at the extent to which we depend on geology in our everyday lives and also at the numerous references to geology in the arts, music, and literature. Many sketches and paintings represent rocks and landscapes realistically. Examples by famous artists include Leonardo da Vinci's *Virgin of the Rocks* and *Virgin and Child with Saint Anne*, Giovanni Bellini's *Saint Francis in Ecstasy* and *Saint Jerome*, and Asher Brown Durand's *Kindred Spirits* (Figure 1.4).

In the field of music, Ferde Grofé's *Grand Canyon Suite* was, no doubt, inspired by the grandeur and timelessness of Arizona's Grand Canyon and its vast rock

"You know, I used to like this hobby ... But shoot! Seems like *everybody's* got a rock collection."

Figure 1.5 *References to geology are frequently found in comics such as this* Far Side *cartoon by Gary Larson.*

exposures. The rocks on the Island of Staffa in the Inner Hebrides provided the inspiration for Felix Mendelssohn's famous *Hebrides* Overture.

References to geology abound in *The German Legends of the Brothers Grimm,* and Jules Verne's *Journey to the Center of the Earth* describes an expedition into Earth's interior. On one level, the poem "Ozymandias" by Percy B. Shelley deals with the fact that nothing lasts forever and even solid rock eventually disintegrates under the ravages of time and weathering. Even comics contain references to geology. Two of the best known are *B.C.* by Johnny Hart and *The Far Side* by Gary Larson (Figure 1.5).

Geology has also played an important role in history. Wars have been fought for the control of such natural resources as oil, gas, gold, silver, diamonds, and other valuable minerals. Empires throughout history have risen and fallen on the distribution and exploitation of natural resources. The configuration of Earth's surface, or its topography, which is shaped by geologic agents, plays a critical role in military tactics. Natural barriers such as mountain ranges and rivers have frequently served as political boundaries.

How Does Geology Affect Our Everyday Lives?

Most readers of this book will not become professional geologists. Everyone, however, should have a basic understanding of the geologic processes that ultimately affect all of us. We can trace many connections between geology and various aspects of our lives. Natural events or disasters, by their sheer magnitude, perhaps provide the most obvious connection. Less apparent, but equally significant, is the connection between geology and economic and political issues. We will likewise find connections between geology and our roles as decision makers in our work lives, as well as our lives as individual consumers and citizens.

Natural Events

Events such as destructive volcanic eruptions, devastating earthquakes, disastrous landslides, gigantic sea waves, floods, and droughts make headlines and affect many people in quite obvious ways (Figure 1.6). Although we cannot prevent most of these natural disasters from happening, the more knowledge we have about what causes them, the better we will be able to predict, and possibly control, the severity of their impact.

Economics and Politics

Equally important, but not always as well understood or appreciated, is the connection between geology and economic and political power. Mineral and energy resources are not equally distributed and no country is self-sufficient in all of them. Throughout history, people have fought wars to secure these resources. We need look no further than 1990–1991 to see that the United States was involved in the Gulf War largely because it needed to protect its oil interests in that region. Mineral and energy availability and needs in many cases help shape foreign policy. The sanctions imposed by the United States on South Africa in 1986, for example, did not include most of the important minerals we had been importing and needed for our industrialized society such as platinum-group minerals. There are many other examples of how foreign policy and treaties are based on the need to acquire and maintain adequate supplies of mineral and energy resources.

Our Role as Decision Makers

You may become involved in geologic decisions in various ways, for instance, as a member of a planning board or as a property owner with mineral rights. In such cases, you will not be able to make informed decisions without a knowledge of geology. Furthermore, many professionals must

Figure 1.6 *As these headlines from various newspapers indicate, geologic events affect our everyday lives.*

deal with geologic issues as part of their jobs. For example, lawyers are becoming more involved in issues ranging from ownership of natural resources to how development activities impact the environment. As government takes a greater role in environmental issues and regulations, members of Congress have increased the number of staff devoted to studying environment- and geology-related issues.

Consumers and Citizens

If issues like nonrenewable energy resources, waste disposal, and pollution are simply too far removed or too complex to be fully appreciated, consider for a moment just how dependent we are on geology in our daily routines.

Much of the electricity for our appliances comes from the burning of coal, oil, or natural gas or from uranium consumed in nuclear-generated plants. It is geologists who locate the coal, petroleum, and uranium. The copper or other metal wires through which electricity travels are manufactured from materials found as the result of mineral exploration. The buildings we live and work in owe their very existence to geologic resources. A few examples include the concrete foundation (concrete is a mixture of clay, sand, or gravel, and limestone), the drywall (made largely from the mineral gypsum), the windows (the mineral quartz is the principal ingredient in the manufacture of glass), and the metal or plastic plumbing fixtures inside the building (the metals are from ore deposits, and the plastics are most likely manufactured from petroleum distillates of crude oil).

When we go to work, the car or public transportation we use is powered and lubricated by some type of petroleum by-product and is constructed of metal alloys and plastics. And the roads or rails we ride over come from geologic materials, such as gravel, asphalt, concrete, or steel. All these items are the result of processing geologic resources.

As individuals and societies, we enjoy a standard of living that is obviously directly dependent on the consumption of geologic materials. Therefore, we need to be aware of geology and of how our use and misuse of geologic resources may affect the delicate balance of nature and irrevocably alter our culture as well as our environment.

Sustainable Development

The concept of *sustainable development* has received increasing attention, particularly since the United Nations Conference on Environment and Development met in Rio de Janeiro, Brazil, during the summer of 1992. This important concept puts satisfying basic human needs side by side with safeguarding our environment to ensure continued economic development. By redefining *wealth* to include such natural capital as clean air and water, as well as productive land, we can take appropriate measures to ensure that future generations have sufficient natural resources to maintain and improve their standard of living.

If we are to have a world in which poverty is not widespread, then we must develop policies that encourage management of our natural resources along with continuing economic development. A growing global population

How Does Geology Affect Our Everyday Lives?

will mean increased demand for food, water, and natural resources, particularly nonrenewable mineral and energy resources. Geologists will play an important role in meeting these demands by locating the needed resources and ensuring protection of the environment for the benefit of future generations.

Global Geologic and Environmental Issues Facing Humankind

Most scientists would argue that the greatest environmental problem facing the world today is overpopulation. With the world's population surpassing 6 billion in 1999, projections indicate that this number will grow by at least another billion people during the next two decades, bringing Earth's human population to more than 7 billion. Although this may not seem to be a geologic problem, we must remember that these people must be fed, housed, and clothed, and all with the least possible impact on the environment. Some of this population growth will be in areas that are already at risk from such geologic hazards as earthquakes, volcanic eruptions, and landslides. Safe and adequate water supplies must be found and kept from being polluted. More oil, gas, coal, and alternative energy resources must be discovered and utilized to provide the energy to fuel the economies of nations with ever-increasing populations. New mineral resources must be located. In addition, ways to reduce usage and to reuse materials must be devised so as to decrease dependency on new sources of these materials.

The problem of overpopulation and how it affects the global ecosystem vary. For many of the poor and non-industrialized countries, the problem is too many people and not enough food. For the more developed and industrialized countries, it is too many people rapidly depleting both the nonrenewable and renewable natural resource base. And in the most industrially developed countries, it is people producing more pollutants than the environment can safely recycle on a human time scale. The common thread tying all of these varied situations together is environmental imbalance created by a human population exceeding Earth's carrying capacity.

Another environmental issue that is currently being debated is global warming. Carbon dioxide is produced as a by-product of respiration and the burning of organic material. As such, it is a component of the global ecosystem and is constantly being recycled as part of the carbon cycle. The concern in recent years over the increase in atmospheric carbon dioxide has to do with its role in the greenhouse effect. The recycling of carbon dioxide between the crust and atmosphere is an important climatic regulator because carbon dioxide and other gases, such as methane, nitrous oxide, chlorofluorocarbons, and water vapor, allow sunlight to pass through them but trap the heat reflected back from Earth's surface. Heat is thus retained, causing the temperature of Earth's surface and, more important, the atmosphere to increase, producing the greenhouse effect.

Until the Industrial Revolution began during the mid-18th century, humans' contribution to the global temperature pattern was negligible. With industrialization and its accompanying burning of tremendous amounts of fossil fuels, carbon dioxide levels in the atmosphere have been steadily increasing since about 1880. In fact, atmospheric levels of carbon dioxide are currently almost 30% higher than a century ago, while nitrous oxide levels have climbed 15% and methane has increased 100% since the Industrial Revolution. At their current yearly rate of increase, greenhouse gas concentrations will probably triple from their present concentrations within the next 100 years, unless something is done to reduce their production.

In 1997 in Kyoto, Japan, a group of nations took the first steps in reducing greenhouse-causing emissions by negotiating a treaty calling for the United States and 37 other industrialized nations to reduce their greenhouse emissions by 2008. The Kyoto Protocol required the United States to cut carbon-dioxide emissions to 7% below 1990 levels, and the European Union and Japan to 8% and 6%, respectively, below previous levels. However, only 13 of the 84 nations signing the Kyoto Protocol have ratified it, and most of the major industrialized nations are unlikely to do so until at least 2001.

Research also indicates that deforestation of large areas, particularly in the tropics, is another cause of increased levels of carbon dioxide. Plants use carbon dioxide in photosynthesis and thus remove it from the atmosphere. With a decrease in the global vegetation cover, less carbon dioxide is removed from the atmosphere.

Because of the increase in human-produced greenhouse gases during the last 200 years, many scientists are concerned that a global warming trend has already begun and will result in severe global climatic shifts. Most computer models based on the current rate of increase in greenhouse gases show Earth warming as a whole by about 3°C during the next hundred years. Such a temperature change will be uneven, however, with the greatest warming occurring in the higher latitudes.

As a consequence of this warming, rainfall patterns and growing conditions will shift dramatically. Under such a climate change, countries in high and midlatitudes such as Canada and Europe will see an increase in crop yields, while those in lower latitudes will generally see a decrease.

With continued global warming, mean sea level will also rise as ice caps and glaciers melt and contribute their water to the world's oceans. It is predicted that by the 2050s, sea level will rise 21 cm, increasing the number of people at risk from flooding in coastal areas by approximately 20 million.

We cannot leave the subject of global warming without pointing out that many scientists are not convinced that the global warming trend is the direct result of increased human activity related to industrialization. They point out that while the amount of greenhouse gases has increased, we are still uncertain about their rate of generation and rate of removal, and whether the rise in global temperature during the past century resulted from normal climatic variations through time or from human activity. Furthermore, they point out that even if there is a general global warming during the next hundred years, it is not certain that the dire predictions made by proponents of global warming will come true.

WHAT WOULD YOU DO?

An important environmental issue facing the world today is global warming. An attempt to address this problem on a global scale was taken in 1997 with the signing of the Kyoto Protocol, which calls for the industrialized nations to reduce their greenhouse-causing gas emissions by 2008. Discuss some of the ways this might be accomplished. Besides reducing industrial emissions, what other steps can individuals, industries, and nations take to help reduce global warming? Do you think reducing industrial emissions will affect our standard of living? What effect might it have on the global economy?

Earth, as we know, is a remarkably complex system, with many feedback mechanisms and interconnections throughout its various subsystems and cycles. It is very difficult to predict all of the consequences that global warming would have for atmospheric and oceanic circulation patterns.

The Origin of the Solar System and the Differentiation of Early Earth

According to the currently accepted theory for the origin of the solar system (Figure 1.7), interstellar material in a spiral arm of the Milky Way Galaxy condensed and began collapsing. As this cloud gradually collapsed under the influence of gravity, it flattened and began rotating counterclockwise, with about 90% of its mass concentrated in the central part of the cloud. The rotation and concentration of material continued, and an embryonic Sun formed, surrounded by a turbulent, rotating cloud of material called a *solar nebula*.

The turbulence in this solar nebula formed localized eddies where gas and solid particles condensed. During the condensation process, gaseous, liquid, and solid particles began accreting into ever-larger masses called *planetesimals* that eventually became true planetary bodies. While the planets were accreting, material that had been pulled into the center of the nebula also condensed, collapsed, and was heated to several million degrees by gravitational compression. The result was the birth of a star, our Sun.

Some 4.6 billion years ago, enough material eventually gathered together in one of the turbulent eddies that swirled around the early Sun to form the planet Earth. Scientists think that this early Earth was rather cool, so the accreting elements and nebular rock fragments were solids rather than gases or liquids. This early Earth is also thought to have been of generally uniform composition and density throughout (Figure 1.8a). It was composed mostly of compounds of silicon, iron, magnesium, oxygen, aluminum, and smaller amounts of all the other chemical elements. Subsequently, when Earth underwent heating, because of meteorite impacts, gravitational compression, and the decay of radioactive elements, this homogeneous composition disappeared (Figure 1.8b) and was replaced by a series of concentric layers of differing composition and density (Figure 1.8c). This differentiation into a layered planet is probably the most significant event in Earth history. Not only did it lead to the formation of a crust and eventually to continents, but it was also probably responsible for the emission of gases from the interior, which eventually led to the formation of the oceans and the atmosphere.

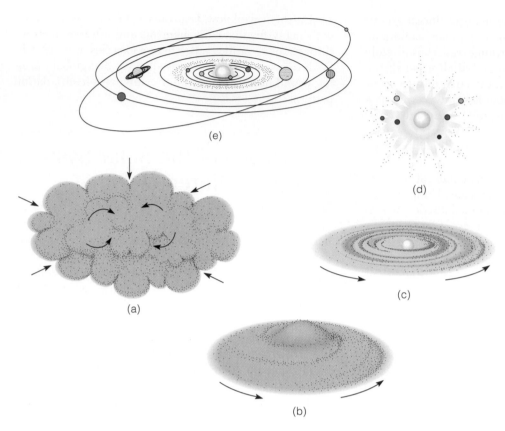

(e)

(d)

(c)

(a)

(b)

Figure 1.7 *The currently accepted theory for the origin of our solar system involves (a) a huge nebula condensing under its own gravitational attraction, then (b) contracting, rotating, and (c) flattening into a disk, with the Sun forming in the center and eddies gathering up material to form planets. As the Sun contracted and began to shine, (d) intense solar radiation blew away unaccreted gas and dust until finally, (e) the Sun began burning hydrogen and the planets completed their formation.*

Why Is Earth a Dynamic Planet?

Earth is a dynamic planet that has continuously changed during its 4.6-billion-year existence. The size, shape, and geographic distribution of continents and ocean basins have changed through time, the composition of the atmosphere has evolved, and life-forms existing today differ from those that lived during the past. We can easily visualize how mountains and hills are worn down by erosion and how landscapes are changed by the forces of wind, water, and ice. Volcanic eruptions and earthquakes reveal an active interior, and folded and fractured rocks indicate the tremendous power of Earth's internal forces.

Earth consists of three concentric layers: the core, the mantle, and the crust (Figure 1.9). This orderly division results from density differences between the layers as a function of variations in composition, temperature, and pressure.

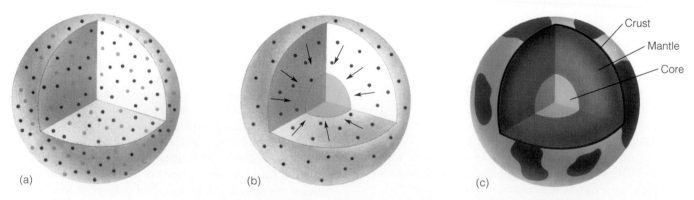

(a)

(b)

(c)

Crust

Mantle

Core

Figure 1.8 *(a) Early Earth was probably of uniform composition and density throughout. (b) Heating of early Earth reached the melting point of iron and nickel which, being denser than silicate minerals, settled to Earth's center. At the same time, the lighter silicate minerals flowed upward to form the mantle and the crust. (c) In this way, a differentiated Earth formed, consisting of a dense iron–nickel core, an iron-rich silicate mantle, and a silicate crust with continents and ocean basins.*

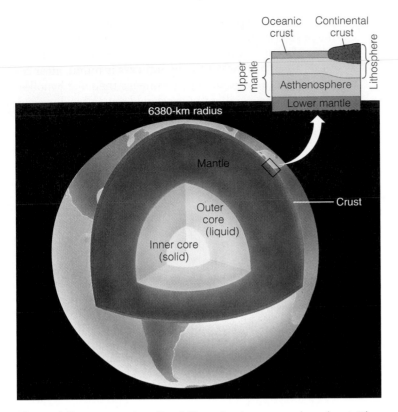

Figure 1.9 *A cross section of Earth illustrating the core, mantle, and crust. The enlarged portion shows the relationship between the lithosphere, composed of the continental crust, oceanic crust, and solid upper mantle, and the underlying asthenosphere and lower mantle.*

The **core** has a calculated density of 10 to 13 grams per cubic centimeter (g/cm³) and occupies about 16% of Earth's total volume. Seismic (earthquake) data indicate that the core consists of a small, solid inner part and a larger, apparently liquid, outer portion. Both are thought to consist largely of iron and a small amount of nickel.

The **mantle** surrounds the core and comprises about 83% of Earth's volume. It is less dense than the core (3.3–5.7 g/cm³) and is thought to be composed largely of *peridotite,* a dark, dense igneous rock containing abundant iron and magnesium. The mantle can be divided into three distinct zones based on physical characteristics. The lower mantle is solid and forms most of the volume of Earth's interior. The **asthenosphere** surrounds the lower mantle. It has the same composition as the lower mantle but behaves plastically and slowly flows. Partial melting within the asthenosphere generates *magma* (molten material), some of which rises to Earth's surface because it is less dense than the rock from which it was derived. The upper mantle consists of the asthenosphere and the overlying solid mantle rocks up to the base of the crust. The solid portion of the upper mantle and the overlying crust constitute the **lithosphere,** which is broken into numerous individual pieces called **plates** that move over the asthenosphere as a result of underlying *convection cells* (Figure 1.10). Interactions of these plates are responsible for such phenomena as earthquakes, volcanic eruptions, and the formation of mountain ranges and ocean basins.

The **crust,** Earth's outermost layer, consists of two types: continental and oceanic. *Continental crust* is thick (20–90 km), has an average density of 2.7 g/cm³, and contains considerable silicon and aluminum. In contrast, *oceanic crust* is thin (5–10 km), denser than continental crust (3.0 g/cm³), and is composed of the igneous rock *basalt.*

Since the widespread acceptance of plate tectonic theory more than 25 years ago, geologists have viewed Earth from a global perspective in which all of its subsystems and

Figure 1.10 *Earth's plates are thought to move as a result of underlying mantle convection cells in which warm material from deep within Earth rises toward the surface, cools, and then, upon losing heat, descends back into the interior. The movement of these convection cells is thought to be the mechanism responsible for the movement of Earth's plates, as shown in this diagrammatic cross section.*

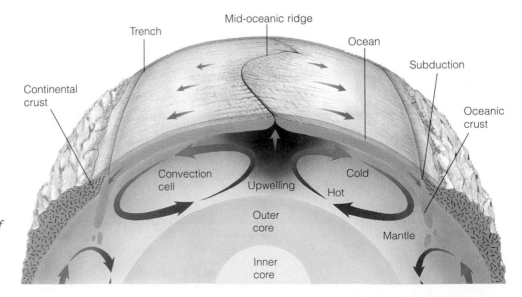

cyles are interconnected. Thus, the distribution of mountain chains, major fault systems, volcanoes, and earthquakes, the origin of new ocean basins, the movement of continents, and several other geological processes and features are all interrelated.

Geology and the Formulation of Theories

The term **theory** has various meanings. In colloquial usage, it means a speculative or conjectural view of something—hence the widespread belief that scientific theories are little more than unsubstantiated wild guesses. In scientific usage, however, a theory is a coherent explanation for one or several related natural phenomena supported by a large body of objective evidence. From a theory are derived predictive statements that can be tested by observations and/or experiments so that their validity can be assessed. The law of universal gravitation is an example of a theory describing the attraction between masses (an apple and the Earth in the popularized account of Newton and his discovery).

Theories are formulated through the process known as the **scientific method.** This method is an orderly, logical approach that involves gathering and analyzing the facts or data about the problem under consideration. Tentative ex-

planations or **hypotheses** are then formulated to explain the observed phenomena. Next, the hypotheses are tested to see if what they predicted actually occurs in a given situation. Finally, if one of the hypotheses is found, after repeated tests, to explain the phenomena, then that hypothesis is proposed as a theory. One should remember, however, that in science, even a theory is still subject to further testing and refinement as new data become available.

The fact that a scientific theory can be tested and is subject to such testing separates science from other forms of human inquiry. Because scientific theories can be tested, they have the potential of being supported or even proved wrong. Accordingly, science must proceed without any appeal to beliefs or supernatural explanations, not because such beliefs or explanations are necessarily untrue, but because we have no way to investigate them. For this reason, science makes no claim about the existence or nonexistence of a supernatural or spiritual realm.

Each scientific discipline has certain theories that are of particular importance for that discipline. In geology, one such theory is plate tectonic theory. This theory has changed the way geologists view Earth. Geologists now view Earth history in terms of interrelated events that are part of a global pattern of change. Before the acceptance of plate tectonic theory by most geologists, numerous hypotheses were proposed and tested. Thus, the evolution of this theory illustrates the scientific method at work (see Perspective 1.2).

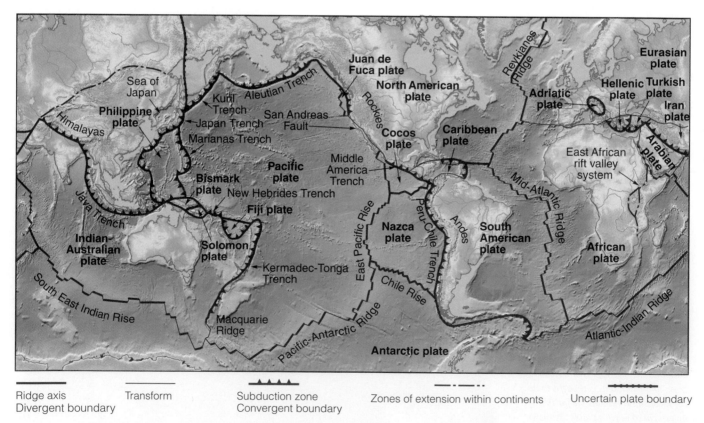

Ridge axis
Divergent boundary

Transform

Subduction zone
Convergent boundary

Zones of extension within continents

Uncertain plate boundary

Figure 1.11 *Earth's lithosphere is divided into rigid plates of various sizes that move over the asthenosphere.*

The Formulation of Plate Tectonic Theory

The idea that continents moved during the past goes back to the time when people first noticed that the margins of eastern South America and western Africa looked as if they fit together. Geologists also noticed that similar or identical fossils occur on widely separated continents, that the same types of rocks from the same time period are found on different continents, and that ancient rocks and features indicating former glacial conditions occur in today's tropical areas. As more and more facts were gathered, hypotheses were proposed to explain them. In 1912, Alfred Wegener, a German meteorologist, amassed a tremendous amount of geological, paleontological, and climatological data that indicated continents moved through time; he proposed the hypothesis of *continental drift* to explain and synthesize this myriad of facts.

Wegener stated that at one time all of the continents were united into one single supercontinent that he named *Pangaea*. Pangaea later broke apart, and the individual continents drifted to their current locations. The continental drift hypothesis explained why the shorelines of different continents fit together, how different mountain ranges were once part of a larger continuous mountain range, why the same fossil animals and plants are found on different continents, and why rocks indicating glacial conditions are now found on continents located in the tropics.

Wegener's hypothesis and its predictions could be tested by asking what types of rocks or fossils one would expect to find at a given location on a continent if that continent was in the tropics 180 million years ago. To test the hypothesis of continental drift, all researchers had to do was to go into the field and examine the rocks and fossils for a particular time period on any continent to see if they supported what the hypothesis predicted for the proposed location of that continent. In almost all cases, the data fit the hypothesis. However, there was one problem with Wegener's hypothesis: It did not explain how continents moved over oceanic crust and what the mechanism of continental movement was.

During the late 1950s and early 1960s, new data about the seafloor emerged that enabled geologists to propose the hypothesis of *seafloor spreading*. This hypothesis suggested that the continents and segments of oceanic crust move together as single units, and that some type of thermal convection cell system operating within Earth was the mechanism responsible for plate movements.

Seafloor spreading and continental drift were then combined into a single hypothesis that proposed that moving rigid plates are composed of continental and/or oceanic crust and the underlying upper mantle. These plates are bounded by mid-oceanic ridges, oceanic trenches, faults, and mountain belts. In this hypothesis, plates move away from mid-oceanic ridges and toward oceanic trenches. Furthermore, new crust is added along the mid-oceanic ridges and consumed or destroyed along oceanic trenches, and mountain chains are formed adjacent to the oceanic trenches.

According to this later hypothesis, Europe and North America should be steadily moving away from each other at a rate of up to several centimeters per year. Precise measurements of continental positions by satellites have verified this, thus confirming the validity of the plate movement hypothesis.

Furthermore, if plates are moving away from mid-oceanic ridges as predicted by the plate tectonic hypothesis, then rocks of the oceanic crust should become progressively older with increasing distance from the mid-oceanic ridges. To test this prediction, deep-sea sediment and oceanic crust were drilled as part of a massive scientific study of the ocean basins called the *Deep Sea Drilling Project*. Analysis of the oceanic crust and the layer of sediment immediately above it showed that the age of the oceanic crust does indeed increase with distance from the mid-oceanic ridges, and that the oldest oceanic crust is adjacent to the continental margins.

With the confirmation of these and other predictions of the plate tectonic hypothesis, most geologists accept that the hypothesis is correct and therefore call it *plate tectonic theory*. Its acceptance has been widespread because of the overwhelming evidence supporting it and also because it explains the relationships between many seemingly unrelated geologic features and events.

Plate Tectonic Theory

The acceptance of **plate tectonic theory** is recognized as a major milestone in the geological sciences. It is comparable to the revolution caused by Darwin's theory of evolution in biology. Plate tectonics has provided a framework for interpreting the composition, structure, and internal processes of Earth on a global scale. It has led to the realization that the continents and ocean basins are part of a lithosphere-atmosphere-hydrosphere (water portion of the planet) system that evolved together with Earth's interior (Table 1.3).

According to plate tectonic theory, the lithosphere is divided into plates that move over the asthenosphere (Figure 1.11). Zones of volcanic activity, earthquake activity, or both, mark most plate boundaries. Along these

Table 1.3

Plate Tectonics and Earth Systems

Solid Earth	Plate tectonics is driven by convection in the mantle and in turn drives mountain-building and associated igneous and metamorphic activity.
Atmosphere	Arrangement of continents affect solar heating and cooling, and thus winds and weather systems. Rapid plate spreading and hot-spot activity may release volcanic carbon dioxide and affect global climate.
Hydrosphere	Continental arrangement affects ocean currents. Rate of spreading affects volume of mid-oceanic ridges and hence sea level. Placement of continents may contribute to onset of ice ages.
Biosphere	Movement of continents creates corridors or barriers to migration, the creation of ecological niches, and transport of habitats into more or less favorable climates.
Extraterrestrial	Arrangement of continents affects free circulation of ocean tides and influences tidal slowing of Earth's rotation.

SOURCE: Adapted by permission from Stephen Dutch, James S. Monroe, and Joseph Moran, *Earth Science* (Minneapolis/St. Paul: West Publishing Co.).

boundaries, plates diverge, converge, or slide sideways past each other.

At **divergent plate boundaries,** plates move apart as magma rises to the surface from the asthenosphere (Figure 1.12). The magma solidifies to form rock, which attaches to the moving plates. The margins of divergent plate boundaries are marked by mid-oceanic ridges in oceanic crust, such as the Mid-Atlantic Ridge (Figure 1.11), and are recognized by linear rift valleys where newly forming divergent boundaries occur beneath continental crust.

Plates move toward one another along **convergent plate boundaries,** where one plate sinks beneath another plate along what is known as a **subduction zone** (Figure 1.12). As the plate descends into Earth, it becomes hotter until it melts, or partially melts, thus generating magma. As this magma rises, it may erupt at Earth's surface, forming a chain of volcanoes. The Andes Mountains on the west coast of South America are a good example of a volcanic mountain range formed as a result of subduction along a convergent plate boundary (Figure 1.11).

Transform plate boundaries are sites where plates slide sideways past each other (Figure 1.12). The San Andreas fault in California is a transform plate boundary separating the Pacific plate from the North American plate (Figure 1.11). The earthquake activity along the San Andreas fault results from the Pacific plate moving northward relative to the North American plate.

A revolutionary concept when it was first proposed in the 1960s, plate tectonic theory has had significant and

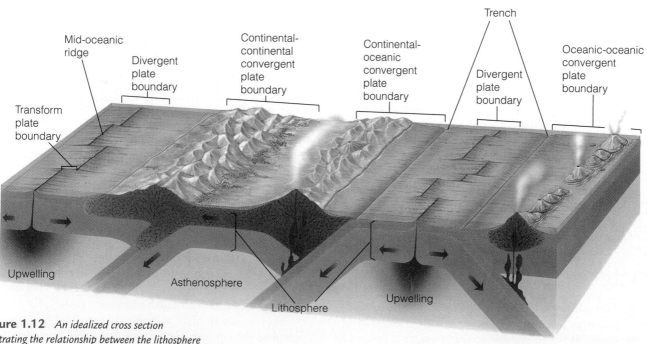

Figure 1.12 *An idealized cross section illustrating the relationship between the lithosphere and the underlying asthenosphere and the three principal types of plate boundaries: divergent, convergent, and transform.*

far-reaching consequences in all fields of geology because it provides the basis for relating many seemingly unrelated geologic phenomena. Its impact has been particularly notable in the interpretation of Earth history. For example, the Appalachian Mountains in eastern North America and the mountain ranges of Greenland, Scotland, Norway, and Sweden are not the result of unrelated mountain-building episodes but, rather, are part of a larger mountain-building event that involved the closing of an ancient "Atlantic Ocean" and the formation of the supercontinent Pangaea about 245 million years ago.

Besides being responsible for the major features of Earth's crust, plate movements also affect the formation and distribution of Earth's natural resources, as well as influence the distribution and evolution of the world's biota.

The Rock Cycle

A **rock** is an aggregate of **minerals,** which are naturally occurring, inorganic, crystalline solids with definite physi-cal and chemical properties. Minerals are composed of elements such as oxygen, silicon, and aluminum, and elements are made up of atoms, the smallest particles of matter that still retain the characteristics of an element. More than 3500 minerals have been identi-fied and described, but only about a dozen make up the bulk of the rocks in the crust (see Table 2.7).

Geologists recognize three major groups of rocks—*igneous, sedimentary,* and *metamorphic*—each of which is characterized by its mode of formation. Each group contains a variety of individual rock types that differ from one another on the basis of composition or texture (the size, shape, and arrangement of mineral grains).

The **rock cycle** provides a way of viewing the interre-lationships between Earth's internal and external processes (Figure 1.13). It relates the three rock groups to each other; to surficial processes such as weathering, trans-portation, and deposition; and to internal processes such as magma generation and metamorphism. Plate movement is the mechanism responsible for recycling rock materials and therefore drives the rock cycle.

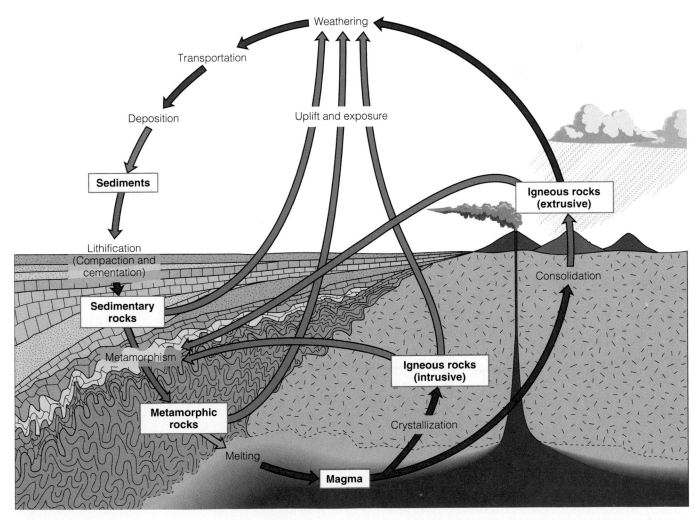

Figure 1.13 *The rock cycle showing the interrelationships between Earth's internal and external processes and how each of the three major rock groups is related to the others.*

Igneous rocks result when magma crystallizes or volcanic ejecta such as ash accumulates and consolidates. As a magma cools, minerals crystallize, and the resulting rock is characterized by interlocking mineral grains. Magma that cools slowly beneath the surface produces *intrusive igneous rocks* (Figure 1.14a); magma that cools at the surface produces *extrusive igneous rocks* (Figure 1.14b).

Rocks exposed at Earth's surface are broken into particles and dissolved by various weathering processes. The particles and dissolved material may be transported by wind, water, or ice and eventually deposited as *sediment.* This sediment may then be compacted or cemented (lithified) into sedimentary rock.

Sedimentary rocks originate by consolidation of mineral or rock fragments, precipitation of mineral matter from solution, or compaction of plant or animal remains (Figure 1.14c and d). Because sedimentary rocks form at or near Earth's surface, geologists can make inferences about the environment in which they were deposited, the type of transporting agent, and perhaps even something about the source from which the sediments were derived.

Accordingly, sedimentary rocks are especially useful for interpreting Earth history.

Metamorphic rocks result from the alteration of other rocks, usually beneath the surface, by heat, pressure, and the chemical activity of fluids. For example, marble, a rock preferred by many sculptors and builders, is a metamorphic rock produced when the agents of metamorphism are applied to the sedimentary rocks limestone or dolostone. Metamorphic rocks are either *foliated* (Figure 1.14e) or *nonfoliated* (Figure 1.14f). Foliation, the parallel alignment of minerals due to pressure, gives the rock a layered or banded appearance.

How Are the Rock Cycle and Plate Tectonics Related?

Interactions among plates determine, to a certain extent, which of the three rock groups will form (Figure 1.15). For example, when plates converge, heat and pressure generated along the plate boundary may lead to igneous

(a) (b) (c)

(d) (e) (f)

Figure 1.14 *Hand specimens of common igneous (a, b), sedimentary (c, d), and metamorphic (e, f) rocks. (a) Granite, an intrusive igneous rock. (b) Basalt, an extrusive igneous rock. (c) Conglomerate, a sedimentary rock formed by the consolidation of rock fragments. (d) Limestone, a sedimentary rock formed by the extraction of mineral matter from seawater by organisms or by the inorganic precipitation of the mineral calcite from seawater. (e) Gneiss, a foliated metamorphic rock. (f) Quartzite, a nonfoliated metamorphic rock.*

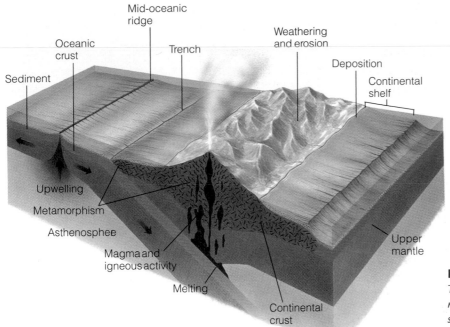

Mid-oceanic ridge

Oceanic crust

Trench

Weathering and erosion

Deposition

Continental shelf

Sediment

Upwelling

Metamorphism

Asthenosphee

Magma and igneous activity

Melting

Continental crust

Upper mantle

Figure 1.15 *Plate tectonics and the rock cycle. The cross section shows how the three major rock groups—igneous, metamorphic, and sedimentary—are recycled through both the continental and oceanic regions.*

activity and metamorphism within the descending oceanic plate, thus producing various igneous and metamorphic rocks.

Some of the sediments and sedimentary rocks on the descending plate are subducted and melt; while other sediments and sedimentary rocks along the boundary of the nonsubducted plate are metamorphosed by the heat and pressure generated along the converging plate boundary. Later, the mountain range or chain of volcanic islands formed along the convergent plate boundary will be weathered and eroded, and the new sediments will be transported by agents such as running water from the continents to the oceans, where they are deposited and accumulate. These sediments, some of which may be lithified and become sedimentary rock, become part of a moving plate along with the underlying oceanic crust.

The interrelationship between the rock cycle and plate tectonics is just one example of how Earth's various subsystems and cycles are all interrelated. We see the same interrelatedness when we expand our view to include other systems and subsystems, both big and small. Heating within Earth's interior results in convection cells that power the movement of plates, and also in magma, which forms intrusive and extrusive igneous rocks. Movement along plate boundaries may result in volcanic activity, earthquakes, and in some cases, mountain building. The interaction between the atmosphere, hydrosphere, and biosphere contributes to the weathering of rocks exposed on Earth's surface. Plates descending back into Earth's interior are subjected to increasing heat and pressure which may lead to metamorphism as well as the generation of magma and yet another recycling of materials.

Organic Evolution

Plate tectonic theory provides us with a model for understanding the internal workings of Earth and its effect on Earth's surface. The theory of **organic evolution** provides the conceptual framework for understanding the history of life. Together, the theories of plate tectonics and organic evolution have changed the way we view our planet, and we should not be surprised at the intimate association between them. While the relationship between plate tectonic processes and the evolution of life is incredibly complex, paleontological data provide indisputable evidence of the influence of plate movement on the distribution of organisms.

The publication in 1859 of Darwin's *On the Origin of Species by Means of Natural Selection* revolutionized biology and marked the beginning of modern evolutionary biology. With its publication, most naturalists recognized that evolution provided a unifying theory that explained an otherwise encyclopedic collection of biologic facts.

The central thesis of organic evolution is that all present-day organisms are related, and that they have descended with modifications from organisms that lived during the past. When Darwin proposed his theory of organic evolution he cited a wealth of supporting evidence, including the way organisms are classified, embryology, comparative anatomy, the geographic distribution of organisms, and, to a limited extent, the fossil record. Furthermore, Darwin proposed that *natural selection*, which results in the survival to reproductive age of those organisms best adapted to their environment, is the mechanism that accounts for evolution.

Perhaps the most compelling evidence in favor of evolution can be found in the fossil record. Just as the rock record allows geologists to interpret physical events and conditions in the geologic past, **fossils,** which are the remains or traces of once-living organisms, not only provide evidence that evolution has occurred, but also demonstrate that Earth has a history extending beyond that recorded by humans.

Geologic Time and Uniformitarianism

An appreciation of the immensity of geologic time is central to understanding the evolution of Earth and its biota. Indeed, time is one of the main aspects that set geology apart from the other sciences, except astronomy. Most people have difficulty comprehending geologic time because they tend to think in terms of the human perspective—seconds, hours, days, and years. Ancient history is what occurred hundreds or even thousands of years ago. When geologists talk of ancient geologic history, however, they are referring to events that happened hundreds of millions or even billions of years ago. To a geologist, recent geologic events are those that occurred within the last million years or so.

As we mentioned earlier in this chapter, it is important to remember that Earth goes through cycles of much longer duration than the human perspective of time. Because of their geologic perspective on time and how the various Earth subsystems and cycles are interrelated, geologists can make valuable contributions to many of the current environmental debates such as those on global warming and sea level changes.

The **geologic time scale** resulted from the work of many 19th-century geologists who pieced together information from numerous rock exposures and constructed a sequential chronology based on changes in Earth's biota through time. Subsequently, with the discovery of radioactivity in 1895 and the development of various radiometric dating techniques, geologists have been able to assign absolute ages in years to the subdivisions of the geologic time scale (Figure 1.16).

One of the cornerstones of geology is the **principle of uniformitarianism,** which is based on the premise that present-day processes have operated throughout geologic time. Therefore, in order to understand and interpret geologic events from evidence preserved in rocks, we must first understand present-day processes and their results.

We should keep in mind that uniformitarianism does not exclude sudden or catastrophic events such as volcanic eruptions, earthquakes, landslides, or flooding. These are processes that shape our modern world, and, in fact, some geologists view the history of Earth as a series of such short-

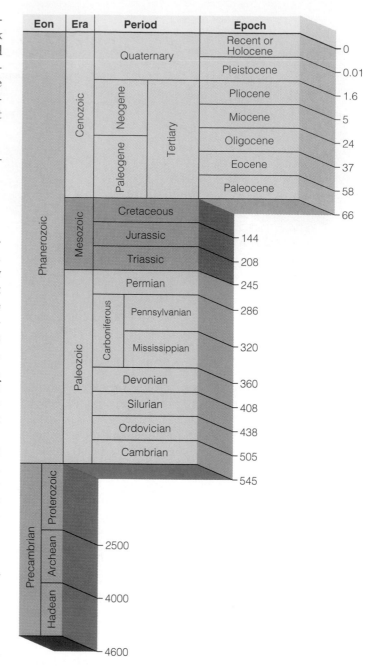

Figure 1.16 *The geologic time scale. Numbers to the right of the columns are ages in millions of years before the present.*

term or punctuated events. Such a view is certainly in keeping with the modern principle of uniformitarianism.

Furthermore, uniformitarianism does not require that the rates and intensities of geological processes be constant through time. We know that volcanic activity was more intense in North America 5 to 10 million years ago than it is today, and that glaciation has been more prevalent during the last several million years than in the previous 300 million years.

What uniformitarianism means is that even though the rates and intensities of geological processes have varied during the past, the physical and chemical laws of nature have remained the same. Although Earth is in a dynamic state of change and has been ever since it formed, the processes that shaped it during the past are the same ones that operate today.

How Does the Study of Geology Benefit Us?

The most meaningful lesson to learn from the study of geology is that Earth is an extremely complex planet in which interactions are taking place between its various subsystems and have been for the past 4.6 billion years. If we want to ensure the survival of the human species, we must understand how the various subsystems work and interact with each other, and more importantly, how our actions affect the delicate balance between these systems.

The study of geology goes beyond learning numerous facts about Earth. In fact, we don't just study geology—we "live" it. Geology is an integral part of our lives. Our standard of living depends directly on our consumption of natural resources and our interaction with the environment. However, the way we consume natural resources and interact with the environment, as individuals and as a society, also determine our ability to pass on this standard of living to the next generation.

As you study the various topics discussed in this book, come back to the themes discussed in this chapter and keep in mind how, like a system, they are all connected. View each chapter's topic in the context of its place in the entire Earth system, and trace the myriad geological links back to your own day-to-day life.

Chapter Summary

1. Geology, the study of Earth, is divided into two broad areas: physical geology is the study of Earth materials as well as the processes that operate within and upon Earth's surface; historical geology examines the origin and evolution of Earth, its continents, oceans, atmosphere, and life.

2. Earth can be viewed as a system of interconnected components that interact and affect each other. The principal subsystems of Earth are the atmosphere, hydrosphere, biosphere, lithosphere, mantle, and core. Earth is considered a dynamic planet that continually changes because of the interaction among its various subsystems and cycles.

3. Geology is part of the human experience. We can find examples of it in the arts, music, and literature. A basic understanding of geology is important for dealing with the many environmental problems and issues facing society.

4. Geologists engage in a variety of occupations, the main one being exploration for mineral and energy resources. They are also becoming increasingly involved in environmental issues and making short- and long-range predictions of the potential dangers from such natural disasters as volcanic eruptions and earthquakes.

5. About 4.6 billion years ago, the solar system formed from a rotating cloud of interstellar matter. Eventually, as this cloud condensed, it collapsed under the influence of gravity and flattened into a rotating disk. Within this rotating disk, the Sun, planets, and moons formed from the turbulent eddies of nebular gases and solids.

6. Earth is differentiated into layers. The outermost layer is the crust, which is divided into continental and oceanic portions. Below the crust is the solid portion of the upper mantle. The crust and solid part of the upper mantle, or lithosphere, overlie the asthenosphere, a zone that slowly flows. The asthenosphere is underlain by the solid lower mantle. Earth's core consists of an outer liquid portion and an inner solid portion.

7. The lithosphere is broken into a series of plates that diverge, converge, and slide sideways past one another.

8. The scientific method is an orderly, logical approach that involves gathering and analyzing facts about a particular phenomenon, formulating hypotheses to explain the phenomenon, testing the hypotheses, and finally proposing a theory. A theory is a testable explanation for some natural phenomenon and has a large body of supporting evidence.

9. Plate tectonic theory provides a unifying explanation for many geological features and events. The interaction between plates is responsible for volcanic eruptions, earthquakes, the formation of mountain ranges and ocean basins, and the recycling of rock materials.

10. Igneous, sedimentary, and metamorphic rocks comprise the three major groups of rocks. Igneous rocks result from the crystallization of magma or consolidation of volcanic ejecta. Sedimentary rocks are formed mostly by the consolidation of minerals or rock fragments, precipitation of mineral matter from solution, or compaction of plant or animal remains. Metamorphic rocks are produced from other rocks, generally beneath the Earth's surface, by heat, pressure, and chemically active fluids.

11. The rock cycle illustrates the interactions between internal and external Earth processes and shows how the three rock groups are interrelated.

12. The central thesis of the theory of organic evolution is that all living organisms evolved (descended with modifications) from organisms that existed in the past.

13. Time sets geology apart from the other sciences, except astronomy, and an appreciation of the immensity of geologic time is central to understanding Earth's evolution. The geologic time scale is the calendar geologists use to date past events.

14. The principle of uniformitarianism is basic to the interpretation of Earth history. This principle holds that the laws of nature have been constant through time and that the same processes operating today have operated in the past, although at different rates.

Important Terms

asthenosphere
convergent plate boundary
core
crust
divergent plate boundary
fossil
geologic time scale

geology
hypothesis
igneous rock
lithosphere
mantle
metamorphic rock
mineral

organic evolution
plate
plate tectonic theory
principle of
   uniformitarianism
rock
rock cycle

scientific method
sedimentary rock
subduction zone
system
theory
transform plate boundary

Review Questions

1. A combination of related parts interacting in an organized fashion is
   a. _____ a theory; b. _____ a cycle; c. _____ uniformitarianism; d. _____ a system; e. _____ a hypothesis.
2. The study of Earth materials is
   a. _____ environmental geology; b. _____ physical geology; c. _____ historical geology; d. _____ structural geology; e. _____ economic geology.
3. The interaction between the atmosphere, hydrosphere, and biosphere is a major contributor to
   a. _____ the generation of magma; b. _____ mountain building; c. _____ weathering of Earth materials; d. _____ metamorphism; e. _____ plate movement.
4. Plates are composed of
   a. _____ the crust and upper mantle; b. _____ the asthenosphere and upper mantle; c. _____ the crust and asthenosphere; d. _____ continental and oceanic crust only; e. _____ the core and mantle.
5. Which of the following is not a subdivision of geology?
   a. _____ paleontology; b. _____ astronomy; c. _____ mineralogy; d. _____ petrology; e. _____ stratigraphy
6. Plate movement is thought to be the result of
   a. _____ density differences between the mantle and core; b. _____ rotation of the mantle around the core; c. _____ gravitational forces; d. _____ the Coriolis effect; e. _____ convection cells.
7. Which of the following statements about a scientific theory is not true?
   a. _____ It is an explanation for some natural phenomenon; b. _____ It is a conjecture or guess; c. _____ It has a large body of supporting evidence; d. _____ It is testable; e. _____ Predictive statements can be derived from it.
8. Rocks resulting from the crystallization of magma are
   a. _____ igneous; b. _____ metamorphic; c. _____ sedimentary; d. _____ migmatites; e. _____ answers (a) and (d).
9. Which of the following is an example of a divergent plate boundary?
   a. _____ the Andes Mountains of South America; b. _____ the San Andreas fault; c. _____ the Mid-Atlantic Ridge; d. _____ the Himalayas; e. _____ the Hawaiian Islands.
10. The premise that present-day processes have operated throughout geologic time is the principle of
    a. _____ continental drift; b. _____ uniformitarianism; c. _____ Earth systems; d. _____ plate tectonics; e. _____ scientific deduction.
11. According to the currently accepted theory for the origin of the solar system,
    a _____ a huge nebula collapsed under its own gravitational attraction; b. _____ the nebula formed a disc with the Sun in the center; c. _____ planetesimals accreted from gaseous, liquid, and solid particles; d. _____ all of the previous; e. _____ none of the previous
12. The driving force of the rock cycle is
    a. _____ uniformitarianism; b. _____ plate tectonics; c. _____ gravitational forces; d. _____ density differences between Earth's interior and surface; e. _____ volcanism.
13. That all living organisms are the descendants of different life-forms that existed in the past is the central claim of
    a. _____ the principle of fossil succession; b. _____ the principle of uniformitarianism; c. _____ plate tectonics; d. _____ organic evolution; e. _____ none of the previous.
14. Why is it important that everyone have a basic understanding of geology, even if he or she is not going to become a geologist?
15. Describe how you would use the scientific method to formulate a hypothesis explaining the similarity of mountain ranges on the east coast of North America and those in England, Scotland, and the Scandinavian countries. How would you test your hypothesis?
16. Discuss what is meant by the statement, "The health and well-being of the world's economy is completely dependent on geologic resources."
17. Discuss how the three major layers of Earth differ from each other and why the differentiation into a layered planet is probably the most significant event in Earth history.
18. Describe the various ways in which geology affects our everyday lives.
19. Explain how the principle of uniformitarianism allows for catastrophic events?
20. Discuss why plate tectonic theory is a unifying theory of geology.
21. Discuss the relationship between overpopulation and geology and why they are related.
22. Explain the advantage of using a systems approach to the study of Earth.
23. Why is an accurate geologic time scale particularly important for geologists in examining changes in global temperatures during the past?

 # World Wide Web Activities

For the following Web site addresses, along with current updates and exercises, log on to **http://www.brookscole.com/geo**

Earth Science Resources on the Internet

This site, maintained by the Geology Department at the University of North Carolina at Chapel Hill, is an excellent starting place to learn how to find geoscience items on the World Wide Web. As you will see, a tremendous amount of information, graphics, and maps is available, and this site will get you started finding many of them and practicing your Net skills.

1. Read the *Starting up* and *How to find things* sections.

2. Visit several of the geology sites mentioned to see the variety of information, videos, and images available.

3. Visit the home page of several of the different geological societies and organizations to see what the societies do and what information is available from them.

4. Check the home page of several of the different schools and universities to find information on course offerings, faculty, and requirements to earn a degree. Compare the home pages and information available from different universities around the world.

West's Geology Directory

This comprehensive directory of geologic Internet links is an excellent place to find links on virtually any geologic topic. The Contents–Index lists numerous topics. Just click on a topic and a list of sites, which you can then click on, will appear. The links are regularly updated, and sites are checked, reviewed, and rated where possible. In addition, there is also a link to field-trip guides and bibliographies.

 # CD-ROM Exploration

Exploring your *Earth Systems Today* CD-ROM will add to your understanding of the material in this chapter.

Topic: Current News

Click on the "Earth Today" button.

Explore activities in this part of the CD to see if you can discover the following for yourself:

Go to the Update Categories and select "Geology" or one of the other categories and examine at least two of the world's "current hot spots" of geological activity. Comment on the geological aspects of these current news stories.

To what extent was each disaster you investigated predictable by geologists, and to what extent was each unpredictable?

If you were a planner or engineer with some geology background, what would you suggest for future mitigation of damage and possible loss of life in such disasters?

What other places on Earth are susceptible to disasters like the ones you just studied? What is it about the geology of those places that is related to their common risk of disaster?

Minerals—The Building Blocks of Rocks

A spectacular example of the mineral tourmaline and quartz (colorless) from the Himalaya Mine, San Diego County, California. Many museums in the United States and several other countries display mineral specimens from this mine.

PROLOGUE

The color, brilliance, and other properties of some minerals, rocks, and fossils have fascinated people for thousands of years. Several of these objects have served as religious symbols and talismans, or have been worn, carried, applied externally, or ingested for their presumed mystical or curative powers. Archaeological evidence indicates that people in Spain and France were carving objects from bone, ivory, horn, and various stones at least 75,000 years ago. By 3400 B.C., the ancient Egyptians were making ornaments from rock crystal (colorless quartz), amethyst (purple quartz), lapis lazuli (a rock composed of a variety of minerals), and several other stones. In fact, they mined turquoise more than 5000 years ago, and by 1650 B.C. they were mining emeralds from near the Red Sea.

Most gemstones are minerals—more rarely, rocks—that are cut and/or polished for jewelry. To qualify as a gemstone, a mineral or rock must be appealing for some reason. The most desirable quality is beauty, but brilliance, durability, and scarcity are important as well. Intense color is preferred in many gemstones, but most gem-quality diamonds are colorless and valued in part because of their brilliance. Gemstones are even more desirable when some kind of lore is associated with them. According to one legend, diamond wards off evil spirits, sickness, and floods. Topaz was once thought to avert mental disorders; ruby was believed to preserve its owner's health and warn of imminent bad luck. A symbol of immortality, a preserver of chastity, and a cure for dysentery are qualities attributed to emerald. Relating gemstones to birth month gives them even more appeal to many people (Table 2.1).

Gemstones are actually the raw materials for gems. That is, gems have been fashioned in some way from gemstones by shaping, polishing, or cutting, the notable exception being pearl which is ready to use when found. Many minerals and rocks are attractive but fail to qualify as gemstones. Cut and polished fluorite is beautiful, but it is fairly common and too soft to be durable. Diamond, in contrast, meets most of the criteria for a gemstone, although not all diamonds are of gem quality; most, in fact, are used in industrial applications. Only about two dozen minerals and rocks are used as gemstones. Some of them are considered *precious;* others are referred to as *semiprecious.* Diamond (Figure 2.1), ruby and sapphire (red and blue varieties of the mineral corundum), and emerald (a bright green variety of the mineral beryl) are the precious gemstones; precious opal is also sometimes included in this group. Several of the semiprecious gemstones are garnet, jade, tourmaline (see chapter-opening photo), topaz, peridot (olivine), aquamarine (light bluish-green beryl), turquoise (Figure 2.2a), and several varieties of quartz, such as amethyst, rose quartz, agate, and tiger's eye.

Also included among the semiprecious gemstones is an organic substance known as amber, and pearl, which is produced by organisms. Amber is a fossil resin from coniferous trees that commonly contains insects. Even though amber is neither a mineral nor a rock, people nevertheless consider it a semiprecious gemstone. Best known from the Baltic Sea region of Europe, sun-worshiping cultures believed that amber, with its golden translucence resembling the Sun's brilliance, had mystical powers. A mollusk such as a clam or oyster forms pearl by depositing successive layers of tiny mineral crystals around some irritant. Most pearls are lustrous white, but some are silver gray, green, or black (Figure 2.2b).

Transparent gemstones are most often cut to yield small, polished plane surfaces known as facets that enhance the quality of reflected light. Gem

OBJECTIVES

At the end of this chapter, you will have learned that

- Minerals are naturally occurring, crystalline solids with certain physical and chemical properties.

- Atoms bond to form elements and compounds, and that most minerals are compounds.

- Geologists draw a distinction between minerals and rocks, most of the latter being composed of minerals.

- Minerals are incredibly varied, but only a few types are particularly common.

- Most minerals are classified as silicates, but several other mineral groups such as carbonates are also important.

- Atomic structure and chemical composition determine the physical properties of minerals.

- Minerals originate in various ways and under varied conditions.

- Various minerals and rocks are important resources essential to industrialized societies.

Table 2.1

Birthstones

Month	Birthstone	Mineral	Symbolizes
January	Garnet	Garnet	Constancy
February	Amethyst	Purple quartz	Sincerity
March	Aquamarine	Light green, blue-green beryl	Courage
	Bloodstone	Greenish chalcedony	
April	Diamond	Diamond	Innocence
May	Emerald	Beryl and several other green minerals	Love, success
June	Pearl	Dense, spherical white or light-colored calcareous concretion produced by organisms, especially mollusks	Health, longevity
	Alexandrite	Greenish chrysoberyl	
	Moonstone	Type of potassium feldspar	
July	Ruby	Red corundum	Contentment
August	Peridot	Pale, clear yellowish-green olivine	Married happiness
	Sardonyx	Gem variety of chalcedony	
September	Sapphire	Blue transparent corundum	Clear thinking
October	Opal	Opal	Hope
	Tourmaline	Tourmaline	
November	Topaz	Topaz	Fidelity
December	Turquoise	Turquoise	Prosperity
	Zircon	Zircon	

cutters maximize the brilliance of diamond by faceting the gem so that it reflects as much light as possible (Figure 2.1). Opaque and translucent gemstones are rarely faceted. Rather, they are cut and polished into dome-shaped stones known as cabochons to emphasize their most interesting features, or they are simply polished by tumbling.

Most people are surprised to learn that diamond, the most sought-after gemstone, and the mineral graphite consist of the same chemical element, carbon. Other than composition, though, they share little in common; gem-quality diamond is colorless and the hardest mineral, whereas graphite is gray, unattractive, and so soft that a fingernail can scratch it. The way carbon atoms are arranged in these two minerals accounts for the differences. In diamond, the atoms bond tightly to one another in a three-dimensional framework, but in graphite the atoms form sheets that are held together by weak bonds. Incidentally, many people think that the true test of a diamond is whether it will scratch glass. Diamond will indeed scratch glass, but so will several other minerals, including common quartz.

"Cutting" a diamond, the hardest substance known, is actually done by several processes. Diamond possesses four internal planes of weakness, or what geologists call cleavage planes, so that if cleaved perfectly along these planes a diamond yields a "stone" shaped like two pyramids placed base to base. Large diamonds are commonly preshaped by cleaving them into smaller pieces that are then further shaped by sawing and grinding with diamond dust. A diamond's value is determined by evaluating its color (colorless specimens are most desirable), clarity (lack of internal flaws), and carat, which is the measure of a precious stone's weight (1 carat = 200 milligrams). The Victoria Transvaal diamond in Figure 2.1 weighs 68 carats, and the world's most famous blue diamond, the Hope diamond, is 45.5 carats.

Gemstones and the gems derived from them have been used for many purposes, especially for personal adornment and as symbols of wealth and power, in addition to their uses for their presumed mystical powers as good luck charms and cures for various ailments. Many people own small gems, but some of the truly magnificent ones are in museums, are parts of collections of crown jewels, or are kept as religious objects.

Figure 2.1 *Their hardness, brilliance, beauty, durability, and scarcity make diamonds the most sought-after gemstones. The pendant in this necklace, housed in the Smithsonian Institution, is the Victoria Transvaal diamond from South Africa.*

(a)

(b)

Figure 2.2 *(a) Turquoise, a sky blue, blue-green, or light green mineral, is a semiprecious gemstone used for jewelry and as a decorative stone. (b) These black pearls valued at about $13,000 are on display at Maui Pearls, Avarua, Rarotonga, Cook Islands in the South Pacific.*

Introduction

In the Prologue we used the term *mineral* without really defining it. The term is used in a variety of ways, as, for example, a designation for dietary substances essential for good nutrition, such as calcium, iron, potassium, and magnesium. These are actually chemical elements and not minerals, at least in the geologic sense. Substances that are neither animal nor vegetable are also commonly called minerals. This usage implies that minerals are inorganic, which is correct, but not all inorganic substances are minerals. For instance, water and water vapor are not minerals even though they are composed of the same chemical elements as ice, which is a mineral because minerals are solids rather than liquids or gases. Geologists define a **mineral** as a naturally occurring, inorganic, crystalline solid. *Crystalline* means that minerals have an ordered internal arrangement of atoms. Minerals also have a narrowly defined chemical composition and characteristic physical properties such as color, hardness, and density. We will examine each part of this definition later in the chapter.

Why Should You Study Minerals?

Why should you study minerals? To understand the rocks of which Earth is composed, you need to be familiar with minerals, the basic building blocks of most rocks. Furthermore, many ore deposits are natural concentrations of minerals that can be profitably extracted and used for such purposes as the manufacture of steel, aluminum, cement, ceramics, glass, wallboard, and fertilizers. The United States and Canada and all other industrialized countries depend on finding and using a variety of mineral resources or importing them from elsewhere. So the distribution of mineral wealth has political as well as economic consequences. Some minerals are quite attractive or desirable for some reason and used as gemstones (see the Prologue). Many minerals are eagerly sought out by private collectors and by museums for attractive and informative displays (Figure 2.3).

Matter—What Is It and What Is It Composed Of?

Matter is defined as anything that has mass and occupies space. Accordingly, water, plants, animals, the atmosphere, and minerals and rocks are composed of matter.

Figure 2.3 *Mineral display at the museum of the California Academy of Sciences in San Francisco.*

Three phases or states of matter are recognized*—*solids, liquids,* and *gases,* all of which are important in geology (Table 2.2). Liquids, such as surface water and groundwater, as well as atmospheric gases are important geologic agents that will be discussed in later chapters. Here, though, we are interested mostly in solids, because by definition minerals are solids.

Elements and Atoms

Our discussion of minerals begins with a consideration of elements and atoms because all matter is made up of chemical **elements,** each of which is composed of tiny particles known as **atoms.** Atoms are the smallest units of matter that retain the characteristics of a particular element. That is, they cannot be split into substances of different composition. Ninety-two naturally occurring elements have been discovered, some of which are listed in Table 2.3, and several more have been made in laboratories (see the Periodic Table of Elements in Appendix A). All naturally occurring elements and most artificially produced ones have a name and a symbol, such as oxygen (O), aluminum (Al), and potassium(K).

At the center of an atom is a tiny **nucleus** made up of one or more particles known as **protons** with a positive electrical charge, and **neutrons,** which are electrically neutral (Figure 2.4). The nucleus is only about 1/100,000 of the diameter of an atom, yet it contains virtually all of the atom's mass. **Electrons,** particles with a negative electrical charge orbiting rapidly around the nucleus, are concentrated within one or more **electron shells** at specific distances from the nucleus. The electrons determine how an atom interacts with other atoms, but the nucleus deter-

---

*Actually, scientists recognize a fourth state of matter known as *plasma,* an ionized gas as in fluorescent and neon lights, and matter in the Sun and stars.

Table 2.2

Phases or States of Matter

Phase	Characteristics	Examples
Solid	Rigid substance that retains its shape unless distorted by a force	Minerals, rocks, iron, wood
Liquid	Flows easily and conforms to the shape of the containing vessel; has a well-defined upper surface and greater density than a gas	Water, lava, wine, blood, gasoline
Gas	Flows easily and expands to fill all parts of a containing vessel; lacks a well-defined upper surface; is compressible	Helium, nitrogen, air, water vapor

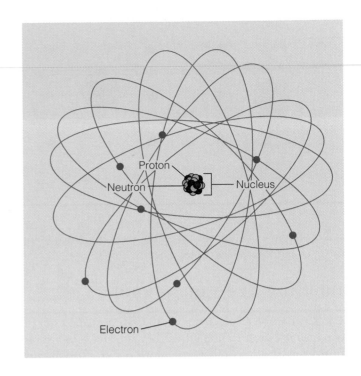

Figure 2.4 *The structure of an atom. The dense nucleus consisting of protons and neutrons is surrounded by a cloud of orbiting electrons.*

Table 2.3

Symbols, Atomic Numbers, and Electron Configurations for Some of the Naturally Occurring Elements

Element	Symbol	Atomic Number	Number of Electrons in Each Shell			
			1	2	3	4
Hydrogen	H	1	1			
Helium	He	2	2			
Lithium	Li	3	2	1		
Beryllium	Be	4	2	2		
Boron	B	5	2	3		
Carbon	C	6	2	4		
Nitrogen	N	7	2	5		
Oxygen	O	8	2	6		
Fluorine	F	9	2	7		
Neon	Ne	10	2	8		
Sodium	Na	11	2	8	1	
Magnesium	Mg	12	2	8	2	
Aluminum	Al	13	2	8	3	
Silicon	Si	14	2	8	4	
Phosphorus	P	15	2	8	5	
Sulfur	S	16	2	8	6	
Chlorine	Cl	17	2	8	7	
Argon	Ar	18	2	8	8	
Potassium	K	19	2	8	8	1
Calcium	Ca	20	2	8	8	2

mines how many electrons an atom has, because the positively charged protons attract and hold negatively charged electrons in their orbits.

The number of protons in its nucleus determines an atom's identity and its **atomic number.** Hydrogen (H), for instance, has 1 proton in its nucleus and thus has an atomic number of 1. The nuclei of helium (He) atoms possess 2 protons, whereas those of carbon (C) have 6, and uranium (U) have 92, so their atomic numbers are 2, 6, and 92, respectively. Atoms are also characterized by their **atomic mass number,** which is the sum of protons and neutrons in the nucleus (electrons contribute negligible mass to atoms). However, atoms of the same chemical element might have different atomic mass numbers because the number of neutrons can vary. All carbon (C) atoms have 6 protons—otherwise they would not be carbon—but the number of neutrons can be 12, 13, or 14. Thus, we recognize three types of carbon, or what are known as *isotopes,* each with a different atomic mass number (Figure 2.5).

These isotopes of carbon, or those of any other element, behave the same chemically—carbon 12 and carbon 14 are both present in carbon dioxide (CO_2), for example. However, some isotopes are radioactive, meaning that they spontaneously decay or change to other stable elements. Carbon 14 is radioactive, whereas both carbon 12 and 13 are

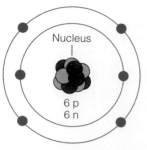

^{12}C (Carbon 12)

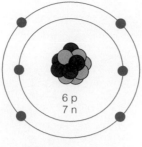

^{13}C (Carbon 13)

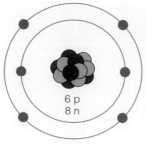

^{14}C (Carbon 14)

Figure 2.5 *Schematic representation of the isotopes of carbon. Carbon has an atomic number of 6 and an atomic mass number of 12, 13, or 14 depending on the number of neutrons in its nucleus.*

stable. Radioactive isotopes are important in determining the ages of rocks (see Chapter 17).

Bonding and Compounds

The interaction of electrons around atoms can result in two or more atoms joining together, a process known as **bonding.** If atoms of two or more different elements bond, the resulting substance is a **compound.** Gaseous oxygen consists only of oxygen atoms and is thus an element, whereas the mineral quartz, consisting of silicon and oxygen atoms, is a compound. Most minerals are compounds, although gold, silver, and several others are important exceptions.

To understand bonding, you must delve deeper into the structure of atoms. Remember that negatively charged electrons orbit the nuclei of atoms in electron shells. With the exception of hydrogen, which has only one proton and one electron, the innermost electron shell of an atom contains no more than two electrons. The other shells contain various numbers of electrons, but the outermost shell never contains more than eight (Table 2.3). The electrons in the outermost shell are those that are usually involved in chemical bonding.

Two types of chemical bonds are particularly important in minerals, *ionic* and *covalent*, and many minerals contain both types of bonds. Two other types of chemical bonds, *metallic* and *van der Waals*, are much less common, but are extremely important in determining the properties of some useful minerals.

Ionic Bonding

Notice in Table 2.3 that most atoms have fewer than eight electrons in their outermost electron shell. However, some elements, including neon and argon, have complete outer shells containing eight electrons. These elements, because of this electron configuration, are known as the *noble gases;* they do not react readily with other elements to form compounds. Interactions among atoms tend to produce electron configurations similar to those of the noble gases. That is, atoms interact so that their outermost

electron shell is filled with eight electrons, unless the first shell (with two electrons) is also the outermost electron shell, as in helium.

One way to attain the noble gas configuration is by the transfer of one or more electrons from one atom to another. Common salt is composed of the elements sodium (Na) and chlorine (Cl), each of which is poisonous, but when combined chemically, they form the compound sodium chloride (NaCl), better known as the mineral halite. Notice in Figure 2.6a that sodium has 11 protons and 11 electrons; thus, the positive electrical charges of the protons are exactly balanced by the negative charges of the electrons, and the atom is electrically neutral. Likewise, chlorine with 17 protons and 17 electrons is electrically neutral (Figure 2.6a), but neither sodium nor chlorine has eight electrons in its outermost electron shell; sodium has only one whereas chlorine has seven. In order to attain a stable configuration, sodium loses the electron in its outermost electron shell, leaving its next shell with eight electrons as the outermost one (Figure 2.6a). However, sodium now has one fewer electron (negative charge) than it has protons (positive charge) so it is an electrically charged particle. Such a particle is an **ion** and, in the case of sodium, is symbolized Na^{+1}.

The electron lost by sodium is transferred to the outermost electron shell of chlorine, which had seven electrons to begin with. Thus, the addition of one more electron gives chlorine an outermost electron shell of eight electrons, the configuration of a noble gas. Its total number of electrons, though, is now 18, which exceeds by one the number of protons. Accordingly, chlorine also becomes an ion, but it is negatively charged (Cl^{-1}). An **ionic bond** forms between sodium and chlorine because of the attractive force between the positively charged sodium ion and the negatively charged chlorine ion (Figure 2.6a).

In ionic compounds, such as sodium chloride (the mineral halite), the ions are arranged in a three-dimensional framework that results in overall electrical neutrality. In halite, sodium ions bond to chlorine ions on all sides, and chlorine ions are surrounded by sodium ions (Figure 2.6b).

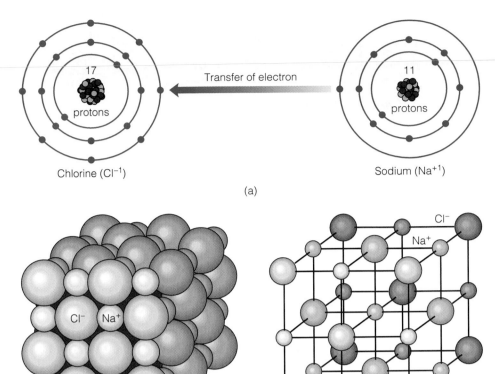

Figure 2.6 *(a) Ionic bonding. The electron in the outermost shell of sodium is transferred to the outermost electron shell of chlorine. Once the transfer has occurred, sodium and chlorine are positively and negatively charged ions, respectively. (b) The crystal structure of sodium chloride, the mineral halite. The diagram on the left shows the relative sizes of sodium and chlorine ions, and the diagram on the right shows the locations of the ions in the crystal structure.*

Covalent Bonding

Covalent bonds form between atoms when their electron shells overlap and electrons are shared. For example, atoms of the same element, such as oxygen in oxygen gas, cannot bond by transferring electrons from one atom to another. And carbon (C), which forms the minerals graphite and diamond, has four electrons in its outermost electron shell (Figure 2.7a). If these four electrons were transferred to another carbon atom, the atom receiving the electrons would have the noble gas configuration of eight electrons in its outermost electron shell, but the atom contributing the electrons would not.

In such situations, adjacent atoms share electrons by overlapping their electron shells. For instance, a carbon atom in diamond shares all four of its outermost electrons with a neighbor to produce a stable noble gas configuration (Figure 2.7a).

Covalent bonds are not restricted to substances composed of atoms of a single kind. Among the most common minerals, the silicates (discussed later in this chapter), the element silicon forms partly covalent and partly ionic bonds with oxygen.

Metallic and van der Waals Bonds

Metallic bonding results from an extreme type of electron sharing. The electrons of the outermost electron shell of metals such as gold, silver, and copper are readily lost and move about from one atom to another. This electron mobility accounts for metals' metallic luster (their appearance in reflected light), their good electrical and thermal conductivity, and their malleability. Only a few minerals possess metallic bonds, but those that do are very useful; copper, for example, is used for electrical wiring because of its high electrical conductivity.

Some electrically neutral atoms and molecules* have no electrons available for ionic, covalent, or metallic bonding. They nevertheless have a weak attractive force known as *van der Waals* or *residual bond* between them when in proximity. The carbon atoms in the mineral graphite covalently bond to form sheets, but the sheets are weakly held together by van der Waals bonds (Figure 2.7b). This type of bonding in graphite makes it useful for pencil leads; when a pencil is moved across a piece of paper, small pieces of graphite flake off along the planes held together by van der Waals bonds and adhere to the paper.

---

*A molecule is the smallest unit of a substance having the properties of that substance. A water molecule (H_2O), for example, possesses two hydrogen atoms and one oxygen atom.

Matter—What Is It and What Is It Composed Of?

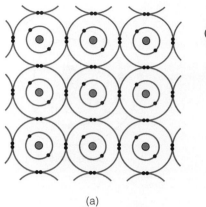 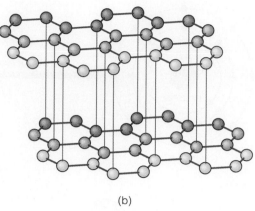

(a)   (b)

Figure 2.7 *(a) Covalent bonds formed by adjacent atoms sharing electrons in diamond. (b) Covalent bonding also occurs in graphite, but here the carbon atoms are bonded together to form sheets that are held to one another by van der Waals bonds. The sheets themselves are strong, but the bonds between sheets are weak.*

What Are Minerals?

In the preceding section, Bonding and Compounds, we noted that minerals are made up of atoms bonded together. In most cases, two or more elements are involved, thus forming compounds; but some minerals, known as *native elements,* consist of bonded atoms of only one element. As we turn to a more detailed discussion of minerals, let us recall our definition: A mineral is a naturally occurring, crystalline solid, with a narrowly defined chemical composition and characteristic physical properties. The next sections examine each part of this definition.

Naturally Occurring Inorganic Substances

The criterion, "naturally occurring," excludes from minerals all substances manufactured by humans. Accord-

ingly, most geologists do not regard synthetic diamonds and rubies and a number of other artificially synthesized substances as minerals.

Some geologists think the term *inorganic* in the mineral definition is superfluous. It does remind us that animal matter and vegetable matter are not minerals. Nevertheless, some organisms, such as corals and clams, construct their shells of the compound calcium carbonate ($CaCO_3$), which is either aragonite or calcite, both minerals.

Mineral Crystals

By definition minerals are **crystalline solids** in which the constituent atoms are arranged in a regular, three-dimensional framework (Figure 2.6b). Under ideal conditions, such as in a cavity, mineral crystals can grow and form perfect crystals that possess planar surfaces (crystal faces), sharp corners, and straight edges (Figure 2.8). In other words, the regular geometric shape of a well-formed mineral crystal is the exterior manifestation of an ordered internal atomic arrangement. Not all rigid substances are crystalline solids; natural and manufactured glass both lack the ordered arrangement of atoms and are said to be *amorphous,* meaning without form.

As early as 1669, a well-known Danish scientist, Nicholas Steno, determined that the angles of intersection of equivalent crystal faces on different specimens of quartz are identical. Since then this *constancy of interfacial angles* has been demonstrated for many other minerals, regardless of their size, shape, age, or geographic occurrence (Figure 2.9). Steno postulated that mineral crystals are composed of small, identical building blocks and that the arrangement of these blocks determines the external form of the crystals. Such regularity of the external form

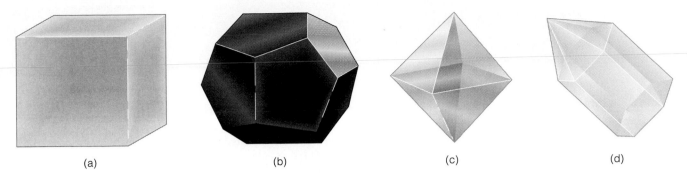

Figure 2.8 *Mineral crystals occur in a variety of shapes, several of which are shown here. (a) Cubic crystals typically develop in the minerals halite, galena, and pyrite. (b) Dodecahedron crystals such as those of garnet have 12 sides. (c) Diamond has octahedral or 8-sided crystals. (d) A prism terminated by pyramids is found in quartz.*

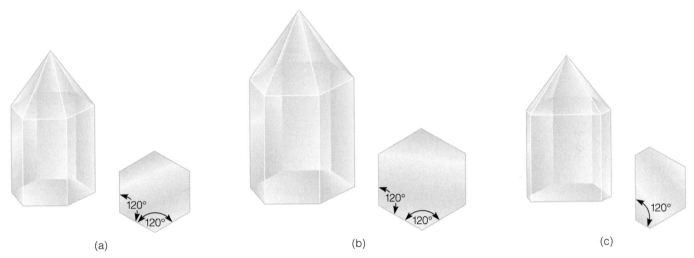

Figure 2.9 *Side views and cross sections of three quartz crystals showing the constancy of interfacial angles: (a) a well-shaped crystal; (b) a larger crystal; and (c) a poorly shaped crystal. The angles formed between equivalent crystal faces on different specimens of the same mineral are the same regardless of the size or shape of the specimens.*

of minerals must surely mean that external crystal form is controlled by internal structure.

Crystalline structure can be demonstrated even in minerals lacking obvious crystals. For example, many minerals possess a property called *cleavage*, meaning that they break or split along closely spaced, smooth planes. The fact that these minerals split along such smooth planar surfaces indicates that the mineral's internal structure controls such breakage. The behavior of light and X-ray beams transmitted through minerals also provides compelling evidence for an orderly arrangement of atoms within minerals.

Chemical Composition of Minerals

Mineral composition is generally shown by a chemical formula, which is a shorthand way of indicating the num-

bers of atoms of different elements composing a mineral. The mineral quartz consists of one silicon (Si) atom for every two oxygen (O) atoms, and thus has the formula SiO_2; the subscript number indicates the number of atoms. Orthoclase is composed of one potassium, one aluminum, three silicon, and eight oxygen atoms so its formula is $KAlSi_3O_8$. Some minerals, known as **native elements,** are composed of a single element; they include gold (Au), silver (Ag), platinum (Pt) (see Perspective 2.1), and graphite and diamond, both of which are composed of carbon (C).

The definition of a mineral contains the phrase "a narrowly defined chemical composition," because some minerals actually have a range of compositions. For many minerals the chemical composition is constant: Quartz is always composed of silicon and oxygen (SiO_2), and halite contains only sodium and chlorine (NaCl). Other minerals have a range of compositions because one element may

As one would expect from their name, the precious metals are desirable for some reason. The best known is gold, which is valued for its pleasing appearance, the ease with which it can be reshaped, and its durability and scarcity. Silver and platinum-group metals are also included among the precious metals, and these too are sought out and used in several ways, including a number of industrial applications. Gold and platinum rarely combine with other elements, so they are usually found as native elements. Silver, however, occurs as either a native element or a compound such as the mineral argentite (Ag_2S).

Among the hundreds of minerals used by humans, gold is certainly one of the most highly prized and sought after. This deep yellow mineral (Figure 1a) has been mined for at least 6000 years, and archaeological evidence indicates that people in Spain possessed small quantities of gold 40,000 years ago. It has been the cause of feuds and wars and one incentive for European exploration of the Americas, especially Central and South America. Gold is too heavy for use in tools or weapons and too soft to hold a cutting edge, so it has been prized for jewelry, for ritual objects, as a symbol of wealth, and as a monetary standard. Much of the gold used in North America still goes into jewelry, but it is also used in the chemical industry, gold plating, electrical circuitry, dentistry, and glass making. South Africa was the 1998 world leader in gold production, with the United States a distant second, followed by Australia, China, and Canada. Most U.S. production comes from mines in Nevada, California, South Dakota, and Alaska; mines in Ontario, followed by Quebec and Alberta, account for most Canadian production.

In the United States, gold was first mined profitably in North Carolina in 1801 and in Georgia in 1829, but the truly spectacular discovery was made in California in 1848, which culminated in the gold rush of 1849. Thousands of people flocked to California to find riches, but only a few found what they sought. Nevertheless, more than $200 million in gold was mined during the five years of the gold rush proper (1849–1853). Another gold rush took place in 1876 when Lt. Col. George Armstrong Custer reported that "gold in satisfactory quantities can be obtained in the Black Hills" of South Dakota. The flood of miners into the Black Hills, the Holy Wilderness of the Sioux Indians, resulted in the Indian War, during which Custer and some 260 of his men were annihilated at the Battle of the Little Bighorn in Montana in June 1876.

Canada, too, has had its gold rushes. The first discovery came in 1850 in the Queen Charlotte Islands along the Pacific Coast, and by 1858 about 10,000 people were panning for gold there. However, the greatest Canadian gold rush was between 1897 and 1899 when about 35,000 people traveled to the remote, hostile Klondike region of the Yukon Territory. Dawson City grew so rapidly that during the winter of 1897 hundreds of people had to be evacuated because of food shortages. As in other gold rushes, merchants were more successful than miners, most of whom barely eked out a living.

Certainly the lure of gold is not new. Spanish explorers during the 1500s and 1600s believed stories of a fabled city of gold in South America known as El Dorado, "the Gilded One," and the explorer Francisco Vasquez de Coronado (1510–1554?) went as far north as Kansas in quest of Gran Quivera, where people reportedly ate from plates made of gold. In the southwestern United States one still hears rumors of lost gold mines in the Superstition Mountains. Probably the most famous is the *Lost Dutchman's Mine*, which was actually "discovered" by a German immigrant, Jacob Waltz, between 1863 and 1886. Waltz did find gold, but he died in 1891 without revealing where it came from. Whether the legend of the Lost Dutchman's Mine is fantasy or based on fact remains unresolved, but rumors of its existence have enticed many to search for it.

The photography industry uses most of the silver (Figure 1b) consumed in North America for silver halide film, a market that will surely diminish as digital imaging becomes more common.

Figure 1 *(a) Specimen of gold from Grass Valley, California. (b) Silver. (c) Platinum.*

(a)

(b)

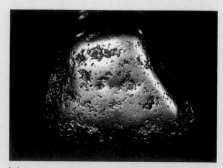

(c)

(a)

(b)

Figure 2 *(a)The Yellowjacket Mine in the Comstock Lode at Gold Hill, Nevada. The tall, wooden structure is the mine's headframe where equipment and miners were lowered into a mine shaft. (b) Hydrothermal (hot water) solutions invaded fractured igneous rocks (gray) where silver- and gold-bearing quartz (milky white) formed. Even though the Comstock Lode was a remarkably rich deposit, ore was found at fewer than 100 of about 16,000 mining claims made there.*

Silver also has uses in jewelry, flatware, surgical instruments, and backing for mirrors. The United States, second only to Mexico in silver production, must nevertheless import more than half the silver it needs, most from Canada where British Columbia is the largest producer. Most silver in the United States is mined in Nevada followed by Alaska, Arizona, and Idaho.

Probably the most famous area of silver mining in North America is the now defunct Comstock Lode at Virginia City, Gold Hill, and Silver City, Nevada, which was active from 1859 to 1898 (Figure 2). Of these three communities, Virginia City is best known: It had a population of more than 30,000 at its peak, but its economy is now based entirely on tourism. One of its better-known residents, at least for a time, was the author Samuel Clemens, better known as Mark Twain.

The Comstock Lode was discovered in 1857 when miners Pat McLaughlin and Peter O'Reilly found gold. But the mine was named for Henry Comstock, who convinced McLaughlin and O'Reilly that the gold was on his property. In any case, the deposit yielded more than $300 million during the first 20 years of mining. It was the richest silver deposit ever discovered on this continent and was largely responsible for bringing Nevada into the Union in 1864, even though this area had too few people to qualify for statehood. This fact was overlooked because silver and gold from the Comstock Lode helped maintain the U.S. economy during the Civil War (1861–1865).

Silver continues to be an important mineral commodity, although its value is considerably less than that of gold; in March 2000, gold was selling for $288 per ounce, whereas silver commanded a price of only $5.11 per ounce. Silver production from U.S. mines in 1998 was valued at $338 million as opposed to the $4 billion in gold mined in the same year.

The platinum-group metals include *platinum, palladium, rhodium, ruthenium, iridium,* and *osmium* (Figure 1c). However, only palladium and platinum are used in comparatively large quantities, although the others have various uses as well. In the United States, only one mine in Montana produced platinum-group metals during 1998, but some metals were derived as by-products of copper refining in Texas and Utah. The world's largest producer by far is South Africa, with Russia a distant second, followed by the United States and Canada.

The automotive industry is the largest user of platinum-group metals, as oxidation catalysts in catalytic converters. These metals are also used in the chemical industry and for cancer chemotherapy; some platinum alloys are used for jewelry. It is interesting to note that many scientists think a thin iridium-rich layer at the boundary between the Cretaceous and Tertiary periods resulted from a meteorite impact that was also responsible for dinosaur extinctions.

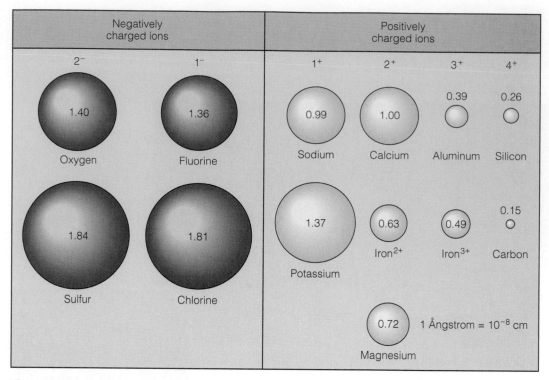

Figure 2.10 *Electrical charges and relative sizes of ions common in minerals. The numbers within the ions are the radii shown in Ångstrom units.*

substitute for another if the atoms of two or more elements are nearly the same size and the same charge. Notice in Figure 2.10 that iron and magnesium atoms are about the same size and electrical charge, and therefore they can substitute for one another. The chemical formula for the mineral olivine is $(Mg,Fe)_2SiO_4$, meaning that, in addition to silicon and oxygen, it may contain only magnesium, only iron, or a combination of both. As a matter of fact, the term *olivine* is usually applied to minerals containing both iron and magnesium, whereas forsterite is olivine with only magnesium (Mg_2SiO_4), and olivine with only iron is fayalite (Fe_2SiO_4). A number of other minerals also have ranges of compositions, so these are actually mineral groups with several members.

Physical Properties of Minerals

The last criterion in our definition of a mineral, "characteristic physical properties," refers to such properties as hardness, color, and crystal form. Composition and structure control these properties. We shall have more to say about physical properties of minerals later in this chapter.

How Many Minerals Are Known?

More than 3500 minerals have been identified and described, but only a few—perhaps two dozen—are particularly common. As we previously mentioned, 92 naturally occurring elements have been discovered. One might think that an extremely large number of minerals could be formed from so many elements, but several factors limit the number of possible minerals. For one thing, many combinations of elements are chemically impossible; no compounds are composed of only potassium and sodium or of silicon and iron, for example. Another important factor restricting the number of common minerals is that only eight chemical elements make up the bulk of Earth's crust (Table 2.4). Oxygen and silicon constitute more than 74% (by weight) of Earth's crust and nearly 84% of the atoms available to form compounds. By far the most common minerals in the crust consist of silicon and oxygen, combined with one or more of the other elements in Table 2.4.

Table 2.4

Common Elements in Earth's Crust

Element	Symbol	Percentage of Crust (by weight)	Percentage of Crust (by atoms)
Oxygen	O	46.6%	62.6%
Silicon	Si	27.7	21.2
Aluminum	Al	8.1	6.5
Iron	Fe	5.0	1.9
Calcium	Ca	3.6	1.9
Sodium	Na	2.8	2.6
Potassium	K	2.6	1.4
Magnesium	Mg	2.1	1.8
All others		1.5	0.1

Table 2.5

Some of the Mineral Groups Recognized by Geologists

Mineral Group	Negatively Charged Ion or Ion Group	Examples	Composition
Carbonate	$(CO_3)^{-2}$	Calcite	$CaCO_3$
		Dolomite	$CaMg(CO_3)_2$
Halide	Cl^{-1}, F^{-1}	Halite	$NaCl$
		Fluorite	CaF_2
Native element	—	Gold	Au
		Silver	Ag
		Diamond	C
		Graphite	C
Oxide	O^{-2}	Hematite	Fe_2O_3
		Magnetite	Fe_3O_4
Silicate	$(SiO_4)^{-4}$	Quartz	SiO_2
		Potassium feldspar	$KAlSi_3O_8$
		Olivine	$(Mg,Fe)_2SiO_4$
Sulfate	$(SO_4)^{-2}$	Anhydrite	$CaSO_4$
		Gypsum	$CaSO_4 \cdot 2H_2O$
Sulfide	S^{-2}	Galena	PbS
		Pyrite	FeS_2

Mineral Groups Recognized by Geologists

Geologists recognize mineral classes or groups, each with members sharing the same negatively charged ion or ion group (Table 2.5). We mentioned in a previous section that ions are atoms having either a positive or negative electrical charge resulting from the loss or gain of electrons in their outermost shell. In addition to ions, some minerals contain tightly bonded, complex groups of different atoms known as *radicals* that act as single units within minerals. A good example is the carbonate radical consisting of a carbon atom bonded to three oxygen atoms, thus having the formula CO_3 and a -2 electrical charge. Other common radicals and their charges are sulfate (SO_4, -2), hydroxyl (OH_2, -1), and silicate (SiO_2, -4).

The Silicate Minerals

Because silicon and oxygen are the two most abundant elements in the crust, it is not surprising that many minerals contain these elements. A combination of silicon and oxygen is known as **silica,** and the minerals containing silica are **silicates.** Quartz (SiO_2) is pure silica because it is composed entirely of silicon and oxygen. But most silicates have one or more additional elements, as in orthoclase ($KAlSi_3O_8$) and olivine [$(Mg,Fe)_2SiO_4$]. Silicate minerals include about one-third of all known minerals, but their abundance is even more impressive when one considers that they make up perhaps 95% of Earth's crust.

The basic building block of all silicate minerals is the **silica tetrahedron,** which consists of one silicon atom and four oxygen atoms (Figure 2.11). These atoms are arranged so that the four oxygen atoms surround a silicon atom, which occupies the space between the oxygen atoms; thus, a four-faced pyramidal structure is formed. The silicon atom has a positive charge of 4, and each of the four oxygen atoms has a negative charge of 2, resulting in a radical with a total negative charge of 4, $(SiO_4)^{-4}$.

Because the silica tetrahedron has a negative charge, it does not exist in nature as an isolated radical; rather it combines with positively charged ions or shares its oxygen atoms with other silica tetrahedra. In the simplest silicate minerals, the silica tetrahedra exist as single units bonded to positively charged ions. In minerals containing isolated tetrahedra, the silicon to oxygen ratio is 1:4, and the negative charge of the silica ion is balanced by positive ions (Figure 2.12a). Olivine [$(Mg,Fe)_2SiO_4$], for example, has either two magnesium (Mg^{+2}) ions, two iron (Fe^{+2}) ions, or one of each to offset the -4 charge of the silica ion.

Silica tetrahedra may also be arranged so that they join together to form chains of indefinite length (Figure 2.12b). Single chains, as in the pyroxene minerals, form when each

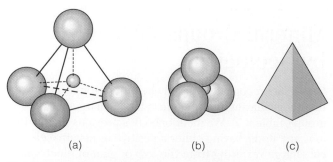

Figure 2.11 *The silica tetrahedron. (a) Expanded view showing oxygen atoms at the corners of a tetrahedron and a small silicon atom at the center. (b) View of the silica tetrahedron as it really exists with the oxygen atoms touching. (c) The silica tetrahedron represented diagrammatically; the oxygen atoms are at the four points of the tetrahedron.*

tetrahedron shares two of its oxygens with adjacent tetrahedra; the result is a silicon to oxygen ratio of 1:3. Enstatite, a pyroxene group mineral, reflects this ratio in its chemical formula, $MgSiO_3$. Individual chains, however, possess a net -2 electrical charge, so they are balanced by positive ions, such as Mg^{+2}, that link parallel chains together (Figure 2.12b).

The amphibole group of minerals is characterized by a double-chain structure in which alternate tetrahedra in two parallel rows are cross-linked (Figure 2.12b). The formation of double chains results in a silicon to oxygen ratio of 4:11, so that each double chain possesses a -6 electrical charge. Mg^{+2}, Fe^{+2}, and Al^{+2} are usually involved in linking the double chains together.

In sheet structure silicates, three oxygens of each tetrahedron are shared by adjacent tetrahedra (Figure 2.12c). Such structures result in continuous sheets of silica tetrahedra with silicon to oxygen ratios of 2:5. Continuous sheets also possess a negative electrical charge satisfied by positive ions located between the sheets. This particular structure accounts for the characteristic sheet structure of the *micas,* such as biotite and muscovite, and the *clay minerals.*

Three-dimensional frameworks of silica tetrahedra form when all four oxygens of the silica tetrahedron are shared by adjacent tetrahedra (Figure 2.12d). Such sharing of oxygen atoms results in a silicon to oxygen ratio of 1:2, which is electrically neutral. Quartz is a common framework silicate (see Perspective 2.2).

Two subgroups of silicates are recognized: ferromagnesian and nonferromagnesian silicates. The

			Formula of negatively charged ion group	Silicon to oxygen ratio	Example
(a)	Isolated tetrahedra		$(SiO_4)^{-4}$	1:4	Olivine
(b)	Continuous chains of tetrahedra	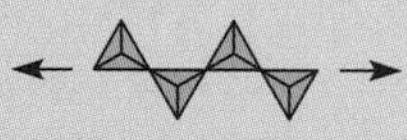 Single chain	$(SiO_3)^{-2}$	1:3	Pyroxene group
		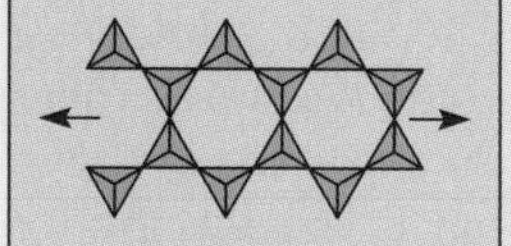 Double chain	$(Si_4O_{11})^{-6}$	4:11	Amphibole group
(c)	Continuous sheets	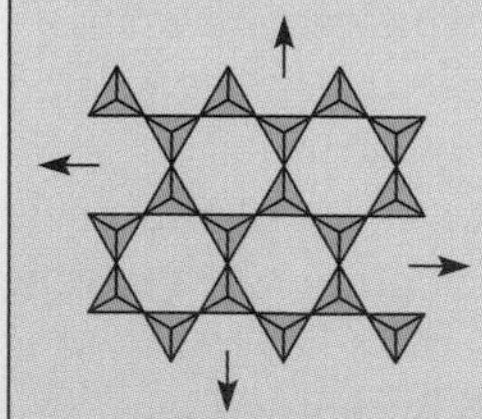	$(Si_4O_{10})^{-4}$	2:5	Micas
(d)	Three-dimensional networks	Too complex to be shown by a simple two-dimensional drawing	$(SiO_2)^0$	1:2	Quartz

Figure 2.12 *Structures of some of the common silicate minerals shown by various arrangements of silica tetrahedra: (a) Isolated tetrahedra. (b) Continuous chains. (c) Continuous sheets. (d) Networks. The arrows adjacent to single-chain, double-chain, and sheet silicates indicate that these structures continue indefinitely in the directions shown.*

ferromagnesian silicates are those containing iron (Fe), magnesium (Mg), or both. These minerals are commonly dark colored and more dense than nonferromagnesian silicates. Some of the common ferromagnesian silicate minerals are olivine, the pyroxenes, the amphiboles, and biotite (Figure 2.13a).

The **nonferromagnesian silicates** lack iron and magnesium, are generally light colored, and are less dense than ferromagnesian silicates (Figure 2.13b). The most common minerals in Earth's crust are nonferromagnesian silicates known as *feldspars*. Feldspar is a general name, however, and two distinct groups are recognized, each of which includes several species. The *potassium feldspars* are represented by microcline and orthoclase ($KAlSi_3O_8$). The second group of feldspars, the *plagioclase feldspars*, range from calcium-rich ($CaAl_2Si_2O_8$) to sodium-rich ($NaAlSi_3O_8$) varieties.

Quartz (SiO_2) is another common nonferromagnesian silicate. It is a framework silicate that can usually be recognized by its glassy appearance and hardness (Figure 2.13b). Another fairly common nonferromagnesian silicate is muscovite, which is a mica (Figure 2.13b).

Carbonate Minerals

Silicates are certainly the most abundant minerals in Earth's crust, but **carbonate minerals** with the negatively charged carbonate radical $(CO_3)^{-2}$ are common at or near the surface. An excellent example is calcium carbonate ($CaCO_3$), the mineral *calcite* which is the main constituent of the sedimentary rock *limestone* (Figure 2.14a). Several other carbonate minerals are known, but only *dolomite* $[CaMg(CO_3)_2]$ need concern us. It forms when calcite is altered by the addition of magnesium, thus forming the sedimentary rock *dolostone* (see Chapter 5).

Other Mineral Groups

In addition to silicates and carbonates, geologists recognize several other mineral groups (Table 2.5). The oxides consist of an element combined with oxygen as in *hematite* (Fe_2O_3). Hematite and another iron oxide known as *magnetite* (Fe_3O_4) are both commonly present in small quantities in a variety of rocks. Rocks containing high concentrations of hematite and magnetite, such as those in the Lake Superior region of Canada and the United States, are important sources of iron ores for the manufacture of steel.

The sulfides have a positively charged ion combined with sulfur (S^{-2}), such as in the mineral galena (PbS), which contains lead (Pb) and sulfur (Figure 2.14b). Sulfates contain an element combined with the sulfate radical $(SO_4)^{-2}$; gypsum ($CaSO_4 \cdot 2H_2O$) is a good example (Figure 2.14c). The halides contain halogen elements such as chlorine (Cl^{-1}) and fluorine (F^{-1}); examples include the minerals halite (NaCl) and fluorite (CaF_2) (Figure 2.14d).

Olivine

Pyroxene

Amphibole

Biotite

Quartz

Potassium feldspar

Plagioclase feldspar

Muscovite

(a)

(b)

Figure 2.13 *(a) Examples of the common ferromagnesian silicates: olivine, the pyroxene group mineral augite, the amphibole group mineral hornblende, and biotite. (b) Examples of the nonferromagnesian silicates: quartz, the potassium feldspar orthoclase, plagioclase feldspar, and muscovite.*

According to one estimate, Earth's crust consists of 51% feldspars, 24% ferromagnesian silicates (olivine, pyroxenes, amphiboles, and biotite), 12% quartz, and 13% other minerals. So quartz is abundant to begin with, but it is much more common than this list indicates. Indeed, quartz is by far the most common mineral in sand on beaches, in stream channels, in desert sand dunes, and on barrier islands adjacent to seashores. Three factors account for this abundance: (1) It is abundant in common rocks such as granite and gneiss, (2) quartz is resistant to mechanical breakdown, and (3) it is chemically stable, meaning that it does not decompose easily (see Chapter 5). In any event, in most people's experience it will be the most commonly encountered mineral.

Given that sand is so common—and anyone who has looked closely at it knows that individual sand grains are not particularly impressive—what is so interesting about quartz? For one thing, it has several practical uses, but its interest lies in its variety—from colorless, well-formed crystals to numerous color and textural types. Some kinds of quartz are used as semiprecious gems in jewelry and as decorative stones. Sand deposits composed mostly of quartz are called *silica sands* and are used in the manufacture of glass, optical equipment, abrasives such as sandpaper, and steel alloys, and for a variety of other industrial uses.

During the Middle Ages (late 5th century A.D. to about 1350), transparent quartz crystals were believed to be ice frozen so solidly that it would not melt (Figure 1a). In fact, the term *crystal* is derived from a Greek word meaning "ice." Even today, crystal refers not only to mineral crystals, but also to clear, colorless glass of high quality such as crystal ware, crystal chandeliers, or the transparent glass or plastic cover of a watch or clock dial. Colorless quartz in particular has been used as a semiprecious stone in jewelry. The term *rhinestone* originally referred to trans-

parent quartz crystals used for jewelry in Germany, and *Herkimer "diamonds"* are simply colorless quartz crystals from Herkimer County, New York. In the past, large, transparent quartz crystals were shaped into spheres for use as fortune-tellers' crystal balls.

Color varieties of quartz include milky quartz, which is commonly found as well-formed crystals. A milky quartz crystal 3.5 m long, 1.7 m in diameter, and weighing 11.8 metric tons was discovered in Siberia. Other color varieties of quartz include amethyst (purple), citrine (yellow to orange), rose quartz (pale pink to deep rose), and smoky quartz (smoky brown to black) (Figure 1). Some types of quartz are referred to as cryptocrystalline, meaning they have such tiny crystals that they can be detected only when highly magnified. These varieties include agate, which shows color banding (Figure 1g), flint (usually brownish, gray, or black), jasper (red, yellow, or brown), and several others.

In addition to the pleasing appearance of some quartz, the mineral has the property of *piezoelectricity* ("pressure" electricity), which enables it to be an accurate timekeeper. When pressure is applied to a quartz crystal, an electric current is generated. If an electric current is applied to a quartz crystal, as by a watch's battery, the crystal expands and contracts extremely rapidly and regularly (about 100,000 times per second). Quartz crystal clocks were first developed in 1928, and now quartz watches and clocks are commonplace. Even inexpensive quartz timepieces are very accurate, and precision-manufactured quartz clocks used in astronomical observations do not gain or lose more than 1 second in 10 years.

An interesting historical note is that during World War II (1939–1945) the United States had difficulty obtaining Brazilian quartz crystals needed for making radios. This shortage prompted the development of artificially synthesized quartz, and now most quartz used in watches and clocks is synthetic.

How Are Minerals Identified?

All minerals possess characteristic physical properties determined by their internal structure and chemical composition. Many physical properties are remarkably constant for a given mineral species, but some, especially color, may vary. Though a professional geologist may use sophisticated techniques in studying and identifying minerals, most common minerals can be identified by using the following physical properties (see Appendix C).

Color and Luster

For some minerals, especially those that have the appearance of metals, color is rather consistent, but for many others it varies because of minute amounts of impurities. Although the color of many minerals varies, some generalizations can be made. Ferromagnesian silicates are typically black, brown, or dark green, although olivine is olive green (Figure 2.13a). Nonferromagnesian silicates, on the other hand, can vary considerably in color, but are only rarely dark (Figure 2.13b).

(a) Colorless quartz

Figure 1 *Several varieties of quartz: (a) colorless quartz crystals; (b) smoky quartz; (c) amethyst; (d) citrine; (e) milky quartz; (f) rose quartz; (g) agate.*

(b) Smoky quartz

(c) Amethyst

(d) Citrine

(e) Milky quartz

(f) Rose quartz

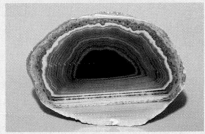

(g) Agate

Luster (not to be confused with color) is the appearance of a mineral in reflected light. Two major types of luster are recognized: *metallic* and *nonmetallic* (Figure 2.15). They are distinguished by the quality of light reflected from a mineral; the observer determines whether it has the appearance of a metal or a nonmetal. Several types of nonmetallic luster are recognized, including glassy or vitreous, greasy, waxy, brilliant (as in diamond), and dull or earthy.

Crystal Form

Many minerals do not show the perfect crystal form typical of that mineral species because they commonly form under less than ideal conditions. Nevertheless, some minerals are typically found as crystals, such as 12-sided crystals of garnet, and 6- and 12-sided crystals of pyrite (Figures 2.8 and 2.16). Minerals that grow in cavities or are precipitated from circulating hot water (hydrothermal solutions) in cracks and crevices in rocks also commonly occur as well-formed crystals.

(a) Calcite

(b) Galena

(c) Gypsum

(d) Halite

Figure 2.14 *(a) Calcite ($CaCO_3$) is the most common carbonate mineral. (b) The sulfide mineral galena (PbS) is the ore of lead. (c) Gypsum ($CaSO_4 \cdot 2H_2O$) is a common sulfate mineral. (d) Halite (NaCl) is a good example of a halide mineral.*

Crystal form can be a useful characteristic for mineral identification, but a number of minerals have the same crystal form. Pyrite (FeS_2), galena (PbS), and halite (NaCl) all form cubic crystals, but they can easily be identified by other properties such as color, luster, hardness, and density. In other words, several mineral properties are usually used for purposes of identification.

(a)

(b)

Figure 2.15 *Hematite (a) has the appearance of a metal and is said to have a metallic luster, whereas orthoclase (b) has a nonmetallic luster.*

Cleavage and Fracture

Not all minerals possess **cleavage,** but those that do tend to break, or split, along a smooth plane or planes of weakness determined by the strength of the bonds within individual mineral crystals. Cleavage is characterized in terms of quality (perfect, good, poor), direction, and angles of intersection of cleavage planes. Biotite, a common ferromagnesian silicate, has perfect cleavage in one direction (Figure 2.17a). The fact that biotite preferentially cleaves along a number of closely spaced, parallel planes is related to its structure; it is a sheet silicate with the sheets of silica tetrahedra weakly bonded to one another by iron and magnesium ions (Figure 2.12c).

Feldspars possess two directions of cleavage that intersect at right angles (Figure 2.17b), and the mineral halite has three directions of cleavage, all of which intersect at right angles (Figure 2.17c). Calcite also possesses three directions of cleavage, but none of the intersection angles is a right angle, so cleavage fragments of calcite are rhombohedrons (Figure 2.17a and Figure 2.17d). Minerals with four directions of cleavage include fluorite and

Figure 2.16 *Mineral crystals. (a) Cubic crystals of fluorite (CaF₂). (b) Calcite (CaCO₃) crystal. (c) Blade-shaped crystals of barite (BaSO₄).*

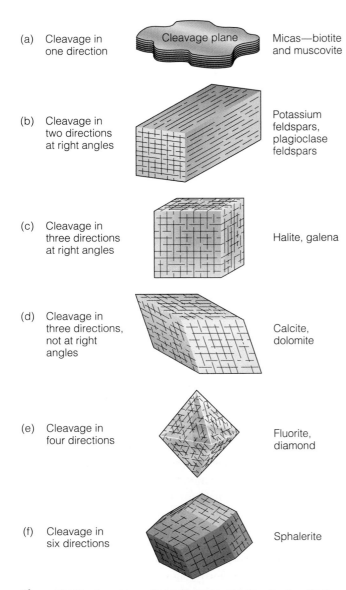

(a)	Cleavage in one direction	Micas—biotite and muscovite
(b)	Cleavage in two directions at right angles	Potassium feldspars, plagioclase feldspars
(c)	Cleavage in three directions at right angles	Halite, galena
(d)	Cleavage in three directions, not at right angles	Calcite, dolomite
(e)	Cleavage in four directions	Fluorite, diamond
(f)	Cleavage in six directions	Sphalerite

Figure 2.17 *Several types of mineral cleavage. (a) One direction. (b) Two directions at right angles. (c) Three directions at right angles. (d) Three directions, not at right angles. (e) Four directions. (f) Six directions.*

diamond (Figure 2.17e). Ironically, diamond, the hardest mineral, can be easily cleaved (see the Prologue). A few minerals such as sphalerite, an ore of zinc, have six directions of cleavage (Figure 2.17f).

Cleavage is an important diagnostic property of minerals, and its recognition is essential in distinguishing between some minerals. For instance, the pyroxene mineral augite and the amphibole mineral hornblende look much alike: both are generally dark green to black, have the same hardness, and possess two directions of cleavage. However, the cleavage planes of augite intersect at about 90°, whereas the cleavage planes of hornblende intersect at angles of 56° and 124° (Figure 2.18). In contrast to cleavage, *fracture* is mineral breakage along irregular surfaces. Any mineral will fracture if enough force is applied, but the fracture surfaces will not be smooth.

Hardness

An Austrian geologist, Friedrich Mohs, devised a relative hardness scale for 10 minerals. He arbitrarily assigned a hardness value of 10 to diamond, the hardest mineral known, and lesser values to the other minerals. Relative hardness can be determined easily by the use of Mohs hardness scale (Table 2.6). Quartz will scratch fluorite but cannot be scratched by fluorite, gypsum can be scratched by a fingernail, and so on. So *hardness* is defined as a mineral's resistance to abrasion. Hardness is controlled mostly by internal structure. For example, both graphite and diamond are composed of carbon, but the former has a hardness of 1 to 2, whereas the latter has a hardness of 10.

Specific Gravity

The *specific gravity* of a mineral is the ratio of its weight to the weight of an equal volume of water. A mineral with a

How Are Minerals Identified? **45**

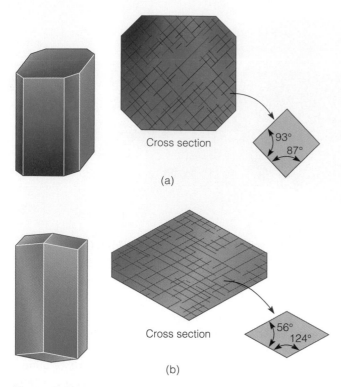

Cross section

(a)

Cross section

(b)

Figure 2.18 *Cleavage in augite and hornblende. (a) Augite crystal and cross section of crystal showing cleavage. (b) Hornblende crystal and cross section of crystal showing cleavage.*

Table 2.6

Mohs Hardness Scale

Hardness	Mineral	Hardness of Some Common Objects
10	Diamond	
9	Corundum	
8	Topaz	
7	Quartz	
		Steel file (6 1/2)
6	Orthoclase	
		Glass (5 1/2–6)
5	Apatite	
4	Fluorite	
3	Calcite	
		Copper penny (3)
		Fingernail (2 1/2)
2	Gypsum	
1	Talc	

specific gravity of 3.0 is three times as heavy as water. Like all ratios, specific gravity is not expressed in units such as grams per cubic centimeter—it is a dimensionless number.

Specific gravity varies in minerals depending upon their composition and structure. Among the common silicates, for example, the ferromagnesian silicates have specific gravities ranging from 2.7 to 4.3, whereas the nonferromagnesian silicates vary from 2.6 to 2.9. Obviously, the ranges of values overlap somewhat, but for the most part ferromagnesian silicates have greater specific gravities than nonferromagnesian silicates. In general, the specific gravity of metallic minerals, such as galena (7.58) and hematite (5.26), is greater than that of nonmetals. Pure gold has a specific gravity of 19.3, making it about 2 1/2 times more dense than lead. Structure as a control of specific gravity is illustrated by the native element carbon (C): the specific gravity of graphite varies from 2.09 to 2.33; that of diamond is 3.5.

Other Useful Mineral Properties

A number of other physical properties characterize some minerals. Talc has a distinctive soapy feel, graphite writes on paper, halite tastes salty, and magnetite is magnetic (Figure 2.19). Calcite possesses the property of *double refraction*, meaning that an object when viewed through a transparent piece of calcite will have a double image (Figure 2.19c). Some minerals are plastic and, when bent into a new shape, will retain that shape, whereas others are flexible and, if bent, will return to their original position when the forces that bent them are removed.

A simple chemical test to identify the minerals calcite and dolomite involves applying a drop of dilute hydrochloric acid to the mineral specimen. If the mineral is calcite, it will react vigorously with the acid and release carbon dioxide, which causes the acid to bubble or effervesce. Dolomite, in contrast, will not react with hydrochloric acid unless it is powdered.

Where and How Do Minerals Originate?

Thus far we have discussed the composition, structure, and physical properties of minerals but have not fully addressed how they originate. A common phenomenon accounting for the origin of some minerals is the cooling of molten rock material known as magma (magma that reaches Earth's surface is called lava). As magma or lava cools, minerals crystallize and grow, thereby determining the mineral composition of various igneous rocks such as basalt (dominated by ferromagnesian silicates) and granite (dominated by nonferromagnesian silicates) (see Chapter 3). Hot water solutions derived from magma commonly invade cracks and crevasses in adjacent rock, and from

(a) Graphite

(b) Magnetite

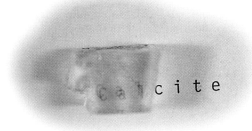

(c) Calcite

Figure 2.19 *Various mineral properties. (a) Graphite, the mineral from which pencil leads are made, writes on paper. (b) Magnetite, an important iron ore, is magnetic. (c) Transparent pieces of calcite show double refraction.*

these solutions several types of minerals might crystallize, some of economic importance. Many well-formed mineral crystals are found in these so-called hydrothermal veins. Minerals also originate when water in hot springs cools (see Chapter 13), and when hot, mineral-rich water discharges onto the seafloor at hot springs known as "black smokers" (see Chapter 8).

Dissolved materials in seawater, more rarely lake water, commonly combine to form minerals such as halite (NaCl), gypsum ($CaSO_4 \cdot 2H_2O$), and others when the water evaporates. Aragonite and/or calcite, both varieties of calcium carbonate ($CaCO_3$), might also form from evaporating water, but most originate when organisms such as clams, oysters, corals, and some floating microorganisms use this compound to construct their shells. And a few plants and animals use silicon dioxide (SiO_2) for their skeletons, which accumulate as mineral deposits on the seafloor when the organisms die (see Chapter 5).

Some clay minerals form when chemical processes alter the composition and structure of other minerals, such as feldspars (see Chapter 5), and others originate when rocks are changed during metamorphism (see Chapter 6). In fact, the agents causing metamorphism—heat, pressure, and chemically active fluids—are responsible for the origin of many minerals. A few minerals even originate from gases when, for example, hydrogen sulfide (H_2S) and sulfur dioxide (SO_2) react at volcanic vents to produce sulfur (Figure 2.20).

Figure 2.20 *Sulfur crystals forming around a gas vent on Kilauea Volcano, Hawaii.*

What Are Rock-Forming Minerals?

The term **rock** has a variety of meanings, but to geologists it refers to a solid aggregate of one or more minerals. Sand on a beach is made up of minerals but is not rock unless the sand grains are compacted and held together by some kind of chemical cement (see Chapter 5). And likewise lava is not rock until it cools and crystallizes, thus forming an interlocking mosaic of mineral crystals (see Chapter 3). Many minerals are present in various kinds of rocks, but only a few varieties are common enough to be designated

as **rock-forming minerals.** Most of the others are present in such small quantities that they can be disregarded in the identification and classification of rocks; these are generally called *accessory minerals.*

Most rocks are composed of silicate minerals, but only a few of the hundreds of known silicates are common in rocks, although many are present as accessories. The common igneous rock basalt, for instance, is made up mostly of ferromagnesian silicates such as pyroxene and olivine, and plagioclase feldspar, a nonferromagnesian silicate. Granite, another igneous rock, is composed predominantly of potassium feldspar and quartz, both nonferromagnesian silicates (Figure 2.21). Both of these rocks contain a variety of accessory minerals, most of which are silicates as well. The minerals just mentioned are also common in metamorphic rocks, and many sedimentary rocks are composed of quartz, feldspars, and various clay minerals. The common rock-forming silicates are summarized in Table 2.7.

(a)

(b)

Figure 2.21 *(a) Granite (top) is made up mostly of potassium feldspar (left) and quartz (middle), and a small amount of biotite (right). (b) Basalt (upper left) contains mostly pyroxene (left) and calcium-rich plagioclase (right), and small amounts of amphibole (center) and other minerals. The minerals are clearly visible in granite but too small to be seen without magnification in basalt (see Chapter 3).*

Table 2.7

Rock-Forming Minerals

Mineral	Composition	Primary Occurrence
Ferromagnesian Silicates		
Olivine	$(Mg,Fe)_2SiO_4$	Igneous, metamorphic rocks
Pyroxene group Augite most common	Ca, Mg, Fe, Al silicate	Igneous, metamorphic rocks
Amphibole group Hornblende most common	Hydrous* Na, Ca, Mg, Fe, Al silicate	Igneous, metamorphic rocks
Biotite	Hydrous K, Mg, Fe silicate	All rock types
Nonferromagnesian Silicates		
Quartz	SiO_2	All rock types
Potassium feldspar group Orthoclase, microcline	$KAlSi_3O_8$	All rock types
Plagioclase feldspar group	Varies from $CaAl_2Si_2O_8$ to $NaAlSi_3O_8$	All rock types
Muscovite	Hydrous K, Al silicate	All rock types
Clay mineral group	Varies	Soils and sedimentary rocks
Carbonates		
Calcite	$CaCO_3$	Sedimentary rocks
Dolomite	$CaMg(CO_3)_2$	Sedimentary rocks
Sulfates		
Anhydrite	$CaSO_4$	Sedimentary rocks
Gypsum	$CaSO_4 \cdot 2H_2O$	Sedimentary rocks
Halides		
Halite	$NaCl$	Sedimentary rocks

*Contains elements of water in some kind of union.

The most common nonsilicate rock-forming minerals are the two carbonates calcite ($CaCO_3$) and dolomite [$CaMg(CO_3)_2$], the primary constituents of the sedimentary rocks limestone and dolostone, respectively. Among the sulfates and halides, gypsum ($CaSO_4 \cdot 2H_2O$) and halite ($NaCl$) are the only rock-forming minerals of any importance.

Mineral Resources and Reserves

As noted previously, the United States and Canada, both highly industrialized nations, enjoy considerable economic success because they have abundant natural resources. But what is a resource? Geologists at the U.S. Geological Survey define a **resource** as "a concentration of naturally occurring solid, liquid, or gaseous material in or on Earth's crust in such form and amount that economic extraction of a commodity from the concentration is currently or poten-

tially feasible." Resources consist of metals or *metallic resources,* sand, gravel, crushed stone, sulfur, salt, and a variety of other *nonmetallic resources,* as well as *energy resources,* such as oil, natural gas, coal, and uranium. An important distinction must be made between a resource, the total amount of a commodity whether discovered or undiscovered, and a **reserve,** that part of the resource base that can be economically extracted. Liquid oil can be extracted from rock known as oil shale, for instance, so it is part of the resource base, but at present it cannot be recovered economically.

What constitutes a resource as opposed to a reserve depends on various factors. Iron-bearing minerals are present in many rocks, but in quantities or ways that make their recovery uneconomical. Geographic location might determine what is considered a reserve. A mineral concentration in a remote region may not be mined because transportation costs are too high, and what might be a resource in the United States and Canada may be mined in a developing nation where labor costs are low.

The market price of a commodity is, of course, important in evaluating a potential resource. From 1935 to 1968, the U.S. government maintained the price of gold at $35 per troy ounce (31.3 g). When this restriction was removed and the price of gold became subject to supply and demand, the price rose, reaching an all-time high of $843 per troy ounce during January 1980. As a result, many marginal deposits became reserves, and many abandoned mines were reopened.

Technological innovations can also change the status of a resource. By the time of World War II (1939–1945), the rich iron ores of the Great Lakes region in the United States and Canada had been depleted. However, the development of a process for separating the iron from unusable rock and shaping it into pellets ideal for use in blast furnaces made it feasible to mine rocks containing less iron (Figure 2.22). Most of the iron ore now mined in North America comes from mines in Newfoundland and Quebec, Canada, and from Minnesota and Michigan. Nearly all of it is shaped into pellets before it is shipped to steel mills.

Most people are aware that numerous mineral commodities are extracted or mined and refined for various uses but have little knowledge about their occurrence, methods of recovery, and economics. Geologists are instrumental in finding recoverable mineral commodities, and engineers and chemists are involved in the extraction and refining processes, but ultimately people with training in business and economics make the decision of whether a commodity is worth extracting. In short, extraction must yield a profit. More than $40 billion in mineral commodities, other than oil, natural gas, and coal, were mined in the United States in 1997, and Canadian mineral production totaled about $17 billion (Canadian dollars).

Everyone knows that metals such as gold, silver, and iron are resources, as are oil and natural gas. However, some quite common minerals are also important to the economies of industrialized nations. Quartz, gypsum, clay minerals, and feldspars are good examples. Pure quartz sand is used for glass and optical instruments as well as sandpaper and steel alloys (see Perspective 2.2). Wallboard, or drywall, is made of gypsum, clay minerals are needed to make ceramics and paper, and feldspars such as orthoclase are used for porcelain, ceramics, enamel, and glass.

Direct access to mineral resources is essential for industrialization and the high standard of living enjoyed by many nations. The United States and Canada are fortunate to have abundant resources, but the amount of resources used in North America since Europeans settled this continent has steadily increased. Each resident of

(a)

Figure 2.22 *(a) Shovels like this one at the Tilden Mine in northern Michigan load trucks with as much as 75 tons of iron ore. The ore is crushed and processed to produce iron pellets (b) measuring about 1 cm for shipment to steel mills.*

(b)

Your country is running short of a natural resource essential to its industrial productivity. To remedy this situation, your choices are to (1) find more deposits of the resource in your country, (2) develop a technology that can use a substitute, or (3) import the commodity from a politically unstable nation. What factors would you evaluate to determine which of these alternatives is the most cost effective? Might factors other than cost be a consideration?

North America now uses about 14 metric tons of resources per year, a large part of which is bulk items such as sand and gravel, cement, and crushed stone. It is no exaggeration to say that industrialized societies are totally dependent on mineral resources, but, unfortunately, most resources are used much faster than they form. Thus, mineral resources are *nonrenewable*, meaning that once a reserve has been depleted, new deposits or suitable substitutes, if available, must be found.

Adequate supplies of some mineral resources are available for the indefinite future (sand and gravel and crushed stone, for example), but many others are either limited or must be imported. For some essential metals, the United States depends totally on imports. No cobalt or nickel was mined in this country during 1998, and all manganese, an essential element in manufacturing steel, is imported from Gabon, Australia, Mexico, and Brazil. More than half the crude oil used in the United States is imported, much from the Middle East where more than 50% of the proven reserves exist. A pointed reminder of our dependence on the availability of resources was the United States' response to the takeover of Kuwait by Iraq during August 1990.

Even though the United States is resource rich, its large population and high standard of living make it dependent on imports of a number of mineral commodities for part or all of its needs. Canada, in contrast, is more self-reliant, meeting most of its own domestic mineral and energy needs. Nevertheless, it must import phosphate, chromium, manganese, and bauxite, the ore of aluminum. Canada also produces more crude oil and natural gas than it uses, and it is among the world leaders in producing and exporting uranium.

Most of the largest and richest mineral deposits have probably already been discovered, and in some cases depleted. To ensure continued supplies of essential minerals and energy resources, geologists and other scientists, government agencies, and leaders of business and industry continually assess the status of resources in view of changing economic and political conditions and developments in science and technology. In the following chapters, we will discuss the origin and distribution of various mineral and energy resources and reserves.

Chapter Summary

1. All matter is composed of chemical elements, each of which consists of atoms. Individual atoms have a nucleus, containing protons and neutrons, and electrons circling the nucleus in electron shells.

2. Atoms are characterized by their atomic number (the number of protons in the nucleus) and their atomic mass number (the number of protons plus the number of neutrons in the nucleus).

3. Bonding is the process whereby atoms are joined to other atoms. If atoms of different elements are bonded, they form a compound. Ionic and covalent bonds are most common in minerals, but metallic and van der Waals bonds also occur in a few.

4. Most minerals are compounds, but a few, including gold and platinum, are composed of a single element and are called native elements.

5. All minerals are crystalline solids, meaning that they possess an orderly internal arrangement of atoms.

6. Some minerals vary in chemical composition because atoms of different elements can substitute for one another provided that the electrical charge is balanced and the atoms are about the same size.

7. Of the more than 3500 known minerals, most are silicates. Ferromagnesian silicates contain iron (Fe) and magnesium (Mg), and nonferromagnesian silicates lack these elements.

8. In addition to silicates, several other mineral groups are recognized, including carbonates, oxides, sulfides, sulfates, and halides.

9. The physical properties of minerals such as color, hardness, cleavage, and crystal form are controlled by composition and structure.

10. A few minerals are common enough constituents of rocks to be designated rock-forming minerals.

11. Any commodity that is or has the potential to be extracted for some use is a resource. Metals constitute metallic resources, whereas commodities such as sand, gravel, and salt are nonmetallic resources, and oil, natural gas, coal, and uranium are energy resources.

12. Reserves are that part of the resource base that can be extracted economically. The status of a resource versus a reserve depends on several factors, including market price, geographic location, and developments in science and technology.

13. The United States, although rich in resources, must import part or all of some essential mineral and energy commodities. Canada is more self-reliant, but it too must import some commodities such as aluminum ore.

Important Terms

atom
atomic mass number
atomic number
bonding
carbonate mineral
cleavage
compound

covalent bond
crystalline solid
electron
electron shell
element
ferromagnesian silicate
ion

ionic bond
mineral
native element
neutron
nonferromagnesian silicate
nucleus
proton

reserve
resource
rock
rock-forming mineral
silica
silica tetrahedron
silicate

Review Questions

1. The term *crystalline* in the definition of a mineral means that a mineral
   a. _____ has an atomic number of at least 92; b. _____ possesses more protons than neutrons; c. _____ is characterized by physical properties such as hardness and cleavage; d. _____ has an orderly internal arrangement of atoms; e. _____ is an essential constituent of granite.
2. Minerals that break along smooth internal planes of weakness have the property known as
   a. _____ cleavage; b. _____ double refraction; c. _____ specific gravity; d. _____ ionic bonding; e. _____ streak.
3. The two common rock-forming minerals from the carbonate group are
   a. _____ hematite and magnetite; b. _____ calcite and dolomite; c. _____ clay and feldspar; d. _____ halite and gypsum; e. _____ coal and obsidian.
4. The atoms of the noble gases do not react to form compounds because they have
   a. _____ a positive electrical charge of 4; b. _____ too many neutrons; c. _____ eight electrons in their outermost electron shell; d. _____ two directions of cleavage that intersect at right angles; e. _____ a deficiency of silica and an excess of iron and magnesium.
5. The number of protons plus neutrons in an atom's nucleus determines its
   a. _____ resistance to abrasion; b. _____ value as a resource; c. _____ frequency of bonding with other atoms; d. _____ chemical formula; e. _____ atomic mass number.
6. The two most abundant elements in Earth's crust are
   a. _____ iron and calcium; b. _____ iridium and platinum; c. _____ oxygen and silicon; d. _____ potassium and magnesium; e. _____ hydrogen and nitrogen.
7. The weight of a mineral compared to that of an equal volume of water is known as its
   a. _____ specific gravity; b. _____ hardness; c. _____ atomic number; d. _____ electrical charge; e. _____ cleavage.
8. Which one group of minerals in the following list contains only ferromagnesian silicates?
   a. _____ quartz-calcite-iron; b. _____ potassium-feldspar-mica-biotite; c. _____ olivine-pyroxene-amphibole; d. _____ halite-gypsum-coal; e. _____ granite-hematite-galena.
9. The ferromagnesian silicate olivine has the chemical formula $(Mg,Fe)_2SiO_4$, which means that

   a. _____ magnesium is heavier than iron; b. _____ magnesium and iron must be present in equal amounts; c. _____ it is a sheet silicate with weak bonds between the sheets; d. _____ magnesium and iron can substitute for one another; e. _____ magnesium is more abundant than iron in Earth's crust.
10. In what type of bonding are electrons transferred from one atom to another?
    a. _____ ionic; b. _____ tetrahedral; c. _____ van der Waals; d. _____ silicate; e. _____ carbonate.
11. A pyramidal structure consisting of four oxygen atoms and one silicon atom is known as the
    a. _____ oxygen-silicon ion; b. _____ oxide building block; c. _____ silica tetrahedron; d. _____ basic amphibole unit; e. _____ carbonate pyramid.
12. The silicon ion has a positive charge of 4, and oxygen has a negative charge of 2. Thus, the ion group (SiO_4) has a
    a. _____ positive charge of 2; b. _____ negative charge of 2; c. _____ negative charge of 1; d. _____ positive charge of 4; e. _____ negative charge of 4.
13. Why is a distinction made between rock-forming minerals and accessory minerals? Also, what are the most common silicate and carbonate rock-forming minerals.
14. How does the fact that all specimens of a given mineral species have the same angles between the same crystal faces indicate that these minerals are crystalline?
15. Why do some minerals, such as plagioclase feldspars, have a range of chemical compositions? Give an example from the ferromagnesian silicates.
16. A hypothetical atom has 19 protons, 14 neutrons, and 12 electrons. What is this atom's atomic number, atomic mass number, and electrical charge?
17. Some minerals possess the physical property known as cleavage. Describe cleavage and explain what controls it. Also, explain why cleavage fragments of calcite and halite, both with three directions of cleavage, are shaped differently.
18. How do compounds and native elements differ? Give two examples along with chemical formulas of minerals that are compounds and two minerals that are native elements.
19. How does a crystalline solid differ from a liquid and a gas?
20. Diamond and graphite are composed of carbon (C), but differ considerably in all physical properties. How and why do they differ?

21. Considering the possible combinations of naturally occurring elements, why are so few minerals actually common components of rocks?
22. How do silicate minerals differ from minerals in other major mineral groups? What are the two subgroups of silicates, and what are the criteria for assigning minerals to either subgroup?

23. Why must the United States, a natural resource-rich nation, import most or all of some of the resources it needs? What are some resources that are in especially short supply in this country, and what problems does dependence on imports create?
24. Minerals most often have irregular shapes unlike their ideal crystal form. How then is it possible to demonstrate that they possess a crystalline structure?

 # World Wide Web Activities

For these Web site addresses, along with current updates and exercises, log on to **http://www.brookscole.com/geo**

Amethyst Galleries

Amethyst Galleries maintains this site, which has a wealth of information on and images of hundreds of minerals. It lists minerals alphabetically by name and class and has minerals in interesting groups such as gemstones and birthstones. A full text search for mineral identification by keyword is also available.

West's Geology Directory

This site was introduced in Chapter 1 as a comprehensive directory of geologic Internet links and an excellent place to find links on virtually any geologic topic. Scroll down and click *Minerals* and then click on *Minerals and Mineralogy.* Here you can find information on all common minerals and many others, including gemstones and jewelry. Click on any of the mineral group names listed.

The Mineral Gallery

This site has images of dozens of minerals, along with information on their chemistry, physical properties, uses, associations with other minerals, and locales where they are found. Click on *sphalerite.* What is its chemical formula, its hardness, and uses? Click on *malachite* and see where it is found, the type of rock it is found in, and its uses. Click on *topaz* and give its chemical formula, specific gravity, and uses. Also, what is the origin of the name *topaz?*

U.S. Geological Survey Mineral Resource Surveys Program

Maintained by the U.S. Geological Survey (USGS), this site contains information on the USGS Mineral Resource Survey Program, Fact Sheets, Contacts, and numerous links to a variety of sources on mineral resources.

 # CD-ROM Exploration

Exploring your *Earth Systems Today* CD-ROM will add to your understanding of the material in this chapter.

Topic: Earth's Materials

Module: Atoms and Crystals

Explore activities in this module to see if you can discover the following for yourself:

Use the animation in this module to gain an understanding of how atoms in magma behave and ultimately bond to make minerals. What bonds are the first to form and why? What mineral forms first? Why?

Use the graphics on mineral structures to observe how silicate minerals are assembled out of separate atoms. How are silica tetrahedra linked to each structure?

Igneous Rocks and Intrusive Igneous Activity

Light-colored granitic rocks exposed along the shoreline in Acadia National Park, Maine.

Nearly everyone is familiar with the term *granite,* although few know that geologists define granite as an igneous rock with specific amounts of quartz, potassium feldspars, and plagioclase feldspars (see Figure 2.21a). Granite and several similar appearing rocks, collectively referred to as *granitic,* are common, especially in mountain ranges and a vast area of North America mostly in Canada known as the Canadian Shield. Granitic rocks are used for building stone, tombstones, and pedestals for statues; they are also crushed and used for aggregate in cement. Several varieties of cut and polished granitic rocks are attractive and used as facing stones on buildings, or for decorative stones in table tops, kitchen counters, and mantelpieces. In addition, the rocks adjacent to granitic rocks in some areas contain important minerals such as copper.

Granitic rocks originate when magma (molten rock material) cools and crystallizes far below the surface. Subsequent uplift and deep erosion has yielded vast exposures of these rocks in many parts of the world. On this continent, one of the most extensive areas of granitic rocks is in British Columbia, Canada; smaller yet still vast areas of similar rocks are found in Alaska, California, the Rocky Mountain states, Texas, the Great Lakes region, some of the Appalachian States, and the New England States. In Acadia National Park, Maine, for instance, magma cooled within the crust, forming granitic rocks that now constitute much of Mount Desert Island, which makes up most of the park. Later magma of a different composition was injected into these granitic rocks, where it formed dark-colored bands (see chapter opening photo). Since then other geologic processes have shaped the area, especially deformation, and erosion by glaciers.

Granitic rocks are also common in the Black Hills of South Dakota. Probably the most frequently visited site in this area is Mount Rushmore National Memorial, where the images of four presidents were carved into rocks of the 1.7 billion-year-old Harney Peak Granite (Figure 3.1a). A memorial to the Native American Crazy Horse is now in progress in these rocks near Custer, South Dakota. The Harney Peak Granite forms the core of the Black Hills, an oval area of crust that was uplifted beginning about 80 million years ago. Chemical and mechanical alteration and erosion by running water continue to modify these rocks.

Between 210 and 70 million years ago, during the Jurassic and Cretaceous periods, huge bodies of granitic rock formed in what are now the Rocky Mountain states, especially in Idaho, Colorado, Wyoming, and Montana, and in California and adjacent parts of Oregon and Nevada. Visitors to Yosemite, Kings Canyon, and Sequoia National Parks in California can see vast rock exposures of little but granitic rocks (Figure 3.1b), as well as giant trees, wildlife, waterfalls, and broad, grass-covered meadows. Presently, chemical and mechanical alteration and erosion by running water modify these rocks, but glaciers were important in shaping them during the past.

Denali (formerly known as Mount McKinley) in Alaska is composed predominantly of granitic rocks (Figure 3.1c). Native Alaskans named the mountain Denali, meaning "The High One," which at more than 6100 m above sea level is North America's highest peak. Snow, ice, and glaciers cover much of the mountain so many of its rocks are hidden from view. Erosion by glaciers is responsible for the mountain's present scenery. Denali and nearby North Peak (5935 m high) are the most conspicuous features in Denali National Park.

Granitic rocks figure importantly in some of our following discussions. In this chapter the types of granitic rocks

OBJECTIVES

At the end of this chapter, you will have learned that

■ With few exceptions magma is composed of silicon and oxygen and lesser amounts of several other chemical elements.

■ Temperature and especially composition control the mobility of magma and lava.

■ Most magma originates within the upper mantle or lower crust at divergent and convergent plate boundaries.

■ Several processes can bring about chemical changes in magma, so magma can evolve from one kind into another.

■ All igneous rocks result from cooling magma or the consolidation of particles erupted during explosive eruptions.

■ Geologists use composition and texture to classify igneous rocks.

■ The size of the minerals in an igneous rock generally indicates whether it cooled from magma below the surface or from lava at the surface.

■ Intrusive igneous bodies known as plutons form when magma cools below the surface, and that the types of plutons are identified by their geometry and relationship to preexisting rocks.

■ Geologists understand how most plutons originated, but for some of the very large ones the mechanism(s) whereby they form is not fully understood.

(a)

(c)

(b)

Figure 3.1 *(a) The presidents' images at Mount Rushmore in the Black Hills of South Dakota are in the Harney Peak Granite. (b) Vast exposures of granite and related rocks in Yosemite National Park, California. (c) Denali, formerly known as Mount McKinley, in Alaska's Denali National Park, is composed largely of granitic rocks.*

and the igneous bodies they form are considered, as well as resources in adjacent rocks. And in Chapter 5, the Mount Airy Granite of North Carolina is important in our discussion of a particular kind of mechanical alteration process. Knowledge

about the origin of granitic rocks will also help you understand the nature of rock alteration by heat and fluids (Chapter 6), interactions among tectonic plates (Chapter 9), and episodes of mountain building (Chapter 10).

Introduction

Rocks that originate from lava flows or particulate matter erupted from volcanoes are common, but they represent only a tiny fraction of the rocks formed by cooling and crystallization of molten rock material known as *magma*. Most magma cools below Earth's surface, where it forms bodies of rock known as *plutons;* these vary in size, shape, and relationships to previously existing rocks adjacent to them. Many of the rocks forming in plutons are granitic (see the Prologue), but a variety of other types are known as well. In this chapter, we are concerned (1) with the ori-

gin, composition, textures, and classification of igneous rocks resulting from either volcanic activity or the formation of plutons; and (2) with the origin, significance, and types of plutons. In the following chapter, we consider volcanism, volcanoes, and associated phenomena, which are produced by magma reaching the surface.

Even though plutons and volcanism are discussed in separate chapters, they are nevertheless related. The same types of magmas are involved in both processes, but some magmas are more mobile than others and more commonly reach the surface. Plutons typically lie beneath areas of volcanism and, in fact, serve as the source for the overlying lavas and fragmental materials ejected during eruptions. Furthermore, present-day plutons and volcanoes are found mostly at or near plate boundaries, indicating that they share a common origin. Thus ancient plutons and rocks resulting from volcanic eruptions serve as some of the criteria used to recognize plate boundaries that existed during the past.

Why Should You Study Igneous Rocks and Intrusive Igneous Activity?

Why should you study igneous rocks and intrusive igneous activity? Igneous rocks, constituting one of the three rock families, are common, making up a large part of the continental crust and virtually all of the oceanic crust. Evaluating the physical and chemical properties of various igneous rocks allows us to make inferences about the processes responsible for their origin, including the plate tectonic setting in which they formed. Intrusive igneous activity—that is, the emplacement and cooling of magma within Earth's crust—accounts for one of the most common igneous rocks, granite (see the Prologue), as well as the origin of several types of mineral deposits.

The Properties and Behavior of Magma and Lava

In Chapter 2, we noted that one process leading to the formation of minerals, and thus of rocks, was the cooling and crystallization of the molten rock material known as magma and lava. **Magma** is simply molten rock below the surface; the same material at the surface is **lava.** No other distinction is necessary, so the two terms simply tell us the location of the molten rock material. Any magma is less dense than the rock from which it was derived, so it tends to move up toward the surface. However, most of it cools and solidifies far below the surface, thus accounting for the origin of various plutons. Some magma does reach the surface either as **lava flows,** or as forcefully ejected particles known as **pyroclastic materials** (from the Greek *pyro,* meaning fire and *klastos,* meaning broken). Lava flows and eruptions of pyroclastic materials are the most awe-inspiring manifestations of all processes related to magma, but as noted, represent only a small percentage of all magma that forms.

All **igneous rocks** derive from magma, but two separate processes account for their origin. They form when (1) magma or lava cools and crystallizes to form minerals, or (2) when pyroclastic materials such as volcanic ash are consolidated, forming solid masses from the previously separate particles. Igneous rocks resulting from cooling lava flows and consolidation of pyroclastic materials are further characterized as **volcanic rocks** or **extrusive igneous rocks,** whereas those forming when magma cools and crystallizes below the surface are **plutonic rocks** or **intrusive igneous rocks.**

This brief introduction to magma and lava is enough to clarify what these substances are and what kinds of rocks are derived from them. However, let's explore magma, the source of all igneous rocks, a bit further and consider its composition, temperature, and resistance to flow, or what is called viscosity.

Composition of Magma

In Chapter 2, we noted that by far the most abundant minerals in Earth's crust are silicates such as quartz, various feldspars, and several ferromagnesian silicates, all composed mostly of silicon, oxygen, and other elements listed in Table 2.4. As a result, melting of crustal rocks usually yields silica-rich magmas that also contain considerable aluminum, calcium, sodium, iron, magnesium, potassium, and several other elements in lesser quantities. However, magma derived by melting rocks of the upper mantle, which are composed predominantly of ferromagnesian silicates, contains comparatively less silica and more iron and magnesium.

Silica is the primary constituent of most magma, but it varies enough to distinguish magmas characterized as felsic, intermediate, and mafic (Table 3.1). **Felsic magma** with more than 65% silica is silica rich and contains considerable sodium, potassium, and aluminum, but little calcium, iron, and magnesium. **Mafic magma,** in contrast, with 45% to 52% silica, is silica poor, but contains proportionately more calcium, iron, and magnesium. As its name implies, **intermediate magma** has a composition between the limits of felsic and mafic magmas (Table 3.1).

How Hot Are Magma and Lava?

Whether you have witnessed a lava flow or not, it is common knowledge that lava is very hot. But how hot is hot? Erupting lavas generally have temperatures in the range of 1000° to 1200°C, although a temperature of 1350°C has been recorded above Hawaiian lava lakes where volcanic gases reacted with the atmosphere. Magma, that is molten rock material below the surface, must be even hotter, but no direct measurements of magma temperatures have been made.

Most temperature measurements are taken where volcanoes show little or no explosive activity, so our best information comes from mafic lava flows such as those issuing from the Hawaiian volcanoes (Figure 3.2). In contrast, eruptions of felsic lava flows are not common, and, in fact, volcanoes erupting silica-rich lavas tend to be explosive and thus cannot be approached safely. Nevertheless,

Table 3.1

The Most Common Types of Magmas and Their Characteristics

Type of Magma	Silica Content (%)	Sodium, Potassium, & Aluminum	Calcium, Iron, & Magnesium
Mafic	45–52	↓	Increase
Intermediate	53–65	↓	↑
Felsic	>65	Increase	↑

the temperatures of some bulbous masses of felsic lava in lava domes have been measured at a distance with an instrument called an optical pyrometer. The surfaces of these domes are as hot as 900°C, but their interiors must surely be even hotter.

When Mount St. Helens erupted in 1980, it ejected felsic magma as particulate matter in pyroclastic flows. Two weeks later, these flows still had temperatures between 300° and 420°C, and a steam explosion took place more than a year later when water encountered some of the still-hot deposits. The reason that magma or lava retains heat so well is that rock is such a poor conductor of heat. Accordingly, the interiors of thick lava flows may remain hot for months or years, whereas plutons, depending on their size and depth, may not completely cool for thousands to millions of years.

Viscosity—Resistance to Flow

All liquids possess the property of **viscosity,** or simply their resistance to flow. For many liquids, such as water, the viscosity is very low so they are highly fluid and flow readily. For other liquids, viscosity is so high that they flow much more slowly. Good examples are cold motor oil and cold syrup, both of which are quite viscous and thus flow only with difficulty. But when these same liquids are heated, their viscosity is much lower and they flow more easily. That is, they become more fluid with increasing temperature. Accordingly, you might suspect that temperature controls the viscosity of magma and lava and this inference is partly correct. We can generalize and say that hot magma or lava moves more readily than cooler magma or lava, but we must qualify this statement by noting that composition is an even more important control of viscosity.

Silica content strongly controls magma and lava viscosity. With increasing silica content, numerous networks of silica tetrahedra form and retard flow, because for flow to take place the strong bonds of the networks must be severed. Mafic magma and lava with only 45% to 52% silica have fewer silica tetrahedra networks and as a result are more mobile than felsic magma and lava flows. One mafic flow in 1783 in Iceland flowed about 80 km, and some ancient flows in Washington State can be traced for more than 500 km. Felsic magma in contrast, because of its higher viscosity, does not reach the surface as commonly as mafic magma. And when felsic lava flows do occur, they tend to be slow moving and thick, and to move only short distances. A thick, pasty lava flow that erupted in 1915 from Lassen Peak in California flowed only about 300 m before it ceased moving.

How Does Magma Originate and Change?

Most people are familiar with magma that reaches Earth's surface as either lava flows or pyroclastic materials ejected during explosive volcanic eruptions. The lava flows issuing from Kilauea in Hawaii fascinate many observers, and the activity at Mount St. Helens, Washington, in 1980 and Mount Pinatubo in the Philippines in 1991 remind us of the violence of some eruptions. Yet most people have little understanding of how magma originates in the first place, how it can change or evolve, or how it

Figure 3.2 *Geologist using a thermocouple to determine the temperature of a lava flow in Hawaii.*

reaches the surface. Indeed, many believe the misconceptions that lava comes from a continuous layer of molten material beneath the crust, or that it comes from Earth's molten core.

Some magma rises from depths of as much as 100 to 300 km, but most of it forms at much shallower depths in the upper mantle or lower crust and accumulates in reservoirs known as **magma chambers.** Beneath spreading ridges in the ocean basins where the crust is thin, magma chambers exist at a depth of only a few kilometers. Along convergent plate boundaries where an oceanic plate is subducted beneath another oceanic plate or beneath a continental plate, magma chambers are a few tens of kilometers deep. The volume of a magma chamber may be several cubic kilometers of molten rock within the otherwise solid lithosphere. This magma might simply cool and crystallize in place, thus forming various intrusive igneous rock bodies, except that some migrates to the surface.

Bowen's Reaction Series

During the early part of this century, N. L. Bowen hypothesized that mafic, intermediate, and felsic magmas could all derive from a parent mafic magma. He knew that minerals do not all crystallize simultaneously from a cooling magma, but rather crystallize in a predictable sequence. Based on his observations and laboratory experiments, Bowen proposed a mechanism, now called **Bowen's reaction series,** to account for the derivation of intermediate and felsic magmas from a mafic magma. Bowen's reaction series consists of two branches: a *discontinuous branch* and a *continuous branch* (Figure 3.3). As the temperature of a magma decreases, crystallization of minerals occurs along both branches simultaneously, but for convenience we will discuss them separately.

In the discontinuous branch, which contains only ferromagnesian silicates, one mineral changes to another over specific temperature ranges (Figure 3.3). As the temperature decreases, a temperature range is reached in which a given mineral begins to crystallize. A previously formed mineral reacts with the remaining liquid magma (the melt) so that it forms the next mineral in the sequence. For example, olivine $[(Mg,Fe)_2SiO_3]$ is the first ferromagnesian silicate to crystallize. As the magma continues to cool, it reaches the temperature range at which pyroxene is stable; a reaction occurs between the olivine and the remaining melt, and pyroxene forms.

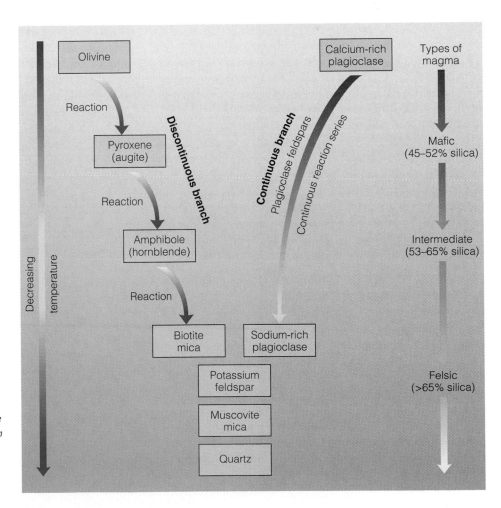

Figure 3.3 *Bowen's reaction series. It consists of a discontinuous branch along which a succession of ferromagnesian silicates crystallize as the magma cools, and a continuous branch along which plagioclase feldspars with increasing amounts of sodium crystallize. Notice also that the composition of the initial mafic magma changes as crystallization takes place along the two branches.*

How Does Magma Originate and Change?

With continued cooling, a similar reaction takes place between pyroxene and the melt, and the pyroxene structure is rearranged to form amphibole. Further cooling causes a reaction between the amphibole and the melt, and its structure is rearranged so that the sheet structure typical of biotite mica forms. Although the reactions just described tend to convert one mineral to the next in the series, the reactions are not always complete. Olivine, for example, might have a rim of pyroxene, indicating an incomplete reaction. If a magma cools rapidly enough, the early-formed minerals do not have time to react with the melt, and thus all the ferromagnesian silicates in the discontinuous branch can be in one rock. In any case, by the time biotite has crystallized, essentially all magnesium and iron present in the original magma have been used up.

Plagioclase feldspars, which are nonferromagnesian silicates, are the only minerals in the continuous branch of Bowen's reaction series (Figure 3.3). Calcium-rich plagioclase crystallizes first. As the magma continues to cool, calcium-rich plagioclase reacts with the melt, and plagioclase containing proportionately more sodium crystallizes until all of the calcium and sodium are used up. In many cases cooling is too rapid for a complete transformation from calcium-rich to sodium-rich plagioclase to take place. Plagioclase forming under these conditions is *zoned,* meaning that it has a calcium-rich core surrounded by zones progressively richer in sodium.

As minerals crystallize along the two branches of Bowen's reaction series, iron and magnesium are depleted, because they are used in ferromagnesian silicates, whereas calcium and sodium are used up in plagioclase feldspars. At this point, any leftover magma will be enriched in potassium, aluminum, and silicon, which combine to form the potassium feldspar orthoclase ($KAlSi_3O_8$), and if water pressure is high, the sheet silicate muscovite forms. Any remaining magma is further enriched in silicon and oxygen (silica) and forms the mineral quartz (SiO_2). The crystallization of orthoclase and quartz is not a true reaction series as is the crystallization of ferromagnesian silicates and plagioclase feldspars, because they form independently rather than by a reaction of orthoclase with the melt.

The Origin of Magma at Spreading Ridges

One fundamental observation we can make regarding the origin of magma is that Earth's temperature increases with depth. This temperature increase, known as the *geothermal gradient,* averages about 25°C/km. Accordingly, rocks only a few kilometers below the surface are hot but remain solid because their melting temperature rises with increasing pressure (Figure 3.4). However, beneath spreading ridges, the temperature locally exceeds the melting temperature, at least in part, because pressure decreases. That is, plate separation at ridges probably causes a decrease in

pressure on the already hot rocks at depth, thus initiating melting (Figure 3.4a). In addition, the presence of water can also decrease the melting temperature beneath spreading ridges because water aids thermal energy in breaking the chemical bonds in minerals (Figure 3.4b).

Another explanation for the origin of spreading-ridge magmas is that localized, cylindrical plumes of hot mantle material, called *mantle plumes,* rise beneath the ridges and spread out in all directions. Perhaps localized concentrations of radioactive minerals within the crust and upper mantle decay and provide the heat necessary to melt rocks and thus generate magma.

The magmas formed beneath spreading ridges are invariably mafic (45–52% silica). But the upper mantle rocks from which these magmas are derived are characterized as ultramafic (<45% silica), consisting largely of ferromagnesian silicates and lesser amounts of nonferromagnesian silicates. To explain how mafic magma originates from ultramafic rock, geologists propose that the magma forms from source rock that only partially melts. This phenomenon of partial melting takes place because various minerals have different melting temperatures.

Recall the sequence of minerals in Bowen's reaction series (Figure 3.3). The order in which these minerals melt is the opposite of their order of crystallization. Accordingly, quartz, potassium feldspar, and sodium-rich plagioclase melt before most of the ferromagnesian silicates and the calcic varieties of plagioclase. So when ultramafic rock begins to melt, the minerals richest in silica melt first, followed by those containing less silica. Therefore, if melting is not complete, a mafic magma containing proportionately more silica than the source rock results. Once this mafic magma forms, some of it rises to the surface where it is erupted as lava flows and pyroclastic materials, and some simply cools beneath the surface thus forming various intrusive igneous bodies.

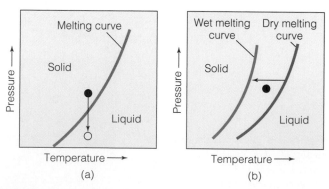

Figure 3.4 *The effects of pressure and water on melting. (a) Melting temperature rises with increasing pressure, so a pressure decrease on already hot rocks can initiate melting. (b) When water is present the melting curve shifts to the left, because water provides an additional agent to break chemical bonds. Accordingly, rocks melt at a lower temperature if water is present.*

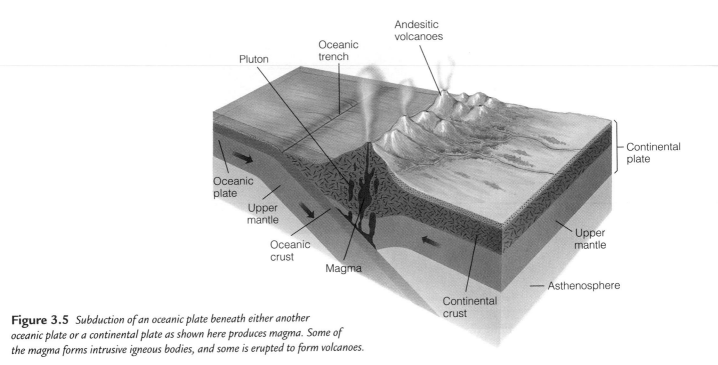

Figure 3.5 *Subduction of an oceanic plate beneath either another oceanic plate or a continental plate as shown here produces magma. Some of the magma forms intrusive igneous bodies, and some is erupted to form volcanoes.*

Subduction Zones and the Origin of Magma

Another basic observation we can make regarding magma is that where an oceanic plate is subducted beneath either a continental plate or another oceanic plate, a belt of volcanoes and plutons is found near the leading edge of the overriding plate (Figure 3.5). It would seem, then, that subduction and the origin of magma must be related in some way, and indeed they are. Furthermore, magma at these convergent plate boundaries is mostly intermediate (53–65% silica) and felsic (>65% silica).

Once again, geologists invoke the phenomenon of partial melting to explain the origin and composition of magma at subduction zones. As a subducted plate descends toward the asthenosphere, it eventually reaches the depth where the temperature is high enough to initiate partial melting. Moreover, the wet oceanic crust descends to a depth at which dewatering takes place, and as the water rises into the overlying mantle, it enhances melting and magma forms (Figure 3.4b).

Recall that partial melting of ultramafic rock at spreading ridges yields mafic magma. Similarly, partial melting of mafic rocks of the oceanic crust yields intermediate (53–65% silica) and felsic (>65% silica) magmas, both of which are richer in silica than the source rock. Additionally, some of the silica-rich sediments and sedimentary rocks of continental margins are probably carried downward with the subducted plate and contribute their silica to the magma. Also, mafic magma rising through the lower continental crust must be contaminated with silica-rich materials, which changes its composition.

Processes Resulting in Compositional Changes in Magma

Once magma forms, its composition may change by **crystal settling,** which involves the physical separation of minerals by crystallization and gravitational settling (Figure 3.6). Olivine, the first ferromagnesian silicate to form in the discontinuous branch of Bowen's reaction series, has a specific gravity greater than that of the remaining magma and tends to sink down in the melt. Accordingly, the remaining melt becomes comparatively rich in silica, sodium, and potassium, because much of the iron and

Figure 3.6 *(a) Crystal settling and assimilation can bring about changes in the composition of magma. Early-formed ferromagnesian silicates have a specific gravity greater than that of the magma so they settle and accumulate in the lower part of the magma chamber. As magma moves upward, fragments of country rock are dislodged. They might melt and be incorporated into the magma by assimilation, or remain as incompletely melted inclusions.(b) Dark-colored inclusions in granitic rock in California.*

magnesium were removed when minerals containing these elements crystallized.

Although crystal settling does take place in magmas, it does not do so on the scale envisioned by Bowen to explain the origin of intermediate and felsic magma from mafic magma. In some thick, sheetlike, intrusive igneous bodies called *sills,* the first-formed minerals in the reaction series are indeed concentrated. The lower parts of these bodies contain more olivine and pyroxene than the upper parts, which are less mafic. But even in these bodies, crystal settling has yielded little felsic magma from an original mafic magma.

If felsic magma could be derived on a large scale from mafic magma as Bowen thought, there should be far more mafic magma than felsic magma. To yield a particular volume of granite (a felsic igneous rock), about 10 times as much mafic magma would have to be present initially for crystal settling to yield the volume of granite in question. If this were so, then mafic intrusive igneous rocks should be much more common than felsic ones. However, just the opposite is the case, so it appears that mechanisms other than crystal settling must account for the large volume of felsic magma. Partial melting of mafic oceanic crust and silica-rich sediments of continental margins during subduction yields magma richer in silica than the source rock. Furthermore, magma rising through the continental crust can absorb some felsic materials and become more enriched in silica.

The composition of a magma can also change by **assimilation,** a process whereby magma reacts with preex-

isting rock, called **country rock,** with which it comes in contact (Figure 3.6). The walls of a volcanic conduit or magma chamber are, of course, heated by the adjacent magma, which may reach temperatures of 1300°C. Some of these rocks can be partly or completely melted, provided their melting temperature is less than that of the magma. Because the assimilated rocks seldom have the same composition as the magma, the composition of the magma is changed.

The fact that assimilation occurs can be demonstrated by *inclusions,* incompletely melted pieces of rock that are fairly common within igneous rocks. Many inclusions were simply wedged loose from the country rock as the magma forced its way into preexisting fractures (Figure 3.6b).

No one doubts that assimilation takes place, but its effect on the bulk composition of most magmas must be slight. The reason is that the heat for melting must come from the magma itself, and this would have the effect of cooling the magma. Only a limited amount of rock can be assimilated by a magma, and that amount is usually insufficient to bring about a major compositional change.

Neither crystal settling nor assimilation can produce a significant amount of felsic magma from a mafic one. But both processes, if operating concurrently, can change the composition of a mafic magma much more than either process acting alone. Some geologists think that this is one way many intermediate magmas form where oceanic lithosphere is subducted beneath continental lithosphere.

The fact that a single volcano can erupt lavas of different composition indicates that magmas of differing com-

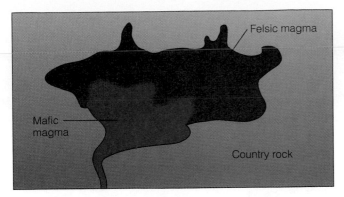

Figure 3.7 *Magma mixing. Two magmas mix and produce a magma with a composition different from either of the parent magmas.*

position must be present. It seems likely that some of these magmas would come into contact and mix with one another. If this is the case, we would expect that the composition of the magma resulting from **magma mixing** would be a modified version of the parent magmas. Suppose a rising mafic magma mixes with a felsic magma of about the same volume (Figure 3.7). The resulting "new" magma would have a more intermediate composition.

Igneous Rocks—What Are They and What Are Their Characteristics?

In the Introduction we briefly defined *plutonic* or *intrusive igneous rocks* and *volcanic* or *extrusive igneous rocks*. Here we will have considerably more to say about the texture, composition, and classification of these rocks, which together constitute one of the three major rock families depicted in the rock cycle (see Figure 1.13).

Igneous Rock Textures

The term *texture* refers to the size, shape, and arrangement of mineral grains composing igneous rocks. Size is the most important because grain size is related to the cooling history of a magma or lava, and generally indicates whether an igneous rock is intrusive or extrusive. The atoms in magma and lava are in constant motion, but when cooling begins, some atoms bond to form small nuclei. As other atoms in the liquid chemically bond to these nuclei, they do so in an orderly geometric arrangement and the nuclei grow into crystalline *mineral grains*, the individual particles that compose igneous rocks.

During rapid cooling, as takes place in lava flows and some shallow plutons, the rate at which mineral nuclei

form exceeds the rate of growth and an aggregate of many small mineral grains results. The result is a fine-grained texture termed **aphanitic** in which individual minerals are too small to be seen without magnification (Figure 3.8a and b). With slow cooling, the rate of growth exceeds the rate of nuclei formation, and relatively large mineral grains form, thus yielding a coarse-grained or **phaneritic** texture in which minerals are clearly visible (Figure 3.8c and d). Aphanitic textures generally indicate an extrusive origin, whereas rocks with phaneritic textures are mostly intrusive. However, rocks that formed in shallow plutons might have an aphanitic texture, and rocks forming in the interiors of some thick lava flows might be phaneritic.

Another common igneous texture is one termed **porphyritic,** in which minerals of markedly different sizes are present in the same rock. The larger minerals are *phenocrysts* and the smaller ones constitute the *groundmass,* which is made up of the grains between phenocrysts (Figure 3.8e and f). The groundmass can be either aphanitic or phaneritic; the only requirement for a porphyritic texture is that the phenocrysts be considerably larger than the minerals in the groundmass. A porphyritic texture might form when magma begins cooling slowly below the surface where some mineral-crystal nuclei form and grow. But if this magma is then extruded onto the surface before crystallization is complete, the remaining liquid phase cools much more rapidly resulting in an aphanitic texture. The resulting igneous rock would have large mineral crystals (phenocrysts) suspended in an aphanitic groundmass and the rock is characterized as a *porphyry*—basalt porphyry, for example.

Lava may cool so rapidly that its constituent atoms do not have time to become arranged in the ordered, three-dimensional frameworks typical of minerals. As a consequence of such rapid cooling, *obsidian* forms, which is a *natural glass* (Figure 3.9a). Even though obsidian is not composed of minerals, geologists still include it with the igneous rocks.

Some lavas contain large amounts of water vapor and other gases that may be trapped in cooling lava where they form numerous small holes or cavities known as **vesicles;** rocks possessing numerous vesicles are termed *vesicular,* as in vesicular basalt (Figure 3.9b).

A **pyroclastic** or **fragmental texture** characterizes igneous rocks formed by explosive volcanic activity. Ash may be discharged high into the atmosphere and eventually settle to the surface where it accumulates; if it is consolidated and becomes solid, it is a pyroclastic igneous rock (Figure 3.9c).

The Composition of Igneous Rocks

Most igneous rocks, just like the magma or lava from which they originate, are characterized as mafic (45–52%

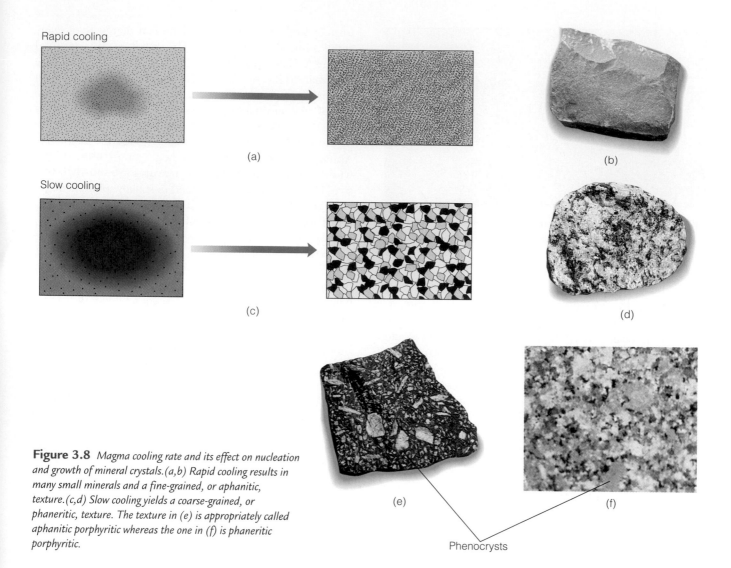

Figure 3.8 *Magma cooling rate and its effect on nucleation and growth of mineral crystals. (a,b) Rapid cooling results in many small minerals and a fine-grained, or aphanitic, texture. (c,d) Slow cooling yields a coarse-grained, or phaneritic, texture. The texture in (e) is appropriately called aphanitic porphyritic whereas the one in (f) is phaneritic porphyritic.*

Rapid cooling

(a)

(b)

Slow cooling

(c)

(d)

(e)

(f)

Phenocrysts

silica), intermediate (53–65% silica), or felsic (>65% silica). A few are referred to as ultramafic (<45% silica), but these probably derived from mafic magma by a process discussed later. The parent magma plays an important role in determining the mineral composition of igneous rocks, yet it is possible for the same magma to yield a variety of igneous rocks, because its composition can change as a result of crystal settling, assimilation, magma mixing, and the sequence in which minerals crystallize.

Classifying Igneous Rocks

With few exceptions, geologists classify igneous rocks on the basis of textural features and composition (Figure 3.10). Notice in Figure 3.10 that all of the rocks, except peridotite, constitute pairs; the members of a pair have the same composition but different textures. Thus, basalt and gabbro, andesite and diorite, and rhyolite and granite are compositional (mineralogical) equivalents, but basalt, andesite, and rhyolite are aphanitic and most commonly

extrusive, whereas gabbro, diorite, and granite have phaneritic textures that generally indicate an intrusive origin.

The igneous rocks shown in Figure 3.10 are also differentiated by composition. Reading across the chart from rhyolite to andesite, to basalt, for example, the relative proportions of nonferromagnesian and ferromagnesian silicates differ. The differences in composition, however, are gradual so that a compositional continuum exists. In other words, rocks exist with compositions intermediate between rhyolite and andesite, and so on.

Ultramafic Rocks

Ultramafic rocks (< 45% silica) are composed largely of ferromagnesian silicate minerals (Figure 3.11). *Peridotite* contains mostly olivine, lesser amounts of pyroxene, and generally a little plagioclase feldspar (Figure 3.10), whereas pyroxenite is composed predominantly of pyroxene. Because these minerals are dark colored, the rocks are generally black or dark green. Peridotite is probably

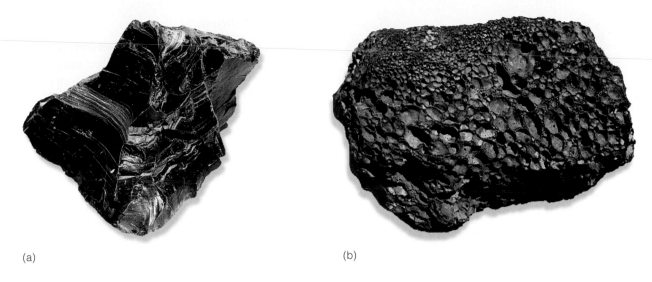

(a)

(b)

Figure 3.9 *(a) Obsidian has a glassy texture because its parent lava cooled too quickly for mineral crystals to form. (b) Vesicles develop when gases expand in cooling lava giving the resulting rocks a vesicular texture. (c) View under a microscope of an igneous rock with a pyroclastic or fragmental texture. The colorless, angular objects are pieces of volcanic glass measuring up to 2 mm.*

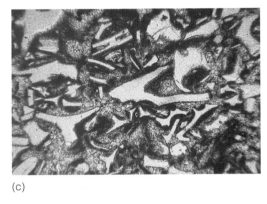

(c)

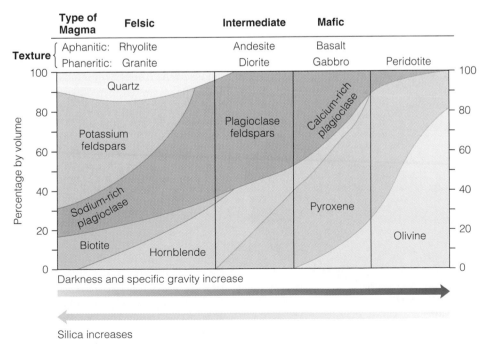

Type of Magma	Felsic		Intermediate	Mafic	
Texture { Aphanitic:	Rhyolite		Andesite	Basalt	
Phaneritic:	Granite		Diorite	Gabbro	Peridotite

Darkness and specific gravity increase

Silica increases

Figure 3.10 *Classification of igneous rocks. This diagram shows the relative proportions of the chief mineral components and the textures of common igneous rocks. For instance, an aphanitic (fine-grained) rock composed mostly of calcium-rich plagioclase and pyroxene is basalt.*

Igneous Rocks—What Are They and What Are Their Characteristics? **65**

Figure 3.11 *The ultramafic igneous rock peridotite. Notice in Figure 3.10 that peridotite is the only phaneritic (coarse-grained) igneous rock not having an aphanitic (fine-grained) equivalent. Peridotite is rare at the surface but is thought to be the rock making up Earth's mantle.*

the rock type composing the upper mantle (see Chapter 7). Ultramafic rocks are generally thought to have originated by concentration of the early formed ferromagnesian silicates that separated from mafic magmas.

Ultramafic lava flows are known in rocks older than 2.5 billion years, but younger ones are rare or absent. The reason is that to erupt, ultramafic lava must have a near-surface temperature of about 1600°C; the surface temperatures of present-day mafic lava flows are generally between 1000° and 1200°C. During early Earth history, though, more radioactive decay heated the mantle to as much as 300°C hotter than now and ultramafic lavas could erupt onto the surface. Because the amount of heat has decreased through time, Earth has cooled, and eruptions of ultramafic lava flows ceased.

Basalt-Gabbro

Basalt and *gabbro* are the fine-grained and coarse-grained rocks, respectively, that crystallize from mafic magmas (45–52% silica) (Figure 3.12a and b). Thus, both have the same composition—mostly calcium-rich plagioclase and pyroxene, with smaller amounts of olivine and amphibole (Figure 3.10). Because they contain a large proportion of ferromagnesian silicates, basalt and gabbro are dark colored; those that are porphyritic typically contain calcium plagioclase or olivine phenocrysts.

Basalt is a very common extrusive igneous rock. Extensive basalt lava flows were erupted in vast areas in Washington, Oregon, Idaho, and northern California (see Chapter 4). Oceanic islands such as Iceland, the Galapagos, the Azores, and the Hawaiian Islands are composed mostly of basalt, and basalt makes up the upper part of the oceanic crust.

Gabbro is much less common than basalt, at least in the continental crust or where it can be easily observed. Small intrusive bodies of gabbro are present in the continental crust, but intermediate and felsic intrusive rocks such as diorite and granite are much more common. However, the lower part of the oceanic crust is composed of gabbro.

Andesite-Diorite

Magmas intermediate in composition (53–65% silica) crystallize to form *andesite* and *diorite,* which are compositionally equivalent fine- and coarse-grained igneous rocks (Figure 3.13). Both are composed predominantly of plagioclase feldspar, with the typical ferromagnesian component being amphibole or biotite (Figure 3.10). Andesite is generally medium to dark gray, but diorite has a salt and pepper appearance because of its white to light gray plagioclase and dark ferromagnesian silicates (Figure 3.13).

Andesite is a common extrusive igneous rock formed from lavas erupted in volcanic chains at convergent plate boundaries. The volcanoes of the Andes Mountains of South America and the Cascade Range in western North America

(a) Basalt

(b) Gabbro

Figure 3.12 *Mafic igneous rocks. (a) Basalt has an aphanitic (fine-grained) texture. Notice the small vesicles. (b) Gabbro is phaneritic (coarse-grained), but has the same composition as basalt. Shiny crystal faces are visible in this specimen.*

(a) Andesite

(b) Diorite

Figure 3.13 *Igneous rocks with an intermediate composition. Both are composed of the same minerals, but andesite (a) is aphanitic whereas diorite (b) has a phaneritic texture. Specimen (a) is actually an andesite porphyry because it contains phenocrysts.*

are composed in part of andesite. Intrusive bodies composed of diorite are fairly common in the continental crust.

Rhyolite-Granite

Rhyolite and *granite* crystallize from felsic magmas (> 65% silica) and are therefore silica-rich rocks (Figure 3.14). They consist largely of potassium feldspar, sodium-rich plagioclase, and quartz, with perhaps some biotite and rarely amphibole (Figure 3.10). Because nonferromagnesian silicates predominate, rhyolite and granite are generally light colored. Rhyolite is fine grained, although most often it contains phenocrysts of potassium feldspar or quartz, and granite is coarse grained. Granite porphyry is also fairly common.

Rhyolite lava flows are much less common than andesite and basalt flows. Recall that the greatest control of viscosity in a magma is silica content. Thus, if a felsic magma rises to the surface, it begins to cool, the pressure

on it decreases, and gases are released explosively, usually yielding rhyolitic pyroclastic materials. The rhyolitic lava flows that do occur are thick and highly viscous and thus move only short distances.

Among geologists, granite has come to mean any coarsely crystalline igneous rock with a composition corresponding to that of the field shown in Figure 3.10. Strictly speaking, not all rocks in this field are granites. For example, a rock with a composition close to the line separating granite and diorite is usually called *granodiorite*. To avoid the confusion that might result from introducing more rock names, we will follow the practice of referring to rocks to the left of the granite-diorite line in Figure 3.10 as *granitic*.

Granitic rocks are by far the most common intrusive igneous rocks, although they are restricted to the continents. Most granitic rocks were intruded at or near convergent plate boundaries during episodes of mountain building. When these mountainous regions are uplifted and eroded,

(a) Rhyolite

(b) Granite

Figure 3.14 *Felsic igneous rocks. (a) Rhyolite (aphanitic) and (b) granite (phaneritic) are typically light-colored because they contain mostly nonferromagnesian silicates. The dark spots in the granite specimen are biotite mica.*

the vast bodies of granitic rocks forming their cores are exposed. The granitic rocks of the Sierra Nevada of California form a composite body measuring about 640 km long and 110 km wide, and the granitic rocks of the Coast Ranges of British Columbia, Canada, are even more voluminous.

Pegmatite

The term *pegmatite* refers to a particular texture rather than a specific composition, but most pegmatites are composed largely of quartz, potassium feldspars, and sodium-rich plagioclase, thus corresponding closely to granite. Their most remarkable feature is the size of their minerals, which measure at least 1 cm across and in some pegmatites are measured in meters or even tens of meters (Figure 3.15). A few pegmatites are mafic or intermediate in composition and are appropriately called gabbro pegmatite and diorite pegmatite.

Many pegmatites are associated with granite plutons and are composed of minerals that formed from the remaining fluid and vapor phases that existed after most of the granite crystallized.

The water-rich vapor phase that exists after most of a magma has crystallized as granite has properties that differ from the magma from which it separated. It has a lower density and viscosity, and thus commonly invades the country rock where it crystallizes as sheetlike igneous bodies known as dikes and, more rarely, as sills. This water-rich vapor phase contains a number of elements that rarely enter into the common minerals that form granite. Pegmatites crystallizing to form coarsely crystalline granite are *simple pegmatites*, whereas those with minerals containing elements such as lithium, beryllium, cesium, tin, and several others are *complex pegmatites* (see Perspective 3.1).

The formation and growth of mineral–crystal nuclei in pegmatites are similar to those processes in magma, but with one critical difference: the vapor phase from which pegmatites crystallize inhibits the formation of nuclei. However, some nuclei do form, and because the appropriate ions in the liquid can move easily and attach themselves to a growing crystal, individual mineral grains have the opportunity to grow to very large sizes.

Other Igneous Rocks

A few igneous rocks, including tuff, volcanic breccia, obsidian, pumice, and scoria, are identified solely by their textures. Much of the fragmental material erupted by volcanoes is *ash*, a designation for pyroclastic materials less than 2 mm in diameter, most of which is broken pieces or shards of volcanic glass. The consolidation of ash forms the pyroclastic rock *tuff* (Figure 3.16). Most tuff is silica-rich and light colored and is appropriately called *rhyolite tuff* (Figure 3.17a). Some ash flows are so hot that as they come to

Figure 3.15 *(a) This pegmatite, the light-colored rocks, is exposed in the Black Hills of South Dakota. (b) Close-up view of a rock from a pegmatite. Most pegmatites have a composition similar to that of granite. The minerals in this specimen measure 2 to 3 cm.*

(a)

(b)

Composition	Felsic ←——————————→ Mafic		
Texture	Vesicular	Pumice	Scoria
	Glassy	Obsidian	
	Pyroclastic or Fragmental	←—————— Volcanic Breccia —————→ Tuff/welded tuff	

Figure 3.16 *Classification of igneous rocks for which texture is the main consideration.*

Pumice is a variety of volcanic glass containing numerous bubble-shaped vesicles that develop when gas escapes through lava and forms a froth (Figure 3.17c). Some pumice forms as crusts on lava flows, and some forms as particles erupted from explosive volcanoes. If pumice falls into water, it can be carried great distances because it is so porous and light that it floats. *Scoria* is dark-colored igneous rock with numerous vesicles (Figure 3.17d) found as a crust on lava flows and as the cinders making up volcanoes known as cinder cones (see Chapter 4). It is more crystalline and denser than pumice, but because of its vesicles it has a lower density than most other igneous rocks.

rest, the ash particles fuse together and form a *welded tuff.* Consolidated deposits of larger pyroclast materials, such as cinders, blocks, and bombs, are *volcanic breccia.*

Both *obsidian* and *pumice* are varieties of volcanic glass (Fig. 3.17b and c). Obsidian may be black, dark gray, red, or brown, with the color depending on the presence of tiny particles of iron minerals. Analyses of numerous samples indicate that most obsidian has a high silica content and is compositionally similar to rhyolite.

(b) Obsidian

(c) Pumice

(a) Tuff

Figure 3.17 *Igneous rocks classified primarily by their textures. (a) Outcrop of tuff in Colorado. It is composed of pyroclastic materials much like those shown in Figure 3.9c. (b) The natural glass obsidian forms when lava cools too quickly for minerals to develop. (c) Pumice is glassy and extremely vesicular. (d) Scoria is also extremely vesicular, but unlike pumice it is darker, more crystalline, and heavier.*

(d) Scoria

Igneous Rocks—What Are They and What Are Their Characteristics? **69**

Complex Pegmatites

Simple pegmatites consist of coarsely crystalline rocks similar to granite (Figure 3.15), but complex pegmatites possess minerals having lithium, cesium, tin, and several other elements that are not normally found in minerals elsewhere. Although complex pegmatites vary considerably, they share some features in common. For one thing, many are characterized by four zones, which are usually designated as border, wall, intermediate, and core zones, the last of which is usually composed entirely of quartz.

The origin of these zones is the subject of continuing debate between proponents of two hypotheses. One group holds that crystallization took place from the border zone inward, with the fluid phase changing as successive layers of minerals crystallized. According to the opposing group, a simple pegmatite formed first and was later partially or completely replaced as hydrothermal fluids circulated through it. No compelling evidence exists that supports one hypothesis to the exclusion of the other.

Another feature shared by complex pegmatites is the occurrence of giant mineral crystals in the inner zone. Crystals of muscovite measuring 2.44 m across have been recovered from pegmatites in Canada, and huge feldspar and quartz crystals are found in pegmatites in many areas. In fact, almost all of the truly giant mineral crystals come from pegmatites.

Pegmatites have been the source of a variety of mineral commodities. In the United States, pegmatites were mined in New England soon after European colonists settled there. In 1803, large transparent sheets of muscovite were recovered from the Ruggles Mine in New Hampshire for use in windows. And by about 1825, feldspar minerals were mined from pegmatites in Connecticut. Gemstones have also been recovered from a number of these pegmatites in the Northeast, especially the Dunton pegmatite in Maine which has yielded hundreds of gem-quality tourmaline crystals (Figure 1).

Anyone familiar with gemstones probably knows that many gem-quality minerals come from pegmatites, but pegmatites are also the sources of many minerals used in various industrial applications. Feldspars, quartz, and muscovite as well as minerals containing elements such as tin, cesium, lithium, rubidium, and beryllium are mined from pegmatites. Pegmatites of the tin-spodumene belt of North Carolina contain vast deposits of lithium, and pegmatites are the only known source of some of the rare earth elements.

Figure 1 *Tourmaline from the Dunton pegmatite mine in Maine.*

Pegmatites are particularly common in the Black Hills of South Dakota, where more than 20,000 have been identified in the country rock adjacent to the Harney Peak Granite; only about 1% of these are complex pegmatites. The stone images of Presidents Washington, Lincoln, Jefferson, and Theodore Roosevelt on Mount Rushmore were carved into rocks of the Harney Peak Granite, which was designated Mount Rushmore National Memorial in 1927 (Figure 3.1a). These pegmatites formed about 1.7 billion years ago when the granite was emplaced as a composite pluton consisting of numerous dikes and sills. More recent uplift and erosion of the area exposed the granite and its associated pegmatites. A well-known pegmatite in this area is the Etta pegmatite, with spodumene ($LiAlSi_2O_6$) crystals more than 12 m long (Figure 2). Lithium from spodumene was mined from the Etta Pegmatite until 1960, but since then lithium has been derived from more economical sources such as dry lakebeds in arid regions.

Complex pegmatites worldwide have yielded numerous gem minerals, but one of the most famous in North America is the Himalaya Mine in the Mesa Grande pegmatite district of San Diego County, California. It is one of the most productive gem pegmatites anywhere in the world; among other minerals it has yielded some of the best specimens of gem-quality tourmaline found anywhere (see Chapter 2 opening photo). And this is only one pegmatite in one of about 20 pegmatite districts in the Peninsular Ranges batholith in southern California. Most major mineral collections in North American museums include specimens from the Himalaya Mine.

A particularly well-exposed and accessible pegmatite is at the Harding Pegmatite Mine in Taos County, New Mexico, which is now owned by the Earth and Planetary Sciences Department of the University of New Mexico. It was mined from 1919 to 1930, but now serves as a textbook example for geology students as well as rock hounds (amateur mineral and rock collectors). Visitors to the mine are allowed to collect and keep small mineral specimens.

Figure 2 *The crystal indicated in the photo measures more than 12 m long and is in the Black Hills of South Dakota.*

Spodumene crystal

Intrusive Igneous Bodies: Plutons—Their Characteristics and Origins

Unlike volcanism and the origin of volcanic rocks, both of which can be observed, intrusive igneous activity can be studied only indirectly. Intrusive igneous bodies known as **plutons** form when magma cools and crystallizes within the crust (Figure 3.18), so plutons can be observed only after uplift and erosion has exposed them at the surface. And geologists cannot duplicate the conditions under which they form except in small laboratory experiments. Accordingly, geologists face a much greater challenge in interpreting the mechanisms whereby plutons are created. The magma that cools to form plutons is emplaced in Earth's crust mostly at divergent and convergent plate boundaries, which are also areas of active volcanism.

Several types of plutons are recognized, all of which are defined by their geometry (three-dimensional shape) and relationships to the country rock (Figure 3.18). Geometrically, plutons may be characterized as massive or irregular, tabular or sheetlike, cylindrical, or mushroom shaped. Plutons are also described as concordant or discordant. A **concordant** pluton, such as a sill, has boundaries parallel to the layering in the country rock. A **discordant** pluton, such as a dike, has boundaries that cut across the layering of the country rock (Figure 3.18).

Dikes and Sills

Both dikes and sills are tabular or sheetlike plutons, but dikes are discordant whereas sills are concordant (Figure 3.18). **Dikes** are common intrusive features (Figure 3.19), most of which are small bodies measuring 1 or 2 m across, but they range from a few centimeters to more than 100 m thick. Dikes are emplaced within fractures or where the fluid pressure is great enough for them to form their own fractures during emplacement.

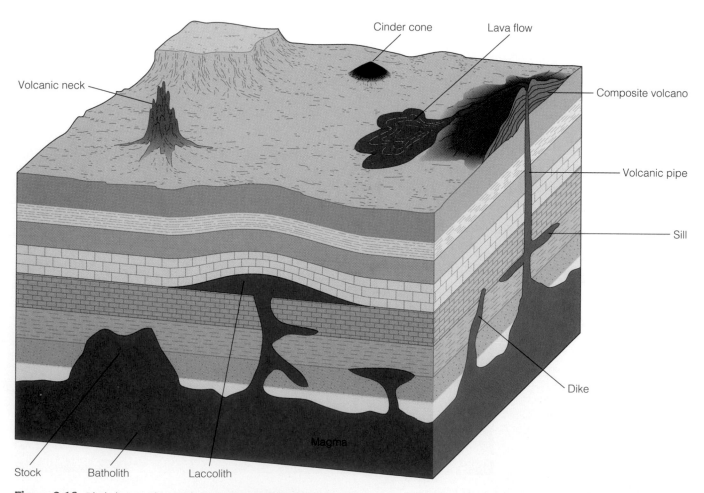

Volcanic neck

Cinder cone

Lava flow

Composite volcano

Volcanic pipe

Sill

Dike

Stock   Batholith   Laccolith

Magma

Figure 3.18 *Block diagram showing the various types of plutons. Notice that some plutons cut across the layering in the country rock and are thus discordant, whereas others parallel the layering and are concordant.*

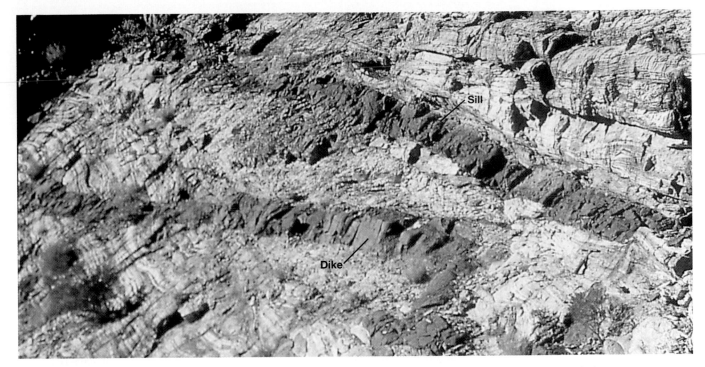

Figure 3.19 *The dark layer cutting diagonally across the rock layers is a dike. The other dark layer is a sill because it parallels the layering.*

Erosion of the Hawaiian volcanoes exposes dikes in rift zones, the large fractures that cut across these volcanoes. The Columbia River basalts in Washington issued from long fissures, and the magma that cooled in the fissures formed dikes. Some of the large historic fissure eruptions are underlain by dikes; for example, dikes underlie both the Laki fissure eruption of 1783 in Iceland and the Eldgja fissure, also in Iceland, where eruptions occurred in A.D. 950 from a fissure 300 km long.

Concordant, sheetlike plutons are **sills,** many of which are a meter or less thick, although some are much thicker (Figure 3.19). A well-known sill in the United States is the Palisades sill that forms the Palisades along the west side of the Hudson River in New York and New Jersey. It is exposed for 60 km along the river and is up to 300 m thick.

Most sills have been intruded into sedimentary rocks, but eroded volcanoes also reveal that sills are commonly injected into piles of volcanic rocks. In fact, some of the inflation of volcanoes preceding eruptions may be caused by the injection of sills (see Chapter 4).

In contrast to dikes, which follow zones of weakness, sills are emplaced when the fluid pressure is so great that the intruding magma actually lifts the overlying rocks. Because emplacement requires fluid pressure exceeding the force exerted by the weight of the overlying rocks, sills are typically shallow intrusive bodies.

Laccoliths

Laccoliths are similar to sills in that they are concordant, but instead of being tabular, they have a mushroomlike geometry (Figure 3.18). They tend to have a flat floor and are domed up in their central part. Like sills, laccoliths are rather shallow intrusive bodies that actually lift up the overlying strata when the magma is intruded. In this case, however, the strata are arched upward over the pluton (Figure 3.18). Most laccoliths are rather small bodies. The best-known laccoliths in North America are in the Henry Mountains of southeastern Utah.

Volcanic Pipes and Necks

A volcano has a cylindrical conduit known as a **volcanic pipe** connecting its crater with an underlying magma chamber (Figure 3.18). Through this structure magma and associated gases rise to the surface. When a volcano ceases to erupt, its slopes are attacked by water, gases, and acids and it erodes, but the magma that solidified in the pipe is commonly more resistant to alteration and erosion. Consequently, much of the volcano is eroded but the pipe remains as a monolithic remnant known as a **volcanic neck** (Figure 3.18). Several volcanic necks are found in the southwestern United States, especially in Arizona and New Mexico, and others are recognized elsewhere (see Perspective 3.2).

Some Remarkable Volcanic Necks

Lava flows and layers of pyroclastic materials composing the flanks of volcanoes are not usually as resistant to chemical and mechanical alteration (weathering—see Chapter 5) as are the rocks that solidified in a volcanic pipe. Accordingly, when extinct volcanoes weather and erode, the volcanic pipe commonly remains as an erosional remnant—a volcanic neck. The geologic origin of volcanic necks is well known, but the necks themselves, rising as monolithic structures above otherwise rather flat land, are awe inspiring and the subject of legends.

Volcanic necks are found in many areas of recently active volcanism, such as the rather small one in LePuy, France, that rises only 79 m above the surrounding countryside. Even though small compared to many other volcanic necks, it is not only scenic but also the site upon which the 11th-century chapel of Saint Michel d'Aiguilhe was built (Figure 1). This monolith is so steep that materials and tools used in the chapel's construction had to be hauled up in baskets.

Many volcanic necks are found in the United States, especially in Arizona and New Mexico. The most impressive one is Shiprock in northwestern New Mexico, which rises nearly 500 m above the surrounding plain and is visible from 160 km. Radiating outward from this conical mass are three dikes that stand like walls above the adjacent plain (Figure 2). The Navajo call Shiprock "Tsae-bidahi," which means "Winged Rock" or "Rock with Wings." According to Navajo legend it represents a giant bird that brought the Navajo people from the north, and the dikes are snakes that have turned to stone.

Figure 1 *A volcanic neck rising 79 m above the surface in LePuy, France. Workers on the chapel had to haul building materials and their tools up in baskets.*

(a)

(b)

Figure 2 *(a) Shiprock, a volcanic neck in northwestern New Mexico, rises more than 550 m above the surrounding plain. (b) View of one of the dikes radiating from Shiprock.*

you might recall that Devil's Tower was featured in the 1977 movie *Close Encounters of the Third Kind.* The Cheyenne and Sioux Indians call Devil's Tower Mateo Teepee, meaning "Grizzly Bear Lodge." It was also called the "Bad God's Tower," and reportedly "Devil's Tower" is a translation of this phrase. The tower's most conspicuous features are the near-vertical lines that according to Sioux legend are scratch marks made by a gigantic grizzly bear. In one legend the bear made the marks while pursuing a group of children. Another tells of six brothers and a woman who were also pursued by a grizzly bear. One brother carried a small rock, and when he sang a song it grew into the present size of Devil's Tower.

The "scratch marks" are actually lines formed by the intersections of columnar joints, features that form in response to cooling and contraction in some lava flows and plutons (see Chapter 4). Many of the columns formed by this kind of fracturing are six-sided, but columns with four, five, and seven sides are present as well. The larger columns are 2.5 m across, and the pile of rubble at the tower's base is simply an accumulation of collapsed columns.

Absolute dating (see Chapter 17) of one of the dikes indicates that Shiprock is about 27 million years old. The material composing Shiprock is known as tuff-breccia, consisting of fragmental volcanic debris along with pieces of various metamorphic, igneous, and sedimentary rocks that the magma penetrated. Apparently the magma was emplaced during explosive eruptions. Shiprock was a favorite with rock climbers for many years until the Navajos put an end to all climbing on their reservation, and because Shiprock itself is sacred, they discourage visits to the rock.

Devil's Tower in northeastern Wyoming is likely another volcanic neck (Figure 3), but another interpretation is possible. Geologists agree that it represents magma that cooled to form a small pluton, and that erosion has exposed it in its present form. However, opinion is divided on whether it is truly a volcanic neck, or an eroded laccolith. In either case, it along with other similar features in the area formed when magma intruded into the crust about 45 to 50 million years ago. President Theodore Roosevelt established Devil's Tower as our first national monument in 1906.

At 260 m high, Devil's Tower is visible from 48 km away, and served as a landmark for early travelers in the area. Some of

Figure 3 *Devil's Tower in northeastern Wyoming. It rises about 260 m above its base and can be seen from 48 km away. It might be a volcanic neck or an eroded laccolith. The vertical lines result from the intersections of fractures known as* columnar joints. *According to Cheyenne legend, however, the lines are deep scratches made by a gigantic grizzly bear.*

Batholiths and Stocks

By definition a **batholith,** the largest of all plutons, must have at least 100 km² of surface area, and most are far larger (Figure 3.20a). A **stock,** in contrast, is similar but smaller, although some stocks are simply the exposed parts of large plutons that once more fully exposed by erosion are batholiths (Figure 3.20b). Both batholiths and stocks are mostly discordant, although locally they may be concordant, and batholiths, especially, consist of multiple intrusions. In other words, a batholith is a large composite body produced by repeated, voluminous intrusions of magma in the same region. The coastal batholith of Peru, for instance, was emplaced during a period of 60 to 70 million years and is made up of as many as 800 individual plutons.

The igneous rocks composing batholiths are mostly granitic, although diorite may also be present. Most batholiths are emplaced along convergent plate boundaries. Examples include the Coast Range batholith of British Columbia, Canada, the Idaho batholith, and the Sierra Nevada batholith in California.

A number of mineral resources are found in rocks of batholiths and stocks and in the adjacent country rocks. Granitic rocks are the primary source of gold, which forms from mineral-rich solutions moving through cracks and fractures of the igneous body. The copper deposits at Butte, Montana, are in rocks near the margins of the granitic rocks of the Boulder batholith. Near Salt Lake City, Utah, copper is mined from the mineralized rocks adjacent to the Bingham stock, a composite pluton composed of granite and granite porphyry.

How Are Batholiths Emplaced in Earth's Crust?

Geologists realized long ago that the emplacement of batholiths posed a space problem; that is, what happened to the rock that formerly occupied the space now occupied by a batholith? One solution to this space problem was to propose that no displacement had occurred, but rather that granite batholiths had been formed in place by alteration of the country rock through a process called *granitization.* According to this view, granite did not originate as a magma, but rather from hot, ion-rich solutions that simply altered the country rock and transformed it into granite.

(a)

(b)

Figure 3.20 *(a) View of granitic rocks of the Sierra Nevada batholith in Yosemite National Park, California. The near vertical cliff is El Capitan, meaning "The Chief." It rises more than 900 m above the valley floor, making it the highest unbroken cliff in the world. (b) Granitic rocks in a small stock at Castle Crags State Park, California.*

Granitization is a solid-state phenomenon so it is essentially an extreme type of metamorphism (see Chapter 6).

Granitization is no doubt a real phenomenon, but most granitic rocks show clear evidence of an igneous origin. For one thing, if granitization had taken place one would expect the change from country rock to granite to take place gradually over some distance. However, in almost all cases no such gradual change can be detected. Thus, granitic rocks most commonly have what geologists refer to as a sharp contact with adjacent rocks. Another feature indicating an igneous origin for granitic rocks is the alignment of elongate minerals parallel with their contacts, which must have occurred when magma was injected. A few granitic rocks lack sharp contacts, and gradually change in character until they resemble the adjacent country rock. These probably did originate by granitization. In the opinion of most geologists, however, only small quantities of granitic rock could form by this process, so it cannot account for the huge volume of granitic rocks of batholiths. These geologists think that an igneous origin for almost all granite is clear, but they still must deal with the space problem.

One solution is that these large igneous bodies melted their way into the crust. In other words, they simply assimilated the country rock as they moved upward (Figure 3.6). The presence of inclusions, especially near the tops of such intrusive bodies, indicates that assimilation does occur. Nevertheless, as we noted previously, assimilation is a limited process because magma is cooled as country rock is assimilated; calculations indicate that far too little heat is available in a magma to assimilate the huge quantities of country rock necessary to make room for a batholith.

Geologists now generally agree that batholiths were emplaced as magma and that the magma, being less dense than the rock from which it was derived, moved upward toward the surface. Recall, however, that granite is derived from viscous felsic magma and, thus, it rises slowly. It appears that the magma deforms and shoulders aside the country rock, and as it rises further, some of the country rock fills the space beneath the magma (Figure 3.21). A somewhat analogous situation was discovered in which large masses of sedimentary rock known as rock salt rise through the overlying rocks to form salt domes.

Salt domes are recognized in several areas of the world, including the Gulf Coast of the United States. Layers of rock salt exist at some depth, but salt is less dense than most other types of rock materials. Thus, when under pressure, it rises toward the surface even though it remains solid, and as it moves upward, it pushes aside and deforms the country rock (Figure 3.22). Natural examples of rock salt flowage are known, and it can easily be demonstrated experimentally. For example, in the arid Middle East, salt moving upward in the manner described actually flows out at the surface.

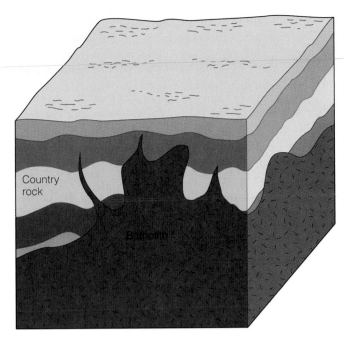

Figure 3.21 *Emplacement of a hypothetical batholith. As the magma rises, it shoulders aside and deforms the country rock.*

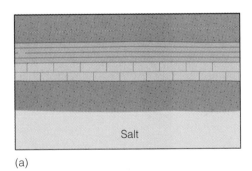

(a)

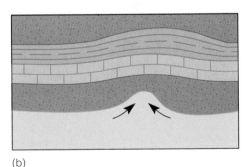

(b)

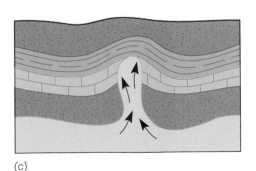

(c)

Figure 3.22 *Three stages in the origin of a salt dome. Rock salt is a low-density sedimentary rock that (a) when deeply buried (b) tends to rise toward the surface, (c) pushing aside and deforming the country rock and forming a dome. Salt domes are thought to rise in much the same manner as batholiths are intruded into the crust.*

Some batholiths do indeed show evidence of having been emplaced forcefully by shouldering aside and deforming the country rock. This mechanism probably occurs in the deeper parts of the crust where temperature and pressure are high and the country rocks are easily deformed in the manner described. At shallower depths, however, the crust is more rigid and tends to deform by fracturing. In this environment, batholiths may be emplaced by **stoping,** a process involving detaching and engulfing pieces of country rock by rising magma (Figure 3.23). According to this concept, magma moves upward along fractures and the planes separating layers of country rock. Eventually, pieces of country rock are detached and settle into the magma. No new room is created during stoping; the magma simply fills the space formerly occupied by country rock.

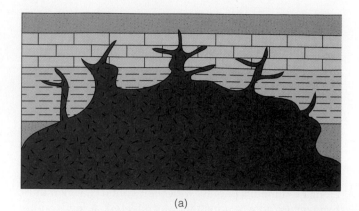

(a)

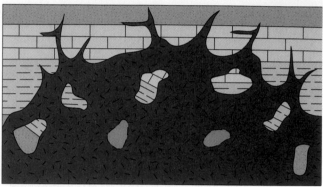

(b)

Figure 3.23 *Emplacement of a batholith by stoping. (a) Magma is injected into fractures and planes between layers in the country rock. (b) Blocks of country rock are detached and engulfed in the magma, thus making room for the magma to rise further. Some of the engulfed blocks might be assimilated, and some might remain as inclusions (Figure 3.6).*

Chapter Summary

1. Magma is molten rock material below Earth's surface, whereas lava is magma that reaches the surface. The silica content of magmas varies and serves to differentiate felsic, intermediate, and mafic magmas.

2. The viscosity of magma and lava depends mostly on their temperature and composition. Silica-rich (felsic) lava is more viscous than silica-poor (mafic) lava.

3. Minerals crystallize from magma and lava when small crystal nuclei form and grow.

4. Volcanic rocks generally have aphanitic textures because of rapid cooling, whereas slow cooling and phaneritic textures characterize plutonic rocks. Igneous rocks with a porphyritic texture have mineral crystals of markedly different sizes. Other igneous rock textures include vesicular, glassy, and pyroclastic.

5. The composition of igneous rocks is determined largely by the composition of the parent magma. It is possible, however, for an individual magma to yield igneous rocks of different compositions.

6. Under ideal cooling conditions, a mafic magma yields a sequence of different minerals that are stable within specific temperature ranges. This sequence, called Bowen's reaction series, consists of a discontinuous branch and a continuous branch.

   a. The discontinuous branch contains only ferromagnesian minerals, each of which reacts with the melt to form the next mineral in the sequence.

   b. The continuous branch involves changes only in plagioclase feldspar as sodium replaces calcium in the crystal structure.

7. The ferromagnesian minerals that form first in Bowen's reaction series can settle and become concentrated near the base of a magma chamber or intrusive body. Such settling of iron- and magnesium-rich minerals causes a chemical change in the remaining melt.

8. A magma can be changed compositionally when it assimilates country rock, but this process usually has only a limited effect. Magma mixing may also bring about compositional changes in magmas.

9. Most igneous rocks are classified on the basis of their textures and composition. Two fundamental groups of igneous rocks are recognized: volcanic or extrusive rocks, and plutonic or intrusive rocks.

   a. Common volcanic rocks include tuff, rhyolite, andesite, and basalt.

   b. Common plutonic rocks include granite, diorite, and gabbro.

10. Pegmatites are coarse-grained igneous rocks, most of which have an overall composition similar to that of granite. Crystallization from a vapor-rich phase left over after the crystallization of granite accounts for the large mineral crystals in pegmatites.

11. Plutons are igneous bodies that formed in place or were intruded into Earth's crust. Various types of plutons are classified by their geometry and whether they are concordant or discordant.

12. Common plutons include dikes (tabular geometry, discordant); sills (tabular geometry, concordant); volcanic necks (cylindrical geometry, discordant); laccoliths (mushroom shaped, concordant); and batholiths and stocks (irregular geometry, mostly discordant).

13. By definition batholiths must have at least 100 km^2 of surface area; stocks are similar but smaller. Batholiths are large composite bodies consisting of many plutons emplaced over a long period of time.

14. Most batholiths appear to have formed in the cores of mountain ranges during episodes of mountain building.

15. Some geologists think that granite batholiths are emplaced when felsic magma moves upward and shoulders aside and deforms the country rock. The upward movement of rock salt and the formation of salt domes provide a somewhat analogous situation.

Important Terms

aphanitic texture	felsic magma	magma mixing	sill
assimilation	igneous rock	phaneritic texture	stock
batholith	intermediate magma	pluton	stoping
Bowen's reaction series	laccolith	plutonic (intrusive igneous)	vesicle
concordant pluton	lava	rock	viscosity
country rock	lava flow	porphyritic texture	volcanic neck
crystal settling	mafic magma	pyroclastic materials	volcanic pipe
dike	magma	pyroclastic (fragmental)	volcanic (extrusive igneous)
discordant pluton	magma chamber	texture	rock

Review Questions

1. An igneous rock having the composition of granite and very large mineral crystals is
   a. ____ basalt; b. ____ pegmatite; c. ____ obsidian; d. ____ tuff; e. ____ rhyolite.

2. The first mineral to crystallize in the continuous branch of Bowen's reaction series is
   a. ____ iron-deficient pyroxene; b. ____ calcium-rich plagioclase; c. ____ sodium-rich amphibole; d. ____ muscovite and biotite; e. ____ silica-rich quartz.

3. The size of the mineral grains composing an igneous rock is a useful criterion for determining whether the rock is____ or ____
   a. ____ discordant/concordant; b. ____ vesicular/fragmental; c. ____ porphyritic/felsic; d. ____ assimilated/ultramafic; e. ____ plutonic/volcanic.

4. Which of the following pairs of igneous rocks have the same mineral composition?
   a. ____ granite-scoria; b. ____ andesite-rhyolite; c. ____ basalt-gabbro; d. ____ pumice-diorite; e. ____ peridotite-rhyolite.

5. Which one of the following is a discordant pluton?
   a. ____ sill; b. ____ laccolith; c. ____ lava flow; d. ____ dike; e. ____ porphyritic andesite

6. An extrusive igneous rock composed mostly of pyroxene and calcium plagioclase is
   a. ____ basalt; b .____ pumice; c. ____ gabbro; d. ____ diorite; e. ____ granite.

7. Any igneous rock having minerals large enough to be seen without magnification is said to have a(an)____texture and is probably____
   a. ____ phaneritic/intrusive; b. ____ laccolith/a batholith; c. ____ assimilated/volcanic; d. ____ aphanitic/a lava flow; e. ____ fragmental/obsidian.

8. The process whereby magma reacts with and incorporates country rock is
   a. ____ crystal differentiation; b .____ granitization; c. ____ plutonism; d. ____ magma mixing; e. ____ assimilation.

9. An intermediate magma is one
   a. ____ with more magnesium than calcium; b. ____ having 53% to 65% silica; c. ____ that crystallizes to form gabbro and basalt; d.____ in which early-formed quartz crystals settle; e.____that formed only during early Earth history.

10. Which of the following regarding batholiths is true?
    a. ____ They form as a series of overlapping lava flows; b.____ Most are composed of granitic rock and consist of multiple intrusions; c.____ The only ones known are small cylindrical bodies that were emplaced explosively; d.____ The Cascade Range volcanoes in the Pacific Northwest are eroded batholiths; e.____ They are found in areas where salt domes are common.

11. The phenomenon whereby a rising body of magma detaches and engulfs pieces of country rock thereby making room for itself is known as
    a. ____ stoping; b.____ granitization; c.____ magma mixing; d.____ crystal settling; e.____ magmatic doming.

12. An igneous rock possessing a combination of minerals of markedly different sizes is
    a. ____ a natural glass; b.____ formed by explosive volcanism; c.____ a porphyry; d.____ muscovite; e.____ produced by very rapid cooling.

13. Describe the process whereby minerals form and grow in a cooling magma.

14. Explain why volcanic rocks are generally aphanitic whereas most plutonic rocks are phaneritic. Give an example of an igneous rock with each texture.

15. Why are felsic lava flows so much more viscous than mafic lava flows? Also, how does viscosity control the shape of a lava flow?

16. Describe the sequence of events leading to the origin of a volcanic neck.

17. How do crystal settling and assimilation bring about changes in the composition of magma? Cite evidence indicating that both of these processes actually take place.

18. Compare the continuous and discontinuous branches of Bowen's reaction series. Why are potassium feldspar and quartz not considered part of either branch?

19. Sills and dikes have the same three-dimensional shape or geometry, so how is it possible to distinguish one from the other? Explain fully.

20. What are pyroclastic materials and what kinds of igneous rock is formed from them?

21. Use Figure 3.10 to classify an aphanitic igneous rock composed mostly of potassium feldspars, quartz, plagioclase feldspars, and biotite.

22. What does the term pegmatite mean, and why are such large mineral crystals found in pegmatites?

23. How might you resolve whether a granite batholith was emplaced as magma rather than by granitization?

24. Two aphanitic igneous rocks have the following compositions: Specimen 1: 15% biotite, 15% sodium-rich plagioclase, 60% potassium feldspar, and 10% quartz. Specimen 2: 10% olivine, 55% pyroxene, 5% hornblende, and 30% calcium-rich plagioclase. Use Figure 3.10 and classify these rocks.

World Wide Web Activities

For these Web site addresses, along with current updates and exercises, log on to **http://www.brookscole.com/geo**

Rob's Granite Page

This site is the work of Robert M. Reed, a Ph.D. candidate in the Department of Geological Sciences, University of Texas at Austin. As you might guess, it is concerned mainly with granite. It has many links to other web sites with information about granite, as well as information about Reed's own research. Check out the various links from his home page.

University of Tulsa Department of Geosciences Igneous Rocks and Processes

This web site contains much information about igneous rocks and processes in general, as well as a few links to other sites.

Igneous Rocks

This site is part of the Georgia Science On-Line project. It was created by Dr. Pamela J. W. Gore for Geology 101 at DeKalb College, Clarkston, Georgia. It has good images of all common igneous rocks, along with descriptions of their textures and compositions. It also has a classification chart similar to the one in the text. Scroll down and read about the various igneous rocks shown.

 # CD-ROM Exploration

Exploring your *Earth Systems Today* CD-ROM will add to your understanding of the material in this chapter.

Topic: Earth's Materials

Module: Rocks and the Rock Cycle

Explore activities in this module to see if you can discover the following for yourself:

Using the "rock laboratory" portion of this module, observe the formation of both a phaneritic and a porphyritic igneous rock through the process of crystallization. Based on what you see in the "rock laboratory" write a short description of how a porphyritic rock crystallizes.

What is the origin of "interlocking texture" among crystals in an igneous rock? How does forming from a liquid contribute to development of this texture?

CHAPTER 4

Volcanism and Volcanoes

A lava fountain and lava flows issuing from Pu'u O'o at Kilauea Volcano, Hawaii.

OBJECTIVES

At the end of this chapter, you will have learned that

■ In addition to lava, volcanoes erupt large quantities of gases and solid particles such as ash.

■ Even though every volcano is unique, most can be classified as one of only a few basic types.

■ The shapes of several types of volcanoes are determined by their eruptive style.

■ Some volcanoes erupt explosively whereas others erupt rather quietly and pose little danger to humans.

■ Eruptions in some areas yield vast overlapping lava flows or layers of ash and other particles.

■ Geologists use several methods to monitor some volcanoes in an effort to predict future eruptions.

■ A semi-quantitative scale is used to express the size of an eruption.

■ Most volcanism takes place in well-defined belts at or near divergent and convergent plate boundaries.

■ The few active volcanoes far from any plate boundary probably result from localized melting of rock at hot spots.

Thousands of volcanoes have erupted during historic time, but few have received as much attention as the eruptions of August 24–25, A.D. 79, that destroyed the cities of Pompeii, Herculaneum, and Stabiae. All were thriving Roman communities along the shore of the Bay of Naples in what is now Italy (Figure 4.1). Fortunately for us, Pliny the Younger, whose uncle, Pliny the Elder, died while trying to investigate the eruption, recorded the event in some detail.

When the eruption began, both Plinys were about 30 km away in the town of Misenum across the Bay of Naples. At first they were not particularly alarmed when a large cloud appeared over Mount Vesuvius, but soon it became apparent that an eruption was beginning. Pliny the Elder, an admiral in the Roman Navy and naturalist, set sail to investigate and perhaps to evacuate friends from the communities closest to the volcano. When be arrived, pyroclastic materials ejected by the volcano had already accumulated along the shore, preventing him from landing, so he changed course and made landfall at Stabiae. The next morning while fleeing from noxious fumes emitted by the volcano, the elder Pliny died, probably of a heart attack.

Pliny the Younger's account of Mount Vesuvius's eruption is so vivid that similar eruptions during which large quantities of pumice are blasted into the air are now referred to as *plinian*. Pompeii, a city of about 20,000 people only 9 km downwind from the volcano, was buried by nearly 3 m of pyroclastic materials, which covered all but the tallest buildings. Pompeii was so utterly destroyed that even though its location was known, it was largely forgotten until 1595 when some of the city was uncovered during the construction of an aqueduct. During the seventeenth and eighteenth centuries, the city was ravaged for arti-facts to grace the homes of wealthy Europeans.

Systematic excavations beginning in the 1800s have exposed much of Pompeii, now a popular tourist attraction (Figure 4.1b). Probably the most famous attractions are the molds of human bodies that formed when the volcanic debris hardened before the bodies decayed. About 2000 victims have been discovered in the city, including a dog still chained to a post, but what happened to the city's other residents is not known. Some probably escaped by sea or overland, but many may remain buried in the debris beyond the city.

Pompeii was the largest city destroyed by Mount Vesuvius, but it was not the only one to suffer such a fate. A number of others, such as Stabiae, still remain buried beneath a vast blanket of debris. Herculaneum was just as close to the volcano as Pompeii, but—as opposed to Pompeii, which was buried rather gradually—it was overwhelmed in minutes by surges of incandescent pyroclastic materials in what are known as nuée ardentes. The debris covered the town to a depth of about 20 m. Prior to 1982, only about a dozen skeletons had been discovered, so it was thought that many of Herculaneum's citizens had escaped; excavations of the city's waterfront, however, revealed hundreds of human skeletons. Many of these were discovered in chambers that probably housed fishing boats, and many more skeletons have since been discovered along the ancient beach.

Before the A.D. 79 eruptions began, Mount Vesuvius was probably not considered a threat although residents of that area were certainly aware of volcanoes. However, this particular volcano had been quiet for at least 300 years, so it no doubt was of little concern. But it is only one in a chain of volcanoes along Italy's south coast where the African plate is subducted beneath the Euro-

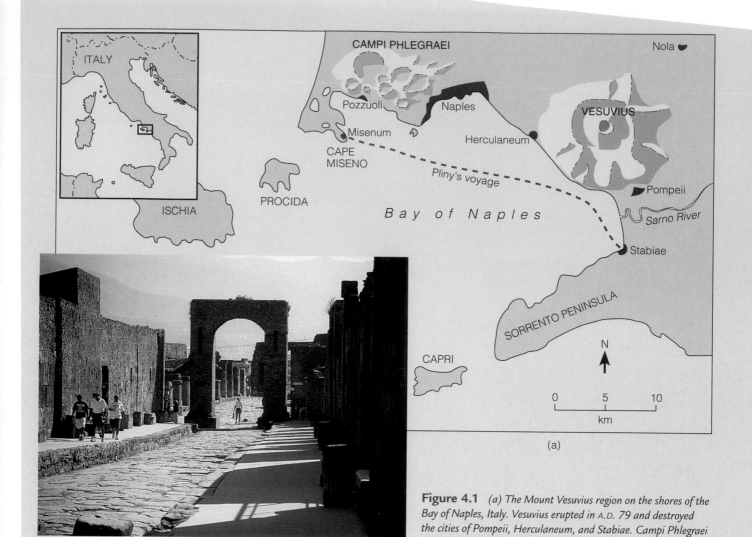

Figure 4.1 *(a) The Mount Vesuvius region on the shores of the Bay of Naples, Italy. Vesuvius erupted in A.D. 79 and destroyed the cities of Pompeii, Herculaneum, and Stabiae. Campi Phlegraei northwest of Vesuvius is another area of active volcanism. (b) The excavated ruins of Pompeii are now a popular tourist attraction.*

pean plate. Since A.D. 79, Mount Vesuvius has erupted 80 times, most violently in 1631 and 1906; it last erupted in 1944.

Mount Vesuvius lies just to the south of a 13-km-diameter volcanic depression, or caldera, known as Campi Phlegraei (Figure 4.1a). The last eruption within the caldera took place in 1538, but movement of magma beneath the surface has generated so many earthquakes that about half the population of Pozzuoli, a city within the caldera, has moved away. Furthermore, the city has experienced vertical ground movements also related to moving masses of magma. In Roman times, Pozzuoli stood high, but by A.D. 1000 it had sunk to about 11 m below sea level, only to rise 12 m between 1000 and 1538. In fact, dur-

ing a 1538 volcanic outburst from nearby Monte Nuovo, Pozzuoli rose 4 m in just two days! Another episode of uplift began in 1982 when the city rose 1.8 m in about a year and a half.

Numerous communities on the shores of the Bay of Naples are within easy reach of eruptions from Mount Vesuvius or Campi Phlegraei caldera. Indeed, the city of Naples is situated on the eastern margin of Campi Phlegraei and only a short distance west of Mount Vesuvius (Figure 4.1). Probably no other major population center in the world is in more danger from devastation by a volcano. The fact that the region remains geologically active was tragically demonstrated in 1980 when an earthquake killed 3000 people in the Naples area.

Introduction

Most magma cools and crystallizes to form plutonic rocks (see Chapter 3), but some rises to the surface where it is erupted as lava flows or pyroclastic materials. Truly, erupting volcanoes are the most impressive manifestation of Earth's dynamic internal processes. Glowing streams of lava and lava fountains are impressive, especially at night (see chapter-opening photo), and explosive eruptions of pyroclastic materials remind us of the violence of some eruptions. Usu in Japan, for instance, began erupting violently during March 2000, and even though it has caused no fatalities or injuries, thousands of people were evacuated from towns and villages near the volcano.

Although most of us will never experience an eruption, such events are commonplace in some parts of the world. Residents of the island of Hawaii, the Philippines, Japan, and Iceland are well aware of volcanic eruptions, but eruptions in the United States are not very common except in Hawaii and Alaska. In the mainland United States only California and Washington have had eruptions during this century, both in the Cascade Range, which stretches from northern California through Oregon and Washington, and into southern British Columbia, Canada. Canada has had no eruptions during historic time.

The fact that lava flows and explosive volcanism can cause property damage, injuries, fatalities, and at least short-term climatic changes indicates that some eruptions are catastrophic events. And indeed some are, from the human perspective, but, ironically, when considered in the context of Earth history, volcanism is actually a constructive process. Oceanic islands such as the Azores, the Galapagos, Iceland, and the Hawaiian Islands owe their existence to volcanism. Oceanic crust is continually produced by volcanism at spreading ridges, and gases released at volcanoes during Earth's early history probably formed the atmosphere and surface waters. Furthermore, chemical and mechanical alteration of lava flows and pyroclastic materials in tropical areas such as Indonesia convert them to productive soils.

Why Should You Study Volcanism and Volcanoes?

Why should you study volcanism and volcanoes? The complex interactions among Earth's major systems are well demonstrated by volcanic eruptions. Volcanism, especially the eruption of gases and pyroclastic materials, has an immediate and profound effect on the atmosphere, hydrosphere, and biosphere, at least in the vicinity of an eruption. And in some cases, the effects are felt worldwide as they were following the eruptions of Tambora (1815), Krakatau (1883), and Pinatubo (1991). Lava flows, in con-trast, commonly depicted in movies as a great threat, actually have little impact on humans, although they might destroy croplands and homes. They are, however, responsible in part for the origin and growth of a number of oceanic islands.

Probably no other geologic phenomenon has captured the public imagination more than erupting volcanoes. Volcanism has yielded some spectacular scenery and interesting features in several of our western national parks and monuments. Craters of the Moon National Monument, Idaho; Lassen Volcanic National Park, California; Crater Lake National Park, Oregon; and Mount Rainier National Park, Washington, provide visitors with opportunities to see the effects of several types of volcanic activity.

Volcanism

Volcanism includes those processes whereby lava and its contained gases and pyroclastic materials are extruded onto the surface or into the atmosphere, as well as the origin of volcanoes and volcanic or extrusive igneous rocks. At present, about 550 volcanoes are *active*—that is, they have erupted during historic time. At any one time about 12 volcanoes are erupting somewhere in the world; although much of this activity is minor, large eruptions are not uncommon. Mauna Loa and Kilauea on the island of Hawaii; Mount Fuji, Japan; Mount Vesuvius, Italy; Mount St. Helens, Washington; and Mount Pinatubo in the Philippines are active volcanoes.

Volcanoes have been active on several other planets and moons in the solar system during the past, but so far, only two of these bodies appear to have active volcanoes now (see Chapter 19). Io, a moon of Jupiter about the size and density of Earth's moon, is by far the most volcanically active body in the solar system. Many of its hundred or so volcanoes are erupting at any given time. The Neptunian moon Triton also appears to have active volcanoes.

Besides active volcanoes, numerous *dormant* volcanoes exist that have not erupted recently but may do so again. Mount Vesuvius in Italy showed no signs of activity in human memory until A.D. 79, when it erupted violently and destroyed the cites of Herculaneum, Pompeii, and Stabiae (the Prologue). Mount Pinatubo in the Philippines lay dormant for 600 years until 1991 when it produced what was probably the largest volcanic outburst in 50 years. Some volcanoes have not erupted during historic time and show no evidence of doing so again. Thousands of these *extinct* or *inactive* volcanoes are known.

Volcanic Gases

Samples taken from present-day volcanoes indicate that 50% to 80% of all volcanic gases are water vapor. Lesser amounts of carbon dioxide, nitrogen, sulfur gases, especially sulfur dioxide and hydrogen sulfide, and very small amounts of carbon monoxide, hydrogen, and chlorine are also commonly emitted. In areas of recent volcanism, such as Lassen Volcanic National Park in California, emission of gases continues, and one cannot help noticing the rotten-egg odor of hydrogen sulfide gas (Figure 4.2a).

When magma rises toward the surface, the pressure is reduced and the contained gases begin to expand. In highly viscous, felsic magmas, expansion is inhibited and gas pressure increases. Eventually, the pressure may become great enough to cause an explosion and produce pyroclastic materials such as ash. In contrast, low-viscosity mafic magmas allow gases to expand and escape easily. Accordingly, mafic magmas generally erupt rather quietly.

The amount of gases contained in magmas varies, but is rarely more than a few percent by weight. Even though volcanic gases constitute a small proportion of a magma, they can be dangerous, and in some cases have had far-reaching climatic effects.

Most volcanic gases quickly dissipate in the atmosphere and pose little danger to humans, but on several occasions they have caused numerous fatalities. In 1783, toxic gases, probably sulfur dioxide, erupting from Laki fissure in Iceland had devastating effects. About 75% of the nation's livestock died, and the haze resulting from the gas caused lower temperatures and crop failures; about 24% of Iceland's population died as a result of the ensuing Blue Haze Famine. The country suffered its coldest winter in 225 years in 1783–1784, with temperatures 4.8°C below the long-term average. The eruption also produced what Benjamin Franklin called a "dry fog" that was responsible for dimming the intensity of sunlight in Europe. The severe winter of 1783–1784 in Europe and eastern North America is attributed to the presence of this "dry fog" in the upper atmosphere.

The particularly cold spring and summer of 1816 are attributed to the 1815 eruption of Tambora in Indonesia, the largest and most deadly eruption during historic time. The eruption of Mayon volcano in the Philippines during the previous year may have contributed to the cool spring and summer of 1816 as well. Another large historic eruption that had widespread climatic effects was the eruption of Krakatau in 1883.

In 1986, in the African nation of Cameroon, 1746 people died when a cloud of carbon dioxide engulfed them. The gas accumulated in the waters of Lake Nyos, which occupies a volcanic crater. Scientists disagree about what caused the gas to suddenly burst forth from the lake, but once it did, it flowed downhill along the surface because it was denser than air. In fact, the density and velocity of the gas cloud were great enough to flatten vegetation, including trees, a few kilometers from the lake. Unfortunately, thousands of animals and many people, some as far as 23 km from the lake, were asphyxiated.

In 1990, forest rangers notice that trees began dying in an area of about 170 acres on the south side of Mammoth Mountain volcano in eastern California (Figure 4.2b). Previously the trees had been thriving, but a swarm of earthquakes in 1991 apparently triggered the release of carbon

(a)

(b)

Figure 4.2 *(a) Gases emitted at the Sulfur Works in Lassen Volcanic National Park, California. (b) Trees killed by carbon dioxide rising from beneath Mammoth Mountain volcano, California.*

dioxide (CO_2) gas from beneath the volcano. Investigations showed that CO_2 gas accounted for 20% to 90% of the gas content of the soil. During photosynthesis, plant leaves produce oxygen (O_2) from CO_2 but their roots must absorb O_2. Diminished quantities of O_2 in the soil were responsible for the tree deaths.

Residents of the island of Hawaii have coined the term *vog* for volcanic smog. Kilauea Volcano has been erupting continuously since 1983, releasing small amounts of lava, copious quantities of carbon dioxide, and about 1000 tons of sulfur dioxide per day. Carbon dioxide has been no problem because it dissipates quickly in the atmosphere, but sulfur dioxide produces a haze and the unpleasant odor of sulfur. As long as Kilauea Volcano erupts, Hawaii will have a vog problem. Vog probably poses little or no health problem for tourists, but a long-term threat exists for residents of the west side of the island where vog is most common.

Lava Flows

Lava flows are frequently portrayed in movies and on televisions as fiery streams of incandescent rock material posing a great danger to humans. Actually, lava flows are the least dangerous manifestation of volcanism, although they may destroy buildings and cover agricultural land. Most lava flows do not move particularly fast, and because they are fluid, they follow existing low areas. Thus, once a flow erupts from a volcano, determining the path it will take is fairly easy, and anyone in areas likely to be affected can be evacuated.

Even low-viscosity lava flows generally do not move very rapidly. One of the fastest flows ever measured in Hawaii had an average speed of about 9.5 km/hr. Flows can move much faster, though, when a flow's margins cool thus forming a channel, and especially when insulated on all sides as in a *lava tube,* where a speed of more than 50 km/hr has been recorded. A conduit known as a **lava tube** beneath the surface of a lava flow forms when the margins and upper surface of the flow solidify. Thus confined and insulated, the flow can move quite rapidly and over great distances. As an eruption ceases, the tube drains leaving an empty tunnel-like structure (Figure 4.3a). Part of the roof of a lava tube may collapse, forming a *skylight* though which an active flow can be observed (Figure 4.3b), or access can be gained to an inactive lava tube. Good examples of lava tubes can be found in several western states, and in Hawaii, lava flows through lava tubes many kilometers long and in some cases discharges into the sea.

Two types of lava flows, both of which were named for Hawaiian flows, are generally recognized. A **pahoehoe** (pronounced *pah-hoy-hoy*) flow has a ropy surface much like taffy (Figure 4.4a). The surface of an **aa** (pronounced *ah-ah*) flow is characterized by rough jagged angular blocks and fragments (Figure 4.4b). Pahoehoe flows are

(a)

(b)

Figure 4.3 *(a) Lava tube. (b) Skylight.*

(a)

(a)

(b)

Figure 4.5 *(a) Pressure ridge on a 1982 lava flow in Hawaii. (b) A spatter cone on a lava flow in Hawaii.*

(b)

Figure 4.4 *(a) Pahoehoe erupted from the Pu'u O'o vent on Kilauea volcano's southeastern flank. (b) An aa flow in the east rift zone of Kilauea volcano in 1983. The flow front is about 2.5 m high.*

less viscous than aa flows; indeed, the latter are viscous enough to break up into blocks and move forward as a wall of rubble.

Such features as pressure ridges and spatter cones may mark the surface of lava flows. Pressure on the partly so-

lidified crust of still-moving flow causes the surface to buckle into **pressure ridges** (Figure 4.5a). Gases escaping from a flow hurl globs of lava into the air, these fall back to the surface and adhere to one another, thus forming small, steep-sided **spatter cones** (Figure 4.5b). Spatter cones a few meters high are common on lava flows in Hawaii, and ancient ones can be seen in Craters of the Moon National Monument, Idaho.

Columnar joints are common in many lava flows, especially mafic flows, but they also occur in other kinds of flows and in some intrusive igneous rocks (Figure 4.6). Once a lava flow ceases moving it contracts as it cools,

Figure 4.6 *Columnar joints in a lava flow at John Day Fossil Beds National Monument, Oregon.*

(a)

(b)

Figure 4.7 *(a) These bulbous masses of pillow lava form when magma is erupted underwater. (b) Ancient pillow lava on land in Marin County, California.*

thus producing forces that cause fractures called *joints* to open. On the surface of a flow, these joints commonly form polygonal (often six-sided) cracks. These cracks extend downward into the flow, thus forming parallel columns with their long axes perpendicular to the principal cooling surface. Excellent examples of columnar joints can be seen at Devil's Postpile National Monument in California, Devil's Tower National Monument in Wyoming, the Giant's Causeway in Ireland, and many other areas (see Perspective 4.1).

Much of the igneous rock in the upper part of the oceanic crust is of a distinctive type, consisting of bulbous masses of basalt resembling pillows, hence the name **pillow lava.** It was long recognized that pillow lava forms when lava is rapidly chilled beneath water, but its formation was not observed until 1971. Divers near Hawaii saw pillows form when a blob of lava broke through the crust of an underwater lava flow and cooled almost instantly, forming a pillow-shaped structure with a glassy exterior. Remaining fluid inside then broke through the crust of the pillow, resulting in an accumulation of interconnected pillows (Figure 4.7).

Pyroclastic Materials

In addition to lava flows, erupting volcanoes may eject pyroclastic materials, especially *ash,* a designation for pyroclastic particles measuring less than 2.0 mm (Figure 4.8). In some cases ash is ejected into the atmosphere and settles to the surface as an *ash fall* that might cover a large area. In 1947, ash erupted from Mount Hekla in Iceland fell 3800 km away on Helsinki, Finland.

Columnar Jointing

Peculiar marks, shapes, patterns, and structures in rocks are not unusual. Some are merely oddities of little or no importance although they might be attractive or interesting. Among some of the more interesting structures are columns yielded by the phenomenon known as *columnar jointing*. They are found in numerous areas of current and past volcanism (Figure 4.6) and in some intrusive igneous rocks as well, such as in Devil's Tower, Wyoming (see Figure 3, Perspective 3.2).

Columnar joints form most often in basalt and andesite lava flows, but they are also known in other volcanic rocks, including some ash flow tuffs, as well as shallow intrusive bodies such as dikes, sills, and volcanic necks. As noted in the text, columnar joints form when lava or magma cools and contracts, and the columns so formed tend to have their long axes perpendicular to the cooling surface—the surface of a lava flow, for example. In the upper part of a cooling flow or igneous body, the columns tend to be irregular and poorly developed, whereas in the lower part of the same flow they are much better developed.

Many columns are remarkably vertical and straight. However, most igneous bodies are irregularly shaped and not homogeneous throughout, nor do all parts of them cool at the same rate. As a result, curved columns such as those at Devil's Tower National Monument, Wyoming, or Devil's Postpile National Monument, California, are not uncommon. In addition, the degree of development of columns varies considerably. In some igneous rocks they are difficult to discern, whereas in others they are remarkably prominent features. The columns outlined by columnar joints are typically six-sided but those with three, four, five, and seven sides are also found. At Devil's Postpile National Monument, California, for instance, six-sided columns ac-

count for 55% of those present, and most of the rest (37%) are five-sided.

In his 1968 book *Chariots of the Gods,* Erich von Daniken claimed that the colossal structures (Figure 1) on the island of Pohnpie in the Federated States of Micronesia were evidence for his thesis that astronauts from elsewhere had visited Earth long ago. According to him, the hexagonal stones used in the structures were so regularly shaped that they could not be natural, and thus must have been shaped and moved by beings with a much more advanced technology than the native islanders possessed. Although the structures represent a phenomenal engineering feat, the stones are nothing more than columns from an ancient lava flow. There is no reason to suppose that the natives could not move the stones and pile them into the remarkable structures. However, nothing is known of why the 500-year-old structures were erected in the first place.

Certainly one of the most impressive examples of columnar jointing is at the Giant's Causeway, a popular tourist attraction in Northern Ireland (Figure 2). The columns are particularly straight and vertical and in several locations have been given fanciful names such as the Honeycomb and the Wishing Chair. According to one legend, a Scottish giant named Benendoner built the causeway from Scotland to Ireland. But when threatened by the Irish giant Finn McCool, he fled back to Scotland and destroyed the causeway except at both ends, one in Ireland and the other on the island of Staffa

Figure 1 *These structures on the island of Pohnpei in the Federated States of Micronesia are made up of long, six-sided (hexagonal) stones. The stones were derived from a lava flow in which columnar joints formed as it cooled.*

(a)

(b)

Figure 2 *Excellent examples of columnar jointing. (a) The Giant's Causeway in Northern Ireland. (b) Devil's Postpile National Monument, California.*

at Fingal's Cave, another place to see well-developed columnar joints. Another legend holds that Finn McCool fell in love with a giant lady on the island of Staffa and built the causeway to bring her to Ireland.

Among the many places in North America where columnar joints can be observed, those at Devil's Postpile National Monument, California, are some of the best developed. Here, columns formed in a lava flow no more than 100,000 years ago show the typical hexagonal shape in both side and top views. However, in some parts of the flow the columns are strongly curved, indicating complex cooling (Figure 2b). Long after the lava flow cooled and solidified, glaciers covered the area and scratched and pol-

ished the surface of the columns. Not far from Devil's Postpile are vast exposures of the 760,000-year-old Bishop Tuff with its radial or rosette columnar joints. These unusual structures appear to have formed around gas vents in the cooling tuff deposit.

Hundreds of localities are known where columns in lava flows and shallow plutons are present. Visitors to Yellowstone National Park, Wyoming, and North Cascades National Park, Washington, can see these structures there as well as at many other places not encompassed by our national park and monument system. Less well known are the columns in the Palisades sill along the Hudson River in New Jersey.

Figure 4.8 *Pyroclastic materials. The large object on the left is a volcanic bomb; it is about 20 cm long. The streamlined shape of bombs indicates they were erupted as globs of magma that cooled and solidified as they descended. The granular objects in the upper right are pyroclastic materials known as lapilli. The pile of gray-white material on the lower right is ash.*

And about 10 million years ago, in what is now northeastern Nebraska, numerous rhinoceroses, horses, camels, and other animals were buried in ash that was apparently erupted more than 1000 km away in New Mexico. As opposed to an ash fall, an *ash flow* is a coherent cloud of ash and gas that flows along or close to the land surface. Ash flows can move at more than 100 km/hr, and some of them cover vast areas.

In populated areas adjacent to volcanoes, ash falls and ash flows can pose some serious problems. Furthermore, volcanic ash in the atmosphere is a serious hazard to aviation. Since 1980, about 80 aircraft have been damaged when they encountered clouds of volcanic ash. The most serious incident took place in 1989 when ash from Redoubt Volcano in Alaska caused all four jet engines to fail on KLM Flight 867. The plane carrying 231 passengers nearly crashed when it fell more than 3 km before the crew could restart the engines. The plane landed safely in Anchorage, Alaska, but it required $80 million in repairs.

In addition to ash (particles measuring less than 2 mm), volcanoes also erupt *lapilli,* consisting of pyroclastic materials measuring from 2 to 64 mm, and *blocks* and *bombs,* both of which are larger than 64 mm (Figure 4.8). Bombs have a twisted, streamlined shape indicating they were erupted as globs of magma that cooled and solidified during their flight through the air. Blocks, in contrast, are angular pieces of rock ripped from a volcanic conduit or pieces of a solidified crust of a lava flow. Because of their size, lapilli, bombs, and blocks are not found over such large areas as ash; instead they are confined to the immediate area of an eruption.

What Are Volcanoes?

According to one definition, a **volcano** is a conical mountain formed around a vent where lava, pyroclastic materials, and gases are erupted. Even though many volcanoes are conical, some are simply solidified bulbous masses of magma at the surface, and others are dome-shaped or resemble an inverted shield lying on the ground. In all cases, though, they have a conduit or conduits leading to a magma chamber beneath the surface. Vulcan, the Roman deity of fire, was the inspiration for calling these mountains volcanoes, and because of their danger and obvious connection to Earth's interior they have been held in awe by many cultures.

Figure 4.9 *Mount Shasta in northern California is a large volcano. The peak on the left is Shastina, a parasitic cone that developed on the flank of Mount Shasta.*

Probably no other geologic process, except perhaps earthquakes, has more lore associated with it. Native Americans of the Northwest tell of a titanic battle between the volcano gods Skel and Llao to account for huge eruptions that took place about 6600 years ago in Oregon, and in Hawaiian legends the volcano goddess Pele resides in the crater of Kilauea on Hawaii. In one of her frequent rages, Pele causes earthquakes and lava flows and she may hurl flaming boulders at those who offend her.

Most volcanoes have a circular depression known as a **crater** at their summit. Craters form as a result of the extrusion of gases and lava from a volcano and are connected via a conduit to a magma chamber below the surface. It is not unusual, however, for magma to erupt from vents on the flanks of large volcanoes where smaller parasitic cones with their own craters develop. For instance, Shastina is a large parasitic cone on the flank of Mount Shasta in California (Figure 4.9), and Mount Etna on Sicily has some 200 smaller vents on its flanks.

Some volcanoes are characterized by a **caldera** rather than a crater. Craters are generally less than 1 km in diameter, whereas calderas exceed this dimension and commonly have steep sides. One of the best-known calderas in this country is the misnamed Crater Lake in Oregon—Crater Lake is actually a caldera (Figure 4.10). It formed about 6600 years ago after voluminous eruptions partially drained its magma chamber. This drainage left the summit of the mountain, Mount Mazama, unsupported, and it collapsed into the magma chamber, forming a caldera more than 1200 m deep and measuring 9.7 by 6.5 km. Most calderas formed when a summit collapsed during particularly large, explosive eruptions as in the case of Crater Lake.

Shield Volcanoes

Shield volcanoes resemble the outer surface of a shield lying on the ground with the convex side up (Figure 4.11). They have low, rounded profiles with gentle slopes ranging from about 2 to 10 degrees. Their slopes indicate that they are composed mostly of mafic flows that had low viscosity, so the flows spread out and formed thin, gently sloping layers. Eruptions from shield volcanoes, sometimes called *Hawaiian-type volcanoes*, are quiet compared to those of volcanoes such as Mount St. Helens; lavas most commonly rise to the surface with little explosive activity, so they usually pose little danger to humans. Lava fountains, some up to 400 m high, contribute some pyroclastic materials to shield volcanoes (see chapter opening photo), but otherwise they are composed largely of basalt lava flows; flows comprise more than 99% of the Hawaiian volcanoes above sea level.

Although eruptions of shield volcanoes tend to be rather quiet, some of the Hawaiian volcanoes have, on occasion, produced sizable explosions. These usually result when groundwater instantly vaporizes as it comes in contact with magma. One such explosion in 1790 killed about 80 warriors in a party headed by Chief Keoua who was leading them across the summit of Kilauea volcano. The current activity at Kilauea is impressive for another reason as well; it has been erupting continuously since January 3, 1983, making it the longest recorded eruption.

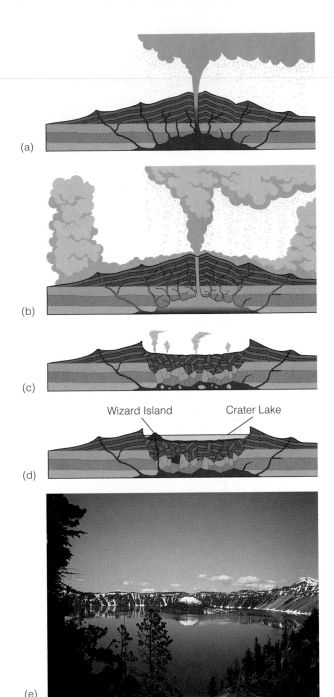

Figure 4.10 *The sequence of events leading to the origin of Crater Lake, Oregon. (a-b) Ash clouds and ash flows partly drain the magma chamber beneath Mount Mazama. (c) The collapse of the summit and formation of the caldera. (d) Postcaldera eruptions partly cover the caldera floor, and the small volcano known as Wizard Island forms. (e) View from the rim of Crater Lake showing Wizard Island.*

What Are Volcanoes? **93**

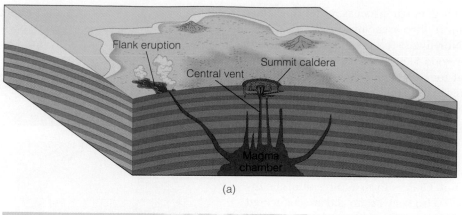

(a)

(b)

(c)

Figure 4.11 *(a) A shield volcano. Each layer shown consists of numerous, thin basalt lava flows. (b) View of Mauna Loa, a large, active shield volcano on the island of Hawaii. (c) An extinct shield volcano in Lassen County, California. Note in the illustration and the images that the flanks of shield volcanoes are not steep.*

Shield volcanoes such as those of the Hawaiian Islands and Iceland, are most common in the ocean basins, but some are also present on the continents—in East Africa, for instance. The island of Hawaii is made up of five huge shield volcanoes, two of which, Kilauea and Mauna Loa, are active much of the time. Mauna Loa is nearly 100 km across its base and stands more than 9.5 km above the surrounding seafloor; it has a volume estimated at 50,000 km^3, making it the world's largest volcano. By contrast, the largest volcano in the continental United States, Mount Shasta in California, has a volume of only about 350 km^3.

Shield volcanoes have a summit crater or caldera, and a number of smaller cones on their flanks through which lava erupts (Figure 4.11). A vent opened on the flank of Kilauea and grew 250 m high between June 1983 and September 1986.

Cinder Cones

Volcanic peaks composed of pyroclastic materials resembling cinders are known as **cinder cones** (Figure 4.12). They form when pyroclastic materials are ejected into the atmosphere and fall back to the surface to accumulate around the vent, thus forming small, steep-sided cones. The slope angles may be as much as 33 degrees, depending on the slope that can be maintained by the irregularly shaped pyroclastic materials. Cinder cones are rarely more than 400 m high, and many have a large, bowl-shaped crater (Figure 4.12b).

Many cinder cones form on the flanks or within the calderas of larger volcanic mountains and appear to represent the final stages of activity, particularly in areas of basaltic volcanism. Wizard Island in Crater Lake, Oregon, is a small cinder cone that formed after the summit of

(a)

(b)

(c)

Figure 4.12 *(a) A 230-m-high cinder cone in Lassen Volcanic National Park, California. It was last active during the 1650s. (b) The large, bowl-shaped crater at the summit of the cinder cone shown in (a). (c) Vestmannaeyjar, Iceland, was threatened by a lava flow from Eldfell, a cinder cone that formed in 1973. Within two days the new volcano had grown to 100 m high. Another cinder cone known as Helgafel is also visible.*

What Are Volcanoes? **95**

Mount Mazama collapsed to form a caldera (Figure 4.10). Cinder cones are common in the southern Rocky Mountain states, particularly New Mexico and Arizona, and many others are in California, Oregon, and Washington.

In 1973, on the Icelandic island of Heimaey, the town of Vestmannaeyjar was threatened by a new cinder cone. The initial eruption began on January 23, and within two days a cinder cone, later named Eldfell, rose to about 100 m above the surrounding area (Figure 4.12c). Pyroclastic materials from the volcano buried parts of the town, and by February a massive aa lava flow was advancing toward the town. The flow's leading edge ranged from 10 to 20 m thick, and its central part was as much as 100 m thick. By spraying the leading edge of the flow with sea water, which caused it to cool and solidify, the residents of Vestmannaeyjar successfully diverted the flow before it did much damage to the town.

Composite Volcanoes

Composite volcanoes, also called *stratovolcanoes,* are composed of both pyroclastic layers and lava flows (Figure 4.13). Typically, both materials have an intermediate composition, and the flows cool to form andesite. Recall that lava of intermediate composition is more viscous than mafic lava. In addition to lava flows and pyroclastic layers, a significant proportion of a composite volcano is made up of **lahars** (volcanic mudflows). Some lahars form when rain falls on pyroclastic materials and creates a muddy slurry that moves downslope. On November 13, 1985, mudflows resulting from a rather minor eruption of Nevado del Ruiz in Colombia killed about 23,000 people. In the Philippines, 83 of the 722 victims of the June 1991 eruptions of Mount Pinatubo were killed by lahars (Figure 4.14; Table 4.1).

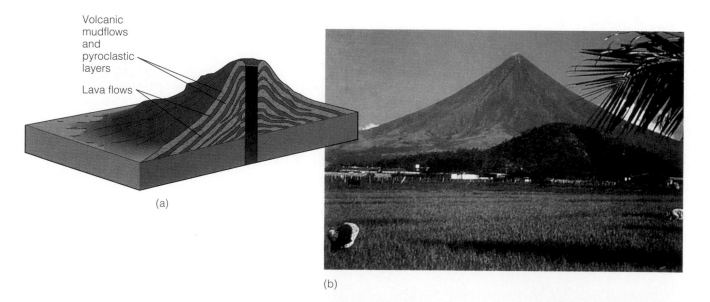

(a)

(b)

Figure 4.13 *(a) Composite volcanoes, also known as stratovolcanoes, are composed of lava flows, layers of pyroclastic materials, and volcanic mudflows or lahars. Note that the slope is steep near the summit but decreases toward the base. (b) Mayon volcano in the Philippines is one of the most nearly symmetrical composite volcanoes in the world. It erupted during 1999 for the 13th time this century. (c) Mount St. Helens, Washington, from the southwest in 1978.*

(c)

(a)

(b)

Figure 4.14 *(a) Mount Pinatubo in the Philippines is one of many volcanoes in a belt nearly encircling the Pacific Ocean basin. It is shown here erupting on June 12, 1991. A huge, thick cloud of ash and steam rises above Clark Air Force Base, from which about 15,000 people had already been evacuated to Subic Bay Naval Base. Following this eruption, the remaining 900 people at the base were also evacuated. (b) Homes partly buried by a volcanic mudflow (lahar) on June 15, 1991. Note that the roof at the far right is still partly covered by pyroclastic materials.*

Composite volcanoes are steep sided near their summits, perhaps as much as 30 degrees, but the slope decreases toward the base where it is generally less than 5 degrees. Mayon volcano in the Philippines is one of the most perfectly symmetrical composite volcanoes on Earth (Figure 4.13b). Its concave slopes rise ever steeper to the summit with its central vent through which lava and pyroclastic materials are periodically erupted. Mayon Volcano erupted for the thirteenth time this century beginning in June 1999.

Composite volcanoes are the typical large volcanoes of the continents and island arcs. Familiar examples include Mount Fuji in Japan and Mount Vesuvius in Italy as well as Mount St. Helens and many of the other volcanic peaks in the Cascade Range of western North America.

Soufrière Hills, a composite volcano on Montserrat in the West Indies, began erupting in 1995 and continues. At least 19 fatalities have been reported, and more than half the island's approximately 12,000 residents have been evacuated (see Chapter 1 Prologue). During February and March 2000, eruptions in the Philippines and Japan caused considerable damage and large-scale evacuations. After several years of eruptions, the activity at Popocatepetl Volcano near Mexico City increased during December 2000 prompting authorities to evacuate more than 40,000 people.

Lava Domes

If the upward pressure in a volcanic conduit is great enough, the most viscous magmas move upward and form bulbous, steep-sided **lava domes** (Figure 4.15). Lava domes are generally composed of felsic lavas although some are of intermediate composition. Because felsic magma is so viscous, it moves upward very slowly; the lava dome that formed in Santa María volcano in Guatemala in 1922 took two years to grow to 500 m high and 1200 m across. Lava domes contribute significantly to many composite volcanoes. Beginning in 1980, a number of lava

Table 4.1

Some Notable Volcanic Eruptions

Date	Volcano	Deaths
Aug. 24, 79	Mt. Vesuvius, Italy	At least 3360 killed in Pompeii and Herculaneum.
1586	Kelut, Java	Mudflows kill 10,000.
Dec. 16, 1631	Mt. Vesuvius, Italy	3500 killed.
Aug. 4, 1672	Merapi, Java	3000 killed by mudflows and pyroclastic flows.
Dec. 10, 1711	Awu, Indonesia	3000 killed by pyroclastic flows.
Sept. 22, 1760	Makian, Indonesia	Eruption kills 2000; island evacuated for seven years.
May 21, 1782	Unzen, Japan	14,500 die in debris avalanche and tsunami.
June 8, 1783	Lakagigar, Iceland	Largest historic lava flows: 12 km^3; 9350 die.
July 26, 1783	Asama, Japan	Pyroclastic flows and floods kill 1200+.
Apr. 10, 1815	Tambora, Indonesia	92,000 killed; another 80,000 reported to have died from famine and disease.
Oct. 8, 1822	Galunggung, Java	4011 die in pyroclastic flows and mudflows.
Mar. 2, 1856	Awu, Indonesia	Pyroclastic flows kill 2806.
Aug. 27, 1883	Krakatau, Indonesia	36,417 die; most killed by tsunami.
June 7, 1892	Awu, Indonesia	1532 die in pyroclastic flows.
May 8, 1902	Mt. Pelée, Martinique	St. Pierre destroyed by pyroclastic flow; 28,000 killed.
Oct. 24, 1902	Santa María, Guatemala	5000 killed.
June 6, 1912	Novarupta, Alaska	Largest 20th-century eruption: about 33 km^3 of pyroclastic materials erupted; no fatalities.
May 19, 1919	Kelut, Java	Mudflows kill 5110, devastate 104 villages.
Jan. 21, 1951	Lamington, New Guinea	2942 killed by pyroclastic flows.
Mar. 17, 1963	Agung, Indonesia	1148 killed.
Aug. 12, 1976	Soufrière, Guadeloupe	74,000 residents evacuated.
May 18, 1980	Mount St. Helens, Washington	57 killed; 600 km^2 of forest devastated.
Mar. 28, 1982	El Chichón, Mexico	Pyroclastic flows kill 1877.
Nov. 13, 1985	Nevado del Ruiz, Colombia	Mudflows kill 23,000.
Aug. 21, 1986	Oku volcanic field, Cameroon	1746 asphyxiated by cloud of CO_2 released from Lake Nyos.
June 1991	Unzen, Japan	43 killed; at least 8500 fled.
June 1991	Mt. Pinatubo, Philippines	~281 killed during initial eruption; 83 killed by later mudflows; 358 died of illness; 200,000 evacuated.
Feb. 2, 1993	Mayon, Philippines	At least 70 killed; 60,000 evacuated.
Nov. 22, 1994	Mt. Morapis, Indonesia	Pyroclastic flows kill 60; more than 6000 evacuated.
July 1999	Soufriére Hills, Montserrat	19 killed; 12,000 evacuated.

SOURCE: American Geological Institute Data Sheets, except for last five entries.

domes were emplaced in the crater of Mount St. Helens; most of these were destroyed during subsequent eruptions. Since 1983, Mount St. Helens has been characterized by sporadic dome growth (see Perspective 4.2).

In June 1991, a dome in Japan's Unzen volcano collapsed, causing a flow of debris and hot ash that killed 43 people in a nearby town. Lava domes are also often responsible for extremely explosive eruptions. In 1902,

Figure 4.15 *Lava domes form when viscous magma, generally of felsic composition, is forced up through a volcanic conduit. The 2 km³ of material composing Lassen Peak in California makes it the world's largest lava dome. It formed about 27,000 years ago, but erupted most recently from 1914 to 1917.*

viscous magma accumulated beneath the summit of Mount Pelée on the island of Martinique. Eventually, the pressure within the mountain increased to the point that it could no longer be contained, and the side of the mountain blew out in a tremendous explosion. When this occurred, a mobile, dense cloud of pyroclastic materials and gases known as a **nuée ardente** (French for "glowing cloud") was ejected and raced downhill at about 100 km/hr, engulfing the city of St. Pierre (Figure 4.16).

A tremendous blast hit St. Pierre leveling buildings; hurling boulders, trees, and pieces of masonry down the streets; and moving a 3-ton statue 16 m. Accompanying the blast was a swirling cloud of incandescent ash and gases with an internal temperature of 700°C that incinerated everything in its path. The nuée ardente passed through St. Pierre in two or three minutes, only to be followed by a firestorm as combustible materials burned and casks of rum exploded. But by then most of the 28,000

(a)

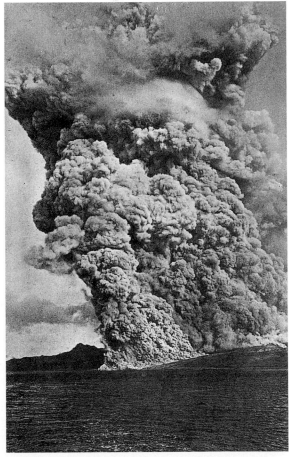

(b)

Figure 4.16 *(a) St. Pierre, Martinique, after it was destroyed by a nuée ardente erupted from Mount Pelée in 1902. Only 2 of the city's 28,000 inhabitants survived. (b) A large nuée ardente from Mount Pelée a few months after the one that destroyed St. Pierre.*

What Are Volcanoes?

Eruptions of Cascade Range Volcanoes

During the summer of 1914, Lassen Peak in northern California began erupting without warning and culminated with the "Great Hot Blast," a huge steam explosion on May 22, 1915. Fortunately, the area was sparsely settled, and little property damage and no deaths resulted, even though a large area of forest on the volcano's eastern and northeastern flanks was devastated. Activity largely ceased by 1917, but hot springs, boiling mud pots, and gas vents known as *fumaroles* remind us that a source of heat is present beneath the surface (Figure 4.2a).

Lassen Peak is the world's largest lava dome. It formed about 27,000 years ago when a bulbous mass of viscous lava was injected into the flank of an older, eroded composite volcano known as Mount Tehama. It is one of 15 large volcanoes in the Cascade Range of northern California, Oregon, Washington, and southern British Columbia, Canada. After Lassen Peak's 1914–1917 eruptions, the Cascade volcanoes remained quiet for 63 years. Then, on March 16, 1980, Mount St. Helens in southern Washington showed signs of renewed activity, and on May 18 it erupted violently, causing the worst volcanic disaster in U.S. history (Figure 1).

The awakening of Mount St. Helens, its first eruption in 123 years, came as no surprise to geologists of the U.S. Geological Survey (USGS) who warned in a 1978 report that Mount St. Helens is ". . . an especially dangerous volcano because [of] its past behavior and [its] relatively high frequency of eruptions during the last 4,500 years."[*] Although no one could predict precisely when Mount St. Helens would erupt, the USGS report

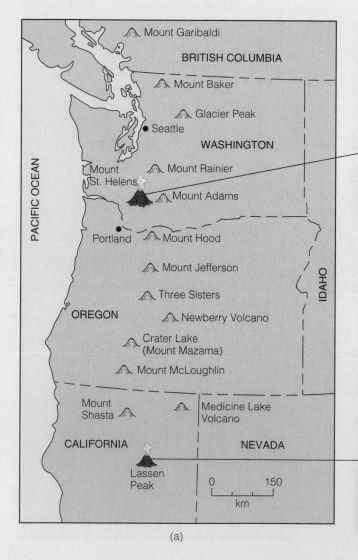

(a)

(b)

(c)

Figure 1. *(a) The major volcanoes of the Cascade Range, several of which have been active during the last 200 years. (b) The eruption of Mount St. Helens, Washington, on May 18, 1980. A lateral blast occurred when a bulge on the mountain's north face collapsed and reduced the pressure on the gas-charged magma within the mountain. (c) Lassen Peak in northern California erupted numerous times from 1914 to 1917. This eruption took place in 1915.*

included maps showing areas in which damage from an eruption could be expected. Forewarned with such data, local officials were better prepared to formulate policies when the eruption did occur.

On March 27, 1980, Mount St. Helens began erupting steam and ash and continued to do so during the rest of March and most of April. By late March, a visible bulge had developed on its north face as molten rock was injected into the mountain, and the bulge continued to expand at about 1.5 m per day. On May 18, an earthquake shook the area, the unstable bulge collapsed, and the pent-up volcanic gases below expanded rapidly, creating a tremendous northward-directed lateral blast that blew out the north side of the mountain. The lateral blast accelerated from 350 to 1080 km/hr, obliterating virtually everything in its path. Some 600 km^2 of forest were completely destroyed; trees were snapped off at their bases and strewn about the countryside, and trees as far as 30 km from the bulge were seared by the intense heat. Tens of thousands of animals were killed; roads, bridges, and buildings were destroyed; and 57 people perished.

Shortly after the lateral blast, volcanic ash and steam erupted, forming a 19-km-high cloud above the volcano. The ash cloud drifted east-northeast, and the resulting ash fall at Yakima, Washington, 130 km to the east, caused almost total darkness at midday. Detectable amounts of ash were deposited over a huge area. Flows of hot gases and volcanic ash raced down the north flank of the mountain, causing steam explosions when they encountered bodies of water or moist ground. Steam explosions continued for weeks, and at least one occurred a year later.

Snow and glacial ice on the upper slopes of Mount St. Helens melted and mixed with ash and other surface debris to form thick, pasty volcanic mudflows. The largest and most destructive mudflow surged down the valley of the North Fork of the Toutle River. Ash and mudflows displaced water in lakes and streams and flooded downstream areas. Ash and other particles carried by the floodwaters were deposited in stream channels; many kilometers from Mount St. Helens, the navigation channel of the Columbia River was reduced from 12 m to less than 4 m as a result of such deposition.

Although the damage resulting from the eruption of Mount St. Helens was significant and the deaths were tragic, it was not a particularly large or deadly eruption compared with some historic eruptions (Table 4.1). The eruption of Tambora in 1815 is the greatest volcanic eruption in recorded history in terms of both casualties and the amount of material erupted; it produced at least 80 times more ash than the 0.9 km^3 that spewed forth from Mount St. Helens.

(continued)

Figure 2. *Mount Rainier as seen from the waterfront of Tacoma, Washington. Its summit elevation of 4392 m makes Mount Rainier the highest peak in the Cascade Range. The summit is only about 80 km from where this picture was taken.*

What Are Volcanoes? **101**

Several other currently dormant Cascade Range volcanoes also pose threats to populated areas. Eruptions of Mount Shasta in northern California would cause damage and perhaps fatalities in several nearby communities, and Mount Hood, Oregon, lies less than 65 km from the densely populated Portland area. But the most dangerous is probably Mount Rainier, Washington (Figure 2).

Rather than lava flows, ash falls, or even a colossal explosion as in the case of Mount St. Helens, the greatest threat from Mount Rainier is volcanic mudflows or lahars. During the last 10,000 years, there have been at least 60 such flows. The largest

consisted of nearly 4 km^3 of debris, and it covered an area now occupied by more than 120,000 people. No one can predict when another mudflow will take place, but at least one community has taken the threat seriously enough to formulate an emergency evacuation plan. Unfortunately, they would have only one or two hours to carry out an evacuation.

* D. R. Crandell and D. R. Mullineaux, "Potential Hazards from Future Eruptions of Mt. St. Helens Volcano, Washington," *United States Geological Survey Bulletin 1383-C* (1978): C1.

residents of the city were already dead. In fact, in the area covered by the nuée ardente, only 2 survived![1] One survivor was on the outer edge of the nuée ardente, but even there he was terribly burned and his family and neighbors were all killed. The other survivor, a stevedore incarcerated the night before for disorderly conduct, was in a windowless cell partly below ground level. He remained in his cell badly burned for four days after the eruption until rescue workers heard his cries for help. He later became an attraction in the Barnum and Bailey

[1] Although reports commonly claim that only two people survived the eruption, at least 69 and possibly as many as 111 people survived beyond the extreme margins of the nuée ardente and on ships in the harbor. Many, however, were badly injured.

Circus where he was advertised as "The only living object that survived in the 'Silent City of Death' where 40,000 beings were suffocated, burned or buried by one belching blast of Mont Pelee's terrible volcanic eruption."[2]

Do All Eruptions Build Volcanoes?

Nearly everyone is aware of magnificent composite volcanoes such as those in Figure 4.13. Indeed, when the term *volcano* is mentioned, mountains such as these immediately come to mind. However, as noted earlier, even though some volcanoes are the typical conical mountains commonly envisioned, numerous volcanoes with other shapes are found in many areas (Figures 4.11 and 4.15). In fact, in some areas of volcanism volcanoes fail to develop at all. For instance, during *fissure eruptions* fluid lava pours out and simply builds up rather flat-lying areas, whereas huge explosive eruptions might yield *pyroclastic sheet deposits*, which, as their name implies, have a sheetlike geometry.

Fissure Eruptions and Basalt Plateaus

Some 164,000 km^2 of eastern Washington and parts of Oregon and Idaho were covered by overlapping basalt lava flows during the Miocene and Pliocene epochs (between 17 and 5 million years ago). Now known as the Columbia River basalts, they are well exposed in canyons eroded by the Snake and Columbia Rivers (Fig-

[2] Quoted from Scarth, A. (1999), *Vulcan's Fury: Man against the volcano*, Yale University Press, p. 177.

(a)

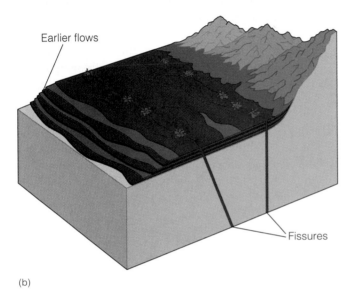

Earlier flows

Fissures

(b)

Figure 4.17 *(a) The Columbia River basalts. (b) A block diagram showing fissure eruptions and the origin of a basalt plateau.*

ure 4.17a). Rather than being erupted from a central vent, these flows issued from long cracks or fissures and are thus known as **fissure eruptions.** Lava erupted from these fissures was so fluid (had such low viscosity) that it simply spread out, covering vast areas and built up a **basalt plateau,** a broad, flat, elevated area underlain by lava flows (Figure 4.17b).

The Columbia River basalt flows have an aggregate thickness of about 1000 m, and some individual flows cover huge areas—for example, the Roza flow, which is 30 m thick, advanced along a front about 100 km wide and covered 40,000 km^2.

Fissure eruptions and basalt plateaus are not common, although several large areas of such features are known. The only area where this kind of activity is currently occurring is in Iceland. A number of volcanic mountains are present in Iceland, but the bulk of the island is composed of basalt flows erupted from fissures. Two major fissure eruptions, one in A.D. 930 and the other in 1783, account for about half of the magma erupted in Iceland during historic time. The 1783 eruption issued from the Laki fissure, which is 25 km long; lava flowed several tens of kilometers from the fissure and in one place filled a valley to a depth of about 200 m.

Pyroclastic Sheet Deposits

More than 100 years ago, geologists were aware of vast areas covered by felsic volcanic rocks a few meters to hundreds of meters thick. It seemed improbable that these could be vast lava flows, but it also seemed equally unlikely that they were ash fall deposits. Based on observations of historic pyroclastic flows, such as the nuée ardente erupted by Mount Pelée in 1902, it now seems probable that these ancient rocks originated as pyroclastic flows, hence the name **pyroclastic sheet deposits** (Figure

Figure 4.18 *Pyroclastic flow deposits in the Marella River valley in the Philippines. The flows that filled this valley to depths of 50 to 200 m were erupted from Mount Pinatubo in June 15, 1991. Some of the flows moved as much as 16 km from the volcano.*

4.18). They cover far greater areas than any observed during historic time, however, and apparently erupted from long fissures rather than from a central vent. The pyroclastic materials of many of these flows were so hot that they fused together to form *welded tuff*.

It now appears that major pyroclastic flows issue from fissures formed during the origin of calderas. For instance, pyroclastic flows were erupted during the formation of a large caldera now occupied by Crater Lake, Oregon.

Similarly, the Bishop Tuff of eastern California appears to have been erupted shortly before the formation of the Long Valley caldera. Interestingly, earthquake activity in the Long Valley caldera and nearby areas beginning in 1978 may indicate that magma is moving upward beneath part of the caldera. Thus, the possibility of future eruptions in that area cannot be discounted.

How Large Is an Eruption and How Long Do Eruptions Last?

Most people are aware of magnitude as a measure expressing the size of an earthquake. The August 1999 earthquake in Turkey was 7.4 on the Richter Magnitude Scale, and because it was centered in a densely populated area, thousands of fatalities and injuries resulted (see Chapter 7). Geologists have several ways of expressing the size of a volcanic eruption. One of these, called the destructiveness index, is based on the area covered by lava or pyroclastic materials during an eruption. They also rank eruptions in terms of their intensity and magnitude, but these have both been incorporated into the more widely used **volcanic explosivity index (VEI)** (Table 4.2). Unlike the Richter Magnitude Scale, though, the VEI is only semiquantitative, being based partly on subjective criteria.

The volcanic explosivity index (VEI) ranges from 0 (nonexplosive) to 8 (megacolossal) and is based on several aspects of a particular eruption such as the volume of material explosively ejected and the height of the eruption plume. However, the volume of lava, fatalities, and property damage are not considered when making a VEI determination. For instance, the 1985 eruption of Nevado del Ruiz in Colombia killed 23,000 people, yet has a VEI of only 3. In contrast, the colossal eruption (VEI = 6) of Katmai in Alaska in 1911 caused no fatalities or injuries. Since A.D. 1500, only the 1815 eruption of Tambora had a value of 7; it was both large and deadly (Table 4.1). Nearly 5700 eruptions during the last 10,000 years have been assigned VEI numbers, but none has exceeded 7, and most (62%) were assigned a value of 2.

The Volcanic Explosivity Index (VEI)

VEI	Description	Plume Height	Volume of Tephra[1]	Duration of Continuous Blast (hours)	Classification[2]	Frequency
0	Nonexplosive	0 – 100 m	1000s m^3	<1	Hawaiian	Daily
1	Gentle	100 – 1000 m	10,000s m^3	<1	Hawaiian/Strombolian	Daily
2	Explosive	1 – 5 km	1,000,000s m^3	1 – 6	Strombolian/Vulcanian	Weekly
3	Severe	3 – 15 km	10,000,000s m^3	1 – 12	Vulcanian	Yearly
4	Cataclysmic	10 – 25 km	100,000,000s m^3	1 – 12	Vulcanian/Plinian	10s of years
5	Paroxysmal	> 25 km	1 km^3	6 – 12	Plinian	100s of years
6	Colossal	> 25 km	10s km^3	> 12	Plinian/Ultra-Plinian	100s of years
7	Super-colossal	> 25 km	100s km^3	> 12	Ultra-Plinian	1000s of years
8	Mega-colossal	> 25 km	1000s km^3	> 12	Ultra-Plinian	10,000s of years

[1] Tephra is a collective term for all pyroclastic materials erupted by a volcano.

[2] The terms in this column refer to types of eruptions: Hawaiian—lava flows but little explosive activity; Strombolian—eruption in which fluid basaltic lava is "jetted" from a central vent; Vulcanian—an eruption of explosively ejected incandescent lava; Plinian—high velocity eruption of streams of fragmented magma and gases from a vent.

SOURCE: Table 4.2, *Physical*, 4/e

Y ou are an enthusiast of natural history and would like to share your interests with your family. Accordingly, you plan a vacation to see some of the volcanic features of our national parks and monuments. Let's assume you plan a route that takes you through California, Oregon, Washington, and then to Hawaii. What specific areas might you visit, and what kinds of volcanic features would you see in these areas? Are there other parts of the United States that you might visit in the future to see additional evidence of volcanism?

The duration of eruptions varies considerably. Fully 42% of about 3300 historic eruptions lasted less than one month, most (about 33%) erupted for one to six months, but some 16 volcanoes have been active more or less continuously for more than 20 years. Stromboli and Mount Etna in Italy and Erta Ale in Ethiopia are good examples. In the case of some explosive volcanoes, the time from the onset of their eruptions to the climactic event is weeks or months. A case in point is the colossal explosive eruption of Mount St. Helens on May 18, 1980, that occurred two months after eruptive activity began. Unfortunately, many volcanoes give little or no warning of such large-scale events; of 252 explosive eruptions, 42% erupted most violently during their first day of activity. As one might imagine, predicting eruptions is complicated by those volcanoes that give so little warning of impending activity (see the next section).

Is It Possible to Predict Eruptions?

Numerous volcanoes have erupted explosively during historic time and many have the potential to do so again. Most dangerous volcanoes are at or near the margins of tectonic plates, especially at convergent plate boundaries. At any one time about a dozen volcanoes are erupting, but most eruptions cause little or no property damage, injuries, or fatalities. Unfortunately, some do. The 1815 eruption of Tambora in Indonesia directly or indirectly resulted in more than 100,000 deaths, and a rather minor eruption of Nevado del Ruiz in Colombia in 1985 triggered volcanic mudflows that killed 23,000 people (Table 4.1).

Only a few of Earth's potentially dangerous volcanoes are monitored, including some in Japan, Italy, Russia, New Zealand, and the United States. Two facilities in the United States are devoted to volcano monitoring: Hawaiian Volcano Observatory on Kilauea Volcano and the David A. Johnston Cascades Volcano Observatory in Vancouver, Washington. Many of the methods now used to monitor volcanoes were developed at the Hawaiian Volcano Observatory.

Volcano monitoring involves recording and analyzing various physical and chemical changes at volcanoes. Tiltmeters are used to detect changes in the slopes of a volcano as it inflates when magma rises beneath it, whereas a geodimeter uses a laser beam to measure horizontal distances, which also change as a volcano inflates (Figure 4.19). Geologists also monitor gas emissions, changes in groundwater level and temperature, hot springs activity, and changes in the local magnetic and electrical fields. Even the accumulating snow and ice, if any, is evaluated to anticipate hazards from floods should an eruption take place. Of critical importance in volcano monitoring and warning of an imminent eruption is the detection of **volcanic tremor,** the continuous ground motion lasting for minutes to hours as opposed to the sudden, sharp jolts produced by most earthquakes. Volcanic tremor, also known as *harmonic tremor,* indicates that magma is moving beneath the surface.

To more fully anticipate the future activity of a particular volcano, its eruptive history must also be known. Accordingly, geologists study the record of past eruptions preserved in rocks. Detailed studies before 1980 indicated that Mount St. Helens, Washington, had erupted explosively 14 or 15 times during the last 4500 years, so geologists concluded that it was one of the most likely Cascade Range volcanoes to erupt again. In fact, maps they prepared showing areas in which damage from an eruption could be expected were helpful in determining which areas should have restricted access and evacuations once an eruption did take place.

Geologists successfully gave timely warnings of impending eruptions of Mount St. Helens, Washington, and Mount Pinatubo in the Philippines, but in both cases the climactic eruptions were preceded by eruptive activity of lesser intensity. In some cases, however, the warning signs are much more subtle and difficult to interpret. Numerous small earthquakes and other warning signs indicated to geologists of the U.S. Geological Survey (USGS) that magma was moving beneath the surface of the Long Valley caldera in eastern California, so in 1987 they issued a low-level warning, and then nothing happened.

Volcanic activity in the Long Valley caldera occurred as recently as 250 years ago, and there is every reason to think it will occur again, but when renewed activity will take place remains an unanswered question. Unfortunately, the local populace was largely unaware of the geologic history of the region, the USGS did a poor job in communicating its concerns, and premature news

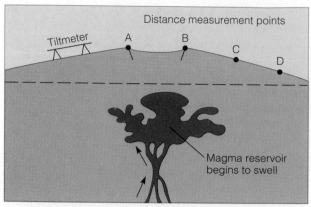

Distance measurement points

Tiltmeter
A   B   C   D

Magma reservoir
begins to swell

(a) Stage 1

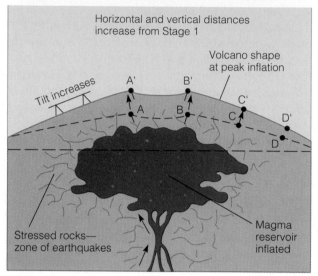

Horizontal and vertical distances
increase from Stage 1

Volcano shape
at peak inflation

Tilt increases
A'   B'   C'
A   B   C   D'
D

Stressed rocks—
zone of earthquakes

Magma
reservoir
inflated

(b) Stage 2

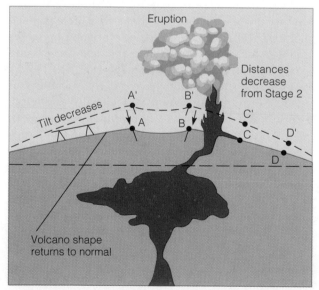

Eruption

Distances
decrease
from Stage 2

Tilt decreases
A'   B'   C'
A   B   C   D'
D

Volcano shape
returns to normal

(c) Stage 3

Figure 4.19 *Volcano monitoring. These illustrations show three stages in the eruption of a Hawaiian volcano, and how tilt and distance measurements are made. (a) The volcano begins to inflate as the magma chamber grows larger. (b) Inflation reaches its peak. (c) The volcano erupts and then deflates, returning to its original shape.*

releases caused more concern than was justified. In any case, local residents were outraged because the warnings caused a decrease in tourism (Mammoth Mountain on the margins of the caldera is the second largest ski area in the country) and property values plummeted. Monitoring continues in the Long Valley caldera, and the signs of renewed volcanism including earthquake swarms, trees being killed by carbon dioxide gas apparently emanating from magma (Figure 4.2b), and hot spring activity, cannot be ignored.

For the better monitored volcanoes, such as those in Hawaii, it is now possible to make accurate short-term predictions of eruptions. But for many volcanoes, little or no information is available. For instance, on January 14, 1993, Colombia's Galeras volcano erupted without much warning, killing 6 of 10 volcanologists on a field trip and 3 Colombian tourists. Ironically, the volcanologists were attending a conference on improving methods for predicting volcanic eruptions.

Distribution of Volcanoes

Rather than being randomly distributed around the globe, most active volcanoes are in well-defined zones or belts. More than 60% of all active volcanoes are in the **circum-Pacific belt,** which includes those in the Andes of South America; the volcanoes of Central America, Mexico, and the Cascade Range of North America; the Alaskan volcanoes and those in Japan, the Philippines, Indonesia, and New Zealand (Figure 4.20). The southernmost active volcanoes at Mount Erebus in Antarctica and a large caldera at Deception Island that erupted most recently in 1970 are also included in the circum-Pacific belt. In fact, this belt nearly encircling the Pacific Ocean basin is popularly referred to as the Ring of Fire, alluding to the many active volcanoes present there.

Another belt of volcanism coincides with the convergent boundary between the African and European plates in the Mediterranean basin (Figure 4.20). This **Mediterranean belt,** with about 20% of all active volcanoes, includes the famous ones in Italy such as Mount Etna and Vesuvius, and the Greek volcano Santorini. Mount Etna has issued lava flows more than 150 times since 1500 B.C. when activity was first recorded. Its most recent episode of eruptive activity began in February 2000. A particularly violent eruption of Santorini in 1390 B.C. might be responsible for the myth of the lost continent of Atlantis (see Chapter 8), and in A.D. 79 eruptions of Vesuvius destroyed Pompeii and other nearby cities (see the Prologue).

The third main area of active volcanism is at or near mid-oceanic ridges or the extensions of these ridges onto land (Figure 4.20). Volcanoes are found along the East Pacific Rise and the longest of all mid-oceanic ridges, the Mid-Atlantic Ridge. The latter is situated near the center of the Atlantic Ocean basin with volcanoes in Iceland, Ascension Island, Tristan da Cunha, and elsewhere. It con-

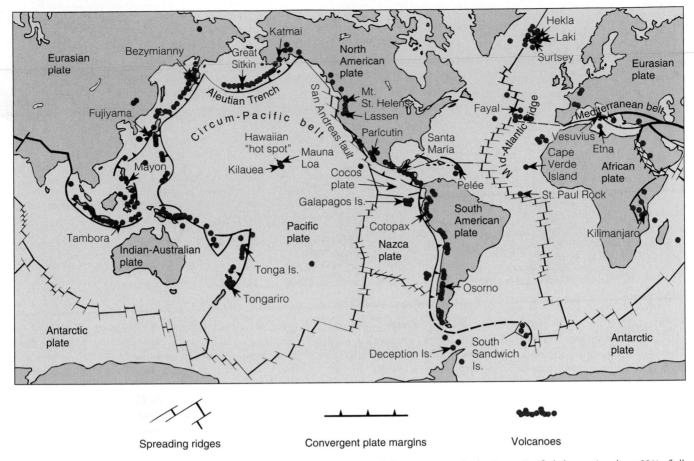

Figure 4.20 *Most volcanoes are at or near plate boundaries. Two major volcano belts are recognized: The circum-Pacific belt contains about 60% of all active volcanoes, about 20% are in the Mediterranean belt, and most of the rest are located along mid-oceanic ridges.*

tinues around the southern tip of Africa, where it continues as the Indian Ridge. Branches of the Indian Ridge extend into the Red Sea and East Africa, where such volcanoes as Kilamanjaro in Tanzania and Erta Ale in Ethiopia with its continuously active lava lake are found.

Anyone familiar with volcanoes will have noticed that so far this discussion has made no mention of the Hawaiian volcanoes. This is not an oversight. Mauna Loa and Kilauea on the island of Hawaii, and Loihi seamount just to the south, are the most notable exceptions to the distribution of volcanoes in well-defined belts at either divergent or convergent plate boundaries. Their location and significance are discussed in the following section.

Plate Tectonics, Volcanoes, and Plutons

IN Chapter 3 we discussed the origin and evolution of magma and concluded that (1) mafic magmas are generated beneath spreading ridges and (2) intermediate and felsic magmas form where an oceanic plate is subducted beneath another oceanic plate or a continental plate. Accordingly, most of Earth's volcanism and emplacement of plutons takes place at divergent and convergent plate boundaries.

Divergent Plate Boundaries and Igneous Activity

Much of the mafic magma that forms beneath spreading ridges is simply emplaced at depth as vertical dikes and gabbro plutons (Figure 4.21). But some of this magma rises to the surface where it forms submarine lava flows and pillow lavas (Figure 4.7). Indeed, the oceanic crust is composed largely of gabbro and basalt. Much of this submarine volcanism goes undetected, but researchers in submersible craft have observed the results of these eruptions.

Pyroclastic materials are not common in this environment because mafic lava is very fluid, allowing gases to escape easily; and at great depth, water pressure prevents gases from expanding. Accordingly, the explosive eruptions that yield pyroclastic materials are not common. Should an eruptive center along a ridge build above sea level, however, pyroclastic materials might be erupted at lava fountains, but most of the magma issues forth as fluid lava flows that form shield volcanoes.

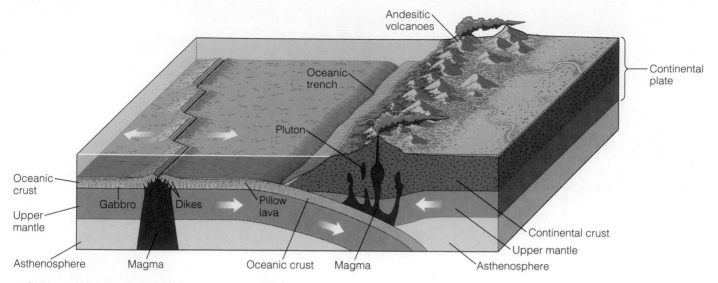

Figure 4.21 *Both intrusive and extrusive igneous activity take place at divergent plate boundaries (spreading ridges) and where plates are subducted at convergent plate boundaries. Oceanic crust is composed largely of vertical dikes and gabbro, and its upper part consists of submarine lavas, especially pillow lavas. Magma forms where an oceanic plate is subducted beneath another ocean plate or beneath a continent plate as shown here. Some of the magma forms plutons, especially batholiths, but some is erupted to form volcanoes.*

Excellent examples of divergent plate boundary volcanism are found along the Mid-Atlantic Ridge, particularly where it takes place above sea level as in Iceland. In November 1963, a new volcanic island, later name Surtsey, rose from the sea just south of Iceland. The East Pacific Rise and the Indian Ridge are also areas of similar volcanism. Not all divergent plate boundaries are beneath sea level as in the previous examples. For instance, divergence and igneous activity are taking place in Africa at the East African Rift system.

Igneous Activity at Convergent Plate Boundaries

Nearly all of the large active volcanoes in both the circum-Pacific and Mediterranean belts are composite volcanoes near the leading edges of overriding plates at convergent plate boundaries (Figure 4.21). A conspicuous exception is Mount Etna on Sicily, a shield volcano that has erupted repeatedly during historic time. The overriding plate, with its chain of volcanoes, may be oceanic as in the case of the Aleutian Islands, or it may be continental as is, for instance, the South American plate with its chain of volcanoes along its western edge.

As previously noted, these volcanoes at convergent plate boundaries consist largely of lava flows and pyroclastic materials of intermediate to felsic composition. Remember that when mafic oceanic crust partially melts, some of the magma generated is emplaced near plate boundaries as plutons and some is erupted to build up composite volcanoes. More viscous magmas, usually of felsic composition, are emplaced as lava domes, thus accounting for the explosive eruptions that typically occur at convergent plate boundaries.

In previous sections of this chapter, we alluded to several eruptions at convergent plate boundaries. Good examples are the explosive eruptions of Mount Pinatubo and Mayon volcano in the Philippines, both of which are situated near a plate boundary beneath which an oceanic plate is subducted. Mount St. Helens, Washington (see Perspective 4.2), is similarly situated, but it is on a continental rather than an oceanic plate. Mount Vesuvius in Italy, one of several active volcanoes in that region, lies on a plate that the northern margin of the African plate is subducted beneath (see the Prologue).

Intraplate Volcanism

Mauna Loa and Kilauea on the island of Hawaii and Loihi just 32 km to the south are within the interior of a rigid plate far from any divergent or convergent plate boundary (Figure 4.20). It is postulated that a mantle plume creates a local "hot spot" beneath Hawaii. However, the magma is derived from the upper mantle, as it is at spreading ridges, and accordingly is mafic so it builds up shield volcanoes. Loihi is particularly interesting because it represents an early stage in the origin of a new Hawaiian island. It is a submarine volcano that rises more than 3000 m above the adjacent seafloor, but its summit is still about 940 m below sea level.

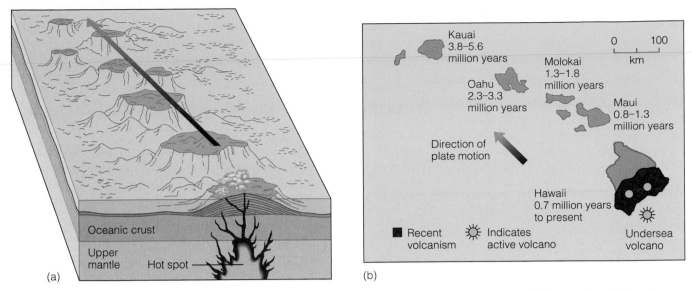

Figure 4.22 *(a) Generalized diagram showing the origin of the Hawaiian Islands. As a lithospheric plate moves over the hot spot beneath Hawaii, a succession of volcanoes form. Present-day volcanism occurs only on Hawaii and beneath the sea just to the south. (b) Map showing the age of the islands in the Hawaiian chain.*

Even though the Hawaiian volcanoes are unrelated to spreading ridges or subduction zones, their evolution is related to plate movements. Notice in Figure 4.22 that the ages of the rocks composing the various Hawaiian Islands increase toward the northwest; Kauai formed 3.8 to 5.6 million years ago, whereas Hawaii began forming less than 1 million years ago, and Loihi began forming even more recently. Continuous movement of the Pacific plate over the hot spot, now beneath Hawaii and Loihi, has formed the islands in succession.

Mantle plumes and hot spots have also been proposed to explain volcanism in several other areas. A mantle plume may exist beneath Yellowstone National Park in Wyoming, for instance. Some source of heat at depth is responsible for the present-day hot springs and geysers such as Old Faithful, but many geologists think the heat source is a body of intruded magma that has not yet completely cooled, rather than a mantle plume.

Chapter Summary

1. Volcanism is the process whereby magma and its associated gases erupt at the surface. Some magma erupts as lava flows, and some is ejected explosively as pyroclastic materials.

2. Only a few percent by weight of a magma consists of gases, most of which is water vapor. Sulfur gases emitted during large eruptions can have far-reaching climatic effects.

3. Many lava flows are characterized by pressure ridges and spatter cones. Columnar joints form in some lava flows when they cool. Pillow lavas are erupted underwater and consist of interconnected bulbous masses.

4. Volcanoes are mountains built up around a vent where lava flows and/or pyroclastic materials are erupted.

5. The summits of volcanoes are characterized by a circular or oval crater or a much larger caldera. Many calderas form by summit collapse when an underlying magma chamber is partly drained.

6. Shield volcanoes have low, rounded profiles and are composed mostly of mafic flows that cooled to form basalt. Cinder cones form where pyroclastic materials that resemble cinders are erupted and accumulate as small, steep-sided cones. Composite volcanoes are composed of lava flows of intermediate composition, layers of pyroclastic materials, and volcanic mudflows.

7. Viscous masses of lava, generally of felsic composition, are forced up through the conduits of some volcanoes and form bulbous, steep-sided lava domes. Volcanoes with lava domes are dangerous because they erupt explosively and frequently eject nuée ardentes.

8. Fluid mafic lava erupted from long fissures (fissure eruptions) spread over large areas to form basalt plateaus.

9. Pyroclastic flows erupted from fissures formed during the origin of calderas cover vast areas. These eruptions of pyroclastic materials form sheetlike deposits.

10. Most active volcanoes are distributed in linear belts. The circum-Pacific belt and Mediterranean belt contain more than 80% of all active volcanoes.

11. Volcanism in the circum-Pacific and Mediterranean belts is at convergent plate margins where subduction occurs. Partial melting of the subducted plate generates intermediate and felsic magmas.

12. Magma derived by partial melting of the upper mantle beneath spreading ridges accounts for the mafic lavas of ocean basins. Melting in these areas may be caused by reduction in pressure and/or hot mantle plumes.

13. The two active volcanoes on the island of Hawaii and one just to the south are thought to lie above a hot mantle plume. The Hawaiian Islands developed as a series of volcanoes formed on the Pacific plate as it moved over the mantle plume.

Important Terms

aa
basalt plateau
caldera
cinder cone
circum-Pacific belt
columnar joint
composite volcano
  (stratovolcano)

crater
fissure eruption
lahar
lava dome
lava tube
Mediterranean belt

nuée ardente
pahoehoe
pillow lava
pressure ridge
pyroclastic sheet deposit
shield volcano

spatter cone
volcanic explosivity index (VEI)
volcanic tremor
volcanism
volcano

Review Questions

1. Much of the upper part of the oceanic crust is made up of interconnected bulbous masses of igneous rock known as
   a. ____ pyroclastic materials; b. ____ volcanic bombs;
   c. ____ parasitic cones; d. ____ lapilli; e. ____ pillow lava.

2. Basalt plateaus form as a result of
   a. ____ repeated eruptions of cinder cones; b. ____ accumulation of thick layers of pyroclastic materials;
   c. ____ eruptions of fluid lava from long fissures; d. ____ deposition by volcanic mudflows; e. ____ erosion of composite volcanoes.

3. Two Cascade Range volcanoes have erupted in the United States during this century. They are ____ and ____.
   a. ____ Mount Rainier, Washington/Crater Lake, Oregon;
   b. ____ Mount St. Helens, Washington/Lassen Peak, California; c. ____ Newberry Volcano, Oregon/Mount Garibaldi, British Columbia; d. ____ Mount Shasta, California/Mount Hood, Oregon; e. ____ Glacier Peak, Washington/Mount Mazama, Oregon.

4. The volcanoes on the island of Hawaii and at Loihi Seamount just to the south
   a. ____ are composite volcanoes; b. ____ have not erupted during historic time; c. ____ are particularly dangerous;
   d. ____ erupt mostly pyroclastic materials; e. ____ are not located at a plate boundary.

5. The fact that ____ have slopes rarely steeper than 10° is because they are composed mostly of low viscosity lava flows
   a. ____ shield volcanoes; b. ____ lava tubes; c. ____ pressure ridges; d. ____ pyroclastic sheet deposits; e. ____ cinder cones.

6. One of the warning signs of an impending volcanic eruption is volcanic tremor, which is
   a. ____ changes in ground water levels; b. ____ inflation of a volcano as magma rises; c. ____ ground shaking lasting for minutes or hours; d. ____ emission of large quantities of gases; e. ____ a marked temperature increase.

7. Which volcanic gas was responsible for the deaths at Lake Nyos in Africa and the dead trees at Mammoth Lakes, California?

   a. ____ methane; b. ____ carbon dioxide; c. ____ hydrogen sulfide; d. ____ water vapor; e. ____ chlorine.

8. A lava flow with a smooth, ropy surface is termed
   a. ____ spatter cone; b. ____ lapilli; c. ____ pahoehoe;
   d. ____ pillow lava; e. ____ vesicular.

9. A nuée ardente is a(an)
   a. ____ lava flow of especially high viscosity; b. ____ thick accumulation of volcanic bombs; c. ____ type of eruption likely to produce a basalt plateau; d. ____ underwater eruption that yields bulbous masses of lava;
   e. ____ incandescent cloud of gases and particles.

10. Most active volcanoes are found in the
    a. ____ mid-oceanic ridge volcanic zone; b. ____ Sierra Nevada–Cascade volcanic province; c. ____ Mediterranean divergence boundary; d. ____ circum-Pacific belt; e. ____ eastern Atlantic subduction zone.

11. The summits of some volcanoes are characterized by very wide, steep-sided depressions known as ____, most of which form by ____.
    a. ____ explosion pits/voluminous eruptions; b. ____ calderas/summit collapse; c. ____ basalt plateaus/fissure eruptions; d. ____ lava domes/forceful injection; e. ____ parasitic cones/lava flows.

12. In addition to lava flows and pyroclastic materials, composite volcanoes are also made up of a significant amount of
    a. ____ lahars; b. ____ pressure ridges; c. ____ spatter cones; d. ____ columnar joints; e. ____ pahoehoe.

13. Why are most composite volcanoes at convergent plate boundaries, whereas most shield volcanoes are along divergent plate boundaries?

14. How do aa and pahoehoe lava flows differ, and what accounts for the difference(s)?

15. What kinds of materials erupted by volcanoes are a hazard to commercial aviation? Explain fully.

16. Explain how a nuée ardente originates and why nuée ardentes are particularly dangerous.

17. Explain why the Hawaiian Islands are progressively older toward the northwest, and why active volcanoes are present only on the island of Hawaii and just to the south?

18. How does a crater differ from a caldera, and how does the latter form? Give an example of a well-known caldera.
19. Why are most volcanoes found in distinct belts rather than being randomly distributed around the world? Where are these belts?
20. What are lava domes? Of what kind(s) of magma are they composed? Why are they especially dangerous?
21. Explain why eruptions of mafic magma are rather quiet, whereas eruptions of felsic magma are commonly explosive.
22. How do columnar joints and spatter cones form? Where are good places to see examples of each?

23. During this century, two Cascade Range volcanoes have erupted, and others will certainly do so in the future. What kinds of evidence would indicate that some of these other volcanoes are showing signs of renewed activity? Should any of them be monitored closely? If so, which ones?
24. In North America active volcanoes are found only in the Pacific Northwest and in Alaska. What geologic events would have to take place for similar volcanism to occur along the East Coast of the United States and Canada?

 World Wide Web Activities

For these Web Site addresses, along with current updates and exercises, log on to **http://www.brookscole.com/geo**

Volcanoes and Global Climate Change

This is a NASA Facts site and contains information on the relationship between volcanoes and global cooling and ozone depletion. Read about how volcanic eruptions affect global temperatures and how scientists monitor the various components erupted from a volcano.

Volcano World

This excellent volcano site is maintained and supported by NASA's Public Use of Earth and Space Science Data Over the Internet program. It contains a wealth of information about volcanoes and volcanic parks and monuments.

U.S. Geological Survey Cascades Volcano Observatory

This site contains information about volcanoes and other natural hazards in the western United States and elsewhere in the world. It is an excellent source of information on the hazards of volcanic eruptions and contains specific information about Mount St. Helens, as well as other volcanoes in the western United States and elsewhere.

Volcano Watch

This is a weekly newsletter mostly about the volcanoes on the island of Hawaii. It is written for the general public by scientists at the USGS's Hawaiian Volcano Observatory. The site contains information regarding the current status of erupting volcanoes in Hawaii.

Global Volcanism Program

This database at the Smithsonian Institution contains information on eruptions that have occurred during the last 10,000 years. The site features *Volcanoes of the World*, *Volcano Net Links*, and the *Bulletin of the Global Volcanism Network*, a weekly newsletter compiled from data submitted by more than 1000 correspondents.

Other Sites

Web sites about volcanoes and volcanism are numerous. Ones to visit include these:

USGS Photoglossary of Volcano Terms. One can read definitions and see images of many volcanic features.

Cenozoic-Mesozoic Volcanism . . . Eastern Sierra Nevada. Take a Field Trip to Owens Valley/Mammoth Lakes in the east-central part of the Sierra Nevada and see images and descriptions of volcanic features. Also, this is an area of potential future volcanism.

USGS Volcano & Hydrologic Monitoring. Great site to see what methods are used to monitor active volcanoes.

Hawaii Center for Volcanology. Devoted to Loihi Volcano, the newest volcano in the Hawaiian chain.

Soufriere Hills, Montserrat, West Indies. This site has images and updates on the continuing volcanism on this Caribbean island.

 CD-ROM Exploration

Exploring your *Earth Systems Today* CD-ROM will add to your understanding of the material in this chapter.

Topic: Earth's Disasters

Module: Volcanism

Explore activities in this module to see if you can discover the following for yourself:

Using aspects of this activity, examine the relationship of magma chemistry and eruptive style.

Using aspects of this activity, examine the global distribution of volcanoes and the nature of volcanic landforms.

Select a volcano from among those available in this activity and study its geologic setting, geologic hazards, geologic history, and current eruptive activity.

Weathering, Soil, and Sedimentary Rocks

Weathering and erosion are responsible for the spectacular scenery in Bryce Canyon National Park, Utah.

PROLOGUE

Geologic processes such as volcanism, glaciation, and shoreline erosion have yielded many areas of exceptional scenery in North America, but some of the most striking examples of scenery produced by the combined effects of weathering and erosion are found in several western states and parts of Canada. All rocks at or near the surface are constantly attacked by physical forces such as the wedging action of water freezing in cracks and by a variety of chemical processes as well as the activities of organisms from bacteria to large, land-dwelling animals. However, the effects of these processes are not the same everywhere, because rocks vary in their resistance to change. As a result, weathering and erosion yield complexly dissected surfaces and many oddly shaped features that go by such fanciful names as spires, splinters, monuments, arches, and hoodoos.

A particularly striking example of the effects of weathering and erosion is found in Bryce Canyon National Park, Utah, where brilliantly colored sedimentary rocks have been intricately sculpted to form mazes of interconnected gullies and pillars and spires of rock (chapter opening photo). Scattered localities of similar landscapes, known as badlands, are present in many other areas from Canada to Arizona and New Mexico. The badlands in Dinosaur National Park, Alberta, Canada, and in Badlands National Park, South Dakota, are excellent examples.

Badlands develop in dry areas with sparse vegetation and nearly impermeable yet easily eroded rocks. Rain falling on such unprotected rocks rapidly runs off and dissects the surface by forming numerous closely spaced, small gullies and steep ravines separated by sharp angular slopes, fluted ridges, arches, and pinnacles. The sedimentary rocks at Bryce Canyon National Park are described as limy siltstones, meaning they are composed of silt-sized (1/256–1/16 mm) particles with calcite ($CaCO_3$) acting as a cement to hold the particles together. These rocks formed 40 to 50 million years ago in a lake and were subsequently uplifted along a large fracture and now form the Pink Cliffs of the park. Incidentally, Bryce Canyon is not a canyon—rather, it is the eroded eastern margin of a high, fairly flat area known as the Paunsaugunt Plateau. The native Paiute called the area "red rocks standing like men in a bend-shaped canyon."

Another area where the remarkable effects of weathering and erosion are evident is in Arches National Park, Utah. Here sedimentary rocks called sandstone have yielded to weathering and erosion to produce a number of isolated spires and balanced rocks (Figure 5.1), as well as the picturesque arches for which the park was named. The term *arch* has been used to describe several features with different origins, but here we restrict it to mean an opening through a wall of rock that formed in response to weathering and erosion.

Arches form when weathering and erosion along parallel fractures leave a slender fin of rock. Some parts of these fins are more susceptible to mechanical and chemical changes than others, and as the sides are altered, a recess may form. If it does, pieces of the unsupported rock at the top of the recess eventually fall away, forming an arch as the original recess is enlarged. Baby Arch shows the initial stage in arch formation,

OBJECTIVES

At the end of this chapter, you will have learned that

■ Weathering yields the raw materials for both soils and for sedimentary rocks.

■ Some weathering processes bring about physical changes in Earth materials with no change in composition, whereas others result in compositional changes.

■ A variety of factors are important in the origin and evolution of soils.

■ Sediments are deposited as aggregates of loose solids that may become sedimentary rocks if they are compacted and/or cemented together.

■ Texture and composition are used to classify sedimentary rocks.

■ A variety of features preserved in sedimentary rocks are good indicators of how the original sediment was deposited.

■ Most evidence of prehistoric life in the form of fossils is found in sedimentary rocks.

■ Weathering is important in the origin and concentration of some resources, and sediments and sedimentary rocks are resources themselves or contain resources such as petroleum and natural gas.

whereas the park's most famous arch, Delicate Arch, represents an advanced stage (Figure 5.2). The arches continue to evolve, as is apparent from historic observations. In 1940, for instance, Skyline Arch was enlarged when a large block fell from its underside. The park also has a number of examples of collapsed arches, which commonly leave pinnacles and spires when they fall.

Figure 5.1 *This isolated spire and the rounded rocks on the right in Arches National Park, Utah are called hoodoos and goblins. They resulted from differential weathering and erosion of sandstone.*

(a)

(b)

Figure 5.2 *Differential weathering and erosion in Arches National Park, Utah. (a) Hole-in-the-Wall or Baby Arch shows an early stage in the development of an arch in a fin of rock. It measures 7.6 m wide and 4.5 m high. (b) An advanced stage of arch formation is shown by Delicate Arch. It is 9.7 m wide and 14 m high. What will this look like when the central span collapses?*

Introduction

Almost everyone is aware that rocklike substances such as pavement and cement in sidewalks, bridges, and foundations decay and crumble with age. Likewise, Earth materials at or near the surface suffer the same fate when exposed to the atmosphere, water, and the activities of organisms. In short, they experience **weathering,** defined as the physical breakdown and chemical alteration of Earth materials at or near the surface. Geologists are interested in weathering because it is an essential part of the rock cycle (see Figure 1.13).

During weathering, **parent material,** which is simply the rock acted upon by weathering processes, is disaggregated to form smaller pieces, and some of its constituent minerals are altered or dissolved. **Erosion** involves the wearing away of soil and rock by various geologic agents, and running water, wind, glaciers, and marine currents **transport** the solid, altered, and dissolved materials elsewhere. Thus erosion involves all processes that wear away soil and rock, whereas transport refers only to the phenomenon of carrying eroded materials elsewhere. For instance, a stream erodes a canyon, but the eroded materials are transported as solids or dissolved substances in the stream.

Some materials derived by weathering simply accumulate in place, that is, they are not eroded, whereas others are transported some distance and eventually deposited as *sediment.* In either case, the weathered materials may be further modified to form soil, or sediment may be transformed into sedimentary rock such as sandstone or limestone. So weathering provides the raw materials for both soils and sedimentary rocks. Furthermore, weathering is responsible for the origin of and enrichment of some natural resources.

Why Should You Study Weathering, Soil, and Sedimentary Rocks?

Why should you study weathering, soil, and sedimentary rocks? A good reason to study these subjects is that they clearly illustrate the interactions of Earth's materials with the atmosphere, hydrosphere, and biosphere. Furthermore, weathering is one aspect of the rock cycle leading to the origin of sedimentary rocks (see Figure 1.13), some of which are economically important. Weathering is also responsible for the origin of and enrichment of some natural resources. All rocks are important in studies of Earth history, but sedimentary rocks have a special place in this endeavor because they preserve features that reveal how they formed in the first place.

How Are Earth Materials Altered?

Even though weathering takes place continuously, its effects commonly vary from area to area and even within the same body of rock. Rocks are not structurally and chemically homogeneous throughout, so **differential weathering** occurs, which yields irregular surfaces. And coupled with *differential erosion,* that is, variable rates of erosion, some unusual and even bizarre features may develop. Indeed, some of the picturesque and peculiar features seen in such areas as Bryce Canyon National Park, Utah (see the Prologue), and Dinosaur National Park, Alberta, Canada, resulted from differential weathering and erosion.

The two recognized types of weathering, *mechanical* and *chemical,* both proceed simultaneously on parent material as well as during erosion and transport, and even in the areas where weathered materials are deposited as sediment. In short, all surface materials are subjected to weathering, although one type of weathering might predominate over the other. Mechanical and chemical weathering are discussed separately in the following sections simply for convenience.

Mechanical Weathering

Mechanical weathering takes place when physical forces break Earth materials into smaller pieces that retain the composition of the parent material. Granite, for instance, might be mechanically weathered and yield smaller pieces of granite or individual grains of quartz, potassium feldspars, plagioclase feldspars, and various accessory minerals such as biotite (Figure 5.3). The physical processes responsible for mechanical weathering in-

Figure 5.3 *Exposure of granitic rock altered mostly by mechanical weathering. The cones of material in the foreground are made up of individual mineral grains and rock fragments. In fact, most of the exposed rocks have been so thoroughly weathered that only a few, rounded masses of rock appear unaltered.*

clude frost action, pressure release, thermal expansion and contraction, salt crystal growth, and organic activity.

Frost action involving water repeatedly freezing and thawing in cracks and pores in rocks is particularly effective where temperatures commonly fluctuate above and below freezing. In the high mountains of the western United States and Canada, frost action is effective even during summer months. But, as one would expect, it is of little or no importance in the tropics or where water is permanently frozen. The reason frost action is so effective is that water expands by about 9% when it freezes, thus exerting great force on the walls of a crack, widening and extending it by *frost wedging* (Figure 5.4a). Repeated freezing and thawing dislodges angular pieces of rock from the parent material that tumble downslope and accumulate as **talus** (Figure 5.4b).

The mechanical weathering process called **pressure release** is especially evident in rocks that formed as plutons such as batholiths, but it also occurs in other rock types. Batholiths crystallize under tremendous pressure, and the rock so formed is stable under these pressure conditions. If uplifted and exposed by erosion, the pressure is reduced and *sheet joints* more or less paralleling the rock surface form as the energy is released by outward expansion (Figure 5.5a). Sheet-joint-

bounded slabs of rock slip or slide off the parent rock leaving large rounded masses known as **exfoliation domes** (Figure 5.5b).

That solid rock expands and produces fractures might be counterintuitive, but is nevertheless a well-known phenomenon. In deep mines, masses of rock detach from the sides of the excavation, often explosively. Spectacular examples of these *rock bursts* and less violent *popping* pose a danger to mine workers, and in South Africa are responsible for about 20 deaths per year. And in some quarries for building stone, excavations to only 7 or 8 m exposed rocks in which sheet joints formed (Figure 5.6), in some cases with enough force to throw quarrying machines weighing more than a ton from their tracks.

During **thermal expansion and contraction** the volume of rocks changes as they heat up and then cool down. For instance, the temperature may vary as much as 30°C in a day in a desert, and rock, being a poor conductor of heat, heats and expands on its outside more than its inside. Even dark minerals absorb heat faster than light-colored ones, so differential expansion takes place between minerals. Surface expansion might generate enough stress to cause fracturing, but experiments in which rocks are heated and cooled repeatedly to simu-

(a)

(b)

Figure 5.4 *(a) When water seeps into cracks and expands as it freezes it causes frost wedging. Repeated freezing and thawing pry angular pieces of rock loose. (b) Talus has accumulated at the base of this slope. Notice that the source rocks are highly fractured.*

(a)

(b)

Figure 5.5 *(a) Slabs of granitic rock bounded by sheet joints in the Sierra Nevada of California. (b) Stone Mountain, Georgia is a good example of an exfoliation dome.*

late years of such activity indicate that thermal expansion and contraction is of minor importance in mechanical weathering.

The formation of salt crystals can exert enough force to widen cracks and dislodge particles in porous, granular

rocks such as sandstone. And even in rocks with an interlocking mosaic of crystals, as in granite, **salt crystal growth** pries loose individual minerals. It takes place mostly in hot, arid regions, but also probably affects rocks in some coastal areas as well.

Animals, plants, and bacteria all participate in the mechanical and chemical alteration of rocks. Burrowing animals, such as worms, reptiles, rodents, termites, and ants, constantly mix soil and sediment particles and bring material from depth to the surface where further weathering occurs. The roots of plants, especially large bushes and trees, wedge themselves into cracks in rocks and further widen them (Figure 5.7).

Chemical Weathering

Chemical weathering includes those processes whereby rock materials are decomposed by alteration of parent material. Several clay minerals, which are sheet silicates, form by the chemical and structural alternation of other minerals such as potassium feldspars and plagioclase feldspars, both of which are framework silicates (see Figure 2.12). Other minerals are completely decomposed during chemical weathering as their ions are taken into solution, but minerals resistant to chemical change might simply be liberated from the parent material.

Important agents of chemical weathering include atmospheric gases, especially oxygen, water, and acids. Organisms also play an important role. Rocks with lichens (composite organisms made up of fungi and algae) on their surfaces undergo more rapid chemical alteration than lichen-free rocks. In addition, plants remove ions from soil water and reduce the chemical stability of soil minerals, and plant roots release organic acids. Other chemical weathering processes include solution, oxidation, and hydrolysis.

During **solution** the ions of a substance separate in a liquid, and the solid substance dissolves. Water is a remarkable solvent because its molecules have an asymmetric shape, consisting of one oxygen atom with two hydrogen atoms arranged so that the angle between the two hydrogens is about 104 degrees (Figure 5.8). Because of this asymmetry, the oxygen end of the molecule retains a slight negative electrical charge, whereas the hydrogen ends retain a slight positive charge. When a soluble substance such as the mineral halite (NaCl) comes in contact with a water molecule, the positively charged sodium ions are attracted to the negative end of the water molecule, and the negatively charged chloride ions are attracted to the positively charged end of the molecule (Figure 5.8). Thus, ions are liberated from the crystal structure, and the solid dissolves.

Most minerals are not very soluble in pure water because the attractive forces of water molecules are not

Figure 5.6 *Sheet-joints formed by expansion in the Mount Airy Granite in North Carolina.*

Figure 5.7 *This small tree is growing in the crack between columns in a lava flow. Its roots will enlarge the crack, thus contributing to mechanical weathering.*

sufficient to overcome the forces between particles in minerals. For instance, the mineral calcite ($CaCO_3$), the major constituent of the sedimentary rock limestone and the metamorphic rock marble, is practically insoluble in pure water, but it rapidly dissolves if a small amount of acid is present. An easy way to make water acidic is by dissociating the ions of carbonic acid as follows:

$$H_2O \; + \; CO_2 \; \rightleftharpoons \; H_2CO_3 \; \rightleftharpoons \; H^+ \; + \; HCO_3^-$$

water   carbon   carbonic hydrogen bicarbonate
      dioxide   acid   ion   ion

According to this chemical equation, water and carbon dioxide combine to form *carbonic acid,* a small amount of which dissociates to yield hydrogen and bicarbonate ions. The concentration of hydrogen ions determines the acidity of a solution; the more hydrogen ions present, the stronger the acid.

Carbon dioxide from several sources may combine with water and react to form acid solutions. The atmosphere is mostly nitrogen and oxygen, but about 0.03% is carbon dioxide, causing rain to be slightly acidic. Decaying organic matter and the respiration of organisms produces carbon dioxide in soils, so groundwater is also generally slightly acidic. Climate affects the acidity, however, with arid regions tending to have alkaline groundwater (that is, it has a low concentration of hydrogen ions).

Whatever the source of carbon dioxide, once an acidic solution is present, calcite rapidly dissolves according to the following reaction:

$$CaCO_3 \; + \; H_2O \; + \; CO_2 \rightleftharpoons Ca^{++} \; + \; 2HCO_3^-$$

calcite   water   carbon calcium bicarbonate
           dioxide   ion   ion

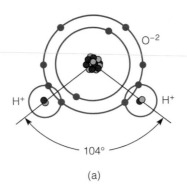

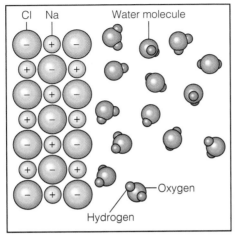
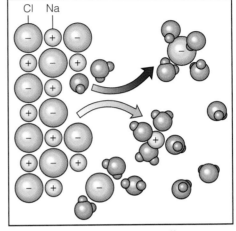

(a)

Figure 5.8 *(a) The structure of a water molecule. The asymmetric arrangement of hydrogen atoms causes the molecule to have a slight positive electrical charge at its hydrogen end and a slight negative charge at its oxygen end. (b) Solution of sodium chloride (NaCl), the mineral halite, in water. Note that the sodium atoms are attracted to the oxygen end of a water molecule, whereas chloride ions are attracted to the hydrogen end of the molecule.*

(b)

The term **oxidation** has a variety of meanings for chemists, but in chemical weathering it refers to reactions with oxygen to form an oxide (one or more metallic elements combined with oxygen), or, if water is present, a hydroxide (a metallic element or radical combined with OH). For example, iron rusts when it combines with oxygen to form the iron oxide hematite:

$$4Fe + 3O_2 \rightarrow 2Fe_2O_3$$
$$\text{iron} \quad \text{oxygen} \quad \text{iron oxide}$$
$$\text{(hematite)}$$

Of course, atmospheric oxygen is abundantly available for oxidation reactions, but oxidation is generally a slow process unless water is present. Thus, most oxidation is carried out by oxygen dissolved in water.

Oxidation is important in the alteration of ferromagnesian silicates such as olivine, pyroxenes, amphiboles, and biotite. Iron in these minerals combines with oxygen to form the reddish iron oxide hematite (Fe_2O_3) or the yellowish or brown hydroxide limonite [$FeO(OH)\cdot nH_2O$]. The yellow, brown, and red colors of many soils and sedimentary rocks are caused by the presence of small amounts of hematite or limonite.

The chemical reaction between the hydrogen (H^+) ions and hydroxyl (OH^-) ions of water and a mineral's ions is known as **hydrolysis.** In hydrolysis, hydrogen ions actually replace positive ions in minerals. Such replacement changes the composition of minerals and liberates iron that then may be oxidized.

The chemical alteration of the potassium feldspar orthoclase provides a good example of hydrolysis. All feldspars are framework silicates, but when altered, they yield soluble salts and clay minerals, such as kaolinite, which are sheet silicates. The chemical weathering of orthoclase by hydrolysis occurs as follows:

$$2KAlSi_3O_8 + 2H^+ + 2HCO_3^- + H_2O \rightarrow$$
$$\text{orthoclase} \quad \text{hydrogen} \quad \text{bicarbonate} \quad \text{water}$$
$$\text{ion} \quad \text{ion}$$

$$Al_2Si_2O_5(OH)_4 + 2K^+ + 2HCO_3^- + 4SiO_2$$
$$\text{clay(kaolinite)} \quad \text{potassium} \quad \text{bicarbonate} \quad \text{silica}$$
$$\text{ion} \quad \text{ion}$$

In this reaction hydrogen ions attack the ions in the orthoclase structure, and some liberated ions are incorporated in a developing clay mineral. The potassium and bicarbonate ions go into solution and combine to form a soluble salt. On the right side of the equation is excess silica that would not fit into the crystal structure of the clay mineral.

Factors Controlling the Rate of Chemical Weathering

Chemical weathering processes operate on the surface of particles so they alter rocks and minerals from the outside inward. In fact, if you break open a weathered stone it is not at all uncommon to see a rind of weathering at and near the surface, but the stone is completely unaltered inside. The rate at which chemical weathering proceeds depends on several factors. One is simply the presence or absence of fractures, because fluids seep along fractures, accounting for more intense chemical weathering along these surfaces (Figure 5.9). Thus, given the same rock type under similar conditions, the more fractures, the more rapid will be the chemical weathering. Of course other factors also control chemi-

Figure 5.9 *Fractured rocks are more susceptible to chemical weathering because fluids seep along fractures. Notice that the fractures in these granitic rocks have been accentuated by weathering.*

cal weathering, including particle size, climate, and parent material.

Because chemical weathering affects particle surfaces, the greater the surface area, the more effective is the weathering. It is important to realize that small particles have larger surface areas compared to their volume than do large particles. Notice in Figure 5.10 that a block measuring 1 m on a side has a total surface area of 6 m^2, but when the block is broken into particles measuring 0.5 m on a side, the total surface area increases to 12 m^2. And if these particles are all reduced to 0.25 m on a side, the total surface area increases to 24 m^2. Note that while the surface area in this example increases, the total volume remains the same at 1 m^3.

The fact that small objects have proportionately more surface area compared to volume than do large objects has profound implications. We can conclude that because mechanical weathering reduces the size of particles, it contributes to chemical weathering by exposing more surface area.

It is not surprising that chemical weathering is more effective in the tropics than in arid and arctic regions because temperatures and rainfall are high and evaporation rates are low. In addition, vegetation and animal life are much more abundant in the tropics. Consequently, the effects of weathering extend to depths of several tens of meters, but commonly extend only centimeters to a few meters deep in arid and arctic regions.

Some rocks are more resistant to chemical alteration than others and thus are not altered as rapidly, so parent material is another control on how rapidly chemical weathering proceeds. For example, the metamorphic rock quartzite, composed of quartz, is an extremely stable substance that alters slowly compared to most other rock types. In contrast, rocks such as basalt, which contain large amounts of calcium-rich plagioclase and pyroxene minerals, decompose rapidly because these minerals are chemically unstable. In fact, the stability of common minerals is just the opposite of their order of

Surface area = 6 m^2

Surface area = 12 m^2

Surface area = 24 m^2

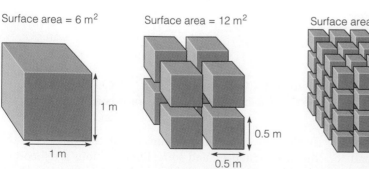

1 m
1 m
(a)

0.5 m
0.5 m
(b)

0.25 m
0.25 m
(c)

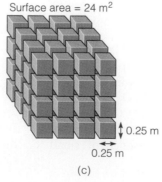

Figure 5.10 *Particle size and chemical weathering. As a rock is reduced into smaller and smaller particles, its surface area increases but its volume remains the same. Thus, in (a) the surface area is 6 m^2, in (b) it is 12 m^2, and in (c) 24 m^2, but the volume remains the same at 1 m^3. Accordingly, small particles have more surface area in proportion to their volume than do large particles.*

Table 5.1

Stability of Silicate Materials

Ferromagnesian Silicates	Nonferromagnesian Silicates
Olivine	Calcium plagioclase
Pyroxene	
Amphibole	Sodium plagioclase
Biotite	Potassium feldspar
	Muscovite
	Quartz

Increasing Stability (arrow pointing downward on left side)

WHAT WOULD YOU DO?

Acid rain is one consequence of industrialization, but it is a problem that can be solved. In your community, a local copper smelter is obviously pouring out gases, thereby contributing to the problem both locally and regionally. As chairperson of a committee of concerned citizens, you make some recommendations to the smelter owners to clean up their emissions. What specific recommendations would you favor? Suppose further that the smelter owners say that the remedies are too expensive and if they are forced to implement them, they will close the smelter. Would you still press forward even though your efforts might mean economic disaster for your community?

crystallization in Bowen's reaction series (Table 5.1, also see Figure 3.3): The minerals that form last in this series are chemically stable, whereas those that form early are more easily altered by chemical processes because they are most out of equilibrium with their conditions of formation.

One manifestation of chemical weathering is **spheroidal weathering** (Figure 5.11). In spheroidal weathering, a stone, even one that is rectangular to begin with, weathers to form a more spherical shape because that is the most stable shape it can assume. The reason is that on a rectangular stone the corners are attacked by weathering processes from three sides, and the edges are attacked from two sides, but the flat surfaces are weathered more or less uniformly (Figure 5.11). Consequently, the corners and edges are altered more rapidly, the material sloughs off them, and a more spherical shape develops (Figure 5.11d). Once a spherical shape is present, all surfaces weather at the same rate.

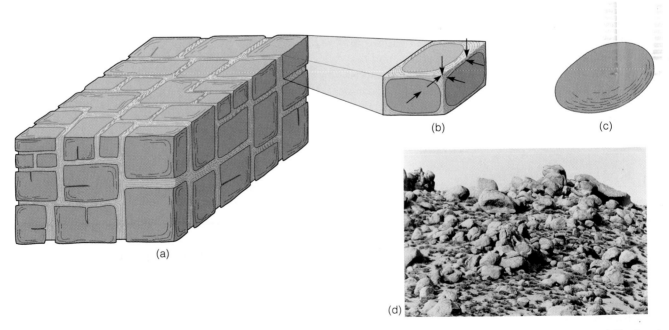

(a)   (b)   (c)   (d)

Figure 5.11 *Spheroidal weathering. (a) The rectangular blocks outlined by fractures are attacked by chemical weathering processes, much like those in Figure 5.9, but (b) the corners and edges are weathered most rapidly. (c) When a block has weathered so that its shape is more nearly spherical, its surface is weathered evenly, and no further change in shape takes place. (d) Exposure of granitic rocks reduced to spherical boulders.*

What Is Soil and How Does It Form?

A layer of **regolith,** a collective term for sediment regardless of how it was deposited, as well as layers of pyroclastic materials and the residue formed in place by weathering, covers most of Earth's land surface. Some regolith consisting of weathered materials, air, water, and organic matter supports vegetation and is called **soil.** Almost all land-dwelling organisms depend directly or indirectly on soil for their existence. Plants grow in soil from which they derive their nutrients and most of their water, whereas many land-dwelling animals depend on plants for nutrients.

About 45% of a good soil for farming and gardening is composed of weathered material, mostly sand, silt, and clay, with much of the remaining volume simply void spaces or pores filled with air and/or water. In addition, a small but important amount of humus is also generally present. *Humus* is carbon derived by bacterial decay of organic matter and is highly resistant to further decay. Even a fertile soil might have as little as 5% humus, but it is nevertheless important as a source of plant nutrients and it enhances a soil's capacity for moisture retention.

Some weathered materials in soils are simply sand- and silt-sized mineral grains, especially quartz, but other minerals may be present as well. These solid particles hold soil particles apart, allowing oxygen and water to circulate more freely. Clay minerals are also important constituents of soils and aid in the retention of water as well as supplying nutrients to plants. Soils with excess clay minerals, however, drain poorly and are sticky when wet and hard when dry.

Residual soils form when parent material weathers in place. For example, if a body of granite weathers, and the weathering residue accumulates over the granite and is converted to soil, the soil thus formed is residual. In contrast, *transported soils* develop on weathered material that was eroded and transported from the weathering site to a new location, where it is altered to soil.

The Soil Profile

Observed in vertical cross section, a soil consists of distinct layers or **soil horizons** that differ from one another in texture, structure, composition, and color (Figure 5.12). Starting from the top, the soil horizons are designated O, A, B, and C, but the boundaries between horizons are transitional rather than sharp. Because soil-forming processes begin at the surface and work downward, horizon A is more altered from the parent material than the layers below.

Horizon O, which is generally only a few centimeters thick, consists of organic matter. The remains of plant materials are clearly recognizable in the upper part of horizon O, but its lower part consists of humus.

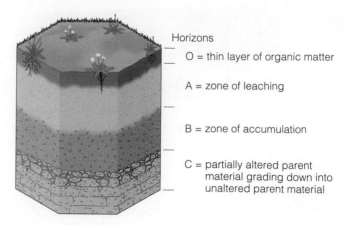

Figure 5.12 *The soil horizons in a fully developed or mature soil.*

Horizons
O = thin layer of organic matter
A = zone of leaching
B = zone of accumulation
C = partially altered parent material grading down into unaltered parent material

Horizon A, called *top soil,* contains more organic matter than horizons B and C below. It is also characterized by intense biological activity because plant roots, bacteria, fungi, and animals such as worms are abundant. Thread-like soil bacteria give freshly plowed soil its earthy aroma. In soils developed over a long period of time, horizon A consists mostly of clays and chemically stable minerals such as quartz. Water percolating down through horizon A dissolves soluble minerals and carries them away or downward to lower levels in the soil by a process called *leaching.* Accordingly, horizon A is also referred to as the *zone of leaching.*

Horizon B, or *subsoil,* contains fewer organisms and less organic matter than horizon A (Figure 5.12). Horizon B is also known as the *zone of accumulation,* because soluble minerals leached from horizon A accumulate as irregular masses. If horizon A is eroded leaving horizon B exposed, plants do not grow as well, and if horizon B is clayey, it is harder when dry and stickier when wet than other soil horizons.

Horizon C, the lowest soil layer, has little organic matter and consists of partially altered parent material grading down into unaltered parent material (Figure 5.12). In horizons A and B, the composition and texture of the parent material have been so thoroughly altered that it is no longer recognizable. In contrast, rock fragments and mineral grains of the parent material retain their identity in horizon C.

Factors Controlling Soil Formation

Soil scientists acknowledge that climate is the single most important factor in soil origins, but complex interactions among several factors account for soil type, thickness, and fertility (Figure 5.13). A very general classification recognizes three major soil types characteristic of different climatic settings. Soils that develop in humid regions such as the eastern United States and much of Canada are **pedalfers,** a name derived from the Greek word *pedon,* meaning soil, and from the chemical symbols for aluminum (Al) and iron (Fe). Because these soils form where abundant moisture is present, most of the soluble minerals have been leached from horizon A. Although it may be gray, horizon A is generally dark colored because of abundant organic matter, and aluminum-rich clays and iron oxides tend to accumulate in horizon B.

Soils found in much of the arid and semiarid western United States, especially the southwest, are **pedocals.** Pedocal derives its name in part from the first three letters of calcite. These soils contain less organic matter than pedalfers, so horizon A is generally lighter colored and contains more unstable minerals because of less intense chemical weathering. As soil water evaporates, calcium carbonate leached from above commonly precipitates in horizon B where it forms irregular masses of *caliche.* Precipitation of sodium salts in some desert areas where soil water evaporation is intense yields *alkali soils* that are so alkaline they cannot support plants.

Laterite forms in the tropics where chemical weathering is intense and leaching of soluble minerals is complete. These soils are red, commonly extend to depths of several tens of meters, and are composed largely of aluminum hydroxides, iron oxides, and clay minerals; even quartz, a chemically stable mineral, is generally leached out (Figure 5.14).

Although laterites support lush vegetation, they are not very fertile. The native vegetation is sustained by nutrients derived mostly from the surface layer of organic matter, but little humus is present in the soil itself because bacterial action destroys it. When laterites are cleared of their native vegetation, the surface accumulation of organic matter is rapidly oxidized, and there is little to replace it. Consequently, when societies practicing slash-and-burn agriculture clear these soils, they can raise crops for only a few years at best. Then the soil is depleted of plant nutrients, the clay-rich laterite bakes brick hard in the tropical sun, and the farmers move on to another area where the process is repeated.

The same rock type can yield different soils in different climatic regimes, and in the same climatic regime the same soils can develop on different rock types. Thus, it seems that climate is more important than parent material in determining the type of soil. Nevertheless, rock type does exert some control. For example, the metamorphic rock quartzite will have a thin soil over it because it is chemically stable, whereas an adjacent body of granite will have a much deeper soil.

Soils depend on organisms for their fertility, and in return they provide a suitable habitat for many organisms.

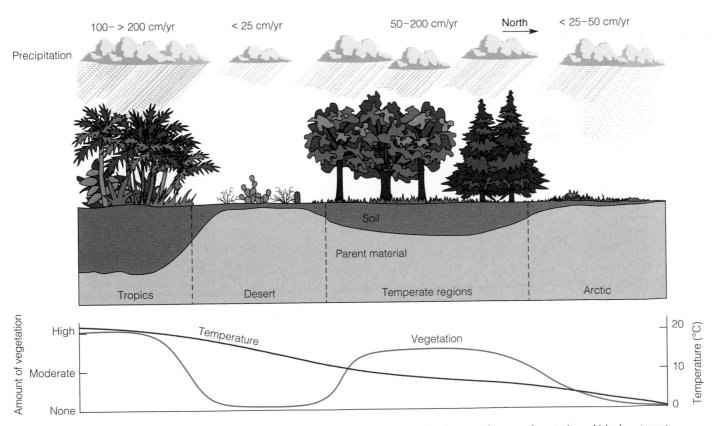

Figure 5.13 *Schematic representation showing soil formation as a function of the relationship between climate and vegetation, which alters parent material over time. Soil-forming processes operate most vigorously where precipitation and temperatures are high.*

Figure 5.14 *Laterite, shown here in Madagascar, is a deep, red soil that forms in response to intense chemical weathering in the tropics.*

Earthworms—as many as one million per acre—ants, sowbugs, termites, centipedes, millipedes, and nematodes, along with various types of fungi, algae, and single-celled animals, make their homes in soil. All contribute to the formation of soils and provide humus when they die and are decomposed by bacterial action.

Much of the humus in soils is provided by grasses or leaf litter that microorganisms decompose to obtain food. In so doing, they break down organic compounds within plants and release nutrients back into the soil. In addition, organic acids produced by decaying soil organisms are important in further weathering of parent materials and soil particles.

Burrowing animals constantly churn and mix soils, and their burrows provide avenues for gases and water. Soil organisms, especially some types of bacteria, are extremely important in changing atmospheric nitrogen into a form of soil nitrogen suitable for use by plants.

The difference in elevation between high and low points in a region is called *relief.* And because climate is such an important factor in soil formation and climate changes with elevation, areas with considerable relief have different soils in mountains and adjacent lowlands. *Slope* also is an important control, but it actually influences soil formation in two ways. One is simply *slope angle;* steep slopes have little or no soil, because weathered materials are eroded faster than soil-forming processes can operate. The other factor is *slope direction.* In the Northern Hemisphere, north-facing slopes receive less sunlight than south-facing slopes, and have cooler internal temperatures, support different vegetation, and if in a cold climate, remain snow covered or frozen longer.

Soil formation begins at the surface and works downward, so horizon A has been altered longer than the other horizons, and thus parent material is no longer recognizable. Even in horizon B parent material is usually not discernible, but it is in horizon C. In fact, soil prop-

erties are determined by climate and organisms altering parent material through time (Figure 5.13), so the longer the processes have operated the more fully developed a soil will be.

How much time is needed to develop a centimeter of soil or a fully developed soil a meter or so deep? No definite answer can be given because weathering proceeds at vastly different rates depending on climate and parent material, but an overall average might be about 2.5 cm per century. Nevertheless, a lava flow a few centuries old in Hawaii may have a well-developed soil on it, whereas a flow the same age in Iceland will have considerably less soil. Given the same climatic conditions, soil develops faster on unconsolidated sediment than it does on bedrock.

Under optimum conditions, soil-forming processes operate rapidly in the context of geologic time. From the human perspective, though, soil formation is a slow process; consequently, soil is a nonrenewable resource.

Soil Degradation

From the human perspective soils are nonrenewable, so soil losses that exceed the rate of formation are viewed with alarm. Likewise any reduction in soil fertility and productivity is cause for concern, especially in areas where soils already provide only a marginal existence. Erosion and chemical and physical deterioration are all forms of **soil degradation,** and are serious problems in many parts of the world.

Erosion, an ongoing natural process, is usually slow enough for soil formation to keep pace, but unfortunately, some human practices add to the problem. Human activities such as removing natural vegetation by plowing, overgrazing, overexploitation for fire wood, and deforestation all contribute to erosion by wind and running water. The Dust Bowl that developed in several Great Plains states during the 1930s is a poignant example of just how effective wind erosion is on soil pulverized and exposed by plowing (see Perspective 5.1).

Although wind has caused considerable soil erosion in some areas, running water is much more effective. Some soil is removed by *sheet erosion,* which involves removal of thin layers of soil more or less evenly over a broad, sloping surface. *Rill erosion,* in contrast, takes place when running water scours small, troughlike channels. Channels shallow enough to be eliminated by plowing are *rills,* but those too deep (about 30 cm) to be plowed over are *gullies* (Figure 5.15). Where gullying is extensive, croplands can no longer be tilled and must be abandoned.

(a)

(b)

Figure 5.15 *(a) Rill erosion in a field in Michigan during a rainstorm. The rill was later plowed over. (b) A large gully in the upper basin of the Rio Reventado in Costa Rica.*

A soil undergoes chemical deterioration when its nutrients are depleted and its productivity decreases. Loss of soil nutrients is most notable in many of the populous developing nations where soils are overused to maintain high levels of agricultural productivity. Chemical deterioration is also caused by insufficient use of fertilizers and by clearing soils of their natural vegetation. Examples of chemical deterioration can be found everywhere, but it is most prevalent in South America, where it accounts for nearly 30% of all soil degradation.

Other types of chemical deterioration are pollution and *salinization,* which occurs when the concentration of salts increases in a soil, making it unfit for agriculture. Improper disposal of domestic and industrial wastes, oil and chemical spills, and the concentration of insecticides and pesticides in soils all cause pollution. Soil pollution is a particularly serious problem in some parts of Eastern Europe.

Physical deterioration of soils results when soil is compacted by the weight of heavy machinery and livestock, especially cattle. Compacted soils are more costly to plow, and plants have a more difficult time emerging from them. Furthermore, water does not readily infiltrate, so more runoff occurs; this in turn accelerates the rate of water erosion.

In North America, the rich prairie soils of the midwestern United States and the Great Plains of the United States and Canada are suffering significant soil degradation. Nevertheless, this degradation, is moderate and less serious than in many other parts of the world. Problems experienced during the past have stimulated the development of methods to minimize soil erosion on agricultural lands. Crop rotation, contour plowing, and the construction of terraces have all proved helpful (Figure 5.16). So has no-till planting in which the residue from the harvested crop is left on the ground to protect the surface from the ravages of wind and water.

Figure 5.16 *One soil conservation practice is contour plowing, which involves plowing parallel to the contours of the land. The furrows and ridges are perpendicular to the direction that water would otherwise flow downhill and thus inhibit erosion.*

What Is Soil and How Does It Form? **125**

The Dust Bowl—An American Tragedy

The stock market crash of 1929 ushered in the Great Depression, a time when millions of people were unemployed and many had no means to acquire food and shelter. Urban areas were affected most severely by the depression, but rural areas suffered as well, especially during the great drought of the 1930s. Prior to the 1930s, farmers had enjoyed a degree of success unparalleled in U.S. history. During World War I (1914–1918), the price of wheat soared, and after the war when Europe was recovering, the government subsidized wheat prices. High prices and mechanized farming resulted in more and more land being tilled. Even the weather cooperated, and land in the western United States that would otherwise have been marginally productive was plowed. Deep-rooted prairie grasses that held the soil in place were replaced by shallow-rooted wheat.

Beginning in about 1930, drought conditions prevailed throughout the country; only two states—Maine and Vermont—were not drought-stricken. Drought conditions varied from moderate to severe but the consequences were particularly severe in the southern Great Plains. Some rain fell, but not enough to maintain agricultural production. And because the land, even marginal land, had been tilled, the native vegetation was no longer available to keep the topsoil from blowing away. And blow away it did—in huge quantities.

A large region in the southern Great Plains that was particularly hard hit by the drought, dust storms, and soil erosion came to be known as the Dust Bowl. Although its boundaries were not well defined, it included parts of Kansas, Colorado, and New Mexico, as well as the panhandles of Oklahoma and Texas (Figure 1a); together the Dust Bowl and its less affected fringe area covered more than 400,000 km^2!

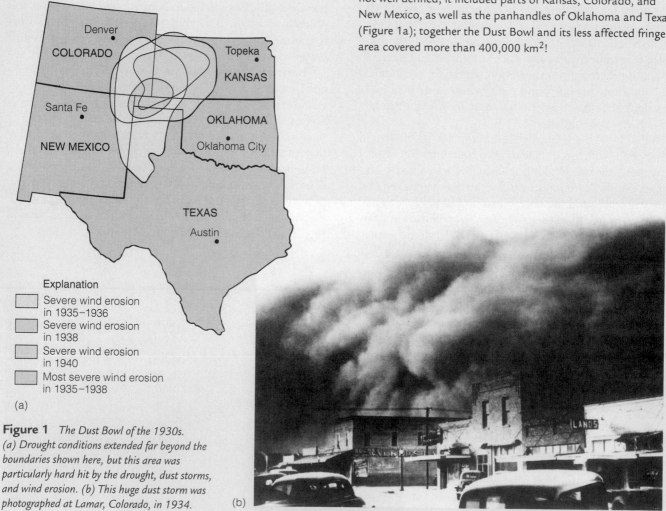

Explanation

- Severe wind erosion in 1935–1936
- Severe wind erosion in 1938
- Severe wind erosion in 1940
- Most severe wind erosion in 1935–1938

(a)

Figure 1 *The Dust Bowl of the 1930s. (a) Drought conditions extended far beyond the boundaries shown here, but this area was particularly hard hit by the drought, dust storms, and wind erosion. (b) This huge dust storm was photographed at Lamar, Colorado, in 1934.*

(b)

Dust storms were common during the 1930s, and some reached phenomenal sizes (Figure 1b). One of the largest storms occurred in 1934 and covered more than 3.5 million km². It lifted dust nearly 5 km into the air, obscured the sky over large parts of six states, and blew hundreds of millions of tons of soil eastward where it settled on New York City, Washington, D.C., and other eastern cities, as well as on ships as far as 480 km out in the Atlantic Ocean. The Soil Conservation Service reported dust storms of regional extent on 140 occasions during 1936 and 1937. Dust was everywhere. It seeped into houses, suffocated wild animals and livestock, and adversely affected human health.

The dust was, of course, the material derived from the tilled lands; in other words, much of the topsoil simply blew away. Blowing dust was not the only problem; sand piled up along fences, drifted against houses and farm machinery, and covered what otherwise might have been productive soils. Agricultural production fell precipitously in the Dust Bowl, farmers could not meet their mortgage payments, and by 1935 tens of thousands were homeless, on relief, or leaving (Figure 2). Many of

these people went west to California and became the migrant farm workers immortalized in John Steinbeck's novel *The Grapes of Wrath*.

The Dust Bowl was an economic disaster of great magnitude. Droughts had stricken the southern Great Plains before, and have done so since—from August 1995 well into the summer of 1996, for instance—but the drought of the 1930s was especially severe. Political and economic factors also contributed to the disaster. Due in part to the artificially inflated wheat prices, many farmers were deeply in debt—mostly because they had purchased farm machinery in order to produce more and benefit from the high prices. Feeling economic pressure because of their huge debts, they tilled marginal land and employed few, if any, soil conservation measures.

If the Dust Bowl has a bright side, it is that the government, farmers, and the public in general no longer take soil for granted or regard it as a substance that needs no nurturing. In addition, a number of soil conservation methods developed then have now become standard practices.

Figure 2 *By the mid 1930s, tens of thousands of people were on relief, homeless, or had left the Dust Bowl. In 1939, Dorthea Lange photographed this family of seven in Pittsburgh County, Oklahoma. Many of these people became the migrant farm workers immortalized in John Steinbeck's novel* The Grapes of Wrath.

What Is Soil and How Does It Form? **127**

You have inherited a piece of property ideally located for everything you consider important. Unfortunately, as you prepare to have a house built, your contractor tells you that the soil is rich in clay that expands when wet and contracts when it dries. You nevertheless go ahead with construction, but must now decide on what measures to take to prevent damage to the structure. Make several proposals that might solve this problem. Which one or ones do you think would be most cost effective?

Weathering and Natural Resources

In a preceding section, we discussed intense chemical weathering in the tropics and the origin of laterite, which is composed largely of aluminum hydroxides, iron oxides, and clay minerals. Laterites are not very productive for agriculture, but one of their qualities is of great economic importance. If the parent material is rich in aluminum, the ore of aluminum known as *bauxite* might accumulate in horizon B. Because such intense chemical weathering does not presently take place in North America, the United States and Canada depend on foreign sources for bauxite. Some aluminum ore is present in Arkansas, Alabama, and Georgia, but importing the ore is cheaper than mining these deposits.

Accumulations of valuable minerals formed by the selective removal of soluble substances are *residual concentrations.* In addition to bauxite, a number of other residual concentrations are important, including ore deposits of iron, manganese, clays, nickel, phosphate, tin, diamonds, and gold. Some of the sedimentary iron deposits of the Lake Superior region were enriched by chemical weathering when the soluble constituents that were originally present were carried away.

A number of kaolinite deposits in the southern United States were formed by the chemical weathering of feldspars in pegmatites and of clay-bearing limestones and dolostones. Kaolinite is a type of clay mineral used in the manufacture of paper and ceramics.

A gossan is a yellow to reddish deposit composed largely of iron hydroxides that formed by the alteration of iron- and sulfur-bearing minerals such as pyrite (FeS_2). The dissolution of such minerals forms sulfuric acid, which causes other metallic minerals to dissolve, and these tend to be carried downward toward the water table. Gossans have been used occasionally as sources of iron, but they are far more important as indicators of underlying ore deposits. One of the oldest known underground mines exploited such ores about 3400 years ago in what is now southern Israel.

Weathering, Sediment, and Sedimentary Rocks

In the Introduction we noted that weathering of parent materials yields the raw materials for both soils and sedimentary rocks, the second major family of rocks (see Figure 1.13). All **sedimentary rocks** are composed of **sediment,** which includes all solid particles derived by mechanical and chemical weathering as well as minerals precipitated from solution by chemical processes and minerals used by organisms to build their skeletons. So all sediment is derived by one weathering process or another, eroded from the weathering site, transported elsewhere, and deposited as a loose aggregate of particles. Sand and gravel in stream channels and mud on the seafloor are examples of sedimentary deposits.

One important criterion for classifying sedimentary particles is their size (Table 5.2), particularly for solid particles, or *detrital sediment,* derived by weathering as opposed to *chemical sediment* consisting of minerals extracted from solution by inorganic chemical processes or the activities of organisms. Particles described as *gravel* measure more than 2 mm, whereas sand measures 1/16–2 mm, and silt is any particle between 1/256 and 1/16 mm. None of these designations implies anything about composition; most gravel is made up of rock fragments—that is, small pieces of granite, basalt, or any other rock type—but sand and silt grains are usually single minerals, especially quartz. Particles smaller than 1/256 mm are termed clay, but clay has two meanings. One is simply a size designation, but the term also refers to certain types of sheet silicates known as *clay minerals.* However, most clay minerals are also clay sized.

Earth's crust is composed mostly of *crystalline rocks,* a term referring loosely to metamorphic rocks and most igneous rocks. Nevertheless, sediment and sedimentary rocks are by far the most common at or near the surface of both the continents and seafloor. All rock types are important in deciphering Earth history, but sedimentary rocks have a special significance in this endeavor because they preserve physical evidence and perhaps fossils useful for determining how they were deposited in the first place. Accordingly, geologists can make inferences from these rocks about the past distribution of streams, deserts, glaciers, lakes, and shorelines, and determine where resources might be found in these rocks.

(a)

(b)

Figure 5.17 *Rounding and sorting in sediments. (a) A deposit of well-sorted, and well rounded gravel. These particles measure about 5 cm across. (b) Angular, poorly sorted gravel.*

Sediment Transport and Deposition

Weathering and erosion are fundamental to the origin of sediment and sedimentary rocks, and so is *transport* and *deposition*, that is, the movement of sediment by natural processes and its accumulation in some area. Because they themselves are solids that are moving, glaciers can carry sediment of any size, whereas wind transports only sand and smaller sediment. Waves and marine currents transport sediment along shorelines, but by far the most common way to transport sediment from its source to other locations is by running water. Even weak currents can move clay- and silt-sized particles and ions in solution, which constitute an unseen but important part of the sediment load in streams, but more vigorous currents are necessary to transport sand and gravel.

During transport, *abrasion* reduces the size of particles, and the sharp corners and edges are worn smooth, a process known as *rounding,* as pieces of sand and gravel collide with one another (Figure 5.17a). Transport and processes operating where sediment accumulates also result in *sorting,* which refers to the particle-size distribution in a sedimentary deposit. Sediment is characterized as well sorted if all particles are about the same size, and poorly sorted if a wide range of particle sizes is present (Figure 5.17b). Both rounding and sorting have important implications for other aspects of sediment and sedimentary rocks, such as how readily fluids move through them, and they also help geologists decipher the history of a deposit.

Regardless of how sediment is transported, it is eventually deposited in some geographic area known as a **depositional environment.** Deposition might take place on a floodplain, in a stream channel, on a beach, or the seafloor, or a variety of other depositional environments where physical, chemical, and biological processes impart various characteristics to the accumulating sediment. Geologists generally recognize three major depositional settings: continental (on the land), transitional (on or near seashores), and marine, each with several specific depositional environments (Figure 5.18).

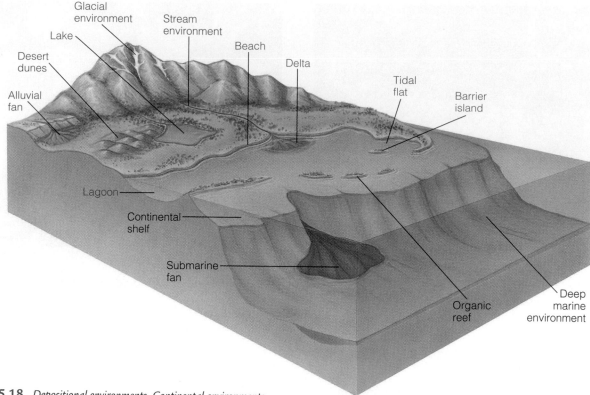

Figure 5.18 *Depositional environments. Continental environments are shown in red type. The environments along the shoreline, shown in blue type, are transitional from continental to marine. The others, shown in black type, are marine environments.*

Lithification—Transforming Sediment into Sedimentary Rock

Chemical sediments such as calcium carbonate ($CaCO_3$) mud are presently accumulating in the shallow waters of Florida Bay, and detrital gravel, sand, and mud are deposited in or adjacent to steams and elsewhere. These sediments consist of loose aggregates of mineral grains and rock fragments. **Lithification,** involving compaction and/or cementation, is the process whereby these sediments are converted into sedimentary rock (see Figure 1.13). Only a few sedimentary rocks skip the unconsolidated sediment stage in their origin—the limestone that makes up coral reefs, for instance.

To illustrate the relative importance of compaction and cementation let us consider two deposits of detrital sediment, one consisting of mud, the other of sand. In both cases, the newly deposited sediment consists of particles and *pore spaces,* the voids between particles. As sediment continues to accumulate, sediment layers are subjected to *compaction* from the weight of the overlying sediments, and the amount of pore space decreases, thus reducing the volume of the deposit. Our hypothetical deposit of mud might have as much as 80% water-filled pore space, but during compaction the water is squeezed out and the volume of the deposit can be reduced by up to 40% (Figure 5.19). The

sand deposit might initially have 50% pore space, although it is usually somewhat less, and it, too, can be compacted so that the sand grains fit more tightly together.

Compaction alone is generally sufficient for lithification of mud, but for larger particles such as sand and gravel, *cementation* is necessary to convert sediment into sedimentary rock. Cement consists of chemically precipitated minerals in pore spaces that bind the particles together. The most common cements in detrital sedimentary rocks are calcium carbonate ($CaCO_3$) and silicon dioxide (SiO_2), both of which are derived by chemical weathering of various minerals and rocks. Iron oxides and hydroxides, such as hematite (Fe_2O_3) and limonite [$FeO(OH) \cdot nH_2O$], respectively, are the chemical cements in some rocks. Exposures of red, yellow, and brown sedimentary rocks in many areas, such as the southwestern United States, owe their color to small amounts of iron oxide or hydroxide cement (Figures 5.1 and 5.2).

By far the most common chemical sediment is calcium carbonate mud and accumulations of larger particles such as seashells that when lithified form limestone. Compaction and cementation are also responsible for lithification. The cement binding these sediments into sedimentary rocks is calcium carbonate ($CaCO_3$) derived by partial solution of some of the particles in the deposit.

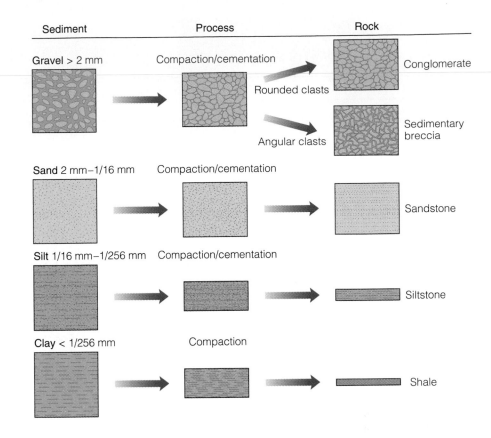

Sediment	Process		Rock
Gravel > 2 mm	Compaction/cementation	Rounded clasts	Conglomerate
		Angular clasts	Sedimentary breccia
Sand 2 mm–1/16 mm	Compaction/cementation		Sandstone
Silt 1/16 mm–1/256 mm	Compaction/cementation		Siltstone
Clay < 1/256 mm	Compaction		Shale

Figure 5.19 *Lithification of detrital sediments by compaction and cementation. Notice that little compaction takes place in sand and gravel.*

What Kinds of Sedimentary Rocks Do Geologists Recognize?

Thus far, we have considered the origin of sediment, its transport, deposition, and lithification, so we now turn to the types of sedimentary rocks and how they are classified. The two broad classes or types of sedimentary rocks are *detrital* and *chemical*, although the latter has a subcategory known as *biochemical* (Table 5.3).

Detrital Sedimentary Rocks

Detrital sedimentary rocks are made up of *detritus*, the solid particles such as sand and gravel derived from parent material by mechanical and chemical weathering. All detrital sedimentary rocks have a *clastic texture*, meaning they are composed of particles or fragments known as *clasts*. The several varieties in this broad category are classified by the size of their constituent particles, although composition is used to modify some rock names (Table 5.3).

Both *conglomerate* and *sedimentary breccia* are composed of gravel-sized particles (Table 5.3, Figure 5.20a,b). The only difference between the two rocks is the shape of their particles; conglomerate has rounded gravel, whereas sedimentary breccia has angular gravel. Conglomerate is common, but sedimentary breccia is rare because gravel-sized particles become rounded very quickly during transport. Thus, if you encounter a sedimentary breccia, you can conclude that its angular gravel has experienced little transport, probably less than a kilometer. Considerable energy is needed to transport gravel, so conglomerate is usually found in environments such as stream channels and beaches.

Sand is simply a size designation for particles between 1/16 and 2 mm, so any mineral or rock fragment can be in *sandstone*. Although composition is not considered in sandstone classification, geologists recognize varieties of sandstone based on mineral content (Table 5.3; Figure 5.20c). *Quartz sandstone* is the most common, and, as the name implies, is made up mostly of quartz grains. Another variety of sandstone called *arkose* contains at least 25% feldspar minerals. Sandstone is found in a number of depositional environments including stream channels, sand dunes, beaches, barrier islands, deltas, and the continental shelf (Figure 5.18).

Mudrock is a general term that encompasses all detrital sedimentary rocks composed of silt- and clay-size particles (Table 5.3). Varieties include *siltstone*, composed mostly of silt-sized particles, *mudstone*, a mixture of silt and clay, and *claystone*, composed primarily of clay-sized particles. Some mudstones and claystones are designated

Table 5.3

Classification of Sedimentary Rocks

DETRITAL SEDIMENTARY ROCKS

Sediment Name and Size	Description	Rock Name	
Gravel (>2 mm)	Rounded gravel particles	Conglomerate	
	Angular gravel particles	Sedimentary breccia	
Sand (1/16–2 mm)	Mostly quartz sand	Quartz sandstone	] Sandstones
	Quartz with >25% feldspar	Arkose	
Mud (<1/16 mm)	Mostly silt	Siltstone	]
	Silt and clay	Mudstone*	] Mudrocks
	Mostly clay	Claystone*	]

CHEMICAL SEDIMENTARY ROCKS

Texture	Composition	Rock Name	
Crystalline	Calcite ($CaCO_3$)	Limestone	] Carbonates
Crystalline	Dolomite [$CaMg(CO_3)_2$]	Dolostone	]
Crystalline	Gypsum ($CaSO_4 \cdot 2H_2O$)	Rock gypsum	] Evaporites
Crystalline	Halite (NaCl)	Rock salt	]

BIOCHEMICAL SEDIMENTARY ROCKS

Texture	Composition	Rock Name
Clastic	Calcium carbonate ($CaCO_3$) shells	Limestone (various types such as chalk and coquina)
Usually crystalline	Altered microscopic shells of silicon dioxide (SiO_2)	Chert
—	Mostly carbon from altered plant remains	Coal

*Mudrocks possessing the property of fissility, meaning they break along closely spaced planes, are commonly called *shale*.

shale if they are fissile, meaning that they break along closely spaced parallel planes (Figure 5.20d). Even weak currents can transport silt- and clay-sized particles, and deposition takes place only where currents and fluid turbulence are minimal, as in the quiet offshore waters of lakes or in lagoons.

Chemical and Biochemical Sedimentary Rocks

Several compounds and ions taken into solution during chemical weathering are the raw materials for **chemical sedimentary rocks.** They are so named because chemical processes are responsible for their origin, as when minerals form either as a result of inorganic chemical reactions or the chemical activities of organisms. Some of these rocks have a *crystalline texture,* meaning they are composed of a mosaic of inter-locking mineral crystals. Others, though, have a clastic texture; some limestones, for instance, are composed of fragmented seashells. Organisms play an important role in the origin of chemical sedimentary rocks designated as **biochemical sedimentary rocks.**

Limestone and dolostone, the most abundant chemical sedimentary rocks, are known as **carbonate rocks** because each is made up of minerals containing the carbonate radical (CO_3). Limestone consists of calcite ($CaCO_3$) and dolostone is made up of dolomite [$CaMg(CO_3)_2$] (see Chapter 2). Recall that calcite rapidly dissolves in acidic water, but the chemical reaction leading to dissolution is reversible, so calcite can precipitate from solution under some circumstances. Thus, some limestone, although probably not very much, forms by inorganic chemical precipitation. Most limestone is biochemical because organisms are so important in its origin—the rock in coral reefs and limestone composed of seashells, for instance (Figure 5.21a).

(a) Conglomerate

(b) Sedimentary breccia

(c) Quartz sandstone

(d)

Figure 5.20 *Detrital sedimentary rocks. (a) Conglomerate is made up of rounded gravel, whereas sedimentary breccia (b) has angular gravel. (c) Quartz sandstone is the most common variety of sandstone. (d) Exposure of the mudrock shale. Notice that the rock splits along closely spaced planes so it is fissile.*

A type of limestone composed almost entirely of fragmented seashells is known as *coquina* (Figure 5.21b), whereas chalk is a soft variety of limestone made up largely of microscopic shells. One distinctive variety of limestone contains small spherical grains called *ooids* that have a small nucleus around which concentric layers of calcite precipitated (Figure 5.21c). Lithified deposits of ooids form *oolitic limestones*.

Dolostone is similar to limestone but most or all of it probably formed secondarily by the alteration of limestone. The consensus among geologists is that dolostone originates when magnesium replaces some of the calcium in calcite, thereby converting calcite to dolomite.

Some of the dissolved substances derived by chemical weathering precipitate from evaporating water and thus form chemical sedimentary rocks known as **evaporites** (Table 5.3). *Rock salt*, composed of halite (NaCl), and *rock gypsum* ($CaSO_4 \cdot 2H_2O$) are the most common (Figures 5.22a, b), although several others are known and some are important resources. Compared with mudrocks, sandstone, and limestone, evaporites are not very common but

nevertheless are significant deposits in some areas such as Michigan, Ohio, New York, and the Gulf Coast region.

Chert is a hard rock composed of microscopic crystals of quartz (Table 5.3; Figure 5.22c). Some of the color varieties of chert are *flint*, which is black because of inclusions of organic matter, and *jasper*, which is colored red or brown by iron oxides. Because chert is hard and lacks cleavage, it can be shaped to form sharp cutting edges, so many cultures have used it to manufacture tools, spear points, and arrowheads. Chert is found as irregular masses or *nodules* in other rocks, especially limestone, and as distinct layers of *bedded chert* made up of tiny shells of silica-secreting organisms.

Coal consists of compressed, altered remains of land plants, but it is nevertheless a biochemical sedimentary rock (Figure 5.22d). It forms in swamps and bogs where the water is oxygen deficient or where organic matter accumulates faster than it decomposes. In oxygen-deficient swamps and bogs, the bacteria that decompose vegetation can live without oxygen, but their wastes must be oxidized, and because little or no oxygen is present, wastes accumulate and kill the bacteria. Bacterial decay ceases and the

What Kinds of Sedimentary Rocks Do Geologists Recognize? **133**

(a) Fossilerous limestone

(b) Coquina

(c) Ooids

Figure 5.21 *(a) Fossiliferous limestone, and (b) coquina composed of broken seashells are biochemical sedimentary rocks. (c) Present-day ooids measuring 1–2 mm across, from the Bahamas. Oolitic limestone is made up of ooids.*

(a) Rock salt

(b) Rock gypsum

(c) Chert

(d) Bituminous coal

Figure 5.22 *Chemical and biochemical sedimentary rocks. (a) Core of rock salt from a well in Michigan. (b) Rock gypsum. (c) Chert, a dense, hard rock composed of microscopic crystals of quartz. (d) Bituminous coal.*

vegetation is not completely decomposed and forms organic muck. When buried and compressed, the muck becomes *peat,* which looks rather like coarse pipe tobacco. Where peat is abundant, as in Ireland and Scotland, it is used for fuel.

Peat represents the first step in forming coal. If peat is more deeply buried and compressed, and especially if it is heated too, it is converted to dull black coal called *lignite.* During this change, the easily vaporized or volatile elements are driven off, enriching the residue in carbon; lignite has about 70% carbon whereas only about 50% is present in peat. *Bituminous coal,* with about 80% carbon, is dense, black, and so thoroughly altered that plant remains are rarely seen. It burns more efficiently than lignite, but the highest-grade coal is *anthracite,* a metamorphic type of coal (see Chapter 6), with up to 98% carbon.

Sedimentary Facies

If a layer of sediment or sedimentary rock is traced laterally, it generally changes in composition, texture, or both. It changes by lateral gradation resulting from the simultaneous operation of different processes in adjacent depositional environments. For example, sand may be deposited in a high-energy nearshore marine environment while mud and carbonate sediments accumulate simultaneously in the laterally adjacent low-energy offshore environments (Figure 5.23). Deposition in each of these environments produces **sedimentary facies,** bodies of sediment each possessing distinctive physical, chemical, and biological attributes. Figure 5.23 illustrates three sedimentary facies: a sand facies, a mud facies, and a carbonate facies. If these sediments become lithified, they are called sandstone, mudstone (or shale), and limestone facies, respectively.

Many sedimentary rocks in the interiors of continents show clear evidence of deposition in marine environments.

The rock layers in Figure 5.24d, for example, consist of a sandstone facies that was deposited in a nearshore marine environment overlain by shale and limestone facies deposited in offshore environments. This vertical sequence of facies can be explained by deposition occurring during a time when sea level rose with respect to the continents. As sea level rises the shoreline moves inland, giving rise to a **marine transgression** (Figure 5.24), and the depositional environments parallel to the shoreline migrate landward. Remember that each laterally adjacent environment in Figure 5.24 is the depositional site of a different sedimentary facies. As a result of a marine transgression, offshore facies are superimposed over nearshore facies, thus accounting for the vertical succession of sedimentary facies. Even though the nearshore environment is long and narrow at any particular time, deposition takes place continuously as the environment migrates landward. The sand deposit may be tens to hundreds of meters thick, but have horizontal dimensions of length and width measured in hundreds of kilometers.

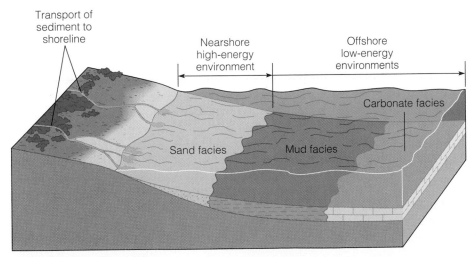

Figure 5.23 *Deposition in adjacent environments yields distinct bodies of sediment, each of which is designated as a sedimentary facies.*

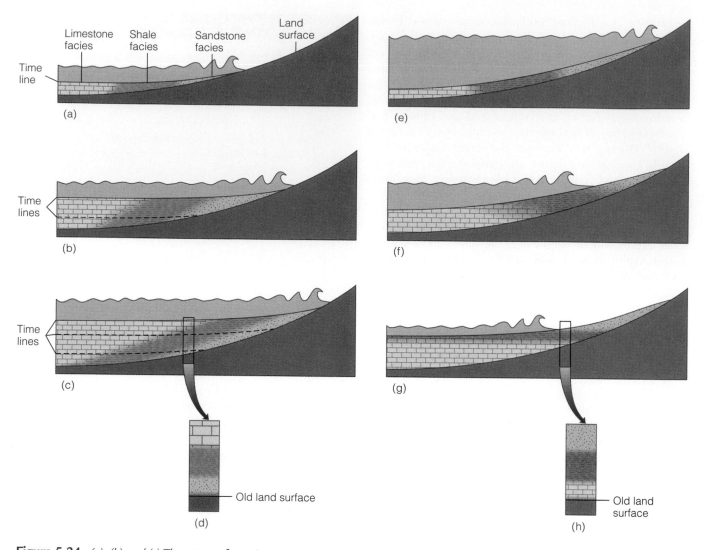

Figure 5.24 *(a), (b), and (c) Three stages of a marine transgression. (d) Diagrammatic representation of the vertical sequence of facies resulting from the transgression. (e), (f), and (g) Three stages of a marine regression. (h) Diagrammatic representation of the vertical sequence of facies resulting from the regression.*

The opposite of a marine transgression is a **marine regression.** If sea level falls with respect to a continent, the shoreline and environments that parallel the shoreline move seaward. The vertical sequence produced by a marine regression has facies of the nearshore environment superposed over facies of offshore environments. Marine regressions also account for the deposition of a facies over a large geographic area.

Reading the Story in Sedimentary Rocks

We mentioned in the Introduction that sedimentary rocks preserve a record of the conditions under which they formed. However, no one was present when ancient sediments were deposited, so geologists must evaluate those aspects of sedimentary rocks that allow them to make inferences about the original depositional environment. And making such determinations is of more than academic interest. For instance, barrier island sand deposits make good reservoirs for hydrocarbons, so knowing the environment of deposition and geometry of such deposits is helpful in exploration for these resources.

Sedimentary textures such as sorting and rounding can give clues to depositional processes. Windblown dune sands tend to be well sorted and well rounded, but poor sorting is typical of glacial deposits. The geometry or three-dimensional shape is another important aspect of sedimentary rock bodies. Marine transgressions and regressions yield sediment bodies with a blanket or sheetlike geometry, but sand deposits in stream channels are long and narrow and described as having a shoestring geometry. Sedimentary textures and geometry alone are usually insufficient to determine depositional environment, but when considered with other sedimentary rock properties, especially *sedimentary structures* and *fossils,* geologists can most often reliably determine the history of a deposit.

Sedimentary Structures

Physical and biological processes operating in depositional environments are responsible for a variety of features known as **sedimentary structures.** One of the most common of these structures is distinct layers known as **strata** or **beds** (Figure 5.25a), with individual layers less than a millimeter up to many meters thick. These strata or beds are separated from one another by surfaces above and below in which the rocks differ in composition, texture, color, or a combination of features. Layering of some kind is present in almost all sedimentary rocks, but a few, such as limestone that formed in coral reefs, lack this feature.

Some individual sedimentary rock layers show an upward decrease in grain size, termed **graded bedding,** most of which forms by turbidity current deposition. A *turbidity current* is an underwater flow of sediment and water with a greater density than sediment-free water. Because of its greater density, a turbidity current flows downslope until it reaches the relatively flat seafloor, or lakefloor,

where it slows and begins depositing large particles followed by progressively smaller ones (Figure 5.26). Some graded bedding also forms in stream channels during the waning stages of floods.

Many sedimentary rocks are characterized by **cross-bedding,** in which layers are arranged at an angle to the surface on which they are deposited (Figure 5.25b). Cross-beds are found in many depositional environments such as sand dunes in deserts and along shorelines, as well as in stream-channel deposits and shallow marine sediments. Invariably, cross-beds result from transport and deposition by wind or water currents, and the cross-beds are inclined downward in the same direction the current flowed. So, ancient deposits with cross-beds inclined down toward the south, for example, indicate that the currents responsible for them flowed from north to south.

The surfaces separating layers in sand deposits commonly have small ridges with intervening troughs known as **ripple marks** giving them a somewhat corrugated appearance. Some ripple marks are asymmetrical in cross section, having a gentle slope on one side and a steep slope on the other. Currents flowing in one direction as in stream channels generate these so-called *current ripple marks* (Figure 5.27a,b). And because the steep slope of these ripples is on the downstream side, they are good indications of ancient current directions. In contrast, *wave-formed ripple marks* tend to be symmetrical in

(a)

(b)

Figure 5.25 *(a) Bedding or stratification is shown in these rocks by alternating layers of shale and sandstone. (b) The inclined layers in the center of this image are cross-beds. Notice that the cross-bedding is bounded above and below by nearly horizontal layers. Because the cross-beds are inclined downward to the left, we know that the currents responsible for them flowed from right to left.*

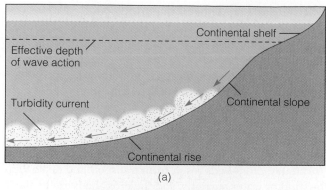

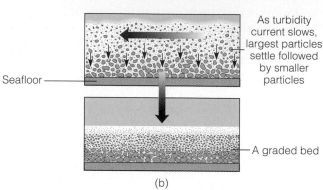

Figure 5.26 *(a) Turbidity current flows downslope along the seafloor (or lake bottom) because it is denser than sediment-free water. (b) Graded bedding forms as a turbidity current slows and deposits large particles followed by smaller ones.*

cross section, and, as their name implies, are generated by the to-and-fro motion of waves (Figure 5.27c,d).

When clay-rich sediment dries it shrinks and develops intersecting fractures called **mud cracks** (Figure 5.28). Mud cracks in ancient sedimentary rocks indicate that the sediment was deposited in an environment where periodic drying took place, such as on a river floodplain, near a lakeshore, or where muddy deposits are exposed along seacoasts at low tide.

The sedimentary structures discussed in the preceding paragraphs are the most common ones, and all are useful for determining ancient environments of deposition. Many others are known, though, some of which form long after deposition (Perspective 5.2).

Fossils—Remains and Traces of Ancient Life

Fossils, the remains or traces of ancient organisms, are interesting as evidence of prehistoric life (Figure 5.29), and are also important in determining depositional environments. Most people are familiar with fossil dinosaurs and some other land-dwelling animals, but are unaware that fossils of invertebrates, animals lacking a seg-

mented vertebral column, such as corals, clams, oysters, and a variety of microorganisms are much more useful because they are so common. It is true that the remains of land-dwelling creatures and plants can be washed into marine environments, but most are preserved in rocks deposited on land or perhaps transitional environments such as deltas. In contrast, fossils of corals tell us the rocks in which they are preserved were deposited in the ocean.

Clams with heavily constructed shells typically live in shallow turbulent seawater, whereas organisms living in low-energy environments commonly have thin, fragile shells. Marine organisms that carry on photosynthesis are restricted to the zone of sunlight penetration, which is generally less than 200 m. The amount of sediment is also a limiting factor on the distribution of organisms. Many corals live in shallow, clear seawater, because suspended sediment clogs their respiratory and food-gathering organs, and some have photosynthesizing algae living in their tissues.

Microfossils are particularly useful for environmental studies because hundreds or even thousands can be recovered from small rock samples. In oil-drilling operations small rock chips known as *well cuttings* are brought to the surface. These samples may contain numerous microfossils, but rarely have entire fossils of larger organisms. These fossils are routinely used to determine depositional environments and to match up rocks of the same relative age (see Chapter 17).

Determining the Environment of Deposition

Ancient sedimentary rocks possess a variety of features that they acquired as a result of physical, chemical, and biological processes that operated in the depositional environment. Geologists have the task of examining these rock features and deciphering the history of the deposit. During initial field studies, they commonly make some preliminary interpretations. Sedimentary particles such as ooids in limestones form in shallow marine environments with vigorous currents. Large-scale cross-bedding is typical of, but not restricted to, desert dunes. Fossils of land plants and animals are most often preserved in rocks that were deposited in continental environments, whereas fossils of corals and oysters obviously indicate a marine environment.

Laboratory analyses of sedimentary rocks collected during field studies might include microscopic and chemical analysis of rocks, fossil identification, and graphic representations showing three-dimensional shapes of rock units and their relationships to other rock units. In addition, the features of sedimentary rocks are compared with those of sediments of present-day depositional environments. In other words, geologists are employing the principle of uniformitarianism (see Chapter 1) by assuming that cross-beds, mud cracks, and ripple

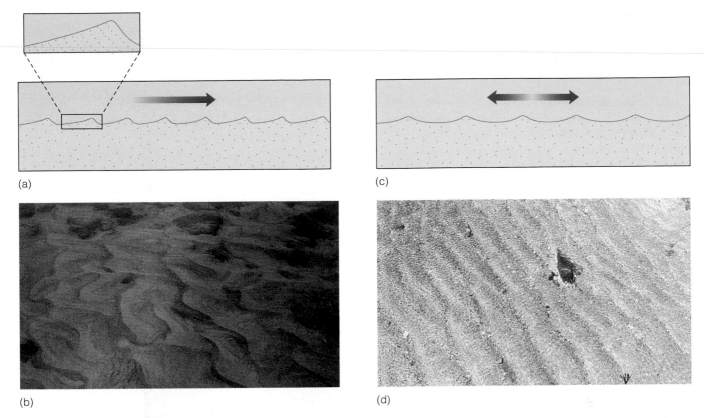

Figure 5.27 *Ripple marks (a) Current ripple marks form in response to flow in one direction as in a stream channel. The enlargement of one ripple shows its internal structure. Note that individual layers within the ripple are inclined, showing an example of cross-bedding. (b) Current ripples that formed in a small stream channel; flow was from right to left. (c) The to-and-fro currents of waves in shallow water deform the surface of the sand layer into wave-formed ripple marks. (d) Wave-formed ripple marks on ancient rocks.*

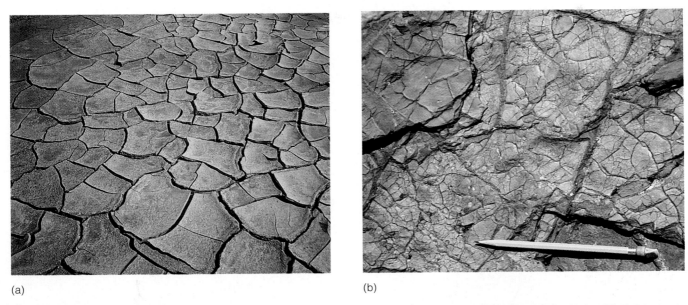

Figure 5.28 *(a) Mud cracks form in clay-rich sediment when they dry and shrink. (b) Mud cracks in ancient rocks in Glacier National Park, Montana. Notice that these cracks have been filled by sediment.*

Concretions, Geodes, and Thunder Eggs

Ripple marks, cross-bedding, and mud cracks are designated *primary sedimentary structures* because they form at the time of sediment deposition or shortly thereafter. All are useful in determining the environment of deposition, but sedimentary rocks also contain a variety of features known as *secondary sedimentary structures*. These formed long after deposition and tell nothing about deposition, but they are interesting and some are quite attractive. Many secondary sedimentary structures are known, but only concretions, geodes, and thunder eggs are discussed here.

Concretions consist of any irregular to spherical mass of material that can easily be separated from the enclosing rock (Figure 1a). Most are small, measuring a few centimeters across, but some are several meters in diameter. Several types of

sedimentary rocks might have concretions, but those that formed by more thorough cementation in parts of sand deposits are common in many sandstones. Because concretions are more thoroughly cemented and thus harder than the surrounding rock, they accumulate at the surface during weathering and erosion of their host rock (Figure 1b).

Imatra stones, or *marlekor,* from glacial lake deposits, are interesting disc-shaped concretions composed of calcium carbonate ($CaCO_3$) and silt (Figure 1c). Most are simple discs, only 2 or 3 cm across, but some unusual shapes develop when discs grow together or where outgrowths develop on the margins of discs. The lower imatra stone in Figure 1c, for instance, has such a regular geometric shape that it and similar ones have been mistaken for fossils or, in some cases, for human artifacts.

(a)

(b)

(c)

Figure 1 *(a) These spherical concretions measuring 6 to 8 cm across have grown together to form a composite concretion. (b) Spherical concretions from the Pumpkin Patch in the Ocotillo Wells State Vehicular Recreation Area, California. (c) Imatra stones, or marlekor, from glacial lake deposits in Connecticut.*

Not only are some imarta stones mistaken for fossils or human artifacts, but amateur fossil collectors have misidentified concretions in Oklahoma, and elsewhere, as human shoe prints. However, even a superficial examination of the rocks reveals the "shoe prints" are nothing more than concretions exposed by differential weathering and erosion. Their resemblance to shoe prints, or anything else, is purely coincidental.

Most concretions have little appeal to casual observers, although oddly shaped ones might elicit some attention. However, concretions known as *septaria,* or *septarian nodules,* are quite attractive. These large, spherical concretions have a series of cracks that widen toward the concretion's center. Smaller cracks more or less parallel the margin of the concretion and intersect the larger cracks (Figure 2a). Apparently, the cracks formed when the original concretion shrank during dehydration; later the cracks were filled with mineral crystals, most commonly calcite. If septeria are released from their host rock and weathered so that the interior cracks are visible, they have the appearance of a turtle shell (Figure 2b).

Sawed and polished septaria are available in any rock and mineral shop as are *geodes,* one of the most popular decorative stones. Most geodes are small, measuring 30 cm across, but some are much larger. They form when minerals grow along the margins of a cavity, partially filling it (Figure 3a). The outermost layer of minerals is a thin, sometimes discontinuous layer of agate, a variety of color-banded, compact, microcrystalline quartz. Next inward toward the cavity's center are inward-pointing crystals of quartz (SiO_2) or calcite ($CaCO_3$), or, more rarely, crystals of sulfate minerals such as barite ($BaSO_4$) and celestite ($SrSO_4$), all of which precipitated from solution.

Geodes or at least geodelike objects are not restricted to sedimentary rocks. *Thunder eggs* resemble geodes, and they too consist of minerals that grew in a cavity. However, thunder eggs are found in igneous rocks such as welded tuffs and rhyolite (Figure 3b). According to one legend, the spirits occupying neighboring volcanoes in Oregon hurled spherical rocks at one another which they had stolen from the thunder bird, hence the name thunder egg. The knobby, ribbed exterior of a thunder egg is not particularly impressive, but inside they are filled with agate, chalcedony, jasper, or opal, all of which are varieties of silicon dioxide (SiO_2). Like geodes and septeria, sawed and polished thunder eggs are popular decorative stones.

(a)

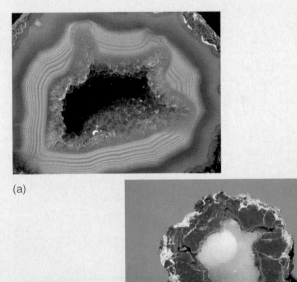

(a)

(b)

(b)

Figure 2 *(a) Sawed specimen of a septaria showing its internal structure. (b) Surface view of a septaria. Notice its resemblance to a turtle shell.*

Figure 3 *(a) A geode formed by the partial filling of a cavity by color-banded agate, which conforms to the walls of the cavity, and by inward-pointing quartz crystals. (b) A thunder egg from Oregon.*

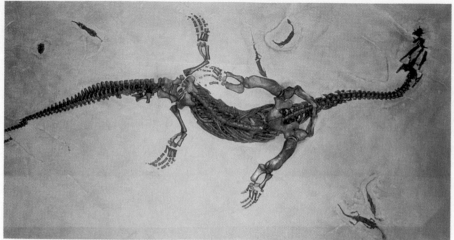
(a)

marks formed during the past just as they do now. Finally, when all data have been analyzed, an environmental interpretation is made.

Perhaps some examples of how environment of deposition is determined will be helpful. The Navajo Sandstone of the southwestern United States has an irregular sheet geometry and consists mostly of well-sorted, well-rounded sand grains .2 to .5 mm in diameter. Some of the sandstone beds have tracks of dinosaurs and other land-dwelling animals clearly indicating deposition in some kind of continental environment. These features along with cross-beds up to 30 m high and ripple marks, both of which look much like those in present-day sand dunes, lead to the conclusion that this is an ancient desert dune deposit (Figure 5.30). The cross-beds are inclined downward to the southwest, indicating that the wind blew mostly from the northeast.

In Arizona, a vertical sequence of three rock units designated the Tapeats Sandstone, Bright Angel Shale, and Muav Limestone is well exposed in the lower part of the Grand Canyon (Figure 5.31). The features preserved in these rocks, including fossils, clearly show that they were deposited in transitional and marine environments, much like the ones shown in Figure 5.23. All three units were being deposited simultaneously in adjacent envi-

(b)

Figure 5.29 *(a) Fossil bones of a 2.3-m-long Mesozoic-aged reptile in the museum at the Glacier Garden, Lucerne, Switzerland. (b) Shells of extinct Mesozoic organisms known as ammonites. Specimen in the Comstock Rock Shop, Virginia City, Nevada.*

Figure 5.30 *These sedimentary rocks making up Checkerboard Mesa in Zion National Park, Utah, belong to the Navajo Sandstone, an ancient desert dune deposit. Vertical fractures intersect the cross-beds, giving this cliff its checkerboard appearance.*

Figure 5.31 *View of the Tapeats Sandstone, Bright Angel Shale, and Muav Limestone in the Grand Canyon in Arizona. These formations were deposited during a widespread marine transgression. Compare this vertical sequence of rocks with the one shown in Figure 5.24d.*

No one was present millions of years ago to record data about the climate, geography, and geologic processes. So how is it possible to decipher unobserved past events? In other words, what features in rocks, and especially sedimentary rocks, would you look for to determine what happened in the far distant past? Can you think of any economic reasons to decipher Earth history from the record preserved in rocks?

ronments, but during a marine transgression they came to be superposed in the order now seen, with offshore facies overlying nearshore facies (Figure 5.31). Similar rock sequences of about the same age in Utah, Colorado, Wyoming, Montana, and South Dakota indicate this marine transgression covered much of what is now the western United States.

Are There Important Resources in Sedimentary Rocks?

The uses of sediments and sedimentary rocks or the materials they contain vary considerably. Sand and gravel are essential to the construction industry, pure clay deposits are used for ceramics, and limestone is used in the manufacture of cement and in blast furnaces where iron ore is refined to make steel. Evaporites are the source of table salt as well as a number of chemical compounds, and rock gypsum is used to manufacture wallboard. The tiny island nation of Nauru, with one of the highest per capita incomes in the world, has an economy based on mining and exporting phosphate-bearing sedimentary rock used in fertilizers.

Some valuable sedimentary deposits are found in streams and on beaches where minerals were concentrated during transport and deposition. These *placer deposits,* as they are called, are surface accumulations resulting from the separation and concentration of materials of greater density from those of lesser density. Much of the gold recovered during the initial stages of the California gold rush (1849–1853) was mined from placer deposits, and placers of a variety of other minerals such as diamonds and tin are important.

Historically, most coal mined in the United States has been bituminous coal from the Appalachian region that formed in coastal swamps during the Pennsylvanian Period (286 and 320 million years ago). Huge lignite and sub-bituminous coal deposits in the western United States are becoming increasingly important. During 1995, more than a billion tons of coal were mined in this country, more than half of it coming from mines in Wyoming, West Virginia, and Kentucky.

Anthracite coal (see Chapter 6) is especially desirable because it burns more efficiently than other types of coal. Unfortunately, it is the least common variety, so most coal used for heating buildings and for generating electricity is bituminous (Figure 5.22d). *Coke,* a hard gray substance consisting of the fused ash of bituminous coal, is used in blast furnaces where steel is produced. Synthetic oil and gas and a number of other products are also made from bituminous coal and lignite.

Petroleum and Natural Gas

Petroleum and natural gas are both *hydrocarbons,* meaning that they are composed of hydrogen and carbon. The remains of microscopic organisms settle to the seafloor, or lakefloor in some cases, where little oxygen is present to decompose them. If buried beneath layers of sediment, they are heated and transformed into petroleum and natural gas. The rock in which hydrocarbons form is known as *source rock,* but for them to accumulate in economic quantities, they must migrate from the source rock into some kind of *reservoir rock*. And finally, the reservoir rock must have some overlying *cap rock,* otherwise the hydrocarbons would eventually reach the surface and escape (Figure 5.32). Effective reservoir rocks must have appreciable pore space and good permeability, the capacity to transmit fluids; otherwise hydrocarbons cannot be extracted from them in reasonable quantities.

Many hydrocarbon reservoirs consist of nearshore marine sandstones with nearby fine-grained, organic-rich source rocks. Such oil and gas traps are called *stratigraphic traps* because they owe their existence to variations in the strata (Figure 5.32a). Ancient coral reefs are also good stratigraphic traps. Indeed, some of the oil in the Persian Gulf region and Michigan is trapped in ancient reefs. *Structural traps* result when rocks are deformed by folding, fracturing, or both. In sedimentary rocks that have been deformed into a series of folds, hydrocarbons migrate to the high parts of these structures (Figure 5.32b). Displacement of rocks along faults (fractures along which movement has occurred) also yields situations conducive to trapping hydrocarbons (Figure 5.32).

Other sources of petroleum that will probably become increasingly important in the future include *oil shales* and *tar sands*. The United States has about two-thirds of all known oil shales, although large deposits are known in South America, and all continents have some oil shale. The richest deposits in the United States are in the Green River Formation of Colorado, Utah, and Wyoming. When the appropriate extraction processes are used, liquid oil and combustible gases can be produced from an organic substance called *kerogen* of oil shale. Oil shales in the Green River Formation yield between 10 and 140 gallons of oil per ton of rock processed, and the total amount of oil recoverable with present processes is estimated at 80 billion barrels. Currently, no oil is produced from oil shale in the United States, because conventional drilling and pumping is less expensive.

Tar sand is a type of sandstone in which viscous, as-phaltlike hydrocarbons fill the pore spaces. This substance is the sticky residue of once-liquid petroleum from which the volatile constituents have been lost. Liquid petroleum can be recovered from tar sand, but for this to happen, large quantities of rock must be mined and processed. Since the United States has few tar sand deposits, it cannot look to this source as a significant future energy resource. The Athabaska tar sands in Alberta, Canada, however, are one of the largest deposits of this type. These deposits are currently being mined, and it is estimated that they contain several hundred billion barrels of recoverable petroleum.

Uranium

Most of the uranium used in nuclear reactors in North America comes from the complex potassium-, uranium-, vanadium-bearing mineral *carnotite* found in some sedimentary rocks. Some uranium is also derived from *uraninite* (UO_2), a uranium oxide in granitic rocks and hydrothermal veins. Uraninite is easily oxidized and dissolved

Figure 5.32 *Oil and natural gas traps. The arrows in the diagrams indicate the migration of hydrocarbons. (a) Two examples of stratigraphic traps. (b) Two examples of structural traps, one formed by folding, the other by faulting.*

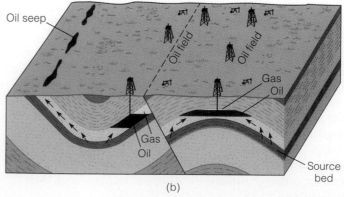

in groundwater, transported elsewhere, and chemically reduced and precipitated in the presence of organic matter.

The richest uranium ores in the United States are widespread in the Colorado Plateau area of Colorado and adjoining parts of Wyoming, Utah, Arizona, and New Mexico. These ores, consisting of fairly pure masses and encrustations of carnotite, are associated with plant remains in sandstones that formed in ancient stream channels. Although most of these ores are associated with fragmentary plant remains, some petrified trees also contain large quantities of uranium.

Large reserves of low-grade uranium ore also are found in the Chattanooga Shale. The uranium is finely disseminated in this black, organic-rich mudrock that underlies large parts of several states including Illinois, Indiana,

Ohio, Kentucky, and Tennessee. Canada remains the world's largest producer and exporter of uranium.

Banded Iron Formation

The chemical sedimentary rock known as *banded iron formation* is of great economic importance. These rocks consist of alternating thin layers of chert and iron minerals, mostly the iron oxides hematite and magnetite. Banded iron formations are present on all the continents and account for most of the iron ore mined in the world today. Vast banded iron formations are present in the Lake Superior region of the United States and Canada and in the Labrador Trough of eastern Canada. The origin of banded iron formations is considered in Chapter 20.

Chapter Summary

1. Mechanical and chemical weathering disintegrate and decompose parent material so that it is more nearly in equilibrium with new physical and chemical conditions. The products of weathering include solid particles and substances in solution.

2. Mechanical weathering includes such processes as frost action, pressure release, salt crystal growth, thermal expansion and contraction, and the activities of organisms. Particles liberated by mechanical weathering retain the chemical composition of the parent material.

3. The chemical weathering processes of solution, oxidation, and hydrolysis bring about chemical changes of parent material. Clay minerals and substances in solution form during chemical weathering.

4. Mechanical weathering aids chemical weathering by breaking parent material into smaller pieces, thereby exposing more surface area.

5. Mechanical and chemical weathering produce regolith, some of which is soil if it consists of solids, air, water, and humus and supports plant growth.

6. Soils are characterized by horizons that are designated, in descending order, as O, A, B, and C; soil horizons differ from one another in texture, structure, composition, and color.

7. Soils called pedalfers develop in humid regions, whereas arid and semiarid region soils are pedocals. Laterite is a soil resulting from intense chemical weathering in the tropics. Laterites are deep and red, and are sources of aluminum ores if derived from aluminum-rich parent material.

8. Soil erosion, caused mostly by sheet and rill erosion, is a problem in some areas. Human practices such as construction, agriculture, and deforestation can accelerate losses of soil to erosion.

9. Sedimentary particles are designated in order of decreasing size as gravel, sand, silt, and clay.

10. Sedimentary particles are rounded and sorted during transport although the degree of rounding and sorting depends on particle size, transport distance, and depositional process.

11. Any area in which sediment is deposited is a depositional environment. Major depositional settings are continental, transitional, and marine, each of which includes several specific depositional environments.

12. Lithification involves compaction and cementation which convert sediment into sedimentary rock. Silica and calcium carbonate are the most common chemical cements, but iron oxide and iron hydroxide cements are important in some rocks.

13. Detrital sedimentary rocks consist of solid particles derived from preexisting rocks. Chemical sedimentary rocks are derived from substances in solution by inorganic chemical processes or the biochemical activities or organisms. A subcategory called biochemical sedimentary rocks is recognized.

14. Sedimentary facies are bodies of sediment or sedimentary rock that are recognizably different from adjacent sediments or rocks.

15. Some sedimentary facies are geographically widespread because they were deposited during marine transgressions or marine regressions.

16. Sedimentary structures such as bedding, cross-bedding, and ripple marks commonly form in sediments when, or shortly after, they are deposited.

17. Depositional environments of ancient sedimentary rocks are determined by studying sedimentary textures and structures, examining fossils, and making comparisons with present-day depositional processes.

18. Intense chemical weathering is responsible for the origin of residual concentrations, many of which contain valuable minerals such as iron, lead, copper, and clay.

19. Many sediments and sedimentary rocks including sand, gravel, evaporites, coal, and banded iron formations are important natural resources. Most oil and natural gas are found in sedimentary rocks.

Important Terms

bed
biochemical sedimentary rock
carbonate rock
chemical sedimentary rock
chemical weathering
cross-bedding
depositional environment
detrital sedimentary rock
differential weathering
erosion
evaporite
exfoliation dome

fossil
frost action
graded bedding
hydrolysis
laterite
lithification
marine regression
marine transgression
mechanical weathering
mud crack
oxidation
parent material

pedalfer
pedocal
pressure release
regolith
ripple mark
salt crystal growth
sediment
sedimentary facies
sedimentary rock
sedimentary structure
soil
soil degradation

soil horizon
solution
spheroidal weathering
strata
talus
thermal expansion and
    contraction
transport
weathering

Review Questions

1. Most of Earth's land surface is covered by soil and unconsolidated rock materials called
   a. _____ regolith; b. _____ horizon A; c. _____ talus;
   d. _____ parent material; e. _____ humus.
2. Limestone, composed of calcite ($CaCO_3$), is nearly insoluble in pure water but dissolves rapidly if _____ is present.
   a. _____ silicon dioxide; b. _____ sodium sulfate;
   c. _____ residual manganese d. _____ carbonic acid
   e. _____ clay
3. The process whereby hydrogen and hydroxyl ions of water replace ions in minerals is
   a. _____ carbonization; b. _____ supergene enrichment;
   c. _____ hydrolysis; d. _____ exfoliation; e. _____ solution.
4. Horizon C differs from other soil horizons in that it
   a. _____ is more fertile; b. _____ contains the most humus;
   c. _____ has pieces of partly altered parent material;
   d. _____ is made up of caliche; e. _____ has been weathered the longest.
5. Horizon B of a soil is also known as the
   a. _____ alkali zonel; b. _____ top soil; c. _____ zone of accumulation; d. _____ talus layer; e. _____ organic-rich bed.
6. The primary weathering process responsible for the origin of exfoliation domes is
   a. _____ pressure release; b. _____ oxidation-solution;
   c. _____ frost heaving; d. _____ soil degradation;
   e. _____ spheroidal weathering.
7. Spheroidal weathering takes place because
   a. _____ the corners and edges of stones weather more rapidly than flat surfaces; b. _____ most naturally occurring rocks are spherical to begin with; c. _____ oxidation changes limestone to clay; d. _____ pedalfers are composed mostly of iron, aluminum, and clay; e. _____ thermal expansion and contraction is such an effective chemical weathering process.
8. Sedimentay breccia is a rare type of rock because
   a. _____ graded bedding forms in the deep sea;
   b. _____ gravel is rounded quickly during transport;
   c. _____ clay is less common than other detrital particles;
   d. _____ sand deposits are typically well sorted;
   e. _____ sodium replaces chlorine in rock salt.
9. Graded bedding is characterized by

   a. _____ beds deposited at an angle to the surface on which they accumulate; b. _____ biochemical precipitation of minerals. c. _____ angular sand and poorly sorted gravel;
   d. _____ carbonate rocks and evaporites; e. _____ an upward decrease in grain size.
10. A vertical sequence of facies in which nearshore deposits overlie offshore deposits results from
    a. _____ a marine regression; b. _____ deposition in a stream channel and its floodplain; c. _____ deposition of evaporites in a desert; d. _____ drying out and cracking of clay-rich sediment; e. _____ compaction of mud.
11. When magnesium replaces some of the calcium in calcite, limestone is converted to
    a. _____ dolostone; b. _____ sedimentary breccia; c. _____ arkose; d. _____ rock gypsum; e. _____ bituminous coal.
12. Traps for petroleum and natural gas formed by deformation such as folding and fracturing of rocks are known as _____ traps.
    a. _____ lithification; b. _____ reservoir; c. _____ structural;
    d. _____ compaction; e. _____ stratigraphic.
13. Which of the following can be used to determine ancient current direction?
    a. _____ mud cracks; b. _____ worm burrows; c. _____ current ripple marks; d. _____ graded bedding; e. _____ evaporites.
14. Rock salt and rock gypsum
    a. _____ form when silica replaces limestone; b. _____ are the most common evaporites; c. _____ are an alternate source of oil and natural gas; d. _____ are the main ores of iron; e. _____ are composed of angular sand.
15. How does mechanical weathering differ from and contribute to chemical weathering?
16. Why are most minerals not very soluble in pure water? Give an example of how a soluble mineral goes into solution.
17. Describe the types of soil degradation. What practices can be used to prevent or at least decrease soil erosion?
18. Draw soil profiles and list the characteristics of each profile for arid- and humid-area soils.
19. Explain fully how an exfoliation dome forms. In what kinds of rocks do these features commonly develop, and where would you go to see examples?

20. What are marine transgressions and regression? Illustrate or explain the vertical sequence of facies that might develop during each.
21. How does coal originate, and what are the varieties of coal? Which variety makes the best fuel?
22. Illustrate and describe two sedimentary structures that can be used to determine ancient current directions.
23. Describe the processes responsible for lithification of detrital sediments. Are these processes equally important in all detrital sediments? Explain.
24. How is it possible to use knowledge of sorting, rounding, and sedimentary structures to determine the depositional environment of ancient sandstone?
25. Consider the following: A soil is 1.5 m thick, new soil forms at the rate of 2.5 cm per century, and the erosion rate is 4 mm per year. Assuming that these rates remain constant, how much soil will remain at this location in 100 years?
26. According to one estimate, Earth's crust is made up of 51% feldspars, 24% ferromagnesian silicates (olivine, pyroxenes, amphiboles, and biotite), 12% quartz, and 13% other minerals. Given this crustal composition, how can you explain the fact that quartz is by far the most common mineral in sandstones?
27. The United States has a total coal reserve of 243 billion metric tons and uses about 860 million metric tons of coal annually. Assuming that all of this coal can be mined, how long will it last at the current rate of consumption? Is there any reason to believe that the current rate of consumption will remain constant, and why is it improbable that all of the reserved can be mined?

World Wide Web Activities

For these Web site addresses, along with current updates and exercises, log on to **http://www.brookscole.com/geo**

Sedimentary Rocks

This web site contains images and information on eleven common sedimentary rocks. It is part of the Soil Science 223 Rocks and Minerals Reference web site. Click on any of the *rock names* and compare the information and images to the information in this chapter.

Devils Marbles

This web site is maintained by Patrick Jennings and contains various images of Devils Marbles from Australia. Devils Marbles are examples of spheroidal weathering. Click on any of the images on this page. They will take you to larger images and some information about the origin and location of those "marbles."

Geology 110 Web Pages of Sedimentary Structures

This web site contains links to two slide collections of sedimentary structures that were used in the Geology 110 class at Duke University. Click on either *Sedimentary Structures, Part 1* or *Sedimentary Structures, Part 2* headings to view slides of different sedimentary structures. Use this collection to review the different sedimentary structures discussed in this chapter.

Rob's Granite Page

This site, which is obviously mostly about granite, is maintained by Robert M. Reed, of the Department of Geological Sciences, University of Texas at Austin. Click on *Llano Uplift* and then click *Enchanted Rock State Park*. Here you will see images of the Town Mountain Granite. Note especially the well-developed exfoliation domes, which are some of the finest examples seen anywhere. Also click on *Turkey Peak* and see an excellent example of differential weathering and erosion.

CD-ROM Exploration

Exploring your *Earth Systems Today* CD-ROM will add to your understanding of the material in this chapter.

Topic: Earth's Materials

Module: Rocks and the Rock Cycle

Explore activities in this module to see if you can discover the following for yourself:

Using the "rock laboratory" portion of this module, observe the formation of a clastic sedimentary rock (sandstone) through the process of lithification. What role does water play in the lithification process? Based on what you see in the "rock laboratory" write a short description of how a sandstone is lithified.

Compaction and cementation must work together to form sedimentary rock. Observe how this happens using the animation and make notes on what you see.

How are igneous rocks and sedimentary rocks different in their texture and mode of formation?

Metamorphism and Metamorphic Rocks

Open-pit asbestos mine, Quebec, Canada.

148

PROLOGUE

Its homogeneity, softness, and various textures have made marble, a metamorphic rock formed from limestone or dolostone, a favorite rock of sculptors throughout history. As the value of authentic marble sculptures has increased through the years, the number of forgeries has also increased. With the price of some marble sculptures in the millions of dollars, private collectors and museums need some means of assuring the authenticity of the works they are buying. Aside from the monetary considerations, it is important that such forgeries not become part of the historical and artistic legacy of human endeavor.

Experts have traditionally relied on the artistic style and weathering characteristics to determine whether a marble sculpture is authentic or a forgery. Because marble is not very resistant to weathering, forgers have resorted to a variety of methods to produce the weathered appearance of an authentic ancient work. Now, however, using new techniques, geologists can distinguish a naturally weathered marble surface from one that has been artificially altered.

Although marble results when the agents of metamorphism (heat, pressure, and fluid activity) are applied to carbonate rocks, the type of marble formed depends, in part, on the composition of the original carbonate rock as well as the type and intensity of metamorphism. Therefore, one way to authenticate a marble sculpture is to determine the origin of the marble itself. During the Preclassical, Greek, and Roman periods, the islands of Naxos, Thasos, and Paros in the Aegean Sea—as well as the Greek mainland, Turkey, and Italy—were all sites of major marble quarries.

To identify the source of the marble in various sculptures, geologists employ a wide variety of analytical techniques. These include hand specimen and thin-section analysis of the marble, trace element analysis by X-ray fluorescence, stable isotopic ratio analysis for carbon and oxygen, and other more esoteric techniques.

The J. Paul Getty Museum in Malibu, California, employed some of these techniques to help authenticate an ancient Greek kouros (a sculptured figure of a Greek youth) thought to have been carved around 530 B.C. (Figure 6.1). The kouros was offered to the Getty Museum in 1984 for a reported price of $7 million. Some of its stylistic features, however, caused some experts to question its authenticity. Consequently, the museum performed a variety of geochemical and mineralogical tests in an effort to determine the authenticity of the kouros.

Isotopic analysis of the weathered surface and fresh interior of the kouros confirmed that the marble probably came from the Cape Vathy quarries on the island of Thasos in the Aegean Sea. But these results did not prove the age of the kouros—it might still have been a forgery carved from marble taken from an archaeological site on the island.

The kouros was carved from dolomitic marble and its surface is covered with a complex thin crust (0.01 to 0.05 mm thick) consisting of whewellite, a calcium oxalate monohydrate mineral. To ensure that the crust is the result of long-term weathering and not a modern forgery, dolomitic marble samples were subjected to a variety of forgery techniques that tried to replicate the surface of the kouros. Samples were soaked or boiled in various mixtures for periods of time ranging from hours to months, and their surfaces treated and retreated to try to match the appearance of the weathered surface of the kouros. Such tests yielded only a few examples that appeared similar to the surface of the kouros. Even those samples, however, were different when examined under high magnification or subjected to geochemical analysis. In fact, all of the samples clearly showed that they were

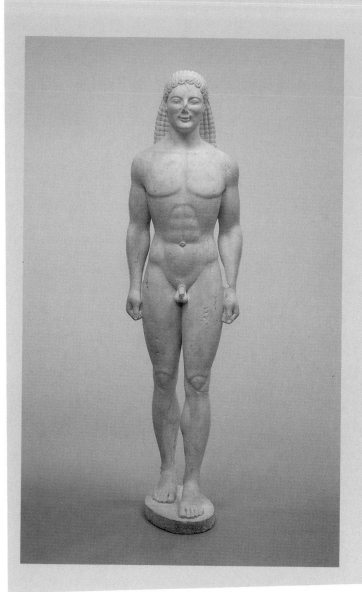

the result of recent alteration and not long-term weathering processes.

Though scientific tests have not unequivocally proved authenticity, they have shown that the weathered surface layer of the kouros bears more similarities to naturally occurring weathered surfaces than to known artificially produced surfaces. Furthermore, no evidence indicates that the surface alteration of the kouros is of modern origin.

Despite intensive study by scientists, archaeologists, and art historians, opinion is still divided as to the authenticity of the Getty kouros. Most scientists accept that the kouros was carved sometime around 530 B.C., but most art historians are doubtful. Pointing to inconsistencies in its style of sculpture for that period, they think it is a modern forgery. Because of the continuing doubts about the statue's authenticity, the J. Paul Getty Center has mounted it in an exhibition listing the evidence for and against its authenticity.

Regardless of the ultimate conclusion on the Getty kouros, geological testing to authenticate marble sculptures is now an important part of many museums' curatorial functions. In addition, a large body of data about the characteristics and origin of marble is being amassed as more sculptures and quarries are analyzed.

Figure 6.1 *This Greek kouros, which stands 206 cm tall, has been the object of an intensive authentication study by the Getty Museum. Using a variety of geological tests, scientists have determined that the kouros was carved from dolomitic marble that probably came from the Cape Vathy quarries on the island of Thasos.*

WHAT WOULD YOU DO?

As the director of a major museum, you have the opportunity to purchase, for a considerable sum of money, a newly discovered marble bust by a famous ancient sculptor. The problem is that you want to be sure it is not a forgery. What would you do to ensure that the bust is authentic and not a clever forgery. After all, you are spending a large sum of the museum's money. As a nonscientist, how would you go about making sure the proper tests are being performed to ensure the bust's authenticity?

Introduction

Metamorphic rocks (from the Greek *meta* meaning change and *morpho* meaning shape) constitute the third major group of rocks. They result from the transformation of other rocks by metamorphic processes that usually occur beneath Earth's surface (see Figure 1.13). During metamorphism, rocks are subjected to sufficient heat, pressure, and fluid activity to change their mineral composition and/or texture, thus forming new rocks. These transformations take place in the solid state, and the type of metamorphic rock formed depends on the composition and texture of the original rock, the agents of metamorphism, and the amount of time the parent rock was subjected to the effects of metamorphism.

A useful analogy for metamorphism is baking. The resulting cake, just like a metamorphic rock, depends on the ingredients, their proportions, how they are mixed together, how much water or milk is added, and the temperature and length of time the cake is baked.

Why Should You Study Metamorphic Rocks?

Why is it important that you study metamorphic rocks? For one thing, a large portion of Earth's continental crust is composed of metamorphic and igneous rocks. Together, they form the crystalline basement rocks that underlie the sedimentary rocks of a continent's surface. This basement rock is widely exposed in regions of the continents known as *shields*, which have been very stable during the past 600 million years (Figure 6.2). Metamorphic rocks also constitute a sizable portion of the crystalline core of large mountain ranges. Some of the oldest known rocks, dated at 3.96 million years from the Canadian Shield, are metamorphic, indicating they formed from even older rocks!

By studying metamorphic rocks, geologists learn about the geologic processes operating within Earth and how these processes have varied through time. The presence of certain minerals in metamorphic rocks allows geologists to determine the approximate temperatures and pressures the original rocks were subjected to during metamorphism, thus providing insights into the physical and chemical changes taking place at different depths within the crust.

Metamorphic rocks such as marble and slate are used as building materials, and certain metamorphic minerals are economically important. Garnets, for example, are used as gemstones or abrasives; talc is used in cosmetics, in the manufacture of paint, and as a lubricant; asbestos is used for insulation and fireproofing; and kyanite is used in the production of heat-resistant materials used in sparkplugs. Therefore a knowledge of metamorphic rocks and processes is of economic value.

Finally, there are different forms of asbestos and they do not all pose the same health hazards. This knowledge would have been useful during the debates over the dangers posed to the public's health by the widespread use of asbestos in buildings and building materials (Perspective 6.1).

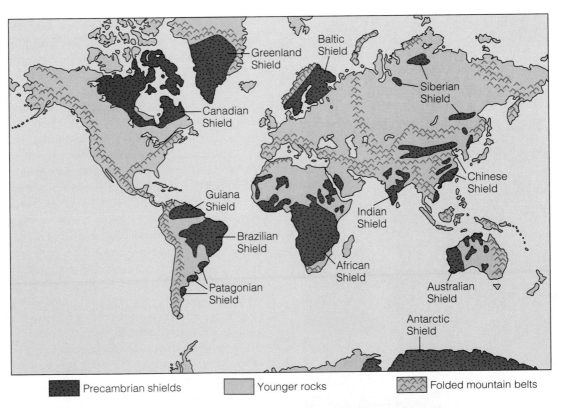

Figure 6.2 *Occurrences of metamorphic rocks. Shields are the exposed portion of the crystalline basement rocks that underlie each continent; these areas have been very stable during the past 600 million years. Metamorphic rocks also constitute the crystalline core of large mountain ranges.*

Why Should You Study Metamorphic Rocks?

Asbestos (from the Latin, meaning unquenchable) is a general term applied to any silicate mineral that easily separates into flexible fibers (Figure 1). The combination of such features as noncombustibility and flexibility makes asbestos an important industrial material of considerable value. In fact, asbestos has more than 3000 known uses, including brake linings, fireproof fabrics, and heat insulators.

Asbestos can be divided into two broad groups, serpentine and amphibole asbestos. *Chrysotile,* which is a hydrous magnesium silicate with the chemical formula $Mg_3Si_2O_5(OH)_4$, is the fibrous form of serpentine asbestos; it is the most valuable type and constitutes the bulk of all commercial asbestos. Chrysotile's strong, silky fibers are easily spun and can withstand temperatures up to 2750°C.

The vast majority of chrysotile asbestos is in serpentine, a type of rock formed by the alteration of ultramafic igneous rocks such as peridotite under low- and medium-grade metamorphic conditions. Serpentine is thought to form from the alteration of olivine by hot, chemically active, residual fluids emanating from cooling magma. The chrysotile asbestos forms veinlets of fiber within the serpentine and may comprise up to 20% of the rock. Other chrysotile results when the metamorphism of magnesium limestone or dolostone produces discontinuous serpentine bands within the carbonate beds.

At least five varieties of amphibole asbestos are known, but *crocidolite,* a sodium-iron amphibole with the chemical formula $Na_2(Fe^{+3})_2(Fe^{+2})_3Si_8O_{22}(OH)_2$, is the most common. Crocidolite, which is also known as blue asbestos, is a long, coarse, spinning fiber that is stronger but more brittle than chrysotile and also less resistant to heat. The other varieties of amphibole asbestos have fewer applications and are used primarily for insulation.

Crocidolite is found in such metamorphic rocks as slates and schists. It is thought that crocidolite forms by the solid-state alteration of other minerals within the high temperature and high pressure environment that results from deep burial. Unlike chrysotile, crocidolite is rarely found associated with igneous intrusions.

Despite the widespread use of asbestos, the federal Environmental Protection Agency (EPA) instituted a gradual ban on all new asbestos products. The ban was imposed because some forms of asbestos can cause lung cancer and scarring of the lungs if its fibers are inhaled. Because the EPA apparently paid little attention to the issue of risks versus benefits when it enacted this rule, the U.S. Fifth Circuit Court of Appeals overturned the EPA ban on asbestos in 1991.

The threat of lung cancer has also resulted in legislation mandating the removal of asbestos already in place in all public buildings, including all public and private schools. Recently, however, important questions have been raised concerning the threat posed by asbestos and the additional potential hazards that may arise from its improper removal.

The current policy of the EPA mandates that all forms of asbestos be treated as identical hazards. Yet studies indicate that only the amphibole forms constitute a known health hazard. Chrysotile, whose fibers tend to be curly, does not become lodged in the lungs. Furthermore, its fibers are generally soluble and disappear in tissue. In contrast, crocidolite has long, straight, thin fibers that penetrate the lungs and stay there. These fibers irritate the lung tissue, and over a long period of time, can lead to lung cancer. Thus, crocidolite, and not chrysotile, is overwhelmingly responsible for asbestos-related lung cancer. Because about 95% of the asbestos in place in the United States is chrysotile, many people are questioning whether the dangers from asbestos have been somewhat exaggerated.

Removing asbestos from buildings where it has been installed might cost as much as $100 billion, and some recent studies have indicated that the air in buildings containing asbestos has essentially the same amount of airborne fibers as the air outdoors. In fact, unless the material containing the asbestos is disturbed, asbestos does not shed fibers. Furthermore, improper removal of asbestos can lead to contamination. In most cases of improper removal, the concentration of airborne asbestos fibers is far higher than if the asbestos had been left in place.

The problem of asbestos contamination is a good example of how geology affects our lives and why a basic knowledge of science is important.

Figure 1 *Hand specimen of chrysotile from Thetford, Quebec, Canada. Chrysotile is the fibrous form of serpentine asbestos.*

What Are the Agents of Metamorphism?

The three agents of metamorphism are heat, pressure, and fluid activity. During metamorphism, the original rock undergoes change so as to achieve equilibrium with its new environment. The changes may result in the formation of new minerals and/or a change in the texture of the rock by the reorientation of the original minerals. In some instances, the change is minor, and features of the original rock can still be recognized. In other cases the rock changes so much that the identity of the original rock can be determined only with great difficulty, if at all.

Besides heat, pressure, and fluid activity, time is also important to the metamorphic process. Chemical reactions proceed at different rates and thus require different amounts of time to complete. Reactions involving silicate compounds are particularly slow, and because most metamorphic rocks are composed of silicate minerals, it is thought that metamorphism is a slow geologic process.

Heat

Heat is an important agent of metamorphism because it increases the rate of chemical reactions that may produce minerals different from those in the original rock. The heat may come from intrusive magmas or result from deep burial in the crust such as occurs during subduction along a convergent plate boundary.

When rocks are intruded by bodies of magma, they are subjected to intense heat that affects the surrounding rock; the most intense heating usually occurs adjacent to the magma body and gradually decreases with distance from the intrusion. The zone of metamorphosed rocks that forms in the country rock adjacent to an intrusive igneous body is usually distinct and easy to recognize.

Recall that temperature increases with depth and that Earth's geothermal gradient averages about 25°C/km. Rocks forming at the surface may be transported to great depths by subduction along a convergent plate boundary and thus subjected to increasing temperature and pressure. During subduction, some minerals may be transformed into other minerals that are more stable under the higher temperature and pressure conditions.

Pressure

When rocks are buried, they are subjected to increasingly greater **lithostatic pressure;** this pressure, which results from the weight of the overlying rocks, is applied equally in all directions (Figure 6.3a). A similar situation occurs when an object is immersed in water. For example, the deeper a Styrofoam cup is submerged in the ocean, the smaller it gets because pressure increases with depth and is exerted on the cup equally in all directions, thereby compressing the Styrofoam (Figure 6.3b).

Just as in the Styrofoam cup example, rocks are subjected to increasing lithostatic pressure with depth such that the mineral grains within a rock may become more closely packed. Under these conditions, the minerals may *recrystallize;* that is, they may form smaller and denser minerals.

Along with lithostatic pressure resulting from burial, rocks may also experience **differential pressures** (Figure 6.4). In this case, the pressures are not equal on all sides, and the rock is consequently distorted. Differential pressures typically occur during deformation associated with mountain building and can produce distinctive metamorphic textures and features.

Fluid Activity

In almost every region of metamorphism, water and carbon dioxide (CO_2) are present in varying amounts along mineral grain boundaries or in the pore spaces of rocks. These fluids, which may contain ions in solution, enhance metamorphism by increasing the rate of chemical reactions. Under dry conditions, most minerals react very slowly, but when even small amounts of fluid are introduced, reaction rates increase, mainly because ions

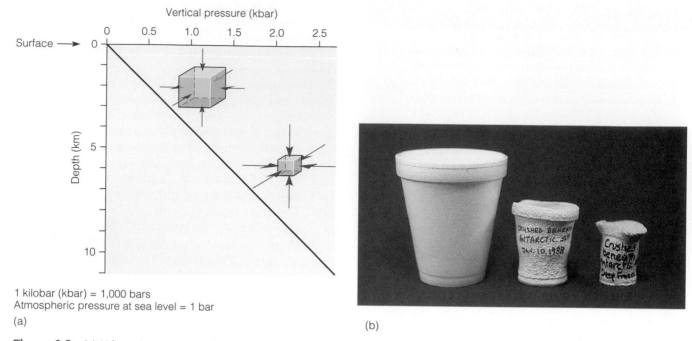

Figure 6.3 *(a) Lithostatic pressure is applied equally in all directions in Earth's crust due to the weight of the overlying rocks. Thus, pressure increases with depth, as indicated by the sloping black line. (b) A similar situation occurs when 200 ml Styrofoam cups are lowered to ocean depths of approximately 750 m and 1500 m. Increased pressure is exerted equally in all directions on the cups, and they consequently decrease in volume, while still maintaining their general shape.*

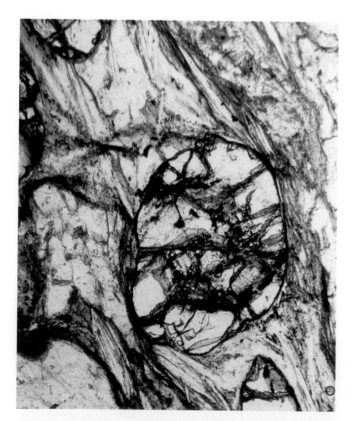

Figure 6.4 *Differential pressure is pressure that is unequally applied to an object. Rotated garnets are a good example of the effects of differential pressure applied to a rock during metamorphism. This rotated garnet (center) comes from a schist in northeast Sardinia.*

can move readily through the fluid and thus enhance chemical reactions and the formation of new minerals.

The following reaction provides a good example of how new minerals can be formed by **fluid activity.** Here, seawater moving through hot basaltic rock of the oceanic crust transforms olivine into the metamorphic mineral serpentine:

$$2Mg_2SiO_4 + 2H_2O \rightarrow Mg_3Si_2O_5(OH)_4 + MgO$$

olivine   water   serpentine   carried away in solution

The chemically active fluids important in metamorphism come primarily from three sources. The first is water trapped in the pore spaces of sedimentary rocks as they form; as these rocks are subjected to heat and pressure, the water is heated, thus accelerating the various chemical reaction rates. A second source is the volatile fluid within magma; as these hot fluids disperse through the surrounding rock, they frequently react with and alter the minerals of the country rock by adding or removing ions. The third source is the dehydration of water-bearing minerals such as gypsum ($CaSO_4 \cdot 2H_2O$) and some clays; when these minerals, which contain water as part of their crystal chemistry, are subjected to heat and pressure, the water may be driven off and enhance metamorphism.

What Are the Three Types of Metamorphism?

Three major types of metamorphism are recognized: *contact metamorphism* in which magmatic heat and fluids act to produce change; *dynamic metamorphism,* which is principally the result of high differential pressures associated with intense deformation; and *regional metamorphism,* which occurs within a large area and is caused primarily by mountain-building forces. Even though we will discuss each type of metamorphism separately, the boundary between them is not always distinct and depends largely on which of the three metamorphic agents was dominant.

Contact Metamorphism

Contact metamorphism takes place when a body of magma alters the surrounding country rock. At shallow depths an intruding magma raises the temperature of the surrounding rock, causing thermal alteration. Furthermore, the release of hot fluids into the country rock by the cooling intrusion can also aid in the formation of new minerals.

Important factors in contact metamorphism are the initial temperature and size of the intrusion as well as the fluid content of the magma and/or country rock. The initial temperature of an intrusion is controlled, in part, by its composition: Mafic magmas are hotter than felsic magmas and hence have a greater thermal effect on the rocks directly surrounding them. The size of the intrusion is also important. In the case of small intrusions, such as dikes and sills, usually only those rocks in immediate contact with the intrusion are affected. Because large intrusions, such as batholiths, take a long time to cool, the increased temperature in the surrounding rock may last long enough for a larger area to be affected.

Temperatures can reach nearly 900°C adjacent to an intrusion, but they gradually decrease with distance. The effects of such heat and the resulting chemical reactions usually occur in concentric zones known as **aureoles** (Figure 6.5). The boundary between an intrusion and its aureole may be either sharp or transitional (Figure 6.6).

Metamorphic aureoles vary in width depending on the size, temperature, and composition of the intrusion as well as the mineralogy of the surrounding country rock. Typically, large intrusive bodies have several metamorphic zones, each characterized by distinctive mineral assemblages indicating the decrease in temperature with distance from the intrusion (Figure 6.5). The zone closest to the intrusion, and hence subject to the highest temperatures, may contain high-temperature metamorphic minerals (that is, minerals in equilibrium with the higher temperature environment) such as sillimanite. The outer zones may be characterized by lower temperature metamorphic minerals such as chlorite, talc, and epidote.

Contact metamorphism can result not only from igneous intrusions, but also from lava flows (Figure 6.7). In this instance, lava flowing over the land may thermally alter the underlying rocks. While recognizing a recent lava flow and the resulting contact metamorphism of the rocks below it is easy, less obvious is whether an igneous body is intrusive or extrusive in a rock outcrop where sedimentary rocks occur above and below the igneous body. The recognition of which sedimentary rock units have been metamorphosed enables geologists to determine whether the igneous body is intrusive (such as a sill or dike) or extrusive (lava flow). Such a determination is critical in reconstructing the geologic history of an area (see Chapter 17) and may have important economic implications as well.

Fluids also play an important role in contact metamorphism. Many magmas are wet and contain hot, chemically active fluids that may emanate into the surrounding rock. These fluids can react with the rock and aid in the formation of new minerals. In addition, the country rock may contain pore fluids that, when heated by magma, also increase reaction rates.

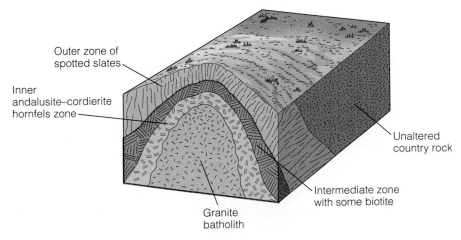

Outer zone of spotted slates

Inner andalusite–cordierite hornfels zone

Granite batholith

Intermediate zone with some biotite

Unaltered country rock

Figure 6.5 *A metamorphic aureole typically surrounds many igneous intrusions. The metamorphic aureole associated with this idealized granite batholith contains three zones of mineral assemblages reflecting the decrease in temperature with distance from the intrusion. An andalusite-cordierite hornfels forms the inner zone adjacent to the batholith. This is followed by an intermediate zone of extensive recrystallization in which some biotite develops, and farthest from the intrusion is the outer zone, which is characterized by spotted slates.*

Igneous rock   Metamorphic rock

Figure 6.6 *A sharp and clearly defined boundary (red line) occurs between the intruding light-colored igneous rock on the left and the dark-colored metamorphosed country rock on the right. The intrusion is part of the Peninsular Ranges Batholith, east of San Diego, California.*

Figure 6.7 *A highly weathered basaltic lava flow near Susanville, California, has altered an underlying rhyolitic volcanic ash by contact metamorphism. The red zone below the lava flow has been baked by the heat of the lava when it flowed over the ash layer. The lava flow displays spheroidal weathering, a type of weathering common in fractured rocks (see Chapter 5).*

The formation of new minerals by contact metamorphism depends not only on proximity to the intrusion, but also on the composition of the country rock. Shales, mudstones, impure limestones, and impure dolostones are particularly susceptible to the formation of new minerals by contact metamorphism, whereas pure sandstones or pure limestones typically are not.

Because heat and fluids are the primary agents of contact metamorphism, two types of contact metamorphic rocks are generally recognized: those resulting from baking of country rock and those altered by hot solutions. Many of the rocks resulting from contact metamorphism have the texture of porcelain; that is, they are hard and fine grained. This is particularly true for rocks with a high clay content, such as shale. Such texture results because the clay minerals in the rock are baked, just as a clay pot is baked when fired in a kiln.

During the final stages of cooling when an intruding magma begins to crystallize, large amounts of hot, watery solutions are often released. These solutions may react with the country rock and produce new metamorphic minerals. This process, which usually occurs near Earth's surface, is called *hydrothermal alteration* (from the Greek *hydro* meaning water and *therme* meaning heat), and may result in valuable mineral deposits. Geologists think that many of the world's ore deposits result from the migration of metallic ions in hydrothermal solutions. Examples include copper, gold, iron ores, tin, and zinc in various localities including Australia, Canada, China, Cyprus, Finland, Russia, and the western United States.

Dynamic Metamorphism

Most **dynamic metamorphism** is associated with fault (fractures along which movement has occurred) zones where rocks are subjected to high differential pressures. The metamorphic rocks resulting from pure dynamic metamorphism are *mylonites* and are typically restricted to narrow zones adjacent to faults. Mylonites are hard, dense, fine-grained rocks, many of which are characterized by thin laminations (layers less than 1 cm thick) (Figure 6.8). Tectonic settings where mylonites occur include the Moine Thrust Zone in northwest Scotland and portions of the San Andreas fault in California (see Chapter 9).

Regional Metamorphism

Most metamorphic rocks result from **regional metamorphism,** which occurs over a large area and is usually caused by tremendous temperatures, pressures, and deformation within the deeper portions of the crust. Regional metamorphism is most obvious along convergent plate margins where rocks are intensely deformed and re-

Figure 6.8 *Mylonite from the Adirondack Highlands, New York. Note the thin laminations.*

crystallized during convergence and subduction. Within these metamorphic rocks, there is usually a gradation of metamorphic intensity from areas that were subjected to the most intense pressures and/or highest temperatures to areas of lower pressures and temperatures. Such a gradation in metamorphism can be recognized by the metamorphic minerals that are present.

Regional metamorphism is not just confined to convergent margins. It also occurs in areas where plates diverge, though usually at much shallower depths because of the high geothermal gradient associated with these areas.

From field studies and laboratory experiments, certain minerals are known to form only within specific temperature and pressure ranges. Such minerals are known as **index minerals** because their presence allows geologists to recognize low-, intermediate-, and high-grade metamorphic zones (Figure 6.9). When a clay-rich rock such as shale is metamorphosed, new minerals form as a result of metamorphic processes. The mineral chlorite, for example, is produced under relatively low temperatures of about 200°C, so its presence indicates low-grade metamorphism. As temperatures and pressures continue to increase, new minerals form that are stable under those conditions. Thus, there is a progression in the appearance of new minerals from chlorite, whose presence indicates low-grade metamorphism, to sillimanite, whose presence indicates high-grade metamorphism and temperatures exceeding 500°C.

Different rock compositions develop different index minerals. When sandy dolomites are metamorphosed, for example, they produce an entirely different set of index minerals. Thus, a specific set of index minerals commonly forms in specific rock types as metamorphism progresses.

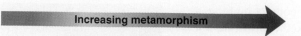

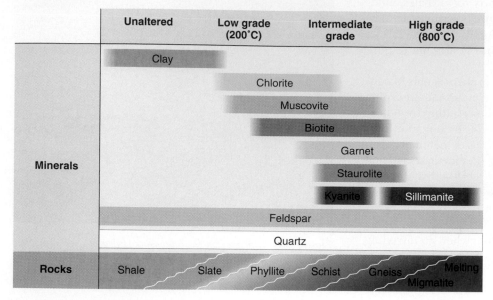

Figure 6.9 *Change in mineral assemblage and rock type with increasing metamorphism of a shale. When a clay-rich rock, such as a shale is subjected to increasing metamorphism, new minerals form as shown by the various colored bars. The progressive appearance of particular minerals allows geologists to recognize low-, intermediate-, and high-grade metamorphic zones.*

Although such common minerals as mica, quartz, and feldspar can occur in both igneous and metamorphic rocks, other minerals such as andalusite, sillimanite, and kyanite generally are found only in metamorphic rocks derived from clay-rich sediments. While these three minerals all have the same chemical formula (Al_2SiO_5), they differ in crystal structure and other physical properties because each forms under a different range of pressures and temperatures. Thus, they are sometimes used as index minerals for metamorphic rocks formed from clay-rich sediments.

How Are Metamorphic Rocks Classified?

For purposes of classification, metamorphic rocks are commonly divided into two groups: those exhibiting a *foliated texture* (from the Latin *folium* meaning leaf) and those with a *nonfoliated texture* (Table 6.1).

Foliated Metamorphic Rocks

Rocks subjected to heat and differential pressure during metamorphism typically have minerals arranged in a parallel fashion that gives them a **foliated texture** (Figure 6.10). The size and shape of the mineral grains determine whether the foliation is fine or coarse. If the foliation is such that the individual grains cannot be recognized without magnification, the rock is said to be slate (Figure 6.11a). A coarse foliation

results when granular minerals such as quartz and feldspar are segregated into roughly parallel and streaky zones that differ in composition and color as in gneiss. Foliated metamorphic rocks can be arranged in order of increasingly coarse grain size and perfection of foliation.

Slate is a fine-grained metamorphic rock that commonly exhibits *slaty cleavage* (Figure 6.11b). Slate is the result of low-grade regional metamorphism of shale or, more rarely, volcanic ash. Because it can easily be split along cleavage planes into flat pieces, slate is an excellent rock for roofing and floor tiles, billiard and pool table tops, and blackboards (Figure 6.11c). The different colors of most slates are caused by minute amounts of graphite (black), iron oxide (red and purple), and/or chlorite (green).

Phyllite is similar in composition to slate, but is coarser grained. The minerals, however, are still too small to be identified without magnification. Phyllite can be distinguished from slate by its glossy or lustrous sheen (Figure 6.12). It represents an intermediate grain size between slate and schist.

Schist is most commonly produced by regional metamorphism. The type of schist formed depends on the intensity of metamorphism and the character of the original rock (Figure 6.13). Metamorphism of many rock types can yield schist, but most schist appears to have formed from clay-rich sedimentary rocks (Table 6.1).

All schists contain more than 50% platy and elongated minerals, all of which are large enough to be clearly visible. Their mineral composition imparts a *schistosity* or *schistose foliation* to the rock that commonly produces a

Table 6.1

Classification of Common Metamorphic Rocks

Texture	Metamorphic Rock	Typical Minerals	Metamorphic Grade	Characteristics of Rocks	Parent Rock
Foliated	Slate	Clays, micas, chlorite	Low	Fine-grained, splits easily into flat pieces	Mudrocks, volcanic ash
	Phyllite	Fine-grained quartz, micas, chlorite	Low to medium	Fine-grained, glossy or lustrous sheen	Mudrocks
	Schist	Micas, chlorite, quartz, talc, hornblende, garnet, staurolite, graphite	Low to high	Distinct foliation, minerals visible	Mudrocks, carbonates, mafic igneous rocks
	Gneiss	Quartz, feldspars, hornblende, micas	High	Segregated light and dark bands visible	Mudrocks, sandstones, felsic igneous rocks
	Amphibolite	Hornblende, plagioclase	Medium to high	Dark-colored, weakly foliated	Mafic igneous rocks
	Migmatite	Quartz, feldspars, hornblende, micas	High	Streaks or lenses of granite intermixed with gneiss	Felsic igneous rocks mixed with sedimentary rocks
Nonfoliated	Marble	Calcite, dolomite	Low to high	Interlocking grains of calcite or dolomite, reacts with HCl	Limestone or dolostone
	Quartzite	Quartz	Medium to high	Interlocking quartz grains, hard, dense	Quartz sandstone
	Greenstone	Chlorite, epidote, hornblende	Low to high	Fine-grained, green	Mafic igneous rocks
	Hornfels	Micas, garnet, andalusite, cordierite, quartz	Low to medium	Fine-grained, equidimensional grains, hard, dense	Mudrocks
	Anthracite	Carbon	High	Black, lustrous, subconcoidal fracture	Coal

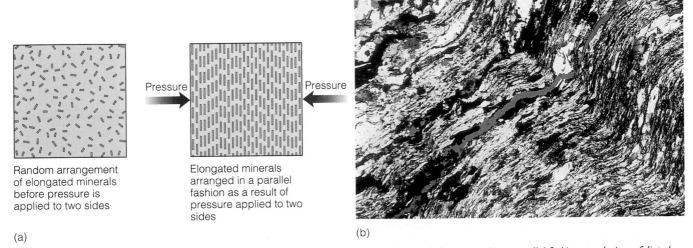

(a) Random arrangement of elongated minerals before pressure is applied to two sides

Pressure → Elongated minerals arranged in a parallel fashion as a result of pressure applied to two sides ← Pressure

(b)

Figure 6.10 (a) When rocks are subjected to differential pressure, the mineral grains are typically arranged in a parallel fashion, producing a foliated texture. (b) Photomicrograph of a metamorphic rock with a foliated texture showing the parallel arrangement of mineral grains.

(a)

Figure 6.12 *Hand specimen of phyllite. Note the lustrous sheen as well as the bedding (upper left to lower right) at an angle to the cleavage of this specimen.*

(b)

(a)

(b)

Figure 6.13 *Schist. (a) Garnet—mica schist. (b) Hornblende—mica—garnet schist.*

(c)

Figure 6.11 *(a) Hand specimen of slate. (b) This panel of Arvonia Slate from Yancy Quarry of the Arvonia-Buckingham Slate Company, Virginia, shows bedding (upper right to lower left) at an angle to the slaty cleavage. (c) Slate roof of Chalet Enzian, Switzerland.*

wavy type of parting when split. Schistosity is common in low- to high-grade metamorphic environments, and each type of schist is known by its most conspicuous mineral or minerals, such as mica schist, chlorite schist, and talc schist.

Gneiss is a metamorphic rock that is streaked or has segregated bands of light and dark minerals. Gneisses are composed mostly of granular minerals such as quartz and/or feldspar, with lesser percentages of platy or elongated minerals such as micas or amphiboles (Figure 6.14). Quartz and feldspar are the principal light-colored minerals, while biotite and hornblende are the typical dark-colored minerals. Gneiss typically breaks in an irregular manner, much like coarsely crystalline nonfoliated rocks.

Most gneiss probably results from recrystallization of clay-rich sedimentary rocks during regional metamorphism (Table 6.1). Gneiss also can form from crystalline igneous rocks such as granite or older metamorphic rocks.

Another fairly common foliated metamorphic rock is *amphibolite*. A dark-colored rock, it is composed mainly of hornblende and plagioclase. The alignment of the hornblende crystals produces a slightly foliated texture. Many amphibolites result from medium- to high-grade metamorphism of such ferromagnesian mineral-rich igneous rocks as basalt.

In some areas of regional metamorphism, exposures of "mixed rocks" having both igneous and high-grade metamorphic characteristics are present. In these rocks, called *migmatites*, streaks or lenses of granite are usually intermixed with high-grade ferromagnesian-rich metamorphic rocks, imparting a wavy appearance to the rock (Figure 6.15).

Figure 6.14 *Gneiss is characterized by segregated bands of light and dark minerals. This folded gneiss is exposed at Wawa, Ontario, Canada.*

Most migmatites are thought to be the product of extremely high-grade metamorphism, and several models for their origin have been proposed. Part of the problem in determining the origin of migmatites is ex-

Figure 6.15 *Migmatites consist of high-grade metamorphic rock intermixed with streaks or lenses of granite. This migmatite is exposed at Thirty Thousand Islands of Georgian Bay, Lake Huron, Ontario, Canada.*

How Are Metamorphic Rocks Classified?

plaining how the granitic component formed. According to one model, the granitic magma formed in place by the partial melting of rock during intense metamorphism. Such an origin is possible provided that the host rocks contained quartz and feldspars and that water was present. Another possibility is that the granitic components formed by the redistribution of minerals by recrystallization in the solid state, that is, by pure metamorphism.

Nonfoliated Metamorphic Rocks

In some metamorphic rocks, the mineral grains do not show a discernible preferred orientation. Instead, these rocks consist of a mosaic of roughly equidimensional minerals and are characterized as having a **nonfoliated texture** (Figure 6.16). Most nonfoliated

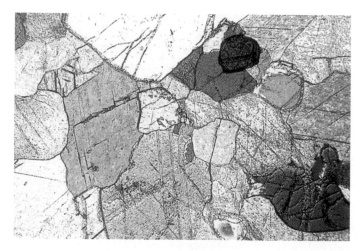

Figure 6.16 *Nonfoliated textures are characterized by a mosaic of roughly equidimensional minerals as in this photomicrograph of marble.*

metamorphic rocks result from contact or regional metamorphism of rocks in which no platy or prismatic minerals are present. Frequently, the only indication that a granular rock has been metamorphosed is the large grain size resulting from recrystallization. Nonfoliated metamorphic rocks are generally of two types: those composed mainly of only one mineral—for example, marble or quartzite—and those in which the different mineral grains are too small to be seen without magnification—such as greenstone and hornfels.

Marble is a well-known metamorphic rock composed predominantly of calcite or dolomite; its grain size ranges from fine to coarsely granular (Figures 6.1 and 6.17). Marble results from either contact or regional metamorphism of limestones or dolostones (Table 6.1). Pure marble is snowy white or bluish, but varieties of many colors exist because of the presence of mineral impurities in the original sedimentary rock. The softness of marble, its uniform texture, and its various colors have made it the favorite rock of builders and sculptors throughout history (see the Prologue).

Quartzite is a hard, compact rock typically formed from quartz sandstone under medium-to-high-grade metamorphic conditions during contact or regional metamorphism (Figure 6.18). Because recrystallization is so complete, metamorphic quartzite is of uniform strength and therefore usually breaks across the component quartz grains rather than around them when it is struck. Pure quartzite is white, but iron and other impurities commonly impart a reddish or other color to it. Quartzite is commonly used as foundation material for road and railway beds.

The name *greenstone* is applied to any compact, dark-green, altered, mafic igneous rock that formed under low-to-high-grade metamorphic conditions. The green color results from the presence of chlorite, epidote, and hornblende.

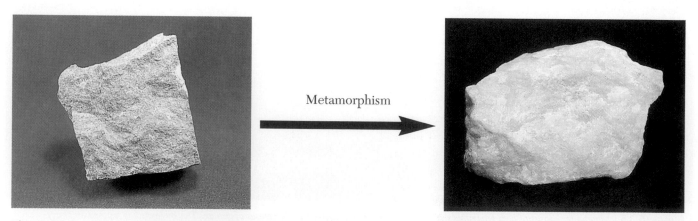

Metamorphism

Figure 6.17 *Marble results from the metamorphism of the sedimentary rock limestone or dolostone.*

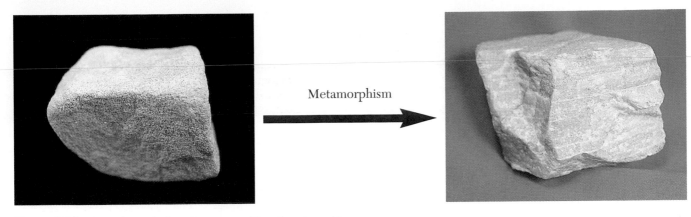

Figure 6.18 *Quartzite results from the metamorphism of quartz sandstone.*

Hornfels, a fine-grained, nonfoliated metamorphic rock resulting from contact metamorphism, is composed of various equidimensional mineral grains. The composition of hornfels directly depends on the composition of the original rock, and many compositional varieties are known. The majority of hornfels, however, are apparently derived from contact metamorphism of clay-rich sedimentary rocks or impure dolostones.

Anthracite is a black, lustrous, hard coal that contains a high percentage of fixed carbon and a low percentage of volatile matter. It usually forms from the metamorphism of lower-grade coals by heat and pressure and is thus considered by many geologists to be a metamorphic rock.

What Are Metamorphic Zones and Facies?

The first systematic study of metamorphic zones was conducted during the late 1800s by George Barrow and other British geologists working in the Dalradian schists of the southwestern Scottish Highlands. Here, clay-rich sedimentary rocks have been subjected to regional metamorphism, and the resulting metamorphic rocks can be divided into different zones based on the presence of distinctive silicate mineral assemblages. These mineral assemblages, each recognized by the presence of one or more index minerals, indicate different degrees of metamorphism. The index minerals Barrow and his associates chose to represent increasing metamorphic intensity were chlorite, biotite, garnet, staurolite, kyanite, and sillimanite (Figure 6.9). Note that these are the metamorphic minerals produced from clay-rich sediments. Other mineral assemblages and index minerals are produced from rocks with different original compositions.

The successive appearance of metamorphic index minerals indicates gradually increasing or decreasing intensity of metamorphism. Going from lower- to higher-grade zones, the first appearance of a particular index mineral indicates the location of the minimum temperature and pressure conditions needed for the formation of that mineral. When the locations of the first appearances of that index mineral are connected on a map, the result is a line of equal metamorphic intensity or an *isograd.* The region between isograds is known as a **metamorphic zone.** By noting the occurrence of metamorphic index minerals, geologists can construct a map showing the metamorphic zones of an entire area (Figure 6.19).

Numerous studies of different metamorphic rocks have demonstrated that while the texture and mineralogy of any rock may be altered by metamorphism, the overall chemical composition may be little changed. Thus, the different mineral assemblages found in increasingly higher grade metamorphic rocks derived from the same parent rock result from changes in temperature and pressure.

A **metamorphic facies** is a group of metamorphic rocks characterized by particular mineral assemblages formed under the same broad temperature-pressure conditions (Figure 6.20). Each facies is named after its most characteristic rock or mineral. For example, the green metamorphic mineral chlorite, which forms under relatively low temperatures and pressures, yields rocks said to belong to the *greenschist facies.* Under increasingly higher temperatures and pressures, other metamorphic facies, such as the *amphibolite* and *granulite facies,* develop.

Although usually applied to areas where the original rocks were clay-rich, the concept of metamorphic facies can be used with modification in other situations. It cannot, however, be used in areas where the original rocks were pure quartz sandstones or pure limestones or dolostones. Such rocks would yield only quartzites and marbles, respectively.

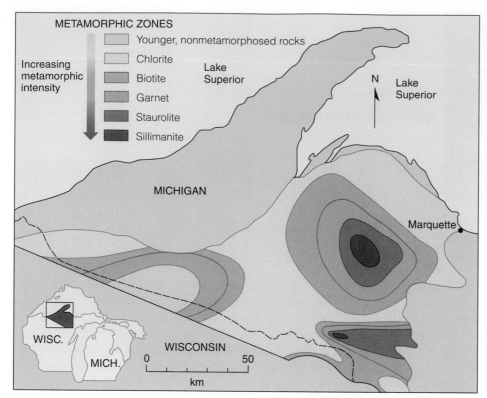

Figure 6.19 *Metamorphic zones in the Upper Peninsula of Michigan. The zones in this region are based on the presence of distinctive silicate mineral assemblages resulting from the metamorphism of sedimentary rocks during an interval of mountain building and minor granitic intrusion during the Proterozoic Eon, about 1.5 billion years ago. The lines separating the different metamorphic zones are isograds.*

How Does Metamorphism Relate to Plate Tectonics?

Although metamorphism is associated with all three types of plate boundaries (see Figure 1.12), it is most common along convergent plate margins. Metamorphic rocks form at convergent plate boundaries because temperature and pressure increase as a result of plate collisions.

Figure 6.21 illustrates the various temperature-pressure regimes produced along an oceanic-continental convergent plate boundary and the type of metamorphic facies and rocks that can result. When an oceanic plate collides with a continental plate, tremendous pressure is generated as the oceanic plate is subducted. Because rock is a poor heat conductor, the cold descending oceanic plate heats slowly, and metamorphism is caused mostly by increasing pressure with depth. Metamorphism in such an environment produces rocks typical of the *blueschist facies* (low temperature, high pressure), which is characterized by the blue-colored amphibole mineral glaucophane (Figure 6.20). Geologists use the occurrence of blueschist facies rocks as evidence of ancient subduction zones. An excellent example of blueschist metamorphism can be found in the California Coast Ranges. Here rocks of the

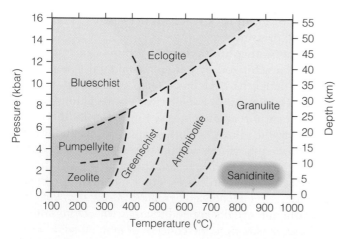

Figure 6.20 *A pressure-temperature diagram showing the conditions under which various metamorphic facies occur. A facies is characterized by a particular mineral assemblage that formed under the same broad temperature-pressure conditions. Each facies is named after its most characteristic rock or mineral.*

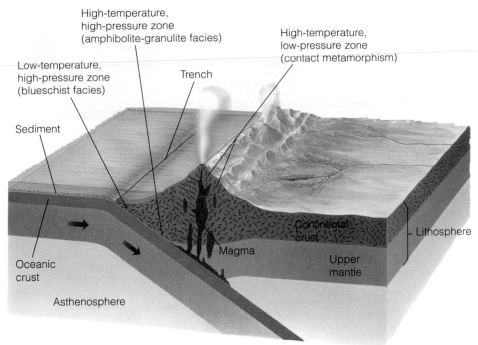

Low-temperature,
high-pressure zone
(blueschist facies)

High-temperature,
high-pressure zone
(amphibolite-granulite facies)

High-temperature,
low-pressure zone
(contact metamorphism)

Trench

Sediment

Oceanic
crust

Asthenosphere

Magma

Continental
crust

Upper
mantle

Lithosphere

Figure 6.21 *Metamorphic facies resulting from various temperature-pressure conditions produced along an oceanic-continental convergent plate boundary.*

Franciscan Complex were metamorphosed under low-temperature, high-pressure conditions that clearly indicate the presence of a former subduction zone (Figure 6.22).

As subduction along the oceanic-continental plate boundary continues, both temperature and pressure increase with depth and can result in high-grade metamorphic rocks. Eventually, the descending plate begins to melt and generates magma that moves upward. This rising magma may alter the surrounding rock by contact metamorphism, producing migmatites in the deeper portions of the crust and hornfels at shallower depths. Such an environment is characterized by high temperatures and low to medium pressures.

While metamorphism is most common along convergent plate margins, many divergent plate boundaries are characterized by contact metamorphism. Rising magma at mid-oceanic ridges heats the adjacent rocks, producing contact metamorphic minerals and textures. Besides contact metamorphism, fluids emanating from the rising magma—and its reaction with seawater—very commonly produce metal-bearing hydrothermal solutions rich in copper, iron, lead, zinc, and other metals, which are precipitated along mid-oceanic ridges. These deposits may eventually be brought to Earth's surface by later tectonic activity. The copper ores of Cyprus are a good example of such hydrothermal activity (see Chapter 9).

Metamorphism and Natural Resources

Many metamorphic rocks and minerals are valuable natural resources. While these resources include various types of ore deposits, the two most familiar and widely used metamorphic rocks, as such, are marble and slate, which, as previously discussed, have been used for centuries in a variety of ways (Figure 6.23).

Many ore deposits result from contact metamorphism during which hot, ion-rich fluids migrate from igneous intrusions into the surrounding rock, thereby producing rich ore deposits. The most common sulfide ore minerals associated with contact metamorphism are bornite, chalcopyrite, galena, pyrite, and sphalerite, while two common oxide ore minerals are hematite and magnetite. Tin and tungsten are also important ores associated with contact metamorphism (Table 6.2).

Other economically important metamorphic minerals include talc for talcum powder; graphite for pencils and dry lubricants (see Perspective 6.2); garnets and corundum, which are used as abrasives or gemstones, depending on their quality; and andalusite, kyanite, and sillimanite, all of which are used in the manufacture of high-temperature porcelains and temperature-resistant minerals for products such as sparkplugs and the linings of furnaces.

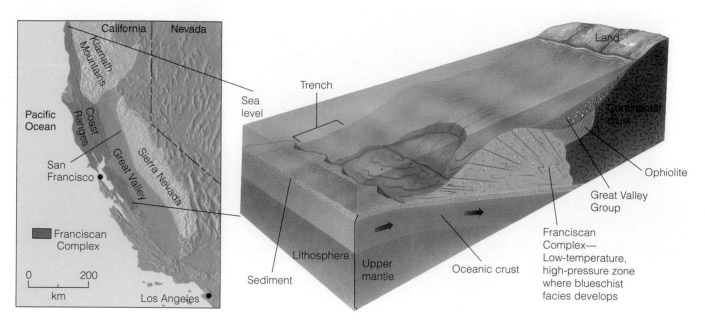

Figure 6.22 *Index map of California showing the location of the Franciscan Complex and a diagrammatic reconstruction of the environment in which it was regionally metamorphosed under low-temperature, high-pressure subduction conditions approximately 150 million years ago. The red line on the index map shows the orientation of the reconstruction to the current geography.*

Figure 6.23 *Slate quarry in Wales. Slate, which has a variety of uses, is the result of low-grade metamorphism of shale. These high quality slates were formed by a mountain-building episode that occurred approximately 400 to 440 million years ago in the present-day countries of Ireland, Scotland, Wales, and Norway.*

Graphite

Graphite (from the Greek *grapho*, meaning write) is a soft gray to black mineral that has a greasy feel and is composed of the element carbon (see Figure 2.19a). Graphite occurs in two varieties; crystalline, which consists of thin, flat, nearly pure black flakes; and massive, an impure variety found in compact masses.

Graphite has the same composition as diamond, but its carbon atoms are strongly bonded together in sheets, with the sheets weakly held together by van der Waals bonds (see Figure 2.7). Because the sheets are loosely held together, they easily slide over one another, giving graphite its ability to mark paper and serve as a dry lubricant.

Graphite occurs mainly in metamorphic rocks produced by contact and regional metamorphism. It is found in marble, quartzite, schist, gneiss, and even in anthracite. Contact metamorphism of impure limestones by igneous intrusions produces some of the graphite found in marbles. The graphite resulting from regional metamorphism of sedimentary rocks probably came from organic matter present in the sediments. Some evidence, however, indicates that the graphite in Precambrian-age rocks (> 545 million years) may be the result of the reduction of calcium carbonate ($CaCO_3$) by an inorganic process. Graphite is also found in igneous rocks, pegmatite dikes, and veins; it is thought to have formed in these environments from the primary constituents of the magma or from the hot fluids and vapors released by the cooling magma.

The major producers of graphite are Mexico, Russia, and South Korea. In the United States, graphite has been mined in 27 states, but production is now generally limited to Alabama and New York.

Graphite is used for many purposes. The oldest use is in pencil leads, where it is finely ground, mixed with clay, and baked. The amount of clay and the baking time give pencil leads their desired hardness. Other important uses include batteries, brake linings, carbon brushes, crucibles, foundry facings, lubricants, refractories, and steelmaking.

Synthetic graphite can be produced from anthracite coal or petroleum coke and now accounts for most graphite production. Its extreme purity (99% to 99.5% pure) makes it especially valuable where high purity is required such as in the rods that slow down the reaction rates in nuclear reactors.

Table 6.2

The Main Ore Deposits Resulting from Contact Metamorphism

Ore Deposit	Major Mineral	Formula	Use
Copper	Bornite Chalcopyrite	Cu_5FeS_4 $CuFeS_2$	Important sources of copper, which is used in various aspects of manufacturing, transportation, communications, and construction
Iron	Hematite Magnetite	Fe_2O_3 Fe_3O_4	Major sources of iron for manufacture of steel, which is used in nearly every form of construction, manufacturing, transportation, and communications
Lead	Galena	PbS	Chief source of lead, which is used in batteries, pipes, solder, and elsewhere where resistance to corrosion is required
Tin	Cassiterite	SnO_2	Principal source of tin, which is used for tin plating, solder, alloys, and chemicals
Tungsten	Scheelite Wolframite	$CaWO_4$ $(Fe,Mn)WO_4$	Chief sources of tungsten, which is used in hardening metals and manufacturing carbides
Zinc	Sphalerite	$(Zn,Fe)S$	Major source of zinc, which is used in batteries and galvanizing iron and making brass

Chapter Summary

1. Metamorphic rocks result from the transformation of other rocks, usually beneath Earth's surface, as a consequence of one or a combination of three agents: heat, pressure, and fluid activity.

2. Heat for metamorphism comes from intrusive or extrusive magmas or deep burial. Pressure is either lithostatic or differential. Fluids trapped in sedimentary rocks or emanating from intruding magmas can enhance chemical changes and the formation of new minerals.

3. The three major types of metamorphism are contact, dynamic, and regional.

4. Metamorphic rocks are primarily classified according to their texture. In a foliated texture, platy and elongate minerals have a preferred orientation. A nonfoliated texture does not exhibit any discernible preferred orientation of the mineral grains.

5. Foliated metamorphic rocks can be arranged in order of grain size and/or perfection of their foliation. Slate is fine grained, followed by phyllite and schist; gneiss displays segregated bands of minerals. Amphibolite is another fairly common foliated metamorphic rock.

6. Marble, quartzite, greenstone, and hornfels are common nonfoliated metamorphic rocks.

7. Metamorphic rocks can be arranged into metamorphic zones based on the conditions of metamorphism. Individual metamorphic facies are characterized by particular minerals that formed under specific metamorphic conditions. Such facies are named for a characteristic rock or mineral.

8. Metamorphism can occur along all three types of plate boundaries, but is most widespread along convergent plate boundaries.

9. Metamorphic rocks formed near Earth's surface along an oceanic-continental plate boundary result from low-temperature, high-pressure conditions. As a subducted oceanic plate descends, it is subjected to increasingly higher temperatures and pressures that result in higher grade metamorphism.

10. Many metamorphic rocks and minerals, such as marble, slate, graphite, talc, and asbestos, are valuable natural resources.

Important Terms

aureole	fluid activity	lithostatic pressure	nonfoliated texture
contact metamorphism	foliated texture	metamorphic facies	regional metamorphism
differential pressure	heat	metamorphic rock	
dynamic metamorphism	index mineral	metamorphic zone	

Review Questions

1. The widely exposed portion of the crystalline basement rocks that underlies each continent is the
   a. _____ mantle; b. _____ craton; c. _____ mountain core; d. _____ shield; e. _____ none of the previous.

2. The study of metamorphic rocks provides geologists with information about
   a. _____ Earth's interior processes; b. _____ the temperature and pressure conditions under which the metamorphic rocks formed; c. _____ economically important minerals; d. _____ the different forms of asbestos; e. _____ all of the previous.

3. Which of the following is not an agent or process of metamorphism?
   a. _____ pressure; b. _____ heat; c. _____ fluid activity; d. _____ time; e. _____ gravity.

4. The nonfoliated metamorphic rock formed from limestone or dolostone is
   a. _____ quartzite; b. _____ marble; c. _____ hornfels; d. _____ greenstone; e. _____ schist.

5. Metamorphic rocks can form from which type of original rock?

   a. _____ igneous; b. _____ sedimentary; c. _____ metamorphic; d. _____ volcanic; e. _____ all of the previous.

6. Pressure resulting from deep burial and applied equally in all directions on a rock is
   a. _____ directional; b. _____ differential; c. _____ lithostatic; d. _____ shear; e. _____ unilateral.

7. Metamorphic rocks resulting from pure dynamic pressure are
   a. _____ hornfels; b. _____ mylonites; c. _____ greenstones; d. _____ quartzites; e. _____ slates.

8. Magmatic heat and fluid activity are the primary agents involved in what type of metamorphism?
   a. _____ dynamic; b. _____ lithostatic; c. _____ contact; d. _____ regional; e. _____ thermodynamic.

9. Along which type of plate boundary is metamorphism most widespread?
   a. _____ divergent; b. _____ transform; c. _____ aseismic; d. _____ convergent; e. _____ lithospheric.

10. Concentric zones surrounding an igneous intrusion and characterized by distinctive mineral assemblages are
a. _____ thermodynamic rings; b. _____ hydrothermal regions; c. _____ metamorphic layers; d. _____ regional facies; e. _____ aureoles.

11. The majority of metamorphic rocks result from which type of metamorphism?
a. _____ lithostatic; b. _____ contact; c. _____ regional; d. _____ local; e. _____ dynamic.

12. Which is the order of increasingly coarser grain size and perfection of foliation?
a. _____ gneiss → schist → phyllite → slate; b. _____ phyllite → slate → schist → gneiss; c. _____ schist → slate → gneiss → phyllite; d. _____ slate → phyllite → schist → gneiss; e. _____ slate → schist → phyllite → gneiss.

13. Metamorphic zones
a. _____ reflect a metamorphic grade; b. _____ are characterized by distinctive mineral assemblages; c. _____ are separated from each other by isograds.; d. _____ all of the previous; e. _____ none of the previous.

14. To which metamorphic facies do metamorphic rocks formed under high-temperature, high-pressure conditions belong?
a. _____ amphibolite-granulite; b. _____ blueshist; c. _____ zeolite; d. _____ greenschist; e. _____ pumpellyite.

15. Discuss the role each of the three agents of metamorphism plays in transforming a rock into a metamorphic rock.

16. Why is metamorphism more common along convergent plate boundaries than any other type of plate boundary?

17. How can aureoles be used to determine the effects of metamorphism?

18. What is a metamorphic zone and how does it differ from a metamorphic facies?

19. Describe the two types of metamorphic texture and discuss how they are produced?

20. Why should the average citizen know anything about metamorphic rocks and how they form?

21. What is regional metamorphism and under what conditions does it form?

22. What specific features about foliated metamorphic rocks make them unsuitable as foundations for dams? Are there any metamorphic rocks that would make good foundations and why?

23. Where does contact metamorphism occur and what types of changes does it produce?

24. Name several economically valuable metamorphic minerals or rocks and discuss why they are valuable.

25. Using Figure 6.20 and a temperature of 400°C, what metamorphic facies occurs at a depth of 10 km? What is the pressure at that depth in kilobars? At the same temperature, what metamorphic facies occurs at 50 km depth and what is the pressure at that depth?

 World Wide Web Activities

For these web site addresses, along with current updates and exercises, log on to **http://www.brookscole.com/geo**

Metamorphic Rocks

This web site contains images and information on metamorphic settings, facies, rock names, and textures. Click on any of the topics and compare the information and images to the information in this chapter.

Big Bend National Park—Metamorphic Processes

This web site explains the process of metamorphism and relates it to Dominguez Mountain, which is the only area in the park where metamorphic rocks are exposed.

Metamorphic Rocks

This web site contains the online lecture notes of Dr. Pamela J. W. Gore, Georgia Perimeter College. These notes are an excellent summary of metamorphic rocks and metamorphic processes and contain many pictures of metamorphic minerals and rocks.

 CD-ROM Exploration

Exploring your *Earth Systems Today* CD-ROM will add to your understanding of the material in this chapter.

Topic: Earth's Materials

Module: Rocks and the Rock Cycle

Explore activities in this module to see if you can discover the following for yourself:

Go to the "rock laboratory" in this activity and observe how shale is transformed into slate. What progressive changes do you see in the process of metamorphism? Based on what you see in the "rock laboratory," write a short description of all the changes that clay in mudstone must experience to become mica in slate.

How are metamorphic rocks different from the other two types of rocks in the way they form? (What would you look for to tell metamorphic rocks from other types of rocks?)

Earthquakes and Earth's Interior

Residents of a collapsed building in Ahmedabad, India, salvage whatever they can before it is demolished by authorities. At least 14,000 people are confirmed dead (although estimates range between 25,000 and 100,000), more than 61,000 injured, and more than 600,000 left homeless from a massive 7.7 magnitude earthquake that struck India's western state of Gujarat on January 26, 2001.

PROLOGUE

At 3:02 A.M. on August 17, 1999, violent shaking from an earthquake awakened millions of people in Turkey. Soon after the earthquake was over, an estimated 17,000 people were dead, at least 50,000 were injured, and tens of thousands of survivors were left homeless. The amount of destruction this earthquake caused is staggering. More than 150,000 buildings were moderately to heavily damaged, and another 90,000 suffered slight damage (Figure 7.1). Collapsed buildings were everywhere, streets were strewn with rubble, and all communications were knocked out. And if this were not enough, the same area was struck again only three months later, on November 12, 1999, by an aftershock nearly as large as the original earthquake; it killed an additional 374 people, while injuring about 3000 more. All in all, this was a disaster of epic proportions. However, it is hardly the first, nor will it be the last in this part of the world (Table 7.1).

Most residents of North America have heard of the San Andreas fault, a huge fracture in Earth's crust, that cuts through coastal California and occasionally spawns large earthquakes (see Figure 9.25). Here, two large plates slide horizontally past each other, generating the energy that causes those destructive earthquakes. The situation is very similar in northern Turkey, where the North Anatolian fault cuts east-west across the country (Figure 7.1a). Movements on this fault have caused several large earthquakes dur-

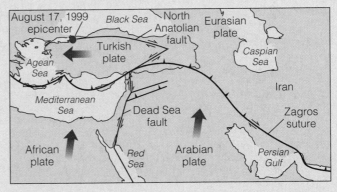
(a)

(b)

Figure 7.1 *The Izmit, Turkey, earthquake of August 17, 1999. (a) Map showing the plate tectonic relationships in the region. The earthquake took place on a large fracture known as the North Anatolian fault. (b) Fatma Tandogan weeps as she searches through the rubble of her collapsed house in Golcuk, Turkey. Ms. Tandogan's house was one of more than 150,000 buildings that were moderately to heavily damaged by the 7.4 magnitude earthquake.*

Table 7.1

Some Significant Earthquakes

Year	Location	Magnitude (Estimated Before 1935)	Deaths (Estimated)
1556	China (Shanxi Province)	8.0	1,000,000
1755	Portugal (Lisbon)	8.6	70,000
1811–12	USA (New Madrid, Missouri)	7.5	20
1886	USA (Charleston, South Carolina)	7.0	60
1906	USA (San Francisco, California)	8.3	700
1923	Japan (Tokyo)	8.3	143,000
1964	USA (Alaska)	8.6	131
1970	China (Yunnan Province)	7.7	15,621
1971	USA (San Fernando, California)	6.6	65
1976	China (Tangshan)	8.0	242,000
1985	Mexico (Mexico City)	8.1	9500
1988	Armenia	7.0	25,000
1989	USA (Loma Prieta, California)	7.1	63
1990	Iran	7.3	40,000
1992	Turkey	6.8	570
1992	Egypt (Cairo)	5.9	550
1993	India	6.4	30,000
1994	USA (Northridge, California)	6.7	61
1995	Japan (Kobe)	7.2	5000+
1995	Russia	7.6	2000+
1996	China (Lijiang)	6.5	304
1997	Iran	5.5	554
1997	Iran	7.3	2400+
1998	Afghanistan	6.1	5000+
1999	Taiwan	7.6	2400
1999	Turkey	7.4	17,000
1999	Turkey	7.1	374
2001	El Salvador	7.7	700
2001	India	7.7	14,000+
2001	El Salvador	6.6	300
2001	USA (Washington)	6.8	0

ing this century, the most recent being the 7.4 magnitude one centered near Izmit, Turkey, and its 7.1 magnitude aftershock centered only a few tens of kilometers to the east.

What makes Turkey such an earthquake-prone area? And why are there so many deaths and so much destruction when earthquakes of similar size occur along the San Andreas fault and usually result in far fewer casualties and much less damage? The answer to the first question can be found in Figure 7.1a,

which shows the plate tectonic regime of the region. Notice that a divergent plate boundary exists in the Red Sea and that the Arabian plate is moving northward against the Eurasian plate. One consequence of these moving plates is that the smaller Turkish plate is being forced west along the North Anatolian fault. Periodic movements along this fault result in earthquakes; eight of nine earthquakes during the last 60 years have had magnitudes greater than 7.

Turkey, of course, does not exist in isolation. The plate motions just described also affect the much larger region extending east-west through the Mediterranean Sea through Turkey, and through northern Iran, Iraq, and India. The details of plate interactions differ in different parts of this extensive earthquake belt. In the case of Turkey, horizontal movement takes place between plates along a transform plate boundary, whereas farther west in Italy and Greece, earthquakes are generated along a convergent plate boundary where oceanic crust is subducted beneath continental crust. And to the east, as in India, earthquakes are caused by convergence and a collision between two continental crustal plates.

Now to the second question posed above: Why are there so many deaths and so much destruction? Since 1988, more than 120,000 people have died in earthquakes in the region just discussed. For example, 40,000 people died as a result of earthquakes in Iran in 1990, and 30,000 died in India in 1993. One factor that affects the number of people killed is population density. Several of these earthquakes, including the most recent ones in Turkey, have struck large population centers. An even more important factor, however, is the types of structures built in different areas. Many of the buildings in this region are made of brick or unreinforced concrete, or they have stone roofs; all of these are especially unstable when shaken. Unfortunately, if an earthquake occurs when most people are inside, as happened in Turkey, the death toll is enormous. We will have more to say about the different types of structures, building practices, and other factors important in determining fatalities and damage later in this chapter.

Introduction

Moving plates and volcanism as well as interactions among the hydrosphere, biosphere, atmosphere, and solid Earth are all manifestations of Earth's dynamic nature. Earthquakes are also an indication that Earth is an internally active planet. As one of nature's most frightening and destructive phenomena, earthquakes have always aroused a sense of fear and as such, have been the subject of numerous myths and legends. What makes an earthquake so frightening is that when it begins, there is no way to tell how long it will last or how violent it will be. About 13 million people have died in earthquakes during the last 4000 years, with about 2.7 million of these deaths occurring during this century alone (Table 7.1). The August 1999 earthquake in Turkey is a case in point in which thousands died and millions of dollars in property damage occurred (see the Prologue).

Geologists define an **earthquake** as the shaking or trembling caused by the sudden release of energy, usually as a result of faulting that involves displacement of rocks along fractures (the different types of faults are discussed in Chapter 10). Following an earthquake, continuing adjustments along a fault might generate a series of earthquakes known as *aftershocks.* Most of these are smaller than the main shock, but they can cause considerable damage to already weakened structures. Indeed, much of the damage and many of the fatalities caused by the 1755 Lisbon, Portugal, earthquake resulted from aftershocks. After a small earthquake, aftershocks usually cease within a few days, but following a large earthquake they might continue for months.

While the geologic definition of an earthquake is accurate, it is not nearly as imaginative or colorful as explanations held by many people during the past. In many cultures the cause of earthquakes was attributed to movements of some kind of animal upon which Earth rested. In Japan, it was a giant catfish; in Mongolia, a giant frog; in China, an ox; in South America, a whale; and to the Algonquin of North America, an immense tortoise. According to a story from India, Earth rests on the backs of four elephants standing on the back of a turtle, which in turn is balanced on a cobra; movement of any of these animals causes earthquakes (Figure 7.2). And a legend from Mexico holds that earthquakes occur when the devil, El Diablo, rips open the crust so he and his friends can reach the surface.

Even in more recent times many people believed that earthquakes were divine retribution or warnings to the unrepentant. This view was strongly reinforced by the earthquake on November 1, 1755 (All Saint's Day), in Lisbon, Portugal, when the churches were crowded with worshipers. So strong was the shaking that it was felt throughout Europe, and chandeliers rattled as far away as North America. A combination of collapsing buildings, huge sea waves that devastated the waterfront, and a fire that swept through the city resulted in 30,000 to 40,000 deaths, and another 20,000 people died in the following months as a direct result of the earthquake.

If earthquakes are not the result of divine retribution, the Devil ripping opening the crust, or animal movement, what does cause earthquakes? The Greek philosopher Aristotle (384–322 B.C.) was the first to offer what he thought was a natural explanation for earthquakes. According to him, atmospheric winds drawn into Earth's interior caused fires and swept through subterranean cavities as they tried to escape. This moving underground air was the cause of earthquakes and occasional volcanic eruptions.

Today, geologists know that most earthquakes result from energy released along divergent, convergent, and transform plate boundaries. It is true that earthquakes occur preceding and during volcanic eruptions as well, but even though these can and do cause damage, at least locally, volcanism is not responsible for most large earthquakes.

Figure 7.2 *A legend from India holds that Earth rests on the backs of four elephants standing on the back of a turtle standing on a cobra. Movements of any of these creatures result in earthquakes.*

Why Should You Study Earthquakes?

Why should you study earthquakes? The obvious answer is because they are destructive and you someday may be caught in one. Even if you don't plan on living in an area prone to earthquakes, you probably will someday travel where there is the threat of earthquakes and you should know what to do if you experience one. In this chapter you will learn how structures are made earthquake resistant, what precautions to take if you live where earthquakes are common, and what to do during and after an earthquake to minimize your chances of serious injury or even death.

Why should you study Earth's interior? One reason is that Earth's interior is part of the whole Earth system. If we are to fully understand and appreciate the relationship between Earth's various subsystems, we must understand the role Earth's interior plays. Therefore, it is important to learn about Earth's interior and how it relates to Earth's surface features and external processes.

What Is the Elastic Rebound Theory?

Based on studies conducted after the 1906 San Francisco earthquake, H. F. Reid of Johns Hopkins University pro-posed the **elastic rebound theory** to explain how energy is released during earthquakes. Reid studied three sets of measurements taken across a portion of the San Andreas fault that had broken during the 1906 earthquake. The measurements revealed that points on opposite sides of the fault had moved 3.2 m during the 50-year period prior to breakage in 1906, with the west side moving northward (Figure 7.3).

According to Reid, any straight line such as a fence or road that crossed the San Andreas fault would gradually be bent, as rocks on one side of the fault moved relative to rocks on the other side. Eventually, the strength of the rocks was exceeded, the rocks on opposite sides of the fault re-bounded or "snapped back" to their former undeformed shape, and the energy stored was released as earthquake waves radiating outward from the break (Figure 7.3). Additional field and laboratory studies conducted by Reid and others have confirmed that elastic rebound is the mechanism by which energy is released during earthquakes.

The energy stored in rocks undergoing elastic deformation is analogous to the energy stored in a tightly wound watch spring. The tighter the spring is wound, the more energy is stored, thus making more energy available for release. If the spring is wound so tightly that it breaks, the stored energy is released as the spring rapidly unwinds and partially regains its original shape. Perhaps an even more meaningful analogy is simply bending a long, straight stick over one's knee. As the stick bends it deforms and eventually reaches the point at which it breaks. When this happens, the two pieces of the original stick snap back into

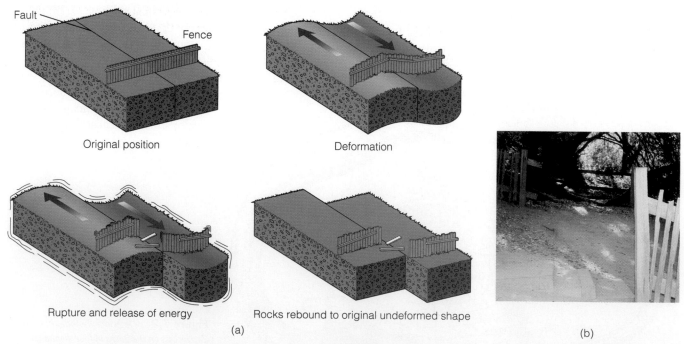

Original position

Deformation

Rupture and release of energy

Rocks rebound to original undeformed shape

(a)

(b)

Figure 7.3 *(a) According to the elastic rebound theory, when rocks are deformed, they store energy and bend. When the internal strength of the rocks is exceeded, they fracture, releasing energy as they rebound to their former undeformed shape. This sudden release of energy causes an earthquake. (b) During the 1906 San Francisco earthquake, this fence in Marin County was displaced nearly 5 m.*

their original straight position. Likewise, rocks subjected to intense forces bend until they break and then return to their original position, releasing energy in the process.

What Is Seismology?

Seismology, the study of earthquakes, emerged as a true science during the 1880s when instruments were developed to effectively record earthquake waves. As early as 132 A.D., the Chinese scholar Chang Heng invented the first earthquake detector, but it only revealed that an earthquake took place in one of two directions from the device (Figure 7.4). By the 1800s, scientists tried using pendulums to record earthquake waves, but they had little success until 1875 when Filippo Cecchi in Italy made the first successful **seismograph,** an instrument that detects, records, and measures the vibrations produced by an earthquake. By 1880, improvements in seismograph sensitivity began yielding meaningful data about earthquakes. Although some seismographs today still use suspended masses to detect earthquake waves (Figure 7.5), most now employ electronic sensors, and of course, computer printouts have largely replaced the strip-charts of earlier seismographs.

Measuring ground motion during an earthquake is a problem because any instrument on Earth moves as the ground moves. Notice in Figure 7.5 that the seismograph employs a weight suspended on a cable, which is essentially a pendulum. Contrary to intuition, ground motion does not make the pendulum swing. Instead, the pendulum, because of its inertia, tends to remain stationary while the ground and instrument attached to it moves. This type of seismograph responds to ground motions at right angles to the pendulum, so a complete seismic station must have at least two such instruments, one sensitive to north-south motions, and one to east-west motions. In addition, a seismograph with a spring-supported weight detects vertical movements (Figure 7.5c).

When an earthquake occurs, energy in the form of *seismic waves* radiates out from the point of release; the record of these seismic waves detected by a seismograph is a *seismogram* (Figure 7.5). These waves are somewhat analogous to the ripples that move out concentrically from the point where a stone is thrown into a pond, but unlike waves on a pond, seismic waves move outward in all directions from their source.

Earthquakes take place because rocks are capable of storing energy, but their strength is limited so if enough force is present, they rupture and thus release their stored energy. In other words, most earthquakes result when movement occurs along fractures (faults), most of which are related to plate movements. Once a fracture begins, it moves along the fault at several kilometers per second for as long as conditions for failure exist. Accordingly, anywhere from a few meters to several hundred kilometers of a fault might experience movement. The longer the fracture along which movement occurs, the more time it takes for the stored energy to be released, and thus the longer the ground will shake. During some very large earthquakes, the ground might shake for up to three minutes, a seemingly brief period under normal conditions, but an interminably long time if you are experiencing the earthquake firsthand!

The Focus and Epicenter of an Earthquake

The point within Earth where fracturing begins—that is, the point at which energy is first released—is an earthquake's **focus,** or *hypocenter.* What we usually hear in news reports, however, is the location of the **epicenter,** the point on Earth's surface directly above the focus (Figure 7.6). For instance, according to a report by the United States Geological Survey, the August 1999 earthquake in Turkey (see

Figure 7.4 *The world's first earthquake detector was invented by Chang Heng sometime around 132 A.D. Movement of the vase dislodges a ball from a dragon's mouth into the waiting mouth of a frog below. For example, if a ball from the dragons on either the east or west sides of the vase were dislodged, then the earthquake must have come from one of those two directions.*

(a)

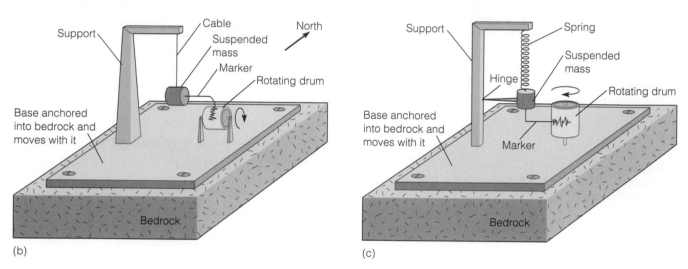

(b)   (c)

Figure 7.5 *Modern seismographs record earthquake waves electronically. (a) Earthquakes recorded by a seismograph. (b) A horizontal-motion seismograph. Because of its inertia, the heavy mass that contains the marker remains stationary while the rest of the structure moves along with the ground during an earthquake. As long as the length of the arm is not parallel to the direction of ground movement, the marker will record the earthquake waves on the rotating drum. (c) A vertical-motion seismograph. This seismograph operates on the same principle as a horizontal-motion instrument and records vertical ground movement.*

the Prologue) had an epicenter about 11 km southeast of the city of Izmit, and its focal depth was about 17 km. The closer structures are to an earthquake's epicenter, the more likely they are to be damaged or destroyed.

Seismologists recognize three categories of earthquakes based on focal depth. *Shallow-focus* earthquakes have focal depths of less than 70 km from the surface, whereas those with focuses between 70 and 300 km are *intermediate-focus,* and the focuses of those characterized as *deep-focus* are more than 300 km deep. However, earthquakes are not evenly distributed among these three categories. Approximately 90% of all earthquake focuses are at depths of less than 100 km, whereas only about 3% of all earthquakes are deep. Shallow-focus earthquakes are, with few exceptions, the most destructive.

An interesting relationship exists between earthquake focuses and plate margins. Earthquakes generated along divergent or transform plate boundaries are invariably shallow focus, while many shallow and nearly all intermediate- and deep-focus earthquakes occur along convergent margins (Figure 7.7). Furthermore, a pattern emerges when the focal depths of earthquakes near island arcs and their adjacent ocean trenches are plotted. Notice in Figure 7.8 that the focal depth increases beneath the Tonga Trench in a narrow, well-defined zone that dips approximately 45 degrees. Dipping seismic zones, called *Benioff zones,* are common to convergent plate boundaries where one plate is subducted beneath another. Such dipping seismic zones indicate the angle of plate descent along a convergent plate boundary.

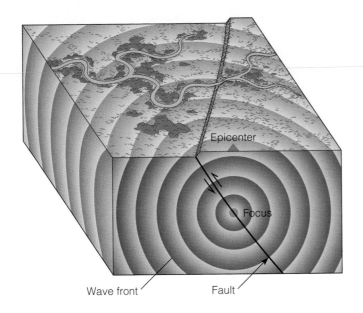

Epicenter

Focus

Wave front

Fault

Figure 7.6 *The focus of an earthquake is the location where rupture begins and energy is released. The place on the surface vertically above the focus is the epicenter. Seismic wave fronts move out in all directions from their source, the focus of an earthquake.*

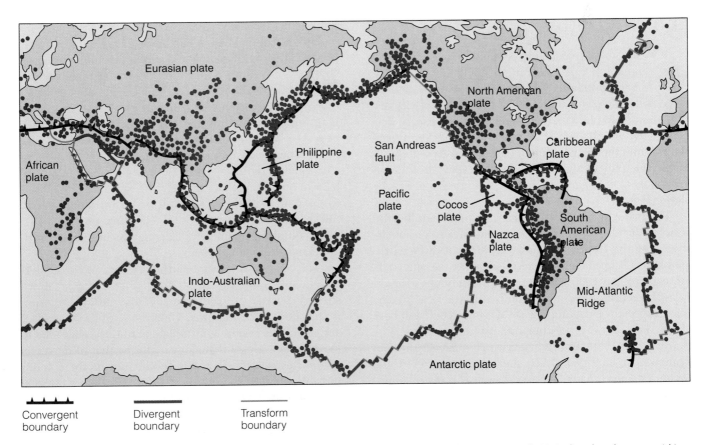

Eurasian plate

North American plate

African plate

Philippine plate

San Andreas fault

Caribbean plate

Pacific plate

Cocos plate

South American plate

Nazca plate

Indo-Australian plate

Mid-Atlantic Ridge

Antarctic plate

Convergent boundary

Divergent boundary

Transform boundary

Figure 7.7 *The relationship between the distribution of earthquake epicenters and plate boundaries. Approximately 80% of earthquakes occur within the circum-Pacific belt, 15% within the Mediterranean-Asiatic belt, and the remaining 5% within plate interiors or along oceanic spreading ridges. Each dot represents a single earthquake epicenter.*

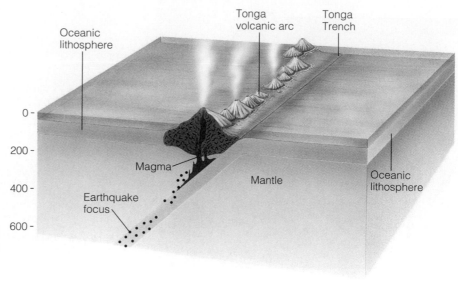

Figure 7.8 *Focal depth increases in a well-defined zone that dips approximately 45 degrees beneath the Tonga volcanic arc in the South Pacific. Dipping seismic zones are called Benioff zones.*

Where Do Earthquakes Occur and How Often?

While no place on Earth is immune to earthquakes, almost 95% take place in seismic belts corresponding to boundaries where plates converge, diverge, and slide past each other. Earthquake activity distant from plate margins is minimal, but can be devastating when it occurs. The relationship between plate margins and the distribution of earthquakes is readily apparent when the locations of earthquake epicenters are superimposed on a map showing the boundaries of Earth's plates (Figure 7.7).

The majority of all earthquakes (approximately 80%) occur in the *circum-Pacific belt*, a zone of seismic activity encircling the Pacific Ocean basin. Most of these earthquakes result from convergence along plate margins, as in the case of the 1995 Kobe, Japan, earthquake (Figure 7.9a). The earthquakes along the North American Pacific coast, especially in California, are also in this belt, but here plates slide past one another rather than converge. The October 17, 1989, Loma Prieta earthquake in the San Francisco area (Figure 7.9b), and the January 17, 1994, Northridge earthquake happened along this plate boundary (see Perspective 7.1).

The second major seismic belt, accounting for 15% of all earthquakes, is the *Mediterranean-Asiatic belt.* This belt extends westerly from Indonesia through the Himalayas, across Iran and Turkey, and westerly through the Mediterranean region of Europe. The devastating 1990 earthquake in Iran that killed 40,000 people, the 1999 Turkey earthquake that killed about 17,000, and the 2001 India earthquake that killed more than 14,000 people, are recent examples of the destructive earthquakes that strike this region (Table 7.1).

The remaining 5% of earthquakes take place mostly in the interiors of plates and along oceanic spreading ridge systems—that is, at divergent plate boundaries. Most of these earthquakes are not strong although several major intraplate earthquakes are worthy of mention. For example, the 1811 and 1812 earthquakes near New Madrid, Missouri, killed approximately 20 people and nearly destroyed the town. So strong were these earthquakes that they were felt from the Rocky Mountains to the Atlantic Ocean and from the Canadian border to the Gulf of Mexico. Another major intraplate earthquake struck Charleston, South Carolina, on August 31, 1886, killing 60 people and causing $23 million in property damage (Figure 7.10).

The cause of intraplate earthquakes is not well understood, but geologists think they arise from localized stresses caused by the compression that most plates experience along their margins. The release of these stresses and hence the resulting intraplate earthquakes are due to local factors. A useful analogy might be that of moving a house. Regardless of how careful the movers are, it is impossible to move something so large without its internal parts shifting slightly. Similarly, plates are not likely to move without some internal stresses that occasionally cause earthquakes. Interestingly, many intraplate earthquakes are associated with very ancient and presumed inactive faults that are reactivated at various intervals.

More than 150,000 earthquakes strong enough to be felt are recorded every year by the worldwide network

(a)

(b)

Figure 7.9 *Earthquake damage in the circum-Pacific belt. (a) Some of the damage in Kobe, Japan, caused by the January 1995 earthquake in which more than 5000 people died. The earthquake in Japan took place at a convergent plate boundary, whereas the Loma Prieta earthquake was generated along a transform plate boundary. (b) Damage in Oakland, California, resulting from the October 1989 Loma Prieta earthquake. The columns supporting the upper deck of Interstate 880 failed, causing the upper deck to collapse onto the lower one.*

Figure 7.10 *Damage done to Charleston, South Carolina, by the earthquake of August 31, 1886. This earthquake is the largest reported in the eastern United States.*

The San Andreas Fault: One Segment of the Circum-Pacific Belt

The circum-Pacific belt is well known for both its volcanic activity and earthquakes. Indeed, about 60% of all volcanic eruptions occur and 80% of all energy released by earthquakes is unleashed in this belt that nearly encircles the Pacific Ocean basin (Figure 7.7). Although it is commonly called the *Ring of Fire,* because of its large number of active volcanoes, you should not draw the conclusion that volcanoes cause earthquakes or vice versa. These two geologic phenomena are common in the circum-Pacific belt because of the interactions of moving plates. Where plates diverge, as along the East Pacific Rise, volcanism and shallow-focus earthquakes occur, whereas convergent plate boundaries, as along the west coast of South America, are characterized by volcanism and shallow-, intermediate-, and deep-focus earthquakes.

One of the best-known and much studied segments of the circum-Pacific belt is the 1300-km-long San Andreas fault extending from the Gulf of California north through coastal California until it terminates at the Mendocino fracture zone off California's north coast (Figure 7.7). In plate tectonic terminology it marks a transform plate boundary with the North American plate sliding horizontally past the Pacific plate (see Chapter 9). In some places, the two plates slide more or less continuously, thus releasing energy in the form of numerous small earthquakes (many of these can be detected only by sensitive instruments). However, some parts of the fault are locked, that is, they are not moving past each other and releasing energy, but are storing energy in the rocks on opposite sides of the fault. Because stored energy is building up in the rocks, these locked segments have the potential to cause large earthquakes when they rupture.

Many people think of a fault as a single crack in the ground, but most faults, and the San Andreas is no exception, consist of a complex zone of fractured rocks anywhere from a few meters to kilometers wide. The San Andreas fault zone, in many areas 1 to 2 km wide, is easily seen on the surface and in aerial photographs (Figure 1). Furthermore, it has numerous subsidiary faults branching from it or paralleling it.

Periodically, rocks on opposite sides of the San Andreas fault lurch past one another, thereby generating large earthquakes. Probably the most famous of these large earthquakes destroyed San Francisco on April 18, 1906. No other earthquake has received such intense study or yielded more scientifically useful information. It resulted when 465 km of the fault ruptured, causing about 6 m of horizontal displacement in some areas (Figure 7.3b). Initial reports put the death toll at 700 to 800, but it was probably closer to 3000, and more than half of the city's approximately 400,000 residents were left homeless. The shaking lasted nearly one minute and caused property damage estimated at $400 million in 1906 dollars!

Some 28,000 buildings were destroyed, many of them by the 3-day fire that raged out of control and devastated about 12 km^2 of the city (Figure 2). In fact, the so-called *San Francisco fire* caused more damage than the earthquake, although the earthquake was responsible for the fire, and, additionally, ruptured water lines so that the fires could not effectively be brought under control. Finally, water was pumped from San Francisco Bay to fight the fires, but by then most of the city was in ruins.

The San Andreas fault and its subsidiary faults have spawned many more earthquakes since 1906, but one of the most tragic was centered at Northridge, California, a community north of Los Angeles. This earthquake actually took place on a buried fault along which rocks were displaced vertically rather than horizontally. During the early morning hours of January 17, 1994, Northridge and surrounding areas were shaken for 40 seconds. When the earthquake was over, 61 people were dead, thousands were

(a)

(b)

Figure 1 *(a) View across the San Andreas fault at Tomales Bay north of San Francisco. The low area occupied by the bay is underlain by shattered rocks of the San Andreas fault zone. Rocks underlying the hills in the distance are on the North American plate, whereas those at the point where the photograph was taken are on the Pacific plate. (b) This shop in Olema, California, is rather whimsically called The Epicenter, alluding to the fact that it is in the San Andreas fault zone.*

Figure 2 *San Francisco following the 1906 earthquake. View along Sacramento Street showing damaged buildings and the approaching fire.*

northern California. Most of the reinforced structures suffered little or no damage during the Northridge earthquake, but several awaiting reinforcement collapsed, including the vital east-west Santa Monica Freeway.

Earthquakes along the San Andreas and related faults have been and will continue to occur. But other segments of the circum-Pacific belt as well as the Mediterranean-Asiatic belt are also quite active, and these areas too will experience earthquakes. Keep in mind as we discuss this topic further that even though the two earthquakes featured in this Perspective were tragic, similar and even smaller earthquakes in China, Japan, Iran, Afghanistan, Turkey, and India have killed more than 150 times as many people and brought about unimaginable human suffering. In no way do we mean to minimize the death and destruction wrought by earthquakes along the San Andreas fault, but we must be aware of and consider why even greater earthquake disasters occur elsewhere.

injured, an oil main and at least 250 gas lines had ruptured igniting numerous fires, nine freeways were destroyed, and thousands of homes and other buildings were damaged or destroyed (Figure 3). So many power lines were knocked down and circuits blown that 3.1 million people were without electricity, and at least 40,000 had no water because of broken water mains.

More than 1000 aftershocks followed the main Northridge earthquake, many of them strong enough to contribute to the already considerable damage. The destruction left in the earthquake's wake was enormous; it amounted to $15 billion to $30 billion in property damage. But considering what had been learned in other large earthquakes, many of the newer structures in this area had been built to stricter standards and generally escaped unscathed or with only minor damage. However, many older unreinforced buildings and more recent wood-frame apartments built over ground-floor garages were destroyed or heavily damaged.

Caltrans, the state transportation department, began a program of reinforcing bridges and freeway overpasses soon after the 1971 Sylmar earthquake in the Los Angeles area and began a second round of reinforcing structures following the 1989 Loma Prieta earthquake in

(a)

(b)

(c)

Figure 3 *Damage resulting from the 1994 Northridge earthquake. (a) Severe damage to the Northridge Meadows apartments. Sixteen died in this building. (b) Fire caused by a gas main explosion. (c) Damage done to Interstate 5 Golden State Freeway.*

Where Do Earthquakes Occur and How Often? **181**

of seismograph stations. In addition, an estimated 900,000 earthquakes are recorded annually by seismographs, but are too small to be individually cataloged. These small earthquakes result from the energy released as continual adjustments take place between the various plates.

What Are Seismic Waves?

Many people have experienced an earthquake, but are probably unaware that the shaking they experience and damage to structures are caused by the arrival of various *seismic waves,* a general term encompassing all waves generated by an earthquake. When movement on a fault takes place, energy is released in the form of two kinds of waves that radiate outward in all directions from an earthquake's focus. *Body waves,* so called because they travel through the solid body of Earth, are somewhat like sound waves, and *surface waves,* which travel along the ground surface, are analogous to undulations or waves on water surfaces.

Body Waves

An earthquake generates two types of body waves: P-waves and S-waves (Figure 7.11). **P-waves** or *primary*

waves are the fastest seismic waves and can travel through solids, liquids, and gases. P-waves are compressional, or push-pull, waves and are similar to sound waves in that they move material forward and backward along a line in the same direction that the waves themselves are moving (Figure 7.11b). Thus, the material through which P-waves travel is expanded and compressed as the waves move through it and returns to its original size and shape after the wave passes by.

S-waves or *secondary waves* are somewhat slower than P-waves and can travel through solids only. S-waves are *shear waves* because they move the material perpendicular to the direction of travel, thereby producing shear stresses in the material they move through (Figure 7.11c). Because liquids (as well as gases) are not rigid, they have no shear strength and S-waves cannot be transmitted through them.

The velocities of P- and S-waves are determined by the density and elasticity of the materials through which they travel. For example, seismic waves travel more slowly through rocks of greater density, but more rapidly through rocks with greater elasticity. *Elasticity* is a property of solids, such as rocks, and means that once they have been deformed by an applied force, they return to their original shape when the force is no longer present. Because P-wave velocity is greater than S-wave velocity in all materials, P-waves always arrive at seismic stations first.

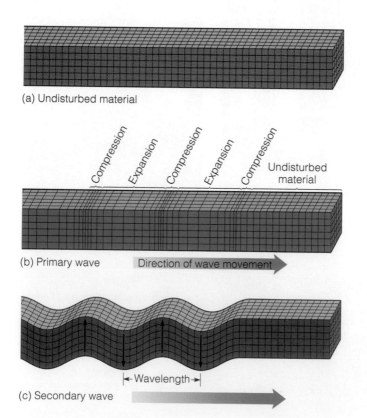

(a) Undisturbed material

Compression Expansion Compression Expansion Compression   Undisturbed material

(b) Primary wave   Direction of wave movement

|←Wavelength→|

(c) Secondary wave

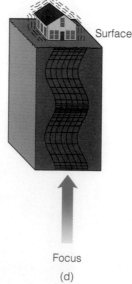

Surface

Focus

(d)

Figure 7.11 *Seismic waves. (a) Undisturbed material for reference. (b) and (c) show how body waves travel through Earth. (b) Primary waves (P-waves) compress and expand material in the same direction they travel. (c) Secondary waves (S-waves) move material perpendicular to the direction of wave movement. (d) P- and S-waves and their effect on surface structures.*

Figure 2 *San Francisco following the 1906 earthquake. View along Sacramento Street showing damaged buildings and the approaching fire.*

northern California. Most of the reinforced structures suffered little or no damage during the Northridge earthquake, but several awaiting reinforcement collapsed, including the vital east-west Santa Monica Freeway.

Earthquakes along the San Andreas and related faults have been and will continue to occur. But other segments of the circum-Pacific belt as well as the Mediterranean-Asiatic belt are also quite active, and these areas too will experience earthquakes. Keep in mind as we discuss this topic further that even though the two earthquakes featured in this Perspective were tragic, similar and even smaller earthquakes in China, Japan, Iran, Afghanistan, Turkey, and India have killed more than 150 times as many people and brought about unimaginable human suffering. In no way do we mean to minimize the death and destruction wrought by earthquakes along the San Andreas fault, but we must be aware of and consider why even greater earthquake disasters occur elsewhere.

injured, an oil main and at least 250 gas lines had ruptured igniting numerous fires, nine freeways were destroyed, and thousands of homes and other buildings were damaged or destroyed (Figure 3). So many power lines were knocked down and circuits blown that 3.1 million people were without electricity, and at least 40,000 had no water because of broken water mains.

More than 1000 aftershocks followed the main Northridge earthquake, many of them strong enough to contribute to the already considerable damage. The destruction left in the earthquake's wake was enormous; it amounted to $15 billion to $30 billion in property damage. But considering what had been learned in other large earthquakes, many of the newer structures in this area had been built to stricter standards and generally escaped unscathed or with only minor damage. However, many older unreinforced buildings and more recent wood-frame apartments built over ground-floor garages were destroyed or heavily damaged.

Caltrans, the state transportation department, began a program of reinforcing bridges and freeway overpasses soon after the 1971 Sylmar earthquake in the Los Angeles area and began a second round of reinforcing structures following the 1989 Loma Prieta earthquake in

(a)

(b)

(c)

Figure 3 *Damage resulting from the 1994 Northridge earthquake. (a) Severe damage to the Northridge Meadows apartments. Sixteen died in this building. (b) Fire caused by a gas main explosion. (c) Damage done to Interstate 5 Golden State Freeway.*

of seismograph stations. In addition, an estimated 900,000 earthquakes are recorded annually by seismographs, but are too small to be individually cataloged. These small earthquakes result from the energy released as continual adjustments take place between the various plates.

What Are Seismic Waves?

Many people have experienced an earthquake, but are probably unaware that the shaking they experience and damage to structures are caused by the arrival of various *seismic waves,* a general term encompassing all waves generated by an earthquake. When movement on a fault takes place, energy is released in the form of two kinds of waves that radiate outward in all directions from an earthquake's focus. *Body waves,* so called because they travel through the solid body of Earth, are somewhat like sound waves, and *surface waves,* which travel along the ground surface, are analogous to undulations or waves on water surfaces.

Body Waves

An earthquake generates two types of body waves: P-waves and S-waves (Figure 7.11). **P-waves** or *primary*

waves are the fastest seismic waves and can travel through solids, liquids, and gases. P-waves are compressional, or push-pull, waves and are similar to sound waves in that they move material forward and backward along a line in the same direction that the waves themselves are moving (Figure 7.11b). Thus, the material through which P-waves travel is expanded and compressed as the waves move through it and returns to its original size and shape after the wave passes by.

S-waves or *secondary waves* are somewhat slower than P-waves and can travel through solids only. S-waves are *shear waves* because they move the material perpendicular to the direction of travel, thereby producing shear stresses in the material they move through (Figure 7.11c). Because liquids (as well as gases) are not rigid, they have no shear strength and S-waves cannot be transmitted through them.

The velocities of P- and S-waves are determined by the density and elasticity of the materials through which they travel. For example, seismic waves travel more slowly through rocks of greater density, but more rapidly through rocks with greater elasticity. *Elasticity* is a property of solids, such as rocks, and means that once they have been deformed by an applied force, they return to their original shape when the force is no longer present. Because P-wave velocity is greater than S-wave velocity in all materials, P-waves always arrive at seismic stations first.

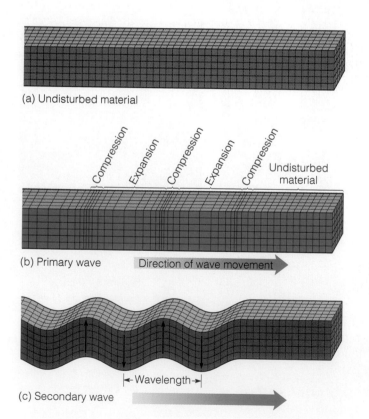

(a) Undisturbed material

Compression Expansion Compression Expansion Compression Undisturbed material

(b) Primary wave   Direction of wave movement

←Wavelength→

(c) Secondary wave

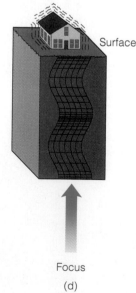

Surface

Focus

(d)

Figure 7.11 *Seismic waves. (a) Undisturbed material for reference. (b) and (c) show how body waves travel through Earth. (b) Primary waves (P-waves) compress and expand material in the same direction they travel. (c) Secondary waves (S-waves) move material perpendicular to the direction of wave movement. (d) P- and S-waves and their effect on surface structures.*

Surface Waves

Surface waves travel along the surface of the ground, or just below it, and are slower than body waves. Unlike the sharp jolting and shaking that body waves cause, surface waves generally produce a rolling or swaying motion, much like the experience of being on a boat.

Several types of surface waves are recognized. The two most important are **Rayleigh waves (R-waves)** and **Love waves (L-waves),** named after the British scientists who discovered them, Lord Rayleigh and A.E.H. Love. Rayleigh waves are generally the slower of the two and behave like water waves in that they move forward while the individual particles of material move in an elliptical path within a vertical plane oriented in the direction of wave movement (Figure 7.12a).

The motion of a Love wave is similar to that of an S-wave, but the individual particles of the material only move back and forth in a horizontal plane perpendicular to the direction of wave travel (Figure 7.12b). This type of lateral motion can be particularly damaging to building foundations.

How Is an Earthquake's Epicenter Located?

Previously we mentioned that news articles commonly report an earthquake's epicenter, but just how is the location of an epicenter determined? Once again, geologists rely on the study of seismic waves. We already know that P-waves travel faster than S-waves, nearly twice as fast in all substances, so P-waves arrive at a seismograph station first followed some time later by S-waves. Both P- and S-waves travel directly from the focus to the seismograph station through Earth's interior, but L- and R-waves arrive last because they are the slowest, and they also travel the longest route along the surface (Figure 7.13). L- and R-waves cause much of the damage during earthquakes, but only P- and S-waves need concern us here because they are the ones important in finding an epicenter.

Seismologists, geologists who study seismology, have accumulated a tremendous amount of data over the years and now know the average speeds of P- and S-waves for any specific distance from their source. These P- and S-wave travel times are published in **time-distance graphs** illustrating that the difference between the arrival times of the two waves is a function of distance between a seismograph and an earthquake's focus (Figure 7.14). That is, the

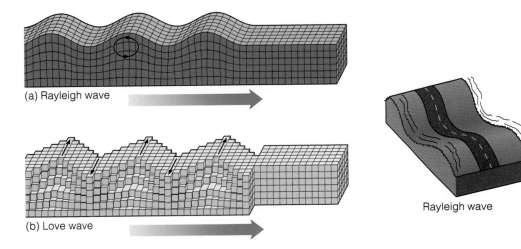

(a) Rayleigh wave

(b) Love wave

Rayleigh wave   Love wave

(c)

Figure 7.12 *Surface waves. (a) Rayleigh waves (R-waves) move material in an elliptical path in a plane oriented parallel to the direction of wave movement. (b) Love waves (L-waves) move material back and forth in a horizontal plane perpendicular to the direction of wave movement. (c) The arrival of R- and L-waves causes the surface to undulate and shake from side to side.*

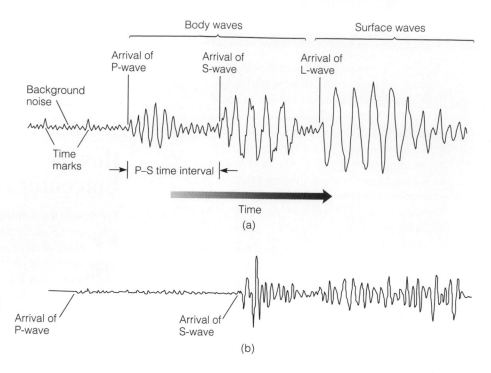

Figure 7.13 *(a) A schematic seismogram showing the arrival order and pattern produced by P-, S-, and L-waves. When an earthquake occurs, body and surface waves radiate out from the focus at the same time. Because P-waves are the fastest, they arrive at a seismograph first, followed by S-waves, and then by surface waves, which are the slowest waves. The difference between the arrival times of the P- and S-waves is the P-S time interval; it is a function of the distance of the seismograph station from the focus. (b) Seismogram for the 1906 San Francisco earthquake, recorded 14,668 km away in Göttingen, Germany. The total record represents about 26 minutes, so considerable time passed between the arrival of the P-waves and the slower moving S-waves. The arrival of surface waves, not shown here, caused the instrument to go off the scale.*

farther the waves travel, the greater the *P-S time interval* or simply the time difference between the arrivals of P- and S-waves (Figure 7.14).

If the P-S time intervals are known from at least three seismograph stations, then the epicenter of any earthquake can be determined (Figure 7.15). Here is how it works. Subtracting the arrival time of the first P-wave from the arrival time of the first S-wave gives the P-S time interval for each seismic station. Each of these time intervals is then plotted on a time-distance graph, and a line is drawn straight down to the distance axis of the graph, thus giving the distance from the focus to each seismic station (Figure 7.14). Next, a circle whose radius equals the distance shown on the time-distance graph from each of the seismic stations is drawn on a map (Figure 7.15). The intersection of the three circles is the location of the earthquake's epicenter. It should be obvious from Figure 7.15 that P-S time intervals from at least three seismic stations are needed. If only one were used, the epicenter could be at any location on the circle drawn around that station, and two stations would give two possible locations for the epicenter.

Determining the focal depth of an earthquake is much more difficult and considerably less precise than finding its epicenter. It is usually found by making computations based on several assumptions, comparisons with the results obtained at other seismic stations, recalculating and approximating the depth as closely as possible. Even so, the results are not highly accurate, but they do tell us that

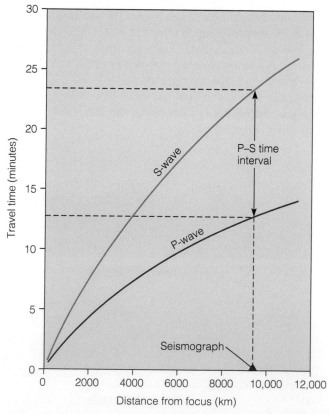

Figure 7.14 *A time-distance graph showing the average travel times for P- and S-waves. The farther away a seismograph station is from the focus of an earthquake, the longer the interval between the arrivals of the P- and S-waves, and hence the greater the distance between the curves on the time-distance graph as indicated by the P-S time interval.*

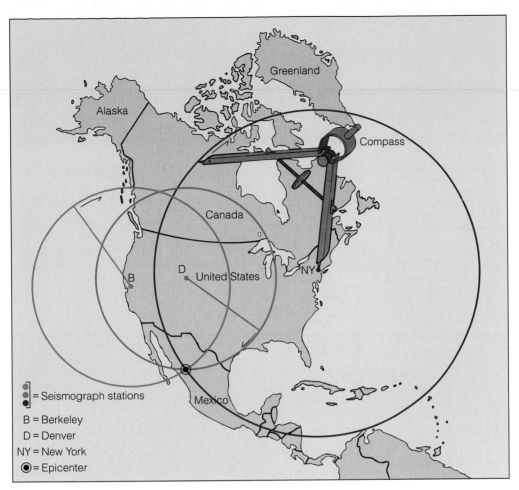

Figure 7.15 *Three seismograph stations are needed to locate the epicenter of an earthquake. The P-S time interval is plotted on a time-distance graph for each seismograph station to determine the distance that station is from the epicenter. A circle with that radius is drawn from each station, and the intersection of the three circles is the epicenter of the earthquake.*

most earthquakes, probably 75% in fact, have focuses no deeper than 10 to 15 km and that a few are as much as 680 km deep.

How Is the Size and Strength of an Earthquake Measured?

Following any earthquake that causes extensive damage, fatalities, and injuries, graphic reports of the quake's violence and human suffering are quite common. Headlines tell us that thousands died, many more were injured or homeless, and property damage is in the millions and possibly billions of dollars. Even though some of the information might be exaggerated, few other natural phenomena account for such tragic consequences. And while descriptions of fatalities and damage give some indication of the size of an earthquake, geologists are interested in more reliable methods of determining an earthquake's size and strength.

Two measures of an earthquakes' strength are commonly used. One is *intensity,* a qualitative assessment of the damage done by an earthquake. The

other, *magnitude,* is a quantitative measure of the amount of energy released during an earthquake. Each method provides important information that can be used to prepare for future earthquakes.

Intensity

Intensity is a subjective measure of the kind of damage done by an earthquake as well as people's reaction to it. Since the mid-19th century, intensity has been used as a rough approximation of the size and strength of an earthquake. The most common intensity scale used in the United States is the **Modified Mercalli Intensity Scale,** which has values ranging from I to XII (Table 7.2).

After an assessment of the earthquake damage is made, *isoseismal lines* (lines of equal intensity) are drawn on a map, dividing the affected region into various intensity zones. The intensity value given for each zone is the maximum intensity that the earthquake produced for that zone. Even though intensity maps are not precise because of the subjective nature of the measurements, they do provide geologists with a rough approximation of the location of the earthquake, the kind and extent of the damage done,

How Is the Size and Strength of an Earthquake Measured?

Table 7.2

Modified Mercalli Intensity Scale

I   Not felt except by a very few under especially favorable circumstances.

II   Felt by only a few people at rest, especially on upper floors of buildings.

III   Felt quite noticeably indoors, especially on upper floors of buildings, but many people do not recognize it as an earthquake. Standing automobiles may rock slightly.

IV   During the day felt indoors by many, outdoors by few. At night some awakened. Sensation like heavy truck striking building, standing automobiles rocked noticeably.

V   Felt by nearly everyone, many awakened. Some dishes, windows, etc., broken, a few instances of cracked plaster. Disturbance of trees, poles, and other tall objects sometimes noticed.

VI   Felt by all, many frightened and run outdoors. Some heavy furniture moved, a few instances of fallen plaster or damaged chimneys. Damages slight.

VII   Everybody runs outdoors. Damage negligible in buildings of good design and construction; slight to moderate in well-built ordinary structures; considerable in poorly built or badly designed structures; some chimneys broken. Noticed by people driving automobiles.

VIII   Damage slight in specially designed structures; considerable in normally constructed buildings with possible partial collapse; great in poorly built structures. Fall of chimneys, monuments, walls. Heavy furniture overturned. Sand and mud ejected in small amounts.

IX   Damage considerable in specially designed structures. Buildings shifted off foundations. Ground noticeably cracked. Underground pipes broken.

X   Some well-built wooden structures destroyed; most masonry and frame structures with foundations destroyed; ground badly cracked. Rails bent. Landslides considerable from river banks and steep slopes. Water splashed over river banks.

XI   Few, if any (masonry) structures remain standing. Bridges destroyed. Broad fissures in ground. Underground pipelines completely out of service.

XII   Damage total. Waves seen on ground surfaces. Objects thrown upward into the air.

SOURCE: United States Geological Survey.

and the effects of local geology on different types of building construction (Figure 7.16). Because intensity is a measure of the kind of damage done by an earthquake, insurance companies still classify earthquakes on the basis of intensity.

Generally, a large earthquake will produce greater intensity values than a small earthquake, but many other factors besides the amount of energy released by an earthquake affect its intensity. These include distance from the epicenter, focal depth of the earthquake, population density and local geology of the area, type of building construction employed, and duration of shaking.

Magnitude

If earthquakes are to be compared quantitatively, we must use a scale that measures the amount of energy released and is independent of intensity. Such a scale was developed in 1935 by Charles F. Richter, a seismologist at the California Institute of Technology. The **Richter Magnitude Scale** measures earthquake **magnitude,** which is the total amount of energy released by an earthquake at its source. It is an open-ended scale with values beginning at 1. The largest magnitude recorded has been 8.6, and though values greater than 9 are theoretically possible, they are highly improbable because rocks are not able to store the energy necessary to generate earthquakes of this magnitude.

The magnitude of an earthquake is determined by measuring the amplitude of the largest seismic wave as recorded on a seismogram (Figure 7.17). To avoid large numbers, Richter used a conventional base-10 logarithmic scale to convert the amplitude of the largest recorded seismic wave to a numerical magnitude value (Figure 7.17). Therefore, each whole-number increase in magnitude represents a 10-fold increase in wave amplitude. For example, the amplitude of the largest seismic wave for an earthquake of magnitude 6 is 10 times that produced by an earthquake of magnitude 5, 100 times as large as a magnitude 4 earthquake, and 1000 times that of an earthquake of magnitude 3 ($10 \times 10 \times 10 = 1000$).

A common misconception about the size of earthquakes is that an increase of one unit on the Richter

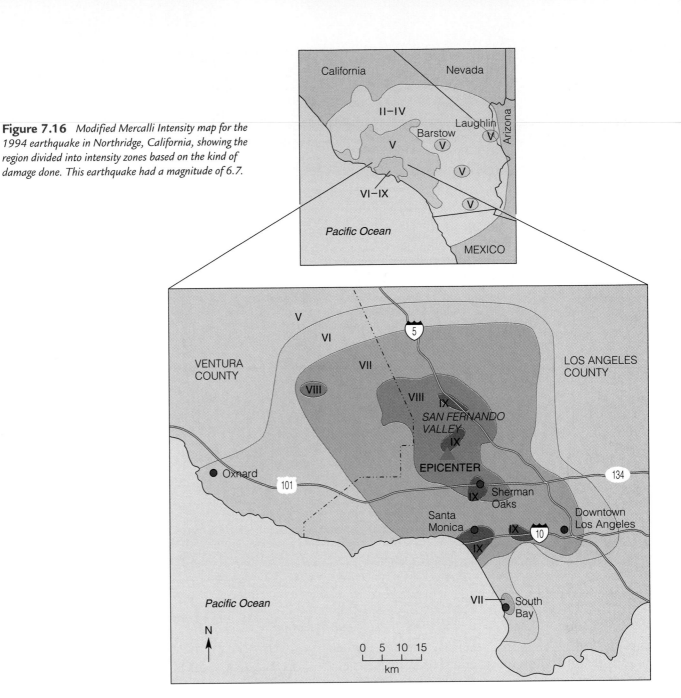

Figure 7.16 *Modified Mercalli Intensity map for the 1994 earthquake in Northridge, California, showing the region divided into intensity zones based on the kind of damage done. This earthquake had a magnitude of 6.7.*

Magnitude Scale, a 7 versus a 6, for instance, means a 10-fold increase in size. It is true that each whole-number increase in magnitude represents a 10-fold increase in the wave amplitude, but each magnitude increase corresponds to a roughly 30-fold increase in the amount of energy released (actually it is 31.5, but 30 is close enough for our purposes). What this means is that it would take about 30 earthquakes of magnitude 6 to equal the energy released in one with a magnitude of 7. The 1964 Alaska earthquake with a magnitude of 8.6 released almost 900 times more energy than the 1994 Northridge, California, earthquake of magnitude 6.7! And if we compare the Alaska earthquake to one with a magnitude of 5.6, it released more than 27,000 times as much energy.

We have already mentioned that more than 900,000 earthquakes are recorded around the world each year. These figures can be placed in better perspective by reference to Table 7.3, which shows that the vast majority of earthquakes have a Richter magnitude of less than 2.5, and that great earthquakes (those with a magnitude greater than 8.0) occur, on average, only once every five years.

The Richter Magnitude Scale was devised to measure earthquake waves on a particular seismograph and a

How Is the Size and Strength of an Earthquake Measured?

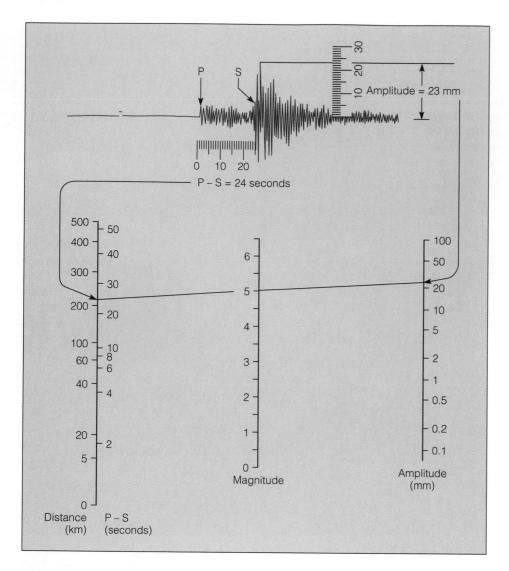

Figure 7.17 *The Richter Magnitude Scale measures magnitude, which is the total amount of energy released by an earthquake at its source. The magnitude is determined by measuring the maximum amplitude of the largest seismic wave and marking it on the right-hand scale. The difference between the arrival times of the P- and S-waves (recorded in seconds) is marked on the left-hand scale. When a line is drawn between the two points, the magnitude of the earthquake is the point at which the line crosses the center scale.*

specific distance from an earthquake. One of its limitations is that it underestimates the energy of very large earthquakes because it measures the highest peak on a seismogram, which represents only an instant during an earthquake. For large earthquakes, though, the energy might be released over several minutes and along hundreds of kilometers of a fault. For example, during the 1857 Fort Tejon, California, earthquake, the ground shook for more than 2 minutes and energy was released for 360 km along the fault. Despite its shortcomings, Richter Magnitude still commonly appears in news releases. More recently, seismologists developed a *Moment Magnitude Scale* that considers the area of a fault along which rupture occurred and the amount of movement of rocks adjacent to the fault. Seismologists are confident that they now have a scale with which they can not only compare different size earthquakes more effectively, but also evaluate the size of earthquakes that occurred before instruments were available to record them.

What Are the Destructive Effects of Earthquakes?

Certainly, earthquakes are one of nature's most destructive phenomena. Little or no warning precedes an earthquake, and once it begins, little or nothing can be done to minimize its effects. Only planning before an earthquake can be very effective, but earthquake prediction might become a reality in the future (discussed in a later section). The destructive effects of earthquakes include ground shaking, fire, seismic sea waves, and landslides, as well as panic, disruption of vital services, and psychological shock. In some cases, rescue attempts are hampered by inadequate resources or planning, existing conditions of civil unrest, or simply the magnitude of the disaster.

The number of deaths and injuries as well as the amount of property damage resulting from an earthquake

Table 7.3

*Average Number of Earthquakes
of Various Magnitudes per Year
Worldwide*

Magnitude	Effects	Average Number per Year
<2.5	Typically not felt but recorded	900,000
2.5–6.0	Usually felt; minor to moderate damage to structures	31,000
6.1–6.9	Potentially destructive, especially in populated areas	100
7.0–7.9	Major earthquakes; serious damage results	20
>8.0	Great earthquakes; usually result in total destruction	1 every 5 years

SOURCE: Modified from Earthquake Information Bulletin and *Seismicity of the Earth and Associated Phenomena,* by B. Gutenberg and C. F. Richter (Princeton, NJ: Princeton University Press, 1949).

depends on several factors. Generally speaking, earthquakes during working hours and school hours in densely populated urban areas are the most destructive and cause most fatalities and injuries. However, magnitude, duration of shaking, distance from the epicenter, geology of the affected region, and type of structures are also important considerations. Given these variables, it should not be surprising that a comparatively small earthquake can have disastrous effects whereas a much larger one might go largely unnoticed, except perhaps by seismologists.

Ground Shaking

Ground shaking is the most obvious and immediate effect of an earthquake, but it varies depending on magnitude, distance from the epicenter, and the type of underlying materials in the area—unconsolidated sediment or fill versus bedrock, for instance. Certainly ground shaking is terrifying, and it might be violent enough for fissures to open in the ground. Nevertheless, contrary to popular myth, fissures do not swallow up people and buildings and then close on them. And although California will no doubt have big earthquakes in the future, rocks cannot store enough energy to displace a landmass this large into the Pacific Ocean as one sometimes reads in the tabloids.

The effects of ground shaking, such as collapsing buildings, falling building facades and window glass, and toppling monuments and statues, cause more damage and result in more loss of life and injuries than any other earthquake hazard. Structures built on bedrock generally suffer less damage than those built on poorly consolidated material such as water-saturated sediments or artificial fill (see Perspective 7.2).

Structures on poorly consolidated or water-saturated material are subjected to ground shaking of longer duration and greater S-wave amplitude than those on bedrock (Figure 7.18). In addition, fill and water-saturated sediments tend to liquefy, or behave as a fluid, a process known as *liquefaction*. When shaken, the individual grains lose cohesion and the ground flows. Dramatic examples of damage resulting from liquefaction include Niigata, Japan, where large apartment buildings were tipped to their sides after the water-saturated soil of the hillside collapsed

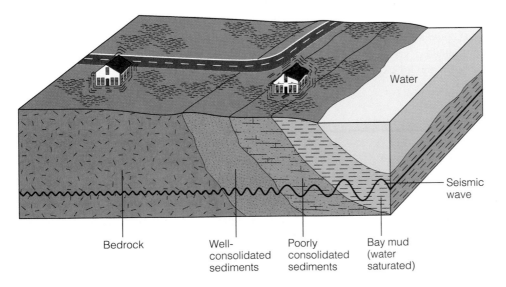

Bedrock Well-consolidated sediments Poorly consolidated sediments Bay mud (water saturated) Water Seismic wave

Figure 7.18 *The amplitude and duration of seismic waves generally increase as they pass from bedrock to poorly consolidated or water-saturated material. Thus, structures built on weaker material typically suffer greater damage than similar structures built on bedrock.*

What Are the Destructive Effects of Earthquakes? **189**

Designing Earthquake-Resistant Structures

One way to reduce property damage, injuries, and loss of life is to design and build structures as earthquake-resistant as possible. Many things can be done to improve the safety of current structures and of new buildings.

California has a Uniform Code that sets minimum standards for building earthquake-resistant structures that is used as a model around the world. The California code is far more stringent than federal earthquake building codes and requires that structures be able to withstand a 25-second main shock. Unfortunately, many earthquakes are of far longer duration. For instance, the main shock of the 1964 Alaskan earthquake lasted approximately three minutes and was followed by numerous aftershocks. While many of the extensively damaged buildings in this earthquake had been built according to the California code, they were not designed to withstand shaking of such long duration. Nevertheless, in California and elsewhere, structures built since the California code went into effect have fared much better during moderate to major earthquakes than those built before its implementation.

To design earthquake-resistant structures, engineers must understand the dynamics and mechanics of earthquakes including the type and duration of the ground motion and how rapidly the ground accelerates during an earthquake. An understanding of the area's geology is also important because certain ground materials such as water-saturated sediments or landfill can lose their strength and cohesiveness during an earthquake (Figure 7.18). Finally, engineers must be aware of how different structures behave under different earthquake conditions.

With the level of technology currently available, a well-designed, properly constructed building should be able to withstand small, short-duration earthquakes of less than 5.5 magnitude with little or no damage. In moderate earthquakes (5.5 to 7.0 magnitude), the damage suffered should not be serious and should be repairable. In a major earthquake of greater than 7.0 magnitude, the building should not collapse, although it may later have to be demolished.

Many factors enter into the design of an earthquake-resistant structure, but the most important is that the building be tied together; that is, the foundation, walls, floors, and roof should all be joined together to create a structure that can withstand both horizontal and vertical shaking (Figure 1). Almost all of the structural failures resulting from earthquake ground movement occur at weak connections, where the various parts of a structure were not securely tied together. Buildings with open or unsupported first stories are particularly susceptible to damage. Some reinforcement must be done or collapse is a distinct possibility (Figure 1).

The size and shape of a building can also affect its resistance to earthquakes (Figure 2). Rectangular box-shaped buildings are inherently stronger than those of irregular size or shape because different parts of an irregular building may sway at different rates, increasing the stress and likelihood of structural failure (Figure 2b).

Tall buildings, such as skyscrapers, must be designed so that a certain amount of swaying or flexing can occur, but not so much that they touch neighboring buildings during swaying (Figure 2d). If a building is brittle and does not give, it will crack and fail. Besides designed flexibility, engineers must make sure that a building does not vibrate at the same frequency as the ground does during an earthquake. When that happens, the force applied by the seismic waves at ground level is multiplied several times by the time they reach the top of the building (Figure 2c). This condition is particularly troublesome in areas of poorly consolidated sediment. Fortunately, buildings can be designed so that they will sway at a different frequency from the ground.

During the Mexico City earthquake of 1985, many buildings between 6 and 15 stories high were either badly damaged or collapsed, even if constructed of reinforced concrete. In contrast, the same kinds of buildings with fewer or more stories suffered comparatively little. The reason? The 6- to 15-story buildings vibrated at the same frequency as the ground and accordingly swayed so much that structural elements failed, and, in some cases, one floor collapsed on another (Figure 3).

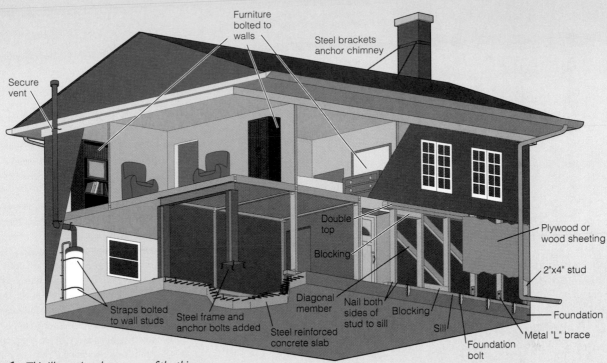

Furniture bolted to walls

Steel brackets anchor chimney

Secure vent

Double top

Blocking

Plywood or wood sheeting

2"x4" stud

Diagonal member

Nail both sides of stud to sill

Blocking

Foundation

Straps bolted to wall studs

Steel frame and anchor bolts added

Steel reinforced concrete slab

Sill

Metal "L" brace

Foundation bolt

Figure 1 *This illustration shows some of the things a homeowner can do to reduce damage to a building because of ground shaking during an earthquake. Notice that the structure must be solidly attached to its foundation, and bracing the walls helps prevent damage from horizontal motion.*

Damage to high-rise structures can also be minimized or prevented by using diagonal steel beams to help prevent swaying. In addition, tall buildings in earthquake-prone areas are now commonly placed on layered steel and rubber structures and devices similar to shock absorbers that help decrease the amount of sway.

What about structures built many years ago? Almost every city and town has older single and multistory structures, constructed of unreinforced brick masonry, poor-quality concrete, and rotting or decaying wood. Just as in new buildings, the most important thing that can be done to increase the stability and safety is to tie the different components of a building together. This can be done by adding a steel frame to unreinforced parts of a building such as a garage, bolting the walls to the foundation, adding reinforced beams to the exterior, and using beam and joist connectors whenever possible. Although such modifications are expensive, they are usually cheaper than having to replace a building that was destroyed by an earthquake.

(continued)

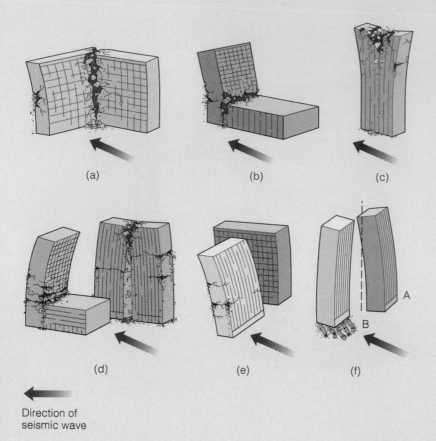

(a)   (b)   (c)

(d)   (e)   (f)

Direction of
seismic wave

Figure 2 *The effects of ground shaking on various tall buildings of differing shapes. (a) Damage will occur if two wings of a building are joined at right angles and experience different motions. (b) Buildings of different heights will sway differently leading to damage at the point of connection. (c) Shaking increases with height and is greatest at the top of a building. (d) Closely spaced buildings may crash into each other due to swaying. (e) A building whose long axis is parallel to the direction of the seismic waves will sway less than a building whose axis is perpendicular. (f) Two buildings of different design will behave differently even when subjected to the same shaking conditions. Building A sways as a unit and remains standing while building B whose first story is composed of only tall columns collapses because most of the swaying takes place in the "soft" first story.*

Figure 3 *This 15-story reinforced concrete building collapsed due to the ground shaking that occurred during the 1985 Mexico City earthquake. The soft lake bed sediments on which Mexico City is built amplified the seismic waves as they passed through.*

(Figure 7.19), and Mexico City, which is built on soft lake bed sediments (see Perspective 7.2, Figure 3).

Besides the magnitude of an earthquake and the underlying geology, the material used and the type of construction also affect the amount of damage done (see Perspective 7.2). Adobe and mud-walled structures are the weakest of all and almost always collapse during an earthquake. Unreinforced brick structures and poorly built concrete structures are also particularly susceptible to collapse. For example, the 1976 earthquake in Tangshan, China, completely leveled the city because hardly any structures were built to resist seismic forces. In fact, most had unrein-

Figure 7.19 *The effects of ground shaking on water-saturated soil are dramatically illustrated by the collapse of these buildings in Niigata, Japan, during a 1964 earthquake. The buildings were designed to be earthquake-resistant and fell over on their sides intact.*

forced brick walls, which have no flexibility, and consequently they collapsed during the shaking (Figure 7.20).

The 6.4 magnitude earthquake that struck India in 1993 killed about 30,000 people whereas the 6.7 magnitude Northridge, California, earthquake one year later, resulted in only 61 deaths. Why such a difference in the death toll? Both earthquakes occurred in densely populated regions, but in India the brick and stone buildings could not withstand the ground shaking and most collapsed entombing their occupants.

Figure 7.20 *Many of the approximately 242,000 people who died in the 1976 earthquake in Tangshan, China, were killed by collapsing structures. Many of the buildings were constructed from unreinforced brick, which has no flexibility, and quickly fell down during the earthquake.*

Fire

In many earthquakes, particularly in urban areas, fire is a major hazard. Almost 90% of the damage done in the 1906 San Francisco earthquake was caused by fire. The shaking severed many of the electrical and gas lines, which touched off flames and started numerous fires all over the city. Because water mains were ruptured by the earthquake, there was no effective way to fight the fires that raged out of control for three days, destroying much of the city.

Eighty-three years later during the 1989 Loma Prieta earthquake, a fire broke out in the Marina district of San Francisco (Figure 7.21). This time, however, the fire was contained within a small area because San Francisco had a system of valves throughout its water and gas pipeline system so that lines could be isolated from breaks.

During the September 1, 1923, earthquake in Japan, fires destroyed 71% of the houses in Tokyo and practically all the houses in Yokohama. In all, a total of 576,262 houses were destroyed by fire, and 143,000 people died, many as a result of fire.

Killer Waves—Tsunami

On April 1, 1946, the residents of Hilo, Hawaii, were completely unaware of an earthquake more than 3500 km away in the Aleutian Islands, but this earthquake ultimately killed 159 people in Hilo and caused $25 million in property damage (Figure 7.22). Thus, earthquakes can kill at a considerable distance, in some cases even much farther than in this example. This earthquake generated what is popularly called a tidal wave but is more correctly termed a *seismic sea wave* or **tsunami,** the latter a Japanese term meaning harbor wave. The term

Figure 7.21 *San Francisco Marina district fire caused by broken gas lines during the 1989 Loma Prieta earthquake.*

tidal wave nevertheless persists in popular literature and some news accounts, but these waves are not caused by or related to tides. Indeed, tsunami are destructive sea waves generated when large amounts of energy are rapidly released into a body of water. Many result from submarine earthquakes, but volcanoes at sea or submarine landslides can also cause them. For example, the 1883 eruption of Krakatau between Java and Sumatra generated a large sea wave that killed 36,000 on nearby islands.

Once a tsunami is generated, it can travel across an entire ocean and cause devastation far from its source. In the open sea tsunami travel at several hundred kilometers per hour, and commonly go unnoticed as they pass beneath ships because they are usually less than 1 m high and the distance between wave crests is typically hundreds of kilometers. When they enter shallow water, however, the wave slows down and water piles up to heights anywhere from a meter or two to many meters high. The 1946 tsunami that struck Hilo, Hawaii, was 16.5 m high! In any case, the tremendous energy possessed by a tsunami is concentrated on a shoreline when it hits either as a large breaking wave or, in some cases, as what appears to be a very rapidly rising tide.

A common popular belief is that a tsunami is a single large wave that crashes onto a shoreline. Any tsunami consists of a series of waves that pour onshore for as much as 30 minutes followed by an equal time during which water rushes back to sea. Furthermore, after the first wave hits, more waves follow at 20 to 60 minute intervals. About 80 minutes after the 1755 earthquake in Lisbon, Portugal, the first of three tsunami, the largest more than 12 m high, destroyed the waterfront area and killed numerous people. Following the arrival of a 2-m-high tsunami in Crescent City, California, in 1964, curious people went to the waterfront to inspect the damage. Unfortunately, 10 were killed shortly thereafter by another 4-m-high wave.

Figure 7.22 *As a tsunami crashes into the street behind them, residents of Hilo, Hawaii, run for their lives. This tsunami was generated by an earthquake in the Aleutian Islands and resulted in massive damage to Hilo and the deaths of 159 people.*

One of nature's warning signs of an approaching tsunami is that some are preceded by a sudden withdrawal of the sea from a coastal region. In fact, the sea might withdraw so far that it cannot even be seen and the seafloor is laid bare over a huge area. On more than one occasion, people have rushed out to inspect exposed reefs or collect fish and shells only to be swept away when the tsunami arrives.

Following the tragic 1946 tsunami that hit Hilo, Hawaii, the U.S. Coast and Geodetic Survey established a Tsunami Early Warning System in Honolulu, Hawaii, in an attempt to minimize tsunami devastation. This system combines seismographs and instruments that can detect earthquake-generated sea waves. Whenever a strong earthquake occurs anywhere within the Pacific basin, its location is determined, and instruments are checked to see if a tsunami has been generated. If it has, a warning is sent out to evacuate people from low-lying areas that may be affected. Nevertheless, tsunami remain a threat to people in coastal areas, especially around the Pacific Ocean (Table 7.4).

Ground Failure

Earthquake-triggered landslides are particularly dangerous in mountainous regions and have been responsible for tremendous amounts of damage and many deaths. The 1959 earthquake in Madison Canyon, Montana, for example, caused a huge rock slide (Figure 7.23), while the 1970 Peru earthquake caused an avalanche that destroyed the town of Yungay and killed an estimated 66,000 people. Most of the 100,000 deaths from the 1920 earthquake in Gansu, China, resulted when cliffs composed of loess (wind-deposited silt) collapsed. More than 20,000 people were killed when two-thirds of the town of Port Royal, Jamaica, slid into the sea following an earthquake on June 7, 1692.

Can Earthquakes Be Predicted?

Can earthquakes be predicted? A successful prediction must include a time frame for the occurrence of the earthquake, its location, and its strength. Despite the tremendous amount of information geologists have gathered about the cause of earthquakes, successful predictions are still rare. Nevertheless, if reliable predictions can be made, they can greatly reduce the number of deaths and injuries.

From an analysis of historic records and the distribution of known faults, *seismic risk maps* can be constructed that indicate the likelihood and potential

Table 7.4

Tsunami Fatalities in the Pacific since 1990

Date	Location	Maximum Wave Height	Fatalities
September 2, 1992	Nicaragua	10 m	170
December 12, 1992	Flores Island	26 m	>1000
July 12, 1993	Okushiri, Japan	31 m	239
June 2, 1994	East Java	14 m	238
November 14, 1994	Mindoro Island	7 m	49
October 9, 1995	Jalisco, Mexico	11 m	1
January 1, 1996	Sulawesi Island	3.4 m	9
February 17, 1996	Irian Jaya	7.7 m	161
February 21, 1996	North coast of Peru	5 m	12
July 17, 1998	Papua New Guinea	15 m	>2200

SOURCE: Gonzales, F. I. 1999. Tsunami! *Scientific American* 280, no. 5, p. 59.

(a)

(b)

Figure 7.23 *On August 17, 1959, an earthquake with a Richter magnitude of 7.3 shook southwestern Montana and a large area in adjacent states. (a) The fault scarp in this image was produced when the block in the background moved up several meters compared to the one in the foreground. (b) The earthquake triggered a landslide (visible in the distance) that blocked the Madison River in Montana and created Earthquake Lake (foreground). The slide entombed about 26 people in a campground at the valley bottom.*

severity of future earthquakes based on the intensity of past earthquakes.

An international effort by scientists from several countries recently resulted in the publication of the first Global Seismic Hazard Assessment Map in December 1999 (Figure 7.24). Although such maps cannot be used to predict when an earthquake will take place in any particular area, they are useful in anticipating future earthquakes and helping people plan and prepare for them.

Earthquake Precursors

Studies conducted during the past several decades indicate that most earthquakes are preceded by both short-term and long-term changes within Earth. Such changes are called *precursors.*

Changes in elevation and tilting of the land surface have frequently preceded earthquakes and may be warnings of impending quakes. Extremely slight changes in the angle of the ground surface can be measured by tiltmeters. Tiltmeters have been placed on both sides of the San Andreas fault to measure tilting of the ground surface that is thought to result from increasing pressure in the rocks. Data from measurements in central California indicate significant tilting immediately preceding small earthquakes. Furthermore, extensive tiltmeter work performed in Japan prior to the 1964 Niigata earthquake clearly showed a relationship between increased tilting and the main shock. Though more research is needed, such precursors appear to be useful in making short-term earthquake predictions.

Other earthquake precursors include fluctuations in the water level of wells and changes in Earth's magnetic field and the electrical resistance of the ground. These fluctuations are thought to result from changes in the amount of pore space in rocks due to increasing pressure.

Many of the precursors just discussed can be related to the *dilatancy model,* which is based on changes occurring in rocks subjected to very high pressures. Laboratory experiments have shown that rocks undergo an increase in volume, known as dilatancy, just before rupturing. As pressure builds in rocks along faults, numerous small cracks are produced that alter the physical properties of the rocks. Water enters the cracks and increases the fluid pressure; this further increases the volume of the rocks and decreases their strength until failure eventually occurs, producing an earthquake.

Besides the various precursors just discussed, one long-range prediction technique used in seismically active areas involves plotting the location of major earthquakes and their aftershocks to detect areas that have had major earthquakes in the past but are currently inactive. Such regions are locked and not releasing energy, making these *seismic gaps* prime locations for future earthquakes. Several seismic gaps along the San Andreas fault have the potential for future major earthquakes.

Earthquake Prediction Programs

Currently, only four nations—the United States, Japan, Russia, and China—have government-sponsored earthquake prediction programs. These programs include laboratory and field studies of rock behavior before, during, and after large earthquakes as well as monitoring activity along major active faults. Most earthquake prediction work in the United States is done by the United

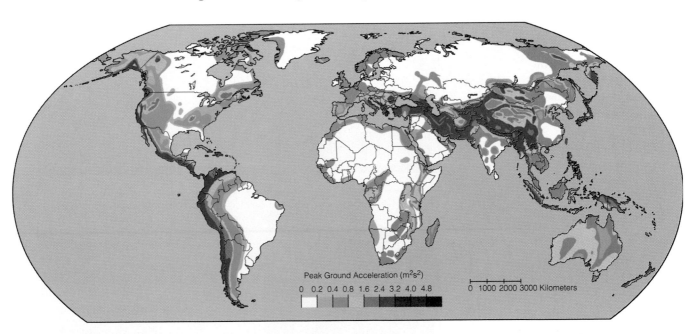

Figure 7.24 *The Global Seismic Hazard Assessment Program recently published this seismic hazard map showing peak ground accelerations. The values are based on a 90% probability that the indicated horizontal ground acceleration during an earthquake is not likely to be exceeded in 50 years. The higher the number, the greater the hazard. As expected, the greatest seismic risks are in the circum-Pacific belt and the Mediterranean-Asiatic belt.*

Can Earthquakes Be Predicted?

States Geological Survey (USGS) and involves research into all aspects of earthquake-related phenomena.

The Chinese have perhaps the most ambitious earthquake prediction program in the world, which is understandable considering their long history of destructive earthquakes. Their earthquake prediction program was initiated soon after two large earthquakes occurred at Xingtai (300 km southwest of Beijing) in 1966. The program includes extensive study and monitoring of all possible earthquake precursors. In addition, the Chinese also emphasize changes in phenomena that can be observed and heard without the use of sophisticated instruments. They successfully predicted the 1975 Haicheng earthquake, but failed to predict the devastating 1976 Tangshan earthquake that killed at least 242,000 people (Figure 7.20).

Progress is being made toward dependable, accurate earthquake predictions, and studies are under way to assess public reactions to long-, medium-, and short-term earthquake warnings. However, unless short-term warnings are actually followed by an earthquake, most people will probably ignore the warnings as they frequently do now for hurricanes, tornadoes, and tsunami. Perhaps the best we can hope for is that people in seismically active areas will take measures to minimize their risk from the next major earthquake.

Can Earthquakes Be Controlled?

Reliable earthquake prediction is still in the future, but can anything be done to control or at least partly control earthquakes? Because of the tremendous energy involved, it seems unlikely that humans will ever be able to prevent earthquakes. However, it might be possible to gradually release the energy stored in rocks, thus decreas-

ing the probability of a large earthquake and extensive damage.

During the early to mid-1960s, Denver, Colorado, experienced numerous small earthquakes. This was surprising because Denver had not been prone to earthquakes in the past. In 1962, geologist David M. Evans suggested that Denver's earthquakes were directly related to the injection of contaminated waste water into a disposal well 3674 m deep at the Rocky Mountain Arsenal, northeast of Denver. The U.S. Army initially denied that a connection existed, but a USGS study concluded that the pumping of waste fluids into the disposal well was the cause of the earthquakes.

Figure 7.25 shows the relationship between the average number of earthquakes in Denver per month and the average amount of contaminated waste fluids injected into the disposal well per month. Obviously, a high degree of correlation between the two exists, and the correlation is particularly convincing considering that during the time when no waste fluids were injected, earthquake activity decreased dramatically. The area beneath the Rocky Mountain Arsenal consists of highly fractured gneiss overlain by sedimentary rocks. When water was pumped into these fractures, it decreased the friction on opposite sides of the fractures and, in essence, lubricated them so that movement occurred, causing the earthquakes that Denver experienced.

Experiments conducted in 1969 at an abandoned oil field near Rangely, Colorado, confirmed the arsenal hypothesis. Water was pumped in and out of abandoned oil wells, the pore-water pressure in these wells was measured, and seismographs were installed in the area to

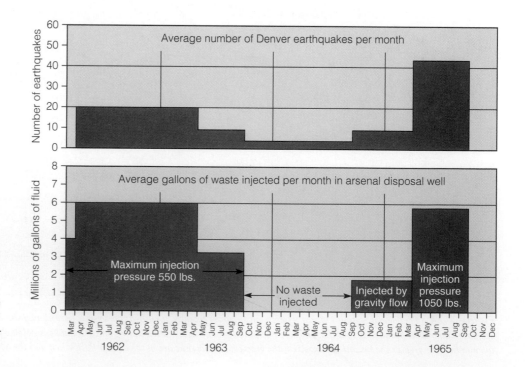

Figure 7.25 *A graph showing the relationship between the amount of waste injected into the arsenal disposal well per month and the average number of Denver earthquakes per month.*

measure any seismic activity. Monitoring showed that small earthquakes were occurring in the area when fluid was injected and that earthquake activity declined when the fluids were pumped out. What the geologists were doing was starting and stopping earthquakes at will, and the relationship between pore-water pressures and earthquakes was established.

Based on these results, some geologists have proposed that fluids be pumped into the locked segments or seismic gaps of active faults to cause small- to moderate-sized earthquakes. They think that this would relieve the pressure on the fault and prevent a major earthquake from occurring. While this plan is intriguing, it also has many potential problems. For instance, there is no guarantee that only a small earthquake might result. Instead a major earthquake might occur, causing tremendous property damage and loss of life. Who would be responsible? Certainly, a great deal more research is needed before such an experiment is performed, even in an area of low population density.

It appears that until such time as earthquakes can be accurately predicted or controlled, the best means of defense is careful planning and preparation (Table 7.5).

WHAT WOULD YOU DO?

Some geologists think that by pumping fluids into locked segments of active faults, small- to moderate-sized earthquakes can be generated. These earthquakes would relieve the buildup of pressure along a fault and thus prevent very large earthquakes from taking place. What do you think of this proposal? What kinds of social, political, and economic consequences would there be, and do you think such an effort will ever actually reduce the threat of earthquakes?

Table 7.5

What You Can Do to Prepare for an Earthquake

Anyone who lives in an area that is subject to earthquakes or who will be visiting or moving to such an area can take certain precautions to reduce the risks and losses resulting from an earthquake.

Before an earthquake:

1. Become familiar with the geologic hazards of the area where you live and work.

2. Make sure your house is securely attached to the foundation by anchor bolts and that the walls, floors, and roof are all firmly connected together.

3. Heavy furniture such as bookcases should be bolted to the walls; semiflexible natural gas lines should be used so that they can give without breaking; water heaters and furnaces should be strapped and the straps bolted to wall studs to prevent gas-line rupture and fire. Brick chimneys should have a bracket or brace that can be anchored to the roof.

4. Maintain a several-day supply of fresh water and canned foods, and keep a fresh supply of flashlight and radio batteries as well as a fire extinguisher.

5. Maintain a basic first-aid kit, and have a working knowledge of first-aid procedures.

6. Learn how to turn off the various utilities at your house.

7. Above all, have a planned course of action for when an earthquake strikes.

During an earthquake:

1. Remain calm and avoid panic.

2. If you are indoors, get under a desk or table if possible, or stand in an interior doorway or room corners as these are the strongest parts of a room; avoid windows and falling debris.

3. In a tall building, do not rush for the stairwells or elevators.

4. In an unreinforced or other hazardous building, it may be better to get out of the building rather than to stay in it. Be on the alert for fallen power lines and the possibility of falling debris.

5. If you are outside, get to an open area away from buildings if possible.

6. If you are in an automobile, stay in the car, and avoid tall buildings, overpasses, and bridges if possible.

After an earthquake:

1. If you are uninjured, remain calm and assess the situation.

2. Help anyone who is injured.

3. Make sure there are no fires or fire hazards.

4. Check for damage to utilities and turn off gas valves if you smell gas.

5. Use your telephone only for emergencies.

6. Do not go sightseeing or move around the streets unnecessarily.

7. Avoid landslide and beach areas.

8. Be prepared for aftershocks.

What Is Earth's Interior Like?

During most of historic time, Earth's interior was perceived as an underground world of vast caverns, heat, and sulfur gases, populated by demons (see Perspective 7.3). By the 1860s, scientists knew what the average density of Earth was and that pressure and temperature increase with depth. And even though Earth's interior is hidden from direct observation, scientists now have a reasonably good idea of its internal structure and composition.

Scientists have known for more than 200 years that Earth's interior is not homogeneous throughout. Sir Isaac Newton (1642–1727) noted in a study of the planets that Earth's average density is 5.0 to 6.0 g/cm^3 (water has a density of 1 g/cm^3). In 1797, Henry Cavendish calculated a density value very close to the 5.5 g/cm^3 now accepted. Earth's average density is considerably greater than that of surface rocks, most of which range from 2.5 to 3.0 g/cm^3. Thus, in order for the average density to be 5.5 g/cm^3, much of the interior must consist of materials with a density greater than Earth's average density.

Earth is generally depicted as consisting of concentric layers that differ in composition and density separated from adjacent layers by rather distinct boundaries (Figure 7.26). Recall that the outermost layer, or the **crust,** is Earth's thin skin. Below the crust and extending about halfway to Earth's center is the **mantle,** which comprises more than 80% of Earth's volume. The central part of Earth consists of a **core,** which is divided into a solid inner portion and a liquid outer part (Figure 7.26).

The behavior and travel times of P- and S-waves provide geologists with much information about Earth's internal structure. Seismic waves travel outward as wave fronts from their source areas, although it is most convenient to depict them as *wave rays,* which are lines showing the direction of movement of small parts of wave fronts (Figure 7.6). Any disturbance, such as a passing train or construction equipment, can cause seismic waves, but only those generated by large earthquakes, explosive volcanism, asteroid impacts, and nuclear explosions can travel completely through Earth.

As we noted earlier, the velocities of P- and S-waves are determined by the density and elasticity of the materials they travel through, both of which increase with depth. Wave velocity is slowed by increasing density but increases in materials with greater elasticity. Because elasticity increases with depth faster than density, a general increase in the velocity of seismic waves takes place as they penetrate to greater depths. P-waves travel faster than S-waves under all circumstances, but unlike P-waves, S-waves cannot be transmitted through a liquid because liquids have no shear strength (rigidity); they simply flow in response to shear stress.

As a seismic wave travels from one material into another of different density and elasticity, its velocity and

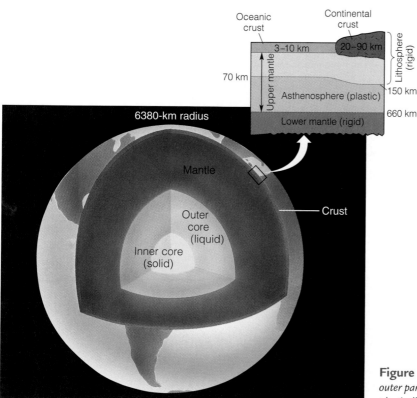

Figure 7.26 *Earth's internal structure. The inset shows Earth's outer part in more detail. The asthenosphere is solid but behaves plastically and flows.*

By the 1800s, scientists had some sketchy ideas about Earth's internal structure, but outside scientific circles, all kinds of bizarre ideas were proposed. In 1869, for example, Cyrus Reed Teed claimed that Earth was hollow and that humans lived on the inside. As recently as 1913, Marshall B. Gardner held that Earth is a large hollow sphere with a 1300-km-thick outer shell surrounding a central sun. In addition, he claimed that Eskimos and mammoths now preserved as fossils originally dwelled within this hollow Earth.

Although making no claim to present a reliable picture of Earth's interior, Jules Verne's 1864 novel *A Journey to the Center of the Earth* described the adventures of Professor Hardwigg, his nephew, and an Icelandic guide as they descended into Earth through the crater of Mount Sneffels in Iceland. During their travels, they followed a labyrinth of passageways until they finally arrived 140 km below the surface. Here, they encountered a vast cavern containing "the central sea" illuminated by some electrical phenomenon related to the northern lights. Along the margins of the sea, they saw forests of prehistoric ferns and palms and a herd of mastodons complete with a gigantic human shepherd. Dwelling in the central sea were Mesozoic-aged marine reptiles and gigantic turtles. Their adventure ended when they were carried upward to the surface on a raft by a rising plume of water.

Scientists in 1864 knew what the average density of Earth was and that pressure and temperature increase with depth. They also knew that the fabled passageways followed by Professor Hardwigg could not exist, but little else was known, even though humans had probed Earth through mines and wells for centuries. Even the deepest mines (the gold mines in South Africa) descend only about 3 km below the surface, barely penetrating the uppermost part of the crust. The deepest drill hole reaches a depth of a little more than 12 km. A drill hole 12 km deep is impressive, but it is less than 0.2% of the distance to Earth's center. Indeed, if Earth were the size of an apple, this drill hole would be roughly equivalent to a pinprick penetrating less than halfway through the apple's skin!

In 1958, an ambitious project to drill through the oceanic crust to the mantle was launched. Known as *Mohole,* the project attracted considerable attention and support from scientists and the public, and in 1962 Congress appropriated more than $40 million to finance it. Mohole encountered technological difficulties, however, and by 1966 funding had dried up; soon thereafter, project Mohole was quickly forgotten.

Because Earth's interior is hidden from direct observation, it is more inaccessible than the surfaces of the Moon and Mars. Nevertheless, scientists have a reasonably good idea of Earth's internal structure and composition. No vast openings or passageways exist as in Jules Verne's story; even the deepest known caverns extend to depths of less than 1500 m. Even at the modest depths to which Professor Hardwigg and his companions are supposed to have descended, the pressure and temperature are so great that rock actually flows even though it remains solid. In deep mines the rock is under such tremendous pressure that rock bursts and popping are constant problems (see Chapter 5). In short, the behavior of solids at depth where the temperature and pressure are great is very different from their brittle behavior at the surface.

---

direction of travel change. That is, the wave is bent, a phenomenon known as **refraction;** in much the same way light waves are refracted as they pass from air into a more dense medium such as water (Figure 7.27). Because seismic waves pass through materials of differing density and elasticity, they are continually refracted so that their paths are curved; wave rays travel only in a straight line and are not refracted when their direction of travel is perpendicular to a boundary (Figure 7.27).

In addition to refraction, seismic rays are also **reflected,** much as light is reflected from a mirror. When seismic rays encounter a boundary separating materials of different density or elasticity, some of the rays' energy is reflected back to the surface (Figure 7.27). If we know the wave velocity and the time required for it to travel from its source to the boundary and back to the surface, we can calculate the depth of the reflecting

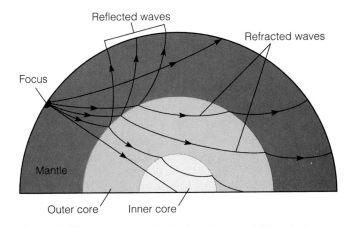

Figure 7.27 *Refraction and reflection of P-waves. When seismic waves pass through a boundary separating Earth materials of different density or elasticity, they are refracted, and some of their energy is reflected back to the surface.*

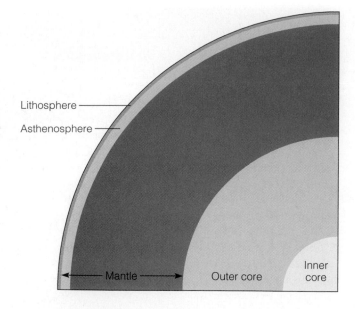

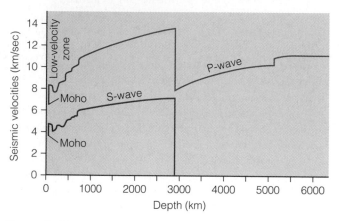

Figure 7.28 *Profiles showing seismic wave velocities versus depth. Several discontinuities are shown across which seismic wave velocities change rapidly.*

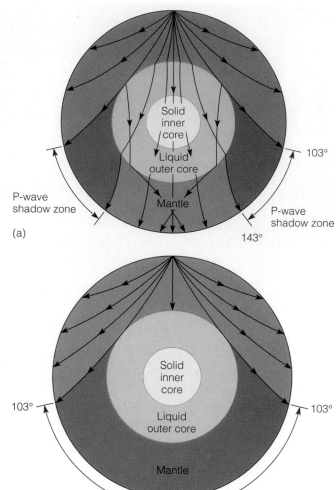

Figure 7.29 *(a) P-waves are refracted so that little P-wave energy reaches the surface in the P-wave shadow zone. (b) The presence of an S-wave shadow zone indicates that S-waves are being blocked within Earth.*

boundary. Such information is useful in determining not only the depths of the various layers but also the depths of sedimentary rocks that may contain petroleum.

Although changes in seismic wave velocity occur continuously with depth, P-wave velocity increases suddenly at the base of the crust and decreases abruptly at a depth of about 2900 km (Figure 7.28). Such marked changes in the velocity of seismic waves indicate a boundary called a **discontinuity** across which a significant change in Earth materials or their properties occurs. These discontinuities are the basis for subdividing Earth's interior into concentric layers.

The Core

In 1906, R. D. Oldham of the Geological Survey of India discovered that seismic waves arrived later than expected

at seismic stations more than 130° from an earthquake focus. He postulated the existence of a core that transmits seismic waves at a slower rate than shallower Earth materials. We now know that P-wave velocity decreases markedly at a depth of 2900 km, thus indicating a major discontinuity now recognized as the core-mantle boundary (Figure 7.28).

The sudden decrease in P-wave velocity at the core-mantle boundary causes P-waves entering the core to be refracted in such a way that little P-wave energy reaches the surface in the area between 103° and 143° from an earthquake focus (Figure 7.29a). This area in which little P-wave energy is recorded by seismographs is a **P-wave shadow zone.**

The P-wave shadow zone is not a perfect shadow zone. That is, some weak P-wave energy reaches the surface within the zone. Several hypotheses were proposed to explain this phenomenon, all of which were rejected by the Danish seismologist Inge Lehmann, who in 1936 postulated that the core is not entirely liquid. She demonstrated that reflection from a solid inner core could account for the arrival of weak P-waves in the P-wave shadow zone. Lehmann's proposal of a solid inner core was quickly accepted by seismologists.

In 1926, the British physicist Harold Jeffreys realized that S-waves were not simply slowed by the core, but were completely blocked by it. So, besides a P-wave shadow zone, a much larger and more complete **S-wave shadow zone** exists (Figure 7.29b). At locations greater than 103° from an earthquake focus, no S-waves are recorded, indicating that S-waves cannot be transmitted through the core. S-waves will not pass through a liquid, so it seems that the outer core must be liquid or behave like a liquid.

Density and Composition of the Core

The core constitutes 16.4% of Earth's volume and nearly one-third of its mass. We can estimate the core's density and composition by using seismic evidence and laboratory experiments. Furthermore, meteorites, which are thought to represent remnants of the material from which the solar system formed, can be used to make estimates of density and composition. For example, meteorites composed of iron and nickel alloys may represent the differentiated interiors of large asteroids and approximate the density and composition of Earth's core. The density of the outer core varies from 9.9 to 12.2 g/cm^3. At Earth's center, the pressure is equivalent to about 3.5 million times normal atmospheric pressure.

The core cannot be composed of the minerals most common at the surface, because even under the tremendous pressures at great depth they would still not be dense enough to yield an average density of 5.5 g/cm^3 for Earth. Both the outer and inner core are thought to be composed largely of iron, but pure iron is too dense to be the sole constituent of the outer core. Thus, it must be "diluted"

with elements of lesser density. Laboratory experiments and comparisons with iron meteorites indicate that about 12% of the outer core may consist of sulfur, and perhaps some silicon and small amounts of nickel and potassium.

In contrast, pure iron is not dense enough to account for the estimated density of the inner core. Many geologists think that perhaps 10 to 20% of the inner core also contains nickel. These metals form an iron-nickel alloy thought to be sufficiently dense under the pressure at that depth to account for the density of the inner core. When the core formed during early Earth history, it was probably molten and has since cooled to the point that its interior has crystallized. Indeed, the inner core continues to grow as Earth slowly cools and liquid of the outer core crystallizes as iron. Recent evidence also indicates that the inner core rotates faster than the outer core, moving about 20 km/yr relative to the outer core.

The Mantle

Another significant discovery about Earth's interior was made in 1909 when the Yugoslavian seismologist Andrija Mohorovičić detected a discontinuity at a depth of about 30 km. While studying arrival times of seismic waves from Balkan earthquakes, Mohorovičić noticed that seismic stations a few hundred kilometers from an earthquake's epicenter were recording two distinct sets of P- and S-waves. He reasoned that one set of waves traveled directly from the epicenter to the seismic station, whereas the other waves had penetrated a deeper layer where they were refracted (Figure 7.30).

From his observations Mohorovičić concluded that a sharp boundary separating rocks with different properties exists at a depth of about 30 km. He postulated that P-waves below this boundary travel at 8 km/sec, whereas those above the boundary travel at 6.75 km/sec. When an earthquake occurs, some waves travel directly from the focus to a seismic station, while others travel through the deeper layer and some of their energy is refracted back to the surface (Figure 7.30). Waves traveling through the

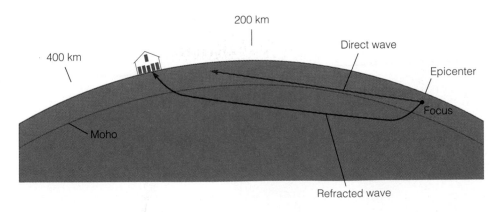

Figure 7.30 Andrija Mohorovičić studied seismic waves and detected a seismic discontinuity at a depth of about 30 km. The deeper, faster seismic waves arrive at seismic stations first, even though they travel farther. This discontinuity, now known as the Moho, is between the crust and mantle.

The Mantle

deeper layer travel farther to a seismic station but they do so more rapidly than those in the shallower layer. The boundary identified by Mohorovičić separates the crust from the mantle and is now called the **Mohorovičić discontinuity,** or simply the **Moho.** It is present everywhere except beneath spreading ridges, but its depth varies: Beneath the continents it ranges from 20 to 90 km with an average of 35 km; beneath the sea floor it is 5 to 10 km deep.

Structure and Composition of the Mantle

Although seismic wave velocity in the mantle generally increases with depth, several discontinuities also exist. Between depths of 100 and 250 km, both P- and S-wave velocities decrease markedly (Figure 7.31). This 100- to 250-km-deep layer is the **low-velocity zone;** it corresponds closely to the *asthenosphere,* a layer in which the rocks are close to their melting point and are less elastic, accounting for the observed decrease in seismic wave velocity. The asthenosphere is an important zone because it may be where some magmas are generated. Furthermore, it lacks strength, flows plastically, and is thought to be the layer over which the outer, rigid *lithosphere* moves.

Other discontinuities have been detected at deeper levels within the mantle; but unlike those between the crust and mantle or between the mantle and core, these probably represent structural changes in minerals rather than compositional changes. In other words, geologists think the mantle is composed of the same material throughout, but the structural states of minerals such as olivine change with depth. At a depth of 410 km, seismic wave velocity increases slightly as a consequence of such changes in mineral structure (Figure 7.31). Another velocity increase occurs at about 660 km where the minerals break down into metal oxides, such as FeO (iron oxide) and MgO (magnesium oxide), and silicon dioxide (SiO_2). These two discontinuities define the top and base of a *transition zone* separating the upper mantle from the lower mantle (Figure 7.31).

Although the mantle's density, which varies from 3.3 to 5.7 g/cm^3, can be inferred rather accurately from seismic waves, its composition is less certain. The igneous rock *peridotite,* containing mostly ferromagnesian silicates, is considered the most likely component. Laboratory experiments indicate that it possesses physical properties that would account for the mantle's density and observed rates of seismic wave transmissions. Peridotite also forms the lower parts of igneous rock sequences thought to be fragments of the oceanic crust and upper mantle emplaced on land. In addition, peridotite occurs as inclusions in volcanic rock bodies such as *kimberlite pipes* that are known to have come from great depths. These inclusions appear to be pieces of the mantle.

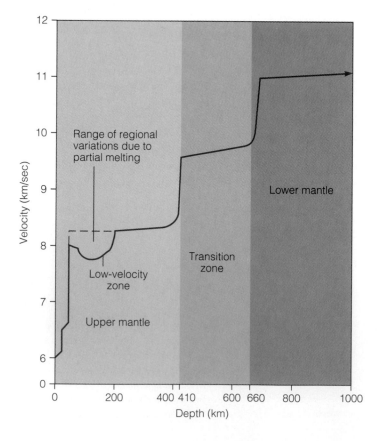

Figure 7.31 *Variations in P-wave velocity in the upper mantle and transition zone.*

Seismic Tomography

The model of Earth's interior consisting of an iron-rich core and a rocky mantle is probably accurate but is also rather imprecise. However, geophysicists have developed a technique called *seismic tomography* that allows them to develop three-dimensional models of Earth's interior. In seismic tomography numerous crossing seismic waves are analyzed in much the same way radiologists analyze CAT (computerized axial tomography) scans. In CAT scans, X-rays penetrate the body, and a two-dimensional image of the inside of a patient is formed. Repeated CAT scans, each from a slightly different angle, are computer analyzed and stacked to produce a three-dimensional picture.

In a similar fashion geophysicists use seismic waves to probe Earth's interior. From the time of arrival and distance traveled, the velocity of a seismic ray is computed at a seismic station. Only average velocity is determined, rather than variations in velocity. In seismic tomography numerous wave rays are analyzed so that "slow" and "fast" areas of wave travel can be detected (Figure 7.32). Recall that seismic wave velocity is controlled partly by elasticity; cold rocks have greater elasticity and therefore transmit seismic waves faster than hot rocks.

Using this technique, geophysicists have detected areas within the mantle at a depth of about 150 km where seismic velocities are slower than expected. These anomalously hot regions lie beneath volcanic areas and beneath mid-oceanic ridges, where convection cells of rising hot mantle rock are thought to exist. In contrast, beneath the older interior parts of continents, where tectonic activity ceased hundreds of millions or billions of years ago, anomalously cold spots are recognized. In effect, tomographic maps and three-dimensional diagrams show heat variations within Earth.

Seismic tomography has also yielded additional and sometimes surprising information about the core. For example, the core-mantle boundary is not a smooth surface, but has broad depressions and rises extending several kilometers into the mantle. Of course, the base of the mantle possesses the same features in reverse; geophysicists have termed these features *anticontinents* and *antimountains*. It appears that the surface of the core is continually deformed by sinking and rising masses of mantle material.

As a result of seismic tomography, a much clearer picture of Earth's interior is emerging. It has already given us a better understanding of complex convection within the mantle, including upwelling convection currents thought to be responsible for the movement of lithospheric plates (see Chapter 9).

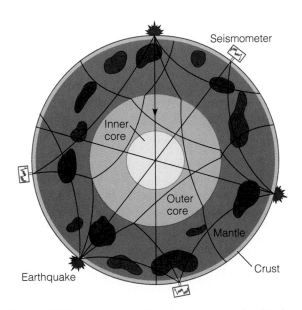

Figure 7.32 *Numerous earthquake waves are analyzed to detect areas within Earth that transmit seismic waves faster or slower than adjacent areas. Areas of fast wave travel correspond to "cold" regions (blue), whereas "hot" regions (red) transmit seismic waves more slowly.*

Earth's Internal Heat

During the nineteenth century, scientists realized that the temperature in deep mines increases with depth. The same trend has been observed in deep drill holes, but even in these we can measure temperatures directly down to a depth of only a few kilometers. The temperature increase with depth, or **geothermal gradient,** near the surface is about 25°C/km, although it varies from area to area. In areas of active or recently active volcanism, the geothermal gradient is greater than in adjacent nonvolcanic areas, and temperature rises faster beneath spreading ridges than elsewhere beneath the sea floor.

Most of Earth's internal heat is generated by radioactive decay, especially the decay of isotopes of uranium and thorium and to a lesser degree of potassium 40. When these isotopes decay, they emit energetic particles and gamma rays, which heat surrounding rocks. And because rock is such a poor conductor of heat, it takes little radioactive decay to build up considerable heat, given enough time.

Unfortunately, the geothermal gradient is not useful for estimating temperatures at great depth. If we were simply to extrapolate from the surface downward, the temperature at 100 km would be so high that despite the great pressure, all known rocks would melt. Yet except for pockets of magma, it appears that the mantle is solid rather than liquid because it transmits S-waves. Accordingly, the geothermal gradient must decrease markedly.

Current estimates of the temperature at the base of the crust are 800° to 1200°C. The latter figure seems to be an upper limit: if it were any higher, melting would be expected. Furthermore, fragments of mantle rock in igneous rocks, thought to have come from depths of about 100 to 300 km, appear to have reached equilibrium at these depths and at a temperature of about 1200°C. At the core-mantle boundary, the temperature is probably between 3500° and 5000°C; the wide spread of values indicates the uncertainties of such estimates. If these figures are reasonably accurate, the geothermal gradient in the mantle is only about 1°C/km.

Because the core is so remote and its composition so uncertain, only general estimates of its temperature can be made. Based on various experiments, the maximum temperature at the center of the core is thought to be about 6500°C, very close to the estimated temperature for the surface of the Sun!

The Crust

Our main concern in the latter part of this chapter is Earth's interior, but to be complete we must briefly discuss the crust, which along with the upper mantle constitutes the lithosphere.

Continental crust is complex, consisting of all rock types but it is usually described as granitic, meaning its overall composition is similar to granite. With the exception of metal-rich rocks such as iron ore deposits, most rocks of the continental crust have densities between 2.0 and 3.0 g/cm^3, with the average density of the crust being about 2.70 g/cm^3. P-wave velocity in continental crust is about 6.75 km/sec. Continental crust averages 35 km thick, but it is much thicker beneath mountain ranges and considerably thinner in some regions such as the Basin and Range Province of the western United States where it is being stretched and thinned. Thicker crust beneath mountain ranges is an important point that will be discussed further in Chapter 10.

In contrast to continental crust, oceanic crust is comparatively simple, consisting of gabbro in its lower part and overlain by basalt. It is thinnest, about 5 km, at spreading ridges, and nowhere is it thicker than 10 km. Its density of 3.0 g/cm^3 accounts for the fact that it transmits P-waves at about 7 km/sec. In fact, this P-wave velocity is what one would expect if oceanic crust is composed of basalt and gabbro. A more detailed description of the oceanic crust's composition and structure appears in Chapter 8.

Chapter Summary

1. Earthquakes are vibrations of Earth, caused by the sudden release of energy, usually along a fault.

2. The elastic rebound theory holds that pressure builds in rocks on opposite sides of a fault until the strength of the rocks is exceeded and rupture occurs. When the rocks rupture, stored energy is released as they snap back to their original position.

3. Seismology is the study of earthquakes. Earthquakes are recorded on seismographs, and the record of an earthquake is a seismogram.

4. The point where energy is released is an earthquake's focus, whereas its epicenter is vertically above the focus on the surface.

5. Approximately 80% of all earthquakes occur in the circum-Pacific belt, 15% within the Mediterranean-Asiatic belt, and the remaining 5% mostly in the interior of the plates or along oceanic spreading ridge systems.

6. The two types of body waves are P-waves and S-waves. Both travel through Earth, although S-waves do not travel through liquids. P-waves are the fastest and are compressional, whereas S-waves are shear. Surface waves travel along or just

below the surface. The two types of surface waves are Rayleigh and Love waves.

7. The epicenter of an earthquake is located by the use of a time-distance graph of the P- and S-waves from any given distance. Three seismographs are needed to locate the epicenter.

8. Intensity is a measure of the kind of damage done by an earthquake and is expressed by values from I to XII in the Modified Mercalli Intensity Scale.

9. Magnitude measures the amount of energy released by an earthquake and is expressed in the Richter Magnitude Scale. Each increase in the magnitude number represents about a 30-fold increase in energy released. The Seismic-Moment Magnitude Scale more accurately estimates the energy released during very large earthquakes.

10. Ground shaking is the most destructive of all earthquake hazards. The amount of damage done by an earthquake depends on the geology of the area, the type of building construction, the magnitude of the earthquake, and the duration of shaking.

11. Tsunami are seismic sea waves that are produced by earthquakes, submarine landslides, and eruptions of volcanoes at sea.

12. Seismic risk maps are helpful in making long-term predictions about the severity of earthquakes based on past occurrences.

13. Earthquake precursors include changes preceding an earthquake that can be used to predict earthquakes, such as seismic gaps, changes in surface elevations, and fluctuations in water-well levels.

14. A variety of earthquake research programs are underway in the United States, Japan, Russia, and China. Studies indicate that most people would probably not heed a short-term earthquake warning.

15. Injecting fluids into locked segments of an active fault holds some promise as a possible means of earthquake control.

16. Earth is concentrically layered into an iron-rich core with a solid inner portion and a liquid outer part, a rocky mantle, and an oceanic and continental crust.

17. Much of the information about Earth's interior has been derived from studies of P- and S-waves that travel through Earth. Laboratory experiments, comparisons with meteorites, and studies of inclusions in volcanic rocks provide additional information.

18. Density and elasticity of Earth materials determine the velocity of seismic waves. Seismic waves are refracted when their direction of travel changes. Wave reflection occurs at boundaries across which the properties of rocks change.

19. The behavior of P- and S-waves within Earth and the presence of P- and S-wave shadow zones allow geologists to estimate the density and composition of Earth's interior and to estimate the size and depth of the core and mantle.

20. Earth's inner core is thought to be composed of iron and nickel, whereas the outer core is probably composed mostly of iron with 10 to 20% sulfur and other substances in smaller quantities. Peridotite is the most likely component of the mantle.

21. Continental crust is complex, with an overall composition similar to granite. Its thickness averages 35 km. Oceanic crust has a thickness between 5 and 10 km and is composed of gabbro and basalt.

22. The geothermal gradient of 25°C/km cannot continue to great depths, otherwise most of Earth would be molten. The geothermal gradient for the mantle and core is probably about 1°C/km. The temperature at Earth's center is estimated to be 6500°C.

Important Terms

core	intensity	Mohorovičić discontinuity	Richter Magnitude Scale
crust	Love wave (L-wave)	(Moho)	seismograph
discontinuity	low-velocity zone	P-wave	seismology
earthquake	magnitude	P-wave shadow zone	S-wave
elastic rebound theory	mantle	Rayleigh wave (R-wave)	S-wave shadow zone
epicenter	Modified Mercalli Intensity	reflection	time-distance graph
focus	Scale	refraction	tsunami
geothermal gradient			

Review Questions

1. An earthquake's epicenter is
   a. _____ usually in the lower part of the mantle; b. _____ a point on the surface directly above the focus; c. _____ determined by analyzing surface wave arrival times at seismic stations; d. _____ a measure of the energy released during an earthquake; e. _____ the damage corresponding to a value of IV on the Modified Mercalli Intensity Scale.

2. With few exceptions, the most damaging earthquakes are
   a. _____ deep focus; b. _____ caused by volcanic eruptions; c. _____ those with Richter magnitudes of about 2; d. _____ shallow focus; e. _____ those occurring along spreading ridges.

3. A tsunami is a(n)
   a. _____ part of a fault with a seismic gap; b. _____ precursor to an earthquake; c. _____ seismic sea wave; d. _____ particularly large and destructive earthquake; e. _____ earthquake with a focal depth exceeding 300 km.

4. A qualitative assessment of the damage done by an earthquake is expressed by

   a. _____ intensity; b. _____ dilatancy; c. _____ seismicity; d. _____ magnitude; e. _____ liquefaction.

5. It would take about _____ earthquakes with a Richter magnitude of 3 to equal the energy released in one earthquake with a magnitude of 6.
   a. _____ 9; b. _____ 2,000,000; c. _____ 27,000; d. _____ 30; e. _____ 250

6. Most earthquakes take place in the
   a. _____ spreading ridge zone; b. _____ Mediterranean-Asiatic belt; c. _____ rifts in continental interiors; d. _____ circum-Pacific belt; e. _____ Appalachian fault zone.

7. In which area would you most likely experience an earthquake?
   a. _____ England; b. _____ Kansas; c. _____ Germany; d. _____ Florida; e. _____ Japan

8. A P-wave is one in which
   a. _____ movement is perpendicular to the direction of wave travel; b. _____ Earth's surface moves as a series of waves; c. _____ materials move forward and back along a line in the same direction that the wave moves; d. _____

large waves crash onto a shoreline following a submarine earthquake; e. _____ movement at the surface is similar to that in water waves.

9. Earth's average density is _____ g/cm^3.
   a. _____ 0.5; b. _____ 2.5; c. _____ 5.5; d. _____ 8.5; e. _____ 12.5

10. Continental crust has an overall composition corresponding closely to that of
    a._____ basalt; b. _____ sandstone; c. _____ peridotite; d. _____ gabbro; e. _____ granite.

11. The geothermal gradient is Earth's
    a. _____ capacity to reflect and refract seismic waves; b. _____ most destructive aspect of earthquakes; c. _____ temperature increase with depth; d. _____ average rate of seismic wave velocity in the mantle; e. _____ elastic rebound potential.

12. Earth's mantle is likely composed of the rock
    a. _____ basalt; b. _____ gneiss; c. _____ peridotite; d. _____ arkose; e. _____ kimberlite.

13. The seismic discontinuity at the base of the crust is known as the
    a. _____ transition zone; b. _____ magnetic reflection point; c. _____ dilatancy level; d. _____ Moho; e. _____ high-velocity zone.

14. The phenomenon in which seismic waves change their direction of travel when they move through materials having different properties is known as
    a. _____ refraction; b. _____ reflection; c. _____ deflection; d. _____ elasticity; e. _____ dispersal.

15. What are precursors and how can they be used to predict earthquakes?

16. Describe how a suspended mass can be used to detect earthquake waves.

17. What are the differences between intensity and magnitude?

18. Why are structures built on bedrock usually damaged less during an earthquake than those sited on unconsolidated material?

19. What plate tectonic settings account for earthquakes along the west coasts of North America and South America? In which of these two areas would you expect deep-focus earthquakes?

20. How does the elastic rebound theory account for energy released during an earthquake?

21. Describe the various ways earthquakes are destructive.

22. How does the density and elasticity of Earth materials affect the velocity and direction of travel of seismic waves?

23. What accounts for the various seismic discontinuities found within the mantle?

24. Describe the composition, density, and depth of Earth's core, mantle, and crust.

25. Explain what the S-wave shadow zone is and what implications it has about Earth's internal structure.

26. From the following arrival times of P- and S-waves and the graph in Figure 7.14, calculate how far away from each seismograph station the earthquake occurred. How would you determine the epicenter of this earthquake?

	Arrival Time of P-Wave	Arrival Time of S-Wave
Station A:	2:59:03 P.M.	3:04:03 P.M.
Station B:	2:51:16 P.M.	3:01:16 P.M.
Station C:	2:48:25 P.M.	2:55:55 P.M.

Distance from the earthquake:

Station A: _____

Station B: _____

Station C: _____

27. Use the graph in Figure 7.17 and the following information to determine the magnitude of this earthquake and how far away from Berkeley it originated. A seismograph in Berkeley, California recorded the arrival time of the earthquake's P-waves at 5:59:54 P.M. and the S-waves at 7:00:02 P.M. The maximum amplitude of the S-waves as recorded on the seismogram is 75 mm.

28. Discuss why insurance companies use the qualitative Modified Mercalli Intensity Scale instead of the quantitative Richter Magnitude Scale to classify earthquakes.

29. How would P- and S-waves behave if Earth were completely solid and had the same composition and density throughout?

30. What factors account for higher-than-average heat flow values at spreading ridges? How is heat flow related to the age of crustal rocks?

 World Wide Web Activities

For these web site addresses, along with current updates and exercises, log on to **http://www.brookscole.com/geo**

The Virtual Earthquake

This interactive web site, run by California State University, Los Angeles, allows you to determine the epicenter of an earthquake and to calculate its magnitude using seismic data from three seismographs. After successfully completing the exercise, you will be awarded a personalized Certificate as a "Virtual Seismologist."

U.S. Geological Survey National Earthquake Information Center

This site provides current and general seismicity information as well as links to other web sites concerned with earthquake activity. From the home page you can choose links to current earthquake information, seismograph station codes and coordinates, and information about recent earthquakes. You can also report an earthquake and click on the current world seismicity map to see the distribution of earthquakes, their depth, and their magnitude.

Seismic Monitor

This interactive educational display of global seismicity allows you to monitor earthquakes from around the world in near real time. The home page world map is updated every 30 minutes using data from the National Earthquake Information Center. By clicking on a particular location, you can view records of seismic activity in that area, see seismograms, and see all the earthquakes that have occurred in that region in the recent past.

1906 San Francisco Earthquake

This site, maintained by the U.S. Geological Survey, contains a variety of topics related to the devastating 1906 San Francisco earthquake, including eyewitness accounts, many photographs, and information about possible future earthquakes.

CD-ROM Exploration

Exploring your *Earth Systems Today* CD-ROM will add to your understanding of the material in this chapter.

Topic: Earth's Structure

Module: Earth's Layers

Explore activities in this module to see if you can discover the following for yourself:

Using activities in this module, study how P- and S-waves travel through the body of Earth. Note the difference between reflection of waves and refraction of waves.

Using activities in this module, study seismic wave propagation in a theoretical homogeneous Earth, a theoretical heterogeneous Earth, and a realistic layered Earth. Using the S-wave shadow zone, compute the depth to the core-mantle boundary.

Topic: Earth's Disasters

Module: Earthquakes and Tsunami

Explore activities in this module to see if you can discover the following for yourself:

Using activities in this module, examine the distribution of earthquakes in space (including depth) and time (i.e., 1975-present). How are earthquakes related to plate boundaries?

Using activities in this module, study those regions of Earth known for generating tsunami and those regions of Earth known to experience the effects of tsunami.

Using activities in this module, find an earthquake epicenter. In addition, observe current seismic activity in Alaska. What regions of the United States are at greatest seismic risk?

The Seafloor

Pillow lava on the Mid-Atlantic Ridge.

Headlines in supermarket tabloids and some of the more sensationalized TV productions allude to a continent that once existed in the Atlantic Ocean basin but disappeared beneath the waves during some kind of catastrophe. Of course, this is the mythical lost continent of Atlantis that most people have heard of, but few are aware of the origin of the myth or the evidence purported to support the idea of its existence. There are only two known sources of the Atlantis story, both written about 350 B.C. by the Greek philosopher Plato. In two of his philosophical dialogues, Plato tells of a large continent that, according to him, was in the Atlantic Ocean west of the Pillars of Hercules, now known as the Strait of Gibraltar (Figure 8.1). Plato was clear on the location of Atlantis, but it has been reported in the popular literature in just about every imaginable place on Earth. One tabloid even gave an account of its discovery in the North Atlantic, and one year later reported it found in the Pacific!

According to Plato's account, Atlantis was technologically advanced and extended its power over a huge area, including parts of southern Europe and through the Mediterranean as far east as Egypt. Yet despite its advantage of wealth, advanced technology, and a large army and navy, Atlantis was defeated in war by Athens. Plato wrote that following the conquest of Atlantis by Athens

> . . . there were violent earthquakes and floods and one terrible day and night came when . . . Atlantis . . . disappeared beneath the sea. And for this reason even now the sea there has become unnavigable and unsearchable, blocked as it is by the mud shallows which the island produced as it sank.*

If one assumes that the destruction of Atlantis was a real event, rather than one conjured up by Plato to make a philosophical point, he nevertheless lived long after it was supposed to have occurred. Plato tells us that Solon, an Athenian who lived about 200 years before Plato, heard the story from Egyptian priests who claimed the event had taken place 9000 years before their time. Solon told the story to his grandson, Critias, who in turn told it to Plato.

Present-day proponents of the Atlantis legend generally cite two types of evidence to support their claim that Atlantis actually existed. First, they point to supposed cultural similarities on opposite sides of the Atlantic Ocean basin, such as the similar shapes of the pyramids of Egypt and Central and South America. They contend that these similarities are due to cultural diffusion from the highly developed civilization of Atlantis. According to archaeologists, however, few similarities actually exist, and those that do can be explained as the

*From the Timaeus. Quoted in E. W. Ramage, ed., *Atlantis: Fact or Fiction?* (Bloomington: Indiana University Press, 1978), p. 13.

OBJECTIVES

At the end of this chapter, you will have learned that

- Scientists employ a variety of techniques to study the largely hidden seafloor.

- Oceanic crust is thinner and compositionally less complex than continental crust.

- The margins of continents consist of several elements, but vary depending on the type of geologic activity that occurs in these marginal areas.

- Although the seafloor is flat and featureless in some places, it also possesses ridges, trenches, seamounts, and other features.

- Geologic activities at or near divergent and convergent plate boundaries account for several distinctive seafloor features.

- Most seafloor sediment comes from weathering and erosion of continents and oceanic islands, and from the shells of tiny marine organisms.

- Organisms in warm, shallow seas build wave-resistant structures known as reefs.

- Several important resources come from seawater or from seafloor sediments and rocks.

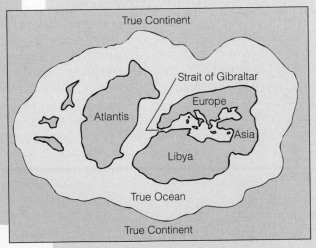

Figure 8.1 *According to Plato, Atlantis was a large continent west of the Pillars of Hercules, which we now call the Strait of Gibraltar. In the* Timaeus, *Plato noted that "[Atlantis] was larger than Libya and Asia put together and was the way to other islands and from these you might pass to the whole of the opposite continent which surrounded the true ocean. . . ."*

independent development of analogous features by different cultures.

Second, supporters of the legend assert that remnants of the sunken continent have been found. No "mud shallows" exist in the Atlantic as Plato claimed, but the Azores, Bermuda, the Bahamas, and the Mid-Atlantic Ridge are alleged to be remnants of Atlantis. If a continent had actually sunk in the Atlantic, though, it could be easily detected by a gravity survey. Continental crust has a granitic composition and a lower density than oceanic crust. Thus, if a continent were actually present beneath the Atlantic Ocean, gravity data should support this contention, but it does not. Furthermore, the crust beneath the Atlantic has been drilled in many places, and all samples recovered indicate that its composition is the same as that of oceanic crust elsewhere rather than continental crust.

In short, no geological or archaeological evidence demonstrates that Atlantis actually existed. Nevertheless, some archaeologists think that the legend might be based on a real event. About 1390 B.C. a huge volcanic eruption destroyed the island of Thera in the Mediterranean Sea—an important center of early Greek civilization (Figure 8.2). (Some researchers now think the eruption was in 1628 B.C.) The eruption was one of the most violent during historic time, and much of the island disappeared when it subsided to form a caldera. Most of the island's inhabitants

Figure 8.2 *An artist's rendition of the volcanic eruption on Thera that destroyed most of the island in about 1390 B.C. Most of the island's inhabitants escaped the devastation.*

escaped, but the eruption probably contributed to the demise of the Minoan culture on Crete. At least 10 cm of ash fell on parts of Crete, and the coastal areas of the island were probably devastated by tsunami. It is possible that Plato used an account of the destruction of Thera, but fictionalized it for his own purposes, thereby giving rise to the Atlantis legend.

Introduction

The Atlantis legend has probably persisted for as long as it has because no one, until recently, had much knowledge of what lay beneath the surface of the oceans. Indeed, the most fundamental observation we can make regarding Earth is that much of its surface is covered by water. Thus, it has vast areas largely hidden from view by oceans, and continents, which at first glance might seem to be nothing more than areas not covered by water. However, considerable differences exist between continents and ocean basins. For one thing, ocean basins are lower than continents; continental crust is thicker and less dense than oceanic crust, so it should stand higher (see Chapter 10).

In addition to their topographic differences, oceanic crust and continental crust are composed of different types of rocks. Oceanic crust is compositionally simple, made up of basalt and gabbro; it is continually generated at spreading ridges and consumed at subduction zones. It also consists of relatively young rocks, geologically speaking. No oceanic crust is more than about 180 million years old, whereas the age of continental crust varies from recent to nearly 4 billion years old. And even though we have characterized continental crust as granitic, it actually consists of all kinds of rocks.

When Earth first formed, it was probably hot and airless and had no surface water. Volcanism was no doubt more pervasive than at present because Earth possessed more internal heat. Gases derived from the interior were released during volcanic eruptions and probably resulted in the origin of the atmosphere and surface waters in a process called *outgassing*. In Chapter 4 we noted that present-day volcanoes emit a variety of gases, the most plentiful being water vapor. Erupting volcanoes surely did the same during Earth's early history, but more frequently. In any case, as Earth cooled, water vapor began condens-

Table 8.1

Numeric Data for the Oceans

Ocean*	Surface Area (million km^2)	Water Volume (million km^3)	Average Depth (km)	Maximum Depth (km)
Pacific	180	700	4.0	11.0
Atlantic	93	335	3.6	9.2
Indian	77	285	3.7	7.5
Arctic	15	17	1.1	5.2

SOURCE: P. R. Pinet, 1992. Oceanography. (St. Paul: West.)

*Excludes adjacent seas, such as the Caribbean Sea and Sea of Japan, which are marginal parts of oceans.

ing and fell as rain; this accumulated to form the surface waters.

Now an interconnected body of salt water we designate as oceans and seas covers 71% of Earth's surface. Because of this interconnection we can refer to a world ocean, but four areas are distinct enough to be recognized as the Pacific, Atlantic, Arctic, and Indian oceans (Figure 8.3). The Pacific is by far the largest, containing almost 53% of all water on Earth (Table 8.1). The term *ocean* applies to these larger areas of salt water, while *sea* designates a smaller body of water, usually a marginal part of an ocean (Figure 8.3). Sea is also used for some bodies of water completely enclosed by land, such as the Dead Sea and Caspian Sea, but these are actually saline lakes on continents.

In previous chapters, we made the point that Earth's internal heat is responsible for geologic activity such as

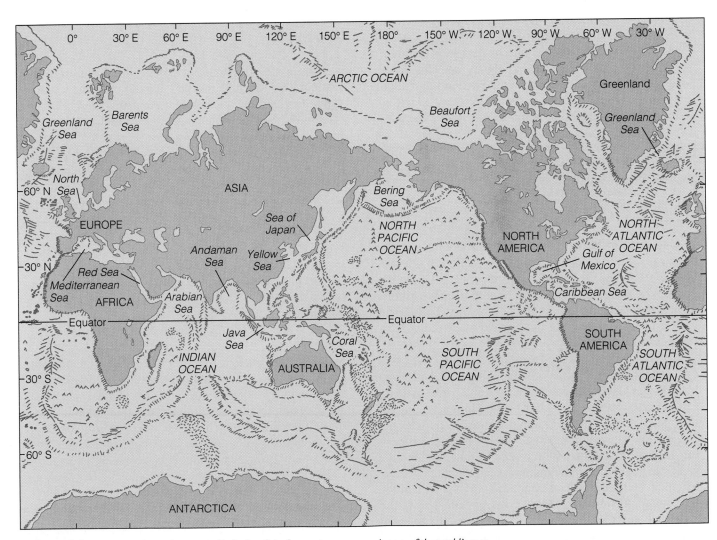

Figure 8.3 *This map shows the geographic limits of the four major oceans and many of the world's seas.*

volcanism, earthquakes, the origin of mountains, and plate movements. Likewise, it is also responsible for the differentiation of the crust into two types: oceanic and continental. Earth's earliest crust is not preserved, so we can only make some reasonable speculations on what it was like. This earliest crust might have been composed of dark colored ultramafic rock (<45% silica), which because of its density would have been subducted and thus not preserved. In any case, many geologists are convinced that plate motions coupled with subduction and collisions of island arcs formed several small continents with intervening oceanic crust at least 4 billion years ago.

Why Should You Study the Seafloor?

Why should you study the seafloor? One reason is that it constitutes the largest part of Earth's surface, although it is hidden from our direct view. And despite the commonly held misconception that the seafloor is flat and featureless, it possesses topography as varied as that of the continents. Many of these seafloor features as well as several aspects of the oceanic crust, such as its age, provide important evidence for plate tectonic theory (see Chapter 9). Several important natural resources are found in seafloor sediments or rocks, and the potential for extracting some other resources, including various metals, shows considerable promise.

Early Exploration of the Oceans

During most of historic time, people knew little of the oceans and, until recently, thought the seafloor was a vast, featureless plain. In fact, through most of this time the seafloor, in one sense, was more remote than the Moon's surface because it could not even be observed. The ancient Greeks had determined Earth's size and shape rather accurately, but western Europeans were not aware of the vastness of the oceans until the 1400s and 1500s, when explorers sought trade routes to the Indies. Even when Christopher Columbus set sail on August 3, 1492, in an effort to find a route to the Indies, he greatly underestimated the width of the Atlantic Ocean. Contrary to popular belief, he was not attempting to demonstrate Earth's spherical shape; its shape was well accepted by then. The controversy was over Earth's circumference and the shortest route to China; on these points, Columbus's critics were correct.

These and similar voyages added considerably to the growing body of knowledge about the oceans, but truly scientific investigations did not begin until the late 1700s. At that time Great Britain was the dominant maritime power,

and to maintain that dominance the British sought to increase their knowledge of the oceans. So, scientific voyages led by Captain James Cook were launched in 1768, 1772, and 1777. From 1831 until 1836, the HMS *Beagle* sailed the seas. Aboard was Charles Darwin who is well known for his views of organic evolution, but who also proposed a theory on the evolution of coral reefs. In 1872 the converted British warship HMS *Challenger* began a four-year voyage to sample seawater, determine oceanic depths, collect samples of seafloor sediment and rock, and name and classify thousands of species of marine organisms.

During these voyages many oceanic islands previously unknown to Europeans were visited. And even though exploration of the oceans was still limited, it was becoming increasingly apparent that the seafloor was not flat and featureless as formerly believed. Indeed, scientists discovered that the seafloor has varied topography just as continents do, and they recognized such features as oceanic trenches, submarine ridges, broad plateaus, hills, and vast plains.

How Are Oceans Explored Today?

Measuring the length of a weighted line lowered to the seafloor was the first method for determining oceanic depths. Now scientists use an instrument called an *echo sounder*, which detects sound waves that travel from a ship to the seafloor and back (Figure 8.4). Depth is calculated by knowing the velocity of sound in water and the time required for the waves to reach the seafloor and return to the ship, thus yielding a continuous profile of seafloor depths along the ship's route. *Seismic profiling* is similar to echo sounding but even more useful. Strong waves from an energy source reflect from the seafloor, and some of the waves penetrate seafloor layers and reflect from various horizons back to the surface (Figure 8.4). Seismic profiling is particularly useful for mapping the structure of the oceanic crust where it is buried beneath seafloor sediments.

The Deep Sea Drilling Project, an international program sponsored by several oceanographic institutions, began in 1968. Its first research vessel, the *Glomar Challenger*, could drill in water more than 6000 m deep and recover long cores of seafloor sediment and oceanic crust. The *Glomar Challenger* drilled more than 1000 holes in the seafloor during the 15 years of the program. The Deep Sea Drilling Project ended in 1983, but beginning in 1985 the Ocean Drilling Program with its research vessel the JOIDES* *Resolution* continued exploring the seafloor (Figure 8.5a). Research vessels also sample the seafloor using *clamshell samplers* and *piston corers* (Figure 8.6).

*JOIDES is an acronym for Joint Oceanographic Institutions for Deep Earth Sampling.

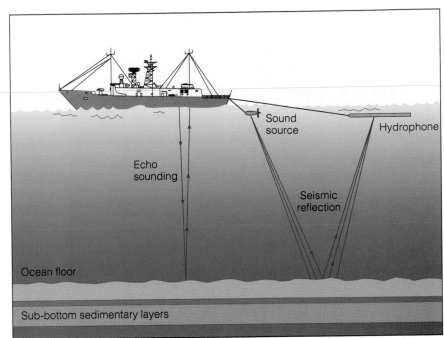

Figure 8.4 *Diagram showing how echo sounding and seismic profiling are used to study the seafloor. Some of the energy generated at the energy source is reflected from various horizons back to the surface where it is detected by hydrophones.*

Sound source

Hydrophone

Echo sounding

Seismic reflection

Ocean floor

Sub-bottom sedimentary layers

(b)

(a)

Figure 8.5 *Research vessels. (a) The* JOIDES *Resolution is capable of drilling the deep seafloor. (b) The submersible* Alvin *is used for observations and sampling of the deep seafloor.*

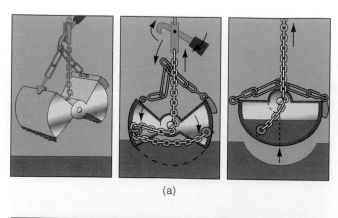

(a)

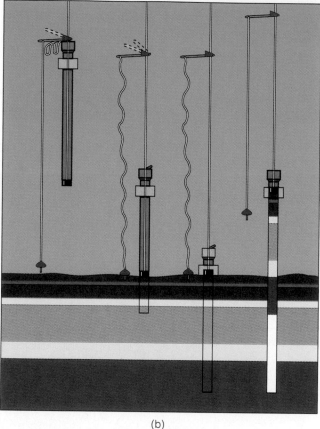

(b)

Figure 8.6 *Sampling the seafloor. (a) A clamshell sampler taking a seafloor sample. (b) A piston corer falls to the seafloor, penetrates the sediment, and then is retrieved.*

In addition to surface vessels, submersibles are now important vehicles for seafloor exploration. Some, such as the *Argo,* are remotely controlled and towed by a surface vessel. In 1985, the *Argo* equipped with sonar and television systems provided the first views of the British ocean liner HMS *Titanic* since it sank in 1912. The U.S. Geological Survey uses a towed device with sonar to produce seafloor images resembling aerial photographs. Scientists aboard submersibles such as *Alvin* (Figure 8.5b) have de-

scended to the seafloor in many areas to make observations and collect samples.

Scientific investigations have yielded important information about the oceans for more than 200 years, but much of our current knowledge has been acquired since World War II (1939–1945). This is particularly true of the seafloor, because only in recent decades has instrumentation been available to study this largely hidden domain. The data amassed during this time are interesting in their own right, but they also provide some of the most compelling evidence to date in support of plate tectonic theory (see Chapter 9).

Oceanic Crust—Its Structure and Composition

We have already mentioned that oceanic crust is composed of basalt and gabbro and is continually generated at spreading ridges. Of course, drilling into the oceanic crust gives some details about its composition and structure, but it has never been completely penetrated and sampled. So how do we know what it is composed of and how it varies with depth? Actually, even before it was sampled and observed, these details about the oceanic crust were known.

Remember that oceanic crust is consumed at subduction zones and thus most of it is recycled, but a small amount is found in mountain ranges on land where it was emplaced by moving along large fractures called thrust faults (faults are discussed more fully in Chapter 10). These preserved slivers of oceanic crust along with part of the underlying upper mantle are known as **ophiolites.** Detailed studies reveal that an ideal ophiolite consists of deep-sea sedimentary rocks underlain by rocks of the upper oceanic crust, especially pillow lava and sheet lava flows (Figure 8.7). Proceeding downward in an ophiolite is a sheeted dike complex, consisting of vertical basaltic dikes, then massive gabbro and layered gabbro that appears to have formed in the upper part of a magma chamber. And finally, the lowermost unit is peridotite representing the upper mantle; this is sometimes altered by metamorphism to a greenish rock known as serpentinite. Thus a complete ophiolite consists of deep-sea sedimentary rocks underlain by rocks of the oceanic crust and upper mantle (Figure 8.7).

Sampling and drilling at oceanic ridges reveal that oceanic crust is indeed made up of pillow lava and sheet lava flows underlain by a sheeted dike complex, just as predicted from studies of ophiolites. But it was not until 1989 that a submersible carrying scientists descended to the walls of a seafloor fracture in the North Atlantic and verified what lay below the sheeted dike complex. Just as expected from ophiolites on land, the lower oceanic crust consists of gabbro and the upper mantle is made up of peridotite. In short, the composition and structure of oceanic crust and upper mantle were inferred from rocks

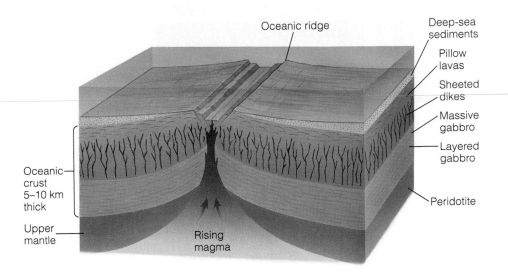

Figure 8.7 *Oceanic crust consisting of the layers shown here forms as magma rises beneath oceanic ridges. Fragments of oceanic crust and upper mantle on land are known as ophiolites.*

on land and then these inferences were verified by sampling, drilling, and observing seafloor rocks.

What Are Continental Margins?

In the Introduction we made the point that continents are not simply areas above sea level, although most people perceive of continents as land areas outlined by the seas. However, the true geologic margin of a continent—that is where granitic continental crust changes to oceanic crust composed of basalt and gabbro—is below sea level. Accordingly, the margins of continents are submerged, and we recognize **continental margins** as separating the part of a continent above sea level from the deep seafloor.

A continental margin is made up of a gently sloping continental shelf, a more steeply inclined continental slope, and in some cases, a deeper, gently sloping continental rise (Figure 8.8). Seaward of the continental margin lies the deep ocean basin. Thus the continental margins extend to increasingly greater depths until they merge with the deep seafloor. The change from continental crust to oceanic crust generally takes places somewhere beneath the continental rise, so part of the continental slope and the continental rise actually rest on oceanic crust.

The Continental Shelf

As one proceeds seaward from the shoreline across the continental margin, the first area encountered is a gently sloping **continental shelf** lying between the shore and the more steeply dipping continental slope (Figure 8.8).

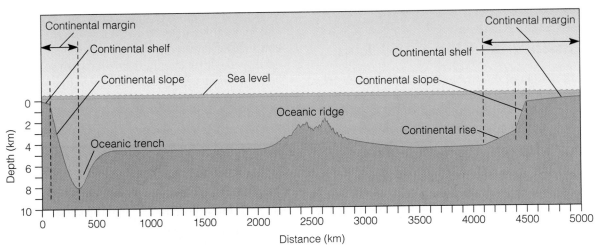

Figure 8.8 *A generalized profile of the seafloor showing features of the continental margins. The vertical dimensions of the features in this profile are greatly exaggerated because the vertical and horizontal scales differ.*

What Are Continental Margins?

The width of the continental shelf varies considerably, ranging from a few tens of meters to more than 1000 km; the shelf terminates where the inclination of the seafloor increases abruptly from 1 degree or less to several degrees. The outer margin of the continental shelf, or simply the *shelf-slope break*, is at an average depth of 135 m, so by oceanic standards the continental shelves are covered by shallow water.

At times during the Pleistocene Epoch (1.6 million to 10,000 years ago), sea level was as much as 130 m lower than it is now. As a result, the continental shelves were above sea level and were areas of stream channel and floodplain deposition. In addition, in many parts of northern Europe and North America, glaciers extended well out onto the continental shelves and deposited gravel, sand, and mud. Since the Pleistocene ended, sea level has risen, submerging these deposits, which are now being reworked by marine processes. Evidence that these sediments were in fact deposited on land includes remains of human settlements and fossils of a variety of land-dwelling animals (see Chapter 24).

The Continental Slope and Rise

The seaward margin of the continental shelf is marked by the *shelf-slope break* (at an average depth of 135 m) where the more steeply inclined **continental slope** begins (Figure 8.8). In most areas around the margins of the Atlantic, the continental slope merges with a more gently sloping **continental rise.** This rise is generally absent around the margins of the Pacific where continental slopes commonly descend directly into an oceanic trench (Figure 8.8).

The shelf-slope break, marking the boundary between the shelf and slope, is an important feature in terms of sediment transport and deposition. Landward of the break— that is, on the shelf—sediments are affected by waves and tidal currents, but these processes have no effect on sedi-ment that is seaward of the break, where gravity is responsible for their transport and deposition on the slope and rise. In fact, much of the land-derived sediment crosses the shelves and is eventually deposited on the continental slopes and rises, where more than 70% of all sediments in the oceans are found. Much of this sediment is transported through submarine canyons by turbidity currents.

Submarine Canyons, Turbidity Currents, and Submarine Fans

In Chapter 5, we discussed the origin of graded bedding, most of which results from **turbidity currents,** underwater flows of sediment-water mixtures with densities greater than sediment-free water. As a turbidity current flows onto the relatively flat seafloor, it slows and begins depositing sediment, the largest particles first, followed by progressively smaller particles, thus forming a layer with graded bedding (see Figure 5.26). Deposition by turbidity currents results in the origin of a series of overlapping **submarine fans,** which constitute a large part of the continental rise (Figure 8.9). Submarine fans are distinctive features, but their outer margins are difficult to discern because they grade into deposits of the deep-ocean basin.

No one has ever observed a turbidity current in progress in the oceans, so for many years some doubted their existence. However, evidence now available removes all doubt. In 1971, for instance, abnormally turbid water was sampled just above the seafloor in the North Atlantic, indicating that a turbidity current had recently occurred. In addition, seafloor samples from many areas show a succession of layers with graded bedding and the remains of shallow-water organisms that were apparently displaced into deeper water by turbidity currents.

Perhaps the most compelling evidence for the existence of turbidity currents is the pattern of trans-Atlantic cable breaks that took place in the North Atlantic near Newfoundland on November 18, 1929. Initially, an earthquake was assumed to have ruptured several telephone and telegraph cables. However, while the breaks on the continental shelf near the epicenter occurred when the earthquake struck, cables farther seaward were broken later and in succession. The last cable to break was 720 km from the source of the earthquake, and it did not snap until 13 hours after the first break. In 1949, geologists realized that an earthquake-generated turbidity current had moved downslope, breaking the cables in succession. The precise time at which each cable broke was known, so calculating the velocity of the turbidity current was simple. It moved at about 80 km/hr on the continental slope, but slowed to about 27 km/hr when it reached the continental rise.

Deep, steep-sided **submarine canyons** are present on continental shelves, but they are best developed on continental slopes (Figure 8.9). Some submarine canyons can be traced across the shelf to rivers on land; they ap-

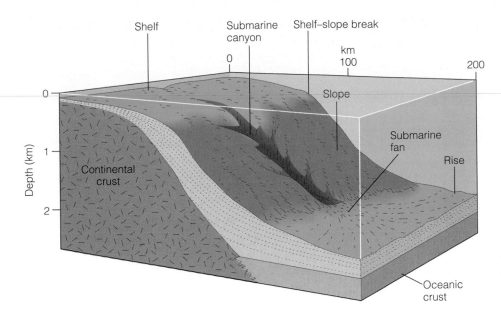

Figure 8.9 *Submarine fans form by deposition of sediments carried down submarine canyons by turbidity currents. Much of the continental rise is composed of overlapping submarine fans.*

parently formed as river valleys when sea level was lower during the Pleistocene. However, many have no such association, and some extend far deeper than can be accounted for by river erosion during times of lower sea level. Scientists know that strong currents move through submarine canyons and perhaps play some role in their origin. Furthermore, turbidity currents periodically move through these canyons and are now thought to be the primary agent responsible for their erosion.

Types of Continental Margins

Continental margins are characterized as either *active* or *passive*, depending on their relationship to plate boundaries. An **active continental margin** develops at the leading edge of a continental plate where oceanic lithosphere is subducted (Figure 8.10). The western margin of South America is a good example. Here, an oceanic plate is subducted beneath the continent, resulting in seismic activity, a geologically young mountain range, and active volcanism. In addition, the continental shelf is narrow, and the continental slope descends directly into the Peru-Chile Trench, so sediment is dumped into the trench and no continental rise develops. The western margin of North America is also considered an active continental margin, although much of it is now bounded by transform faults rather than a subduction zone. However, plate convergence and subduction still take place in the Pacific Northwest along the continental margins of northern California, Oregon, and Washington.

The continental margins of eastern North America and South America differ considerably from their western mar-

gins. For one thing, they possess broad continental shelves as well as a continental slope and rise; also, vast, flat *abyssal plains* are commonly present adjacent to the rises (Figure 8.10). Furthermore, these **passive continental margins** are within a plate rather than at a plate boundary, and they lack the typical volcanic and seismic activity found at active continental margins. Nevertheless, earthquakes do take place occasionally at these margins—the Charleston, South Carolina, earthquake of 1886, for instance (see Figure 7.10).

Active and passive continental margins share some features, but they are notably different in the widths of their continental shelves, and active margins have an oceanic trench but no continental rise. Why the differences? At both types of continental margins, turbidity currents transport sediment into deeper water. At passive margins, the sediment forms a series of overlapping submarine fans and thus develops a continental rise, whereas at an active margin, sediment is simply dumped into the trench and no rise forms. The proximity of a trench to a continent also explains why the continental shelves of active margins are so narrow. In contrast, land-derived sedimentary deposits at passive margins have built a broad platform extending far out into the ocean.

It should be clear from the preceding discussion that continental margins are active or passive depending on their location with respect to a plate boundary. However, just as Earth as a whole has evolved, so have continental margins, and one type can change to another. For instance, eastern North America now has a passive continental margin, but during the Paleozoic Era it was bounded by an active continental margin (see Chapter 21). A change such as this depends on the stage of development of an ocean basin.

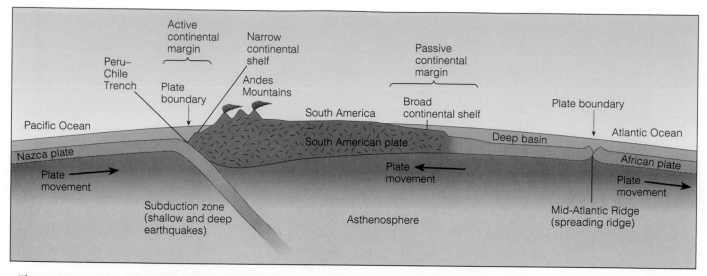

Figure 8.10 *Active and passive continental margins along the west and east coasts of South America. Notice that passive continental margins are much wider than active ones. Seafloor sediment is not shown.*

What Features Are Found in the Deep-Ocean Basins?

Most of the seafloor, with an average depth of 3.8 km, lies far below the depth of sunlight penetration, which is generally less than 100 m. Accordingly, most of the seafloor is completely dark, no plant life exists, the temperature is generally just above freezing, and the pressure varies from 200 to more than 1000 atmospheres depending on depth. In fact, biologic productivity is low on the deep seafloor with the exception of hydrothermal vent communities (see Perspective 8.1).

Scientists have descended to the greatest oceanic depths, submarine ridges, and elsewhere in submersibles, so some of the seafloor has been observed. Nevertheless, much of the seafloor has been studied only by echo sounding, seismic profiling, and sampling of seafloor sediments and oceanic crust, and remote devices have explored oceanic depths in excess of 11,000 m. Oceanographers are developing a more thorough understanding of the oceans and the deep seafloor, and they now know of many deep ocean features such as vast plains, trenches, and ridges.

Abyssal Plains

Beyond the continental rises of passive continental margins are **abyssal plains,** flat surfaces covering vast areas of the seafloor. In some areas they are interrupted by peaks rising more than 1 km, but in general abyssal plains are the flattest, most featureless areas on Earth (Figure 8.11). Their flatness is a result of sediment deposition covering the rugged topography of the oceanic crust.

Abyssal plains are invariably found adjacent to the continental rises, which are composed mostly of overlapping submarine fans that owe their origin to deposition by turbidity currents. Along active continental margins, sediments derived from the shelf and slope are trapped in an oceanic trench, and abyssal plains fail to develop. Accordingly, abyssal plains are common in the Atlantic Ocean basin, but rare in the Pacific Ocean basin (Figure 8.11).

Oceanic Trenches

Long, steep-sided depressions on the seafloor near convergent plate boundaries, or simply **oceanic trenches,** constitute no more than 2% of the seafloor, but they are important features because it is here that lithospheric plates are consumed by subduction (see Chapter 9). Because oceanic trenches are found along active continental margins, they are common in the Pacific Ocean basin but largely lacking in the Atlantic, those in the Caribbean being notable exceptions (Figure 8.11). On the landward sides of oceanic trenches, the continental slope descends into them at up to 25 degrees, and many have thick accumulations of sediments. The greatest oceanic depths are found in these trenches; the Challenger Deep of the Marianas Trench in the Pacific is more than 11,000 m deep!

Sensitive instruments can detect the amount of heat energy escaping from Earth's interior by the phenomenon of *heat flow.* As one might expect, heat flow is greatest in areas of active or recently active volcanism. For instance, higher-than-average heat flow takes place at spreading ridges, but at subduction zones heat flow values are less than the average for Earth as a whole. It seems that oceanic crust at oceanic trenches is cooler and slightly denser than elsewhere.

Seismic activity also takes place at or near oceanic trenches along planes dipping at about 45 degrees. In Chapter 7 we discussed these inclined seismic zones called Benioff zones (see Figure 7.8), where most of Earth's intermediate and deep earthquakes occur. Volcanism does not take place in trenches, but because these are zones where oceanic lithosphere is subducted beneath either oceanic or continental lithosphere, an arcuate chain of volcanoes is found on the overriding plate (Figure 8.10). The Aleutian Islands and the volcanoes along the western margin of South America are good examples of such a chain.

Oceanic Ridges

When the first submarine cable was laid between North America and Europe during the late 1800s, a feature called the Telegraph Plateau was discovered in the North Atlantic. Using data from the 1925–27 voyage of the German research vessel *Meteor,* scientists proposed that the plateau was actually a continuous ridge extending the length of the Atlantic Ocean basin. Subsequent investigations revealed that this conjecture was correct, and we now call this feature the Mid-Atlantic Ridge (Figure 8.11).

We now know that the Mid-Atlantic Ridge is more than 2000 km wide and rises 2 to 2.5 km above the adjacent seafloor. Furthermore, it is part of a much larger **oceanic ridge** system of mostly submarine mountainous topography. This system runs from the Arctic Ocean through the middle of the Atlantic and curves around South Africa, where the Indian Ridge continues into the Indian Ocean; the Atlantic-Pacific Ridge extends eastward and a branch of this, the East Pacific Rise, trends northeast until it reaches the Gulf of California (Figure 8.11). The entire system is at least 65,000 km long, far surpassing the length of any mountain system on land. Oceanic ridges are composed almost entirely of basalt and gabbro and possess features produced by tensional forces. Mountain ranges on land, in contrast, consist of igneous, metamorphic, and sedimentary rocks; these ranges formed when rocks were folded and fractured by compressive forces (see Chapter 10).

Oceanic ridges are mostly below sea level, but they rise above the sea in some places such as Iceland, the Azores, and Easter Island. Of course, oceanic ridges are the sites where new oceanic crust is generated and plates diverge (see Chapter 9). The rate of plate divergence is important because it determines the cross-sectional profile of a ridge. For example, the Mid-Atlantic Ridge has a comparatively steep profile because divergence is slow, allowing the new oceanic crust to cool,

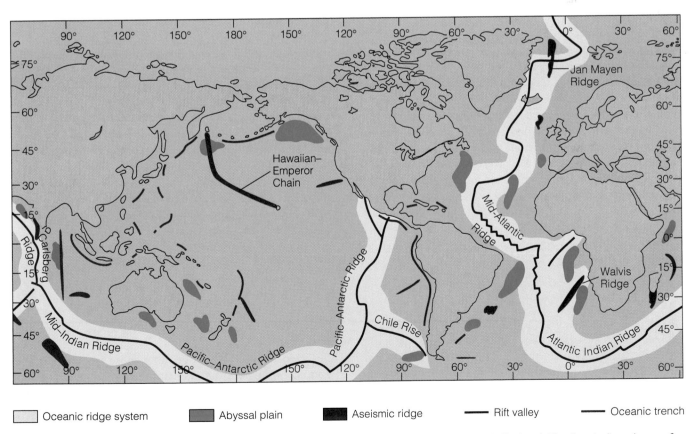

| | Oceanic ridge system | | Abyssal plain | | Aseismic ridge | — Rift valley | — Oceanic trench |

Figure 8.11 *The distribution of oceanic trenches (brown), abyssal plains (green), the oceanic ridge system (yellow) and rift valleys (red), and some of the aseismic ridges (blue).*

What Features Are Found in the Deep-Ocean Basins?

Submarine Hydrothermal Vents

One prediction of plate tectonic theory is that water seeping into seafloor cracks and fissures at or near spreading ridges should be heated and then rise to the surface as hot springs. And indeed such *submarine hydrothermal vents* have been detected or observed at many locations. As early as 1960, hot metal-rich brines derived from seafloor hot springs were detected in the Red Sea. During the 1970s, researchers observed hydrothermal vents on the Mid-Atlantic Ridge about 2900 km east of Miami, Florida, and in 1978 minerals deposited by similar features were sampled from the East Pacific Rise just east of the Gulf of California.

In 1979, the submersible *Alvin* carrying scientists descended about 2900 m to the Galapagos Rift in the Pacific Ocean where they saw chimneylike structures discharging plumes of hot, black water (Figure 1a). These particular submarine hydrothermal vents are now called *black smokers*. The plumes of water contain dissolved minerals that precipitate as the water cools and forms chimneys rising as much as 20 m above the seafloor. When a black smoker ceases activity, the chimney collapses and forms a moundlike accumulation of minerals. Since 1979, scientists have observed submarine hydrothermal vents in several other areas.

Submarine hydrothermal vents are interesting for several reasons. For one thing, near the vents are communities of organisms including bacteria, crabs, mussels, starfish, and tubeworms, many of which had never been seen before (Figure 1b). The life forms in most biological communities are dependent on energy provided by sunlight for photosynthesizing organisms that lie at the base of a food chain. These provide nutrients for herbivores (plant eaters), which in turn are eaten by carnivores and scavengers. In any case, the ultimate source of energy is sunlight. In vent communities, though, no sunlight is available for photosynthesis. Instead, bacteria lie at the base of this food chain. As opposed to photosynthesis, these bacteria practice a metabolic process called chemosynthesis, in which they oxidize sulfur compounds from the hot vent waters, thus providing their own nutrients and the nutrients for other members of the food chain.

Submarine hydrothermal vents are also interesting because of their economic potential. Seawater circulating down into cracks is heated to perhaps 400°C, and it becomes a metal-bearing solution as it reacts with the oceanic crust. When this hot water discharges onto the seafloor, it cools, thereby reducing its ability to keep minerals in solution, so iron, copper, and zinc sulfides precipitate along with several other minerals. The accumulating minerals form a chimney, which apparently grows quite rapidly. After a 10-m high chimney was accidentally knocked over in 1991, it grew about 6 m by the time it was revisited three months later. When activity ceases at a chimney, it eventually collapses and is incorporated into a large moundlike mineral deposit.

Submarine hydrothermal vent deposits have formed throughout geologic time and continue to do so. None on the seafloor are currently being mined, but the technology to exploit some of them exists. And some deposits are huge. For instance, the deposits in the Atlantis II Deep in the Red Sea contain an estimated 100 million tons of metals, including iron, copper, zinc, silver, and gold. As resources on land diminish some of these deposits will no doubt attract more attention than they do now.

And finally, submarine hydrothermal vents provide another example of interactions among Earth's systems. Heat derived from within the planet drives the system in which superheated, metal-bearing brines discharge into seawater, and gases are released that eventually make their way into the atmosphere. Of course, the hot vent waters contain sulfur compounds oxidized by bacteria that support a complex food chain and numerous organisms. In fact, some scientists think that the first self-replicating cells originated near these vents. It is interesting to note that amino acids, the building blocks of more complex molecules typical of organisms, have been detected in vent waters. This hypothesis, however, is controversial and not accepted by all scientists familiar with such matters.

shrink, and subside closer to the ridge crest than it does in areas of faster divergence such as the East Pacific Rise. A ridge may also have a rift along its crest that opens in response to tension (Figure 8.12). A rift is particularly obvious along the Mid-Atlantic Ridge, but appears to be absent along parts of the East Pacific rise. These rifts are commonly 1 to 2 km deep and several kilometers wide. They open as seafloor spreading takes place (discussed in Chapter 9) and are characterized by shallow-focus earthquakes, basaltic volcanism, and high heat flow.

Even though most oceanographic research is still done by echo sounding, seismic profiling, and seafloor sampling, scientists have been making direct observations of oceanic ridges and their rifts since 1974. As part of Project FAMOUS (French-American Mid-Ocean Undersea Study), submersibles have descended to the ridges and into their rifts in several areas. No active vol-

Figure 1 *(a) A hydrothermal vent known as a black smoker on the seafloor. Seawater seeps down through the oceanic crust, becomes heated, and then rises and builds chimneys composed of anhydrite ($CaSO_4$) and sulfides of iron, copper, and zinc. The plume of "black smoke" is simply heated water saturated with dissolved minerals. (b) Several types of organisms including these tubeworms live near black smokers.*

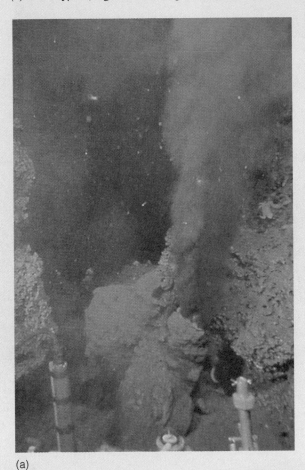

(a)

(b)

canism has been observed, but researchers did see pillow lavas (see Figure 4.7a), lava tubes, and sheet lava flows, some of which seemed to have formed very recently. In fact, on return visits to a site they have seen the effects of volcanism that occurred since their previous visit. And on January 25, 1998, a submarine volcano began erupting along the Juan de Fuca Ridge west of Oregon. Researchers aboard submersibles have also observed hot water being discharged from the seafloor at or near ridges in submarine hydrothermal vents (see Perspective 8.1).

Fractures in the Seafloor

Oceanic ridges are not continuous features winding without interruption around the globe. They abruptly terminate where they are offset along major fractures oriented more or less at right angles to ridge axes (Figure 8.13).

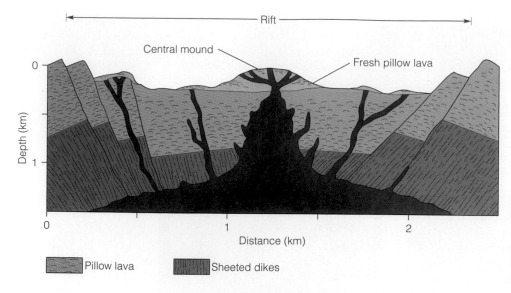

Figure 8.12 *Studies during the early 1970s revealed that recent moundlike accumulations of volcanic rocks, mostly basaltic pillow lavas, occurred in a rift along the axis of the Mid-Atlantic Ridge.*

Such large-scale fractures run for hundreds of kilometers, although they are difficult to trace where they are buried beneath seafloor sediments. Many geologists are convinced that some geologic features on the continents can best be accounted for by the extension of these fractures into continents.

Shallow-focus earthquakes take place along these fractures, but only between the displaced ridge segments. Furthermore, because ridges are higher than the adjacent seafloor, the offset segments yield nearly vertical escarpments 2 or 3 km high (Figure 8.13). The reason oceanic ridges have so many fractures is that plate divergence takes place irregularly on a sphere, resulting in stresses that cause fracturing. We will have more to say about these fractures between offset ridge segments in Chapter 9 where they are termed *transform faults*.

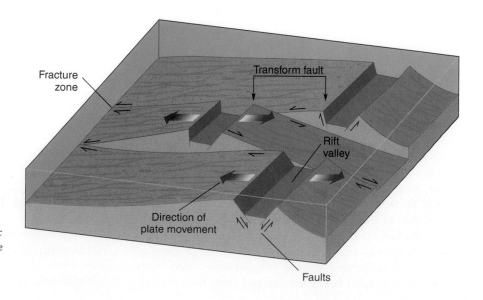

Figure 8.13 *Diagrammatic view of an oceanic ridge offset along fractures. That part of a fracture between displaced segments of the ridge crest is known as a transform fault (see Chapter 9).*

Seamounts, Guyots, and Aseismic Ridges

As noted previously, the seafloor is not a flat, featureless plain, except for the abyssal plains, and even these are underlain by rugged topography. In fact, a large number of volcanic hills, seamounts, and guyots rise above the seafloor in all ocean basins, but they are particularly abundant in the Pacific. All are of volcanic origin and differ from one another mostly in size. **Seamounts** rise more than one kilometer above the seafloor, and if flat topped, they are called **guyots** (Figure 8.14). Guyots are volcanoes that originally extended above sea level. However, as the plate upon which they were situated continued to move, they were carried away from a spreading ridge, and as the oceanic crust cooled it descended to greater depths. Thus, what was once an island slowly sank beneath the sea, and as it did, wave erosion produced the typical flat-topped appearance (Figure 8.14). Many other volcanic features smaller than seamounts exist on the seafloor, but they probably originated in the same way. These so-called *abyssal hills* average only about 250 m high.

Other common features in the ocean basins are long, narrow ridges and broad plateaulike features rising as much as 2 to 3 km above the surrounding seafloor. These **aseismic ridges** are so-called because they lack seismic activity. A few of these ridges are thought to be small fragments separated from continents during rifting and are referred to as *microcontinents*. The Jan Mayen Ridge in the North Atlantic is probably a microcontinent (Figure 8.11).

Most aseismic ridges form as a linear succession of hot spot volcanoes. These may develop at or near an oceanic ridge, but each volcano so formed is carried laterally with the plate upon which it originated. The net result of such activity is a line of seamounts/guyots extending from an oceanic ridge (Figure 8.14); the Walvis Ridge in the South Atlantic is a good example (Figure 8.11). Aseismic ridges also form over hot spots unrelated to ridges. The Hawaiian-Emperor chain in the Pacific formed in such a manner (Figure 8.11).

Sedimentation and Sediments on the Deep Seafloor

Sediments on the deep seafloor are mostly fine grained—that is, composed of silt- and clay-sized particles—because few processes can transport sand and gravel very far from land. Certainly icebergs can carry sand and gravel, and, in fact, a broad band of glacial-marine sediment is adjacent to Antarctica and Greenland (Figure 8.15). Floating vegetation might also carry large particles far out to sea, but it contributes very little sediment to the deep seafloor.

Most of the fine-grained sediment on the deep seafloor is derived from (1) windblown dust and volcanic ash from the continents and volcanic islands, and (2) the shells of microscopic plants and animals that live in the near-surface oceanic waters. Minor sources are chemical reactions in seawater that yield manganese nodules found in all ocean basins (Figure 8.16) and cosmic dust. Researchers think that as many as 40,000 metric tons of cosmic dust fall to Earth each year, but this is a trivial quantity compared to the volume of sediment derived from the two primary sources.

Most sediment on the deep seafloor is *pelagic*, meaning that it settled from suspension far from land. Pelagic sediment is further characterized as pelagic clay and ooze. **Pelagic clay** is generally brown or red, and, as its name implies, is composed of clay-sized particles from

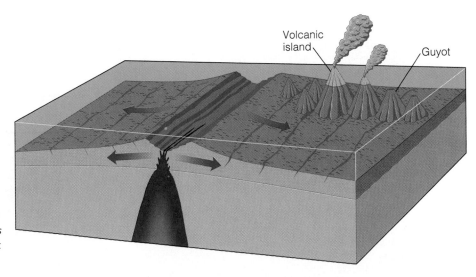

Figure 8.14 *Submarine volcanoes may build up above sea level to form seamounts. As the plate upon which these volcanoes rest moves away from a spreading ridge, the volcanoes sink beneath sea level and become guyots.*

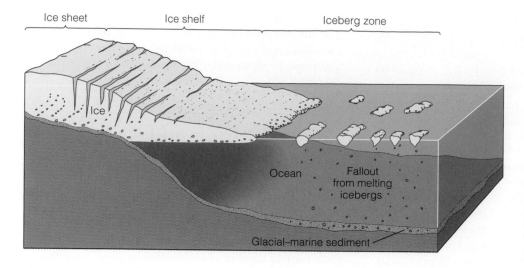

Figure 8.15 *The origin of glacial-marine sediments by ice rafting, a process in which icebergs transport sediment into the ocean basins, releasing it as they melt.*

(a)

(b)

Figure 8.16 *(a) Manganese nodules on the seafloor. (b) Sectioned manganese nodule showing its internal structure. This nodule measures about 6 cm across.*

the continents or oceanic islands. **Ooze** in contrast, is made up mostly of tiny shells of marine organisms. *Calcareous ooze* consists primarily of calcium carbonate ($CaCO_3$) skeletons of marine organisms such as foraminifera, and *siliceous ooze* is composed of the silica (SiO_2) skeletons of such single-celled organisms as radiolarians (animals) and diatoms (plants).

Reefs—Rocks Made by Organisms

The term **reef** has a variety of meanings such as shallowly submerged rocks that pose a hazard to navigation, but here we restrict it to mean a moundlike, wave-resistant structure composed of the skeletons of marine organisms (Figure 8.17). Although commonly called coral reefs, they actually have a solid framework composed of skeletons of corals and various mollusks such as clams, and encrusting organisms including sponges and algae. Reefs grow to a depth of 45 or 50 m and are restricted to shallow, tropical seas where the water is clear and its temperature does not fall below about 20°C. The depth to which reefs grow depends on sunlight penetration because many of the corals rely on symbiotic algae that must have sunlight for energy.

Reefs of many shapes are known, but most can be classified as one of three basic varieties: fringing, barrier, and atoll (Figure 8.18). *Fringing reefs* are solidly attached to the margins of an island or continent. They have a rough, tablelike surface, are as much as one kilometer wide, and on their seaward side slope steeply down to the seafloor. *Barrier reefs*

are similar to fringing reefs, except that they are separated from the mainland by a lagoon (Figure 8.19a). The best-known barrier reef in the world is the 2000-km-long Great Barrier Reef of Australia.

Circular to oval reefs surrounding a lagoon are known as *atolls*. They form around volcanic islands that subside below sea level as the plate on which they rest is carried progressively farther from an oceanic ridge (Figure 8.19b). As subsidence proceeds, the reef organisms construct the reef upward so that the living part of the reef remains in shallow water. However, the island eventually subsides below sea level, leaving a circular lagoon surrounded by a more or less continuous reef. Atolls are particularly common in the western Pacific Ocean basin. Many of these began as fringing reefs, but as the plate they were on was carried into deeper water they evolved first to barrier reefs and finally to atolls (Figure 8.18).

Figure 8.17 *Reefs such as this one in the Red Sea are wave-resistant structures composed of the skeletons of organisms.*

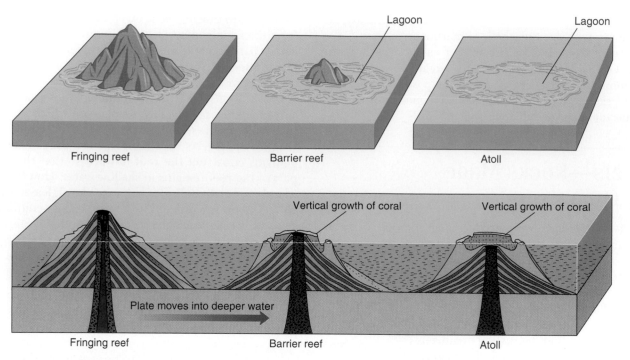

Figure 8.18 *The three types of coral reefs. Fringing reefs grow around the perimeter of an island, but as the island resting on a moving plate is carried into deeper water the reef becomes separated from the island by a lagoon, thus forming a barrier reef. An atoll forms as the island disappears beneath the sea and the reef encircles a lagoon.*

What Resources Come from Seawater and the Seafloor?

Seawater contains many elements in solution, some of which are extracted for various industrial and domestic uses. Sodium chloride (table salt) is produced by the evaporation of seawater, and a large proportion of the world's magnesium is derived from seawater. Numerous other elements and compounds can be extracted from seawater, but for many, such as gold, the cost is prohibitive.

In addition to substances in seawater, deposits on the seafloor or within seafloor sediments are becoming increasingly important. Many of these potential resources lie well beyond continental margins, so their ownership is a political and legal problem that has not yet been resolved.

Most nations bordering the ocean claim those resources within their adjacent continental margin. The United States by a presidential proclamation issued on March 10, 1983, claims sovereign rights over an area designed as the **Exclusive Economic Zone (EEZ)** (Figure 8.20). The EEZ extends seaward 200 nautical miles (371 km) from the coast and includes areas adjacent to U.S. territories such as Guam, American Samoa, Wake Island, and Puerto Rico. In short, the United States claims rights to all resources within an area about 1.7 times larger than its land area. A

number of other nations also make similar claims, and some small Pacific Island nations are considering recovering resources from their disproportionately large EEZs.

Numerous resources are found within the EEZ, some of which have been exploited for many years. Sand and gravel for construction are mined from the continental shelf in several areas, and about 17% of U.S. oil and natural gas production comes from wells on the continental shelf. Ancient shelf deposits in the Persian Gulf region contain the world's largest reserves of oil.

A potential resource within the EEZ is methane hydrate, consisting of single methane molecules bound up in networks formed by frozen water. These methane hydrates are stable at water depths of more than 500 m and near freezing temperatures. According to one estimate, the carbon in these deposits is double that in all coal, oil, and natural gas reserves. However, no one knows yet whether methane hydrates can be effectively recovered and used as an energy source. Additionally, their contribution to global warming must be assessed, because a volume of methane 3000 times greater than in the atmosphere is present in seafloor deposits—and methane is 10 times more effective than carbon dioxide as a greenhouse gas.

The manganese nodules previously discussed constitute another potential seafloor resource (Figure 8.16). These spherical objects are composed mostly of manganese and iron oxides but also contain copper, nickel, and cobalt. The United States, which must import most of the

(a)

(b)

Figure 8.19 *Two reefs in the Pacific Ocean. (a) A barrier reef along the shore of Rarotonga in the Cook Islands. Notice the numerous rounded to oval structures known as patch reefs within the lagoon. This lagoon has a maximum depth of about 4 m and water depth over the reef is generally less than 1 m. (b) View of an atoll. The light-colored material within the lagoon is sand- and gravel-sized sediment derived from the reef.*

manganese and cobalt it uses, is particularly interested in these nodules as a potential resource.

Other seafloor resources of interest include massive sulfide deposits that form by submarine hydrothermal activity at spreading ridges (see Perspective 8.1). These deposits containing iron, copper, zinc, and other metals have been identified within the EEZ at the Gorda Ridge off the coasts of California and Oregon; similar deposits occur at the Juan de Fuca Ridge within the Canadian EEZ.

Within the EEZ, manganese nodules are found near Johnston Island in the Pacific Ocean and on the Blake Plateau off the east coast of South Carolina and Georgia. In addition, seamounts and seamount chains within the EEZ in the Pacific are known to have metalliferous oxide crusts several centimeters thick from which cobalt and manganese could be mined.

Another important resource found in shallow marine deposits is phosphate-rich sedimentary rock known as *phosphorite* (see Perspective 8.2).

Oceanic Circulation and Resources from the Sea

Earth's oceans are in constant motion. Huge quantities of water circulate horizontally in surface currents such as the Gulf Stream and the South Equatorial Current that carry water great distances. Surface currents transfer heat from the equatorial regions toward the poles and thus have a modifying effect on climate. Deep-ocean waters also move horizontally as a result of temperature and salinity differences between adjacent water masses. And vertical circulation takes place in the oceans when *upwelling* slowly transfers cold water from depth to the surface and *downwelling* transfers warm surface water to depth.

Upwelling is of more than academic interest. It not only transfers water from depth to the surface, but it also carries nutrients, especially nitrates and phosphate, into the zone of sunlight penetration. Here these nutrients sustain huge concentrations of floating organisms, which in turn support other organisms. Other than the continental shelves and areas adjacent to hydrothermal vents on the seafloor, areas of upwelling are the only parts of the oceans where biological productivity is very high. In fact, they are so productive that even though constituting less than 1% of the ocean surface they support more than 50% (by weight) of all fishes.

Scientists recognize three types of upwelling, but only *coastal upwelling* need concern us here. Most coastal upwelling takes place along the west coasts of Africa, North America, and South America, although one notable exception is present in the Indian Ocean. Coastal upwelling involves movement of water offshore, which is replaced by water rising from depth (Figure 1). Along the coast of Peru, for example, the winds coupled with the Coriolis effect transport surface water seaward, and cold, nutrient-

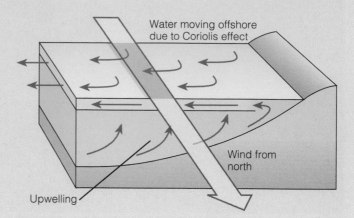

Figure 1 *Wind from the north along the west coast of a continent coupled with the Coriolis effect causes surface water to move offshore resulting in upwelling of cold, nutrient-rich deep water.*

rich water rises to replace it. This area is a major fishery, and changes in the surface-water circulation every three to seven years adjacent to South America are associated with the onset of El Niño, a weather phenomenon with far-reaching consequences.

Our interest here lies in the knowledge that among the nutrients in upwelling oceanic waters is considerable phosphorous, an essential element for animal and plant nutrition. Although present in minute quantities in many sedimentary rocks, most commercial phosphorous is derived from *phosphorite,* a sedimentary rock with such phosphate-rich minerals as fluorapatite ($Ca_5(PO_4)_3F$). Areas of upwelling along the outer margins of continental shelves are the depositional sites of most of the so-

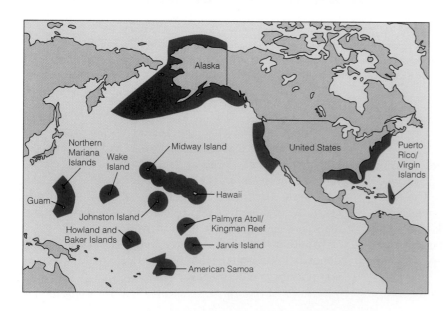

Figure 8.20 *The Exclusive Economic Zone (EEZ), shown in dark blue, includes a vast area adjacent to the United States and its possessions.*

called bedded phosphorites, which are interlayered with carbonate rocks, chert, shale, and sandstone. Vast deposits in the Permian-aged Phosphoria Formation of Montana, Wyoming, and Idaho formed in this manner.

Upwelling accounts for the origin of most of Earth's bedded phosphorites, but other processes are responsible for some phosphate deposits. For instance, in *phosphatization,* carbonate grains such as animal skeletons and ooids are replaced by phosphate, and guano deposits consist of calcium phosphate from bird and bat excrement. Another type of phosphorous-rich deposit is essentially a placer deposit in which the skeletons of vertebrate animals are found in large numbers (vertebrate skeletons and teeth are made up mostly of hydroxyapatite $(Ca_5(PO_4)_3OH)$). A good example is the Miocene to Pliocene (15 to 3 million-year-old) Bone Valley Formation of Florida consisting of a complex of rocks deposited in continental and transitional environments (Figure 2). The phosphorous in this formation comes from skeletons of sharks, manatees, horses, rhinoceroses, and other vertebrate animals.

The United States is the world leader in production and consumption of phosphate rock, most of it (88%) coming from deposits in Florida and North Carolina, but some is also mined in Idaho and Utah. In 1998, 93% of all phosphate mined in this country was used to make chemical fertilizers and animal-feed supplements. However, phosphorus from phosphorite has several other uses, as in metallurgy, food preservation, and production of ceramics and matches.

Figure 2 *Phosphorite of the Bone Valley Formation in the IMC Four Corners Mine, Polk County, Florida.*

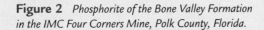

Chapter Summary

1. Scientific investigations of the oceans began during the late 1700s. Present-day research vessels are equipped to investigate the seafloor by sampling, drilling, echo sounding, and seismic profiling.

2. Continental margins separate the continents above sea level from the deep ocean basin. They consist of a continental shelf, continental slope, and in some cases a continental rise.

3. Continental shelves slope gently seaward and vary from a few tens of meters to more than 1000 km wide.

4. The continental slope begins at an average depth of 135 m where the inclination of the seafloor increases rather abruptly from less than 1 degree to several degrees.

5. Submarine canyons are characteristic of the continental slope, but some of them extend well up onto the shelf and lie offshore from large rivers. Some submarine canyons were probably eroded by turbidity currents.

6. Turbidity currents commonly move through submarine canyons and deposit an overlapping series of submarine fans that constitutes a large part of the continental rise.

7. Active continental margins are characterized by a narrow shelf and a slope that descends directly into an oceanic trench with no rise present. These margins are also characterized by seismic activity and volcanism.

8. Passive continental margins lack volcanism and exhibit little seismic activity. The continental shelf along such margins is

Chapter Summary

broad, and the slope merges with a continental rise. Abyssal plains are commonly present seaward of the rise.

9. Oceanic trenches are long, narrow features where oceanic crust is subducted. They are characterized by low heat flow, negative gravity anomalies, and the greatest oceanic depths.

10. Oceanic ridges consisting of mountainous topography are composed of volcanic rocks, and many ridges possess a central rift caused by tensional forces. Basaltic volcanism and shallow focus earthquakes occur at ridges. Oceanic ridges nearly encircle the globe, but they are interrupted and offset by large fractures in the seafloor.

11. Other important features on the seafloor include seamounts that rise more than a kilometer high and guyots, which are flat-topped seamounts. Many aseismic ridges are oriented more or less perpendicular to oceanic ridges and consist of a chain of seamounts and/or guyots.

12. Deep-sea sediments consist mostly of fine-grained particles derived from continents and oceanic islands and the microscopic shells of organisms. The primary types of deep-sea sediments are pelagic clay and ooze.

13. Reefs are wave-resistant structures composed of animal skeletons, particularly corals. Three types of reefs are recognized: fringing, barrier, and atoll.

14. The United States has claimed rights to all resources within 200 nautical miles (371 km) of its shorelines. Numerous resources including various metals are found within this Exclusive Economic Zone.

Important Terms

abyssal plain
active continental margin
aseismic ridge
continental margin
continental rise
continental shelf

continental slope
Exclusive Economic
    Zone (EEZ)
guyot
oceanic ridge
oceanic trench

ooze
ophiolites
passive continental margin
pelagic clay
reef

seamount
submarine canyon
submarine fan
turbidity current

Review Questions

1. Although submarine canyons are best developed on continental slopes, they are also found on (in)
   a. _____ aseismic ridges; b. _____ continental shelves;
   c. _____ oceanic ridges; d. _____ oceanic trenches;
   e. _____ abyssal plains.

2. Submarine fans constitute a large part of
   a. _____ barrier reefs; b. _____ deep seafloor sediments;
   c. _____ the Mid-Atlantic ridge; d. _____ fractures in the seafloor; e. _____ continental rises.

3. The greatest oceanic depths are found at
   a. _____ oceanic trenches; b. _____ abyssal plains;
   c. _____ aseismic ridges; d. _____ guyots; e. _____ passive continental margins.

4. Which one of the following is characteristic of a passive continental margin?
   a. _____ volcanism; b. _____ wide continental shelf;
   c. _____ hydrothermal vents; d. _____ ophiolites; e. _____ oceanic trench.

5. The continental shelf has a slope that averages about
   a. _____ 15 degrees; b. _____ less than 1 degree; c. _____ 4 degrees; d. _____ 10 degrees; e. _____ 20 degrees.

6. Turbidity current deposits typically show or possess
   a. _____ sheeted dikes; b. _____ hydrothermal-vent deposits; c. _____ many coral skeletons; d _____ graded bedding; e _____ a large component of calcareous ooze.

7. Which of the following statements is correct?
   a. _____ Oceanic crust is composed mostly of basalt and gabbro; b. _____ Most passive continental margins are found around the margins of the Pacific Ocean basin; c. _____ The exclusive Economic Zone (EEZ) extends seaward to the shelf-slope break; d. _____ Most

intermediate- and deep-focus earthquakes take place at or near passive continental margins; e. _____ The greatest source of deep-sea sediments is meteorite dust.

8. Sediment that settles from suspension far from land is
   a. _____ abyssal; b. _____ pelagic; c. _____ volcanic;
   d. _____ generally gravel sized; e. _____ sulfide.

9. Outgassing is the process that probably accounts for
   a. _____ the fact that oceanic crust is denser than continental crust; b. _____ the salt content of ocean waters; c. _____ Earth's surface waters; d. _____ fractures in the seafloor; e. _____ the orientation of aseismic ridges perpendicular to oceanic ridges.

10. A circular reef enclosing a lagoon is a(n)
    a. _____ barrier reef; b. _____ atoll; c. _____ seamount;
    d. _____ ophiolite; e. _____ submarine.

11. A flat-topped seamount rising more than 1 km above the seafloor is known as a(n)
    a. _____ guyot; b. _____ seismic profile; c. _____ turbidity-current deposit; d. _____ reef; e. _____ oceanic ridge.

12. Which of the following is a resource derived from seawater or seafloor deposits?
    a. _____ basalt; b. _____ pelagic clay; c. _____ siliceous ooze; d. _____ phosphorite; e. _____ gabbro

13. How do mid-oceanic ridges form, and how do they differ from mountain ranges on land?

14. Explain what calcareous and siliceous oozes are and how they are deposited.

15. Illustrate and label an ideal sequence of rocks in an ophiolite. Also, for each rock type explain how and where it formed.

16. Why are abyssal plains common around the margins of the Atlantic Ocean basin but rare around the margins of the Pacific?

17. Describe the sequence of events leading first to a fringing reef, then a barrier reef, and finally to an atoll.

18. What are the characteristics of an active continental margin? Why are margins of this type so common along the margins of the Pacific but not the Atlantic?

19. What is the Exclusive Economic Zone (EEZ)? What types of metal deposits are found within it?

20. Describe a submarine canyon. Also, explain how scientists think they form, and what supporting evidence they have for their interpretation.

21. What are the characteristics of a passive continental margin?

22. How are echo sounding and seismic profiling used to study the seafloor?

23. The most distant part of an aseismic ridge is 1000 km from an oceanic ridge system, and its age is 30 million years. How fast, on the average, did the plate with this ridge move in centimeters per year?

24. Hydrothermal vents on the seafloor are difficult to locate because they lie far below the surface of the oceans. How would you propose looking for and locating such vents? Also, how can you account for the fact that metal deposits formed at hydrothermal vents can be found hundreds of kilometers from active vents?

 # World Wide Web Activities

For these web site addresses, along with current updates and exercises, log on to **http://www.brookscole.com/geo**

AMNH-Past Day Logs

This site, maintained by the American Museum of Natural History, has an excellent animation of how black smokers work. The first screen one sees is American Museum of Natural History Expeditions. On this screen click *Black Smokers* and see how hydrothermal vent systems operate. Click on *World Ridge System* and then click *Life Forms* to see creatures living near hydrothermal vents. Also, click on *Underwater Tools* to see how underwater research is conducted.

USGS BAMG Seafloor Mapping Server

This site is presented by the U.S. Geological Survey Coastal and Marine Geology Program and the Woods Hole Field Center at Woods Hole, Massachusetts. Click on and see what the *GLORIA Mapping Program* is. Also, click GLORIA II Sidescan Sonar System.

NOVA Online | Join Us/Feedback

When the main page is on your screen, go to search and enter black smokers and then investigate topics such as NOVA Online | Into the Abyss: Billowing black smoker chimney in the Mothra field, and NOVA Online | Teacher's Guide | Volcanoes of the Deep, and many others.

Gas (Methane) Hydrates—A New Frontier

The U.S. Geological Survey Marine and Coastal Geology Program maintains this site that tells of methane hydrates in seafloor sediments. See the recent map by the USGS of the methane hydrates off the North and South Carolina coasts.

 # CD-ROM Exploration

Exploring your *Earth Systems Today* CD-ROM will add to your understanding of the material in this chapter.

Topic: Earth's Processes

Module: Plate Tectonics

Explore activities in this module to see if you can discover the following for yourself:

Explore links within this activity that take you to a triple-junction site study (i.e., junction of the Nazca, Antarctic, and South American plates) and links that show you several kinds of instrumentation being deployed to study underwater plate boundaries like ridges and trenches.

Plate Tectonics: A Unifying Theory

The northwest-southeast trending, snow-covered, linear mountain range is the King Ata Tag Mountains, located in extreme westernmost China and including part of Tajikistan. Highest peaks are just over 6130 m above sea level. The town or village of Muji is located in the larger river valley south of this mountain range.

PROLOGUE

OBJECTIVES

At the end of this chapter, you will have learned that

■ Plate tectonics is the unifying theory of geology and has revolutionized geology.

■ The hypothesis of continental drift was based on considerable geologic, paleontologic, and climatologic evidence.

■ The hypothesis of seafloor spreading accounts for continental movement, and that thermal convection cells provide a mechanism for plate movement.

■ The three types of plate boundaries are divergent, convergent, and transform, and along these boundaries new plates are formed, consumed, or slide past one another.

■ Interaction along plate boundaries accounts for most of Earth's earthquake and volcanic activity.

■ The rate of movement and motion of plates can be calculated in several ways.

■ Some type of convective heat system is involved in plate movement.

■ Plate movement affects the distribution of natural resources.

■ Plate movement affects the distribution of the world's biota and has influenced evolution.

Two tragic events during 1985 serve to remind us of the dangers of living near a convergent plate margin. On September 19, a magnitude 8.1 earthquake killed more than 9000 people in Mexico City. Two months later and 3200 km to the south, a minor eruption of Colombia's Nevado del Ruiz volcano partially melted its summit glacial ice, causing a mudflow that engulfed Armero and several other villages and killed more than 23,000 people. These two tragedies resulted in more than 32,000 deaths, tens of thousands of injuries, and billions of dollars in property damage.

Both events occurred along the eastern portion of the Ring of Fire, a chain of intense seismic and volcanic activity that encircles the Pacific Ocean basin (see Figure 7.7). Some of the world's greatest disasters occur along this ring because of volcanism and earthquakes generated by plate convergence. The Mexico City earthquake resulted from subduction of the Cocos plate at the Middle America Trench (see Figure 1.11). Sudden movement of the Cocos plate beneath Central America generated seismic waves that traveled outward in all directions. The violent shaking experienced in Mexico City, 350 km away, and elsewhere was caused by these seismic waves.

The strata underlying Mexico City consist of unconsolidated sediment deposited in a large ancient lake. Such sediment amplifies the shaking during earthquakes with the unfortunate consequence that buildings constructed there are commonly more heavily damaged than those built on solid bedrock (see Perspective 7.2, Figure 3).

Less than two months after the Mexico City earthquake, Colombia experienced its greatest recorded natural disaster. Nevado del Ruiz is one of several active volcanoes resulting from the rise of magma generated where the Nazca plate is subducted beneath South America (see Figure 1.11). A minor eruption of Nevado del Ruiz partially melted the glacial ice on the mountain; the meltwater rushed down the valleys, mixed with the sediment, and turned it into a deadly viscous mudflow.

The city of Armero, Colombia, lies in the valley of the Lagunilla River, one of several river valleys inundated by mudflows. Of the city's 23,000 inhabitants, 20,000 died, and most of the city was destroyed. Another 3000 people were killed in nearby valleys. A geologic hazard assessment study completed one month before the eruption showed that Armero was in a high-hazard mudflow area!

These two examples vividly illustrate some of the dangers of living in proximity to a convergent plate boundary. Subduction of one plate beneath another repeatedly triggers large earthquakes, the effects of which are frequently felt far from their epicenters. Since 1900, earthquakes have killed more than 120,000 people in Central and South America alone. Even though volcanic eruptions in this region have not caused nearly as many casualties as earthquakes, they have, nevertheless, been responsible for tremendous property damage and have the potential for triggering devastating events such as the 1985 Colombian mudflow.

Because the Ring of Fire is home to millions of people, can anything be done to decrease the devastation that inevitably results from the earthquake and volcanic activity in that region? Given our present state of knowledge, most of the disasters could not have been accurately predicted, but better planning and advance preparations by the nations bordering the Ring of Fire could have prevented much tragic loss of life. As long as people live near convergent plate margins, disasters will continue. By studying and understanding geologic activity along plate margins, however, geologists can help to minimize the destruction.

Introduction

The decade of the 1960s was a time of geological as well as social and cultural revolution. The ramifications of the newly proposed plate tectonic theory radically changed the way geologists viewed our planet. Earth could now be treated as a system in which seemingly unrelated geologic phenomena were related and interconnected (Table 9.1). No longer could Earth be regarded as an unchanging planet on which continents and ocean basins remained fixed through time, but instead, it was recognized as an integrated and dynamically changing planet.

Plate tectonics has been the dominant process affecting the evolution of Earth. The interactions between moving plates determine the locations of continents, ocean basins, and mountain systems, which in turn affect atmospheric and oceanic circulation patterns that ultimately determine global climates. Plate movement has also profoundly influenced the geographic distribution, evolution, and extinction of plants and animals.

Although most people have only a vague notion of what plate tectonic theory is, plate tectonics has had and continues to have a profound effect on all our lives. We now realize that the devastating earthquakes and volcanic eruptions we read about are not random occurrences but take place near plate margins. Furthermore, the formation and distribution of many important geologic resources, such as metallic ores and petroleum, are related to plate tectonic processes. This realization has resulted in geologists incorporating plate tectonic theory into not only their prospecting efforts but virtually all areas of geologic research.

The acceptance of plate tectonic theory has led to a greater understanding of how Earth has evolved and continues to do so. This powerful, unifying theory enables geologists to view Earth history in terms of interrelated events that are part of a global panorama of dynamic change through time.

This chapter begins by reviewing the various hypotheses that preceded plate tectonic theory, examining the evidence that led some people to accept the idea of continental movement and others to reject it. We follow with a discussion of how the seafloor spreading hypothesis provided a mechanism to explain the movement of continents. The rest of the chapter covers plate tectonic theory, its ramifications, and why it is the unifying theory of geology.

Why Should You Study Plate Tectonics?

If you're like most people, you probably have no idea or only a vague notion of what plate tectonic theory is. Yet, plate tectonics has had and continues to have a profound effect on all of our lives. We now realize that the devastating earthquakes and volcanic eruptions we read about in the newspapers are not random occurrences, but take place near plate margins. Furthermore, the formation and distribution of many important geologic resources, such as metallic ores and petroleum, are related to plate tectonic processes.

Plate tectonics thus affects all of us, not only in terms of the destruction caused by volcanoes and earthquakes, but also politically and economically, because the distribution and concentration of natural resources is controlled by plate movement. Therefore, it is important to understand this unifying theory because it not only affects us as individuals and citizens of nation states, but also because it ties together many of the aspects of geology you will be studying.

What Were Some of the Early Ideas about Continental Drift?

The idea that Earth's geography was different during the past is not new. The earliest maps showing the east coast of South America and the west coast of Africa probably provided people with the first evidence that continents may have once been joined together, then broken apart and moved to their present positions.

Table 9.1

Plate Tectonics and Earth Systems

Solid Earth	Plate tectonics is driven by convection in the mantle and in turn drives mountain-building processes and associated igneous and metamorphic activity.
Atmosphere	Arrangement of continents affects solar heating and cooling and thus winds and weather systems. Rapid plate spreading and hot-spot activity may release volcanic carbon dioxide and affect global climate.
Hydrosphere	Continental arrangement affects ocean currents. Rate of spreading affects volume of mid-ocean ridges and hence sea level. Placement of continents may contribute to onset of ice ages.
Biosphere	Movement of continents creates corridors or barriers to migration and thus creates ecological niches. Habitats may be transported into more or less favorable climates.
Extraterrestrial	Arrangement of continents affects free circulation of ocean tides and influences tidal slowing of Earth's rotation.

During the late 19th century, the Austrian geologist Edward Suess noted the similarities between the Late Paleozoic plant fossils of India, Australia, South Africa, and South America as well as evidence of glaciation in the rock sequences of these southern continents. The plant fossils comprise a unique flora that occurs in the coal layers just above the glacial deposits of these southern continents. This flora is very different from the contemporaneous coal swamp flora of the northern continents and is collectively known as the **Glossopteris flora** after its most conspicuous genus (Figure 9.1).

In his book, *The Face of the Earth,* published in 1885, Suess proposed the name *Gondwanaland* (or **Gondwana** as we will use here) for a supercontinent composed of the aforementioned southern continents. Gondwana is a province in India where abundant fossils of the *Glossopteris* flora are found in coal beds. Suess thought these continents were connected by land bridges over which plants and animals migrated. Thus, in his view, the similar fossils of the continents were due to the appearance and disappearance of the connecting land bridges.

The American geologist Frank Taylor published a pamphlet in 1910 presenting his own theory of continental drift. He explained the formation of mountain ranges as a result of the lateral movement of continents. He also envisioned the present-day continents as parts of larger polar continents that eventually broke apart and migrated toward the equator after Earth's rotation supposedly slowed due to gigantic tidal forces. According to Taylor, these tidal forces were generated when Earth captured the Moon about 100 million years ago.

Although we now know that Taylor's mechanism is incorrect, one of his most significant contributions was his suggestion that the Mid-Atlantic Ridge, discovered by the 1872–1876 British HMS *Challenger* expeditions, might mark the site along which an ancient continent broke apart to form the present-day Atlantic Ocean.

Alfred Wegener and the Continental Drift Hypothesis

Alfred Wegener, a German meteorologist (Figure 9.2), is generally credited with developing the hypothesis of **continental drift.** In his monumental book, *The Origin of Continents and Oceans* (first published in 1915), Wegener proposed that all landmasses were originally united into a single supercontinent that he named **Pangaea,** from the Greek meaning "all land." Wegener portrayed his grand concept of continental movement in a series of maps showing the breakup of Pangaea and the movement of the various continents to their present-day locations. Wegener had amassed a tremendous amount of geologic, paleontologic, and climatologic evidence in support of continental drift, but the initial reaction of scientists to his then-heretical ideas can best be described as mixed.

Nevertheless, the eminent South African geologist Alexander du Toit further developed Wegener's arguments and introduced more geologic evidence in support of continental drift. In 1937 du Toit published *Our Wandering Continents,* in which he contrasted the glacial deposits of Gondwana with coal deposits of the same age found in the continents of the Northern Hemisphere. To resolve this apparent climatologic paradox, du Toit placed the southern continents of Gondwana at or near the South Pole and arranged the northern continents together so that the coal deposits were located at the equator. He named this northern landmass, consisting of present-day North America, Greenland, Europe, and Asia (except for India) **Laurasia.**

(a)

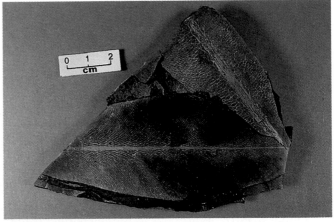

(b)

Figure 9.1 *Representative members of the* Glossopteris *flora. Fossils of these plants are found on all five of the Gondwana continents.* Glossopteris *leaves from (a) the Upper Permian Dunedoo Formation and (b) the Upper Permian Illawarra Coal Measures, Australia.*

Figure 9.2 *Alfred Wegener, a German meteorologist, proposed the continental drift hypothesis in 1912 based on a tremendous amount of geologic, paleontologic, and climatologic evidence. He is shown here waiting out the Arctic winter in an expedition hut.*

Figure 9.3 *The best fit among continents occurs along the continental slope, where erosion would be minimal.*

What Is the Evidence for Continental Drift?

What then was the evidence used by Wegener, du Toit, and others to support the hypothesis of continental drift? It includes the fit of the shorelines of continents, the appearance of the same rock sequences and mountain ranges of the same age on continents now widely separated, the matching of glacial deposits and paleoclimatic zones, and the similarities of many extinct plant and animal groups whose fossil remains are found today on widely separated continents.

Continental Fit

Wegener, like some before him, was impressed by the close resemblance between the coastlines of continents on  opposite sides of the Atlantic Ocean, particularly between South America and Africa. He cited these similarities as partial evidence that the continents were at one time joined together as a supercontinent that subsequently split apart. As his critics pointed out, though, the configuration of coastlines results from erosional and depositional processes and therefore is continually being modified. So even if the continents had separated during the Mesozoic Era, as Wegener proposed, it is not likely that the coastlines would fit exactly.

A more realistic approach is to fit the continents together along the continental slope where erosion would be minimal. Recall from Chapter 8 that the true margin of a continent—that is, where continental crust changes to

oceanic crust—is beneath the continental slope. In 1965 Sir Edward Bullard, an English geophysicist, and two associates showed that the best fit between the continents occurs at a depth of about 2000 m (Figure 9.3). Since then, other reconstructions using the latest ocean basin data have confirmed the close fit between continents when they are reassembled to form Pangaea.

Similarity of Rock Sequences and Mountain Ranges

If the continents were at one time joined together, then the rocks and mountain ranges of the same age in adjoining locations on the opposite continents should closely match. Such is the case for the Gondwana continents. Marine, nonmarine, and glacial rock sequences of Pennsylvanian to Jurassic age are almost identical for all five Gondwana continents, strongly indicating that they were joined together at one time (Figure 9.4).

The trends of several major mountain ranges also support the hypothesis of continental drift. These mountain ranges seemingly end at the coastline of one continent only to apparently continue on another continent across the ocean. The folded Appalachian Mountains of North America, for example, trend northeastward through the eastern United States and Canada and terminate abruptly at the Newfoundland coastline. Mountain ranges of the same age and deformational style occur in eastern Greenland, Ireland, Great Britain, and Norway. Even though these

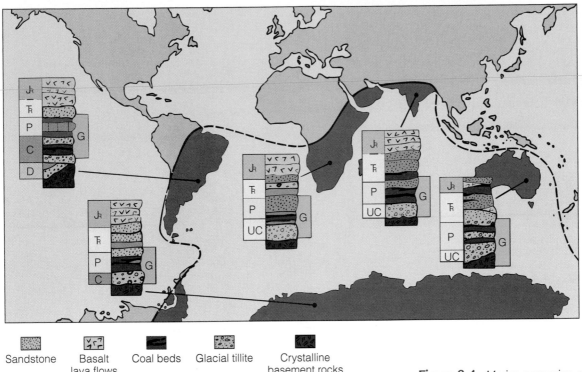

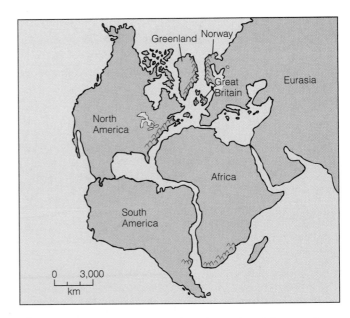

Figure 9.4 *Marine, nonmarine, and glacial rock sequences of Pennsylvanian to Jurassic age are nearly the same for all Gondwana continents. Such close similarity strongly suggests that they were joined together at one time. The range indicated by G is that of the Glossopteris flora.*

Legend for Figure 9.4:

Symbol	Rock type
	Sandstone
	Basalt lava flows
	Coal beds
	Glacial tillite
	Crystalline basement rocks

Abbreviation	Period
Jr	Jurassic
Tr	Triassic
P	Permian
UC	Pennsylvanian
C	Carboniferous (Mississippian and Pennsylvanian)
D	Devonian

mountain ranges are currently separated by the Atlantic Ocean, they form an essentially continuous mountain range when the continents are positioned next to each other (Figure 9.5).

Glacial Evidence

During the Late Paleozoic Era, massive glaciers covered large continental areas of the Southern Hemisphere. Evidence for this glaciation includes layers of till (sediments deposited by glaciers) and striations (scratch marks) in the bedrock beneath the till. Fossils and sedimentary rocks of the same age from the Northern Hemisphere, however, give no indication of glaciation. Fossil plants found in coals indicate that the Northern Hemisphere had a tropical climate during the time that the Southern Hemisphere was glaciated.

All the Gondwana continents except Antarctica are currently located near the equator in subtropical to tropical climates. Mapping of glacial striations in bedrock in Australia (Figure 9.6a), India, and South America indicates that the glaciers moved from the areas of the present-day oceans onto land (Figure 9.6b). This would be highly unlikely because large continental glaciers flow outward from their central area of accumulation toward the sea.

Figure 9.5 *When continents are brought together, their mountain ranges form a single continuous range of the same age and style of deformation throughout. Such evidence indicates the continents were at one time joined together and were subsequently separated.*

What Is the Evidence for Continental Drift? **239**

(a)

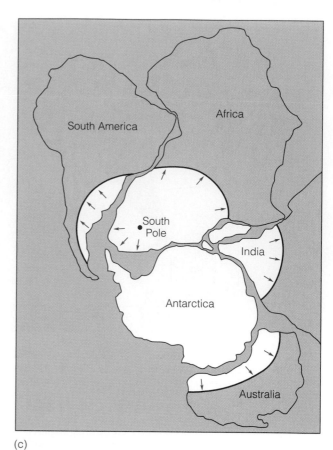

(c)

Figure 9.6 *(a) Permian-aged glacial striations in bedrock exposed at Hallet's Cove, Australia, indicate the direction of glacial movement more than 200 million years ago. (b) If the continents did not move in the past, then Late Paleozoic glacial striations preserved in bedrock in Australia, India, and South America indicate that glacial movement for each continent was from the oceans onto land within a subtropical to tropical climate. Such an occurrence is highly unlikely. (c) If the Gondwana continents are brought together so that South Africa is located at the South Pole, then the glacial movement indicated by the striations makes sense. In this situation, the glacier, located in a polar climate, moved radially outward from a thick central area toward its periphery.*

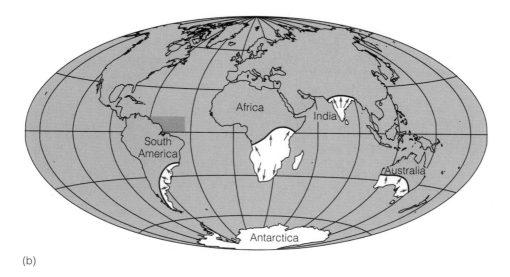

(b)

If the continents did not move during the past, one would have to explain how glaciers moved from the oceans onto land and how large-scale continental glaciers formed near the equator. But if the continents are reassembled as a single landmass with South Africa located at the south pole, the direction of movement of Late Paleozoic continental glaciers makes sense. Furthermore, this geographic arrangement places the northern continents nearer the tropics, which is consistent with the fossil and climatologic evidence from Laurasia.

Fossil Evidence

Some of the most compelling evidence for continental drift comes from the fossil record (Figure 9.7). Fossils of the *Glossopteris* flora are found in equivalent Penn-

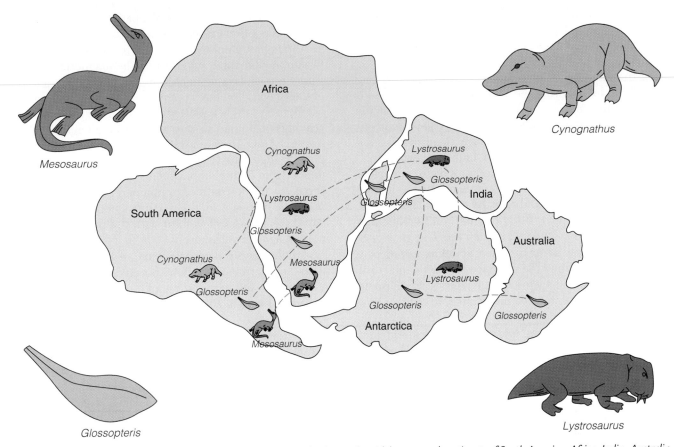

Figure 9.7 *Some of the animals and plants whose fossils are found today on the widely separated continents of South America, Africa, India, Australia, and Antarctica. These continents were joined together during the Late Paleozoic to form Gondwana, the southern landmass of Pangaea.* Glossopteris *and similar plants are found in Pennsylvanian- and Permian-age deposits on all five continents.* Mesosaurus *is a freshwater reptile whose fossils are found in Permian-age rocks in Brazil and South Africa.* Cynognathus *and* Lystrosaurus *are land reptiles who lived during the Early Triassic Period. Fossils of* Cynognathus *are found in South America and Africa, and fossils of* Lystrosaurus *have been recovered from Africa, India, and Antarctica.*

sylvanian- and Permian-aged coal deposits on all five Gondwana continents. The *Glossopteris* flora is characterized by the seed fern *Glossopteris* (Figure 9.1) as well as by many other distinctive and easily identifiable plants. Pollen and spores of plants can be dispersed over great distances by wind, but *Glossopteris*-type plants produced seeds that are too large to have been carried by winds. Even if the seeds had floated across the ocean, they probably would not have remained viable for any length of time in salt water.

The present-day climates of South America, Africa, India, Australia, and Antarctica range from tropical to polar and are much too diverse to support the types of plants that compose the *Glossopteris* flora. Wegener therefore reasoned that these continents must once have been joined so that these widely separated localities were all in the same latitudinal climatic belt (Figure 9.7).

The fossil remains of animals also provide strong evidence for continental drift. One of the best examples is *Mesosaurus*, a freshwater reptile whose fossils are found in Permian-aged rocks in certain regions of Brazil and South Africa and nowhere else in the world (Figure 9.8).

Because the physiology of freshwater and marine animals is completely different, it is hard to imagine how a freshwater reptile could have swum across the Atlantic Ocean and found a freshwater environment nearly identical to its former habitat. Moreover, if *Mesosaurus* could have swum across the ocean, its fossil remains should be widely

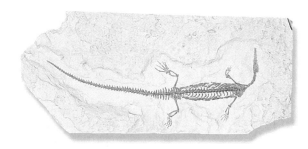

Figure 9.8 Mesosaurus, *a Permian-age freshwater reptile whose fossil remains are found in Brazil and South Africa, indicating these two continents were joined together at the end of the Paleozoic Era.*

dispersed. It is more logical to assume that *Mesosaurus* lived in lakes in what are now adjacent areas of South America and Africa, but were then united into a single continent.

Lystrosaurus and *Cynognathus* are both land-dwelling reptiles that lived during the Triassic Period; their fossils are found only on the present-day continental fragments of Gondwana (Figure 9.7). Because they are both land animals, they certainly could not have swum across the oceans currently separating the Gondwana continents. Therefore, the continents must once have been connected.

Despite what seemed to be overwhelming evidence presented by Wegener and later by du Toit and others, most geologists simply refused to entertain the idea that continents might have moved in the past. The geologists were not necessarily being obstinate about accepting new ideas; rather, they found the evidence for continental drift inadequate and unconvincing. In part, this was because no one could provide a suitable mechanism to explain how continents could move over Earth's surface. Not until new evidence from studies of Earth's magnetic field and oceanographic research showed that the ocean basins were geologically young features did renewed interest in continental drift occur.

Paleomagnetism and Polar Wandering

Interest in continental drift revived during the 1950s as a result of new evidence from paleomagnetic studies. **Paleomagnetism** is the remanent magnetism in ancient rocks recording the direction of Earth's magnetic poles at the time of the rock's formation. Earth can be thought of as a giant dipole magnet in which the magnetic poles essentially coincide with the geographic poles (Figure 9.9). Such an arrangement means that the strength of the magnetic field is not constant, but varies, being weakest at the equator and strongest at the poles. Earth's magnetic field is thought to result from the different rotation speeds of the outer core and mantle.

When magma cools, the magnetic iron-bearing minerals align themselves with Earth's magnetic field, recording both its direction and strength. The temperature at which iron-bearing minerals gain their magnetization is called the **Curie point.** As long as the rock is not subsequently heated above the Curie point, it will preserve that remanent magnetism. Thus, an ancient lava flow provides a record of the orientation and strength of Earth's magnetic field at the time the lava flow cooled.

As paleomagnetic research progressed during the 1950s, some unexpected results emerged. When geologists measured the magnetism of recent rocks, they found it was generally consistent with Earth's current magnetic field. The paleomagnetism of ancient rocks, though, showed dif-

ferent orientations. For example, paleomagnetic studies of Silurian lava flows in North America indicated that the north magnetic pole was located in the western Pacific Ocean at that time, while the paleomagnetic evidence from Permian lava flows indicated a pole in Asia, and that of Cretaceous lava flows pointed to yet another location in northern Asia. When plotted on a map, the paleomagnetic readings of numerous lava flows from all ages in North America trace the apparent movement of the magnetic pole through time (Figure 9.10). This paleomagnetic evidence from a single continent could be interpreted in three ways: The continent remained fixed and the north magnetic pole moved; the north magnetic pole stood still and the continent moved; or both the continent and the north magnetic pole moved.

Upon analysis, magnetic minerals from European Silurian and Permian lava flows pointed to a different magnetic pole location from those of the same age from North America (Figure 9.10). Furthermore, analysis of lava flows from all continents indicated each continent had its own series of magnetic poles. Does this mean there were different north magnetic poles for each continent? That would be highly unlikely and difficult to reconcile with the theory accounting for Earth's magnetic field.

The best explanation for such data is that the magnetic poles have remained at their present locations near the geographic north and south poles and the continents have moved. When the continental margins are fitted together so that the paleomagnetic data point to only one magnetic pole, we find, just as Wegener did, that the rock sequences and glacial deposits match, and that the fossil evidence is consistent with the reconstructed paleogeography (see Perspective 9.1).

What Is the Relationship Between Magnetic Reversals and Seafloor Spreading?

Geologists refer to Earth's present magnetic field as normal, that is, with the north and south magnetic poles located roughly at the north and south geographic poles. As early as 1906, though, rocks were discovered that showed reversed magnetism. Paleomagnetic studies initially conducted on continental lava flows have clearly shown that the magnetic field has completely reversed itself numerous times during the geologic past (Figure 9.11). When these **magnetic reversals** occur, the magnetic polarity is reversed so that the north arrow on a compass would point south rather than north.

Once the existence of magnetic reversals was well established for continental lava flows, magnetic reversals

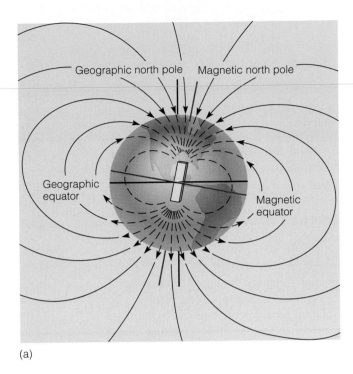

Figure 9.9 *(a) Earth's magnetic field has lines of force just like those of a bar magnet. (b) The strength of the magnetic field changes uniformly from the magnetic equator to the magnetic poles. This change in strength causes a dip needle to parallel Earth's surface only at the magnetic equator, whereas its inclination with respect to the surface increases to 90 degrees at the magnetic poles.*

(a)

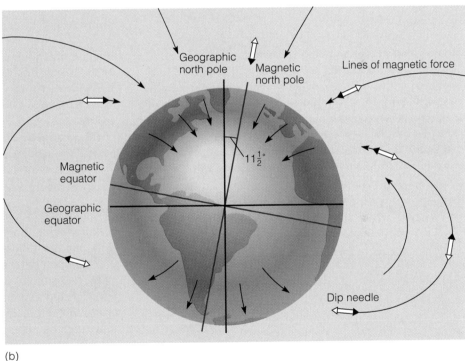

(b)

were also discovered in igneous rocks of the oceanic crust as part of the extensive mapping of the ocean basins during the 1960s (Figure 9.12). Although the cause of magnetic reversals is not completely known, it appears to be related to changes in the intensity of the magnetic field.

Besides the discovery of magnetic reversals, mapping of the ocean basins also revealed a ridge system 65,000 km long, constituting the most extensive mountain range in the world. Perhaps the best-known part of the ridge system is the Mid-Atlantic Ridge, which divides the Atlantic Ocean basin into two nearly equal parts (Figure 9.13).

Paleogeographic Reconstructions and Maps

The key to any reconstruction of world paleogeography is the correct positioning of the continents in terms of latitude and longitude as well as orientation of a paleocontinent relative to the paleonorth pole. The main criteria used for paleogeographic reconstructions are paleomagnetism, biogeography, tectonic patterns, and climatology.

Paleomagnetism provides the only source of quantitative data on the orientations of the continents. For the Paleozoic Era, the paleomagnetic data are often inconsistent and contradictory because of secondary magnetizations acquired through the effects of metamorphism or weathering.

The distribution of faunas and floras provides a useful check on the latitudes determined by paleomagnetism and can provide additional limits on longitudinal separation of continents. As is well known, the distribution of plants and animals is controlled by both climatic and geographic barriers. Such information can be used to position continents and ocean basins in a way that accounts for the biogeographic patterns indicated by fossil evidence.

Tectonic activity is indicated by deformed sediments associated with andesitic volcanics and ophiolites. Such features allow geologists to recognize ancient mountain ranges and zones of subduction. These mountain ranges may subsequently have been separated by plate movement, so the identification of large, continuous mountain ranges provides important information about continental positions in the geologic past.

Climate-sensitive sedimentary rocks are used to interpret past climatic conditions. Desert dunes are typically well sorted and cross-bedded on a large scale and associated with other deposits that indicate an arid environment. Coal forms in freshwater swamps where climatic conditions promote abundant plant growth. Evaporites result when evaporation exceeds precipitation, such as in desert regions or along hot, dry shorelines. Tillites result from glacial activity and indicate cold, wet environments.

Paleogeographic features can be determined by associations of sedimentary rocks and sedimentary structures. For example, large-scale cross-beds may indicate aeolian or wind-blown conditions such as in deserts. Delta complexes and deep-sea fans have characteristic internal features and three-dimensional forms that can be recognized in the geologic record, just as coal and associated deposits usually follow a particular sequence. These features can be used to interpret such geographic features as lakes, streams, swamps, and shallow and deep marine areas.

Former mountain ranges can be recognized by folded and faulted sedimentary rocks associated with metamorphic and igneous rocks. Furthermore, the presence of andesites and ophiolites can also be used as evidence of former mountain building. By combining all relevant geologic, paleontologic, and climatologic information, geologists can construct paleogeographic maps. Such maps are simply interpretations of the geography of an area for a particular time in the geologic past. The majority of paleogeographic maps show the distribution of land and sea, probable climatic regimes, and such geographic features as mountain ranges, swamps, and glaciers (see Figure 21.2).

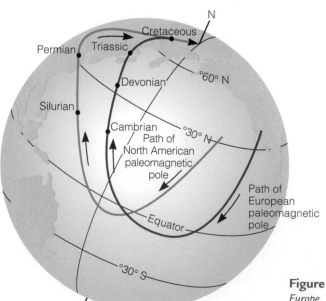

Figure 9.10 *The apparent paths of polar wandering for North America and Europe. The apparent location of the north magnetic pole is shown for different periods on each continent's polar wandering path.*

As a result of oceanographic research conducted during the 1950s, Harry Hess of Princeton University proposed the theory of **seafloor spreading** in 1962 to account for continental movement. Hess suggested that continents do not move across oceanic crust, but rather that the continents and oceanic crust move together. He suggested that the seafloor separates at oceanic ridges where new crust is formed by upwelling magma. As the magma cools, the newly formed oceanic crust moves laterally away from the ridge. As a mechanism to drive this system, Hess revived the idea (proposed in the 1930s and 1940s by Arthur Holmes and others) of **thermal convection cells** in the mantle; that is, hot

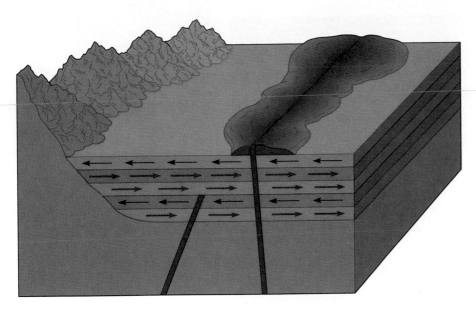

Figure 9.11 *Magnetic reversals recorded in a succession of lava flows are shown diagrammatically by red arrows, and the record of normal polarity events is shown by black arrows.*

magma rises from the mantle, intrudes along fractures defining oceanic ridges and thus forms new crust. Cold crust is subducted back into the mantle at deep-sea trenches, where it is heated and recycled, thus completing a thermal convection cell (see Figure 1.10).

How could Hess's hypothesis be confirmed? Magnetic surveys of the oceanic crust revealed striped **magnetic**

anomalies (deviations from the average strength of Earth's magnetic field) in the rocks that were both parallel to and symmetric with the oceanic ridges (Figure 9.12). Furthermore, the pattern of oceanic magnetic anomalies matched the pattern of magnetic reversals already known from studies of continental lava flows (Figure 9.11). When magma wells up and cools along a ridge

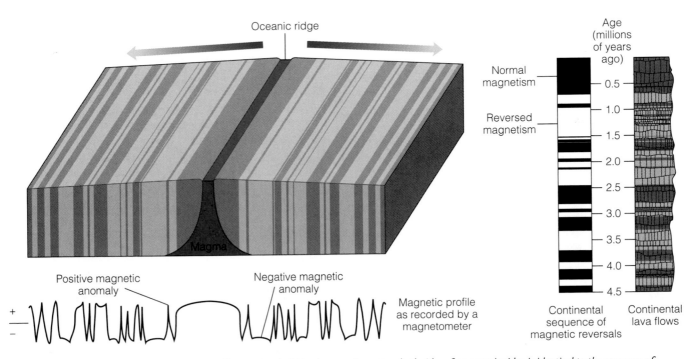

Figure 9.12 *The sequence of magnetic anomalies preserved within the oceanic crust on both sides of an oceanic ridge is identical to the sequence of magnetic reversals already known from continental lava flows. Magnetic anomalies are formed when magma intrudes into oceanic ridges; when the magma cools below the Curie point, it records Earth's magnetic polarity at that time. Seafloor spreading splits the previously formed crust in half, so that it moves laterally away from the oceanic ridge. Repeated intrusions record a symmetrical series of magnetic anomalies that record periods of normal and reversed polarity. The magnetic anomalies are recorded by a magnetometer, which measures the strength of the magnetic field.*

What Is the Relationship Between Magnetic Reversals and Seafloor Spreading?

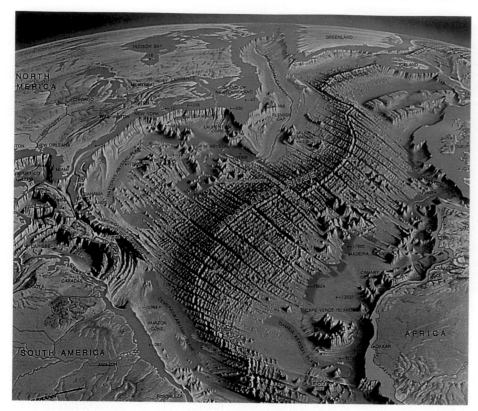

Figure 9.13 *Artistic view of what the Atlantic Ocean basin would look like without water. The major feature is the Mid-Atlantic Ridge.*

summit, it records Earth's magnetic field at that time as either normal or reversed. As new crust forms at the summit, the previously formed crust moves laterally away from the ridge. These magnetic stripes, representing times of normal or reversed polarity, are parallel to and symmetrical around oceanic ridges (where upwelling magma forms new oceanic crust), conclusively confirming Hess's theory of seafloor spreading.

One of the consequences of the seafloor spreading theory is its confirmation that ocean basins are geologically young features whose openings and closings are partially responsible for continental movement (Figure 9.14). Radiometric dating reveals that the oldest oceanic crust is less than 180 million years old, whereas the oldest continental crust is 3.96 billion years old.

Deep-Sea Drilling and the Confirmation of Seafloor Spreading

To many geologists, the paleomagnetic data amassed in support of continental drift and seafloor spreading were convincing. Results from the Deep-Sea Drilling Project (see Chapter 8) have confirmed the interpretations made from earlier paleomagnetic studies.

According to the seafloor spreading hypothesis, oceanic crust is continuously forming at mid-oceanic ridges, moving away from these ridges by seafloor spreading, and being consumed at subduction zones. If this is the case, oceanic crust should be youngest at the ridges and become progressively older with increasing distance away from them. Moreover, the age of the oceanic crust should be symmetrically distributed about the ridges. As we have just noted, paleomagnetic data confirm these statements. Furthermore, fossils from sediments overlying the oceanic crust and radiometric dating of rocks found on oceanic islands substantiate this predicted age distribution.

Sediments in the open ocean accumulate, on average, at a rate of less than 0.3 cm per 1000 years. If the ocean basins were as old as the continents, we would expect deep-sea sediments to be several kilometers thick. However, data from numerous drill holes indicate that deep-sea sediments are at most only a few hundred meters thick and are thin or absent at oceanic ridges. Their near-absence at the ridges should come as no surprise because these are the areas where new crust is continuously produced by volcanism and seafloor spreading. Accordingly, sediments have had little time to accumulate at or very close to spreading ridges where the oceanic crust is young, but their thickness increases with distance away from the ridges (Figure 9.15).

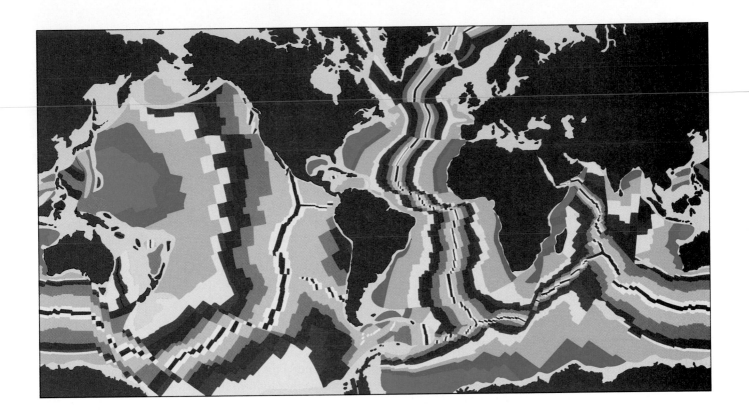

■ Pleistocene to Recent (0–1.6 M.Y.A.)	■ Paleocene (58–66 M.Y.A.)
□ Pliocene (1.6–5 M.Y.A.)	□ Late Cretaceous (66–88 M.Y.A.)
▨ Miocene (5–24 M.Y.A.)	▨ Middle Cretaceous (88–118 M.Y.A.)
▨ Oligocene (24–37 M.Y.A.)	▨ Early Cretaceous (118–144 M.Y.A.)
■ Eocene (37–58 M.Y.A.)	□ Late Jurassic (144–161 M.Y.A.)

Figure 9.14 *The age of the world's ocean basins established from magnetic anomalies demonstrates that the youngest oceanic crust is adjacent to the oceanic ridges and that its age increases away from the ridge axis.*

Figure 9.15 *The total thickness of deep-sea sediments increases away from oceanic ridges. This is because oceanic crust becomes older away from oceanic ridges, and there has been more time for sediment to accumulate.*

Oceanic ridge

Oceanic crust

Deep-sea sediments

Total thickness of sediment increases away from oceanic ridge

Upper mantle

Magma

Increasing age of crust

What Is the Relationship Between Magnetic Reversals and Seafloor Spreading?   **247**

Plate Tectonic Theory

Plate tectonic theory is based on a simple model of Earth. The rigid lithosphere, consisting of both oceanic and continental crust, as well as the underlying upper mantle, consists of numerous variable-sized pieces called **plates** (Figure 9.16). The plates vary in thickness; those composed of upper mantle and continental crust are as much as 250 km thick, whereas those of upper mantle and oceanic crust are up to 100 km thick.

The lithosphere overlies the hotter and weaker semiplastic asthenosphere. It is thought that movement resulting from some type of heat-transfer system within the asthenosphere causes the overlying plates to move. As plates move over the asthenosphere, they separate, mostly at oceanic ridges; in other areas such as at oceanic trenches, they collide and are subducted back into the mantle.

An easy way to visualize plate movement is to think of a conveyer belt moving luggage from an airplane's cargo hold to a baggage cart. The conveyer belt represents convection currents within the mantle, and the luggage represents Earth's lithospheric plates. The luggage is moved along by the conveyer belt until it is dumped into the baggage cart in the same way plates are moved by convection cells until they are subducted into Earth's interior. While this analogy allows one to visualize how the mechanism of plate movement takes place, it must be remembered that there are limitations to this analogy. The major limitation is that unlike the luggage, plates consist of continental and oceanic crust which have different densities, and only oceanic crust is subducted into Earth's interior. Nonetheless, this analogy does provide an easy way to visualize plate movement.

Most geologists accept plate tectonic theory because the evidence for it is overwhelming, and because it ties together many seemingly unrelated geologic features and events and shows how they are interrelated. Consequently, geologists now view such geologic processes as mountain building, seismicity, and volcanism, from the perspective of plate tectonics. Furthermore, because all of the inner planets have had a similar origin and early history, geologists are interested in determining whether plate tectonics is unique to Earth or whether it operates on other planets (see Chapter 19).

The Supercontinent Cycle

By the end of the Paleozoic Era, all of the continents had come together to form the supercontinent Pangaea. Soon thereafter, Pangaea began breaking apart into the familiar continents we know today. The breakup of Pangaea during the Triassic Period and the subsequent movement of plates resulted in the present distribution of continents

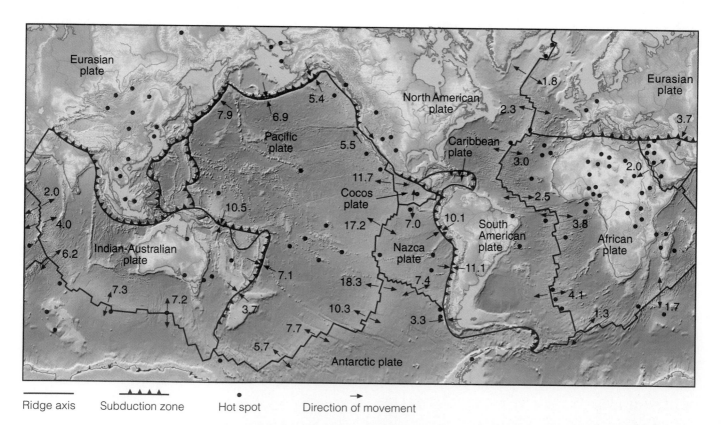

Ridge axis   Subduction zone   Hot spot   Direction of movement

Figure 9.16 *A map of the world showing the plates, their boundaries, relative motion and rates of movement in centimeters per year, and hot spots.*

Y ou've been selected to be part of the first astronaut team to go to Mars. While your two fellow crew members descend to the Martian surface, you'll be staying in the command module and circling the Red Planet. As part of the geologic investigation of Mars, one of the crew members will be mapping the geology around the landing site and deciphering the geologic history of the area. Your job will be to observe and photograph the planet's surface and try to determine whether Mars had an active plate tectonic regime in the past and whether there is current plate movement. What features would you look for and what evidence might reveal current or previous plate activity?

rising from below fills the fractures. As a basalt-filled fracture widens, it begins subsiding and forms a long narrow ocean such as the present-day Red Sea. Continued rifting eventually forms an expansive ocean basin such as the Atlantic Ocean basin.

One of the most convincing arguments for proponents of the supercontinent cycle hypothesis is the "surprising regularity" of mountain building caused by compression during continental collisions. These mountain-building episodes occur about every 400 to 500 million years and are followed by an episode of rifting about 100 million years later. In other words, a supercontinent breaks apart causing individual continental blocks to move away from each other. This results in interior oceans forming between the separating continents, followed by the moving continents colliding, forming mountain ranges and reassembling into another supercontinent.

The supercontinent cycle is one more example of how the various Earth systems and subsystems are interrelated and integrated.

and ocean basins. Some scientists have proposed that supercontinents consisting of all or most of Earth's landmasses form, break up, and re-form in a cycle spanning about 500 million years.

The *supercontinent cycle hypothesis* is an expansion on the ideas of the Canadian geologist J. Tuzo Wilson. During the early 1970s, Wilson proposed a cycle (now known as the Wilson cycle) that includes continental fragmentation, the opening and closing of an ocean basin, and reassembly of the continent. According to the supercontinent cycle hypothesis, heat accumulates beneath a supercontinent because rocks of continents are poor conductors of heat. As a result of the heat accumulation, the supercontinent domes upward and fractures. Basaltic magma

What Are the Three Types of Plate Boundaries?

Because plate tectonics appears to have operated since at least the Proterozoic, it is important that we understand how plates move and interact with each other. After all, the movement of plates has had a profound effect on the geologic and biologic history of this planet.

Geologists recognize three major types of plate boundaries: *divergent, convergent,* and *transform* (Table 9.2). It is along these boundaries that new plates are formed, are consumed, or slide laterally past one another. Interaction

Table 9.2

Types of Plate Boundaries

Type	Example	Landforms	Volcanism
Divergent			
Oceanic	Mid-Atlantic Ridge	Mid-oceanic ridge with axial rift valley	Basalt
Continental	East African Rift Valley	Rift valley	Basalt and rhyolite, no andesite
Convergent			
Oceanic–oceanic	Aleutian Islands	Volcanic island arc, offshore oceanic trench	Andesite
Oceanic–continental	Andes	Offshore oceanic trench, volcanic mountain chain, mountain belt	Andesite
Continental–continental	Himalayas	Mountain belt	Minor
Transform	San Andreas fault	Fault valley	Minor

of plates along their boundaries accounts for most of Earth's volcanic eruptions and earthquakes as well as the formation and evolution of its mountain systems.

Divergent Boundaries

Divergent plate boundaries or *spreading ridges* occur where plates are separating and new oceanic lithosphere is forming. Divergent boundaries are places where the crust is extended, thinned, and fractured as magma, derived from the partial melting of the mantle, rises to the surface. The magma is almost entirely basaltic, some of it intrudes into vertical fractures to form dikes and some is extruded as lava flows (Figure 9.17). As successive injections of magma cool and solidify, they form new oceanic crust and record the intensity and orientation of the magnetic field (Figure 9.12). Divergent boundaries most commonly occur along the crests of oceanic ridges, for example, the Mid-Atlantic Ridge. Oceanic ridges are thus characterized by rugged topography with high relief resulting from displacement of rocks along large fractures, shallow-focus earthquakes, high heat flow, and basaltic flows or pillow lavas (Figure 9.17).

Divergent plate boundaries are also present under continents during the early stages of continental breakup (Figure 9.18). When magma wells up beneath a continent, the crust is initially elevated, stretched, and thinned, producing fractures and rift valleys (Figure 9.18a). During this stage, magma typically intrudes into the fractures and flows onto the valley floor (Figure 9.18b). The East African rift valleys are an excellent example of this stage of continental breakup (Figure 9.19).

As spreading proceeds, some rift valleys will continue to lengthen and deepen until the continental crust eventually breaks, and a narrow linear sea is formed, separating two continental blocks (Figure 9.18c). The Red Sea separating the Arabian Peninsula from Africa is a good example of this stage of rifting (Figure 9.19a).

As a newly formed narrow sea continues enlarging, it may eventually become an expansive ocean basin such as the Atlantic Ocean basin is today, separating North and South America from Europe and Africa by thousands of kilometers (Figure 9.18d). The Mid-Atlantic Ridge is the boundary between these diverging plates; the American plates are moving westward, and the Eurasian and African plates are moving eastward.

An Example of Ancient Rifting

What features in the rock record can geologists use to recognize ancient rifting? Associated with regions of continental rifting are faults, dikes, sills, lava flows, and thick sedimentary sequences within rift valleys. The Triassic fault basins of the eastern United States are a good example of ancient continental rifting (see Figure 23.8). These

Figure 9.17 *Pillow lavas forming along the Mid-Atlantic Ridge. Their distinctive bulbous shape is the result of underwater eruption.*

fault basins mark the zone of rifting that occurred when North America split apart from Africa. They contain thousands of meters of continental sediment and are riddled with dikes and sills (see Chapter 23).

Convergent Boundaries

Whereas new crust forms at divergent plate boundaries, older crust must be destroyed and recycled in order for the entire surface area of Earth to remain constant. Otherwise we would have an expanding Earth. Such plate destruction occurs at **convergent plate boundaries** where two plates collide and the leading edge of one plate is subducted beneath the margin of the other plate and eventually incorporated into the asthenosphere.

Convergent boundaries are characterized by deformation, volcanism, mountain building, metamorphism, earthquake activity, and important mineral deposits. It is important to remember that when we talk about convergent plate boundaries, we are really talking about three different types of boundaries: *oceanic-oceanic, oceanic-continental,* and *continental-continental.* And while the basic processes are the same along all three boundaries, the results are different because different types of crust are involved.

Oceanic-Oceanic Boundaries

When two oceanic plates converge, one of them is subducted beneath the other along an **oceanic-oceanic plate boundary** (Figure 9.20). The subducting plate bends downward to form the outer wall of an oceanic trench. A *subduction complex,*

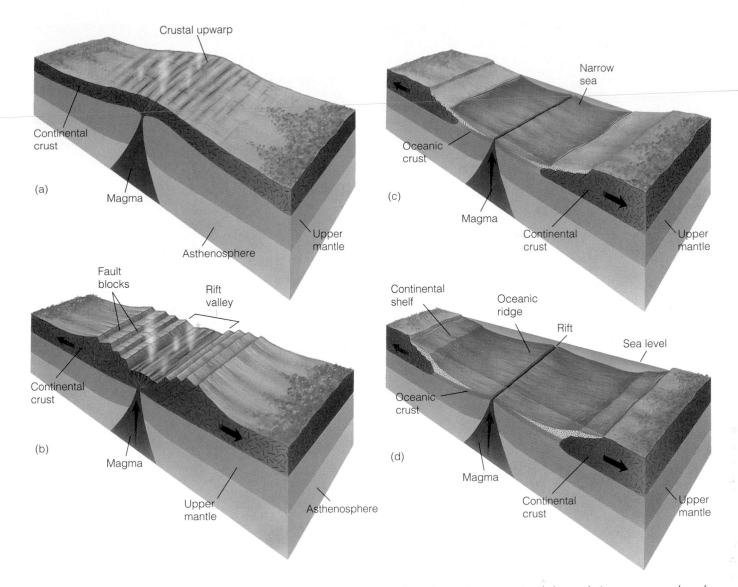

Figure 9.18 *History of a divergent plate boundary. (a) Heat from rising magma beneath a continent causes it to bulge, producing numerous cracks and fractures. (b) As the crust is stretched and thinned, rift valleys develop, and lava flows onto the valley floors. (c) Continued spreading further separates the continent until a narrow seaway develops. (d) As spreading continues, an oceanic ridge system forms, and an ocean basin develops and grows.*

composed of wedge-shaped slices of highly folded and faulted marine sediments and oceanic lithosphere scraped off the descending plate, forms along the inner wall of the oceanic trench. As the subducting plate descends into the mantle, it is heated and partially melted, thus generating magma commonly of andesitic composition. This magma is less dense than the surrounding mantle rocks and rises to the surface of the nonsubducted plate, forming a curved chain of volcanic islands called a **volcanic island arc** (any plane intersecting a sphere makes an arc). This arc is nearly parallel to the oceanic trench and is separated from it by a distance of up to several hundred kilometers—the distance depends on the angle of dip of the subducting plate (Figure 9.20).

In those areas where the rate of subduction is faster than the forward movement of the overriding plate, the lithosphere on the landward side of the volcanic island arc may be subjected to tensional stress and stretched and thinned, resulting in the formation of a *back-arc basin*. This back-arc basin may grow by spreading if magma breaks through the thin crust and forms new oceanic crust (Figure 9.20). A good example of a back-arc basin associated with an oceanic-oceanic plate boundary is the Sea of Japan between the Asian continent and the islands of Japan.

Most present-day active volcanic island arcs are in the Pacific Ocean basin and include the Aleutian Islands, the Kermadec-Tonga arc, and the Japanese and Philippine Islands. The Scotia and Antillean (Caribbean) island arcs are present in the Atlantic Ocean basin.

What Are the Three Types of Plate Boundaries? **251**

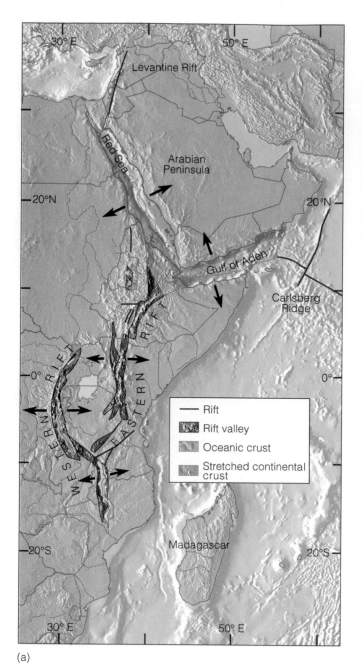

(a)

(b)

Figure 9.19 *(a) The East African Rift Valley is being formed by the separation of eastern Africa from the rest of the continent along a divergent plate boundary. The Red Sea represents a more advanced stage of rifting, in which two continental blocks are separated by a narrow sea. (b) View looking down the Great Rift Valley of Africa. Little Magadi, seen in the background, is one of numerous soda lakes forming in the valley. Because of high evaporation rates and lack of any drainage outlets, these lakes are very saline. The Great Rift Valley is part of the system of rift valleys resulting from stretching of the crust as plates move away from each other in eastern Africa.*

Oceanic-Continental Boundaries

When an oceanic and continental plate converge, the denser oceanic plate is subducted under the continental plate along an **oceanic-continental plate boundary** (Figure 9.21). Just as along oceanic-oceanic plate boundaries, the descending oceanic plate forms the outer wall of an oceanic trench.

As the cold, wet, and slightly denser oceanic plate descends into the hot asthenosphere, melting occurs and magma is generated. This magma rises beneath the overriding continental plate and can erupt at the surface, producing a chain of andesitic volcanoes (also called a volcanic

arc), or intrude into the continental margin as plutons, especially batholiths. An excellent example of an oceanic-continental plate boundary is the Pacific coast of South America where the oceanic Nazca plate is currently being subducted under South America (Figure 9.16). The Peru-Chile Trench marks the site of subduction, and the Andes Mountains are the resulting volcanic mountain chain on the nonsubducting plate.

Continental-Continental Boundaries

Two continents approaching each other will initially be separated by an ocean floor that is being subducted under one

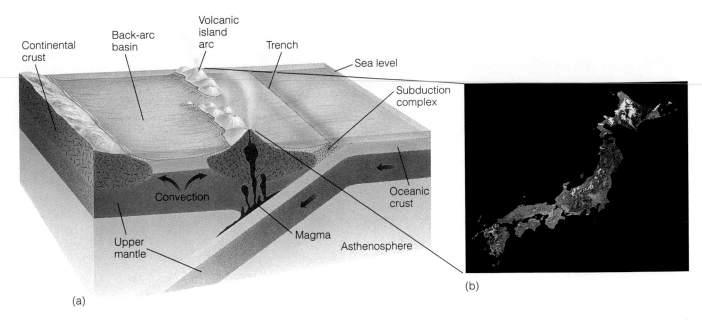

(a)

(b)

Figure 9.20 *Oceanic-oceanic plate boundary. (a) An oceanic trench forms where one oceanic plate is subducted beneath another. On the nonsubducted plate, a volcanic island arc forms from the rising magma generated from the subducting plate. (b) Satellite image of Japan. The Japanese Islands are a volcanic island arc resulting from the subduction of one oceanic plate beneath another oceanic plate.*

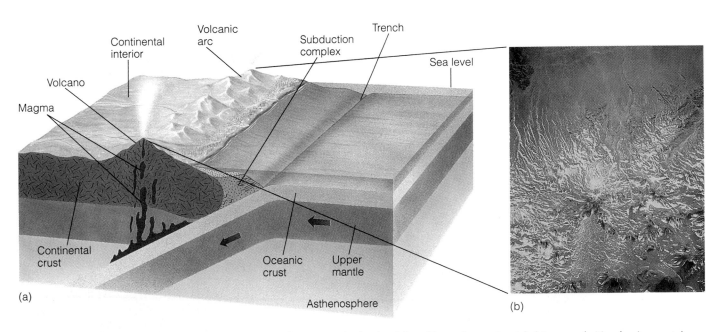

(a)

(b)

Figure 9.21 *Oceanic-continental plate boundary. (a) When an oceanic plate is subducted beneath a continental plate, an andesitic volcanic mountain range is formed on the continental plate as a result of rising magma. (b) A portion of the Andes mountain range in the Potosi Region of Boliva as photographed by the crew of the space shuttle* Atlantis *in 1992. The Andes are one of the best examples of continuing mountain building at an oceanic-continental plate boundary.*

continent. The edge of that continent will display the features characteristic of oceanic-continental convergence. As the ocean floor continues to be subducted, the two continents will come closer together until they eventually collide. Because continental lithosphere, which consists of continental crust and the upper mantle, is less dense than oceanic lithosphere (oceanic crust and upper mantle), it cannot sink into the asthenosphere. Although one continent may partly slide under the other, it cannot be pulled or pushed down into a subduction zone (Figure 9.22).

When two continents collide, they are welded together along a zone marking the former site of subduction. At this **continental-continental plate boundary,** an interior mountain belt is formed consisting of deformed sediments and sedimentary rocks, igneous intrusions, metamorphic rocks, and fragments of oceanic crust. In addition, the entire region is subjected to numerous earthquakes. The Himalayas in central Asia, the world's youngest and highest mountain system, resulted from the collision between India and Asia that began about 40 to 50 million years ago and is still continuing (see Chapter 10).

Recognizing Ancient Convergent Plate Boundaries

How can former subduction zones be recognized in the rock record? One clue is provided by igneous rocks. The magma erupted at the surface, forming island arc volcanoes and continental volcanoes, is of andesitic composition. Another clue can be found in the zone of intensely deformed rocks between the deep-sea trench where subduction is taking place and the area of igneous activity. Here, sediments and submarine rocks are folded, faulted, and metamorphosed into a chaotic mixture of rocks termed a *mélange.*

During subduction, pieces of oceanic lithosphere are sometimes incorporated into the mélange and accreted onto the edge of the continent. Such slices of oceanic crust and upper mantle are called **ophiolites** (Figure 9.23). They consist of a layer of deep-sea sediments that include graywackes (poorly sorted sandstones containing abundant feldspars and rock fragments, usually in a clay-rich matrix), black shales, and cherts. These deep-sea sediments are underlain by pillow basalts, a sheeted dike complex, massive gabbro, and layered gabbro, all of which form the oceanic crust. Beneath the gabbro is peridotite, which probably represents the upper mantle. Ophiolites are key features in recognizing plate convergence along a subduction zone.

Elongate belts of folded and faulted marine sediments, andesites, and ophiolites are found in the Appalachians, Alps, Himalayas, and Andes mountains. The combination of such features is good evidence that these mountain ranges resulted from deformation along convergent plate boundaries.

Transform Boundaries

The third type of plate boundary is a **transform plate boundary.** These usually occur along fractures in the

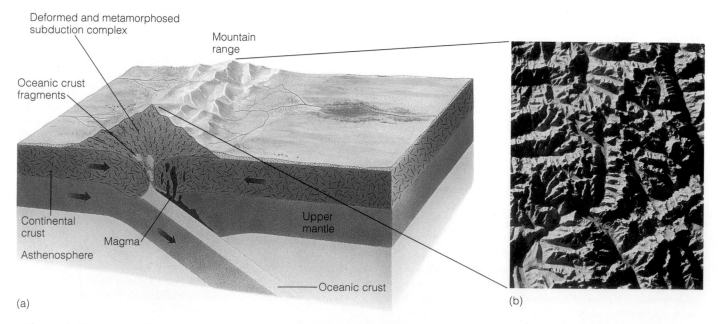

(a)

(b)

Figure 9.22 *Continental-continental plate boundary. (a) When two continental plates converge, neither is subducted because of their great thickness and low and equal densities. As the two continental plates collide, a mountain range is formed in the interior of a new and larger continent. (b) Vertical view of the Himalayas, the youngest and highest mountain system in the world. The Himalayas began forming when India collided with Asia 40 to 50 million years ago.*

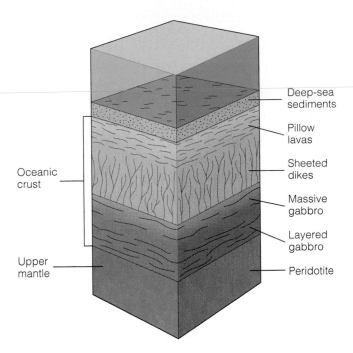

Figure 9.23 *Ophiolites are sequences of rock on land consisting of deep-sea sediments, oceanic crust, and upper mantle.*

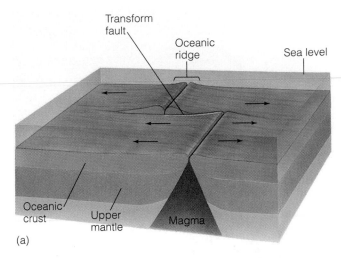

(a)

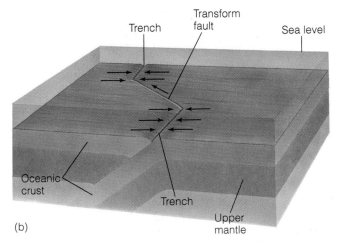

(b)

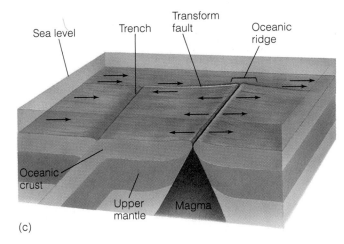

(c)

Figure 9.24 *Horizontal movement between plates takes place along a transform fault. (a) The majority of transform faults connect two oceanic ridge segments. Note that relative motion between the plates only occurs between the two ridges. (b) A transform fault connecting two trenches. (c) A transform fault connecting a ridge and a trench.*

seafloor known as *transform faults* where plates slide laterally past each other, roughly parallel to the direction of plate movement. Although lithosphere is neither created nor destroyed along a transform boundary, the movement between plates results in a zone of intensely shattered rock and numerous shallow-focus earthquakes.

Transform faults are particular types of faults that "transform" or change one type of motion between plates into another type of motion. Most commonly, transform faults connect two oceanic ridge segments, but they can also connect ridges to trenches and trenches to trenches (Figure 9.24). While the majority of transform faults are in oceanic crust and are marked by distinct fracture zones, they may also extend into continents.

One of the best-known transform faults is the San Andreas fault in California. It separates the Pacific plate from the North American plate and connects spreading ridges in the Gulf of California with the ridge separating the Juan de Fuca and Pacific plates off the coast of northern California (Figure 9.25). Many of the earthquakes affecting California are the result of movement along this fault.

Unfortunately, transform faults generally do not leave any characteristic or diagnostic features except for the obvious displacement of the rocks with which they are associated. This displacement is commonly large, on the order of tens to hundreds of kilometers. Such large displacements in ancient rocks can sometimes be related to transform fault systems.

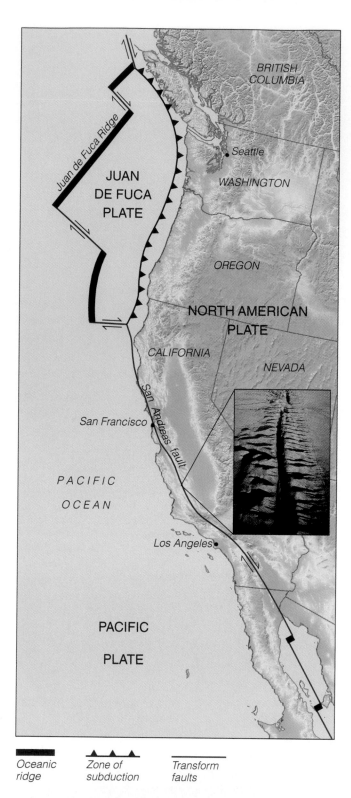

Oceanic ridge · Zone of subduction · Transform faults

Figure 9.25 *Transform plate boundary. The San Andreas fault is a transform fault separating the Pacific plate from the North American plate. Movement along this fault has caused numerous earthquakes. The inset photograph shows a segment of the San Andreas fault as it cuts through the Carrizo Plain, California.*

How Are Plate Movement and Motion Determined?

How fast and in what direction are Earth's various plates moving, and do they all move at the same rate? Rates of plate movement can be calculated in several ways. The least accurate method is to determine the age of the sediments immediately above any portion of the oceanic crust and divide that age by the distance from the spreading ridge. Such calculations give an average rate of movement.

A more accurate method of determining both the average rate of movement and relative motion is by dating the magnetic anomalies in the crust of the seafloor. The distance from an oceanic ridge axis to any magnetic anomaly indicates the width of new seafloor that formed during that time interval. Thus, for a given interval of time, the wider the strip of seafloor, the faster the plate has moved. In this way not only can the present average rate of movement and relative motion be determined (Figure 9.16), but the average rate of movement during the past can also be calculated by dividing the distance between anomalies by the amount of time elapsed between anomalies.

Geologists can not only calculate the average rate of plate movement from magnetic anomalies, but they can also determine plate positions at various times in the past using magnetic anomalies. Because magnetic anomalies are parallel and symmetrical with respect to spreading ridges, all one has to do to determine the position of continents when particular anomalies formed is to move the anomalies back to the spreading ridge, which will also move the continents with them (Figure 9.26). Unfortunately, subduction destroys oceanic crust and the magnetic record it carries. Thus, we have an excellent record of plate movements since the breakup of Pangaea, but not as good an understanding of plate movement before that time.

From the information in Figure 9.16, it is obvious that the rate of movement varies among plates. The southeastern part of the Pacific plate and the Cocos plates are the two fastest moving plates, while the Arabian and southern African plates are the slowest.

The average rate of movement as well as the relative motion between any two plates can also be determined by satellite laser ranging techniques. Laser beams from a station on one plate are bounced off a satellite (in geosynchronous orbit) and returned to a station on a different plate. As the plates move away from each other, the laser beam takes more time to go from the sending station to the stationary satellite and back to the receiving station. This difference in elapsed time is used to calculate the rate of movement and relative motion between plates. In addition, rates of movement and relative motion have also been calculated by measuring the difference between arrival

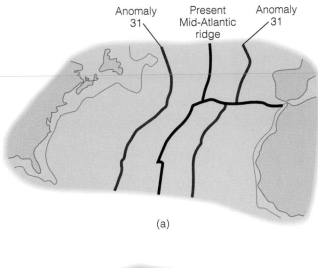

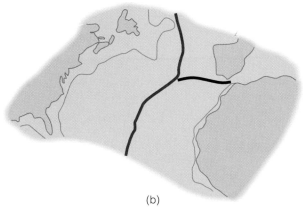

Figure 9.26 *Reconstructing plate positions using magnetic anomalies. (a) The present North Atlantic, showing the present ridge and magnetic anomaly 31, which formed 67 million years ago. (b) The Atlantic 67 million years ago. Anomaly 31 marks the plate boundary 67 million years ago. By moving the anomalies back together, along with the plates they are on, we reconstruct the former positions of the continents.*

times of radio signals from the same quasar to receiving stations on different plates. The rate of plate movement determined by these two techniques correlates closely with those determined from magnetic reversals.

Hot Spots and Absolute Motion

Plate motions determined from magnetic reversals, lasers bounced off satellites, and the difference between arrival times of radio signals from quasars give only the relative motion of one plate with respect to another. To determine absolute motion, we must have a fixed reference from which the rate and direction of plate movement can be determined. **Hot spots,** which may provide such reference points, are locations where stationary columns

of magma, originating deep within the mantle (mantle plumes), slowly rise to the surface and form volcanoes (Figure 9.16). Because the mantle plumes apparently remain stationary (although some evidence suggests that they might not) while the plates move over them, the resulting hot spots leave a trail of extinct and progressively older volcanoes called *aseismic ridges,* which records the movement of the plate.

One of the best examples of aseismic ridges and hot spots is the Emperor Seamount-Hawaiian Island chain (Figure 9.27). This chain of islands and seamounts (structures of volcanic origin rising more than 1 km above the seafloor) extends from the island of Hawaii to the Aleutian Trench off Alaska, a distance of some 6000 km, and consists of more than 80 volcanic structures.

Currently, the only active volcanoes in this island chain are the island of Hawaii and the Loihi Seamount. The rest of the islands are extinct volcanic structures that become progressively older toward the north and northwest. This means that the Emperor Seamount-Hawaiian Island chain records the direction that the Pacific plate traveled as it moved over an apparently stationary mantle plume. In this case, the Pacific plate first moved in a north-northwesterly direction, and then, as indicated by the sharp bend in the chain, changed to a west-northwesterly direction about 43 million years ago. The reason the Pacific plate changed directions is not known, but the shift might be related to the collision of India with the Asian continent at around the same time (see Figure 10.25).

What Is the Driving Mechanism of Plate Tectonics?

As you recall from our discussion earlier in the chapter, a major obstacle to the acceptance of continental drift was the lack of a driving mechanism to explain continental movement. When they saw that continents and ocean floors moved together, not separately, and that new crust formed at spreading ridges by rising magma, most geologists accepted some type of convective heat system as the basic process responsible for plate motion. The question still remains, however, as to what exactly drives the plates.

Two models involving thermal convection cells have been proposed to explain plate movement (Figure 9.28). In one model, thermal convection cells are restricted to the asthenosphere, whereas in the second model the entire mantle is involved. In both models spreading ridges mark the ascending limbs of adjacent convection cells, and trenches occur where convection cells descend back into Earth's interior. The locations of spreading ridges and trenches are therefore determined by the convection cells

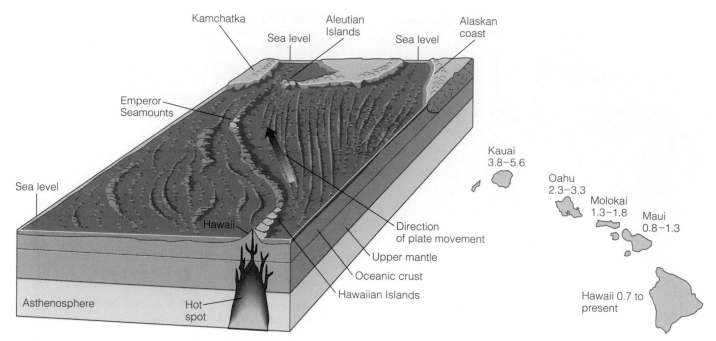

Figure 9.27 *The Emperor Seamount-Hawaiian Island chain formed as a result of movement of the Pacific plate over a hot spot. The line of the volcanic islands traces the direction of plate movement. The numbers indicate the age of the islands in millions of years.*

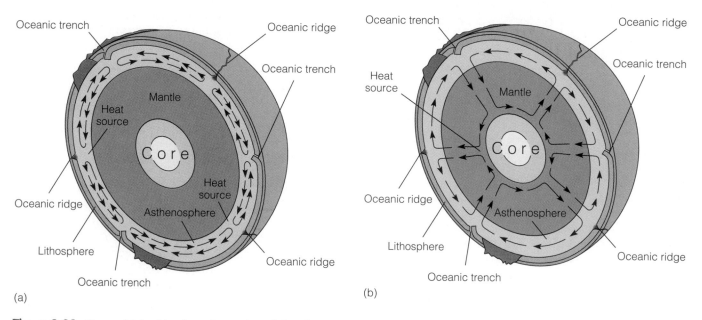

(a)

(b)

Figure 9.28 *Two models involving thermal convection cells have been proposed to explain plate movement. (a) In one model, thermal convection cells are restricted to the asthenosphere. (b) In the other model, thermal convection cells involve the entire mantle.*

themselves, and the lithosphere is considered to be the top of the thermal convection cell.

Although most geologists agree that Earth's internal heat plays an important role in plate movement, there are problems with both models. The major problem associated with the first model is the difficulty in explaining the source of heat for the convection cells and why they are restricted to the asthenosphere. In the second model, the source of heat comes from the outer core, but no one has yet explained how heat is transferred from the outer core to the mantle. Nor is it clear how convection can involve both the lower mantle and the asthenosphere.

In addition to some type of thermal convective system driving plate movement, some geologists think plate

Y ou are part of a mining exploration team that is exploring a promising and remote area of central Asia. You know that former convergent and divergent plate boundaries frequently are sites of ore deposits. What evidence would you look for to determine if the area you're exploring might be an ancient convergent or divergent plate boundary? Is there anything you can do before visiting the area that might help you in determining what the geology of the area is?

movement is also aided by "ridge-push" or "slab-pull" (Figure 9.29). According to the proponents of the ridge-push mechanism, the intrusion of magma into a spreading ridge provides an additional force that pushes the plates apart, thus aiding in the movement of plates. In slab-pull, the subducting cold slab of lithosphere, being denser than the surrounding warmer asthenosphere, pulls the rest of the plate along with it as it descends into the asthenosphere. While many geologists think that the forces associated with subduction, such as slab-pull, play an important role in plate movement, others are not yet convinced.

Geologists are fairly certain that some type of convective system is involved in plate movement, but the extent to which other mechanisms, such as ridge-push and slab-pull, are involved is still unresolved. Presently, none of the proposed mechanisms can explain all aspects of plate movement because the underlying forces are part of Earth's interior and difficult or impossible to measure directly. However, that plates have moved in the past and are still moving today has been proven beyond a doubt. And while a comprehensive theory of plate movement has not yet been developed, more and more of the pieces are falling into place as geologists learn more about Earth's interior.

How Does Plate Tectonics Affect the Distribution of Natural Resources?

Besides being responsible for the major features of Earth's crust, and influencing the distribution and evolution of the world's biota, plate movements also affect the formation and distribution of natural resources (see Perspective 9.2). Consequently, geologists are using plate tec-

tonic theory in their search for petroleum, natural gas, and mineral deposits and in explaining the occurrence of these natural resources.

It is becoming increasingly clear that if we are to keep up with the continuing demands of a global industrialized society, the application of plate tectonic theory to the origin and distribution of natural resources is essential.

Many metallic mineral deposits such as copper, gold, lead, silver, tin, and zinc are related to igneous and associated hydrothermal activity, so it is not surprising that a close relationship exists between plate boundaries and the occurrence of these valuable deposits.

The magma generated by partial melting of a subducting plate rises toward the surface, and as it cools, it precipitates and concentrates various metallic ores. Many of the world's major metallic ore deposits are associated with convergent plate boundaries including those in the Andes of South America, the Coast Ranges and Rockies of North America, Japan, the Philippines, Russia, and a zone extending from the eastern Mediterranean region to Pakistan. In addition, the majority of the world's gold is associated with sulfide deposits located at ancient convergent plate boundaries in such areas as South Africa, Canada, California, Alaska, Venezuela, Brazil, southern India, Russia, and western Australia.

The copper deposits of western North and South America are an excellent example of the relationship between convergent plate boundaries and the distribution, concentration, and exploitation of valuable metallic ores (Figure 9.30). The world's largest copper deposits are found along this belt. The majority of the copper deposits in the Andes and the southwestern United States were formed less than 60 million years ago when oceanic plates were subducted under the North and South American plates. The rising magma and associated hydrothermal fluids carried minute amounts of copper, which was originally widely disseminated but eventually became concentrated in the cracks and fractures of the surrounding andesites. These low-grade copper deposits contain from 0.2% to 2% copper and are extracted from large open-pit mines (Figure 9.30b).

Divergent plate boundaries also yield valuable resources. The island of Cyprus in the Mediterranean is rich in copper and has been supplying all or part of the world's needs for the last 3000 years. The concentration of copper on Cyprus formed as a result of precipitation adjacent to hydrothermal vents. This deposit was brought to the surface when the copper-rich seafloor collided with the European plate, warping the seafloor and forming Cyprus.

Studies indicate that minerals containing such metals as copper, gold, iron, lead, silver, and zinc are currently forming as sulfides in the Red Sea. The Red Sea is opening as a result of plate divergence and represents the earliest stage in the growth of an ocean basin (Figures 9.18c and 9.19).

Oil, Plate Tectonics, and Politics

It is certainly not surprising that oil and politics are closely linked. The Iran-Iraq War of 1980–1989 and the Gulf War of 1990–1991 were both fought over oil. Indeed, many of the conflicts in the Middle East today have as their underlying cause, control of the vast deposits of petroleum in the region. Most people, however, are not aware of why there is so much oil in this part of the world.

Although large concentrations of petroleum occur in many areas of the world, more than 50% of all proven reserves are in the Persian Gulf region. Interestingly, however, this region did not become a significant petroleum-producing area until the economic recovery following World War II (1939–1945). After the war, Western Europe and Japan in particular became dependent on Persian Gulf oil and still rely heavily on this region for most of their supply. The United States is also dependent on imports from the Persian Gulf, but receives significant quantities of petroleum from other sources such as Mexico and Venezuela.

Why is there so much oil in the Persian Gulf region? The answer lies in the paleogeography and plate movements of this region during the Mesozoic and Cenozoic eras (Figure 1). During the Meso-

zoic Era, and particularly the Cretaceous Period when most of the petroleum formed, the Persian Gulf area was a broad, stable marine shelf extending eastward from Africa. This passive continental margin lay near the equator where countless microorganisms lived in the surface waters. The remains of these organisms accumulated with the bottom sediments and were buried, beginning the complex process of petroleum generation and the formation of source beds.

As a consequence of rifting in the Red Sea and Gulf of Aden during the Cenozoic Era, the Arabian plate is moving northeast away from Africa and subducting beneath Iran. As the sediments of the passive continental margin were initially subducted, during the early stages of collision between Arabia and Iran, the heating broke down the organic molecules and led to the formation of petroleum. The tilting of the Arabian block to the northeast allowed the newly formed petroleum to migrate upward into the interior of the Arabian plate. The continued subduction and collision with Iran folded the rocks, creating traps for petroleum to accumulate. Thus, the vast area south of the collision zone (known as the Zagros suture) is oil producing.

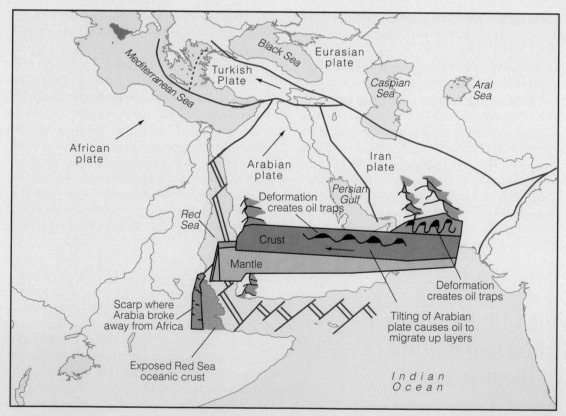

Figure 1 *Petroleum geologists divide oil-producing regions into two categories: the Middle East and everything else. In this oblique view, an imaginary trench has been cut across Arabia and the Persian Gulf. The collision between Arabia and Iran has tilted the Arabian plate and crumpled rocks on the edges of both plates. The tilting of Arabia allows oil to migrate upslope to accumulate in traps created by folding. Along almost the entire length of the plate boundary, conditions are ideal for maximum accumulation of oil. Elsewhere, the convergence of Arabia and the Eurasian plate is squeezing Turkey westward. Also, the opposing coastlines of the Red Sea fit almost perfectly, except at the southern end. There, a portion of Red Sea ocean floor is exposed, one of the few places oceanic crust is exposed on dry land. The true edge of the African crust is marked by a steep scarp, which matches the southwestern corner of the Arabian plate almost perfectly.*

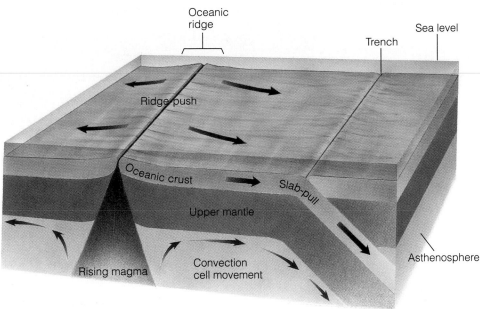

Oceanic ridge

Sea level

Trench

Ridge-push

Oceanic crust

Slab-pull

Upper mantle

Rising magma

Convection cell movement

Asthenosphere

Figure 9.29 *Plate movement is also thought to be aided by "ridge-push" or "slab-pull" mechanisms. In "ridge-push," the intrusion of magma into a spreading ridge provides an additional force that pushes the plates apart. In "slab-pull," the edge of the colder and denser subducting plate descends into Earth's interior, and pulls the rest of the plate downward.*

Figure 9.30 *(a) Important copper deposits are located along the west coasts of North and South America. (b) Bingham Mine in Utah is a huge open-pit copper mine with reserves estimated at 1.7 billion tons. More than 400,000 tons of rock are removed each day.*

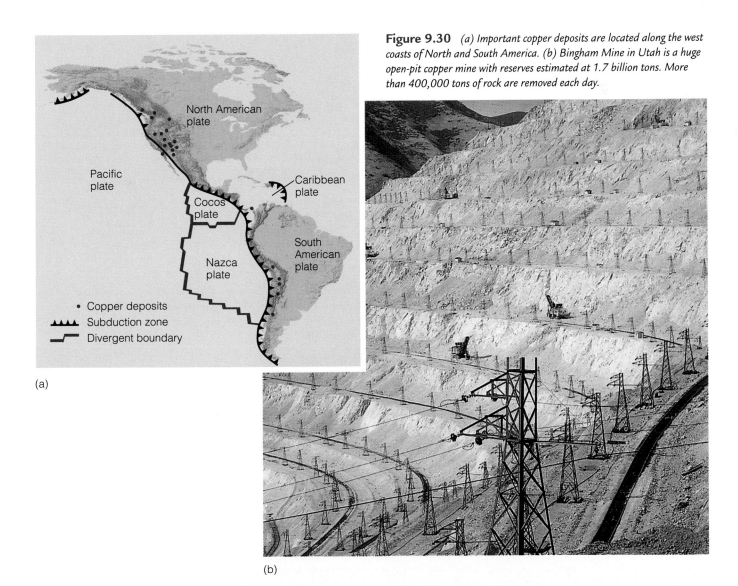

North American plate

Pacific plate

Caribbean plate

Cocos plate

South American plate

Nazca plate

• Copper deposits
▲▲▲ Subduction zone
⌐_⌐ Divergent boundary

(a)

(b)

How Does Plate Tectonics Affect the Distribution of Natural Resources? **261**

How Has Plate Tectonics Influenced Evolution and Affected the Distribution of the World's Biota?

Plate tectonic theory is as revolutionary and far-reaching in its implications for geology as the theory of evolution was for biology when it was proposed. Together, the theories of plate tectonics and evolution have changed the way we view our planet, and we should not be surprised at the intimate association between them. While the relationship between plate tectonic processes and the evolution of life is incredibly complex, paleontological data provide convincing evidence of the influence of plate movement on the distribution of organisms.

The present distribution of plants and animals is not random but is controlled largely by climate and geographic barriers. The world's organisms occupy *biotic provinces*, each of which is a region characterized by a distinctive assemblage of plants and animals. Organisms within a province have similar ecological requirements, and the boundaries separating provinces are therefore natural ecological breaks. Climatic or geographic barriers are the most common province boundaries, and these are largely controlled by plate movements.

Because adjacent provinces usually have less than 20% of their species in common, global diversity is a direct reflection of the number of provinces; the more provinces there are, the greater the global diversity. When continents break up, for example, the opportunity for new provinces to form increases, with a resultant increase in diversity. Just the opposite occurs when continents come together. Plate tectonics thus plays an important role in the distribution of organisms and their evolutionary history.

Complex interactions of wind and ocean currents have a strong influence on the world's climates. These currents are influenced by the number, distribution, topography, and orientation of continents. Temperature is one of the major limiting factors for organisms, and province boundaries often reflect temperature barriers. Because atmospheric and oceanic temperatures decrease from the equator to the poles, most species exhibit a strong climatic zonation. This biotic zonation parallels the world's latitudinal atmospheric and oceanic circulation patterns. Changes in climate thus have a profound effect on the distribution and evolution of organisms.

The distribution of continents and ocean basins not only influences wind and ocean currents, but also affects provinciality by creating physical barriers to, or pathways for, the migration of organisms. Intraplate volcanoes, island arcs, mid-oceanic ridges, mountain ranges, and sub-

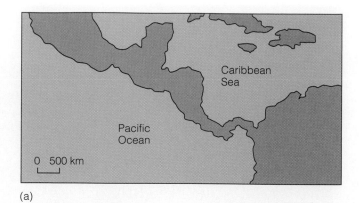

(a)

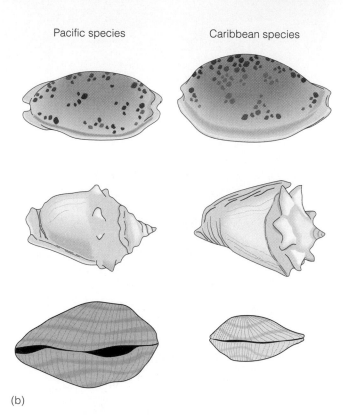

Pacific species   Caribbean species

(b)

Figure 9.31 *(a) The Isthmus of Panama forms a barrier that divides a once-uniform fauna. (b) Divergence of molluscan species after the formation of the Isthmus of Panama. Each pair belongs to the same genus but is a different species.*

duction zones all result from the interaction of plates, and their orientation and distribution strongly influence the number of provinces and hence total global diversity. Thus, provinciality and diversity will be highest when there are numerous small continents spread across many zones of latitude.

When a geographic barrier separates a once-uniform fauna, species may undergo divergence. If conditions on opposite sides of the barrier are sufficiently different, then species must adapt to the new conditions, migrate, or become extinct. Adaptation to the new environment by various species may involve enough change that new species eventually evolve. The marine invertebrates found on opposite sides of the Isthmus of Panama provide an excellent example of divergence caused by the formation of a geographic barrier. Prior to the rise of this land connection between North and South America, a homogeneous population of bottom-dwelling invertebrates inhabited the shallow seas of the area. After the rise of the Isthmus of Panama by subduction of the Pacific plate about 5 million years ago, the original population was divided. In response to the changing environment, new species evolved on opposite sides of the isthmus (Figure 9.31).

The formation of the Isthmus of Panama also had an impact on the evolution of the North and South American mammalian faunas. During most of the Cenozoic Era, South America was an island continent, and its mammalian fauna evolved in isolation from the rest of the world's faunas. When North and South America were connected by the Isthmus of Panama, most of the indigenous South American mammals were replaced by migrants from North America. Surprisingly, only a few South American mammal groups migrated northward.

Chapter Summary

1. The concept of continental movement is not new. The earliest maps showing the similarity between the east coast of South America and the west coast of Africa provided people with the first evidence that continents may once have been united and subsequently separated from each other.

2. Alfred Wegener is generally credited with developing the hypothesis of continental drift. He provided abundant geological and paleontological evidence to show that the continents were once united into one supercontinent he named Pangaea. Unfortunately, Wegener could not explain how the continents moved, and most geologists ignored his ideas.

3. The hypothesis of continental drift was revived during the 1950s when paleomagnetic studies indicated the presence of multiple magnetic north poles instead of just one as there is today. This paradox was resolved by moving the continents into different positions, making the paleomagnetic data consistent with a single magnetic north pole.

4. Magnetic surveys of the oceanic crust revealed magnetic anomalies in the rocks, indicating that Earth's magnetic field has reversed itself in the past. Because the anomalies are parallel and form symmetric belts adjacent to oceanic ridges, new oceanic crust must have formed as the seafloor was spreading.

5. Seafloor spreading has been confirmed by dating the sediments overlying the oceanic crust and by radiometric dating of rocks on oceanic islands. Such dating reveals that the oceanic crust becomes older with distance from spreading ridges.

6. Plate tectonic theory became widely accepted by the 1970s because of the overwhelming evidence supporting it and because it provides geologists with a powerful theory for explaining such phenomena as volcanism, earthquake activity, mountain building, global climatic changes, the distribution of the world's biota, and the distribution of mineral resources.

7. The supercontinent cycle indicates that all or most of Earth's landmasses form, break up, and re-form in a cycle spanning about 500 million years.

8. Three types of plate boundaries are recognized: divergent boundaries, where plates move away from each other; convergent boundaries, where two plates collide; and transform boundaries, where two plates slide past each other.

9. Ancient plate boundaries can be recognized by their associated rock assemblages and geologic structures. For divergent boundaries, these may include rift valleys with thick sedimentary sequences and numerous dikes and sills. For convergent boundaries, ophiolites and andesitic rocks are two characteristic features. Transform faults generally do not leave any characteristic or diagnostic features in the rock record.

10. The average rate of movement and relative motion of plates can be calculated in several ways. The results of these different methods all agree and indicate that the plates move at different average velocities.

11. Absolute motion of plates can be determined by the movement of plates over mantle plumes. A mantle plume is an apparently stationary column of magma that rises to the surface where it becomes a hot spot and forms a volcano.

12. Although a comprehensive theory of plate movement has yet to be developed, geologists think that some type of convective heat system is the major driving force.

13. A close relationship exists between the formation of some mineral deposits and petroleum, and plate boundaries. Furthermore, the formation and distribution of some natural resources are related to plate movements.

14. The relationship between plate tectonic processes and the evolution of life is complex. The distribution of plants and animals is not random, but is controlled largely by climate and geographic barriers, which are controlled, to a large extent, by the movement of plates.

Important Terms

continental-continental plate
 boundary
continental drift
convergent plate boundary
Curie point
divergent plate boundary
Glossopteris flora

Gondwana
hot spot
Laurasia
magnetic anomaly
magnetic reversal
oceanic-continental plate
 boundary

oceanic-oceanic plate boundary
ophiolite
paleomagnetism
Pangaea
plate
plate tectonic theory
seafloor spreading

thermal convection cell
transform plate boundary
transform fault
volcanic island arc

Review Questions

1. The man credited with developing the continental drift hypothesis is
   a. _____ Wilson; b. _____ Wegener; c. _____ Hess;
   d. _____ du Toit; e. _____ Vine.

2. The name of the supercontinent that formed at the end of the Paleozoic Era is
   a. _____ Laurasia; b. _____ Gondwana; c. _____
   Panthalassa; d. _____ Atlantis; e. _____ Pangaea.

3. Magnetic surveys of the ocean basins indicate that
   a. _____ the oceanic crust is youngest adjacent to mid-oceanic ridges; b. _____ the oceanic crust is oldest adjacent to mid-oceanic ridges; c. _____ the oceanic crust is youngest adjacent to the continents; d. _____ the oceanic crust is the same age everywhere; e. _____ answers (b) and (c).

4. The East African rift valleys are good examples of what type of plate boundary?
   a. _____ continental-continental; b. _____ oceanic-oceanic;
   c. _____ oceanic-continental; d. _____ divergent; e. _____ transform

5. Hot spots and aseismic ridges can be used to determine
   a. _____ the location of divergent plate boundaries;
   b. _____ the absolute motion of plates; c. _____ the location of magnetic anomalies in oceanic crust; d. _____ the relative motion of plates; e. _____ the location of convergent plate boundaries.

6. Which of the following lines of evidence was not used by Wegener and du Toit to show that continents had moved during the past?
   a. _____ continental fit; b. _____ similarity of rock sequences; c. _____ trends of mountain ranges on different continents; d. _____ fossil plants and animals;
   e. _____ paleomagnetism

7. The driving mechanism of plate movement is thought to be
   a. _____ isostasy; b. _____ thermal convection cells;
   c. _____ magnetism; d. _____ Earth's rotation; e. _____ polar wandering.

8. The San Andreas fault is an example of what type of plate boundary?
   a. _____ transform; b. _____ oceanic-continental; c. _____ divergent; d. _____ oceanic-oceanic;
   e. _____ continental-continental

9. Subduction occurs along what type of plate boundary?
   a. _____ divergent; b. _____ transform; c. _____ convergent; d. _____ answers (a) and (b);
   e. _____ answers (a) and (c).

10. Iron-bearing minerals in magma gain their magnetism and align themselves with the magnetic field when they cool through the
    a. _____ Curie point; b. _____ magnetic anomaly point;
    c. _____ thermal convection point; d. _____ hot spot point;
    e. _____ isostatic point.

11. Convergent plate boundaries are zones where
    a. _____ new continental lithosphere is forming; b. _____ new oceanic lithosphere is forming; c. _____ two plates come together; d. _____ two plates slide past each other;
    e. _____ two plates move away from each other.

12. The formation and distribution of copper deposits are associated with what type(s) of plate boundaries?
    a. _____ divergent; b. _____ convergent; c. _____ transform; d. _____ answers (a) and (b); e. _____ answers (b) and (c)

13. A distinctive assemblage of deep-sea sediments, oceanic crustal rocks, and upper mantle constitutes a(n)
    a. _____ back-arc basin; b. _____ volcanic island arc;
    c. _____ ophiolite; d. _____ subduction complex;
    e. _____ none of the previous answers.

14. The most common biotic province boundaries are
    a. _____ geographic barriers; b. _____ biologic barriers;
    c. _____ climatic barriers; d. _____ answers (a) and (b);
    e. _____ answers (a) and (c).

15. What evidence convinced Wegener that the continents were once joined together and subsequently broke apart?

16. What is the significance of polar wandering in relation to continental drift?

17. How can magnetic anomalies be used to show that the seafloor has been spreading?

18. What are mantle plumes and hot spots? How can they be used to determine the direction and rate of plate movement?

19. Using the age for each of the Hawaiian Islands in Figure 9.27, calculate the average rate of movement per year for the Pacific plate since each island formed. Is the average rate of movement the same for each island? Would you expect it to be? Explain why it may not be.

20. If movement along the San Andreas fault, which separates the Pacific plate from the North American plate, averages 5.5 cm per year, how long will it take before Los Angeles is opposite San Francisco?

21. How have plate tectonic processes affected the formation and distribution of natural resources?

22. Why is some type of thermal convection system thought to be the major force driving plate movement?
23. Briefly summarize the supercontinent cycle.
24. Explain how plate tectonics affects the evolution of life.
25. Explain why global diversity increases with an increase in biotic provinces. How does plate movement affect the number of biotic provinces?
26. Why is plate tectonics the unifying theory of geology?

27. The average rate of movement away from the Mid-Atlantic Ridge between North America and Africa is 3.0 cm per year (Figure 9.16), and the average sedimentation rate in the open ocean is 0.275 cm per thousand years. Calculate what the age of the oceanic crust is adjacent to Norfolk, Virginia, and what the thickness of the sediment is overlying the oceanic crust at that location. Does your calculation agree with the age of the oceanic crust at that location as shown in Figure 9.14?

 World Wide Web Activities

For these web site addresses, along with current updates and exercises, log on to **http://www.brookscole.com/geo**

Active Tectonics
This site functions mainly to disseminate information and links to other sites related to plate tectonics. Check out the comprehensive listing of web sites related to Active Tectonics as well as the images section.

Global Earth History
Developed at the University of Northern Arizona, this site contains plate and paleogeographic reconstructions for different time periods. See how plates have moved throughout the geologic past and how these movements have affected paleoclimates.

This Dynamic Earth: The Story of Plate Tectonics
This site provides the complete text and graphics from the original book written by W. Jacquelyne Kious and Robert I. Tilling and published in 1996 by the United States Geological Survey. The book covers the history and development of plate tectonic theory, as well as how it relates to people.

 CD-ROM Exploration

Exploring your *Earth Systems Today* CD-ROM will add to your understanding of the material in this chapter.

Topic: Earth's Processes

Module: Plate Tectonics

Explore activities in this module to see if you can discover the following for yourself:

Using the functions in this activity, examine the distribution of volcanoes and earthquakes in relation to plate boundaries. Also, look at the types of plate boundaries and their distribution.

Using this activity, play the Expedition game, which allows you to resolve placement of a plate boundary using an exploration budget. Draw your own plate boundaries based on the data that you collect.

CHAPTER 10

Deformation, Mountain Building, and the Continents

One of many scenic views of the Teton Range in Wyoming. Jenny Lake is in the foreground.

PROLOGUE

At the end of this chapter, you will have learned that

■ Rock deformation results from several forces and is characterized in different ways.

■ Several criteria are used to differentiate among the various types of geologic structures such as folds, joints, and faults.

■ Geologic structures are important to consider in human endeavors such as construction of highways and dams, the placement of power plants, and in finding and extracting some resources.

■ Deformation and the origin of geologic structures is important in the origin of many of Earth's mountains.

■ Most of the large mountain ranges on land formed, and in some cases continue to form, at the three types of convergent plate boundaries.

■ Terranes have special significance in mountain building.

■ Earth's continental crust, and especially mountains, stands higher than adjacent crust because of its composition and thickness.

North America has many scenic mountains, but few compare in grandeur to the Teton Range in northwestern Wyoming (see the chapter opening photo). The Teton Range is one of many mountain ranges in the Rocky Mountains, a complex mountain system stretching from central Mexico through the continental United States and Canada and into Alaska. Higher and larger ranges are present within the Rocky Mountain system, but none rise so abruptly as the Teton Range, which ascends nearly vertically more than 2100 m above Jackson Hole, the valley to the east (Figure 10.1a). The range consists of numerous jagged peaks, the highest of which, the Grand Teton, stands 4190 m above sea level. Native Americans in the region called these mountains Teewinot, meaning many pinnacles, an appropriate name indeed. The range and surrounding area now make up Grand Teton National Park.

The area now occupied by the Teton Range has a history of mountain building going back at least 90 million years. But these earlier mountains differed in several aspects from the ones present now. For one thing, their long axes were oriented northwest-southeast, whereas the Teton Range is oriented north-south. Furthermore, the earlier ranges formed by folding of the crust in response to compression, but the Teton range, which began forming about 10 million years ago, was uplifted along a large fracture known as the Teton fault (Figure 10.1b). This particular kind of fault, known as a normal fault, indicates that tension rather than compression was involved in the range's origin.

Notice in Figure 10.1b that most of the rocks in the Teton Range are meta-morphic and plutonic; both formed at great depth beneath a thick layer of sedimentary rocks. However, movement on the Teton fault resulted in uplift of the range as much as 6100 m relative to the rocks on the east. As the block that was to become the Teton Range rose it was deeply eroded thus exposing the underlying metamorphic and plutonic rocks. Notice also in Figure 10.1b that only the eastern side of the Teton Range is bounded by a fault. As a result of movement on this fault, the uplifted block has been tilted ever more steeply toward the west. Displacement of recent sediments indicates that the Teton fault is still active.

Uplift along the Teton fault was responsible for elevating one block of Earth's crust relative to another, but the range's spectacular, rugged topography resulted from erosion, most of it rather recently by geologic standards. Today, the Teton Range supports about a dozen small glaciers, moving bodies of ice on land. But during the past 200,000 years it was, at times, much more heavily glaciated. As moving solids, glaciers are particularly effective at erosion, transport, and deposition. Indeed, they were responsible for the deeply scoured valleys and intricately sculpted ridges and peaks as well as many deposits in Jackson Hole. For example, a glacial deposit known as a moraine impounds Jenny Lake (see chapter opening photo). Many glacial landforms are easily seen in the area, including the Grand Teton, which is a horn peak (see Chapter 14). Today, glaciers have only a minimal effect in this area, but other geologic processes such as ongoing uplift, weathering, and erosion by running water continue to modify this scenic mountain range.

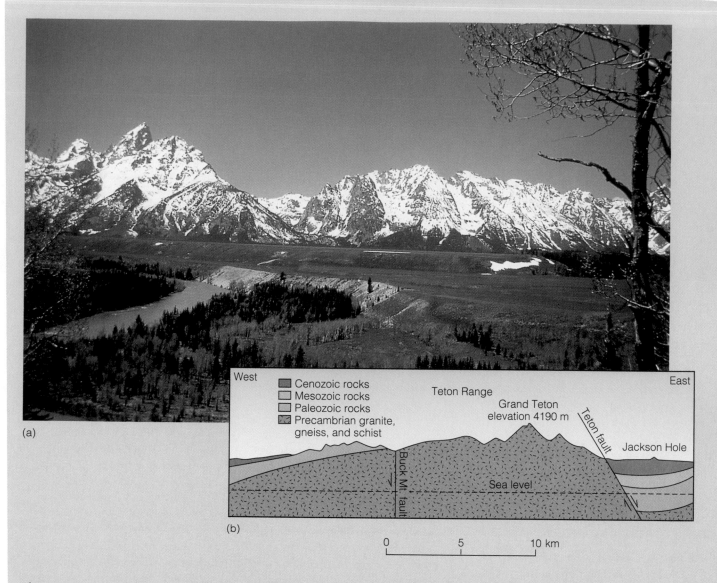

Figure 10.1 *(a) View from the east of the Teton Range. The Grand Teton is the highest peak visible. (b) East-west cross section of the Teton Range showing uplift along the Teton fault. This fault is not visible in (a), but it lies at the base of the mountains.*

Introduction

In previous chapters we noted that rocks are not necessarily as solid as the expression "solid as rock" implies. They decompose in response to chemical activities, and physical agents disaggregate them; heat, pressure, and chemical fluids bring about changes during metamorphism; and at great depth where they are under high temperature and pressure, rocks behave very differently from the way they behave at the surface. Indeed, many rocks show the effects of deformation, meaning that dynamic forces caused fracturing, contortion of rock layers, or both (Figure 10.2). Some manifestations of these dynamic forces at work within Earth are seismic activity and the ongoing evolution of mountains in South America, Asia, and elsewhere. In short, Earth remains an active planet with a variety of processes being driven by internal heat, particularly the movement of lithospheric plates. In fact, most of Earth's present-day seismic activity, volcanism, rock deformation, and mountain building takes place at divergent, convergent, and transform plate boundaries.

Fold

Fold

(a)

(b)

Figure 10.2 *Many rocks show the effects of deformation. (a) Folded rock layers. (b) Rock deformation resulting in numerous closely spaced fractures.*

A number of processes account for the origin of mountains, some of which involve little or no deformation, but in most mountains on continents the rocks have been complexly deformed by compression at convergent plate boundaries. The Alps in Europe, the Appalachians of North America, the Himalayas of Asia, and the Andes of South America owe their existence, and in some cases continuing evolution, to deformation and perhaps volcanism at convergent plate boundaries. In short, rock deformation and mountain building are closely related phenomena and are thus both discussed in this chapter.

Why Should You Study Deformation and Mountain Building?

Why should you study deformation and mountain building? Although much of this chapter is devoted to a review of the various types of geologic structures, their descriptive terminology, and the forces responsible for them, the study of rock deformation has several practical applications. For one thing, features such as folds and fractures resulting from deformation provide a record of the kinds and intensities of forces that operated during the past. Thus, the interpretation of these geologic structures allows us to partly satisfy our curiosity about Earth history. In addition, understanding the nature of geologic structures is essential in a number of engineering endeavors such as choosing sites for dams, large bridges, and nuclear power plants, especially if the sites are in areas of active deformation. And finally, many aspects of mining and exploration for petroleum and natural gas rely on correctly interpreting a variety of geologic structures.

Rock Deformation—What Does It Mean, and How Does It Occur?

Deformation is a general term encompassing all changes in the shape or volume or both of rocks. That is, rocks might be fractured or crumpled into folds as a result of **stress,** which is simply the force applied to a given area of rock. If the intensity of the stress is greater than the rock's internal strength, it will undergo **strain,** which is deformation caused by stress. The terminology is a bit confusing at first, but keep in mind that deformation and strain are synonyms, and stress is the force that causes deformation or strain. The following example will perhaps clarify the meaning of stress and the distinction between stress and strain.

Remember that stress is the force applied to a given area of rock, so it can be expressed in pounds per square inch (lbs/in^2) or kilograms per square centimeter (kg/cm^2), or any other convenient expression of force per unit area. For instance, the stress, or force, exerted by a person walking on an ice-covered pond is a function of the person's weight and the area beneath his or her feet. The ice's internal strength resists the stress unless it is too great, in which case it might begin to bend or crack as it is strained, or deformed. To avoid breaking through the ice and having a very chilly swim, the person might lie down; this does not reduce the total force but does distribute it over a larger area, thus reducing the stress per unit area.

Although stress is force per unit area, it comes in three varieties—*compression, tension,* and *shear*—depending on the directions of the applied forces. In **compression,** rocks are squeezed or compressed by external forces directed toward one another. Rock layers subjected to compression are commonly shortened in the direction of stress by either folding or faulting (Figure 10.3a). **Tension** results from forces acting in opposite directions along the same line and tends to lengthen rocks or pull them apart (Figure 10.3b). In **shear stress,** forces act parallel to one another but in opposite directions, resulting in deformation by displacement of adjacent layers along closely spaced planes (Figure 10.3c).

Three types of strain can result depending on the response of rocks to stress. In **elastic strain,** deformed rocks return to their original shape when the stress is relaxed. A somewhat analogous situation is found in squeezing a rubber ball that regains its original shape when no longer squeezed. Or, in the example of our ice-covered pond, the ice might bend under a person's weight but return to its original shape when no longer under stress. As one might expect, most rocks are not very elastic, but when loaded

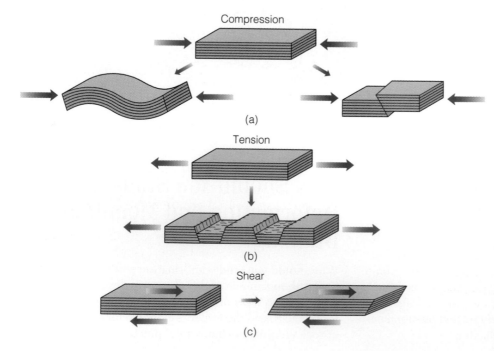

Figure 10.3 *Stress and possible types of resulting deformation. (a) Compression causes shortening of rock layers by folding or faulting. (b) Tension lengthens rock layers and causes faulting. (c) Shear stress causes deformation by displacement along closely spaced planes.*

The types of stresses as well as elastic versus plastic strain might seem rather esoteric, but perhaps understanding these concepts has some practical applications. What relevance do you think knowing about stress and strain has to some professions, other than geology, and what professions might these be? Can you think of stresses and strain that we contend with in our daily lives? As an example, what happens when a car smashes into a tree?

Figure 10.5 *This marble slab in the Rock Creek Cemetery, Washington, D.C., bent under its own weight in about 80 years.*

with glacial ice, Earth's crust responds by elastic strain and is depressed into the mantle.

When stress is applied to rocks they respond first by elastic strain, but when strained beyond their elastic limit, they respond by **plastic strain,** as when they are folded, or they behave as brittle solids and **fracture** (Figure 10.4). In either case, the rocks cannot recover their original shape. So we can summarize by saying that elastic strain results in no permanent deformation, whereas rocks experiencing plastic deformation or fracture are permanently deformed. One more point should be made regarding stress and strain. Rocks are considerably stronger in compression than they are in tension.

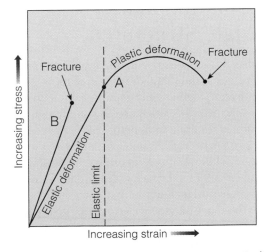

Figure 10.4 *Strain, or deformation, in response to stress. Rocks initially respond to stress by elastic deformation and return to their original shape when the stress is released. If the elastic limit is exceeded as in curve A, rocks deform plastically, which is permanent deformation. The amount of plastic deformation rocks exhibit before fracturing depends on their ductility: if they are ductile, they show considerable plastic deformation (curve A), but if they are brittle, they show little or no plastic deformation before failing by fracture (curve B).*

Many rocks show the effects of deformation, but deformation differs depending on such factors as the kind of stress, pressure and temperature, rock type, and the length of time stress is applied. A small stress applied over a long period, as on the slab in Figure 10.5, will cause plastic deformation, but if struck by a hammer (a large stress applied rapidly), the same slab will fracture. Rock type is important because not all rocks possess the same internal strength and thus respond differently to stress. In fact, geologists characterize rocks as either *ductile* or *brittle* depending on the amount of plastic strain they exhibit. Ductile rocks show much more plastic strain before they fracture than brittle rocks do (Figure 10.4). For instance, mudrocks such as mudstone and shale are generally more ductile than sandstone and limestone, but any rock type at depth where it is under great pressure and high temperature is more ductile than it would be at or near the surface.

Strike and Dip—Determining the Orientation of Rock Layers

During the 1660s, Nicholas Steno, a Danish anatomist, proposed several principles essential for deciphering Earth history from the record preserved in rocks, or what we refer to as the geologic record (see Chapter 17). One is the *principle of original horizontality*, which means that when deposited, sediments accumulate in horizontal or nearly horizontal layers. Thus, if we observe steeply inclined beds of sedimentary rock we are justified in inferring that they were deposited nearly horizontally, lithified, and then tilted into their present position (Figure 10.6). Some igneous rocks also form nearly horizontal layers. Rock layers deformed by folding and/or faulting

(a)

(b)

Figure 10.6 *(a) These sedimentary rock layers in the Valley of the Gods, Utah, are still horizontal as when they were deposited and lithified. (b) We can infer that these sandstone beds in Colorado were deposited horizontally, lithified, and then tilted into their present position.*

are no longer in their original position, so geologists use *strike* and *dip* to describe their orientation with respect to a horizontal plane.

By definition **strike** is the direction of a line formed by the intersection of a horizontal plane with an inclined plane. The surfaces of the rock layers in Figure 10.7 are good examples of inclined planes, whereas the water surface is a horizontal plane. The direction of the line formed at the intersection of these planes is the strike of the rock layers. The strike line's orientation is determined by using a compass to measure its angle with respect to north. **Dip** is a measure of an inclined plane's maximum inclination

from horizontal, so it must be measured at right angles to strike direction (Figure 10.7).

Geologic maps showing the age, aerial distribution, and geologic structures of rocks in an area employ a special symbol to indicate strike and dip. A long line oriented in the appropriate direction indicates strike and a short line perpendicular to the strike line shows the direction of dip (Figure 10.7). Adjacent to the strike and dip symbol is a number corresponding to the dip angle. The usefulness of strike and dip symbols will become apparent in the following sections on folds and faults.

Deformation and Geologic Structures

Remember that deformation and its synonym strain refer to changes in the shape or volume of rocks. During deformation rocks might be crumpled into folds, or they might be fractured, or perhaps folded and fractured. Any of these features resulting from deformation is referred to as a *geologic structure*. Various geologic structures are present almost everywhere that rock exposures can be observed, and many can be detected far below the surface by drilling and several geophysical techniques. In the following sections we examine the types of geologic structures geologists recognize, and we also discuss deformation and the origin of mountains at convergent plate boundaries.

Folded Rock Layers

Geologic structures known as folds, in which planar features are crumpled and bent, are quite common. Compression is responsible for most folding, as when you place your hands on a tablecloth and move

Figure 10.7 *Strike and dip. The intersection of a horizontal plane (the water surface) and an inclined plane (the surface of any of the rock layers) forms a line known as strike. The dip of these layers is their maximum angular deviation from horizontal. Notice the strike and dip symbol with 50 adjacent to it indicating the angle of dip.*

them toward one another, thereby producing a series of up- and down-arches in the fabric. Rock layers in the crust respond similarly to compression, but unlike the tablecloth, folding in rock layers is permanent. That is, plastic strain has taken place—so once folded, the rocks stay folded. Most folding probably takes place deep in the crust where rocks are more ductile than they are at or near the surface. The configuration of folds and the intensity of folding varies considerably, but only three basic types of folds are recognized: *monoclines, anticlines,* and *synclines.*

Monoclines

A simple bend or flexure in otherwise horizontal or uniformly dipping rock layers is a **monocline** (Figure 10.8a). The large monocline in Figure 10.8b formed when the Bighorn Mountains in Wyoming rose vertically along a fracture. The fracture did not penetrate to the surface, so as uplift of the mountains proceeded, the near-surface rocks were bent so that they now appear to be draped over the margin of the uplifted block. In a manner of speaking, a monocline is simply one half of an anticline or syncline (see below).

Anticlines and Synclines

Monoclines are not rare, but they are not nearly as common as anticlines and synclines. An **anticline** is an up-arched or convex upward fold with the oldest rock layers in its core, whereas a **syncline** is a down-arched or concave downward fold in which the youngest rock layers are in its core (Figure 10.9). Anticlines and synclines possess an axial plane connecting the points of maximum curvature of each folded layer (Figure 10.10); the axial plane divides folds into halves, each half being a *limb.* Because folds are most often found in a series of anticlines alternating with synclines, an anticline and adjacent syncline share a limb. It is important to remember that anticlines and synclines are simply folded rock layers and do not necessarily correspond to high and low areas at the surface. In fact, folds might underlie rather flat areas (Figure 10.11).

Folds are commonly exposed to view in areas of deep erosion, but even where eroded, strike and dip and the relative ages of the folded rock layers can easily distinguish anticlines from synclines. Notice in Figure 10.12 that in the surface view of the anticline, each limb dips outward or away from the center of the fold, and the oldest exposed

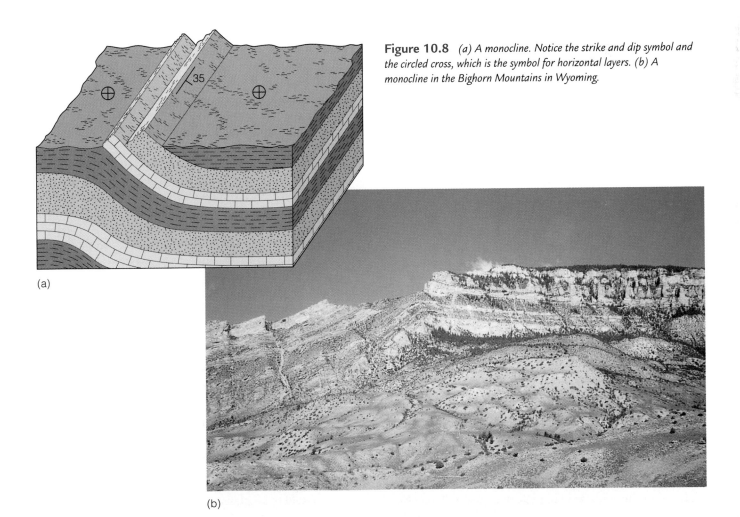

Figure 10.8 (a) A monocline. Notice the strike and dip symbol and the circled cross, which is the symbol for horizontal layers. (b) A monocline in the Bighorn Mountains in Wyoming.

(a)

(b)

Figure 10.9 *Folded rocks in the Calico Mountains of southeastern California. Three folds are visible from left to right and consist of a syncline, an anticline, and another syncline. We can infer that compression was responsible for these folds.*

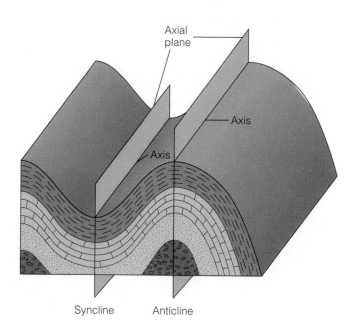

Figure 10.10 *Syncline and anticline showing the axial plane, axis, and fold limbs.*

rocks are in the fold's core. In an eroded syncline, though, each limb dips inward toward the fold's center, where the youngest exposed rocks are found.

The folds described so far are *upright,* meaning that their axial planes are vertical and both fold limbs dip at the same angle (Figure 10.12). In many folds, though, the axial plane is not vertical, the limbs dip at different angles, and the folds are characterized as *inclined* (Fig-

ure 10.13a). If both limbs dip in the same direction, the fold is *overturned.* That is, one limb has been rotated more than 90° from its original position so that it is now upside down (Figure 10.13b). In some areas deformation has been so intense that axial planes of folds are now horizontal, giving rise to what geologists call *recumbent folds* (Figure 10.13c). Overturned and recumbent folds are particularly common in mountains resulting from compression at convergent plate boundaries (discussed later in this chapter).

For upright folds, the distinction between anticlines and synclines is straightforward; but interpreting complex folds that have been tipped on their sides or turned completely upside down is more difficult. Can you determine which of the two fold types shown in Figure 10.13c is an anticline? Even if strike and dip symbols were shown you could still not resolve this question, but knowing the relative ages of the folded rock layers provides a solution. Remember that an anticline has the oldest rock layers in its core, so the fold nearest the surface is an anticline and the lower fold is a syncline.

Plunging Folds

As if upright and inclined folds were not enough, geologists further characterize folds as *nonplunging* or *plunging.* In some folds the fold axis, a line formed by the intersection of the axial plane with the folded layers, is horizontal and the folds are nonplunging (Figure 10.12). Much more commonly, though, fold axes are inclined so that they appear to plunge beneath adjacent rocks and the folds are said to be plunging (Figure 10.14).

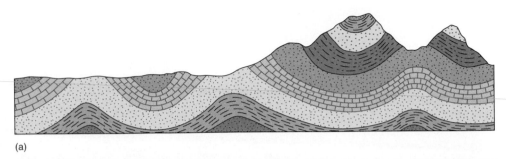

(a)

(b)

Figure 10.11 *Folds and their relationship to topography. (a) Cross section showing that anticlines and synclines do not necessarily correspond with high and low areas. Notice that folds underlie the rather flat area. (c) A syncline is at the peak of this mountain in Kootenay National Park, British Columbia, Canada. Notice the anticline and syncline on the lower left flank of the mountain.*

It might seem that with this additional complication, differentiating plunging anticlines from plunging synclines would be much more difficult, but geologists use exactly the same criteria they use for nonplunging folds. Therefore, all rock layers dip away from the fold axis in plunging anticline and toward the axis in plunging synclines. The oldest exposed rocks are in the core of an eroded plunging anticline, whereas the youngest exposed rock layers are found in the core of an eroded plunging syncline (Figure 10.14b).

In Chapter 5 we noted that anticlines form one type of structural trap in which petroleum and natural gas might accumulate (see Figure 5.32b). As a matter of fact, most

of the world's petroleum production comes from anticlines, although other geologic structures and stratigraphic traps are important, too. Accordingly, geologists and their employers are particularly interested in correctly identifying geologic structures in areas of potential hydrocarbon production.

Domes and Basins

Anticlines and synclines are elongate structures, meaning that their length greatly exceeds their width. In contrast, **domes** and **basins** are circular to oval folds (Figure 10.15). In fact, we can think of domes and basins as the circular to

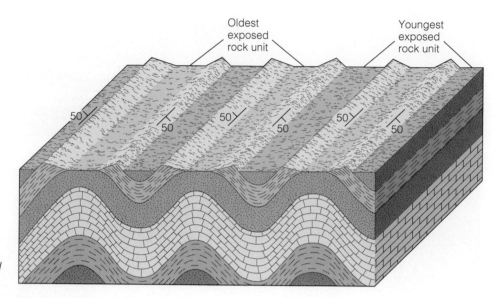

Figure 10.12 *Identifying eroded anticlines and synclines by strike and dip and the relative ages of the folded rocks.*

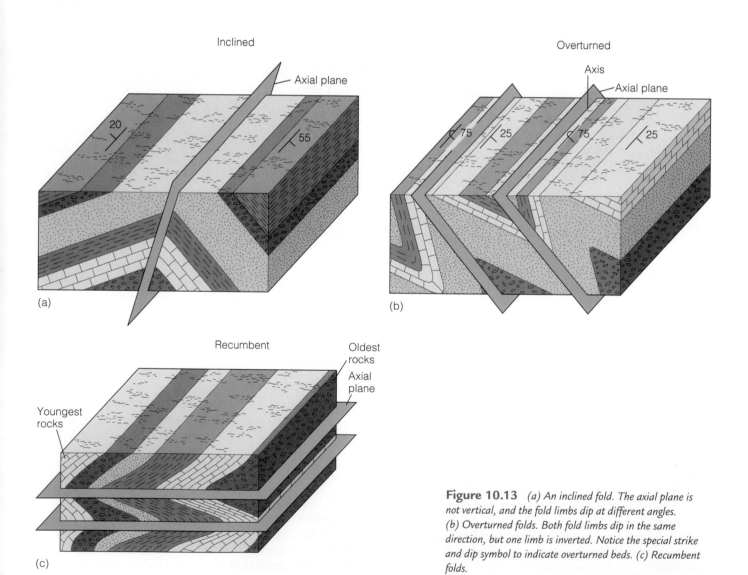

(a)

(b)

(c)

Figure 10.13 *(a) An inclined fold. The axial plane is not vertical, and the fold limbs dip at different angles. (b) Overturned folds. Both fold limbs dip in the same direction, but one limb is inverted. Notice the special strike and dip symbol to indicate overturned beds. (c) Recumbent folds.*

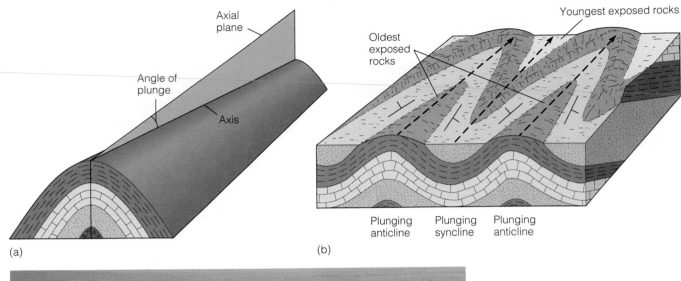

Figure 10.14 *Plunging folds. (a) A schematic illustration of a plunging fold. (b) A block diagram showing surface and cross-sectional views of plunging folds. The long arrow at the center of each fold shows the direction of plunge. (c) Surface view of the eroded, plunging Sheep Mountain anticline in Wyoming.*

oval equivalents of anticlines and synclines, respectively. In an eroded dome, all rock layers dip outward from a central point and the oldest exposed rocks are at the center of the structure. Just the opposite is true in a basin—that is, all rock layers dip inward toward a central point and the youngest exposed rocks are at the center (Figure 10.15).

Many domes and basins are so large that they can be visualized only on geologic maps or aerial photographs. The Black Hills of South Dakota, for example, are a large oval dome. One of the best-known large basins in the United States is the Michigan basin, most of which is buried beneath younger strata so it is not directly observable at the surface. Nevertheless, strike and dip of exposed strata near the basin margin and thousands of drill holes for oil and gas clearly show that the strata are deformed into a large basin.

Unfortunately, the terms *dome* and *basin* are also used to distinguish high and low areas of Earth's surface, but as with anticlines and synclines, domes and basins resulting from deformation do not necessarily correspond with mountains or valleys. In some of the following discussions we will have occasion to use these terms in other contexts, but we will endeavor to be clear when we refer to surface elevations as opposed to geologic structures.

Joints

One does not have to be a geologist to realize that surface rocks of all kinds are riddled with fractures. If the blocks of rock on opposite sides of fractures have not moved parallel to the fracture surface they are

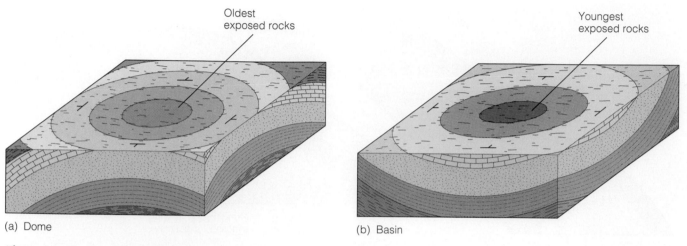

Oldest
exposed rocks

Youngest
exposed rocks

(a) Dome

(b) Basin

Figure 10.15 *In a dome (a) the oldest exposed rocks are in the center and all rock layers dip away from a central point, whereas in a basin (b) the youngest exposed rocks are in the center and all layers dip toward a central point.*

joints (Figure 10.16). The only movement shown on joints is perpendicular to the fracture surfaces; that is, joints might open up. Coal miners originally used the term *joint* long ago for cracks in rocks that appeared to be surfaces where adjacent blocks were "joined" together. Joints are extremely common, but we must distinguish them from *faults,* fractures along which movement has taken place parallel with the fracture surface (see the next section).

Joints form under a variety of conditions from compression, tension, and shear, but in all cases the rocks behaved as brittle solids. When an anticline forms in response to compression, the rock layers are flexed so that in their upper parts they experience tension, and fractures form parallel with the fold's long axis. Joints also form when rocks are stretched as in tension and when subjected to shear stress.

Some joints are simply minute fractures whereas others are structures of regional extent (Figure 10.16a). Additionally, they are commonly arranged in parallel or nearly parallel sets, and two or perhaps three prominent sets may be present in a region. Regional mapping reveals that joint sets are usually related to other geologic structures such as large folds. Weathering and erosion of rocks with prominent joints in Utah has yielded the spectacular scenery in Arches National Park (see Chapter 5 Prologue).

We have already discussed columnar joints that form when lava or magma in some shallow plutons cools and contracts (see Figure 4.6 and Perspective 4.1). Another type of jointing previously discussed is sheet jointing that forms in response to pressure release (see Figures 5.5 and 5.6).

Faults

Another type of fracture known as a **fault** is one along which blocks of rock on opposite sides of the fracture have moved parallel with the fracture surface, and the surface along which movement takes place is a **fault plane** (Figure 10.17a). Not all faults penetrate to the surface, but those that do might show a *fault scarp,* a bluff or cliff formed by vertical movement (Figure 10.17b). Fault scarps are usually quickly eroded and obscured. When movement takes place on a fault plane, the rocks on opposite sides may be scratched and polished (Figure 10.17b), or crushed and shattered into angular block forming a *fault breccia* (Figure 10.17c).

Refer to Figure 10.17a and notice the designations *hanging wall block* and *footwall block*. The **hanging wall block** consists of the rock overlying the fault, whereas the **footwall block** lies beneath the fault plane. You can recognize these two blocks on any fault except a vertical one—that is, one that dips at 90°. To identify some kinds of faults you must not only correctly identify these two blocks but also determine which one moved relatively up or down. We use the phrase *relative movement* because you cannot usually tell which block moved or if both moved. In Figure 10.17a, the footwall block may have moved up, the hanging wall block may have moved down, or both could have moved. Nevertheless, the hanging wall block appears to have moved down relative to the footwall block.

Remember our discussion of strike and dip of rock layers. Fault planes are also inclined planes and they too are characterized by their strike and dip (Figure 10.17a). In fact, the two basic varieties of faults are defined on the basis of whether the blocks on opposite sides of the fault plane moved parallel to the direction of dip (dip-slip faults) or along the direction of strike (strike-slip faults).

Dip-Slip Faults

All movement on dip-slip faults takes place parallel with the fault's dip; that is, all movement is vertical, either up or down the fault plane. In Figures 10.18a and 10.19a, for

(a)

(b)

(c)

Figure 10.16 *(a) Erosion along parallel joints in Arches National Park, Utah. (b) Joints intersecting at right angles yield this rectangular pattern in Wales. (c) Intersecting joints forming an "X" pattern in granite at Marquette, Michigan.*

instance, the hanging wall block moved down relative to the footwall block, giving rise to a **normal fault.** In contrast, in a **reverse fault** the hanging wall block moves up relative to the footwall block (Figures 10.18b and 10.19b). In Figure 10.18c, the hanging wall block also moved up relative to the footwall block, but the fault has a dip of less than 45° and is a special variety of reverse fault known as a **thrust fault.**

If you refer to Figure 10.3b you can see that normal faults result from tension. Numerous normal faults are present along one or both sides of mountain ranges in the Basin and Range Province of the western United States where the crust is being stretched and thinned. The Sierra Nevada at the western margin of the Basin and Range is bounded by normal faults, and the range has risen along

these faults so that it now stands more than 3000 m above the lowlands to the east (see Chapter 24). Also, an active normal fault is found along the eastern margin of the Teton Range in Wyoming, accounting for the 2100 m elevation difference between the valley floor and the highest peaks in the mountains (see the Prologue).

Reverse and thrust faults both result from compression (Figure 10.18b, c, and Figure 10.19b). Large-scale examples of both are found in mountain ranges that formed at convergent plate margins, where one would expect compression (discussed later in this chapter). A very well-known thrust fault is the Lewis overthrust of Montana. (An overthrust is simply a low angle thrust fault with movement measured in kilometers.) On this fault a huge slab of Precambrian-aged rocks moved at least 75 km eastward

Deformation and Geologic Structures **279**

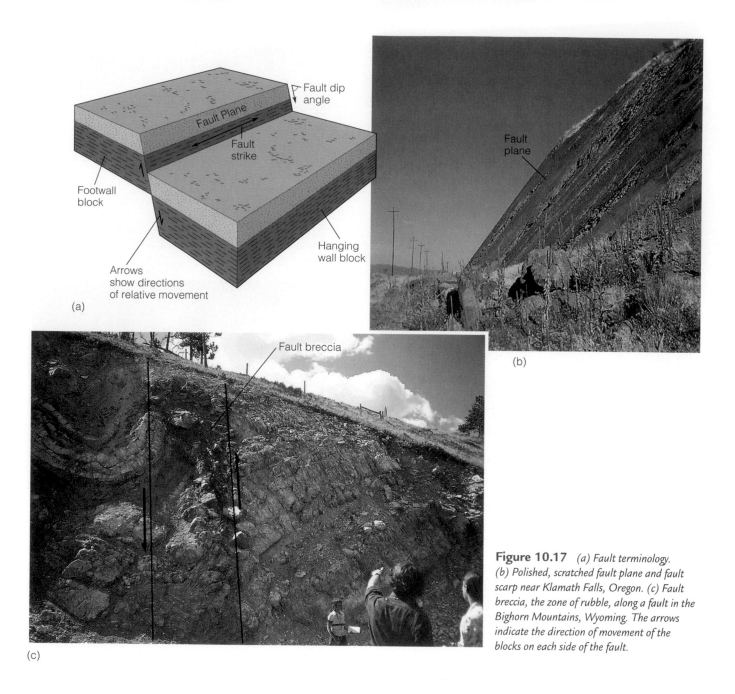

Figure 10.17 *(a) Fault terminology. (b) Polished, scratched fault plane and fault scarp near Klamath Falls, Oregon. (c) Fault breccia, the zone of rubble, along a fault in the Bighorn Mountains, Wyoming. The arrows indicate the direction of movement of the blocks on each side of the fault.*

and now rests upon much younger Cretaceous-aged rocks (see Chapter 24).

Strike-Slip Faults

Shearing forces are responsible for **strike-slip faults,** a type of faulting involving horizontal movement in which blocks on opposite sides of a fault plane slide past one another (Figure 10.18d). In other words, all movement is in the direction of the fault plane's strike, hence the name strike-slip fault. A number of large strike-slip faults are recognized, but one of the best studied is the San Andreas fault that cuts through coastal California. Remem-

ber from Chapter 9 that the San Andreas fault is also called a *transform fault* in plate tectonics terminology.

Strike-slip faults are further characterized as either right-lateral or left-lateral depending on the apparent direction of offset. In Figure 10.18d, for instance, observers looking at the block on the opposite side of the fault from their point of observation notice that it appears to have moved to the left. Accordingly, this is a *left-lateral strike-slip fault.* Had this been a *right-lateral strike-slip fault,* the block across the fault from the observers would appear to have moved to the right. Refer to Figure 7.3b and see if you can determine whether the San Andreas fault is a right-lateral or left-lateral strike-slip fault.

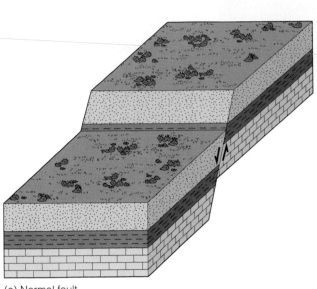

(a) Normal fault

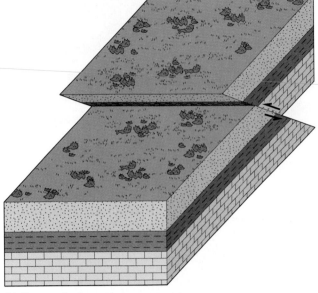

(b) Reverse fault

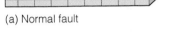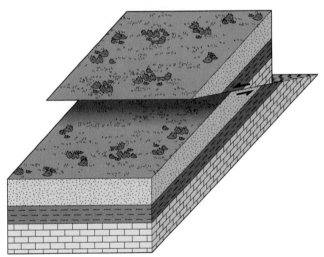

(c) Thrust fault

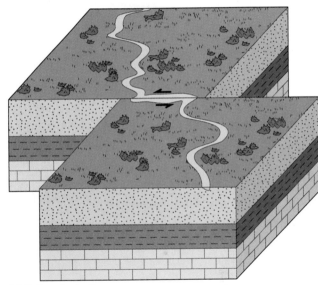

(d) Strike-slip fault

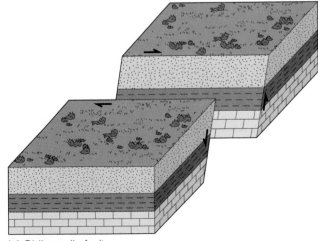

(e) Oblique-slip fault

Figure 10.18 *Types of faults. (a), (b), and (c) are dip-slip faults. (a) Normal fault—hanging wall block moves down relative to footwall block. (b) and (c) Reverse and thrust faults—hanging wall block moves up relative to footwall block. (d) Strike-slip fault—all movement is parallel to the strike of the fault. (e) Oblique-slip fault—combination of dip-slip and strike-slip movements.*

(a)

(b)

Figure 10.19 *(a) Layers of volcanic ash in Oregon cut by two small normal faults. (b) Notice that the sandstone layers to the right of the hammer have been cut by a reverse fault. Compare the sense of movement of the hanging wall blocks in the two images.*

Oblique-Slip Faults

The movement on most faults is primarily dip-slip or strike-slip, but on some, both types of movement take place. These are known as **oblique-slip faults.** Strike-slip movement might be accompanied by a component of dip-slip giving rise to a combined movement that includes left-lateral and reverse, or right-lateral and normal (Figure 10.18e).

Various geologic structures along with colors and symbols for different rock types are portrayed on *geologic maps.* As one would expect, geologists construct and use these maps, but engineers, city and regional planners, and people in several other professions may have occasion to refer to geologic maps (see Perspective 10.1).

Deformation and the Origin of Mountains

The forces necessary to form lofty mountains are difficult to comprehend, yet when compared with Earth's size even the highest mountains are very small features. An analogy to illustrate the size of mountains in relation to the entire planet is to compare Earth with a billiard ball or ball bearing. Both are exceptionally smooth, but if they were expanded to Earth's size, imperceptibly small features would become deep canyons and high mountains. Thus, a globe depicting Earth features at true scale would be quite smooth. Nevertheless, from the human perspective mountains are indeed impressive, towering as much as 8840 m above sea level. And even many of those of more modest heights are quite majestic.

We noted in the Introduction that mountains originate in several ways, but that the truly large mountains on continents result mostly from compression-induced deformation at convergent plate boundaries. Before discussing mountain building, though, we should define what we mean by the term *mountain* and briefly discuss the types of mountains. *Mountain* is a designation for any area of land that stands significantly higher, at least 300 m, than surrounding country and has a restricted summit area. Some mountains are single, isolated peaks, but more commonly they are parts of linear associations of peaks and or ridges known as *mountain ranges* that are related in age and origin. A *mountain system,* a complex linear zone of deformation and crustal thickening, on the other hand, consists of several or many mountain ranges. The Teton Range discussed in the Prologue is one of many ranges in the Rocky Mountains. The Appalachian Mountains of the eastern United States and Canada is another complex mountain system made up of many ranges, such as the Great Smoky Mountains of North Carolina and Tennessee, the Adirondack Mountains of New York, and the Green Mountains of Vermont.

Types of Mountains

Mountains develop in a variety of ways, some involving little or no deformation. For instance, differential weathering and erosion has yielded high areas with adjacent lowlands in the southwestern United States, but these erosional remnants are rather flat topped or pinnacle-shaped and go by the names mesa and butte, respectively (see Chapter 15). A single volcanic mountain might develop over a hot spot, although more commonly a series of volcanoes forms as a plate moves over a hot spot, as in the Hawaiian Islands (see Figure 9.27). And keep in mind that the oceanic ridge system consists of mountains that exceed the size of any mountains on land. But the oceanic ridges form by volcanism at divergent plate boundaries and show features produced by tensional stresses. Large mountains on land, however, are composed of all rock types and show clear indications of compression.

Mountainous topography also forms where the crust has been intruded by batholiths that are subsequently uplifted and eroded (Figure 10.20). The Sweetgrass Hills of northern Montana and the Llano uplift in Texas consist of resistant plutonic rocks exposed following uplift and erosion of the softer overlying sedimentary rocks.

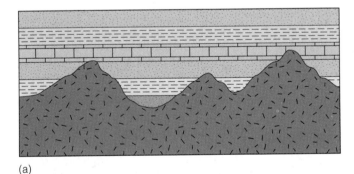

(a)

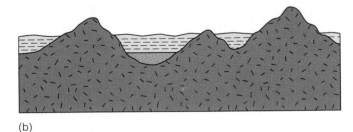

(b)

Figure 10.20 *(a) Pluton intruded into sedimentary rocks.*
(b) Erosion of the softer overlying rocks reveals the pluton and forms small mountains.

Geologic Maps—Their Construction and Uses

Most people are familiar with road maps depicting highways, communities, waterways, and political boundaries such as county and state lines. Maps such as these always have a scale so we can determine distances between points of interest and a legend explaining any symbols or colors used on the map. Maps using lines of equal elevation, or what are called topographic maps, are also used by people in a variety of professions and by backpackers and other outdoor enthusiasts (see Appendix D). On geologic maps, lines, symbols, and colors are used to depict the distribution of various rocks, show age relationships among rocks, and delineate geologic structures such as folds and faults (Figure 1). Geologic maps are generally printed on a base map, that is, a map showing locations and perhaps elevations. In short, geologic maps provide in graphic form considerable information about the composition, structure, and distribution of geologic materials in a given region, and like other maps they have a scale and legend.

The area depicted on a geologic map depends on the purpose of the study that led to the map's construction in the first place. Geologic maps are available for entire continents or countries, but because they display such large areas, they cannot show small-scale details. These maps are useful for determining the regional distribution of rocks of a given age and for delineating very large geologic structures, but geologic maps showing smaller areas as in Figure 1 are generally of more immediate use.

Geologic maps are constructed in a rather straightforward manner, using information gathered during field studies (Figure 2). Surface exposures of rocks are studied and pertinent information recorded, such as strike and dip, composition, and relative age relationships. Notice in Figure 2 that surface exposures, or outcrops of the various rocks are discontinuous, the most common situation encountered by geologists. Nevertheless, the exposures can be used to infer what lies beneath the surface between the outcrops. Thus, the geologic map shows the rocks of this area as if there were no soil cover (Figure 2c).

Geologists, of course, are the primary users of geologic maps. Once geologic structures and the types and relative ages of rocks in an area are known, geologists can interpret geologic processes and the geologic history. Furthermore, because geologic maps use strike and dip symbols and other symbols depicting geologic structures, cross sections of mapped areas can be constructed to illustrate three-dimensional relationships among rock units and geologic structures (Figure 1). This step is particularly important for many economic ventures because cross sections may indicate where an oil well should be drilled or where other resources might be present beneath the surface.

In addition to geologists, people in many other disciplines use geologic maps. You, for example, may one day be a member of a safety-planning board of a city with known active faults. Geologic maps might therefore be useful in developing zoning regulations and

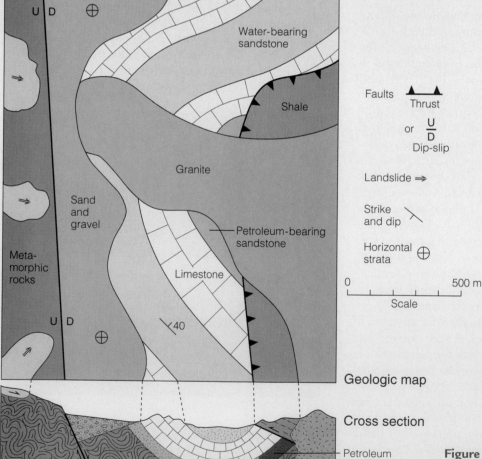

Geologic map

Cross section

Figure 1 *Geologic map and cross section showing the rocks and geologic structures of an area.*

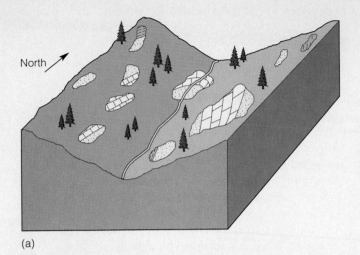

(a)

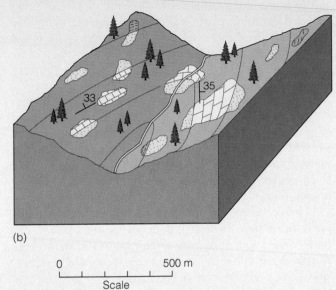

(b)

0   500 m

Scale

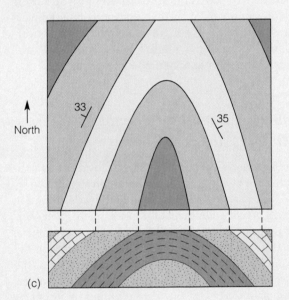

(c)

Figure 2 *Construction of a geologic map and cross section from surface rock exposures (outcrops). (a) Valley with outcrops. In much of the area, the rocks are covered by soil. (b) Data from the outcrops are used to infer what is present in the covered areas between outcrops. The lines shown represent boundaries between different types of rock. (c) A geologic map (top) showing the area as if the soil had been removed. Strike and dip would be recorded at many places, but only two are shown here. The orientation of the rocks as recorded by strike and dip is used to construct the cross section (bottom).*

construction codes. Or perhaps you will become a land-use planner or a member of a county commission charged with selecting a site for a sanitary landfill or securing an adequate supply of groundwater. Geologic maps would almost certainly be important in making these decisions.

Geologic maps are also used extensively in planning and constructing dams, choosing the best routes for highways through mountainous regions, and selecting sites for nuclear reactors. All such structures must be situated on stable foundations. For example, it would be unwise to build a nuclear reactor in an area of active faulting or to construct a dam across a valley with an active fault or on rocks too weak to support such a large structure.

In areas of continuing volcanic activity, geologic maps provide essential information about the kinds of volcanic processes that might occur, such as lava flows, ash falls, or mudflows. Recall from Chapter 4 that one way in which volcanic hazards are as-

sessed is by mapping and determining the geologic history of a region. Studies and maps prepared by geologists of the U.S. Geological Survey two years before the eruption of Mount St. Helens in Washington in 1980 allowed local officials to deal more effectively with the consequences of the eruption when it occurred.

Figure 1, a geologic map of a hypothetical area, depicts not only bedrock and various geologic structures, but also recent landslide and stream deposits. Given the position of the fault, which is still active, and the landslide-prone area, zoning this area for housing developments would not seem prudent, although it might be perfectly satisfactory for agriculture or some other kind of land use. The map also shows the distribution of unconsolidated layers of sand, and these might be important sources of groundwater. The identification of a petroleum-bearing sandstone might be important to the economic development of this region. In short, this map provides a wealth of information that can be used for a variety of purposes.

Block-faulting is yet another way to form mountains but it involves considerable deformation (Figure 10.21). Block-faulting entails movement on normal faults so that one or more blocks are elevated relative to adjacent areas. A classic example is the large-scale active block-faulting in the Basin and Range Province of the western United States, a large area centered on Nevada but extending into several adjacent states and northern Mexico. In the Basin and Range Province, the crust is being stretched in an east-west direction; thus, tensional stresses produce north-south oriented, range-bounding faults. Differential movement on these faults has produced uplifted blocks called *horsts* and down-dropped blocks called *grabens* (Figure 10.21). Horsts and grabens are bounded on both sides by parallel normal faults. Erosion of the horsts has yielded the mountainous topography now present, and the grabens have filled with sediments eroded from the horsts (Figure 10.21).

The processes discussed in this section can certainly yield mountains. However, the truly large mountain systems of the continents, such as the Alps of Europe and the Appalachians in North America, were formed by compression along convergent plate boundaries.

How Does Plate Tectonics Account for the Origin of Mountains?

Geologists define the term **orogeny** as an episode of mountain building during which intense deformation takes place, generally accompanied by metamorphism, the emplacement of plutons, especially batholiths, and thickening of Earth's crust. The processes responsible for an orogeny are still not fully understood, but it is now known that mountain building is related to plate movements. In fact, the advent of plate tectonic theory has completely changed the way geologists view the origin of mountain systems.

Any theory accounting for mountain building must adequately explain the characteristics of mountains, such as their geometry and location; they tend to be long and narrow and at or near plate margins. Mountains also show intense deformation, especially compression-induced overturned and recumbent folds as well as reverse and thrust faults. Furthermore, granitic plutons and regional metamorphism characterize the interiors or cores of mountain ranges. Another feature is sedimentary rocks now far above sea level that were clearly deposited in shallow and deep marine environments.

Deformation and associated activities at convergent plate boundaries are certainly important processes in mountain building. They account for a mountain range's location and geometry, as well as complex geologic structures, plutons, and metamophism. Yet the present-day topographic expression of mountains is also related to several surface processes such as mass wasting (gravity-driven processes including landslides), glaciers, and running water. In other words, erosion also plays an important role in the evolution of mountains.

Plate Boundaries and Mountain Building

Most of Earth's geologically recent and present-day orogenic activity is concentrated in two major zones or belts: the *Alpine-Himalayan orogenic belt* and the *circum-Pacific orogenic belt* (Figure 10.22). Both belts are composed of a number of smaller segments known as *orogens,* each is a zone of deformed rocks, many of which have been metamorphosed and intruded by plutons.

In fact, we can explain most of Earth's past and present orogenies in terms of the geologic activity at convergent plate boundaries. Recall from Chapter 9 that convergent plate boundaries might be oceanic-oceanic, oceanic-continental, or continental-continental.

Orogenies at Oceanic-Oceanic Plate Boundaries

Orogenies occurring where oceanic lithosphere is subducted beneath oceanic lithosphere are characterized by the formation of a volcanic island arc and by deformation and igneous activity. Deformation takes place when sediments from the volcanic island arc are deposited on the adjacent seafloor and in the oceanic trench, and then deformed and scraped off against the landward side of the trench (Figure 10.23). These form a subduction complex, or *accretionary wedge,* of intricately folded rocks cut by numerous thrust faults. In addition, orogenies generated by plate convergence results in low-temperature, high-pressure metamorphism characteristic of the blueschist facies (see Figure 6.20).

Deformation caused largely by the emplacement of plutons also takes place in the island arc system where many rocks show evidence of high-temperature, low-pressure metamorphism. The overall effect of island arc orogenesis is the origin of two more or less parallel orogenic belts consisting of a landward volcanic island arc underlain by batholiths and a seaward belt of deformed trench rocks (Figure 10.23). The Japanese Islands are a good example of this type of deformation.

In the area between the island arc and its nearby continent, the back-arc basin, volcanic rocks and sediments derived from the island arc and the adjacent continent are

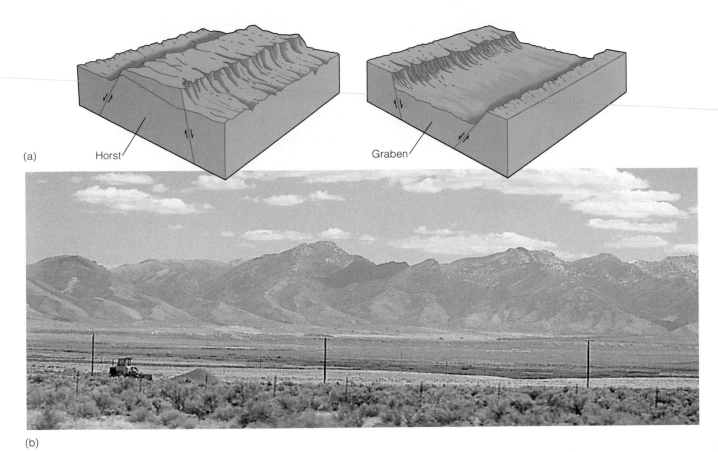

(a)

Horst

Graben

(b)

Figure 10.21 *(a) Block-faulting and the origin of horsts and grabens. (b) View of the Humboldt Range in Nevada, which is a horst bounded by normal faults.*

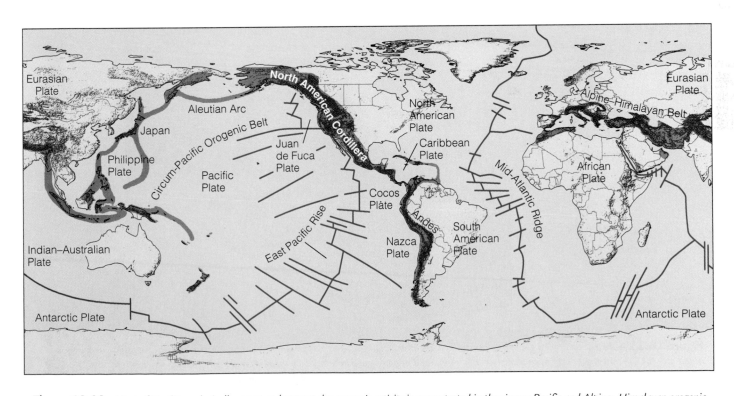

Figure 10.22 *Most of Earth's geologically recent and present-day orogenic activity is concentrated in the circum-Pacific and Alpine–Himalayan orogenic belts.*

How Does Plate Tectonics Account for the Origin of Mountains?

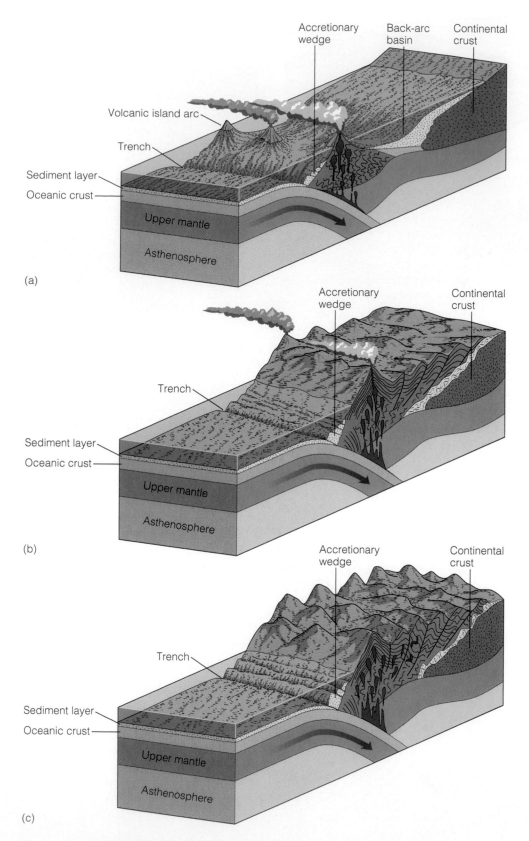

Figure 10.23 *Orogeny and the origin of a volcanic island arc at an oceanic-oceanic plate boundary. (a) Subduction of an oceanic plate beneath an island arc. (b) Continued subduction, emplacement of plutons, and beginning of deformation by thrusting and folding of back-arc basin sediments. (c) Thrusting of back-arc basin sediments onto the adjacent continent and suturing of the island arc to the continent.*

also deformed as the plates continue to converge. The sediments are intensely folded and displaced toward the continent along low-angle thrust faults. Eventually, the entire island arc complex is fused to the edge of the continent, and the back-arc basin sediments are thrust onto the continent forming a thick stack of thrust sheets (Figure 10.23).

Orogenies at Oceanic-Continental Plate Boundaries

Several mountain systems such as the Alps of Europe and the Andes of South America formed at oceanic-continental plate boundaries where oceanic lithosphere is subducted. The Andes are perhaps the best example of such continuing orogeny. Among the ranges of the Andes are the highest mountain peaks in the Americas and many active volcanoes. Furthermore, the west coast of South America is an extremely active segment of the circum-Pacific earthquake belt; and one of Earth's great oceanic trench systems, the Peru-Chile Trench, lies just off the west coast of South America.

Prior to 200 million years ago, the western margin of South America was a site where sediments accumulated on a passive continental margin much as they currently do along the east coast of North America. However, when Pangaea split apart in response to rifting along what is now the Mid-Atlantic Ridge, the South American plate moved westward. As a consequence, the oceanic lithosphere west of South America began subducting beneath the continent (Figure 10.24). Subduction resulted in partial melting of the descending plate, which produced the andesitic volcano arc of composite volcanoes, and the west coast became an active continental margin. More felsic magmas, mostly of granitic composition, were emplaced as large plutons beneath the arc (Figure 10.24).

As a result of the events just described, the Andes Mountains consist of a central core of granitic rocks capped by andesitic volcanoes. To the west of this central core along the coast are the deformed rocks of the accretionary wedge. And to the east of the central core are intensely folded sedimentary rocks that were thrust eastward onto the continent (Figure 10.24). Present-day subduction, volcanism, and seismicity along South America's west coast indicate that the Andes Mountains are still actively forming.

Orogenies at Continental-Continental Plate Boundaries

The best example of an orogeny along a continental-continental plate boundary is the Himalayas of Asia. The Himalayas began forming when India collided with Asia about 40 to 50 million years ago. Prior to that time, India was far south of Asia and separated from it by an ocean basin (Figure 10.25a). As the Indian plate moved northward, a subduction zone formed along the southern margin of Asia where oceanic lithosphere was consumed. Partial melting generated magma, which rose to form a volcanic arc, and large granite plutons were emplaced into what is now Tibet. At this stage, the activity along Asia's southern margin was similar to what is now occurring along the west coast of South America.

The ocean separating India from Asia continued to close, and India eventually collided with Asia (Figure 10.25b). As a result, two continental plates became welded, or sutured, together. Thus, the Himalayas are within a continent rather than along a continental margin. The exact time of India's collision with Asia is uncertain, but between 40 and 50 million years ago, India's rate of northward drift decreased abruptly from 15 to 20 cm per year to about 5 cm per year. Because continental lithosphere is not dense enough to be subducted, this decrease seems to mark the time of collision and India's resistance to subduction. Consequently, the leading margin of India was thrust beneath Asia, causing crustal thickening, thrusting, and uplift. Sedimentary rocks that had been deposited in the sea south of Asia were thrust northward, and two major thrust faults carried rocks of Asian origin onto the Indian plate (Figure 10.25c and d). Rocks deposited in the shallow seas along India's northern margin now form the higher parts of the Himalayas. Since its collision with Asia, India has been underthrust about 2000 km beneath Asia, and now moves north at about 5 cm per year.

A number of other mountain systems also formed as a result of collisions between two continental plates. The Urals in Russia and the Appalachians of North America formed by such collisions. In addition, the Arabian plate is now colliding with Asia along the Zagros Mountains of Iran.

Terranes and the Origin of Mountains

In the preceding sections, we discussed orogenies along convergent plate boundaries resulting in addition of new material to a continent, a process termed *continental accretion*. Much of the material accreted to continents during these events is simply eroded older continental crust, but a significant amount of new material is added to continents as well—igneous rocks that formed during subduction and partial melting and the suturing of an island arc to a continent, for example. Although subduction is the predominant influence on tectonic history, in many regions of orogenies, other processes are also involved in mountain building and continental accretion, especially the accretion of blocks of rock known as *terranes*.

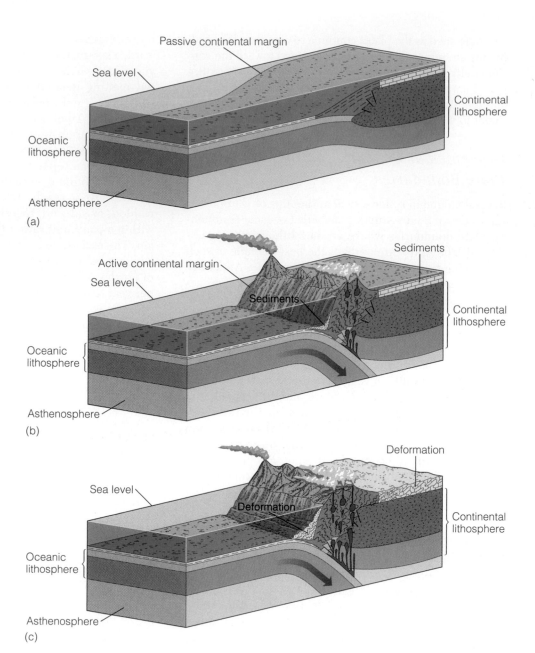

(a)

(b)

Figure 10.24 *Generalized diagrams showing three stages in the development of the Andes of South America. (a) Prior to 200 million years ago, the west coast of South America was a passive continental margin. (b) An orogeny began when the west coast of South America became an active continental margin. (c) Continued deformation, volcanism, and plutonism.*

(c)

During the late 1970s and 1980s, geologists discovered that portions of many mountain systems are composed of small accreted lithospheric blocks that are clearly of foreign origin. These **terranes*** differ completely in their fossil content, structural trends, and paleomagnetic properties from the rocks of the surrounding mountain system. In fact, they are so different from adjacent rocks that most geologists think they formed elsewhere and were carried great distances as parts of other plates until they collided with continents.

*Some geologists prefer the terms *suspect terrane, exotic terrane,* or *displaced terrane.* Notice also the spelling of *terrane* as opposed to the more familiar *terrain,* the later a geographic term indicating a particular area of land.

Geologic evidence indicates that much of the Pacific coast from Alaska to Baja California consists of accreted terranes or igneous intrusions. They are composed of volcanic island arcs, oceanic ridges, seamounts, and small fragments of continents that were scraped off and accreted to the continent's margin as the oceanic plate upon which they were carried was subducted under the continent. And numerous ridges, plateaus, and seamounts in today's oceans are potential terranes, especially in the Pacific Ocean (see Figure 8.11). According to one estimate, more than 100 different-sized terranes have been added to the western margin of North America during the last 200 million years (Figure 10.26). Actually, the terranes illustrated in Figure 10.26 are composed of smaller terranes that cannot be shown at this scale, the Franciscan

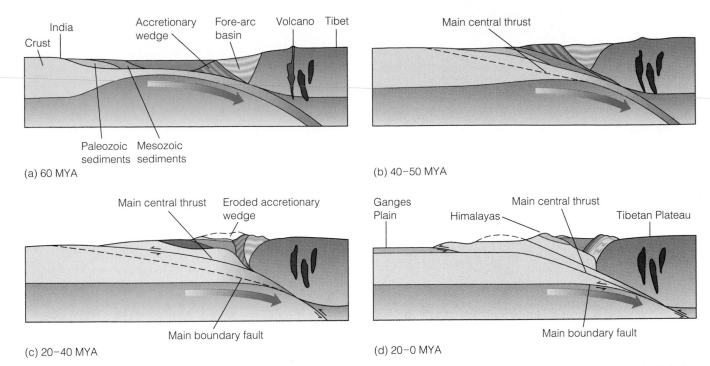

Figure 10.25 *Simplified cross sections showing the collision of India with Asia and the origin of the Himalayas. (a) The northern margin of India before its collision with Asia. Subduction of oceanic lithosphere beneath southern Tibet as India approached Asia. (b) About 40 to 50 million years ago, India collided with Asia, but because India was too light to be subducted, it was underthrust beneath Asia. (c) Continued convergence accompanied by thrusting of rocks of Asian origin onto the Indian subcontinent. (d) Since about 10 million years ago, India has moved beneath Asia along the main boundary fault. Shallow marine sedimentary rocks that were deposited along India's northern margin now form the higher parts of the Himalayas. Sediment eroded from the Himalayas has been deposited on the Ganges Plain.*

complex of the West Coast being an excellent example (see Perspective 10.2).

The basic plate tectonic reconstruction of orogenies and continental accretion remains unchanged, but the details of these reconstructions are decidedly different in view of terrane accretion. For example, growth along active continental margins is faster than along passive continental margins because of the accretion of terranes. Furthermore, these accreted terranes are often new additions to a continent rather than reworked older continental material.

Most of the terranes identified so far are in mountains of the North American Pacific coast region, but a number of such plates are suspected to be present in other mountain systems as well. They are more difficult to recognize in older mountain systems, such as the Appalachians. Nevertheless, about a dozen terrain have been identified though their boundaries are difficult to descern.

Earth's Continental Crust

We noted in Chapter 8 that the crust beneath continents differs from oceanic crust in composition, density, and topographic expression. Obviously oceanic crust is lower

than continental crust, but why is this so? Furthermore, why do mountains stand higher than surrounding continental crust? To answer these questions we must examine continental crust in more detail. As you already know oceanic crust is composed of basalt and gabbro whereas continental crust is characterized as granitic, meaning it has a composition similar to granite. Nevertheless, it contains a wide variety of igneous, sedimentary, and metamorphic rocks, has an average density of 2.7 g/cm^3, and varies from 20 to 90 km thick. In short, it not only differs from oceanic crust in several important aspects, but it is also considerably more complex.

In most places continental crust is about 35 km thick, but it is much thicker beneath the Rocky Mountains, the Appalachians, the Alps in Europe, and the Himalayas of Asia. (Oceanic crust is only 5 to 10 km thick.) As a matter of fact, continental crust is thicker beneath all of Earth's mountain systems. And it is this difference in thickness coupled with the fact that continental crust is less dense than oceanic crust that explains why mountains stand high.

Floating Continents?

Floating immediately brings to mind a ship at sea or some other buoyant object in a fluid, but it certainly does not

Terranes and the Geology of San Francisco, California

S an Francisco, California, is not only among America's most beautiful cities, but it also has an interesting history of geologic events, not the least of which was the disastrous 1906 earthquake. Just offshore to the west of the city lies the famous San Andreas fault, which continually spawns small earthquakes and has the potential to cause large ones. The area, and indeed the entire West Coast of North America, is in a region that has been geologically active for many millions of years.

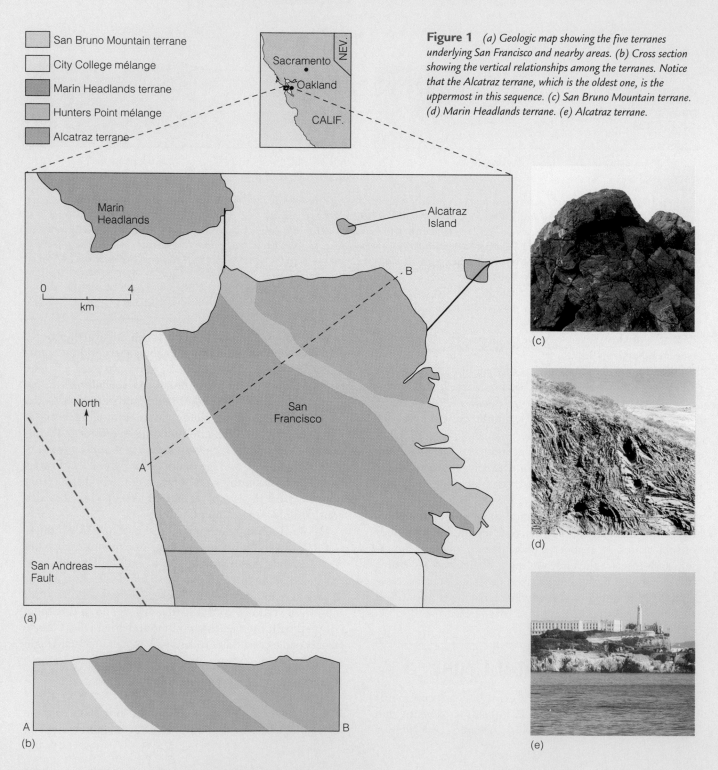

Figure 1 *(a) Geologic map showing the five terranes underlying San Francisco and nearby areas. (b) Cross section showing the vertical relationships among the terranes. Notice that the Alcatraz terrane, which is the oldest one, is the uppermost in this sequence. (c) San Bruno Mountain terrane. (d) Marin Headlands terrane. (e) Alcatraz terrane.*

San Bruno Mountain terrane

City College mélange

Marin Headlands terrane

Hunters Point mélange

Alcatraz terrane

Sacramento

Oakland

CALIF.

NEV.

Marin Headlands

Alcatraz Island

B

0 ——— 4
km

North

San Francisco

A

San Andreas Fault

(a)

A   B
(b)

(c)

(d)

(e)

Notice in Figure 10.26 that a large part of coastal California is made up of the Franciscan Complex, a chaotic jumble of rocks that consists of numerous terranes, five of which are present in San Francisco and nearby areas. Each terrane was accreted to the West Coast as a result of subduction of the Farallon Plate beneath North America during the Jurassic and Cretaceous periods. The geologic map depicted in Figure 1 shows the five terranes, which are progressively older toward the northeast.

At this point it is instructive to point out that the Alcatraz terrane is the oldest and lies above successively younger terranes. However, a basic principle in geology holds that in an undisturbed sequence of rocks the youngest layer is at the top (see Chapter 17). Yet the rocks here are in precisely the reverse order. The word "undisturbed" is the key here. The rocks making up the five terranes were deformed as they were carried beneath one another in succession during subduction of the Farralon plate (Figure 2). Once subduction ceased, uplift and erosion exposed the rocks at the surface in their present form.

Two terranes in Figure 1a are designated as *mélange,* a geologic term for blocks and fragments of rock in a clay matrix formed by intense shearing during subduction. A common rock type in these mélanges is serpentinite, a greenish rock formed by hydrothermal alteration of ultramafic rocks of the mantle. Both mélanges are exposed in San Francisco, but they form subdued outcrops and can be seen clearly in only a few places. However, similar ones are much more easily observed at several locations along the Pacific Coast north of the city.

The other three terranes consist of a variety of rock types and are well exposed in the city and nearby areas. Rocks of the San Bruno Mountain terrane consist mostly of thickly bedded graywacke, a type of sandstone with considerable feldspar minerals, rock fragments, and clay (Figure 1a). It was probably deposited in an oceanic trench, much like the present-day Peru-Chile Trench along the west coast of South America.

Rocks of the Marin Headlands terrane are present in San Francisco, but also easily observed across the Bay in the Marin Headlands (Figure 1). The rocks are pillow lavas, thinly bedded layers of chert, and sandstone, all of which have been intensely deformed. The pillow lavas formed at a spreading ridge—that is, a divergent plate boundary—where they were incorporated into new ocean crust, whereas the chert was deposited on the deep seafloor. The sandstone was derived from land and accumulated on top of the chert beds. Deformation took place during subduction and accretion.

The oldest terrane, the Alcatraz terrane, is present at several locations, especially on Alcatraz Island (Figure 1). The name Alcatraz is well known because from 1934 until 1963 the island was the site of a maximum-security prison where such infamous criminals as Al Capone were incarcerated. The rocks are mostly thick-bedded graywackes (Figure 1a), much like those in the San Bruno Mountains terrane, but with one important difference. These graywackes contain little feldspar, whereas abundant feldspars are found in the San Bruno Mountain terrane graywackes. The reason for this difference is that the rocks of the Alcatraz terrane were deposited before the feldspar-rich granitic rocks of the Sierra Nevada batholith were exposed at the surface. In contrast, the much younger San Bruno Mountain terrane formed when the Sierra granitic rocks were eroding, yielding feldspars and other minerals that were transported to the ocean. One distinctive aspect of San Francisco is its steep hills, many of which are underlain by resistant rock of the Alcatraz terrane.

Following the origin of the five terranes, subduction ceased when North America collided with the Pacific-Farallon ridge, a spreading ridge, and the continent became bounded by a transform fault (see Chapter 24). Since then, several sedimentary rock units were deposited in continental and transitional environments. All of these rocks along with the older terranes have had a subsequent history of uplift, faulting, and erosion.

Figure 2 *According to this model, the Alcatraz terrane (1) was subducted first followed by the Hunters Point mélange (2), the Marin Headlands terrane (3), the City College mélange (4), and finally the San Bruno Mountain terrane (5). Uplift and erosion have exposed the rocks at the surface.*

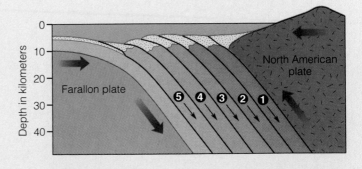

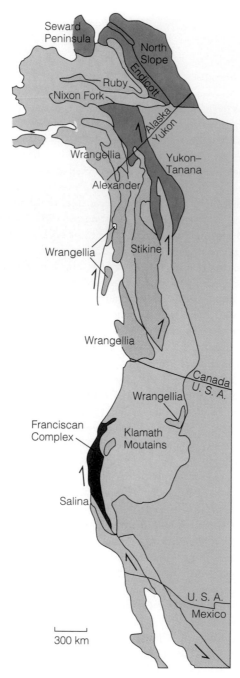

Figure 10.26 *Some of the accreted lithospheric blocks called* terranes *that form the western margin of North America. The light brown blocks probably originated as parts of continents other than North America. The reddish brown blocks are possibly displaced parts of North America. The Franciscan Complex consists of a variety of seafloor rocks.*

evoke an image of a continent buoyed up by some kind of fluid below. Nevertheless, continental crust, and oceanic crust for that matter, is floating in a manner of speaking in a more dense substance. To fully explain why this is so you must be familiar with the concepts of gravity and the principle of isostasy.

Isaac Newton formulated the law of universal gravitation in which the force of gravity (F) between two masses (m_1 and m_2) is directly proportional to the products of their masses and inversely proportional to the square of the distance between their centers of mass. What this means is that an attractive force exists between any two objects, and the magnitude of that force varies depending on the masses of the objects and the distance between their centers. We generally refer to the gravitational force between an object and Earth as its *weight*.

Gravitational attraction would be the same everywhere on the surface if Earth were perfectly spherical, homogeneous throughout, and not rotating. But because Earth varies in all of these aspects, the force of gravity varies from area to area. Geologists use a *gravimeter* to measure gravitational attraction and to detect **gravity anomalies,** that is, departures from the expected force of gravity (Figure 10.27). Gravity anomalies might be *positive,* meaning that an excess of mass is present at some location, or *negative,* when a mass deficiency exists. For instance, a buried iron ore deposit would yield a positive gravity anomaly because of the greater density of these rocks.

The Principle of Isostasy

Suppose that mountains were nothing more than heaps of material piled on the continental crust as shown in Figure 10.27a. If this were so, we would expect a gravity survey across this mountainous area to reveal a huge positive gravity anomaly; that is, an excess of mass between the surface and Earth's center. The fact that no such anomaly exists implies that some of the dense mantle material at depth must be displaced by less dense crustal rocks (Figure 10.27b).

According to the **principle of isostasy,** Earth's crust is in floating equilibrium with the denser mantle below. This phenomenon is easy to understand by analogy to an iceberg. Ice is slightly less dense than water, so it floats. But according to Archimedes' *principle of buoyancy,* an iceberg sinks in the water until it displaces a volume of water equal to the weight of the iceberg. When the iceberg has sunk to its equilibrium position, only about 10% of its volume projects above water level. And if some of the ice above water level melts, the iceberg rises in order to maintain the same proportion of ice above and below water.

Earth's crust is similar to the iceberg in our analogy, in that it sinks into the mantle to its equilibrium level. Where the crust is thickest, as beneath mountain ranges, it sinks further down into the mantle but also rises higher above the equilibrium surface (Figure 10.27). Both continental and oceanic crust are less dense than the upper mantle (its density is 3.3 g/cm³), but continental crust being thicker and less dense than oceanic crust stands higher.

Some of you might realize that crust floating in the mantle raises an apparent contradiction. Remember from Chapter 7 we said that the mantle is a solid because it

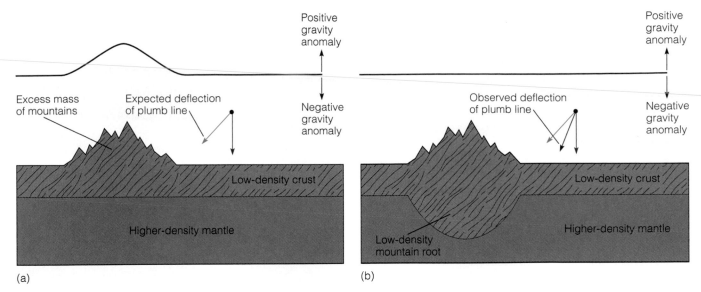

Figure 10.27 *(a) A plumb line (a cord with a suspended weight) is normally vertical, pointing to Earth's center of gravity. Near a mountain range, the plumb line should be deflected as shown if the mountains are simply thicker, low-density material resting on denser material, and a gravity survey across the mountains would indicate a positive gravity anomaly. (b) The actual deflection of the plumb line during the survey in India was less than expected. It was explained by postulating that the Himalayas have a low-density root. A gravity survey in this case would show no anomaly because the mass of the mountains above the surface is compensated for at depth by low-density material displacing denser material.*

transmits S-waves, which will not move through a fluid. But according to the principle of isostasy, the mantle behaves as a fluid. How can this apparent paradox be resolved? When considered in terms of the brief time required for S-waves to pass through it, the mantle is indeed solid. But when subjected to stress over long periods, it yields by flowage and thus at this time scale it can be regarded as a viscous fluid. A familiar substance having the properties of a fluid or a solid depending on how rapidly deforming stress is applied is Silly Putty. If given sufficient time it will flow under its own weight, but shatters as a brittle solid if struck a sharp blow.

Isostatic Rebound

What happens when a ship is loaded with cargo and then later unloaded? Of course it first sinks lower in the water and then rises, but it always finds its equilibrium position. Earth's crust responds similarly to loading and unloading but much more slowly. For example, should the crust be loaded as when widespread glaciers accumulate, the crust responds by sinking further into the mantle to maintain equilibrium. The crust behaves in a similar fashion in areas where huge quantities of sediment accumulate.

If loading by glacial ice or sediment causes Earth's crust to be depressed further into the mantle, it follows that when vast glaciers melt or where deep erosion takes place the crust should rise back up to its equilibrium level. And in fact it does. This phenomenon known as **isostatic**

WHAT WOULD YOU DO?

While teaching high school earth science, you mention that Earth's crust behaves as if it were floating in a more dense mantle below. It is obvious that your students are having a difficult time grasping the concept. After all, how can a solid float in a solid? Can you think of any analogies that might help them understand? Also, what kinds of experiments might you devise to demonstrate that both crustal composition and thickness play a role in isostasy?

rebound is taking place in Scandinavia, which was covered by a thick ice sheet until about 10,000 years ago; it is now rebounding at about 1 m per century. In fact, coastal cities in Scandinavia have rebounded rapidly enough that docks constructed several centuries ago are now far from shore. Isostatic rebound has also occurred in eastern Canada where the crust has risen as much as 100 m in the last 6000 years.

Figure 10.28 shows the response of Earth's continental crust to loading and unloading as mountains form and evolve. Recall that during an orogeny, emplacement of plutons, metamorphism, and general thickening of the crust accompany deformation. Consequently the crust rises higher above and projects further below the equilibrium surface than adjacent thinner crust does. However, as the mountains erode as shown in Figure 10.28, isostatic rebound takes place and the mountains rise whereas adjacent areas of sedimentation subside. If continued long enough, the mountains will disappear and then can be detected only by the plutons and metamorphic rocks that show their former existence.

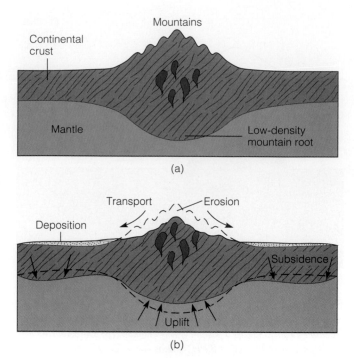

(a)

(b)

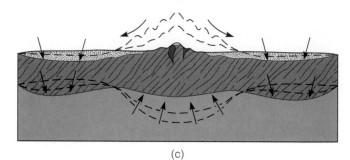

(c)

Figure 10.28 *A diagrammatic representation showing the isostatic response of the crust to erosion (unloading) and widespread deposition (loading). Notice in (b) and (c) that isostatic rebound takes place as the mountain range is eroded, and subsidence of the crust occurs in response to sediment deposition.*

Chapter Summary

1. Contorted and fractured rocks have been deformed or strained by applied stresses.

2. Stresses are characterized as compression, tension, or shear. Elastic strain is not permanent, meaning that when the stress is removed, rocks return to their original shape or volume. Plastic strain and fracture are both permanent types of deformation.

3. The orientation of deformed layers of rock is described by strike and dip.

4. Rock layers that have been buckled into up- and down-arched folds are anticlines and synclines, respectively. They can be identified by the strike and dip of the folded rocks and by the relative age of the rocks in the center of eroded folds.

5. Domes and basins are the circular to oval equivalents of anticlines and synclines, but are commonly much larger structures.

6. Two types of structures resulting from fracturing are recognized: joints, fractures along which the only movement, if any, is perpendicular to the fracture surface; and faults, fractures along which the blocks on opposite sides of the fracture move parallel to the fracture surface.

7. Joints, the commonest geologic structures, form in response to compression, tension, and shear.

8. On dip-slip faults, all movement is in the dip direction of the fault plane. Two varieties of dip-slip faults are recognized: normal faults form in response to tension, while reverse faults are caused by compression.

9. Strike-slip faults are those on which all movement is in the direction of strike of the fault plane. They are characterized as right-lateral or left-lateral depending on the apparent direction of offset of one block relative to the other.

10. Some faults, known as oblique-slip faults, show components of both dip-slip and strike-slip movement.

11. Mountains can form in a variety of ways, some of which involve little or no folding or faulting. Mountain systems consisting of several mountain ranges result from deformation related to plate movements.

12. A volcanic island arc, deformation, igneous activity, and metamorphism characterize orogenies at oceanic-oceanic plate boundaries. Subduction of oceanic lithosphere at an oceanic-continental plate boundary also results in orogeny.

13. Some mountain systems, such as the Himalayas, are within continents far from present-day plate boundary. Such mountains formed when two continental plates collided and became sutured.

14. The oceanic and continental crusts are basaltic and granitic in composition, respectively, and continental crust is much thicker than oceanic crust.

15. According to the principle of isostasy, Earth's crust is floating in equilibrium with the denser mantle below. Continental crust stands higher than oceanic crust because it is thicker and less dense.

Important Terms

anticline	fault plane	normal fault	stress
basin	footwall block	oblique-slip fault	strike
compression	fracture	orogeny	strike-slip fault
deformation	gravity anomaly	plastic strain	syncline
dip	hanging wall block	principle of isostasy	tension
dome	isostatic rebound	reverse fault	terrane
elastic strain	joint	shear stress	thrust fault
fault	monocline	strain	

Review Questions

1. Rocks that show little or no plastic strain are said to be
   a. _____ brittle; b. _____ elastic; c. _____ ductile;
   d. _____ folded; e. _____ compressed.
2. A fault along which the hanging wall block appears to have moved up relative to the footwall block is a _____ fault.
   a. _____ strike-slip; b. _____ transform; c. _____ normal;
   d. _____ transverse; e. _____ reverse.
3. Rocks that have been sheared show
   a. _____ dip-slip movement; b. _____ no deformation;
   c. _____ elastic strain; d. _____ movement along closely spaced planes; e. _____ none of these.
4. Strike is defined as the
   a. _____ number of degrees a plane is inclined from horizontal; b. _____ intensity of folding and faulting in a region; c. _____ line formed by the intersection of a horizontal plane with an inclined plane; d. _____ amount of stress stored in rocks; e. _____ process whereby new materials are accreted to continents.
5. An oval to circular fold with all rocks dipping inward toward a central point is a(n)
   a. _____ basin; b. _____ recumbent syncline;
   c. _____ plunging fold; d. _____ fault scarp; e. _____ compression joint.
6. Most folding of rock layers results from
   a. _____ shear stress; b. _____ compression; c. _____ fracturing; d. _____ convection; e. _____ tension.
7. The fault illustrated in Figure 10.18e shows components of both _____ and _____ faulting.
   a. _____ thrust/reverse; b. _____ reverse dip-slip/left-lateral strike-slip; c. _____ low-angle thrusting/normal;
   d. _____ hanging wall block uplift/footwall block subsidence; e. _____ right-lateral strike-slip/normal.
8. Anticlines and synclines with inclined axes are said to be _____ folds.
   a. _____ overturned; b. _____ recumbent; c. _____ plunging; d. _____ jointed; e. _____ reversed.
9. The geologic term used for an episode of mountain building is
   a. _____ orogeny; b. _____ terrane; c. _____ craton;
   d. _____ shield; e. _____ accretion.
10. The Andes of South America and the Himalayas in Asia are still forming as the results of plates colliding at _____ and _____ plate boundaries.
    a. _____ divergent/transform; b. _____ oceanic-oceanic/continental-continental; c. _____ transform/passive continental margin; d. _____ oceanic-continental/continental-continental; e. _____ strike-slip/compression-tension.

11. An upright fold with all rocks dipping outward away from the fold's center is a(n)
    a. _____ anticline; b. _____ basin; c. _____ dip-slip fault;
    d. _____ platform; e. _____ joint.
12. Which one of the following is correct?
    a. _____ Most mountain building is caused by compression at divergent plate boundaries; b. _____ The San Andreas fault is a large normal fault along the east side of the Sierra Nevada; c. _____ In a recumbent anticline the youngest rock layers are in the core of the fold; d. _____ A monocline is a type of fault along which all movement is in the direction of dip; e. _____ Movement on joints, if any, is perpendicular to the joint surface.
13. Explain what is meant by the term terrane and how terranes are incorporated into continents. Where would one go to see examples of terranes?
14. How do compression, tension, and shear differ from one another? What kinds of deformation does each cause?
15. Discuss the features of mountains formed at an oceanic-oceanic plate boundary and give an example of where such activity is currently taking place.
16. Explain how rock type, time, pressure, and temperature influence the type of deformation in rocks.
17. Draw simple cross sections showing a normal and a reverse fault. Also label the hanging wall and footwall blocks.
18. What is meant by the elastic limit of rocks, and what happens when rocks are strained beyond their elastic limit?
19. How can you determine whether a recumbent fold is an anticline or syncline? Illustrate.
20. What are the similarities and differences between a syncline and a basin?
21. Give an example of how you would explain stress and strain to someone unfamiliar with the concepts.
22. Suppose Earth's crust is deeply eroded in one area and loaded by widespread sedimentary deposits in another. Explain fully how the crust will respond in these two areas.
23. During the Paleozoic Era, eastern North America experienced considerable deformation; during the Mesozoic and Cenozoic eras, deformation has occurred mostly in the western part of the continent. What accounts for this shifting pattern of deformation?
24. Over 5 million years, rocks are displaced 6000 m along a normal fault. What was the average yearly movement on this fault? Is this average likely to represent the actual rate of displacement on this fault? Explain.

World Wide Web Activities

For these web site addresses, along with current updates and exercises, log on to **http://www.brookscole.com/geo**

Structural Geology

This site, maintained by Steven Henry Schimmrich, contains a listing of data sets, bibliographies, organizations, computer software, and online courses in structural geology. Check out the various research projects listed under the Research Information section.

Structural Geology Photo Gallery GS 326, Cornell University

This site, maintained by R. W. Allmendinger, contains a sampling of slides of various geologic structures from field trips and class lectures. Click on any of the designations such as *normal and thrust faults*, then click the thumbnail on the left of the image to see a full-screen photo.

Keck Geology Consortium Structural Geology Slide Set

This slide set was compiled by H. Robert Burger of Smith College with the support of the W. M. Keck Foundation of Los Angeles, California. The database was created at the Department of Earth and Ocean Sciences, University of British Columbia, Vancouver, BC, Canada. The site has a searchable database. Simply enter a keyword such as *anticline, syncline,* or *normal fault* in the "submit query" space. When the image appears, click to enlarge.

USGS Geology in the Parks

The U.S. Geological Survey and National Park Service maintain this site with information on various aspects of the national parks and monuments. Under the menu *The Basics*, click on *Geologic Maps* to see examples of geologic maps and explanations of the colors, symbols, and map key.

 # CD-ROM Exploration

Exploring your *Earth Systems Today* CD-ROM will add to your understanding of the material in this chapter.

Topic: Earth's Processes

Module: Geologic Time

Explore activities in this module to see if you can discover the following for yourself:

Click on "Changing Earth" to see changes in the geography of Earth through time and to examine changes in continental position through time. Note how continents situated at the margins of plates can experience intensive deformation and mountain building during plate tectonic motion. Describe several of these deformation and mountain-building events and their ages. In particular, observe what happens during development of the Appalachian mountains of the eastern United States.

Mass Wasting

Residents of Caracas, Venezuela, clean up the debris from massive flooding and mudslides that devastated large areas of Venezuela during December, 1999.

PROLOGUE

Triggered by relentless torrential rains that began on December 15, 1999, Venezuela was devastated by floods and mudslides that have been described as the country's deadliest natural disaster of the century. And while a reliable death toll is impossible to calculate, it is estimated that 10,000 to 30,000 people were killed, 100,000 to 150,000 were left homeless, 35,000 to 40,000 homes were destroyed or buried by mudslides, and $10 billion to $20 billion in damage was done before the rains and slides abated. Even weeks later, there were still at least 60,000 homes in areas of high risk from new mudslides. It is easy to throw around numbers of dead and homeless, but the human side of the disaster is vividly brought home by a mother who told of standing helplessly by and watching her four small children buried alive in the family car as it was carried away by a raging mudslide.

Sparing no one, mudslides engulfed and buried not only homes, buildings, and roads, but also entire communities, leaving some areas covered with as much as 7 m of mud. In addition, flooding and the accompanying mudslides swept away large parts of many of Venezuela's northern coastal communities, leaving huge areas uninhabitable.

While it was not unexpected there would be mudslides in the areas where they occurred, the magnitude of the disaster surprised many. This disaster was a combination of nature, politics,

and economics. Prior to the flooding and mudslides, Venezuela needed approximately 100,000 new homes annually to keep up with demand, yet only about half that number were built since the early 1980s. Unfortunately, this forced many low- and moderate-income families to find housing in geologically unsuitable areas that were prone to flooding and landslides. Thus, when the rains came, most of these structures and those in the path of the floods and mudslides were destroyed. The government, under newly elected president Hugo Chévez, has promised to build new homes and an economic infrastructure for many of the displaced families in sparsely inhabited inland areas. Whether he and the government will keep this promise remains to be seen.

What makes this terrible tragedy important is how it illustrates the close link between geology and individuals, governments, and society in general, a theme we stressed in Chapter 1. The underlying causes of the mudslides are not unique to Venezuela and can be found anywhere in the world. By being able to recognize and understand these causes and what the ultimate result may be, we can find ways to reduce these hazards and minimize the damage, both in human suffering and in property damage. The important lesson to be learned from this tragedy is how geology impacts all our lives and how interconnected the various systems and subsystems of Earth are.

OBJECTIVES

At the end of this chapter, you will have learned that

■ It is important to understand the different types of mass wasting, because mass wasting affects us all and is economically significant in terms of the destruction it causes.

■ Factors such as slope angle, weathering and climate, water content, vegetation, and overloading are interrelated, and all contribute to mass wasting.

■ Mass movements can be triggered by such factors as overloading, soil saturation, and ground shaking.

■ Mass wasting can be categorized as resulting from either rapid mass movements or slow mass movements.

■ The different types of rapid mass movements are rockfalls, slumps, rock slides, mudflows, debris flows, and quick clays, and each type has recognizable characteristics.

■ The different types of slow mass movements are earthflows, solifluction, and creep, and each type has recognizable characteristics.

■ One can minimize the effects of mass wasting by conducting geologic investigations of an area and stabilizing slopes to prevent and ameliorate movement.

Introduction

The topography of land areas is the result of the interaction among Earth's internal processes, the types of rocks exposed at the surface, the effects of weathering, and the erosional agents of water, ice, and wind. The specific type of landscape developed depends, in part, on which agent of erosion is dominant. Landslides (mass movements), which can be very destructive, are part of the normal adjustment of slopes to changing surface conditions.

Geologists use the term *landslide* in a general sense to cover a wide variety of mass movements that may cause loss of life, property damage, or a general disruption of human activities. In 218 B.C., avalanches in the European Alps buried 18,000 people; an earthquake-generated landslide in Hsian, China, killed an estimated 1,000,000 people in 1556; and 7000 people died when mudflows and avalanches destroyed Huaraz, Peru, in 1941. What makes these mass movements so terrifying, and yet so fascinating, is that they almost always occur with little or no warning and are over in a very short time, leaving behind a legacy of death and destruction (Table 11.1).

Mass wasting (also called *mass movement*) is defined as the downslope movement of material under the direct influence of gravity. Most types of mass wasting are aided by weathering and usually involve surficial material. The material moves at rates ranging from almost imperceptible, as in the case of creep, to extremely fast as in a rockfall or slide. Though water can play an important role, the relentless pull of gravity is the major force behind mass wasting.

Mass wasting is an important geologic process that can occur at any time and almost any place. While most people associate mass wasting with steep and unstable slopes, it can also occur on near-level land, given the right geologic conditions. Furthermore, while the rapid types of mass wasting, such as avalanches and mudflows, typically get the most publicity, the slow, imperceptible types, such as creep, usually do the greatest amount of property damage.

Why Should You Study Mass Wasting?

Why is it important to study mass wasting? Because mass wasting affects all of us, no matter where we live (see the Prologue). In the United States alone, mass wasting occurs in all 50 states and is economically significant in terms of destruction in more than 25 states (Figure 11.1). Furthermore, between 25 and 50 people, on average, are killed each year by landslides in the United States, and the annual cost in damages from them exceeds $1.5 billion. Almost all major landslides have natural causes, yet many

of the smaller ones are the result of human activity and could have been prevented or their damage minimized. In this chapter we will examine the factors that lead to mass wasting and discuss ways to prevent or minimize the damage they cause.

What Factors Influence Mass Wasting?

When the gravitational force acting on a slope exceeds its resisting force, slope failure (mass wasting) occurs. The resisting forces helping to maintain slope stability include the slope material's strength and cohesion, the amount of internal friction between grains, and any external support of the slope (Figure 11.2). These factors collectively define a slope's **shear strength.**

Opposing a slope's shear strength is the force of gravity. Gravity operates vertically but has a component acting parallel to the slope, thereby causing instability (Figure 11.2). The greater a slope's angle, the greater the component of force acting parallel to the slope, and the greater the chance for mass wasting. The steepest angle that a slope can maintain without collapsing is its *angle of repose.* At this angle, the shear strength of the slope's material exactly counterbalances the force of gravity. For unconsolidated material, the angle of repose normally ranges from 25° to 40°. Slopes steeper than 40° usually consist of unweathered solid rock.

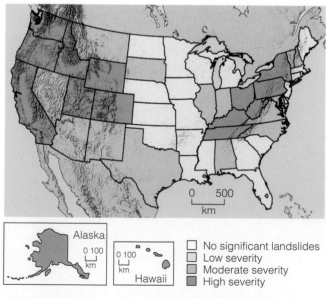

Figure 11.1 *Severity of landslides in the United States. Note that areas of greatest severity occur in the coastal mountain ranges of the West Coast, the Rocky Mountains, and the Appalachian Mountains.*

Table 11.1

Selected Landslides, Their Cause, and the Number of People Killed

Date	Location	Type	Deaths
218 B.C.	Alps (European)	Avalanche—destroyed Hannibal's army	18,000
1512	Alps (Biasco)	Landslide—temporary lake burst	>600
1556	China (Hsian)	Landslides—earthquake triggered	1,000,000
1689	Austria (Montaton Valley)	Avalanche	>300
1806	Switzerland (Goldau)	Rock slide	457
1881	Switzerland (Elm)	Rockfall	115
1892	France (Haute-Savoie)	Icefall, mudflow	150
1903	Canada (Frank, Alberta)	Rock slide	70
1920	China (Kansu)	Landslides—earthquake triggered	~200,000
1936	Norway (Loen)	Rockfall into fiord	73
1941	Peru (Huaraz)	Avalanche and mudflow	7000
1959	USA (Madison Canyon, Montana)	Landslide—earthquake triggered	26
1962	Peru (Mt. Huascarán)	Ice avalanche and mudflow	~4000
1963	Italy (Vaiont Dam)	Landslide—subsequent flood	~2000
1966	Brazil (Rio de Janeiro)	Landslides	279
1966	United Kingdom (Aberfan, South Wales)	Debris flow—collapse of mining-waste tip	144
1970	Peru (Mt. Huascarán)	Rockfall and debris avalanche—earthquake triggered	25,000
1971	Canada (St. Jean-Vianney, Quebec)	Quick clays	31
1972	USA (West Virginia)	Landslide and mudflow—collapse of mining-waste tip	400
1974	Peru (Mayunmarca)	Rockslide and debris flow	430
1978	Japan (Myoko Kogen Machi)	Mudflow	12
1979	Indonesia (Sumatra)	Landslide	80
1980	USA (Washington)	Avalanche and mudflow	63
1981	Indonesia (West Irian)	Landslide—earthquake triggered	261
1981	Indonesia (Java)	Mudflow	252
1983	Iran (Northern area)	Landslide and avalanche	90
1987	El Salvador (San Salvador)	Landslide	1000
1988	Chile (Tupungatito area)	Mudflow	41
1989	Tadzhikistan	Mudflow—earthquake triggered	274
1989	Indonesia (West Irian)	Landslide—earthquake triggered	120
1991	Guatemala (Santa Maria)	Landslide	33
1994	Colombia (Paez River Valley)	Avalanche—earthquake triggered	>300
1995	Brazil (Northeastern area)	Mudflow	15
1996	Brazil (Recife)	Mudflow	49
1997	Peru (Ccocha and Pumaranra)	Mudflow	33
1999	Venezuela	Mudflow	>10,000
2001	El Salvador	Landslide—earthquake triggered	600+

SOURCE: Data from J. Whittow, *Disasters: The Anatomy of Environmental Hazards* (Athens: University of Georgia Press, 1979); *Geotimes;* and *Earth.*

All slopes are in a state of *dynamic equilibrium,* which means that they are constantly adjusting to new conditions. While we tend to view mass wasting as a disruptive and usually destructive event, it is one of the ways that a slope adjusts to new conditions. Whenever a building or road is constructed on a hillside, the equilibrium of that slope is affected. The slope must then adjust, perhaps by mass wasting, to this new set of conditions.

Many factors can cause mass wasting: a change in slope angle, weakening of material by weathering, increased water content, changes in the vegetation cover, and overloading. Although most of these are interrelated, we will

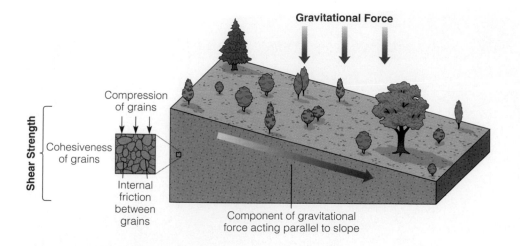

Gravitational Force

Figure 11.2 *A slope's shear strength depends on the slope material's strength and cohesiveness, the amount of internal friction between grains, and any external support of the slope. These factors promote slope stability. The force of gravity operates vertically but has a component acting parallel to the slope. When this force, which promotes instability, exceeds a slope's shear strength, slope failure occurs.*

examine them separately for ease of discussion, but will also show how they individually and collectively affect a slope's equilibrium.

Slope Angle

Slope angle is probably the major cause of mass wasting. Generally speaking, the steeper the slope, the less stable it is. Therefore, steep slopes are more likely to experience mass wasting than gentle ones.

A number of processes can oversteepen a slope. One of the most common is undercutting by stream or wave action (Figure 11.3). This removes the slope's base, increases the slope angle, and thereby increases the gravitational force acting parallel to the slope. Wave action, especially during storms, often results in mass movements along the shores of oceans or large lakes (Figure 11.4).

Excavations for road and hillside building sites are another major cause of slope failure (Figure 11.5). Grading the slope too steeply, or cutting into its side, increases the stress in the rock or soil until it is no longer strong enough to remain at the steeper angle, and mass movement ensues. Such action is analogous to undercutting by streams or waves and has the same result, thus explaining why so many mountain roads are plagued by frequent mass movements.

Weathering and Climate

Mass wasting is more likely to occur in loose or poorly consolidated slope material than in bedrock. As soon as rock is exposed at Earth's surface, weathering begins to disintegrate and decompose it, reducing its shear strength and increasing its susceptibility to mass wasting. The deeper the weathering zone extends, the greater the likelihood of some type of mass movement.

Recall that some rocks are more susceptible to weathering than others and that climate plays an important role

in the rate and type of weathering. In the tropics, where temperatures are high and considerable rainfall occurs, the effects of weathering extend to depths of several tens of meters, and mass movements most commonly occur in the deep weathering zone. In arid and semiarid regions, the weathering zone is usually considerably shallower. Nevertheless, intense, localized cloudbursts can drop large quantities of water on an area in a short time. With little vegetation to absorb this water, runoff is rapid and frequently results in mudflows.

Water Content

The amount of water in rock or soil influences slope stability. Large quantities of water from melting snow or heavy storms greatly increase the likelihood of slope failure. The additional weight that water adds to a slope can be enough to cause mass movement. Furthermore, water percolating through a slope's material helps to decrease friction between grains, contributing to a loss of cohesion. For example, slopes composed of dry clay are usually quite stable, but when wetted, they quickly lose cohesiveness and internal friction and become an unstable slurry. This occurs because clay, which can hold large quantities of water, consists of platy particles that easily slide over each other when wet. For this reason, clay beds are frequently the slippery layer along which overlying rock units slide downslope (see Perspective 11.1).

Vegetation

Vegetation affects slope stability in several ways. By absorbing water from a rainstorm, vegetation decreases water saturation of a slope's material that would otherwise lead to a loss of shear strength. Vegetation's root system also helps stabilize a slope by binding soil particles together and holding the soil to bedrock.

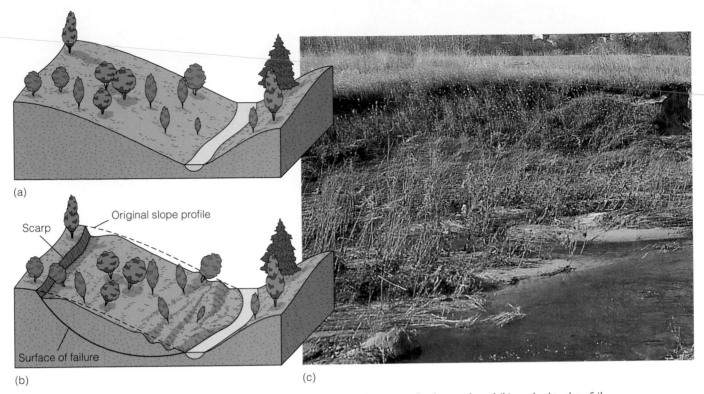

(a)

Original slope profile

Scarp

Surface of failure

(b)   (c)

Figure 11.3 *Undercutting by stream erosion (a) removes a slope's base, which increases the slope angle and (b) can lead to slope failure. (c) Undercutting by stream erosion caused slumping along this stream near Weidman, Michigan.*

The removal of vegetation by either natural or human activity is a major cause of many mass movements. Summer brush and forest fires in southern California frequently leave the hillsides bare of vegetation. Fall rain- storms saturate the ground causing mudslides that do tremendous damage and cost millions of dollars to clean up (Figure 11.6). The soils of many hillsides in New Zealand are sliding because deep-rooted native bushes

Figure 11.4 *This sea cliff north of Bodega Bay, California, was undercut by waves during the winter of 1997–1998. As a result, part of the land slid into the ocean, damaging several houses.*

The Tragedy at Aberfan, Wales

The debris brought out of underground coal mines in southern Wales typically consists of a wet mixture of various sedimentary rock fragments. This material is usually dumped along the nearest valley slope where it builds up into large waste piles called *tips*. A tip is fairly stable as long as the material composing it is relatively dry and its sides are not too steep.

Between 1918 and 1966, seven large tips composed of mine debris were built at various elevations on the valley slopes above the small coal-mining village of Aberfan. Shortly after 9:00 A.M. on October 21, 1966, the 250-m-high, rain-soaked Tip No. 7 collapsed, and a black sludge flowed down the valley with the roar of a loud train (Figure 1). Before it came to a halt 800 m from its starting place, the flow had destroyed two farm cottages, crossed a canal, and buried Pantglas Junior School, suffocating virtually all the children of Aberfan. A total of 144 people died in the flow, among them 116 children who had gathered for morning assembly in the school.

After the disaster, everyone asked, "Why did this tragedy occur and could it have been prevented?" The subsequent investigation revealed that no stability studies had ever been made on the tips and that repeated warnings about potential failure of the tips, as well as previous slides, had all been ignored.

In 1939, 8 km to the south, a tip constructed under conditions almost identical to those of Tip No. 7 collapsed. Luckily no one was injured, but unfortunately the failure was soon forgotten and the Aberfan tips continued to grow. In 1944 Tip No. 4 failed, and again no one was injured.

In 1958 Tip No. 7 was sited solely on the basis of available space, with no regard for the area's geology. In spite of previous tip failures and warnings of slope failure by tip workers and others, mine debris was being piled onto Tip No. 7 until the day of the disaster.

What exactly caused Tip No. 7 and the others to fail? The official investigation revealed that the foundation of the tips had become saturated with water from the springs over which they were built. In the case of the collapsed tips, pore pressure from the water exceeded the friction between grains, and the entire mass liquefied like a "quicksand." Behaving as a liquid, the mass quickly moved downhill spreading out laterally. As it flowed, water escaped from the mass, and the particles regained their cohesion.

Following the inquiry, it was recommended that a National Tip Safety Committee be established to assess the dangers of existing tips and advise on the construction of new tip sites.

Figure 1 *Location map and aerial view of the Aberfan tip disaster in which 144 people died.*

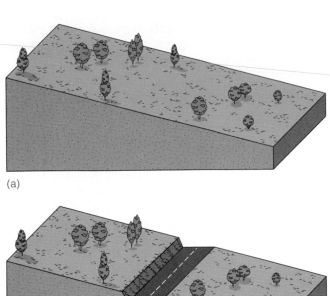

(a)

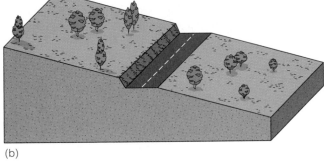

(b)

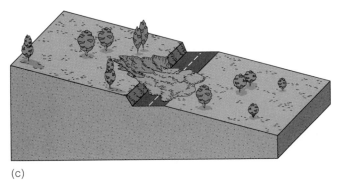

(c)

(d)

Figure 11.5 *(a) Highway excavations disturb the equilibrium of a slope by (b) removing a portion of its support as well as oversteepening it at the point of excavation. (c) Such action can result in frequent landslides. (d) Cutting into the hillside to construct this portion of the Pan American Highway in Mexico resulted in a rockfall that completely blocked the road.*

Figure 11.6 *A California Highway Patrol officer stands on top of a 2-m-high wall of mud that rolled over a patrol car near the Golden State Freeway on October 23, 1987. Flooding and mudslides also trapped other vehicles and closed the freeway.*

What Factors Influence Mass Wasting?

have been replaced by shallow-rooted grasses used for sheep grazing. When heavy rains saturate the soil, the shallow-rooted grasses cannot hold the slope in place, and parts of it slide downhill.

Overloading

Overloading is almost always the result of human activity and typically results from dumping, filling, or piling up of material. Under natural conditions, a material's load is carried by its grain-to-grain contacts, with the friction between the grains maintaining a slope. The additional weight created by overloading increases the water pressure within the material, which in turn decreases its shear strength, thereby weakening the slope material. If enough material is added, the slope will eventually fail, sometimes with tragic consequences.

Geology and Slope Stability

The relationship between topography and the geology of an area is important in determining slope stability (Figure 11.7). If the rocks underlying a slope dip in the same direction as the slope, mass wasting is more likely to occur than if the rocks are horizontal or dip in the opposite direction. When the rocks dip in the same direction as the slope, water can percolate along the various bedding planes and decrease the cohesiveness and friction between adjacent rock units

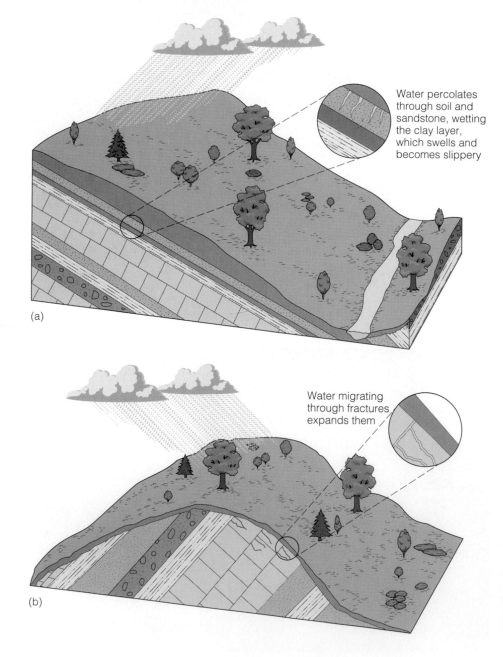

Water percolates through soil and sandstone, wetting the clay layer, which swells and becomes slippery

(a)

Water migrating through fractures expands them

(b)

Figure 11.7 *(a) Rocks dipping in the same direction as a hill's slope are particularly susceptible to mass wasting. Undercutting of the base of the slope by a stream removes support and steepens the slope at the base. Water percolating through the soil and into the underlying rock increases its weight and, if clay layers are present, wets the clay making them slippery. (b) Fractures dipping in the same direction as a slope are enlarged by chemical weathering, which can weaken the rocks and cause mass wasting.*

(Figure 11.7a). This is particularly true when clay layers are present, because clay becomes slippery when wet.

Even if the rocks are horizontal or dip in a direction opposite to that of the slope, joints may dip in the same direction as the slope. Water migrating through them weathers the rock and expands these openings until the weight of the overlying rock causes it to fall (Figure 11.7b).

Triggering Mechanisms

The factors discussed thus far all contribute to slope instability, but most—though not all—rapid mass movements are triggered by a force that temporarily disturbs slope equilibrium. The most common triggering mechanisms are strong vibrations from earthquakes and excessive amounts of water from a winter snow melt or a heavy rainstorm (Figure 11.8).

Volcanic eruptions, explosions, and even loud claps of thunder may also be enough to trigger a landslide if the slope is sufficiently unstable. Many *avalanches*, which are rapid movements of snow and ice down steep mountain slopes, are triggered by the sound of a loud gunshot or, in rare cases, even a person's shout.

What Are the Different Types of Mass Wasting?

Geologists recognize a variety of mass movements (Table 11.2). Some are of one distinct type, while others are a combination of different types. It is not uncommon for one type of mass movement to change into another along its course. Even though many slope failures are combinations of different materials and movements, it is still convenient to classify them according to their dominant behavior.

Mass movements are generally classified on the basis of three major criteria (Table 11.2): (1) rate of movement

Figure 11.8 *Heavy winter rains caused this 200,000 yd³ landslide in March 1995 at La Conchita, California, 75 miles northeast of Los Angeles. While no casualties occurred, nine homes were destroyed or badly damaged.*

What Are the Different Types of Mass Wasting? **309**

Table 11.2

Classification of Mass Movements and Their Characteristics

Type of Movement	Subdivision	Characteristics	Rate of Movement
Falls	Rockfall	Rocks of any size fall through the air from steep cliffs, canyons, and road cuts	Extremely rapid
Slides	Slump	Movement occurs along a curved surface of rupture; most commonly involves unconsolidated or weakly consolidated material	Extremely slow to moderate
	Rock slide	Movement occurs along a generally planar surface	Rapid to very rapid
Flows	Mudflow	Consists of at least 50% silt- and clay-sized particles and up to 30% water	Very rapid
	Debris flow	Contains larger-sized particles and less water than mudflows	Rapid to very rapid
	Earthflow	Thick, viscous, tongue-shaped mass of wet regolith	Slow to moderate
	Quick clays	Composed of fine silt and clay particles saturated with water; when disturbed by a sudden shock, lose their cohesiveness and flow like a liquid	Rapid to very rapid
	Solifluction	Water-saturated surface sediment	Slow
	Creep	Downslope movement of soil and rock	Extremely slow
Complex movements		Combination of different movement types	Slow to extremely rapid

(rapid or slow); (2) type of movement (primarily falling, sliding, or flowing); and (3) type of material involved (rock, soil, or debris).

Rapid mass movements involve a visible movement of material. Such movements usually occur quite suddenly, and the material moves quickly downslope. Rapid mass movements are potentially dangerous and frequently result in loss of life and property damage. Most rapid mass movements occur on relatively steep slopes and can involve rock, soil, or debris.

Slow mass movements advance at an imperceptible rate and are usually only detectable by the effects of their movement such as tilted trees and power poles or cracked foundations. Although rapid mass movements are more dramatic, slow mass movements are responsible for the downslope transport of a much greater volume of weathered material.

Falls

Rockfalls are a common type of extremely rapid mass movement in which rocks of any size fall through the air (Figure 11.9). Rockfalls occur along steep canyons, cliffs, and road cuts and build up accumulations of loose rocks and rock fragments at their base called *talus* (see Figure 5.4b).

Rockfalls result from failure along joints or bedding planes in the bedrock and are commonly triggered by natural or human undercutting of slopes, or by earthquakes. Many rockfalls in cold climates are the result of frost wedging. Chemical weathering caused by water percolating through the fissures in carbonate rocks (limestone,

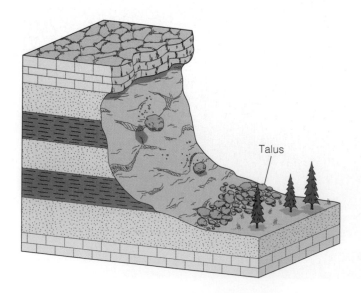

Talus

Figure 11.9 *Rockfalls result from failure along cracks, fractures, or bedding planes in the bedrock and are common features in areas of steep cliffs.*

dolostone, and marble) is also responsible for many rockfalls.

Rockfalls range in size from small rocks falling from a cliff to massive falls involving millions of cubic meters of debris that destroy buildings, bury towns, and block highways (Figure 11.10). Rockfalls are a particularly common hazard in mountainous areas where roads have been built by blasting and grading through steep hillsides of bedrock. Anyone who has ever driven through the Appalachians, the Rocky Mountains, or the Sierra Nevada is familiar with the "Watch for Falling Rocks" signs posted to warn drivers of the danger. Slopes particularly prone to rockfalls are sometimes covered with wire mesh in an effort to prevent dislodged rocks from falling to the road below (Figure 11.11b). Another tactic is to put up wire mesh fences along the base of the slope to catch or slow down bouncing or rolling rocks (Figure 11.11a).

Slides

A **slide** involves movement of material along one or more surfaces of failure. The type of material may be soil, rock, or a combination of the two, and it may break apart during movement or remain intact. A slide's rate of movement can vary from extremely slow to very rapid (Table 11.2).

Two types of slides are generally recognized: (1) slumps or rotational slides, in which movement occurs along a curved surface; and (2) rock or block slides, which move along a more or less planar surface.

A **slump** involves the downward movement of material along a curved surface of a rupture and is characterized by the backward rotation of the slump block (Figure 11.12). Slumps usually occur in unconsolidated or weakly consolidated material and range in size from small individual sets, such as occur along stream banks, to massive, multiple sets that affect large areas and cause considerable damage.

Slumps can be caused by a variety of factors, but the most common is erosion along the base of a slope, which removes support for the overlying material. This local steepening may be caused naturally by stream erosion along its banks (Figure 11.3c) or by wave action at the base of a coastal cliff (Figure 11.13). Slope oversteepening can also be caused by human activity, such as the construction of highways and housing developments. Slumps are particularly prevalent along highway cuts where they are generally the most frequent type of slope failure observed.

While many slumps are merely a nuisance, large-scale slumps involving populated areas and highways can cause extensive damage. Such is the case in coastal southern California where slumping and sliding have been a constant

Figure 11.10 *Rockfall in Jefferson County, Colorado. All eastbound traffic and part of the westbound lane of Interstate 70 was blocked by the rockfall. Heavy rainfall and failure along joints and foliation planes in Precambrian gneiss caused this rockfall.*

What Are the Different Types of Mass Wasting?   **311**

(a)

(b)

Figure 11.11 *Minimizing damage from rockfalls. (a) A wire mesh fence along the base of this hillside of Highway 44 in California has caught many boulders and prevented them from rolling onto the highway. (b) Wire mesh has been used to cover this steep slope in Hawaii. This is a common practice in mountainous areas to prevent rocks from falling on the road.*

problem. Many areas along the coast are underlain by poorly to weakly consolidated silts, sands, and gravels interbedded with clay layers, some of which are weathered ash falls. In addition, southern California is tectonically active so that many of these deposits are cut by faults and joints, which allow the infrequent rains to percolate downward rapidly, wetting and lubricating the clay layers.

Southern California lies in a semiarid climate and is dry most of the year. When it does rain, typically between November and March, large amounts of rain can fall in a

short time. Thus, the ground quickly becomes saturated, leading to landslides along steep canyon walls as well as along coastal cliffs (Figure 11.13). Most of the slope failures along the southern California coast are the result of slumping. These slumps have destroyed many expensive homes and forced numerous roads to be closed and relocated.

A **rock** or *block* **slide** occurs when rocks move downslope along a more or less planar surface. Most rock slides take place because the local slopes and rock layers dip in

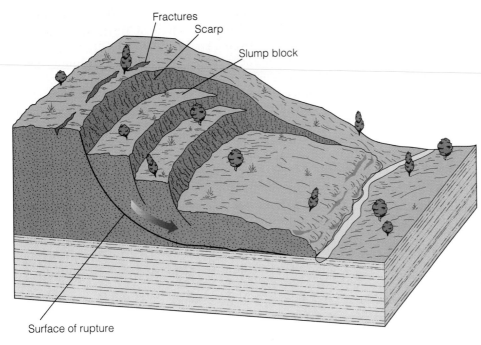

Figure 11.12 *In a slump, material moves downward along the curved surface of a rupture, causing the slump block to rotate backward. Most slumps involve unconsolidated or weakly consolidated material and are typically caused by erosion along the slope's base.*

Courtesy of John S. Shelton

Figure 11.13 *Undercutting of steep sea cliffs by wave action resulted in massive slumping in the Pacific Palisades area of southern California on March 31 and April 3, 1958. Highway 1 was completely blocked. Note the heavy earthmoving equipment for scale.*

What Are the Different Types of Mass Wasting?

the same direction (Figure 11.14), although they can also occur along fractures parallel to a slope. In addition to slumping, rock slides are also common occurrences along the southern California coast. At Point Fermin, seaward-dipping rocks with interbedded slippery clay layers are undercut by waves causing numerous slides (Figure 11.15a).

Farther south in the town of Laguna Beach, startled residents watched as a rock slide destroyed or damaged 50 homes on October 2, 1978 (Figure 11.15b). Just as at Point Fermin, the rocks at Laguna Beach dip about 25° in the same direction as the slope of the canyon walls and contain clay beds that "lubricate" the overlying rock layers, causing the rocks and the houses built on them to slide. Additionally, percolating water from the previous winter's heavy rains wet a subsurface clayey siltstone, thus reducing its shear strength and helping to activate the slide. Although the 1978 slide covered only about five acres, it was part of a larger ancient slide complex.

Not all rock slides are the result of rocks dipping in the same direction as a hill's slope. The rock slide at Frank, Alberta, Canada, on April 29, 1903, illustrates how nature and human activity can combine to create a situation with tragic results (Figure 11.16).

It would appear at first glance that the coal-mining town of Frank, lying at the base of Turtle Mountain, was in no danger from a landslide (Figure 11.16). After all, many of the rocks dipped away from the mining valley. The joints in the massive limestone composing Turtle Mountain, however, dip steeply toward the valley and are essentially parallel with the slope of the mountain itself. Furthermore, Turtle Mountain is supported by weak limestones, shales, and coal layers that underwent slow plastic deformation from the weight of the overlying massive limestone. Coal mining along the base of the valley also contributed to the stress on the rocks by removing some of the underlying support. All of these factors, as well as frost action and chemical weathering that widened the joints, finally resulted in a massive rock slide. Almost 40 million m^3 of rock slid down Turtle Mountain along joint planes, killing 70 people and partially burying the town of Frank.

Flows

Mass movements in which material flows as a viscous fluid or displays plastic movement are termed *flows*. Their rate

Figure 11.14 *Rock slides occur when material moves downslope along a generally planar surface.*

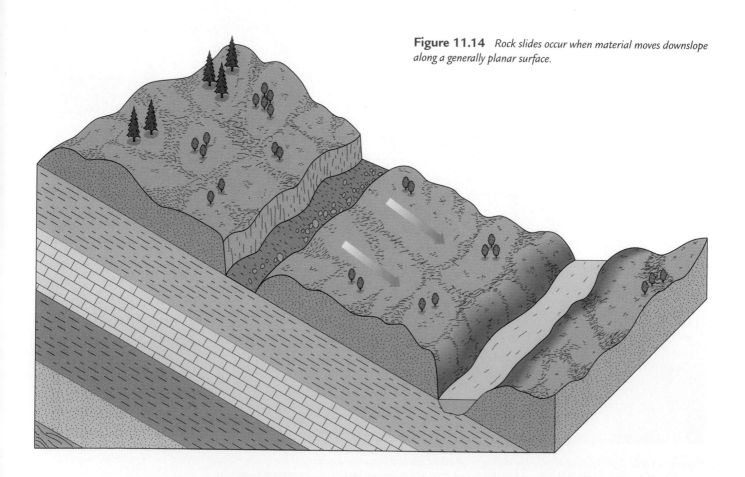

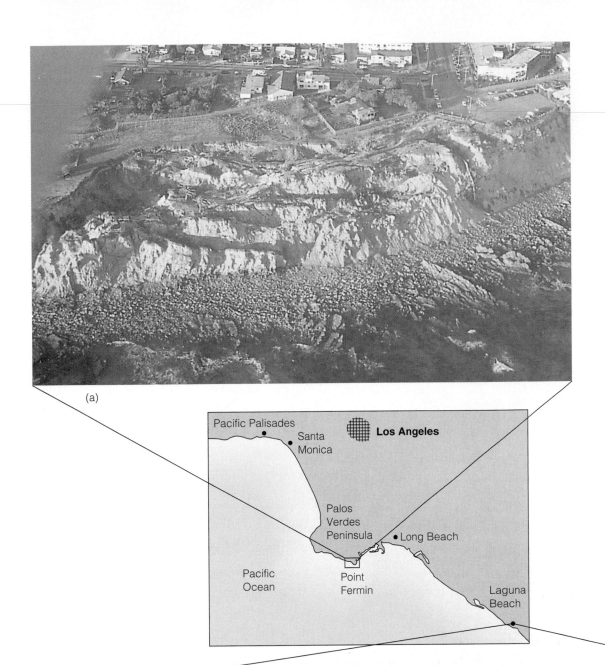

(a)

Pacific Palisades

Santa
Monica

Los Angeles

Palos
Verdes
Peninsula

Long Beach

Pacific
Ocean

Point
Fermin

Laguna
Beach

Figure 11.15 *(a) A combination of interbedded clay beds that become slippery when wet, rocks dipping in the same direction as the slope of the sea cliffs, and undercutting of the sea cliffs by wave action has caused numerous rock slides and slumps at Point Fermin, California. (b) Farther south at Laguna Beach the same combination of factors apparently activated a rock slide that destroyed numerous homes and cars on October 2, 1978.*

(b)

What Are the Different Types of Mass Wasting?

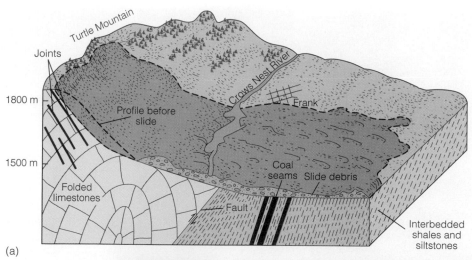

(a)

(b)

Figure 11.16 *(a) The tragic Turtle Mountain rock slide that killed 70 people and partially buried the town of Frank, Alberta, Canada, on April 29, 1903, was caused by a combination of factors. These included joints that dipped in the same direction as the slope of Turtle Mountain, a fault partway down the mountain, weak shale and siltstone beds underlying the base of the mountain, and mined-out coal seams. (b) Results of the 1903 rock slide at Frank.*

of movement ranges from extremely slow to extremely rapid (Table 11.2). In many cases, mass movements begin as falls, slumps, or slides and change into flows further downslope.

Of the major mass movement types, **mudflows** are the most fluid and move most rapidly (at speeds up to 80 km per hour). They consist of at least 50% silt- and clay-sized material combined with a significant amount of water (up to 30%). Mudflows are common in arid and semiarid environments where they are triggered by heavy rainstorms that quickly saturate the regolith, turning it into a raging flow of mud that engulfs everything in its path. Mudflows can also occur in mountain regions (Figure 11.17) and in

areas covered by volcanic ash where they can be particularly destructive (see Chapter 4). Because mudflows are so fluid, they generally follow preexisting channels until the slope decreases or the channel widens, at which point they fan out.

As urban areas in arid and semiarid climates continue to expand, mudflows and the damage they create are becoming problems. Mudflows are common, for example, in the steep hillsides around Los Angeles where they have damaged or destroyed many homes.

Debris flows are composed of larger-sized particles than mudflows and do not contain as much water. Consequently, they are usually more viscous than mudflows, typically do not

Figure 11.17 *Mudflow near Estes Park, Colorado.*

move as rapidly, and rarely are confined to preexisting channels. Debris flows can be just as damaging, though, because they can transport large objects (Figure 11.18).

Earthflows move more slowly than either mudflows or debris flows. An earthflow slumps from the upper part of a hillside, leaving a scarp, and flows slowly downslope as a thick, viscous, tongue-shaped mass of wet regolith (Figure 11.19). Like mudflows and debris flows, earthflows can be of any size, and are frequently destructive. They occur most commonly in humid climates on grassy soil-covered slopes following heavy rains.

Some clays spontaneously liquefy and flow like water when they are disturbed. Such **quick clays** have caused serious damage and loss of lives in Sweden, Norway, eastern Canada (Figure 11.20), and Alaska (Table 11.1). Quick clays are composed of fine silt and clay particles made by the grinding action of glaciers. Geologists think these fine sediments were originally deposited in a marine environment where their pore space was filled with saltwater. The ions in saltwater helped establish strong bonds between the clay particles, thus stabilizing and strengthening the clay. When the clays were subsequently uplifted above sea level, the saltwater was flushed out by fresh groundwater, reducing the effectiveness of the ionic bonds between the clay particles and thereby reducing the overall strength and cohesiveness of the clay. Consequently, when the clay is disturbed by a sudden shock or shaking, it essentially turns to a liquid and flows.

An example of the damage that can be done by quick clays occurred in the Turnagain Heights area of Anchorage, Alaska, in 1964 (Figure 11.21). Underlying most of the Anchorage area is the Bootlegger Cove Clay, a massive clay unit of poor permeability. Because the Bootlegger Cove Clay forms a barrier preventing groundwater from flowing through the adjacent glacial deposits to the sea, considerable hydraulic pressure builds up behind the clay. Some of this water has flushed out the saltwater in the clay and has saturated the lenses of sand and silt associated with the clay beds. When the 8.6-magnitude Good Friday earthquake struck on March 27, 1964, the shaking turned parts

Figure 11.18 *A debris flow and damaged house in lower Ophir Creek, western Nevada. Note the many large boulders that are part of the debris flow.*

What Are the Different Types of Mass Wasting?

(a)

Scarp

(b)

Figure 11.19 (a) Earthflows form tongue-shaped masses of wet regolith that move slowly downslope. They occur most commonly in humid climates on grassy soil-covered slopes. (b) An earthflow near Baraga, Michigan.

Figure 11.20 Quick-clay slide at Nicolet, Quebec, Canada. The house on the slide (to the right of the bridge) traveled several hundred feet with relatively little damage.

Figure 11.21 *(a) Groundshaking by the 1964 Alaska earthquake turned parts of the Bootlegger Cove Clay into a quick clay, causing numerous slides. (b) Low-altitude photograph of the Turnagain Heights subdivision of Anchorage shows some of the numerous landslide fissures that developed as well as the extensive damage to buildings in the area. The remains of the Four Seasons apartment building can be seen in the background.*

(b)

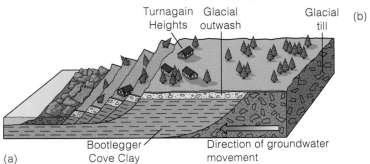

Turnagain Heights · Glacial outwash · Glacial till · Bootlegger Cove Clay · Direction of groundwater movement

(a)

of the Bootlegger Cove Clay into a quick clay and precipitated a series of massive slides in the coastal bluffs that destroyed most of the homes in the Turnagain Heights subdivision (Figure 11.21b).

Solifluction is the slow downslope movement of water-saturated surface sediment. Solifluction can occur in any climate where the ground becomes saturated with water, but is most common in areas of permafrost. **Permafrost,** ground that remains permanently frozen, covers nearly 20% of the world's land surface (Figure 11.22a). During the warmer season when the upper portion of the permafrost thaws, water and surface sediment form a soggy mass that flows by solifluction and produces a characteristic lobate topography (Figure 11.22b).

As might be expected, many problems are associated with construction in a permafrost environment. A good example is what happens when an uninsulated building is constructed directly on permafrost. In this instance, heat escapes through the floor, thaws the ground below, and turns it into a soggy, unstable mush. Because the ground is no longer solid, the building settles unevenly into the ground and numerous structural problems result (Figure 11.23).

Construction of the Alaska pipeline from the oil fields in Prudhoe Bay to the ice-free port of Valdez raised nu-

merous concerns over the effect it might have on the permafrost and the potential for solifluction. Some thought that oil flowing through the pipeline would be warm enough to melt the permafrost, causing the pipeline to sink further into the ground and possibly rupture. After numerous studies were conducted, scientists concluded that the pipeline, completed in 1977, could safely be buried for more than half of its 1280-km length; where melting of the permafrost might cause structural problems to the pipe, it was insulated and installed above ground.

Creep, the slowest type of flow, is the most widespread and significant mass-wasting process in terms of the total amount of material moved downslope and the monetary damage it does annually. Creep involves extremely slow downhill movement of soil or rock. Although it can occur anywhere and in any climate, it is most effective and significant as a geologic agent in humid regions. In fact, it is the most common form of mass wasting in the southeastern United States and the southern Appalachian Mountains.

Because the rate of movement is essentially imperceptible, we are frequently unaware of creep's existence until we notice its effects: tilted trees and power poles, broken streets and sidewalks, or cracked retaining walls or foundations (Figure 11.24). Creep usually involves the

What Are the Different Types of Mass Wasting?

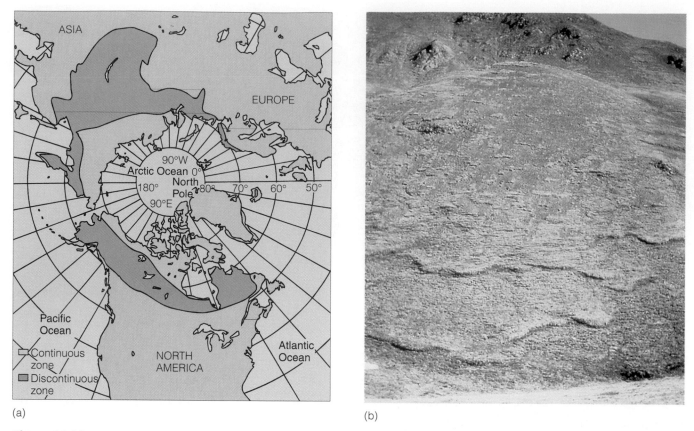

(a)

(b)

Figure 11.22 *(a) Distribution of permafrost areas in the Northern Hemisphere. (b) Solifluction flows near Suslositna Creek, Alaska, show the typical lobate topography that is characteristic of solifluction conditions.*

Figure 11.23 *This house south of Fairbanks, Alaska, has settled unevenly because the underlying permafrost in fine-grained silts and sands has thawed.*

Figure 11.24 *(a) Some evidence of creep: (A) curved tree trunks; (B) displaced monuments; (C) tilted power poles; (D) displaced and tilted fences; (E) roadways moved out of alignment; (F) hummocky surface. (b) Creep has bent these sandstone and shale beds of the Haymond Formation near Marathon, Texas. (c) Trees bent by creep, Wyoming. (d) Stone wall tilted due to creep, Champion, Michigan.*

What Are the Different Types of Mass Wasting?

whole hillside and probably occurs, to some extent, on any weathered or soil-covered, sloping surface.

Creep is not only difficult to recognize but also to control. Although engineers can sometimes slow or stabilize creep, many times the only course of action is to simply avoid the area if at all possible or, if the zone of creep is relatively thin, design structures that can be anchored into the bedrock.

Complex Movements

Recall that many mass movements are combinations of different movement types. When one type is dominant, the movement can be classified as one of those described thus far. If several types are more or less equally involved, however, it is called a **complex movement.**

The most common type of complex movement is the slide-flow in which there is sliding at the head and then some type of flowage farther along its course. Most slide-flow landslides involve well-defined slumping at the head, followed by a debris flow or earthflow (Figure 11.25). Any combination of different mass movement types can, however, be classified as a complex movement.

A *debris avalanche* is a complex movement that often occurs in steep mountain ranges. Debris avalanches typically start out as rockfalls when large quantities of rock, ice, and snow are dislodged from a mountainside, frequently as a result of an earthquake. The material then slides or flows down the mountainside, picking up additional surface material and increasing in speed. The 1970 Peru earthquake (Table 11.1) set in motion the debris avalanche that destroyed the town of Yungay and Ranrahirca, Peru, and killed more than 25,000 people (Figure 11.26).

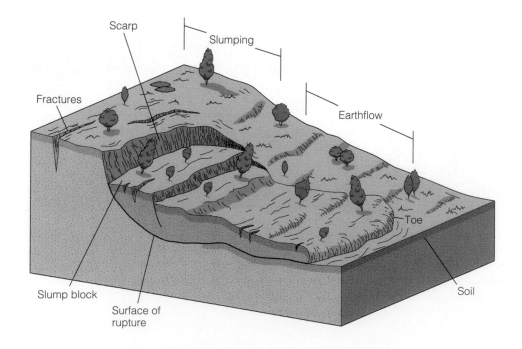

Figure 11.25 *A complex movement in which slumping occurs at the head followed by an earthflow.*

Figure 11.26 *An earthquake 65 km away triggered a landslide on Nevado Huascarán, Peru, that destroyed the towns of Yungay and Ranrahirca and killed more than 25,000 people.*

How Can We Recognize and Minimize the Effects of Mass Movements?

The most important factor in eliminating or minimizing the damaging effects of mass wasting is a thorough geologic investigation of the region in question. In this way, former landslides and areas susceptible to mass movements can be identified and perhaps avoided (see Perspective 11.2). By assessing the risks of possible mass wasting before construction begins, steps can be taken to eliminate or minimize the effects of such events.

Identifying areas with a high potential for slope failure is important in any hazard assessment study; these studies include identifying former landslides as well as sites of potential mass movement. Scarps, open fissures, displaced or tilted objects, a hummocky surface, and sudden changes in vegetation are some of the features indicating former landslides or an area susceptible to slope failure. The effects of weathering, erosion, and vegetation, may, however, obscure the evidence for previous mass wasting.

Soil and bedrock samples are also studied, both in the field and the laboratory, to assess such characteristics as composition, susceptibility to weathering, cohesiveness, and ability to transmit fluids. These studies help geologists and engineers predict slope stability under a variety of conditions.

The information derived from a hazard assessment study can be used to produce *slope stability maps* of the area (Figure 11.27). These maps allow planners and developers to make decisions about where to site roads, utility lines, and housing or industrial developments based on the relative stability or instability of a particular location. In addition, the maps indicate the extent of an area's landslide problem and the type of mass movement that may occur. This information is important for grading slopes or building structures, to prevent or minimize slope-failure damage.

Although most large mass movements usually cannot be prevented, geologists and engineers can employ various methods to minimize the danger and damage resulting from them. Because water plays such an important role in many landslides, one of the most effective and inexpensive ways to reduce the potential for slope failure or to increase existing slope stability is through surface and subsurface drainage of a hillside. Drainage serves two purposes. It reduces the weight of the material likely to slide and increases the shear strength of the slope material by lowering pore pressure.

Surface waters can be drained and diverted by ditches, gutters, or culverts designed to direct water away from slopes. Drainpipes perforated along one surface and driven into a hillside can help remove subsurface water

On October 9, 1963, more than 240 million m³ of rock and soil slid into the Vaiont Reservoir, triggering a destructive flood that killed nearly 3000 people (Figure 1). To fully appreciate the enormity of this catastrophe, consider the following: Within a period of 15 to 30 seconds, the slide filled the reservoir with a mass of debris 2 km long and as high as 175 m above the reservoir level. The impact of the debris created a wave of water that overflowed the dam by 100 m and was still more than 70 m high 1.6 km downstream. The slide also set off a blast of wind that shook houses, broke windows, and even lifted the roof off one house in the town of Casso, which is 260 m above the reservoir on the opposite side of the valley; it also set off shock waves recorded by seismographs throughout Europe. Considering the forces generated by the slide, it is a tribute to the designer and construction engineer that the dam itself survived the disaster (Figure 2)!

The dam was built in a glacial valley underlain by thick layers of folded and faulted limestones and clay layers that were further weakened by jointing (Figure 3). Signs of previous slides in the area were obvious, and the few boreholes in the valley slopes revealed clay layers and small-scale slide planes. In spite of the geological evidence of previous mass wasting in the area and objections to the site by some of the early investigators, construction of the 265-m-high Vaiont Dam began.

A combination of adverse geological features and conditions resulting from the dam construction contributed to the massive landslide. Among the geological causes were the rocks themselves, which were weak to begin with and dipped in the same direction as the valley walls of the reservoir. Fractured

Figure 2 *Aerial view of the Vaiont Dam.*

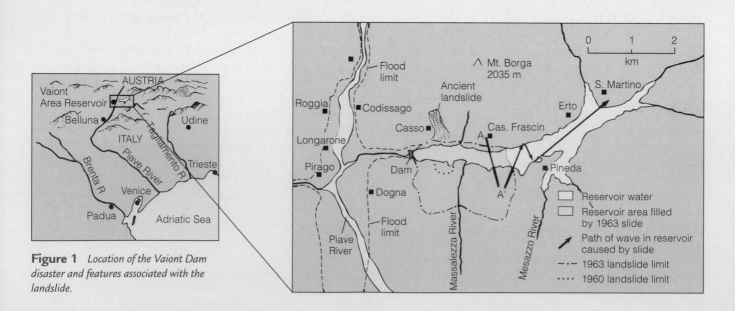

Figure 1 *Location of the Vaiont Dam disaster and features associated with the landslide.*

limestones make up the bulk of the rocks and are interbedded with numerous clay beds that are particularly prone to slippage. Active solution of the limestones by slightly acid groundwater further weakened them by developing and expanding an extensive network of cracks, joints, fissures, and other openings.

During the two weeks before the slide occurred, heavy rains saturated the ground, adding extra weight and reducing the shear strength of the rocks. In addition to water from the rains, water from the reservoir infiltrated the rocks of the lower valley walls, further reducing their strength.

Soon after the dam was completed, a relatively small slide of one million m³ of material occurred on the south side of the reservoir. Following this slide, it was decided to limit the amount of water in the reservoir and to install monitoring devices throughout the potential slide area. Between 1960 and 1963, the eventual slide area moved an average of about 1 cm per week. On September 18, 1963, numerous monitoring stations reported movement had increased to about 1 cm per day. It was assumed that these were individual blocks moving, but it was actually the entire slide area!

Heavy rains fell between September 28 and October 9, increasing the amount of subsurface water. By October 8, the creep rate had increased to almost 39 cm per day. Engineers finally realized that the entire slide area was moving, and quickly began lowering the reservoir level. On October 9, the rate of movement in the slide area had increased still further, in some locations up to 80 cm per day, and there were reports that the reservoir level was actually rising. This was to be expected if the south bank was moving into the reservoir and displacing water. Finally, at 10:41 P.M. that night, during yet another rainstorm, the south bank of the Vaiont valley slid into the reservoir.

The lesson to be learned from this disaster is that before construction on any dam begins, a complete and systematic appraisal of an area must be conducted. Such a study should examine the geology of the area, identify past mass movements, assess their potential for recurrence, and evaluate the effects that the project will have on the rocks, including how it will alter their shear strength over time. Without these precautions, similar disasters will occur and lives will needlessly be lost.

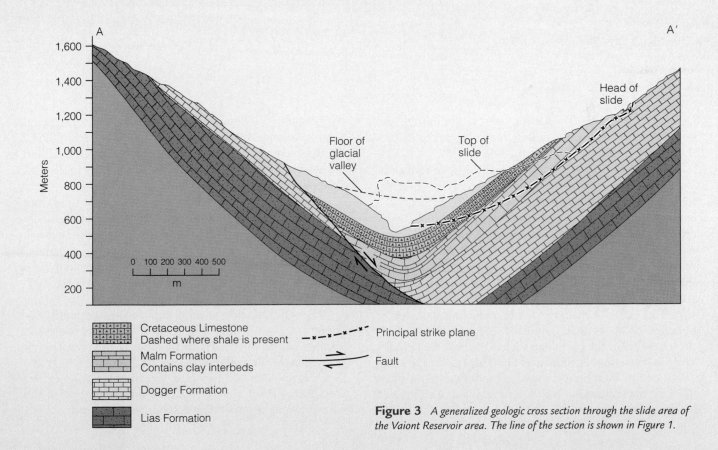

Figure 3 *A generalized geologic cross section through the slide area of the Vaiont Reservoir area. The line of the section is shown in Figure 1.*

How Can We Recognize and Minimize the Effects of Mass Movements?

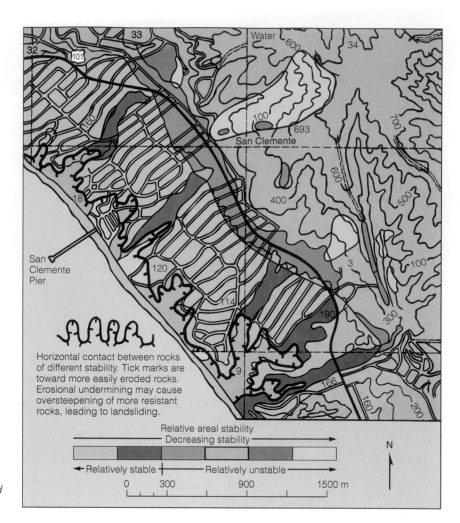

Horizontal contact between rocks of different stability. Tick marks are toward more easily eroded rocks. Erosional undermining may cause oversteepening of more resistant rocks, leading to landsliding.

Relative areal stability
Decreasing stability

← Relatively stable | Relatively unstable →

0   300   900   1500 m

N

Figure 11.27 *Relative slope stability map of part of San Clemente, California, showing areas delineated according to relative stability.*

(Figure 11.28). Finally, planting vegetation on hillsides helps stabilize slopes by holding the soil together and reducing the amount of water in the soil.

Another way to help stabilize a hillside is to reduce its slope. Recall that overloading or oversteepening by grading are common causes of slope failure. By reducing the angle of a hillside, the potential for slope failure is decreased. Two methods are commonly employed to reduce a slope's angle. In the *cut-and-fill* method, material is removed from the upper part of the slope and used as fill at the base, thus providing a flat surface for construction and reducing the slope (Figure 11.29). The second method, which is called *benching* involves cutting a series of benches or steps into a hillside (Figure 11.30). This process reduces the overall average slope, and the benches serve as collecting sites for small landslides or rockfalls that might occur. Benching is most commonly used on steep hillsides in conjunction with a system of surface drains to divert runoff.

In some situations, retaining walls can be constructed to provide support for the base of the slope (Figure 11.31). These are usually anchored well into bedrock, backfilled with crushed rock, and provided with drain holes to prevent the buildup of water pressure in the hillside.

Rock bolts, similar to those employed in tunneling and mining, can sometimes be used to fasten potentially unstable rock masses into the underlying stable bedrock (Figure 11.32). This technique has been used successfully on the hillsides of Rio de Janeiro, Brazil, and to help secure the slopes at the Glen Canyon Dam on the Colorado River.

Recognition, prevention, and control of landslide-prone areas is expensive, but not nearly as expensive as the damage can be when such warning signs are ignored or not recognized. The collapse of Tip No.7 at Aberfan, Wales (see Perspective 11.1) and the Vaiont Dam disaster (see Perspective 11.2) are two tragic examples in which the warning signs of impending disaster were ignored.

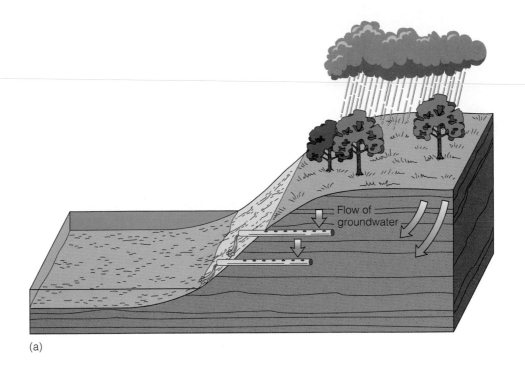

(a)

(b)

Figure 11.28 *(a) Driving drainpipes that are perforated on one side into a hillside, with the perforated side up, can remove some subsurface water and help stabilize a hillside. (b) A drainpipe driven into the hillside at Point Fermin, California, helps remove subsurface water and stabilize the slope.*

How Can We Recognize and Minimize the Effects of Mass Movements?

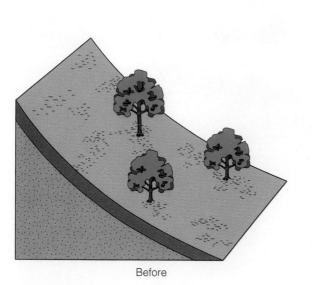

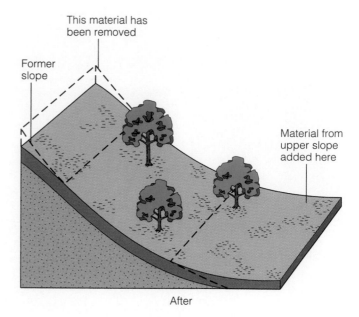

Before

After

Figure 11.29 *One common method used to help stabilize a hillside and reduce its slope is the cut-and-fill method. Here, material from the steeper upper part of the hillside is removed, thereby reducing the slope angle, and is used to fill in the base. This provides some additional support at the base of the slope.*

Figure 11.30 *(a) Another common method used in stabilizing a hillside and reducing its slope is benching. This process involves making several cuts along a hillside to reduce the overall slope. Furthermore, individual slope failures are now limited in size, and the material collects on the benches. (b) Benching is used in nearly all road cuts.*

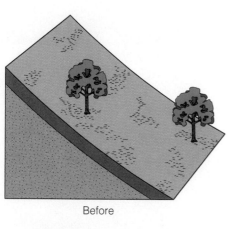

Before

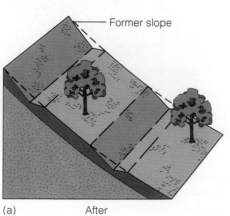

(a)   After

(b)

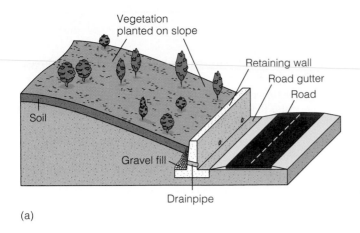

(a)

(b)

Figure 11.31 (a) Retaining walls anchored into bedrock, backfilled with gravel, and provided with drainpipes can support a slope's base and reduce landslides. (b) Steel retaining wall built to stabilize the slope and keep falling and sliding rocks off the highway.

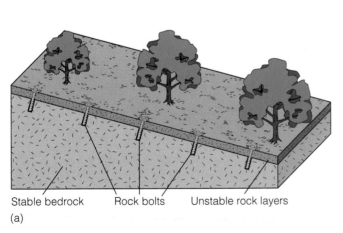

Stable bedrock Rock bolts Unstable rock layers
(a)

(b)

Figure 11.32 (a) Rock bolts secured in bedrock can help stabilize a slope and reduce landslides. (b) Rock bolts and wire mesh are used to help secure rock on a steep hillside in Brisbane, Australia.

How Can We Recognize and Minimize the Effects of Mass Movements?   **329**

Chapter Summary

1. Mass wasting is the downslope movement of material under the influence of gravity. It occurs when the gravitational force acting parallel to a slope exceeds the slope's strength.

2. Mass wasting frequently results in loss of life, as well as causing millions of dollars in damage annually.

3. Mass wasting can be caused by many factors including slope angle, weathering of the slope material, water content, overloading, and removal of vegetation. Usually, several factors in combination contribute to slope failure.

4. Mass movements are generally classified on the basis of their rate of movement (rapid or slow), type of movement (falling, sliding, or flowing), and type of material (rock, soil, or debris).

5. Rockfalls are a common mass movement in which rocks freefall.

6. Two types of slides are recognized. Slumps are rotational slides involving movement along a curved surface; they are most common in poorly consolidated or unconsolidated material. Rock slides occur when movement takes place along a more or less planar surface; they usually involve solid pieces of rock.

7. Several types of flows are recognized on the basis of their rate of movement (rapid or slow), type of material (rock, sediment, or soil), and amount of water.

8. Mudflows consist of mostly clay- and silt-sized particles and contain more than 30% water. They are most common in semiarid and arid environments and generally follow preexisting channels.

9. Debris flows are composed of larger particles and contain less water than mudflows. They are more viscous and do not flow as rapidly as mudflows.

10. Earthflows move more slowly than either debris flows or mudflows; they move downslope as thick, viscous, tongue-shaped masses of wet regolith.

11. Quick clays are clays that spontaneously liquefy and flow like water when they are disturbed.

12. Solifluction is the slow downslope movement of water-saturated surface material and is most common in areas of permafrost.

13. Creep, the slowest type of flow, is the imperceptible downslope movement of soil or rock. Creep is the most widespread of all types of mass wasting.

14. Complex movements are combinations of different types of mass movements in which one type is not dominant. Most complex movements involve sliding and flowing.

15. The most important factor in reducing or eliminating the damaging effects of mass wasting is a thorough geologic investigation of the area to outline areas susceptible to mass movements.

16. Slopes can be stabilized by retaining walls, draining excess water, regrading slopes, and planting vegetation.

Important Terms

complex movement
creep
debris flow
earthflow

mass wasting
mudflow
permafrost
quick clay

rapid mass movement
rockfall
rock slide
shear strength

slide
slow mass movement
slump
solifluction

Review Questions

1. Which of the following are the most fluid of mass movements?
   a. _____ earthflows; b. _____ mudflows; c. _____ debris flows; d. _____ slumps; e. _____ solifluction.

2. Movement of material along a planar surface or surfaces of failure is a(n)
   a. _____ slide; b. _____ fall; c. _____ flow; d. _____ landslide; e. _____ none of these.

3. The force opposing a slope's shear strength is
   a. _____ internal friction; b. _____ cohesion; c. _____ external support; d. _____ internal support; e. _____ gravity.

4. Which of the following is a factor influencing mass wasting?
   a. _____ weathering; b. _____ slope angle; c. _____ water content; d. _____ vegetation; e. _____ all of the previous.

5. The downward movement of material along a curved surface of rupture is a(n)

   a. _____ slump; b. _____ earthflow; c. _____ rockfall; d. _____ rock slide; e. _____ mudflow.

6. Solifluction occurs most commonly in which areas?
   a. _____ deserts; b. _____ tropical forests; c. _____ permafrost; d. _____ beaches; e. _____ none of the previous.

7. The most widespread and costly of all mass wasting processes is
   a. _____ mudflows; b. _____ slumps; c. _____ rockfalls; d. _____ creep; e. _____ quick clay.

8. Which of the following helps increase the shear strength of slope material?
   a. _____ vegetation; b. _____ perforated drainpipes; c. _____ culverts; d. _____ drainage ditches; e. _____ all of the previous.

9. Which of the following helps reduce the slope angle, or provides support at the base, of a hillside?

a. _____ cut and fill; b. _____ benching; c. _____ retaining walls; d. _____ all of the previous; e. _____ none of the previous.

10. Where can mass wasting occur?
   a. _____ only on gentle slopes; b. _____ only on steep slopes; c. _____ only in temperate climates; d. _____ only where bedrock is exposed; e. _____ anywhere.

11. Former landslides and areas currently susceptible to slope failure can be identified by which of the following features?
   a. _____ scarps; b. _____ open fissures; c. _____ hummocky surfaces; d. _____ tilted objects; e. _____ all of the previous.

12. Where in the United States are landslides least likely to be found?
   a. _____ Cascade Ranges; b. _____ Rocky Mountains; c. _____ Appalachian Mountains; d. _____ Midwest; e. _____ Sierra Nevada.

13. The material that accumulates at the base of a rock fall is called
   a. _____ debris; b. _____ quick clay; c. _____ talus; d. _____ till; e. _____ all of the previous.

14. Discuss some of the ways slope stability can be maintained so as to reduce the likelihood of mass movements.

15. Why is creep so prevalent and why does it do so much damage?

16. Discuss some of the ways creep might be controlled.

17. Where are rockfalls most common, and what are some of the methods employed to reduce the damage they cause?

18. Discuss how topography and the underlying geology contribute to slope failure.

19. How can mass wasting be recognized on other planets or moons and what would that tell us about the geology and perhaps atmosphere of the planet or moon on which it has occurred?

20. What precautions must be taken when building in permafrost areas?

21. What features would you look for to determine whether the site on which you want to build your dream house has not been or is not likely to be subject to mass wasting?

22. How can you differentiate between a mudflow, debris flow, and earthflow, and why is it important to be able to do so?

23. Discuss how slope stability maps, used in conjunction with seismic risk maps, can be useful to planners, developers, and the government agencies responsible for overseeing development and growth in an area of increasing population and industry.

24. Discuss how the different factors that influence mass wasting are interconnected.

25. Why is it important to know about the different types of mass wasting?

 # World Wide Web Activities

For these web site addresses, along with current updates and exercises, log on to **http://www.brookscole.com/geo**

U.S. Geological Survey Earth Science in the Public Service: Geologic Hazards

The home page of this site contains links to earthquakes, landslides, geomagnetism, and dynamic maps. Click on *Landslides* to link to the Landslides site. At the Landslides site you can check out information about the National Landslide Hazards Program, the National Landslide Information Center, access the Searchable Bibliographic Database, check on Landslide Program Publications and Current Projects of the U.S. Geological Survey, and learn about Recent Events. This last link gives all sorts of information about recent landslides throughout the world.

Landslides in Nebraska

This site begins with a photographic history of the birth and development of a landslide along Interstate 80, five miles west of Lincoln, Nebraska. It also contains charts and photographs of the five types of landslides found along Nebraska highways, and concludes with a photographic gallery of different types of landslides in Nebraska. Even though this site is limited to Nebraska, the photographs are excellent examples of the different types of mass movements discussed in this chapter.

Mass Wasting in Craters Near the South Pole of Callisto

For an out of this world experience, check out this NASA site of mass wasting in craters near the south pole of Callisto, one of Jupiter's many moons. In some of the craters one can see what appear to be landslide or slump deposits.

Running Water

The South Arm of Rice Creek in northern California cascades about 7 m over volcanic rocks.

At 4:07 P.M. on May 31, 1889, the residents of Johnstown, Pennsylvania, heard "a roar like thunder," and within 10 minutes a flood destroyed the town and killed hundreds of people. A huge wall of water, in places 18 m high, roared down the narrow valley at more than 60 km/hr, sweeping before it debris, houses, and people. A particularly vivid account notes that "thousands of people desperately tried to escape the wave. Those caught by the wave found themselves swept up in a torrent of oily, muddy water, surrounded by tons of grinding debris, which crushed some, provided rafts for others. Many became helplessly entangled in miles of barbed wire from the destroyed wire works."* Eighty people who survived the initial flood by floating on debris died when the debris caught fire after it lodged against a bridge.

The flood killed an estimated 2200 people! One in three bodies recovered was never identified, and many of the flood victims were never found. Ninety-nine of Johnstown's families perished in the flood, while 98 children were left without parents. Survivors made temporary shelters from whatever they could salvage from the wreckage. The town, of course, was in complete ruins and would not fully recover for five years (Figure 12.1). Within days of the tragedy, assistance began pouring in from concerned citizens elsewhere, volunteers, and the Red Cross.

Several factors account for the destruction and tragic loss of life in the Johnstown flood, the most deadly in U.S. history. The city was established in 1784 and began to prosper by the mid-1800s; by the 1880s about 30,000 people were living in Johnstown and nearby communities. Unfortunately, these communities were built on a

floodplain at the junction of the Little Conemaugh River and Stony Creek, and as Johnstown expanded, the rivers were constricted, causing increased problems with flooding. Another factor contributing to the disastrous flood was the 20 to 25 cm of rain that fell in 24 hours. A storm that began in Kansas and Nebraska moved east, causing havoc in its path. In fact, the U.S. Signal Corps issued a storm warning on May 29 for the Mid-Atlantic States.

The large amount of rainfall alone was serious enough, but the most important factor in Johnstown's destruction was the failure of the South Fork Dam about 22 km upstream on the Little Conemaugh River. Measuring about 22 m high and 270 m long, it was not particularly large as dams go, but it impounded a 4.8-km-long reservoir more than 130 m higher than Johnstown. The dam was built during the 1840s, but purchased and repaired by the South Fork Fishing and Hunting Club. Unfortunately, it was poorly maintained thereafter, and its poor condition contributed to its failure when it was filled to overflowing by excess rain.

As the final ingredient in this catastrophe, a warning of the dam's imminent collapse was ignored by residents of one nearby community, apparently because rumors of such a collapse had circulated for years. And the warnings never reached Johnstown because the telegraph lines were down. Although no one doubts that the poor condition of the South Fork Dam was at least partly responsible for the disaster, none of the survivors successfully sued the South Fork Fishing and Hunting Club.

As tragic as the Johnstown flood was, it was certainly not the last time a dam failed resulting in fatalities. Just one week after it was filled in March 1928, the St. Francis Dam in California collapsed (Figure 12.2). A 55-m-high wall of water rushed forth carrying

*National Park Service—U.S. Department of Interior, Johnstown Information Service Online.

At the end of this chapter, you will have learned that

■ Running water, one part of the hydrologic cycle, does considerable geologic work.

■ Water is continually cycled from the oceans to land and back to the oceans.

■ Running water transports large quantities of sediment and deposits sediment in or adjacent to braided and meandering rivers and streams.

■ Alluvial fans (on land) and deltas (in a standing body of water) are deposited when a stream's capacity to transport sediment decreases.

■ Flooding is a natural part of stream activity that takes place when a channel receives more water than it can handle.

■ The several types of structures to control floods are only partly effective.

■ Rivers and streams continually adjust to changes.

■ The concept of a graded stream is an ideal, although many rivers and streams approach the graded condition.

■ Most valleys form and change in response to erosion by running water coupled with other geologic processes such as mass wasting.

(a)

(b)

Figure 12.1 *Johnstown, Pennsylvania, before (a) and after (b) the May 31, 1889, flood that killed about 2200 people.*

pieces of the dam weighing more than 9000 metric tons nearly a kilometer. At least 450 people died because the dam had been built on rock incapable of supporting such a large structure. The death toll was undoubtedly much higher because the flooded canyon was occupied by hundreds of illegal immigrants, many of whom could not be accounted for after the flood.

Both types of dams constructed—those made of concrete, and earth-fill dams consisting of huge piles of gravel, sand, and clayey soil—have failed on occasion. But neither is as prone to collapse as dams constructed of mine tailings—that is, the refuse

rock from mining considered too poor to warrant further processing. So these are earth-fill dams of a kind, but ones with little strength when put under stress. Numerous tailings dams have collapsed, but a particularly notable one in this country occurred on February 26, 1972. Following particularly heavy rains, a dam of coal mining wastes failed and the ensuing flood killed at least 125 people and left 4000 homeless at Buffalo Creek, West Virginia. More than 600 survivors sued the company responsible for the dam and won a settlement of $13.5 million.

(a)

(b)

Figure 12.2 *(a) The St. Francis Dam in southern California in March 1928. (b) On March 13, 1928, only the central part of the dam remained. The catastrophic flooding following the dam's collapse killed more than 450 people.*

Introduction

When considering the solid Earth and the major systems acting on its surface, certainly the hydrosphere has an important impact, although the biosphere-solid Earth and atmosphere-solid Earth interactions are also significant (see Table 1.1). Of course, the hydrosphere consists of several elements including water vapor in the atmosphere; groundwater (see Chapter 13); water frozen in glaciers (see Chapter 14); water in the oceans (see Chapters 8 and 16); and that small but important amount of water confined to channels on land, or simply running water.

The incredible power of running water is well illustrated by the examples in the Prologue, such as the 1889 flood in Johnstown, Pennsylvania, the deadliest in U.S. history. And although floods caused either by human carelessness or purely natural causes continue to be a threat, there are also many benefits from running water. Waterways throughout the world are major avenues of commerce, and when Europeans first explored the interior of North America they did so by following rivers such as the St. Lawrence, Mississippi, Ohio, and Missouri. And even some floods are beneficial. For instance, the agricultural lands along the Nile River in Egypt depend on annual flood-derived deposits to maintain their fertility.

Among the terrestrial planets (Mercury, Venus, Earth, and Mars), Earth is unique in having abundant liquid water. Only Mars has some frozen water and trace amounts of water vapor in its atmosphere, but studies of *Mariner* and *Viking* spacecraft images reveal areas with winding valleys that apparently were eroded by running water during the planet's early history. Recent evidence from the *Mars Global Explorer* indicates that some springs might be present as well. In contrast, 71% of Earth's surface is water covered by oceans and seas (see Figure 8.3), its atmosphere contains a small but important quantity of water, and water is present in streams, lakes, swamps, glaciers, and beneath the land surface as groundwater.

Most of Earth's 1.36 billion km^3 of water (97.2%) is in the oceans, whereas some 2.15% is on land but frozen in glaciers, especially in Antarctica and Greenland, with the remaining 0.65% constituting all the remaining water on the planet (Table 12.1). Thus, only a tiny proportion of all water is in rivers and streams, but running water is nevertheless the most important geologic process modifying Earth's land surface.

No one questions the worldwide importance of running water in erosion, sediment transport, and deposition, but its role is limited in some areas. Antarctica and Greenland are nearly completely ice covered, so in these areas erosion by glaciers is currently important whereas running water has only a minimal effect in some ice-free coastal areas. And parts of some deserts are little affected by running water. Sections of the Atacama Desert of Chile, for example, have had no rain during historic time, but even in most deserts the effects of running water are conspicuous, although channels are dry most of the time (see Chapter 15).

Much of our discussion of running water is necessarily descriptive, but one should always keep in mind that rivers and streams are dynamic systems that continually adjust to natural and human-caused changes. Long-term climatic changes obviously affect the amount of water available for runoff, and paving in urban areas increases surface runoff. In short, as dynamic elements of the hydrosphere, streams and rivers are sensitive to any change that alters them in some way.

Table 12.1
Water on Earth in Km^3

Location	Volume	Percent of Total
Oceans	1,327,500,000	97.20
Icecaps and glaciers	29,315,000	2.15
Groundwater	8,442,580	.625
Freshwater and saline lakes and inland seas	230,325	.017
Atmosphere at sea level	12,982	.001
Average in stream channels	1255	.0001

Why Should You Study Running Water?

Why should you study running water? Running water is the most important geologic agent modifying Earth's land surface in most areas, and it provides one source of fresh water for agriculture, industry, and domestic use. Erosion by running water as well as periodic floods are of course concerns and require considerable effort and cost to manage. Furthermore, running water provides another excellent example of interactions among systems—the hydrosphere, atmosphere, biosphere, and solid Earth. Other reasons for studying this topic are that large waterways—that is, rivers—are important avenues of commerce, and running water is one resource used to generate electricity (Figure 12.3) (see Perspective 12.1).

Figure 12.3 *Hoover Dam on the Colorado River in Nevada where falling water generates electricity. Although now the second highest dam in the United States, at 221 m it was the highest when completed in 1936. In addition to electrical power generation, the reservoir also functions in flood control, irrigation, and recreation.*

The Hydrologic Cycle

The connection between precipitation and clouds is obvious, but where does the moisture for rain and snow come from in the first place? In the Introduction we noted that 97.2% of all water on Earth is in the oceans, so one might immediately suspect that the oceans are the ultimate source of precipitation. In fact, water is continually recycled from the oceans, through the atmosphere, to the continents, and back to the oceans. This **hydrologic cycle,** as it is called, is powered by solar radiation and is possible because water changes easily from liquid to gas (water vapor) under surface conditions (Figure 12.4). A volume of water corresponding to a layer about 1 m thick evaporates from the oceans each year, constituting about 85% of all water entering the atmosphere. The remaining 15% comes from water on land, but almost all of this water originally came from the oceans as well.

Regardless of its source, water vapor rises into the atmosphere where the complex processes of cloud formation and condensation take place. Much of the world's precipitation, about 80%, falls directly back into the oceans, in which case the hydrologic cycle is limited to a three-step process of evaporation, condensation, and precipitation. For the 20% of all precipitation falling on land, the hydrologic cycle is more complex. In this case it involves evaporation, condensation, movement of water vapor from the oceans to land, and precipitation. Although some precipitation evaporates as it falls and reenters the cycle, about 36,000 km³ of the precipitation falling on land returns to the oceans by **runoff,** the surface flow in streams and rivers.

Not all precipitation returns directly to the oceans by runoff, though. Some is temporarily stored in lakes and swamps, snowfields and glaciers, or seeps below the surface where it enters the groundwater system (see Chapter 13). In some of these reservoirs water might remain effectively stored for thousands of years, but eventually glaciers melt, lakes and groundwater feed streams and rivers, and this water returns to the oceans. Even the water used by plants evaporates, a process known as transpiration, and returns to the atmosphere. In short, all water derived from the oceans eventually makes it back to the oceans, and can thus begin the hydrologic cycle again (Figure 12.4). Our concern here is with the comparatively small quantity of water returning to the oceans by surface runoff.

Running Water

Unlike solids, water has no strength so it will flow on any slope no matter how slight. In other words it flows downhill in response to stress, which in this case is generated by that part of the gravity force operating parallel with a slope. Once a fluid begins moving, its flow is characterized as either *laminar* or *turbulent.* In laminar flow, lines of flow known as streamlines parallel one another and no mixing takes place between adjacent layers in the fluid (Figure 12.5a). In true laminar flow, all flow is in one direction only and it remains unchanged through time. When turbulent flow takes place, streamlines intertwine, causing complex mixing within the fluid (Figure 12.5b). So, at any particular location within the fluid a water molecule might move in any direction.

You can easily observe laminar flow in viscous fluids such as cold motor oil or syrup, both of which flow slowly and with difficulty. When heated, however, both flow much more readily. Viscosity is also a consideration in running water, but its viscosity is so low that it most often moves by turbulent flow. Temperature exerts some control on water's viscosity, but the primary controls are velocity and the roughness of the surface over which flow occurs.

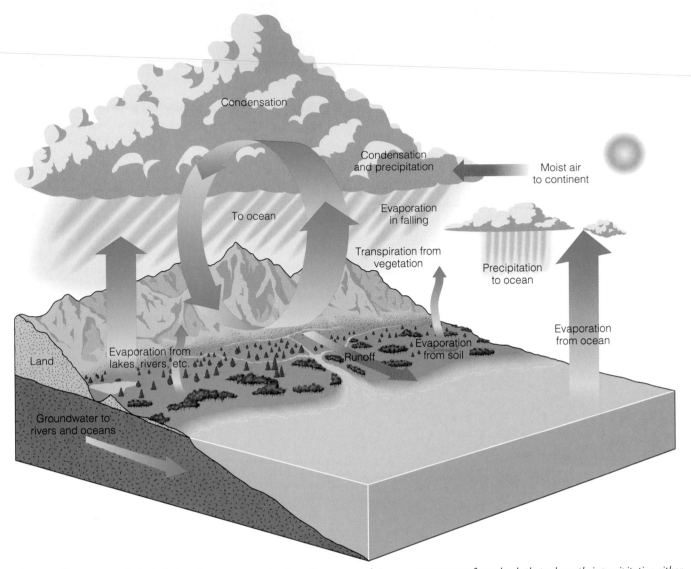

Figure 12.4 *During the hydrologic cycle, water evaporates from the oceans and rises as water vapor to form clouds that release their precipitation either over the oceans or over land. Much of the precipitation falling on land returns to the oceans by surface runoff, thus completing the cycle.*

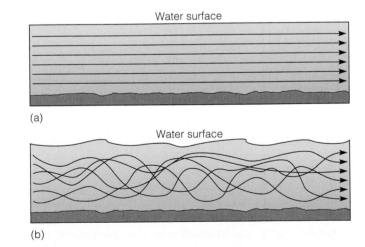

Figure 12.5 *(a) In laminar flow, streamlines are parallel to one another, and little or no mixing takes place between adjacent layers in the fluid. (b) In turbulent flow, streamlines are complexly intertwined, indicating mixing between adjacent layers in the fluid. Most flow in streams is turbulent.*

Dams, Reservoirs, and Hydroelectric Power

Flip a switch and we illuminate our homes, offices, and factories; turn a dial, and we heat our homes or cook our food; and some of our public transportation relies on an unseen energy source—electricity. In fact, about 40% of the total energy in industrialized nations is converted to electricity. Most of it is generated at plants burning fossil fuel (oil, natural gas, and especially coal), but nuclear power plants are important in some countries. Electricity produced by geothermal energy (see Chapter 13), wind, and tides (see Chapter 16) is important in a few areas, but overall they contribute little to the total production.

Hydroelectric power, that is, electricity generated by moving water, accounts for only about 8% of all electricity produced, but its importance varies considerably from area to area. For instance, less than 2% of the electricity generated in Ohio, Texas, and Florida comes from hydroelectric plants, whereas Washington, Oregon, and Idaho derive more than 80% of their electricity from this source.

The same method of generating electricity is used in all hydroelectric plants—moving water spins a turbine connected to a generator containing an electromagnet inside a coil of wire, and electricity is produced. In contrast, in fossil-fuel-burning plants and nuclear power plants, steam is the energy source used to turn turbines. To provide the necessary water, a dam is built impounding a reservoir where the water is higher than the power-generating plant (Figure 12.4). Water moves through a large pipe called a penstock and encounters the blades of a turbine (Figure 1). So the potential energy of the water in the reser-

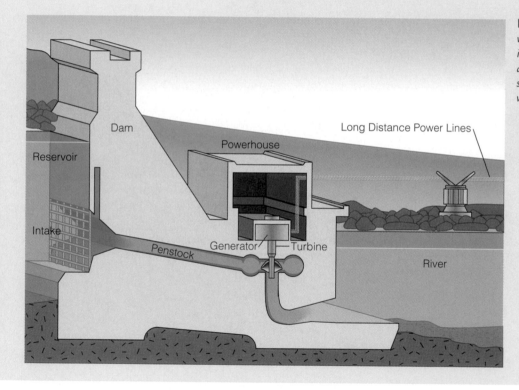

Figure 1 *At a hydroelectric dam, water rushes through the penstock where it spins a turbine connected by a shaft to an electromagnet within a generator. The spinning electromagnet inside a coil of wire generates electricity.*

Although some mudflows and lava flows undoubtedly move by laminar flow, examples of laminar flow in running water are difficult to find. Laminar flow takes place as groundwater moves slowly through the tiny pores in soil, sediment, and rocks (see Chapter 13), but velocity and roughness in almost all surface flow ensures that it is fully turbulent. Even if laminar flow should occur at the surface, it is generally too slow and too shallow to cause any erosion. Turbulent flow, on the other hand, is much more energetic and therefore capable of considerable erosion and sediment transport. We will have occasion to mention laminar flow again, but in this chapter our main concern is with turbulent flow.

Runoff during a rainstorm depends on **infiltration capacity,** the rate at which surface materials can absorb water. Several factors control infiltration capacity, includ-

voir is converted to electrical energy at the power plant. Once generated, electricity is transmitted to areas of use by power lines.

Falling water was first used to generate electricity in 1882 at Appleton, Wisconsin, and since then hydroelectric plants have become common features on the world's waterways. Currently the largest one in terms of generating capacity is on the Parana River on the Paraguay-Brazil border, and an even larger-capacity plant is under construction in the People's Republic of China. A large region in the Pacific Northwest depends on electricity generated at the Grand Coulee Dam in Washington, and hydroelectric plants on the Montreal River in Canada supply power to a vast area. Here, however, we will concentrate on the Tennessee Valley Authority, an agency responsible for supplying electricity to all of Tennessee and parts of six adjacent states (Figure 2).

Established in 1931, the Tennessee Valley Authority (TVA) was part of President Franklin D. Roosevelt's New Deal designed to help restore the economy during the Great Depression that began with the stock market crash of 1929. He directed Congress to create "a corporation clothed with the power of government but possessed of the flexibility and initiative of a private enterprise."* The TVA has several responsibilities such as flood control and river navigation, but generating electricity is its primary function. Currently it produces enough electricity for three cities the size of New York City at 11 coal-burning plants, 29 hydroelectric dams, and 4 nuclear power plants. About 68% of the total electricity generated comes from fossil-fuel-burning plants, whereas 19% comes from plants at hydroelectric dams.

One might be curious about why the TVA and other agencies and governments do not simply increase their hydroelectric output. After all, hydroelectric power generation has several

*A Short History of the TVA at http://www.tva.gov/heritage/ hert_history.htm

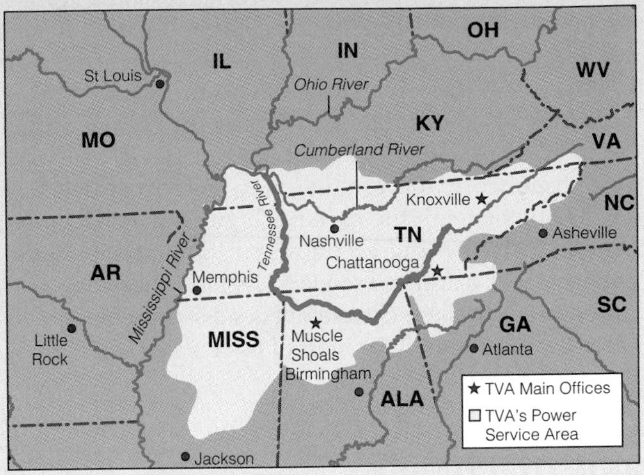

Figure 2 *The Tennessee Valley Authority (TVA) provides electricity to more than 8.3 million people in an area of about 146,000 km².*

appealing aspects, not the least of which is that it is a renewable resource. In fact, some countries, such as New Zealand, have enough hydroelectric generating capacity to meet all their needs, although they use geothermal energy too. However, not all areas have this potential; suitable sites for dams and reservoirs might not be available, for instance. In addition, dams are very expensive to build, reservoirs fill with sediment, and during droughts too little water might be present to keep reservoirs sufficiently full. And of course people must be relocated from areas where reservoirs are impounded and from the discharge areas downstream from dams. So although hydroelectric dams remain an essential element in our total energy production, we cannot realistically look forward to very much additional use of this energy source.

ing intensity and duration of rainfall. If rain is absorbed as fast as it falls, no surface runoff takes place. For instance, loosely packed dry soil absorbs water faster than tightly packed wet soil, and thus more rain must fall on loose dry soil before runoff begins. Regardless of the initial condition of surface materials, once they are saturated, excess water collects on the surface, and if on a slope, it moves downhill.

Sheet Flow and Channel Flow

Even on steep slopes, flow is initially slow and causes little erosion. But as water moves downhill, it accelerates and moves by either *sheet flow* or *channel flow*. In **sheet flow,** a more or less continuous sheet of shallow water moves over the surface and in some cases is responsible for sheet erosion, a problem on some agricultural lands (see Chapter 5). **Channel flow,** in contrast, is confined to long, troughlike depressions that vary from tiny rills with a trickling stream of water to huge rivers. A variety of terms describe channelized flow including rill, brook, creek, stream, and river, most of which are distinguished by size and volume. Here we will use the terms *stream* and *river,* the latter generally designating a larger body of flowing water.

Rivers and streams receive water from several sources, including sheet flow and rain falling directly into their channels. Much more important, though, is water supplied by soil moisture and groundwater, both of which flow downslope beneath the surface (Figure 12.4). In humid regions with plentiful groundwater, flow remains more constant than in arid and semiarid regions where flow fluctuates widely because channels depend on infrequent rainfall and surface runoff for most of their water.

Gradient, Velocity, and Discharge

Water in any channel flows downhill over a slope known as its **gradient.** For example, suppose a river has its headwaters (source) 1000 m above sea level and it flows 500 km to the sea, so it drops vertically 1000 m over a horizontal distance of 500 km. Its gradient is found by dividing the vertical drop by the horizontal distance, which in this example is 1000 m/500 km = 2 m/km (Figure 12.6). We can say that on the average this river drops vertically 2 m for every kilometer along its course.

In the preceding example we calculated the average gradient for a hypothetical river, but gradients vary not only among channels, but even along the course of a single channel. Rivers and streams are steeper in their upper reaches (near their headwaters) where they may have gradients of several tens of meters per kilometer, but have gradients of only a few centimeters per kilometer where they discharge into the sea.

The **velocity** of running water is simply a measure of the downstream distance water travels in a given time. It is usually expressed in meters per second (m/sec) or feet per second (ft/sec), and it varies across a channel's width

as well as along its length. Water moves more slowly and with greater turbulence near a channel's bed and banks because friction is greater there than it is some distance from these boundaries (Figure 12.7a). Channel shape and roughness also influence flow velocity. Broad, shallow channels and narrow, deep channels have proportionately more water in contact with their perimeters than channels with semicircular cross sections (Figure 12.7b). So if other variables are the same, water flows faster in a semicircular channel because of less frictional resistance. As one would expect, rough channels, such as those strewn with boulders, offer more frictional resistance to flow than do channels with a bed and banks composed of sand or mud.

Intuitively you might suspect that gradient is the greatest control on velocity—the steeper the gradient, the greater the velocity. In fact, a channel's average velocity actually increases downstream even though its gradient decreases! Keep in mind that we are talking about average velocity for a long segment of a channel, not velocity at a single point. Three factors account for this general downstream increase in velocity. First, velocity increases even with decreasing gradient in response to the acceleration of gravity unless other factors retard flow. Second, the upstream reaches of channels tend to be boulder-strewn, broad, and shallow, so frictional resistance to flow is high, whereas downstream segments of the same channels are generally more semicircular and have banks composed of finer materials. And finally, the number of smaller tributaries joining a larger channel increases downstream. Thus, the total volume of water (discharge) increases, and increasing discharge results in greater velocity.

We mentioned discharge in the preceding paragraph but noted only that it refers to the volume of water. More specifically, **discharge,** the volume of water passing a par-

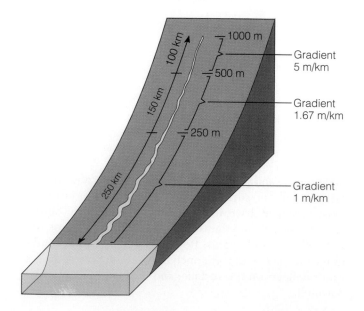

Figure 12.6 *The average gradient of this stream is 2 m/km, but gradient can be calculated for any segment of a stream as shown in this example. Notice that the gradient is steepest in the headwaters area and decreases in a downstream direction.*

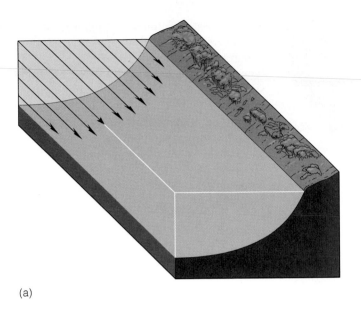

(a)

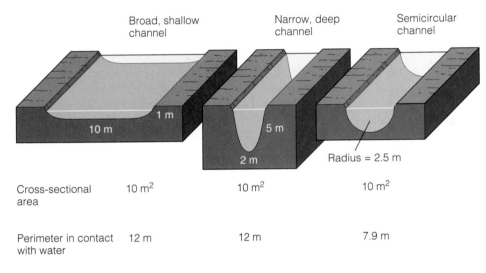

(b)

Figure 12.7 *Flow velocity in rivers and streams varies as a result of friction with their banks and beds. (a) The maximum flow velocity is near the center and top of a straight channel where friction is least. The arrows are proportional to velocity. (b) These three differently shaped channels have the same cross-sectional area. But the semicircular one has less water in contact with its perimeter, and thus less frictional resistance to flow.*

ticular point in a given period of time, is found by knowing the dimensions of a water-filled channel, that is, its cross-sectional area (A) and its flow velocity (V). Discharge (Q) is then calculated with the formula Q = VA, and is expressed in cubic meters per second (m³/sec) or cubic feet per second (ft³/sec) (Figure 12.8). The Mississippi River has an average discharge of 18,000 m³/sec, and average discharge for the Amazon River in South America is 200,000 m³/sec.

In most rivers and streams, discharge increases downstream as more and more water enters a channel. However, there are a few exceptions. Because of high evaporation rates and infiltration, the flow in some desert waterways actually decreases downstream until the streams disappear. And even in perennial rivers

and streams, discharge is obviously highest during times of heavy rainfall and at a minimum during the dry season.

How Does Running Water Erode and Transport Sediment?

We have already mentioned that running water, with few exceptions, is the most important geologic agent modifying Earth's land surface. Its role in erosion cannot be underestimated. Water possesses two kinds of

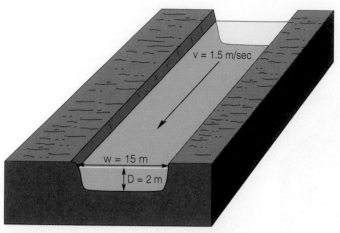

Discharge (Q) = VA = 1.5 m/sec × 30 m² = 45 m³/sec

Figure 12.8 *Discharge (Q) is found by multiplying a river or stream's velocity (V) by its cross-sectional area (A). Cross-sectional area (A) is the product of depth (D) times width (W) of the water-filled part of the channel. The channel in this example is rectangular to make the calculation easier, but natural channels are irregular, so determining (A) is more difficult.*

energy: potential and kinetic. *Potential energy* is the energy of position, as that possessed by water behind a dam or at high elevation. In running water, the energy of position is converted to *kinetic energy,* the energy of motion. Most of this kinetic energy is used up by fluid turbulence, but a small amount is available to erode and transport sediment.

Erosion involves the removal from a source area of dissolved substances as well as loose particles of soil, minerals, and rock. Some of the dissolved materials in running water is acquired from the beds and banks of channels where soluble rocks such as limestone are exposed, but much of it is derived from sheet flow and groundwater. Solid particles from channel perimeters or introduced into channels by mass wasting are set in motion by **hydraulic action,** that is, the direct impart of water on loose materials (Figure 12.9). Running water carrying sand and gravel also erodes by **abrasion,** involving the impact of solid particles with exposed rock surfaces. One obvious manifestation of abrasion is circular to oval *potholes* that form where sand and gravel in eddying currents abrade depressions into rock (Figure 12.10).

Once materials are eroded, they are transported by running water for some distance from their source and eventually deposited. Transport involves moving both a **dissolved load,** consisting of materials taken into solution during chemical weathering, and a *solid load* ranging from clay-sized particles to large boulders (Figure 12.11). This solid load is further divided into a suspended load and a bed load. In the **suspended load,** the smallest particles in transport, such as silt and clay, are kept suspended above the channel bed by fluid turbulence (Figure 12.11). Suspended load in rivers and streams is what

gives the water its murky appearance. Running water also transports a **bed load** of larger particles, especially sand and gravel, that cannot be kept suspended by fluid turbulence. Some of the sand, however, might be temporarily suspended when an eddying current swirls across a channel's bed and lift grains into the water. These grains move forward with the water but also settle and come to rest on the bed where they may be moved again by this same processes of intermittent bouncing and skipping, a phenomenon known as *saltation* (Figure 12.11). Particles too large to be suspended even temporarily move by rolling and sliding.

Deposition by Running Water

Some of the sediment now being deposited in the Gulf of Mexico by the Mississippi River came from such distant sources as Pennsylvania, Minnesota, and Alberta, Canada. In short, transport might be lengthy but deposition eventually takes place. Some deposits accumulate along the way in channels, on adjacent floodplains, or where rivers and streams discharge from mountains onto nearby lowlands or where they flow into lakes or seas.

Rivers and streams constantly erode, transport, and deposit sediment, but most of their geologic work takes place when they flood. Consequently, their deposits, collectively called **alluvium,** do not represent the day-to-day activities of running water, but rather the periodic, large-scale events of sedimentation that take place during floods. Remember from Chapter 5 that sediments accumulate in *depositional environments* characterized as continental, transitional, and marine. Deposits of rivers and streams are found mostly in the first two of these settings; however, much of the detrital sediment found on continental margins is derived from the land and transported to the oceans by running water.

The Deposits of Braided and Meandering Channels

Most rivers and streams possess channels that we characterize as *braided* or *meandering.* A **braided stream** possess an intricate network of dividing and rejoining channels separated from one another by sand and gravel bars (Figure 12.12). Seen from above, the channels resemble the complex strands of a braid. Braided channels develop when sediment supply exceeds the transport capacity of running water resulting in the deposition of sand and gravel bars. During high-water stages, the bars are submerged, but when the water is low, they are exposed and divide a single channel into multiple channels. Braided streams have broad, shallow channels and are characterized as bed-load transport streams because

Figure 12.9 *(a) This stream acquires some of its sediment load by undercutting its banks. (b) Some of the sediment in the Snake River of Idaho comes from these talus cones that accumulated as a result of mass wasting.*

Figure 12.10 *(a) The circular depressions on the bed of the Chippewa River in Ontario, Canada, are potholes. They measure about 1 m across, but two potholes at the top center of the image have merged to form a larger, composite pothole. (b) These stones measuring 7 to 8 cm in diameter from a pothole are remarkably spherical and smooth because of abrasion.*

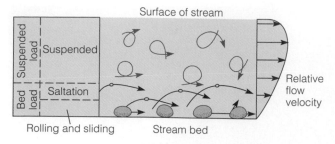

Figure 12.11 *Methods of sediment transport by running water. The velocity profile at the right indicates that the water flows fastest near the surface and slowest along the streambed.*

they transport and deposit mostly sand and gravel (Figure 12.12).

Braided streams are common in arid and semiarid regions where, because of the sparse vegetation, surface materials are unprotected and easily eroded. In fact, contrary to what many people think, erosion rates are higher here than in most humid regions. So much sediment is released from melting glaciers that rivers and streams discharging from them are also commonly braided (see Chapter 14).

Meandering streams have a single sinuous channel with broadly looping curves known as *meanders* (Figure 12.13). Channels of meandering streams are semicircular in cross section along straight reaches, but markedly asymmetric at meanders where they vary from quite shallow to deep across the meander. The deeper side of the channel is known as the *cut bank* because greater velocity and fluid turbulence erodes it. In contrast, flow velocity is at a minimum on the opposite bank, which slopes gently into the channel. As a result of this unequal distribution of flow velocity across meanders, the cut bank erodes and a **point bar** is deposited on the gently sloping inner bank. Most point

(a)

(b)

Figure 12.12 *Braided streams and their deposits. (a) A braided stream with a bed of sand near Santa Fe, New Mexico. (b) A braided stream with gravel bars near Chester, California.*

bars are composed of cross-bedded sand, but some consist of gravel (Figure 12.14).

Meanders commonly become so sinuous that the thin neck of land between adjacent ones gets cut off during a flood. Many of the floors of valleys with meandering channels are marked by crescent-shaped **oxbow lakes,** which are simply cutoff meanders (Figures 12.13 and 12.15). Oxbow lakes may persist for some time, but they eventually fill with organic matter and fine-grained sediments carried by floods.

Floodplain Deposits

Rivers and streams periodically receive more water than their channels can accommodate, so they overflow their banks and spread across adjacent low-lying, relatively flat **floodplains** (Figure 12.13). Floodplain sediments might be sand and gravel that accumulated as meandering streams deposited a succession of point bars as they migrated laterally (Figure 12.16). More commonly, however, fine-grained sediments, mostly mud, are dominant on floodplains. During a flood, a stream overtops its banks and water pours onto the floodplain, but as it does so its velocity and depth rapidly decrease. As a result, ridges of sandy alluvium known as **natural levees** are deposited along the channel margins, and mud is carried beyond the natural levees into the floodplain where it settles from suspension (Figure 12.17).

Figure 12.13 *Aerial view of a meandering stream. The broad, flat area adjacent to the stream channel is the floodplain. Notice the crescent-shaped lakes—these are cut-off meanders, or what are known as oxbow lakes.*

Figure 12.14 *(a) In a meandering river or stream, flow velocity is greatest near the cut bank. The dashed line follows the path of maximum velocity, and the solid arrows are proportional to velocity. The cut bank is eroded whereas deposition takes place on the opposite bank forming a point bar. (b) Two small point bars composed of sand. Notice how the point bars are inclined into the deeper part of the channel. (c) This point bar is composed of gravel.*

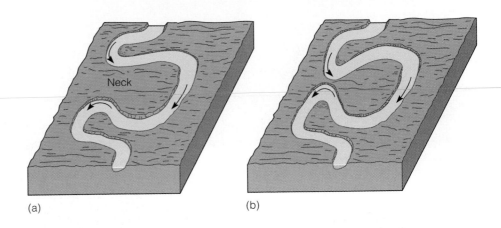

(a)

Neck

(b)

Deposits of
silt and clay

Abandoned
channel

Oxbow
lake

Figure 12.15 *Four stages in the origin of an oxbow lake. In (a) and (b) the meander neck becomes narrower. (c) The meander neck is cut off, and part of the channel is abandoned. (d) When it is completely isolated from the main channel, the abandoned meander is an oxbow lake.*

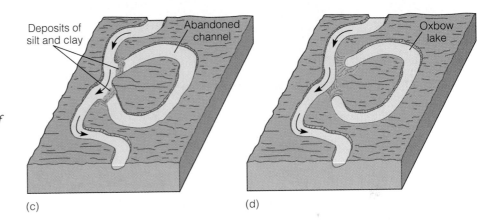

(c)

(d)

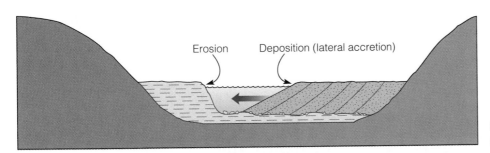

Erosion   Deposition (lateral accretion)

Figure 12.16 *Floodplain deposits forming by lateral growth of point bars.*

Floods cause more than $100 million in property damage annually in the United States. And despite the completion of more and more flood-control projects, the amount for property damage is not decreasing (see Perspective 12.1). As a matter of fact, floods of one kind or another cause more property damage and more fatalities than any other kind of natural disaster.

Deltas

Where a river or stream flows into a standing body of water, such as a lake or the ocean, its flow velocity rap- idly diminishes and any sediment in transport is deposited. Under some circumstances, this deposition leads to the origin of a **delta,** an alluvial deposit that causes the shoreline to build outward into the lake or sea, a process called *progradation.* The simplest prograding deltas have a characteristic vertical sequence of *bottomset beds* overlain successively by *foreset beds* and *topset beds* (Figure 12.18). This vertical sequence develops when a river or stream enters another body of water where the finest sediment (silt and clay) is carried some distance out into the lake or sea; there it settles to form bottomset beds. Nearer shore, foreset beds are

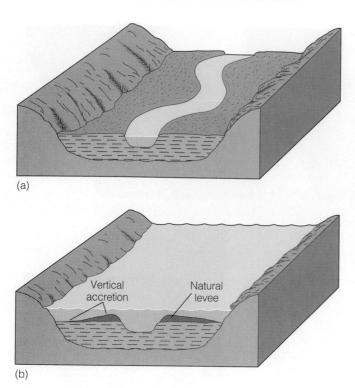

(a)

(b)

Vertical
accretion

Natural
levee

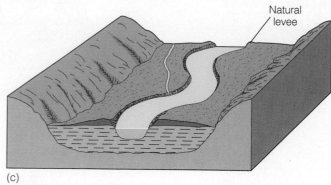

Natural
levee

(c)

Figure 12.17 *Three stages in the formation of deposits on a floodplain. (a) Stream at low-water stage. (b) Flooding stream and deposition. Many such episodes of flooding form natural levees. (c) After flooding.*

deposited as gently inclined layers, and topset beds, consisting of the coarsest sediments are deposited in a network of *distributary channels* traversing the top of the delta (Figure 12.18).

Many small deltas in lakes have the three-part sequence described above, but deltas deposited along seacoasts are much larger, far more complex, and considerably more important as potential areas of natural resources. In fact, depending on the relative importance

of stream (or river), wave, and tide processes, geologists identify three main types of marine deltas (Figure 12.19). *Stream-dominated deltas* have long fingerlike sand bodies, each deposited in a distributary channel that progrades far seaward. The Mississippi delta is a good example. In contrast, the Nile delta in Egypt is *wave-dominated.* It also has distributary channels, but the seaward margin of the delta consists of islands reworked by waves, and the entire margin of the delta progrades. *Tide-dominated deltas* are continually modified into tidal sand bodies that parallel the direction of tidal flow.

Alluvial Fans

Lobate deposits of alluvium on land known as **alluvial fans** form best on lowlands with adjacent highlands in arid and semiarid regions where little vegetation exists to stabilize surface materials (Figure 12.20). During periodic rainstorms, surface materials are quickly saturated and surface runoff is funneled into a mountain canyon leading to adjacent lowlands. In the mountain canyon, the runoff is confined so it cannot spread laterally, but when it discharges onto the lowlands it quickly spreads out, its velocity diminishes, and deposition ensues. Repeated episodes of this kind of sedimentation result in the accumulation of a fan-shaped body of alluvium.

Deposition by running water in the manner just described is responsible for many alluvial fans. In this case

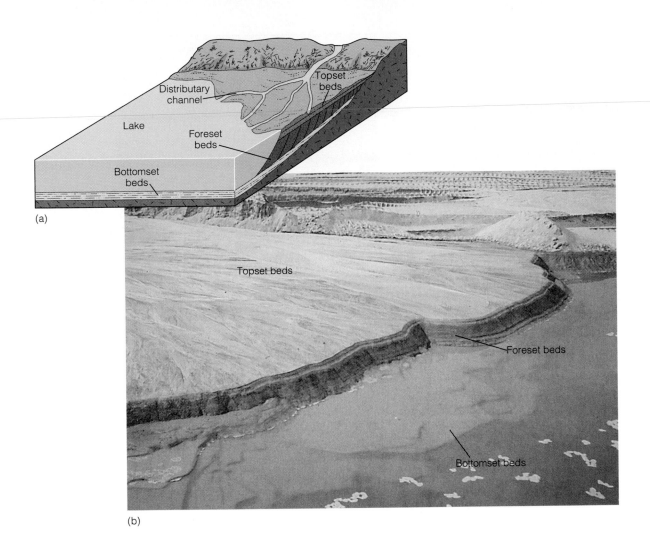

Figure 12.18 *(a) Internal structure of the simplest type of prograding delta. (b) A small delta, measuring about 20m across, in which bottomset, foreset, and topset beds are visible.*

they are composed mostly of sand and gravel, both of which might contain a variety of sedimentary structures. In some cases, though, the water flowing through a canyon picks up so much sediment that it becomes a viscous mudflow. Consequently, some alluvial fans consist mostly of mudflow deposits showing little or no layering. Of course, the dominant type of deposition can change through time, so a particular fan might have both types of deposits.

Can Floods Be Controlled?

When a river or stream receives more water than its channel can handle, it floods, occupying part or all of its floodplain. Indeed, floods are so common that unless they cause considerable property damage or fatalities they rarely rate more than a passing notice in the news. The most recent catastrophic flooding in the United States took place in 1993 (see Perspective 12.2), but since then several other areas in North America and elsewhere have experienced serious flooding (Figure 12.21). One of the most disastrous was during December 1999 when floods and mudslides killed tens of thousands in Venezuela (see Chapter 11).

Flood control has been practiced for thousands of years. Common practices are to construct dams that impound reservoirs (Figure 12.3) and to build levees along stream banks. Levees raise the banks of a stream, thereby restricting flow during floods. Unfortunately, deposition within the channel results in raising the streambed, making the levees useless unless they too are raised. Levees along the banks of the Huang He in China caused the streambed to rise more than 20 m

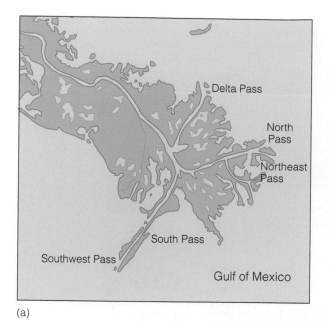

(a)

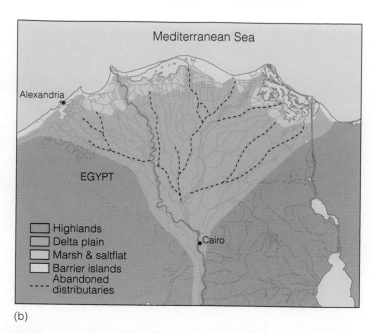

(b)

Figure 12.19 *(a) The Mississippi River delta of the U.S. Gulf Coast is stream-dominated. (b) The Nile Delta of Egypt is wave-dominated. (c) The Ganges-Brahmaputra delta of Bangladesh is tide-dominated.*

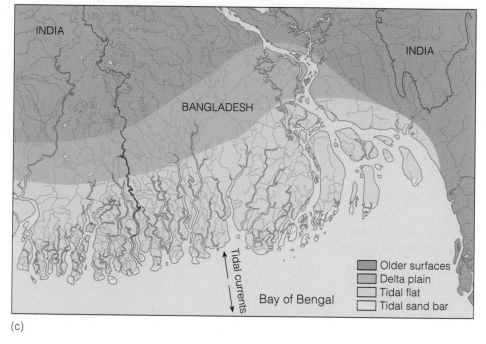

(c)

above its surrounding floodplain in 4000 years. When the Huang He breached its levees in 1887, more than 1 million people were killed.

Dams and levees alone are insufficient to control large floods, so in many areas floodways are also used. These usually consist of a channel constructed to divert part of the excess water in a stream around populated areas or areas of economic importance (Figure 12.22). Reforestation of cleared land also helps reduce the potential for flooding because vegetated soil helps prevent runoff by absorbing more water.

When flood-control projects are well planned and constructed, they are functional. What many people fail to realize is that these projects are designed to contain floods of a given size; should larger floods occur, streams spill onto floodplains anyway. Furthermore, dams occasionally collapse (see the Prologue), and reservoirs eventually fill with sediment unless dredged. In short, flood-control projects are not only initially expensive but also require constant, costly maintenance. Such costs must be weighed against the cost of damage if no control projects were undertaken.

(a)

Alluvial fan

Channels are dry most of the time

(b)

Figure 12.20 *(a) Alluvial fans form where streams discharge from mountain canyons onto adjacent lowlands. (b) Alluvial fans adjacent to the Panamint Range on the margin of Death Valley, California.*

Drainage Basins and Drainage Patterns

Thousands of waterways, which are parts of larger drainage systems, flow directly or indirectly into the oceans. The only exceptions are some rivers and streams that flow into desert basins surrounded by higher areas.

But even these are parts of larger systems consisting of a main channel with all its tributaries, that is, streams that contribute water to another stream. The Mississippi River and its tributaries such as the Ohio, Missouri, Arkansas, and Red Rivers and thousands of smaller ones, or any other drainage system for that matter, carries runoff from an area known as a **drainage basin.** A topographically high area called a **divide** separates a drainage basin from adjoining ones (Figure 12.23). The continental divide

Scarcely a year passes without the Mississippi River or one or more of its tributaries flooding. In 1927, flooding on the Mississippi resulted in 214 deaths and during 1937 floods in the Mississippi and Ohio valleys killed another 250. In both floods, tens of thousands lost their homes and tens to hundreds of millions of dollars of property damage resulted. The death toll in the Flood of '93 was much lower, about 50 people were killed, but at least 70,000 were driven from their homes. And according to one report, the June to August 1993 flooding in several midwestern states caused $15 to $20 billion in property damage.

The Flood of '93 was so vast that more than 200 counties in several states, including every county in Iowa, were declared disaster areas. In addition to Iowa, flooding took place in South Dakota, North Dakota, Minnesota, Wisconsin, Nebraska, Missouri, Illinois, and Kansas. The onset of flooding began in late June and continued into August by which time more than 92,000 km² were underwater. At several cities, such as Davenport, Iowa, Des Moines, Iowa, and Quincy, Illinois, the rivers crested higher than they ever had before.

No shortage of human effort was spared to contain the rising rivers. Nevertheless, despite heroic efforts by the National Guard and thousands of volunteers to stabilize levees by sandbagging them, levees failed everywhere or were simply overtopped by the rising floodwaters. Grafton, Illinois, at the juncture of the Mississippi and Illinois Rivers was 80% underwater. This town, which has no protective levees, has been flooded six times in the last 20 years. In the aftermath of the Flood of '93, the mayor and many of Grafton's residents decided to move the town to a higher, safer location at a cost of $25 million.

A large part of St. Charles County, Missouri, lying near the confluence of the Missouri and Mississippi Rivers, was so extensively flooded that 8000 people were evacuated. In the small town of Portage des Sioux in St. Charles County, only eight homes were not flooded or were only slightly damaged by floodwaters. Even the 5.5-m-high pedestal of a statue near the Mississippi was swamped (Figure 1). In Howard County, Missouri, along the Missouri River, grain silos and other structures were smashed and thick sand and mud deposits covered farmlands.

Obviously the cause of flooding was too much water for the Mississippi and its tributaries to handle. But why so much water, particular at a time when rainfall is generally much less, and why such widespread flooding when so much money and effort has been spent on flood control? The reason for the excess water was the unusual behavior of the *jet stream,* a narrow band of strong winds in the atmosphere. It is usually over the Midwest during the spring and then shifts north into Canada during the summer. In 1993, though, it remained over the Midwest and as hot, moisture-laden air moved north from the Gulf of Mexico, thunderstorms occurred repeatedly over the region resulting in 1 1/2 to 2 times the normal precipitation.

One important lesson learned from the Flood of '93 is that despite our best efforts at flood control, not all floods can be prevented. The 29 dams on the Mississippi and 36 reservoirs on upstream tributaries, as well as about 5800 km of levees, had little effect on holding the floodwaters in check. Reservoirs have a limited capacity, and once filled they are of no further use in flood control. In fact, some reservoirs are kept at high levels for other purposes such as recreation which is inconsistent with their role in flood control.

Levees are effective in protecting many areas during floods, yet in some cases they actually exacerbate the problem by restricting flow that would otherwise spread across a broad floodplain. And in some cases levees simply fail as water seeps through them or flows over their tops. Levees are certainly expensive to build and maintain, and their overall effectiveness is and will continue to be questioned. No one doubts the success in flood control of the 500 or so dams and 16,000 km of levees built during this century, at least within the limits of their design. But in some ways flood control projects actually make the problem of flooding worse. Particularly because development in flood-prone areas commonly follows the completion of flood-control projects, yet nothing can be done to prevent some floods.

---

along the crest of the Rocky Mountains in North America, for instance, separates drainage in opposite directions; drainage to the west goes to the Pacific, whereas that to the east eventually reaches the Gulf of Mexico.

Various arrangements of channels within an area are classified as several types of **drainage patterns.** *Dendritic drainage,* consisting of a network of channels resembling tree branching is the most common (Figure 12.24a). It develops on gently sloping surfaces composed of materials that respond more or less homogeneously to erosion. Areas underlain by nearly horizontal sedimentary rocks and some areas of igneous and metamorphic rock commonly display dendritic drainage.

In dendritic drainage, tributaries join larger channels at various angles, but *rectangular drainage* is characterized by right angle bends and tributaries joining larger channels at right angles (Figure 12.24b). Such regularity in channels is strongly controlled by geologic structures,

(a)

(b)

Figure 1 (a) Portage des Sioux, St. Charles County, Missouri, on July 16, 1993. The channel of the Mississippi River is at the far right. (b) Floodwaters in Portage des Sioux covered the 5.5-m-high pedestal of this statue on the bank of the Mississippi River.

particularly regional joint systems that intersect at right angles.

Trellis drainage consisting of a network of nearly parallel main streams with tributaries joining them at right angles is common in some parts of the eastern United States. In Virginia and Pennsylvania, for example, erosion of folded sedimentary rocks developed a landscape of alternating ridges on resistant rocks and valleys underlain by easily eroded rocks. Main waterways follow the valleys, and short tributaries flowing from the nearby ridges join the main channels at nearly right angles (Figure 12.24c).

In *radial drainage*, streams flow outward in all directions from a central high point such as a large volcano (Figure 12.24d). Many of the volcanoes in the Cascade Range of western North America have radial drainage patterns. In all of the types of drainage mentioned so far some kind of pattern is easily recognized. *Deranged drainage*, however, is characterized by irregularity with streams flowing

Figure 12.21 *Flooding of New Richmond, Ohio, by the Ohio River on March 4, 1997. Heavy thunderstorms on March 1 and 2 caused the Ohio River and several tributaries to flood resulting in 18 Ohio counties being declared federal and state disaster areas. About 20,000 people were evacuated, five died, and property damage was about $180 million.*

Figure 12.22 *This floodway carries excess water from a river (not visible) to reduce the threat of flooding.*

than 1.6 km deep, but the bottom of the canyon is still far above sea level. In fact, the lowest level to which any river or stream can erode is known as **base level.** Sea level is commonly called *ultimate base level,* and theoretically a channel could erode deeply enough that its gradient rose ever so slightly inland from the sea (Figure 12.25). But rise it must, for otherwise no flow would be possible. Ultimate base level applies to an entire river or stream, but channels might also have *local* or *temporary base levels* (Figure 12.25). For instance, local base level for a segment of a stream might be a lake or another stream. And where water flows across particularly resistant rocks, a waterfall might develop, forming a local base level.

Ultimate base level is sea level, but what if sea level should rise or drop with respect to the land, or what if the

into and out of swamps and lakes, streams with only a few short tributaries, and vast swampy areas between channels (Figure 12.24e). The presence of this kind of drainage indicates that it developed recently and has not yet formed a fully organized drainage system. In parts of Minnesota, Wisconsin, and Michigan, where glaciers obliterated the previous drainage, only 10,000 years has elapsed since the glaciers melted. As a result, drainage systems have not fully developed and large areas still remain undrained.

The Significance of Base Level

We have already mentioned that running water is a very effective agent of erosion. But just how deeply can a river or stream erode? The Grand Canyon in Arizona is more

(a)

Figure 12.23 (a) Small drainage basins separated from one another by divides (dashed lines), which are along the crests of the ridges between channels (solid lines). (b) The drainage basin of the Wabash River, which is one of the tributaries of the Ohio River. All tributary streams within the drainage basin, such as the Vermillion River, have their own smaller drainage basins. Divides are shown by dark red lines.

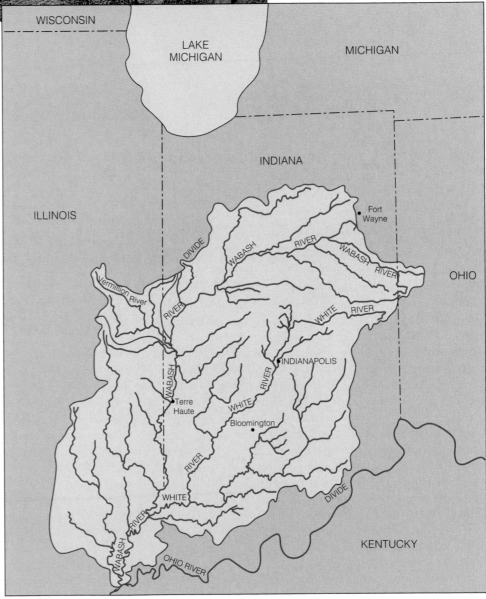

(b)

The Significance of Base Level **355**

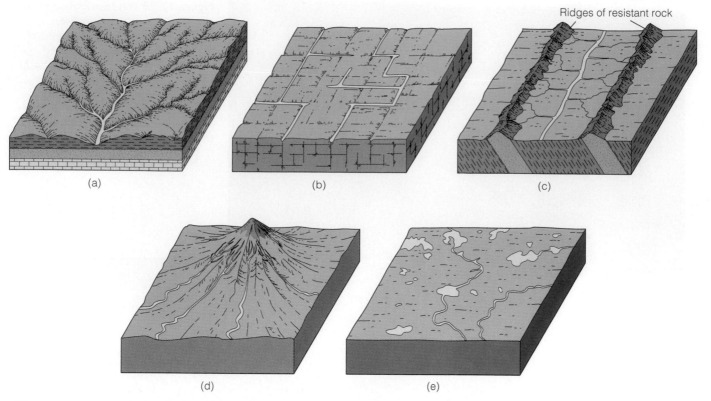

Figure 12.24 *Examples of drainage patterns. (a) Dendritic drainage. (b) Rectangular drainage. (c) Trellis drainage. (d) Radial drainage. (e) Deranged drainage.*

land over which a river flows should be uplifted or subside? In either case changes would occur in base level and the rivers would respond accordingly. During the Pleistocene Epoch, for instance, sea level was as much as 130 m lower than it is now (see Chapter 24), and rivers adjusted to this lower base level by deepening their valleys and extending them well out onto the continental shelves. Rising sea level at the end of the Pleistocene caused base level to rise, thereby flooding the valleys on the continental shelf.

Natural changes, such as fluctuations in sea level during the Pleistocene, alter the dynamics of rivers and streams, but so does human intervention. Geologists and engineers are well aware that building a dam to impound a reservoir creates a local base level (Figure 12.26a). A stream entering a reservoir slows down and deposits sediment, so unless dredged, reservoirs eventually fill with sediment. In addition, the water discharged at a dam is largely sediment-free but still possesses enough energy to carry a sediment load. As a result it is not uncommon for streams to erode vigorously downstream from the dam to acquire a sediment load.

Draining a lake may seem like a small change and well worth the time and expense to expose dry land for agriculture or commercial development. But in doing so a local base level has been eliminated, and a stream that originally

flowed into the lake responds by rapidly eroding a deeper valley as it adjusts to a new base level (Figure 12.26b).

What Is a Graded Stream?

The *longitudinal profile* of any waterway shows the elevations and irregularities, if any, of a channel along its length as viewed in cross section (Figure 12.27). For some rivers and streams the longitudinal profile is smooth, but for others, a number of irregularities such as lakes and waterfalls are present, all of which are local base levels (Figure 12.25). Over time these irregularities tend to be eliminated; deposition takes place where the gradient is insufficient to maintain sediment transport, and erosion decreases the gradient where it is steep. So given enough time, rivers and streams develop a smooth, concave longitudinal profile of equilibrium, meaning that all parts of the system dynamically adjust to one another.

A **graded stream** is one with an equilibrium profile in which a delicate balance exists between gradient, discharge, flow velocity, channel shape, and sediment load so that neither significant erosion nor deposition takes place

Local base
level

Rock resistant
to erosion

Ultimate base
level

(a)

Figure 12.25 *(a) Sea level is ultimate base level and a resistant rock layer forms a local base level. (b) Cumberland Falls on the River in Cumberland Falls State Resort Park, Kentucky. The rock layer the falls plunge over is a local base level. At 38 m wide and 18 m high, Cumberland Falls is the second highest waterfall east of the Rocky Mountains.*

(b)

within its channel. Such a delicate balance is rarely attained, so the concept of a graded stream is an ideal. Nevertheless, the graded condition is closely approached in many streams, although only temporarily and not necessarily along their entire lengths.

Even though the concept of a graded stream is an ideal, we can generally anticipate the response of a graded stream to changes altering its equilibrium. For instance, a change

in base level would cause a stream to adjust as previously discussed. Increased rainfall in a stream's drainage basin would result in greater discharge and flow velocity. In short, the stream would now possess greater energy—energy that must be dissipated within the stream system by, for example, a change from a semicircular to a broad, shallow channel that would dissipate more energy by friction. On the other hand, the stream may respond by eroding a deeper

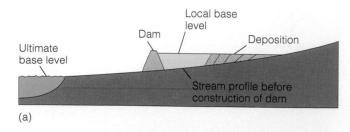

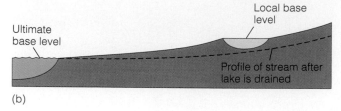

Figure 12.26 *(a) Constructing a dam and impounding a reservoir create a local base level. A stream deposits much of its sediment load where it flows into a reservoir. (b) A stream adjusts to a lower base level when a lake is drained.*

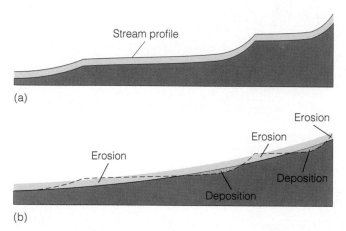

Figure 12.27 *(a) An ungraded stream had irregularities in its longitudinal profile. (b) Erosion and deposition along the course of a stream eliminate irregularities and cause it to develop the smooth, concave profile typical of a graded stream.*

valley and effectively reduce its gradient until it is once again graded.

Vegetation inhibits erosion by having a stabilizing effect on soil and other loose surface materials. So a decrease in vegetation in a drainage basin might lead to higher erosion rates, causing more sediment to be washed into a stream than it can effectively carry. Accordingly, the stream may respond by deposition within its channel, which increases its gradient until it is sufficiently steep to transport the greater sediment load.

How Do Valleys Form and Evolve?

Low areas on the land surface known as **valleys** are bounded by higher land, and most of them have a river or stream running their length, with tributaries draining the nearby high areas. Valleys are common landforms and with few exceptions they form and evolve in response to erosion by running water, although other processes, especially mass wasting, contribute. The shapes and sizes of valleys vary from small, steep-sided *gullies* to those that are broad with gently sloping valley walls. Steep-walled, deep valleys of vast size are *canyons,* and particularly narrow and deep ones are *gorges* (Figure 12.28).

Erosion of a valley might be initiated where runoff has sufficient energy to dislodge surface materials and excavate a small rill. Once formed, a rill collects more runoff and becomes deeper and wider and continues to do so until a full-fledged valley develops. Processes related to running water that contribute to valley formation include downcutting, lateral erosion, headward erosion, and sheet wash. A variety of mass wasting processes are also important.

Downcutting takes place when a river or stream has more energy than it needs to transport sediment, so some of its excess energy is used to deepen its valley. If downcutting were the only process operating, valleys would be narrow and steep sided. In most cases, though, the valley walls are undercut by stream action, a process called *lateral erosion;* this creates unstable slopes that may fail by one or more mass wasting processes. Furthermore, erosion by sheet wash and erosion by tributary streams carry materials from the valley walls into the main stream in the valley.

Valleys not only become deeper and wider but also become longer by *headward erosion,* a phenomenon involving erosion by entering runoff at the upstream end of a valley (Figure 12.29). Continued headward erosion commonly results in *stream piracy,* the breaching of a drainage divide and diversion of part of the drainage of another stream (Figure 12.29). Once stream piracy has taken place, both drainage systems must adjust to these new conditions—one system now has greater discharge and the potential to do more erosion and sediment transport, whereas the other is diminished in its ability to accomplish these tasks.

According to one concept, stream erosion of an area uplifted above sea level yields a distinctive series of landscapes. When erosion begins, streams erode downward; their valleys are deep, narrow, and V-shaped; and a number of irregularities are present in their profiles (Figure 12.30a). As streams cease eroding downward, they start eroding laterally, thereby establishing a meandering pattern and a broad floodplain (Figure 12.30b). Finally, with continued erosion, a vast, rather featureless plain develops (Figure 12.30c).

Many streams do indeed show an association of features typical of these stages. For instance, the Colorado River

(a)

(b)

Figure 12.28 *Valleys of various sizes and shapes form and evolve mostly as a result of erosion by running water coupled with mass wasting processes. (a) Many valleys have gently sloping walls much like this one. (b) Canyon de Cheilly in Arizona is broad and has steep walls.*

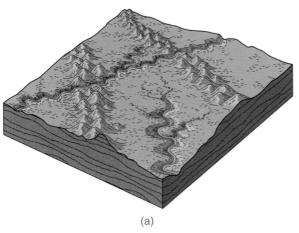

(a)

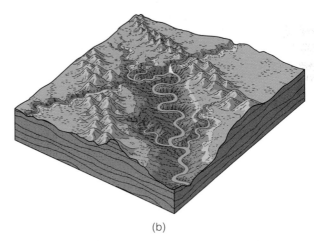

(b)

Figure 12.29 *Two stages in the evolution of a valley. (a) The stream widens its valley by lateral erosion and mass wasting, while simultaneously extending its valley by headward erosion. (b) As the larger stream continues to erode headward, stream piracy takes place when it captures some of the drainage of the smaller stream. Notice also that the larger valley is wider in (b) than it was in (a).*

flows through the Grand Canyon and closely matches the features in the initial stage shown in Figure 12.30a. Streams in many areas approximate the second stage of development, and certainly the lower Mississippi closely resembles the last stage. Nevertheless, the idea of a sequential devel-

opment of stream-eroded landscapes has been largely abandoned because there is no reason to think that streams necessarily follow this idealized cycle. Indeed, a stream on a gently sloping surface near sea level could develop features of the last stage very early in its history. In addition, as long

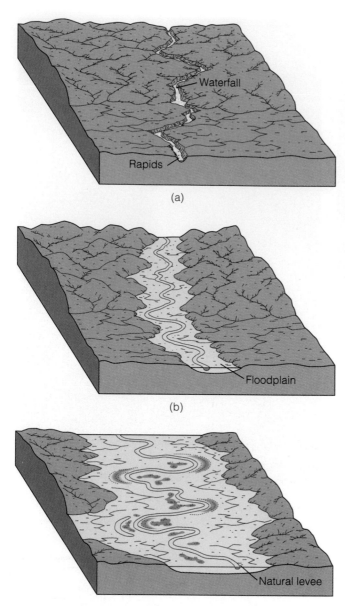

Figure 12.30 *Idealized stages in the development of a stream and its associated landforms.*

Although all stream terraces result from erosion, they are preceded by an episode of floodplain formation and deposition of sediment. Subsequent erosion causes the stream to cut downward until it is once again graded (Figure 12.31). Once the stream again becomes graded, it begins eroding laterally and establishes a new floodplain at a lower level. Several such episodes account for the multiple terrace levels seen adjacent to some streams.

Renewed erosion and the formation of stream terraces are usually attributed to a change in base level. Either uplift of the land a stream flows over or lowering of sea level yields a steeper gradient and increased flow velocity, thus initiating an episode of downcutting. When the stream reaches a level at which it is once again graded, downcutting ceases. Although changes in base level no doubt account for many stream terraces, greater runoff in a stream's drainage basin can also result in the formation of terraces.

Incised Meanders

Some streams are restricted to deep, meandering canyons cut into solid bedrock, where they form features called **incised meanders.** For example, the San Juan River in Utah occupies a meandering canyon more than 390 meters deep (Figure 12.32). Such streams, being restricted by solid rock walls, are generally ineffective in eroding laterally; thus, they lack a floodplain and occupy the entire width of the canyon floor.

It is not difficult to understand how a stream can cut downward into solid rock, but forming a meandering pattern in bedrock is another matter. Because lateral erosion is inhibited once downcutting begins, one must infer that the meandering course was established when the stream flowed across an area covered by alluvium. For example, suppose that a stream near base level has established a meandering pattern. If the land the stream flows over is uplifted, erosion is initiated, and the meanders become incised into the underlying bedrock.

Superposed Streams

Streams flow downhill in response to gravity, so their courses are determined by preexisting topography. Yet a number of streams seem, at first glance, to have defied this fundamental control. For example, the Delaware, Potomac, and Susquehanna Rivers in the eastern United States have valleys that cut directly through ridges lying in their paths. The Madison River in Montana meanders northward through a broad valley, then enters a narrow canyon cut into bedrock that leads to the next valley where the river resumes meandering.

All of these examples are of **superposed streams.** In order to understand superposition, it is necessary to know the geologic histories of these streams. In the case of the Madison River, the valleys it now occupies were once

as the rate of uplift exceeds the rate of downcutting, a stream will continue to erode downward and be confined to a narrow canyon.

Stream Terraces

Adjacent to many streams are erosional remnants of floodplains formed when the streams were flowing at a higher level. These **stream terraces** consist of a fairly flat upper surface and a steep slope descending to the level of the lower, present-day floodplain (Figure 12.31). In some cases, a stream has several steplike surfaces above its present-day floodplain, indicating that stream terraces formed several times.

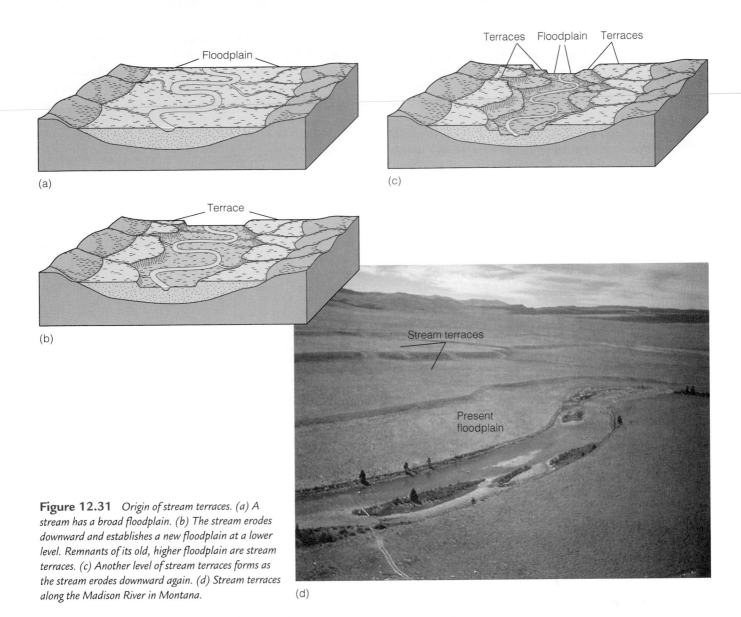

(a)

(b)

(c)

Figure 12.31 *Origin of stream terraces. (a) A stream has a broad floodplain. (b) The stream erodes downward and establishes a new floodplain at a lower level. Remnants of its old, higher floodplain are stream terraces. (c) Another level of stream terraces forms as the stream erodes downward again. (d) Stream terraces along the Madison River in Montana.*

(d)

Figure 12.32 *The Goose Necks of the San Juan River in Utah are incised meanders. This meandering canyon is more than 390 m deep.*

How Do Valleys Form and Evolve? **361**

filled with sedimentary rocks so that the river flowed on a surface at a higher level (Figure 12.33). As the river eroded downward, it was superposed directly upon a pre-existing knob of more resistant rock, and instead of changing its course, it cut a narrow, steep-walled canyon called a *water gap*.

Superposition also accounts for the fact that the Delaware, Potomac, and Susquehanna Rivers flow through water gaps. During the Mesozoic Era, the Appalachian Mountain region was eroded to a sediment-covered plain across which numerous streams flowed generally eastward. During the Cenozoic Era, regional uplift commenced, and as a result of the uplift, the streams began eroding downward and were superposed on resistant strata, thus forming water gaps (Figure 12.33).

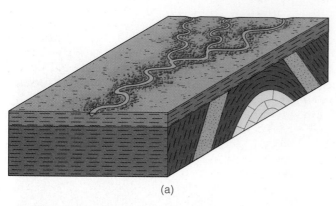

(a)

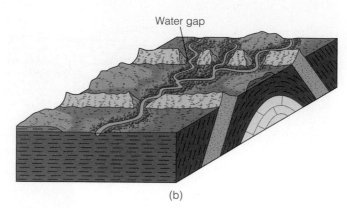

Water gap

(b)

Figure 12.33 *The origin of a superposed stream. (a) A stream begins cutting down into horizontal strata. (b) A horizontal layer is removed by erosion, exposing the underlying structure. The stream flows across resistant beds that form the ridges.*

Chapter Summary

1. Water continuously evaporates from the oceans, rises as water vapor, condenses, and falls as precipitation. About 20% of all precipitation falls on land and eventually returns to the oceans, mostly by surface runoff.

2. Running water moves by either laminar flow or turbulent flow. Laminar flow is shallow and slow and of little importance in erosion and sediment transport, but turbulent flow is much more energetic.

3. Runoff takes place as either sheet flow or channel flow, the latter being confined to long, troughlike depressions, or simply channels.

4. A channel's gradient varies from steep in its upper reaches to gentle in its lower reaches.

5. Flow velocity and discharge are related, so a change in one results in a change in the other. Velocity and discharge increase downstream in most rivers and streams.

6. Rivers and streams along with their tributaries carry runoff from areas known as drainage basins, which are separated from one another by divides.

7. Running water erodes by hydraulic action, abrasion, and dissolution of soluble rocks.

8. The larger particles transported by running water move as bed load, whereas the smallest particles move as suspended load. Rivers and streams also transport a dissolved load of materials in solution.

9. Braided streams have complex multiple, intertwining channels. Their deposits consist mostly of sand and gravel.

10. Meandering streams have a single sinuous channel in which point bars of sand or gravel are deposited. Cut-off meanders known as oxbow lakes eventually fill with fine-grained sediments and organic matter.

11. Floodplain deposits might consist of a succession of point bars deposited by a migrating channel, or mud deposited by water carried into the floodplain during floods.

12. Deltas form where a river or stream enters a standing body of water and deposits its sediment load. Small deltas in lakes commonly have a three-part division of bottomset, foreset, and topset beds, but marine deltas are larger, more complex, and more important economically.

13. In arid and semiarid regions where a river or stream flows from a mountain canyon onto adjacent lowlands a deposit known as an alluvial fan accumulates. Alluvial fans are composed of stream-deposited sand and gravel and/or mudflow deposits.

14. Sea level is ultimate base level, the lowest level to which rivers and streams can erode. Local or temporary base levels are lakes, other rivers or streams, or particularly resistant rocks.

15. Rivers and streams tend to eliminate irregularities in their channels so that they develop a smooth, concave profile of equilibrium. These so-called graded streams approach this ideal condition only temporarily.

16. Valleys develop and evolve by several processes including downcutting, lateral erosion, sheet wash, headward erosion, and mass wasting.

17. Superposed streams once flowed at a higher level, but eroded downward into resistant rocks. They now flow through valleys cut into ridges directly in their paths.

18. The formation of a floodplain followed by renewed downcutting by a stream leaves remnants of the older floodplain at a higher level known as stream terraces.

19. Incised meanders are generally attributed to renewed downcutting by a meandering stream so that it now occupies a deep, meandering valley.

Important Terms

abrasion	discharge	hydraulic action	runoff
alluvial fan	dissolved load	hydrologic cycle	sheet flow
alluvium	divide	incised meander	stream terrace
base level	drainage basin	infiltration capacity	superposed stream
bed load	drainage pattern	meandering stream	suspended load
braided stream	floodplain	natural levee	valley
channel flow	graded stream	oxbow lake	velocity
delta	gradient	point bar	

Review Questions

1. One way streams erode is by hydraulic action, which can be described as the
   a. _____ capacity of water to dissolve minerals; b. _____ direct impact of water; c. _____ amount of bed load in a stream; d. _____ depth of a stream valley; e. _____ velocity of flow.
2. In which one of the following areas would a radial drainage pattern likely develop?
   a. _____ stream terrace; b. _____ delta; c. _____ drainage divide; d. _____ point bar; e. _____ composite volcano.
3. Which one of the following is a temporary or local base level?
   a. _____ ocean; b. _____ lake; c. _____ desert; d. _____ mountain; e. _____ alluvial fan.
4. A meandering stream is one having
   a. _____ a single, sinuous channel; b. _____ long, straight reaches and waterfalls; c. _____ numerous sand and gravel bars; d. _____ a dendritic drainage pattern; e. _____ alluvial fans along its course.
5. The capacity for Earth materials to absorb water is known as
   a. _____ competence; b. _____ infiltration capacity; c. _____ meandering pattern; d. _____ dissolved load; e. _____ base level.
6. The bed load of a stream consists of
   a. _____ dissolved materials and alluvium; b. _____ natural levees; c. _____ sediment deposited during the waning stages of floods; d. _____ particles carried by rolling, sliding, and saltation; e. _____ clay and silt.
7. The quantity of water moving past a given place in a given amount of time is a stream's
   a. _____ gradient; b. _____ velocity; c. _____ capacity; d. _____ runoff; e. _____ discharge.
8. The deposits of braided streams consist mostly of
   a. _____ suspended load; b. _____ terraces; c. _____ sand and gravel; d. _____ point bars; e. _____ dissolved load.
9. An erosional remnant of a floodplain now higher than the current level of a stream is a(n)
   a. _____ delta; b. _____ alluvial fan; c. _____ lake; d. _____ terrace; e. _____ base level.
10. Streams carrying sand and gravel effectively erode by

11. a. _____ abrasion; b. _____ solution; c. _____ deposition; d. _____ stream piracy; e. _____ lateral accretion.
11. An oxbow lake is a(n)
    a. _____ deposit formed where a stream leaves a canyon; b. _____ mound of sediment on the margins of a channel; c. _____ crescent-shaped lake on a floodplain; d. _____ type of valley eroded when meanders are incised; e. _____ type of drainage developed on gently dipping rock layers.
12. A particularly narrow, deep valley is known as a(an)
    a. _____ canyon; b. _____ gully; c. _____ alluvial fan; d. _____ gorge; e. _____ graded stream.
13. Explain how a stream can lengthen its channel at both its upstream and downstream ends.
14. Describe the hydrologic cycle, and explain what makes it operate.
15. How is it possible for a meandering stream to erode laterally and yet maintain a more or less constant channel width?
16. What factors influence infiltration capacity, and why is infiltration capacity important when considering runoff?
17. What are laminar and turbulent flow and what factors determine which will take place? Also, explain why one is more important than the other in erosion and transport.
18. How do alluvial fans and deltas compare and differ?
19. Describe how a point bar and a natural levee develop.
20. Is the following statement correct? "The steeper the gradient, the greater the flow velocity." Explain.
21. Explain the concept of a graded stream, and describe conditions that might upset the graded condition.
22. Describe stream terraces and explain how multiple levels of terraces might form adjacent to a stream.
23. According to one estimate, 10.76 km³ of sediment is eroded from the continents each year, much of it by running water. Given that the volume of the continents above sea level is 92,832,194 km³, they should be eroded to sea level in only 8,627,527 years. Although the calculation is correct, there is something seriously wrong with this line of reasoning. What is it?
24. Calculate the daily discharge for a stream 148 m wide, 2.6 m deep, with a flow velocity of 0.3 m/sec.

World Wide Web Activities

For these web site addresses, along with current updates and exercises, log on to **http://www.brookscole.com/geo**

U.S. Geological Survey Water Resources of the United States

This site contains a wealth of information about streams, flooding, flood forecasting, flood warnings, and groundwater.

Floodplain Management

This site, maintained by the Floodplain Management Association, contains information on all aspects of flooding and floodplain management. Scroll down and click on *Flood Basins,* then go to and read *Basic Facts of Flooding.* Click on *Learning Center* and see Overview of Floods, Cause of Floods, Frequency of Floods, and How Much Risk Is Acceptable.

For Information on Dam Failures and Flooding Go to Any of the Following Sites:

Buffalo Creek Flooding Disaster
The Johnstown Flood 1889 *or* Johnstown Flood National Memorial
St. Francis Dam Disaster Site
Teton Dam Flood

A Short History of TVA

An excellent place to find out about the Tennessee Valley Authority, when and how it developed, and how it supplies power to a huge region. Also has links to other sites that explain how and where electricity is generated and sources of energy.

Groundwater

A variety of cave deposits are present in Blanchard Springs Caverns, Arkansas, one of many caves in the Ozark Plateau.

PROLOGUE

For more than two weeks in February 1925, Floyd Collins, an unknown farmer and cave explorer, became a household word (Figure 13.1). News about attempts to rescue him from a narrow subsurface fissure near Mammoth Cave, Kentucky, captured the attention of the nation. Unfortunately, the story ended in tragedy when 17 days after he became trapped, rescue workers recovered his body.

The saga of Floyd Collins is rooted in what is known as the Great Cave War of Kentucky. The western region of Kentucky is riddled with caves formed by groundwater weathering and erosion. Many of them were developed as tourist attractions to help supplement meager farm earnings. The largest and best known is Mammoth Cave. So spectacular is Mammoth, with its numerous caverns, underground rivers, and dramatic cave deposits, that it soon became the standard by which all other caves were measured.

As Mammoth Cave drew more and more tourists, rival cave owners became increasingly bold in attempting to lure visitors to their caves and curio shops. Signs pointing the way to Mammoth Cave frequently disappeared, while "official" cave information booths redirected unsuspecting tourists away from Mammoth Cave. It was in this environment that Floyd Collins grew up.

Seven years before his tragic death, Collins had discovered Crystal Cave on the family farm and opened it up for visitors. But like most of the caves in the area, Crystal Cave attracted few tourists—they visited Mammoth Cave instead. Perhaps it was the thought of discovering a cave rivaling Mammoth or even connecting to it that drove Collins to his fateful exploration of Sand Cave on January 30, 1925.

As Collins inched his way back up through the narrow fissure he had crawled down, he dislodged from the ceiling a small oblong piece of limestone that immediately pinned his left ankle. Try as he might, he was trapped in total darkness 17 m below ground. As he lay half on his left side, Collins's left arm was partially wedged under him, while his right arm was held fast by an overhanging ledge. During his struggles to free himself, he dislodged enough silt and small rocks to bury his legs, further immobilizing him and adding to his anguish.

The next day several neighbors reached Collins and were able to talk to him, feed him, encourage him, and try to make him more comfortable, but they could not get him out. Word of his plight quickly spread and the area soon swarmed with reporters. Eventually, volunteers were able to excavate an area around Collins's upper body, but could not free his pinned legs. While an anxious country waited, rescue attempts led by Floyd's brother Homer continued.

Three days after he had become trapped, a harness was put around Collins's chest and rescuers tried to pull him free. After numerous attempts to yank him out, workers had to abandon that plan because Collins was unable to bear the pain. Meanwhile at the surface, a carnival-like atmosphere had developed as a horde of up to 20,000 spectators converged on the scene. Moonshiners (people who illegally distill and sell liquor) set up stands and did a booming business. Unfortunately, drunks interfered with the rescue attempts and started fights. Finally, the National Guard had to be called out to maintain order. This atmosphere also led to rumors and exaggerations in some newspapers, when they reported that the 12.2 kg stone trapping Collins's leg weighed more than 1800 kg and that he was trapped 100 m underground rather than 17 m.

Two days after the attempt to pull Collins out of the fissure failed, part of

OBJECTIVES

At the end of this chapter, you will have learned that

- Groundwater is one reservoir of the hydrologic cycle and represents approximately 22% of the world's supply of freshwater.

- Porosity and permeability are largely responsible for the amount, availability, and movement of groundwater.

- The water table separates the zone of aeration from the underlying zone of saturation and is a subdued replica of the overlying land surface.

- Groundwater moves downward due to the force of gravity.

- An artesian system is one in which groundwater is confined and builds up high hydrostatic pressure; there are several important artesian systems in the United States as well as in many other parts of the world.

- Groundwater is an important agent of erosion and is responsible for sinkholes and karst topography in some areas.

- Groundwater also is an important agent of deposition and is responsible for a variety of cave features.

- Modifications of the groundwater system may result in lowering of the water table, saltwater incursion, subsidence, and contamination.

- Hot springs and geysers result when groundwater is heated, typically in regions of recent volcanic activity.

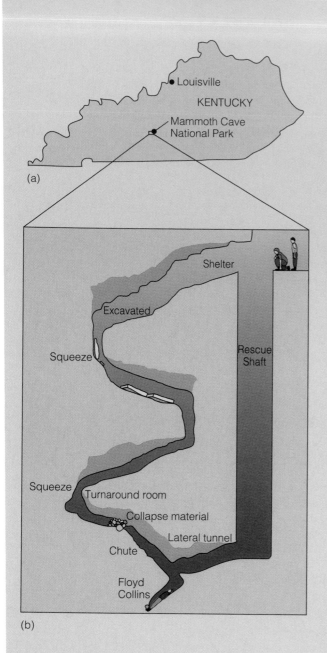

(a)

(b)

the passageway used by the rescuers collapsed, sealing Collins's fate. The only hope now was to dig a vertical relief shaft from which a lateral tunnel could be dug to reach Collins. On February 16, rescuers finally reached the chamber where Collins's lifeless body lay entombed. After his body was brought out, he was buried near Crystal Cave, where his grave is appropriately marked by a beautiful stalagmite and pink granite headstone.

(c)

Figure 13.1 (a) Location of the cave in which Floyd Collins was trapped. (b) Cross section showing the fissure where Collins was trapped, the rescue shaft that was sunk, and the lateral tunnel that finally reached him. (c) Collins looking out of a fissure near the cave where he ultimately died.

Introduction

Even the most casual observations make us aware of some aspects of Earth's hydrosphere. Everyone has seen precipitation in one form or another, and all are aware of running water in channels; many have seen water in the oceans and perhaps even frozen in glaciers. Few, however, have much experience with what is called **groundwater,** water filling open spaces in rocks, sediment, and soil beneath the surface. As a natural resource it is just as important as surface water, and its recovery is essential to many human endeavors.

In North America, as elsewhere in the world, groundwater rights have always been important, especially in semiarid regions where surface water is insufficient to maintain agricultural productivity. Many legal battles have resulted from claims and counterclaims about groundwater rights. Groundwater also played a pivotal role in the westward expansion of people in the United States during the nineteenth century. As the railway system grew, a reliable source of water for the steam locomotives was needed. Much of that water came from groundwater tapped by wells.

Most people probably think of groundwater only in terms of a source of freshwater for domestic, agricultural, and industrial needs. However, as we will see in this chapter, groundwater is also an important erosional agent (caves and caverns) as well as an alternate energy source (geothermal energy) in many parts of the world.

Why Should You Study Groundwater?

The study of groundwater and its movement has become increasingly important as the demand for freshwater by agricultural, industrial, and domestic users has reached an all-time high. More than 65% of the groundwater used in the United States each year goes for irrigation, with industrial use second, followed by domestic needs. These demands have severely depleted the groundwater supply in many areas and led to such problems as ground subsidence and saltwater contamination. In other areas, pollution from landfills, toxic waste, and agriculture has rendered the groundwater supply unsafe.

As the world's population and industrial development expand, the demand for water, particularly groundwater, will increase. Not only must new groundwater sources be located, but, once found, these sources must be protected from pollution and managed properly to ensure that users do not withdraw more water than can be replenished. It is therefore important that people become aware of what a valuable resource groundwater is so that future generations are assured of a clean and adequate supply of this water source.

Groundwater and the Hydrologic Cycle

Groundwater is one reservoir in the hydrologic cycle, representing approximately 22% (8.4 million km^3) of the world's supply of freshwater (see Figure 12.4). Just as for all other water in the hydrologic cycle, the ultimate source of groundwater is the oceans, but its more immediate source is the precipitation that infiltrates the ground and seeps down through the voids in soil, sediment, and rocks. It might also come from water infiltrating from streams, lakes, swamps, artificial recharge ponds, and water treatment systems.

Regardless of its source, groundwater moving through the tiny openings between soil and sediment particles and the spaces in rocks filters out many impurities such as disease-causing microorganisms and many pollutants. Not all soils and rocks are good filters, though, and sometimes so much undesirable material may be present that it contaminates the groundwater. Groundwater movement and its recovery at wells depends on two critical aspects of the materials it moves through—*porosity* and *permeability*.

What Properties of Earth Materials Allow Them to Absorb Water?

Porosity and permeability are important physical properties of Earth materials and are largely responsible for the amount, availability, and movement of groundwater. Water soaks into the ground because the soil, sediment, or rock has open spaces or pores. **Porosity** is simply the percentage of a material's total volume that is pore space. Whereas porosity most often consists of the spaces between particles in soil, sediments, and sedimentary rocks, other types of porosity can include cracks, fractures, faults in any rock type, and vesicles in volcanic rocks (Figure 13.2).

Porosity varies among different rock types and is dependent on the size, shape, and arrangement of the material composing the rock (Table 13.1). Most igneous and metamorphic rocks as well as many limestones and dolostones have very low porosity because they are composed of tightly interlocking crystals. Their porosity can

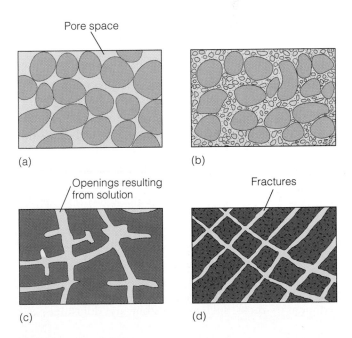

Figure 13.2 *A rock's porosity is dependent on the size, shape, and arrangement of the material composing the rock. (a) A well-sorted sedimentary rock has high porosity whereas (b) a poorly sorted one has lower porosity. (c) In soluble rocks such as limestones, porosity can be increased by solution, whereas (d) crystalline metamorphic and igneous rocks can be rendered porous by fracturing.*

What Properties of Earth Materials Allow Them to Absorb Water? **369**

Table 13.1

Porosity Values for Different Materials

Material	Percentage Porosity
Unconsolidated sediment	
Soil	55
Gravel	20–40
Sand	25–50
Silt	35–50
Clay	50–70
Rocks	
Sandstone	5–30
Shale	0–10
Solution activity in limestone, dolostone	10–30
Fractured basalt	5–40
Fractured granite	10

SOURCE: U.S. Geological Survey, *Water Supply Paper 2220* (1983) and others.

many igneous and metamorphic rocks that are highly fractured can also be very permeable provided that the fractures are interconnected.

The contrasting porosity and permeability of familiar substances is well demonstrated by sand versus clay. Pour some water on sand and it rapidly sinks in; pour water on clay and it remains on the surface. Furthermore, wet sand dries quickly, but once clay absorbs water, it may take days to dry out, because of its low permeability. Neither sand nor clay makes a good substance in which to grow crops or gardens, but a mixture of the two plus some organic matter in the form of humus makes an excellent soil for farming and gardening (see Chapter 5).

A permeable layer transporting groundwater is an **aquifer,** from the Latin *aqua* meaning water. The most effective aquifers are deposits of well-sorted and well-rounded sand and gravel. Limestones in which fractures and bedding planes have been enlarged by solution are also good aquifers. Shales and many igneous and metamorphic rocks make poor aquifers because they are typically impermeable, unless fractured. Rocks such as these and any other materials that prevent the movement of groundwater are **aquicludes.**

be increased, however, if they have been fractured or weathered by groundwater. This is particularly true for massive limestone and dolostone whose fractures can be enlarged by acidic groundwater.

By contrast, detrital sedimentary rocks composed of well-sorted and well-rounded grains can have high porosity because any two grains touch only at a single point, leaving relatively large open spaces between the grains (Figure 13.2a). Poorly sorted sedimentary rocks, on the other hand, typically have low porosity because smaller grains fill in the spaces between the larger grains, further reducing porosity (Figure 13.2b). In addition, the amount of cement between grains can also decrease porosity.

Porosity determines the amount of groundwater Earth materials can hold, but it does not guarantee that the water can be easily extracted as well. So in addition to being porous, Earth materials must also have the capacity to transmit fluids, a property known as **permeability.** Thus, both porosity and permeability play important roles in groundwater movement and recovery. Permeability is dependent not only on porosity, but also on the size of the pores or fractures and their interconnections. For example, deposits of silt or clay are typically more porous than sand or gravel, but they have low permeability because the pores between the clay particles are very small, and the molecular attraction between the particles and water is great, thereby preventing movement of the water. In contrast, pore spaces between grains in sandstone and conglomerate are much larger, and the molecular attraction on the water is therefore low. Chemical and biochemical sedimentary rocks, such as limestone and dolostone, and

What Is the Water Table?

Some of the precipitation on land evaporates, and some enters streams and returns to the oceans by surface runoff; the remainder seeps into the ground. As this water moves down from the surface, a small amount adheres to the material it moves through and halts its downward progress. With the exception of this *suspended water,* however, the rest seeps further downward and collects until it fills all the available pore spaces. Thus, two zones are defined by whether their pore spaces contain mostly air or water: the **zone of aeration,** and the underlying **zone of saturation.** The surface separating these two zones is the **water table** (Figure 13.3).

The base of the zone of saturation varies from place to place, but usually extends to a depth where an impermeable layer is encountered or to a depth where confining pressure closes all open space. Extending irregularly upward a few centimeters to several meters from the zone of saturation is the *capillary fringe.* Water moves upward in this region because of surface tension, much as water moves upward through a paper towel.

In general, the configuration of the water table is a subdued replica of the overlying land surface; that is, it rises beneath hills and has its lowest elevations beneath valleys. Several factors contribute to the surface configuration of a region's water table. These include regional differences in the amount of rainfall, permeability, and the rate of groundwater movement. During periods of high rainfall, groundwater tends to rise beneath hills because it cannot flow fast enough into the adjacent valleys to maintain a level surface. During droughts, the water table falls

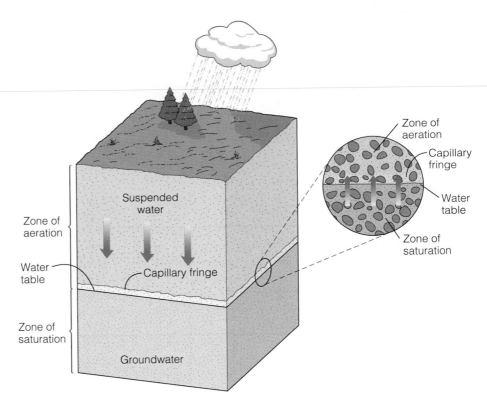

Figure 13.3 *The zone of aeration contains both air and water within its open space, whereas all open spaces in the zone of saturation are filled with groundwater. The water table is the surface separating the zones of aeration and saturation. Within the capillary fringe, water rises by surface tension from the zone of saturation into the zone of aeration.*

and tends to flatten out because it is not being replenished. In arid and semiarid regions, the water table is quite flat regardless of the overlying land surface.

How Does Groundwater Move?

Gravity provides the energy for the downward movement of groundwater. Water entering the ground moves through the zone of aeration to the zone of saturation (Figure 13.4). When water reaches the water table, it continues to move through the zone of saturation from areas where the water table is high toward areas where it is lower, such as streams, lakes, or swamps. Only some of the water follows the direct route along the slope of the water table. Most of it takes longer curving paths down and then enters a stream, lake, or swamp from below, because it moves from areas of high pressure toward areas of lower pressure within the saturated zone.

Groundwater velocity varies greatly and depends on many factors. Velocities range from 250 m per day in some extremely permeable material to less than a few

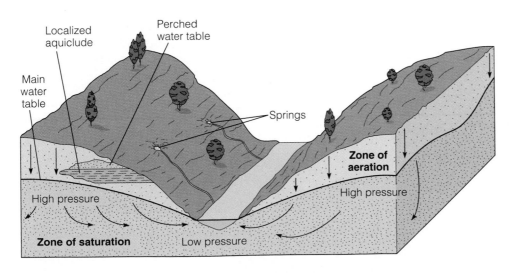

Figure 13.4 *Groundwater moves down through the zone of aeration to the zone of saturation. There some of it moves along the slope of the water table, and the rest moves through the zone of saturation from areas of high pressure toward areas of low pressure. Some water might collect over a local aquiclude, such as a shale layer, thus forming a perched water table.*

centimeters per year in nearly impermeable material. In most ordinary aquifers, the average velocity is a few centimeters per day.

What Are Springs, Water Wells, and Artesian Systems?

We can think of the water in the zone of saturation much like a reservoir whose surface rises or falls depending on additions versus natural and artificial withdrawals. *Recharge*—that is, additions to the zone of saturation—comes from rainfall or melting snow, or water might be added artificially at wastewater-treatment plants or recharge ponds constructed for just this purpose (Figure 13.5). But if groundwater is discharged naturally or withdrawn at wells without sufficient recharge, the water table drops just as a savings account diminishes if withdrawals exceed deposits. Withdrawals from the groundwater system take place where groundwater flows laterally into streams, lakes, or swamps, where it discharges at the surface as *springs,* and where it is withdrawn from the system at water wells.

Springs

Places where groundwater flows or seeps out of the ground as **springs** have always fascinated people. The water flows out of the ground for no apparent reason and from no readily identifiable source. So it is not surprising that springs have long been regarded with superstition and revered for their supposed medicinal value and healing powers. Nevertheless, there is nothing mystical or mysterious about springs.

Although springs can occur under a wide variety of geologic conditions, they all form in basically the same way (Figure 13.6). When percolating water reaches the water table or an impermeable layer, it flows laterally, and if this flow intersects the surface, the water discharges as a spring (Figure 13.7). The Mammoth Cave area in Kentucky is underlain by fractured limestones that have been enlarged into caves by solution activity. In this geologic environment, springs occur where the fractures and caves intersect the ground surface allowing groundwater to exit onto the surface. Most springs are along valley walls where streams have cut valleys below the regional water table.

Springs can also develop wherever a perched water table intersects the surface (Figure 13.4). A *perched water*

Figure 13.5 *Storm Water Basin No. 129, Garden City Park, Long Island, New York, is one of many recharge basins operated by the Nassau County Department of Public Works.*

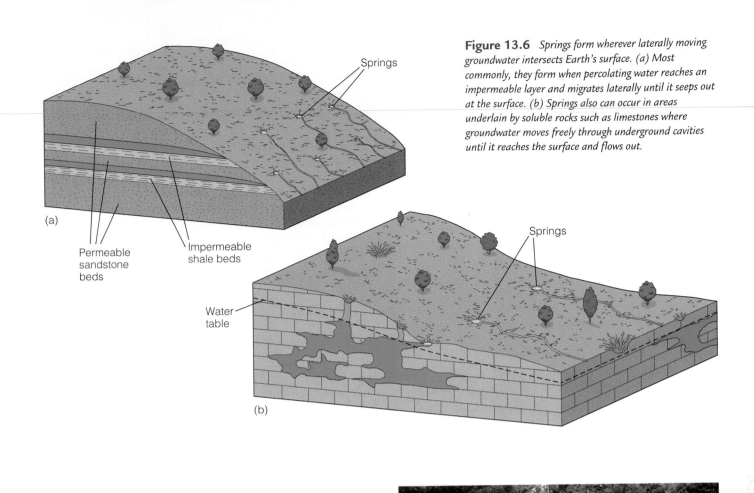

Figure 13.6 *Springs form wherever laterally moving groundwater intersects Earth's surface. (a) Most commonly, they form when percolating water reaches an impermeable layer and migrates laterally until it seeps out at the surface. (b) Springs also can occur in areas underlain by soluble rocks such as limestones where groundwater moves freely through underground cavities until it reaches the surface and flows out.*

(a)

Permeable sandstone beds

Impermeable shale beds

Springs

Water table

Springs

(b)

(a)

Spring

(b)

Figure 13.7 *Spring. (a) This spring issues from rocks at the base of the mountain in the background. (b) A good view of surface runoff and the discharge of groundwater at Burney Falls State Park, California. The water on either side of the falls that discharges from the cliff face flows through a porous zone between lava flows.*

What Are Springs, Water Wells, and Artesian Systems?

table may occur wherever a local aquiclude is present within a larger aquifer, such as a lens of shale within sandstone. As water migrates through the zone of aeration, it is stopped by the local aquiclude, and a localized zone of saturation "perched" above the main water table forms. Water moving laterally along the perched water table may intersect the surface to produce a spring.

Water Wells

Water wells are simply openings made by digging or drilling down into the zone of saturation. Once the zone of saturation has been penetrated, water percolates into the well, filling it to the level of the water table. A few wells are free flowing (see the next section), but for most the water must be brought to the surface. In some parts of the world, water is raised to the surface with nothing more than a bucket on a rope or a hand-operated pump. In many parts of the United States and Canada, one can see windmills from times past that employed wind power to pump water. Most of these are no longer in use, having been replaced by more efficient electric pumps (Figure 13.8b).

When groundwater is pumped from a well, the water table in the area around the well is lowered, forming a **cone of depression** (Figure 13.9). The reason a cone of depression forms is because the rate of water withdrawn from the well is greater than the inflow of water to the well resulting in a lowering of the water table around the well. This lowering of the water table normally does not pose a problem for the average domestic well, provided that the well is drilled sufficiently deep into the zone of saturation. The tremendous amounts of water used by industry and for irrigation, however, may create a large cone of depression that lowers the water table sufficiently to cause shallow wells in the immediate area to go dry (Figure 13.9). This situation is not uncommon and frequently results in lawsuits by the owners of the shallow dry wells. Furthermore, lowering of the regional water table is becoming a serious problem in many areas, particularly in the southwestern United States where rapid growth has placed tremendous demands on the groundwater system. Unrestricted withdrawal of groundwater cannot continue indefinitely, and the rising costs and decreasing supply of groundwater should soon limit the growth of this region of the United States.

Artesian Systems

The word *artesian* comes from the French town and province of Artois (called Artesium during Roman times) near Calais, where the first European artesian well was drilled in A.D. 1126 and is still flowing today. The term **artesian system** can be applied to any system in which groundwater is confined and builds up high hydrostatic (fluid) pressure. Water in such a system is able to rise above the level of the aquifer if a well is drilled through the confining layer, thereby reducing the pressure and forcing the water upward (Figure 13.10). For an artesian system to develop, three geologic conditions must be present (Figure 13.11): (1) The aquifer must be confined above and below by aquicludes to prevent

(a)

(b)

Figure 13.8 *(a) Windmill. (b) Modern electric pump.*

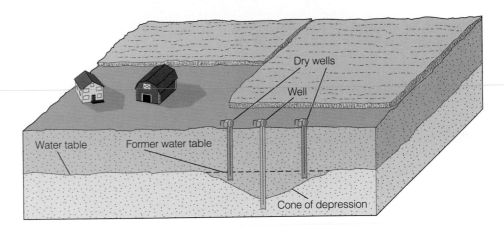

Figure 13.9 *A cone of depression forms whenever water is withdrawn from a well. If the cone of depression is large enough, nearby shallow wells may go dry.*

water from escaping; (2) the rock sequence is usually tilted and exposed at the surface, enabling the aquifer to be recharged; and (3) there is sufficient precipitation in the recharge area to keep the aquifer filled.

The elevation of the water table in the recharge area and the distance of the well from the recharge area determine the height to which artesian water rises in a well. The surface defined by the water table in the recharge area, called the *artesian-pressure surface*, is indicated by the sloping dashed line in Figure 13.11. If there were no friction in the aquifer, well water from an artesian aquifer would rise exactly to the elevation of the artesian-pressure surface. Friction, however, slightly reduces the pressure of the aquifer water and consequently the level to which artesian water rises. This is why the pressure surface slopes.

Figure 13.10 *Artesian well at Deep Well Ranch, South Fork of the Madison River, Gallatin County, Montana.*

What Are Springs, Water Wells, and Artesian Systems? **375**

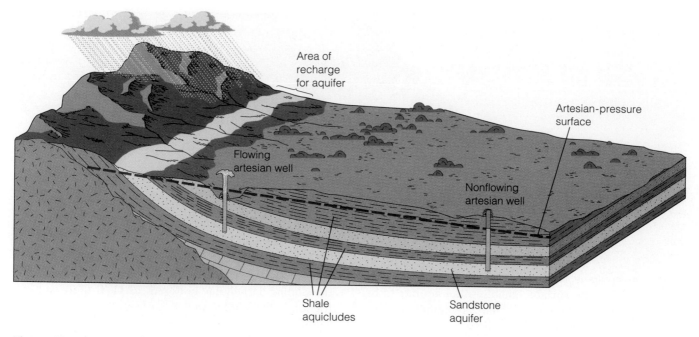

Figure 13.11 *An artesian system must have an aquifer confined above and below by aquicludes, the aquifer must be exposed at the surface, and there must be sufficient precipitation in the recharge area to keep the aquifer filled. The elevation of the water table in the recharge area, which is indicated by a sloping dashed line (the artesian-pressure surface), defines the highest level to which well water can rise. If the elevation of a wellhead is below the elevation of the artesian-pressure surface, the well will be free-flowing because the water will rise toward the artesian-pressure surface, which is at a higher elevation than the wellhead. If the elevation of a wellhead is at or above that of the artesian-pressure surface, the well will be nonflowing.*

An artesian well will flow freely at the ground surface only if the wellhead is at an elevation below the artesian-pressure surface. In this situation, the water flows out of the well because it rises toward the artesian-pressure surface, which is at a higher elevation than the wellhead (Figure 13.10). In a nonflowing artesian well, the wellhead is above the artesian-pressure surface, and thus the water will rise in the well only as high as the artesian-pressure surface.

In addition to artesian wells, many artesian springs also exist. Such springs form if a fault or fracture intersects the confined aquifer, allowing water to rise above the aquifer. Oases in deserts are commonly artesian springs.

Because the geologic conditions necessary for artesian water can occur in a variety of ways, artesian systems are common in many areas of the world underlain by sedimentary rocks. One of the best-known artesian systems in the United States underlies South Dakota and extends southward to central Texas. The majority of the artesian water from this system is used for irrigation. The aquifer of this artesian system, the Dakota Sandstone, is recharged where it is exposed along the margins of the Black Hills of South Dakota. The hydrostatic pressure in this system was originally great enough to produce free-flowing wells and to operate waterwheels. The extensive use of water for irrigation over the years has reduced the pressure in many of the wells so that they are no longer free-flowing and the water must be pumped.

How Does Groundwater Erode and Deposit Material?

When rainwater begins seeping into the ground, it immediately starts to react with the minerals it contacts, weathering them chemically. In areas underlain by soluble rock, groundwater is the principal agent of erosion and is responsible for the formation of many major features of the landscape.

Limestone, a common sedimentary rock composed primarily of the mineral calcite ($CaCO_3$), underlies large areas of Earth's surface (Figure 13.12). Although limestone is practically insoluble in pure water, it readily dissolves if a small amount of acid is present. Carbonic acid (H_2CO_3) is a weak acid that forms when carbon dioxide combines with water ($H_2O + CO_2 \rightarrow H_2CO_3$) (see Chapter 5). Because the atmosphere contains a small amount of carbon dioxide (0.03%), and carbon dioxide is also produced in soil by the decay of organic matter, most groundwater is slightly acidic. When groundwater percolates through the various openings in limestone, the slightly acidic water readily reacts with the calcite to dissolve the rock by forming soluble calcium bicarbonate, which is carried away in solution (see Chapter 5).

Sinkholes and Karst Topography

In regions underlain by soluble rock, the ground surface may be pitted with numerous depressions that vary in size and shape. These depressions, called **sinkholes** or merely *sinks,* mark areas with underlying soluble rock (Figure 13.13). Most sinkholes form in one of two ways. The first is when soluble rock below the soil is dissolved by seeping water and openings in the rock are enlarged and filled in by the overlying soil. As the groundwater continues to dissolve the rock, the soil is eventually removed, leaving shallow depressions with gently sloping sides. When adjacent sinkholes merge, they form a network of larger, irregular, closed depressions called *solution valleys.*

Sinkholes also form when a cave's roof collapses, usually producing a steep-sided crater. Sinkholes formed in this way are a serious hazard, particularly in populated areas. In regions prone to sinkhole formation, the depth and extent of underlying cave systems must be mapped before any development to ensure that the underlying rocks are thick enough to support planned structures.

Karst topography, or simply *karst,* develops largely by groundwater erosion in many areas underlain by soluble rocks (Figure 13.14). The name karst is derived from the plateau region of the border area of Slovenia, Croatia, and northeastern Italy where this type of topography is well developed. In the United States, regions of karst topography include large areas of southwestern Illinois, southern Indiana, Kentucky, Tennessee, northern Missouri, Alabama, and central and northern Florida (Figure 13.12).

Karst topography is characterized by numerous caves, springs, sinkholes, solution valleys, and disappearing streams (Figure 13.14). *Disappearing streams* are so named because they typically flow only a short distance at the surface and then disappear into a sinkhole. The water continues flowing underground through various fractures or caves until it surfaces again at a spring or other stream. Another feature commonly encountered in karst regions is *terra rossa,* a residual red, clayey soil resulting from the solution of limestone.

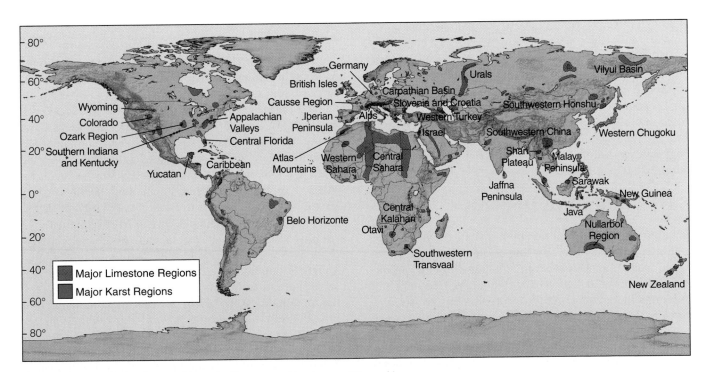

Figure 13.12 *The distribution of the major limestone and karst areas of the world.*

(a)

(b)

Figure 13.13 *(a) This sinkhole formed on May 8 and 9, 1981, in Winter Park, Florida. It formed in previously dissolved limestone following a drop in the water table. The 100-m-wide, 35-m-deep sinkhole destroyed a house, numerous cars, and a municipal swimming pool. (b) A small sinkhole in Montana now occupied by a lake. The water enters the lake from a hot spring so it remains warm year round. In fact, tropical fish have been introduced into the lake and thus live in an otherwise inhospitable climate.*

Karst topography varies from the spectacular high relief landscapes of China to the subdued and pockmarked landforms of Kentucky (Figure 13.15). Common to all karst topography, though, is the presence of thick-bedded, readily soluble rock at the surface or just below the soil, and enough water for solution activity to occur. Karst topography is, therefore, typically restricted to humid and temperate climates.

Caves and Cave Deposits

Caves are perhaps the most spectacular examples of the combined effects of weathering and erosion by groundwater. As groundwater percolates through carbonate rocks, it dissolves and enlarges fractures and openings to form a complex interconnecting system of crevices, caves, caverns, and underground streams. A **cave** is usually defined as a naturally formed subsurface opening that is generally connected to the surface and is large enough for a person to enter. A *cavern* is a very large cave or a system of interconnected caves.

More than 17,000 caves are known in the United States. Most of them are small, but some are quite large and spectacular. Some of the more famous ones are Mammoth Cave, Kentucky; Carlsbad Caverns, New Mexico; Lewis and Clark Caverns, Montana; Wind Cave and Jewel Cave, South Dakota; Lehman Cave, Nevada; and Meramec Caverns, Missouri, which Jesse James and his outlaw band often used as a hideout. The United States has many famous caves, but so has Canada, including the 536 m–deep Arctomys Cave in Mount Robson Provincial Park, British Columbia, the deepest known cave in North America.

Caves and caverns form as a result of the dissolution of carbonate rocks by weakly acidic groundwater (Figure 13.16). Groundwater percolating through the zone of aeration slowly dissolves the carbonate rock and enlarges its fractures and bedding planes. On reaching the water table, the groundwater migrates toward the region's surface streams. As the groundwater moves through the zone of saturation, it continues to dissolve the rock and gradually forms a system of horizontal passageways through which the dissolved rock is carried to the streams. As the surface

Figure 13.14 *Some of the features of karst topography.*

Solution valleys

Springs

Sinkholes

Disappearing streams

Deeply intrenched permanent stream

Cave

(a)

Figure 13.15 *(a) The Stone Forest, 125 km southeast of Kunming, People's Republic of China, is a high relief karst landscape formed by the dissolution of carbonate rocks. (b) Solution valleys, sinkholes, and sinkhole lakes dominate the subdued karst topography east of Bowling Green, Kentucky.*

(b)

How Does Groundwater Erode and Deposit Material? **379**

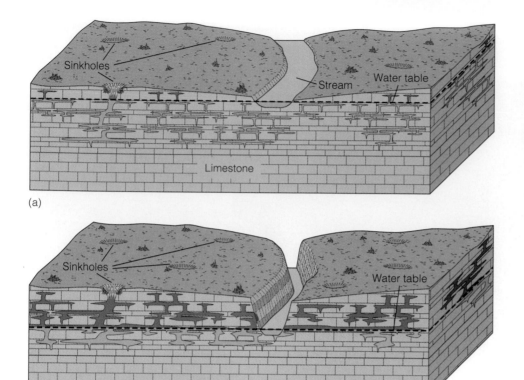

(a)

(b)

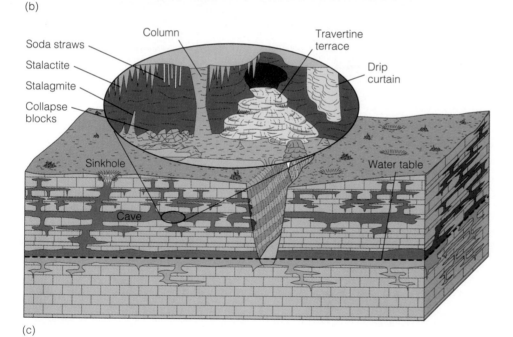

(c)

Figure 13.16 *The formation of caves. (a) As groundwater percolates through the zone of aeration and flows through the zone of saturation, it dissolves the carbonate rocks and gradually forms a system of passageways. (b) Groundwater moves along the surface of the water table, forming a system of horizontal passageways through which dissolved rock is carried to the surface streams, thus enlarging the passageways. (c) As the surface streams erode deeper valleys, the water table drops, and the abandoned channelways form an interconnecting system of caves and caverns.*

streams erode deeper valleys, the water table drops in response to the lower elevation of the streams. The water that flowed through the system of horizontal passageways now percolates to the lower water table where a new system of passageways begins to form. The abandoned channelways now form an interconnecting system of caves and caverns. Caves eventually become unstable and collapse, littering the floor with fallen debris.

When most people think of caves, they think of the seemingly endless variety of colorful and bizarre-shaped deposits found in them. Although a great many different types of cave deposits exist, most form in essentially the same manner and are collectively known as **dripstone.** As water seeps through a cave, some of the dissolved carbon dioxide in the water escapes, and a small amount of calcite is precipitated. In this manner, the various dripstone deposits are formed.

Figure 13.17 *Stalactites are the icicle-shaped structures seen hanging from the cave's ceiling, whereas the upward-pointing structures on the floor are stalagmites. Several columns are present where the stalactites and stalagmites have met in this chamber of Luray Caverns, Virginia.*

Stalactites are icicle-shaped structures hanging from cave ceilings that form as a result of precipitation from dripping water (Figure 13.17). With each drop of water, a thin layer of calcite is deposited over the previous layer, forming a cone-shaped projection that grows down from the ceiling.

The water that drips from a cave's ceiling also precipitates a small amount of calcite when it hits the floor. As additional calcite is deposited, an upward growing projection called a *stalagmite* forms (Figure 13.17). If a stalactite and stalagmite meet, they form a *column*. Groundwater seeping from a crack in a cave's ceiling may form a vertical sheet of rock called a *drip curtain,* while water flowing across a cave's floor may produce *travertine terraces* (Figure 13.16).

How Do Humans Impact the Groundwater System, and What Are the Effects?

Groundwater is a valuable natural resource that is rapidly being exploited with little regard to the effects of overuse and misuse. Currently, about 20% of all water used in the United States is groundwater. This percentage is increasing, and unless this resource is used more wisely, sufficient amounts of clean groundwater will not be available in the future. Modifications of the groundwater system may have many consequences including (1) lowering of the water table, which causes wells to dry up; (2) loss of hydrostatic pressure, which causes once free-flowing wells to require pumping; (3) saltwater incursion; (4) subsidence; and (5) contamination.

Lowering the Water Table

Withdrawing groundwater at a significantly greater rate than it is replaced by either natural or artificial recharge can have serious effects. For example, the High Plains aquifer is one of the most important aquifers in the United States. It underlies more than 450,000 km², including most of Nebraska, large parts of Colorado and Kansas, portions of South Dakota, Wyoming, and New Mexico, as well as the panhandle regions of Oklahoma and Texas, and accounts for approximately 30% of the groundwater used for irrigation in the United States (Figure 13.18). Irrigation from the High Plains aquifer is largely responsible for the region's high agricultural productivity, which includes a significant percentage of the nation's corn, cotton, and wheat, and half of U.S. beef cattle. Large areas of land (about 18 million acres) are currently irrigated with water pumped from the High Plains aquifer. Irrigation is popular because yields from irrigated lands can be triple what they would be without irrigation.

Although the High Plains aquifer has contributed to the high productivity of the region, it cannot continue providing the quantities of water that it has in the past. In some parts of the High Plains, from 2 to 100 times more water is being pumped annually than is being recharged. Consequently, water is being removed from the aquifer faster than it is being replenished, causing the water table to drop significantly in many areas (Figure 13.18).

What will happen to this region's economy if long-term withdrawal of water from the High Plains aquifer greatly exceeds its recharge rate so that it can no longer supply the quantities of water necessary for irrigation? Solutions range from going back to farming without irrigation to diverting water from other regions such as the Great Lakes. Farming without irrigation would result in greatly decreased yields and higher costs and prices for agricultural products, while the diversion of water from elsewhere would cost billions of dollars and the price of agricultural products would still rise.

Another excellent example of what we might call deficit spending with regard to groundwater took place in California during the drought of 1987 to 1992. During that time, the state's aquifers were overdrawn at the rate of 10 million acre/feet per year (an acre/foot is the amount of water that covers one acre one foot deep). In short, each year of the drought California was withdrawing over 12 km^3 of groundwater more than was being replaced. Unfortunately, excessive depletion of the groundwater reservoir has other consequences, such as subsidence involving sinking or settling of the ground surface (discussed in a later section).

Water supply problems certainly exist in many areas, but on the positive side, water use in the United States actually declined during the five years following 1980 and has remained nearly constant since then, even though the population has increased (Figure 13.19). This downturn in demand resulted largely from improved techniques in irrigation, more efficient industrial water use, and a general public awareness of water problems coupled with conservation practices. Nevertheless, the rates of withdrawal of groundwater from some aquifers still exceeds their rate of recharge, and population growth in the arid to semiarid Southwest is putting larger demands on an already limited water supply.

Saltwater Incursion

The excessive pumping of groundwater in coastal areas can result in **saltwater incursion** such as occurred on Long Island, New York, during the 1960s. Along coastlines where permeable rocks or sediments are in con-

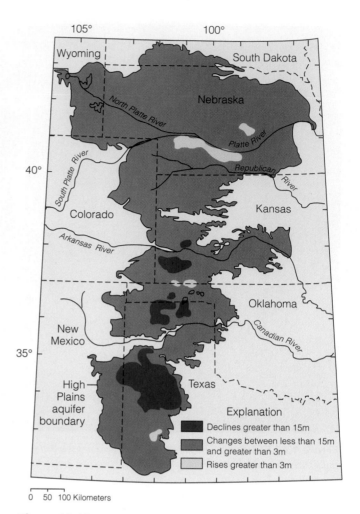

Figure 13.18 *Geographic extent of the High Plains aquifer and changes in water level from predevelopment through 1993.*

tact with the ocean, the fresh groundwater, being less dense than seawater, forms a lens-shaped body above the underlying saltwater (Figure 13.20a). The weight of the freshwater exerts pressure on the underlying saltwater. As long as rates of recharge equal rates of withdrawal, the contact between the fresh groundwater and the seawater will remain the same. If excessive pumping occurs, however, a deep cone of depression forms in the fresh groundwater (Figure 13.20b). Because some of the pressure from the overlying freshwater has been removed, saltwater forms a *cone of ascension* as it rises to fill the pore space that formerly contained freshwater. When this occurs, wells become contaminated with saltwater and remain contaminated until recharge by freshwater restores the former level of the fresh groundwater water table.

Saltwater incursion is a major problem in many rapidly growing coastal communities. As the population in these areas grows, greater demand for groundwater creates an even greater imbalance between recharge and withdrawal.

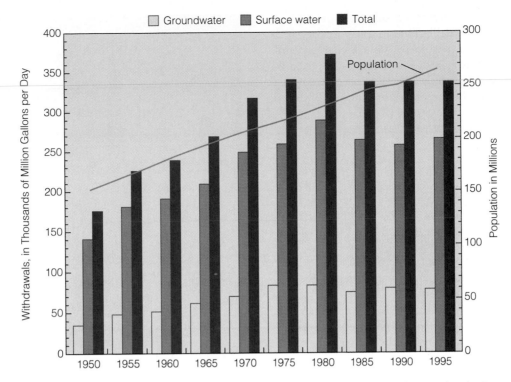

Figure 13.19 *Trends in water use in the United States from 1950 to 1995. Notice the steady increase until 1980, then the decrease to the present level of use.*

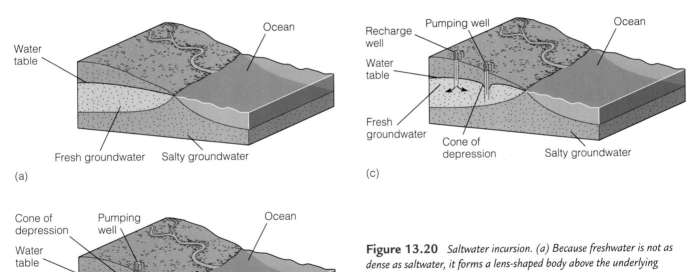

(a)

(b)

(c)

Figure 13.20 *Saltwater incursion. (a) Because freshwater is not as dense as saltwater, it forms a lens-shaped body above the underlying saltwater. (b) If excessive pumping occurs, a cone of depression develops in the fresh groundwater, and a cone of ascension forms in the underlying salty groundwater that may result in saltwater contamination of the well. (c) Pumping water back into the groundwater system through recharge wells can help lower the interface between the fresh groundwater and the salty groundwater and reduce saltwater incursion.*

How Do Humans Impact the Groundwater System, and What Are the Effects?

To counteract the effects of saltwater incursion, recharge wells are often drilled to pump water back into the groundwater system (Figure 13.20c). Recharge ponds that allow large quantities of fresh surface water to infiltrate the groundwater supply may also be constructed (Figure 13.5). Both of these methods are successfully used on Long Island.

Subsidence

As excessive amounts of groundwater are withdrawn from poorly consolidated sediments and sedimentary rocks, the water pressure between grains is reduced, and the weight of the overlying materials causes the grains to pack closer together, resulting in *subsidence* of the ground. As more and more groundwater is pumped to meet the increasing needs of agriculture, industry, and population growth, subsidence is becoming more prevalent.

The San Joaquin Valley of California is a major agricultural region that relies largely on groundwater for irrigation. Between 1925 and 1975, groundwater withdrawals in parts of the valley caused subsidence of almost 9 m (Figure 13.21). Other areas in the United States that have experienced subsidence include New Orleans, Louisiana, and Houston, Texas, both of which have subsided more than 2 m, and Las Vegas, Nevada, where 8.5 m of subsidence has taken place (Table 13.2).

Elsewhere in the world, the tilt of the Leaning Tower of Pisa in Italy is partly due to groundwater withdrawal (Figure 13.22). The tower started tilting soon after construction began in 1173 because of differential compaction of the foundation. During the 1960s, the city of Pisa withdrew ever-larger amounts of groundwater, causing the ground to subside further; as a result, the tilt of the tower increased until it was considered in danger of falling over. Strict control of groundwater withdrawal and stabilization of the foundation have reduced the amount of tilting to about 1 mm per year, thus ensuring that the tower should stand for several more centuries.

A spectacular example of continuing subsidence is taking place in Mexico City, which is built on a former lake bed. As groundwater is removed for the ever-increasing needs of the city's 15.6 million people, the water table has been lowered up to 10 meters. As a result, the fine-grained lake sediments are compacting, and Mexico City is slowly and unevenly subsiding. Its opera house has settled more than 3 m, and half of the first floor is now below ground level. Other parts of the city have subsided more than 7.5 m, creating similar problems for other structures (Figure 13.23).

The extraction of oil can also cause subsidence. Long Beach, California, has subsided 9 m as a result of 34 years of oil production. More than $100 million of damage was done to the pumping, transportation, and harbor facilities

Figure 13.21 *The dates on this power pole dramatically illustrate the amount of subsidence in the San Joaquin Valley, California. Because of groundwater withdrawals and subsequent sediment compaction, the ground subsided nearly 9 meters between 1925 and 1977. For a time, surface water use reduced subsidence, but during the drought of 1987 to 1992 it started again as more groundwater was withdrawn.*

in this area because of subsidence and encroachment of the sea (Figure 13.24). Once water was pumped back into the oil reservoir, thus stabilizing it, subsidence virtually stopped.

Groundwater Contamination

A major problem facing our society is the safe disposal of the numerous pollutant by-products of an industrialized economy. We are becoming increasingly aware that streams, lakes, and oceans are not unlimited reservoirs for waste, and that we must find new safe ways to dispose of pollutants.

The most common sources of contamination are sewage, landfills, toxic-waste disposal sites, and agriculture. Once pollutants get into the groundwater system,

Table 13.2

Subsidence of Cities and Regions Due Primarily to Groundwater Removal

Location	Maximum Subsidence (m)	Area Affected (km²)
Mexico City, Mexico	8.0	25
Long Beach and Los Angeles, California	9.0	50
Taipei Basin, Taiwan	1.0	100
Shanghai, China	2.6	121
Venice, Italy	0.2	150
New Orleans, Louisiana	2.0	175
London, England	0.3	295
Las Vegas, Nevada	8.5	500
Santa Clara Valley, California	4.0	600
Bangkok, Thailand	1.0	800
Osaka and Tokyo, Japan	4.0	3000
San Joaquin Valley, California	9.0	9000
Houston, Texas	2.7	12,100

SOURCE: Data from R. Dolan and H. G. Goodell, "Sinking Cities," *American Scientist* 74 (1986):38–47; and J. Whittow, *Disasters: The Anatomy of Environmental Hazards* (Athens, Ga.: University of Georgia Press, 1979).

Figure 13.22 *The Leaning Tower of Pisa, Italy. The tilting is partly the result of subsidence due to the removal of groundwater.*

they spread wherever groundwater travels, which can make their containment difficult. Furthermore, because groundwater moves so slowly, it takes a long time to cleanse a groundwater reservoir once it has become contaminated.

In many areas, septic tanks are the most common way of disposing of sewage. A septic tank slowly releases sewage into the ground where it is decomposed by oxidation and microorganisms and filtered by the sediment as it percolates through the zone of aeration. In most situations, by the time the water from the sewage reaches the zone of saturation, it has been cleansed of any impurities and is safe to use (Figure 13.25a). If the water table is close to the surface or if the rocks are very permeable, water entering the zone of saturation may still be contaminated and unfit to use.

Landfills are also potential sources of groundwater contamination (Figure 13.25b). Not only does liquid waste seep into the ground, but rainwater also carries dissolved chemicals and other pollutants down into the groundwater

reservoir. Unless the landfill is carefully designed and lined with an impermeable layer such as clay, many toxic compounds such as paints, solvents, cleansers, pesticides, and battery acid will find their way into the groundwater system.

Toxic-waste sites where dangerous chemicals are either buried or pumped underground are an increasing source of groundwater contamination. The United States alone must dispose of several thousand metric tons of hazardous chemical waste per year. Unfortunately, much of this waste has been, and still is, being improperly dumped and is contaminating the surface water, soil, and groundwater.

Examples of indiscriminate dumping of dangerous and toxic chemicals can be found in every state. Perhaps the most famous is the Love Canal, near Niagara Falls, New

Figure 13.23 *Excessive withdrawal of groundwater from beneath Mexico City has resulted in subsidence and uneven settling of buildings. The right side of this church (Our Lady of Guadalupe) in Mexico City has settled slightly more than a meter.*

York. During the 1940s, the Hooker Chemical Company dumped approximately 19,000 tons of chemical waste into Love Canal. In 1953 it covered one of the dump sites with dirt and sold it for one dollar to the Niagara Falls Board of Education, which built an elementary school and playground on the site. Heavy rains and snow during the winter of 1976–1977 raised the water table and turned the area into a muddy swamp in the spring of 1977. Mixed with the mud were thousands of different toxic, noxious chemicals that formed puddles in the playground, oozed into people's basements, and covered gardens and lawns. Trees, lawns, and gardens began to die, and many of the residents of the area suffered from serious illnesses. The cost of cleaning up the Love Canal site and relocating its residents exceeded $100 million, and the site and neighborhood are now vacant.

Groundwater Quality

Finding groundwater is rather easy, because it is present beneath the land surface nearly everywhere, although the depth to the water table varies considerably. But just finding water is not enough. Sufficient amounts in porous and permeable materials must be located if it is to be withdrawn for agricultural, industrial, or domestic use. The availability of groundwater was important in the westward expansion in both Canada and the United States, and now more than one-third of all water for irrigation comes from the groundwater system. More than 90% of the water used for domestic purposes in rural America and the water for a number of large cities comes from groundwater, and, as one would expect, quality is more important here than it is for most other purposes.

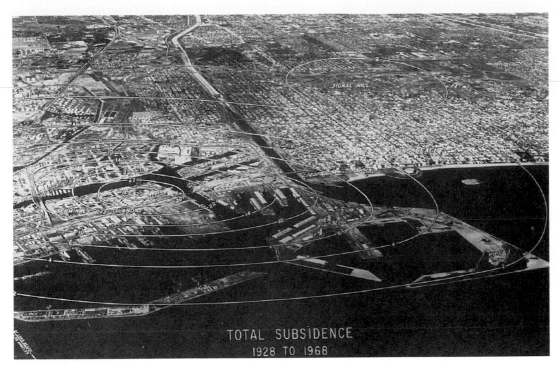

Figure 13.24 *The withdrawal of petroleum from the oil field in Long Beach, California, resulted in up to 9 m of ground subsidence because of sediment compaction. It was not until water was pumped back into the reservoir to replace the petroleum that ground subsidence essentially ceased. (2 to 29 feet = 0.6 to 8.8 meters)*

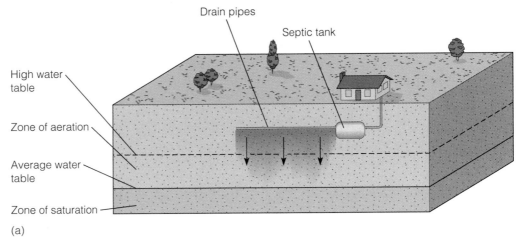

(a)

Figure 13.25 *(a) A septic system slowly releases sewage into the zone of aeration. Oxidation, bacterial degradation, and filtering usually remove natural impurities before they reach the water table. However, if the rocks are very permeable or the water table is too close to the septic system, contamination of the groundwater can result. (b) Unless there is an impermeable barrier between a landfill and the water table, pollutants can be carried into the zone of saturation and contaminate the groundwater supply.*

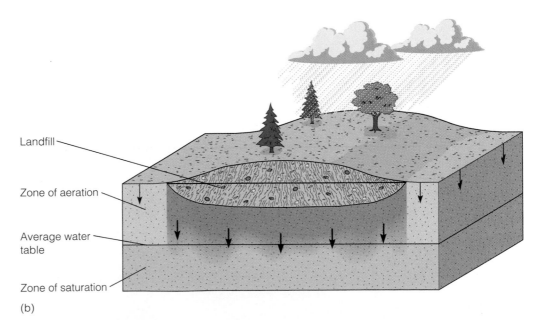

(b)

How Do Humans Impact the Groundwater System, and What Are the Effects?   **387**

Arsenic and Old Lace

Many people probably learned that arsenic is a poison from either reading or seeing the play *Arsenic and Old Lace,* written by Joseph Kesselring. In the play, the elderly Brewster sisters poison lonely old men by adding a small amount of arsenic to their homemade elderberry wine.

Arsenic is a naturally occurring toxic element found in the environment, and several types of cancer have been linked to arsenic in water. In fact, because of its prevalence in the environment and its adverse health effects, arsenic was included in the amendments to the Safe Drinking Water Act by Congress in 1996. Arsenic gets into the groundwater system mainly as a result of arsenic-bearing minerals dissolving as part of the natural weathering process of rocks and soils.

A recently published map by the United States Geological Survey (USGS) shows the extent and concentration of arsenic in the nation's groundwater supply (Figure 1). The highest concentrations of arsenic in groundwater were found throughout the West and in parts of the Midwest and Northeast. Although the map is not intended to provide specific information for individual wells or even a locality within a county, it will aid researchers and policy makers in identifying areas of high concentration so that informed decisions about water use in those areas can be made. It should be pointed out, however, that there can be a high degree of local variability in the amount of arsenic in the groundwater due to the local geology, the type of aquifer, the depth of the well, and

Figure 1 *Arsenic concentrations for 18,850 groundwater samples collected from 1973 to 1997.*

other factors. Testing is the only way to discover the arsenic concentration in any well.

What is considered a safe level of arsenic in drinking water? The current U.S. Environmental Protection Agency (USEPA) standard is 50 micrograms of arsenic per liter. When using that figure, arsenic concentrations in the groundwater samples the USGS analyzed were almost always lower than the USEPA standards. However, that standard is likely to be lowered in 2001 to be the same as the World Health Organization's (WHO) international guideline of 10 micrograms per liter for drinking water. Based on this level, approximately 10 percent of the samples in the USGS study exceed the WHO guideline (Figure 2).

Public water supply systems that exceed the existing USEPA arsenic standard are required to either treat the water to remove the arsenic or find an alternate supply. If the WHO arsenic standard is adopted, then those public water supply systems (currently serving about 91,200,000 people) exceeding the guideline will also have to come into compliance. Although reducing the acceptable level of arsenic in drinking water will surely increase the cost of water to consumers, it will also decrease their exposure to arsenic, and the possible adverse health effects associated with this toxic element.

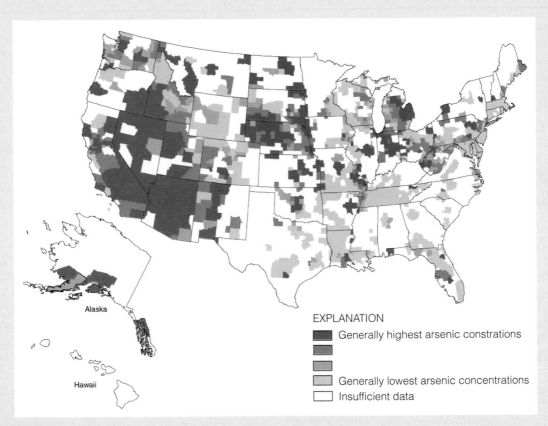

Figure 2 *Counties with arsenic concentrations exceeding possible new guidelines (mg/L = micrograms per liter) in 10 percent or more of groundwater samples.*

Discounting contamination by humans from landfills, septic systems, toxic-waste sites, and industrial effluents, groundwater quality is mostly a function of (1) the kinds of materials making up an aquifer, (2) the residence time of water in an aquifer, and (3) the solubility of rocks and minerals. These factors account for the amount of dissolved materials in groundwater, such as calcium, iron, fluoride, and several others. Most pose no health problems, but some have undesirable effects such as deposition of minerals in water pipes and water heaters as well as offensive taste or smell. Some may stain clothing and fixtures or inhibit the effectiveness of detergents.

Hydrothermal Activity— What Is It, and Where Does It Occur?

Hydrothermal is a term referring to hot water. Some geologists restrict the meaning to encompass only water heated by magma, but here we use it to mean any hot subsurface water and surface activity resulting from its discharge. One manifestation of hydrothermal activity in areas of active or recently active volcanism is the discharge of gases, such as steam, at vents known as *fumeroles* (see Figure 4.2). Of more immediate concern here, however, is the groundwater that rises to the surface as *hot springs* or *geysers*. It may be heated by its proximity to magma or by Earth's geothermal gradient, because it circulates deeply.

Hot Springs

A **hot spring** (also called a *thermal spring* or *warm spring*) is any spring in which the water temperature is greater than 37°C, the temperature of the human body (Figure 13.26a). Some hot springs are much hotter, with temperatures up to the boiling point in many instances (Figure 13.26b). Another type of hot spring, called a *mud pot*, results when chemically altered rocks yield clays that bubble as hot water and steam rise through them (Figure 13.26c). Of the approximately 1100 known hot springs in the United States, more than 1000 of them are in the far West, with the others in the Black Hills of South Dakota, Georgia, the Ouachita region of Arkansas, and the Appalachian region.

Hot springs are also common in other parts of the world. One of the most famous is at Bath, England, where shortly after the Roman conquest of Britain in A.D. 43, numerous bathhouses and a temple were built around the hot springs (Figure 13.27).

The heat for most hot springs comes from magma or cooling igneous rocks. The geologically recent igneous activity in the western United States accounts for the large number of hot springs in that region. The water in some hot springs, however, is circulated deep into Earth, where it is warmed by the normal increase in temperature, the geothermal gradient. For example, the spring water of Warm Springs, Georgia, is heated in this manner. This hot spring was a health and bathing resort long before the Civil War (1861–1865); later with the establishment of the Georgia Warm Springs Foundation, it was used to help treat polio victims.

Geysers

Hot springs that intermittently eject hot water and steam with tremendous force are known as **geysers.** The word comes from the Icelandic *geysir* which means to gush or rush forth. One of the most famous geysers in the world is Old Faithful in Yellowstone National Park in Wyoming (Figure 13.28). With a thunderous roar, it erupts a column of hot water and steam every 30 to 90 minutes. Other well-known geyser areas are found in Iceland and New Zealand.

Geysers are the surface expression of an extensive underground system of interconnected fractures within hot igneous rocks (Figure 13.29). Groundwater percolating into the network of fractures is heated as it comes into contact with the hot rocks. Because the water near the bottom of the fracture system is under greater pressure than that near the top, it must be heated to a higher temperature before it will boil. Thus, when the deeper water is heated to near the boiling point, a slight rise in temperature or a drop in pressure, such as from escaping gas, will cause it to change instantly to steam. The expanding steam quickly pushes the water above it out of the ground and into the air, thereby producing a geyser eruption. After the eruption, relatively cool groundwater starts to seep back into the fracture system where it is heated to near its boiling temperature and the eruption cycle begins again. Such a process explains how geysers can erupt with some regularity.

Hot spring and geyser water typically contains large quantities of dissolved minerals because most minerals dissolve more rapidly in warm water than in cold water. Due to this high mineral content, the waters of many hot springs are believed by some to have medicinal properties. Numerous spas and bathhouses have been built at hot springs throughout the world to take advantage of these supposed healing properties.

When the highly mineralized water of hot springs or geysers cools at the surface, some of the material in solution is precipitated, forming various types of deposits. The amount and type of precipitated mineral depend on the solubility and composition of the material that the groundwater flows through. If the groundwater contains dissolved calcium carbonate ($CaCO_3$), then *travertine* or *calcareous tufa* (both of which are varieties of limestone) are precipitated. Spectacular examples of hot spring travertine

(a)

(b)

(c)

(d)

Figure 13.26 *Hot springs. Hot spring in the West Thumb Geyser Basin, Yellowstone National Park, Wyoming. (b) The water in this hot spring at Bumpass Hell in Lassen Volcanic National Park, California, is boiling. (c) Mud pot at the Sulfur Works, also in Lassen Volcanic National Park. (d) The Park Service warns of the dangers in hydrothermal areas, but some people ignore the warnings and are injured or killed.*

Figure 13.27 *One of the many bathhouses in Bath, England, that were built around hot springs shortly after the Roman conquest in A.D. 43.*

(a)

(b)

Figure 13.28 *Geysers in Yellowstone National Park, Wyoming. (a) Old Faithful Geyser erupts every 30 to 90 minutes spewing water from 32 to 56 m high. (b) A small geyser erupting in Norris Geyser Basin.*

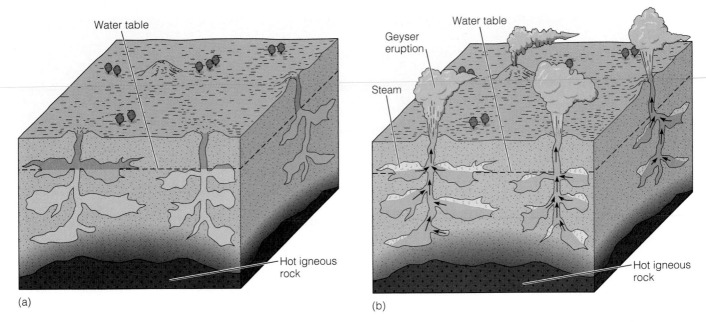

Figure 13.29 *The eruption of a geyser. (a) Groundwater percolates down into a network of interconnected openings and is heated by the hot igneous rocks. The water near the bottom of the fracture system is under greater pressure than that near the top and consequently must be heated to a higher temperature before it will boil. (b) Any rise in temperature of the water above its boiling point or a drop in pressure will cause the water to change to steam, which quickly pushes the water above it up and out of the ground, producing a geyser eruption.*

deposits are found at Pamukhale in Turkey and at Mammoth Hot Springs in Yellowstone National Park (Figure 13.30a). Groundwater containing dissolved silica will, upon reaching the surface, precipitate a soft, white, hydrated mineral called *siliceous sinter* or *geyserite,* which can accumulate around a geyser's opening (Figure 13.30b).

Geothermal Energy

Geothermal energy is defined as any energy produced from Earth's internal heat. In fact, the term *geothermal* comes from *geo,* meaning Earth, and *thermal* meaning heat. Several forms of internal heat are known, such as hot dry rocks and magma, but so far only hot water and steam are used.

Approximately 1 to 2% of the world's current energy needs could be met by geothermal energy. In those areas where it is plentiful, geothermal energy can supply most, if not all, of the energy needs, sometimes at a fraction of the cost of other types of energy. Some of the countries currently using geothermal energy in one form or another include Iceland, the United States, Mexico, Italy, New Zealand, Japan, the Philippines, and Indonesia.

WHAT WOULD YOU DO?

Americans generate tremendous amounts of waste. Some of this waste, such as battery acid, paint, cleaning agents, insecticides, and pesticides can easily contaminate the groundwater system. Your community is planning to construct a city dump to contain waste products, but simply wants to dig a hole, dump waste in, and then bury it. What do you think of this plan? Are you skeptical of this plan's merits, and if so, what would you suggest to remedy any potential problems?

As oil reserves decline, geothermal energy is becoming an attractive alternative, particularly in parts of the western United States, such as the Salton Sea area of southern California, where geothermal exploration and development have begun.

(a)

(b)

Figure 13.30 *Hot spring deposits in Yellowstone National Park, Wyoming. (a) Minerva Terrace formed when calcium-carbonate-rich hot spring water cooled, precipitating travertine. (b) Liberty Cap is a geyserite mound formed by numerous geyser eruptions of silicon-dioxide-rich hot spring water.*

Chapter Summary

1. Groundwater consists of all subsurface water trapped in the pores and other open spaces in rocks, sediment, and soil.

2. About 22% of the world's supply of freshwater is groundwater, which constitutes one reservoir in the hydrologic cycle.

3. For groundwater to move through materials, they must be porous and permeable. Any material that transmits groundwater is an aquifer, whereas materials that prevent groundwater movement are aquicludes.

4. The zone of saturation (in which pores are filled with water) is separated from the zone of aeration (in which pores are filled with air and water) by the water table. The water table is a subdued replica of the overlying land surface in most places.

5. Groundwater moves slowly through the pore spaces in the zone of aeration and moves through the zone of saturation to outlets such as streams, lakes, and swamps.

6. Springs are found wherever the water table intersects the surface. Some springs are the result of a perched water table, that is, a localized aquiclude within an aquifer and above the regional water table.

7. Water wells are made by digging or drilling into the zone of saturation. When water is pumped out of a well, a cone of depression forms.

8. In an artesian system, confined groundwater builds up high hydrostatic pressure. Three conditions must generally be met for an artesian system to form: The aquifer must be confined above and below by aquicludes; the aquifer is usually tilted and exposed at the surface so it can be recharged; and precipitation must be sufficient to keep the aquifer filled.

9. Karst topography results from groundwater weathering and erosion and is characterized by sinkholes, caves, solution valleys, and disappearing streams.

10. Caves form when groundwater in the zone of saturation weathers and erodes soluble rock such as limestone. Cave deposits, called dripstone, result from the precipitation of calcite.

11. Modifications of the groundwater system can cause serious problems. Excessive withdrawal of groundwater might result in dry wells, loss of hydrostatic pressure, saltwater incursion, and ground subsidence.

12. Groundwater contamination is becoming a serious problem and can result from sewage, landfills, and toxic waste.

13. Groundwater might be heated by magma or be heated by the geothermal gradient as it circulates deeply. In either case, the water commonly rises to the surface, thus accounting for hydrothermal activity in the form of hot springs, geysers, and several other features.

14. Geothermal energy comes from the steam and hot water trapped within the crust. It is a relatively nonpolluting form of energy that is used as a source of heat and to generate electricity.

Important Terms

aquiclude	geothermal energy	karst topography	spring
aquifer	geyser	permeability	water table
artesian system	groundwater	porosity	water well
cave	hot spring	saltwater incursion	zone of aeration
cone of depression	hydrothermal	sinkhole	zone of saturation
dripstone			

Review Questions

1. A layer of Earth materials that inhibits the movement of groundwater is a(n)
   a. _____ geyser; b. _____ dripstone; c. _____ cone of depression; d. _____ aquiclude; e. _____ capillary fringe.

2. Two features typical of areas of karst topography are
   a. _____ geysers and hot springs; b. _____ hydrothermal activity and springs; c. _____ saltwater incursion and pollution; d. _____ dripstone and a cone of depression; e. _____ sinkholes and disappearing streams.

3. When water is pumped from wells in some coastal areas, a problem arises known as
   a. _____ dripstone deposition; b. _____ saltwater incursion; c. _____ permeability decrease; d. _____ geothermal depression; e. _____ artesian recharge.

4. Which of the following is a cave deposit?
   a. _____ aquiclude; b. _____ artesian spring; c. _____ sinkhole; d. _____ stalagmite; e. _____ chamber.

5. A hot spring that periodically erupts is known as a
   a. _____ mud pot; b. _____ travertine terrace; c. _____ geyser; d. _____ stalactite; e. _____ cone of ascension.

6. What is the correct order, from highest to lowest, of groundwater usage in the United States?
   a. _____ agricultural, industrial, domestic; b. _____ industrial, domestic, agricultural; c. _____ domestic, agricultural, industrial; d. _____ agricultural, domestic, industrial; e. _____ industrial, agricultural, domestic

7. Which of the following conditions must exist for an artesian system to form?

a. _____ The water table must be at or very near the surface; b. _____ Water must rise very high in the capillary fringe; c. _____ An aquifer must be confined above and below by aquicludes; d. _____ Groundwater must circulate near magma; e. _____ The rocks below the surface must be especially resistant to solution.

8. The porosity of Earth materials is defined as a. _____ their ability to transmit fluids; b. _____ the depth of the zone of saturation; c. _____ the percentage of void spaces; d. _____ their solubility in the presence of weak acids; e. _____ the temperature of groundwater.

9. A cone of depression forms when a. _____ a stream flows into a sinkhole; b. _____ water in the zone of aeration is replaced by water from the zone of saturation; c. _____ the ceiling of a cave collapses, forming a steep-sided crater; d. _____ a spring forms where a perched water table intersects the surface; e. _____ water is withdrawn from a well faster than it can be replaced.

10. Hydrothermal is a term referring to a. _____ groundwater contamination; b. _____ artesian wells; c. _____ hot water; d. _____ sinkhole formation; e. _____ calcareous tufa.

11. Which of the following statements is correct? a. _____ The water from an artesian well is better than water from other wells; b. _____ The water table dips beneath hills and rises beneath valleys; c. _____ In the United States, karst topography is best developed in the Pacific Northwest; d. _____ Groundwater erosion yields geysers and hot springs; e. _____ A perched water table is found above the main water table.

12. In which of the following states is geothermal energy used to generate electricity? a. _____ Texas; b. _____ California; c. _____ New York; d. _____ Idaho; e. _____ Arkansas.

13. Groundwater represents what percentage of the world's freshwater supply? a. _____ 50; b. _____ 45; c. _____ 22; d. _____ 17; e. _____ 6.

14. Describe some of the ways of quantitatively measuring the rate of groundwater movement.

15. Explain how groundwater weathers and erodes.

16. Why does groundwater move so much slower than surface water?

17. One concern geologists have about burying nuclear waste in present-day arid regions, such as Nevada, is that the climate may change during the next several thousand years and become more humid, thus allowing more water to percolate through the zone of aeration. Why is this a concern? What would the average rate of groundwater movement have to be to reach the canisters containing radioactive waste buried at a depth of 400 m?

18. Explain how some Earth materials can be porous, yet not be permeable. Give an example.

19. Explain how saltwater incursion takes place and why it is a problem in coastal areas.

20. Discuss the various effects that excessive groundwater removal may have on a region. Give some examples.

21. Diagram the conditions necessary for an artesian system to form. Do all artesian wells flow freely at the surface? Explain.

22. Discuss the role of groundwater in the hydrologic cycle.

23. Describe the configuration of the water table beneath a humid area and an arid region. Why do they differ?

24. Explain what geothermal energy is, and briefly discuss its potential to decrease our dependence on traditional energy sources.

25. Describe three features you might see in an active hydrothermal area. Where in the United States would you go to see such activity?

 # World Wide Web Activities

For these website addresses, along with current updates and exercises, log on to **http://www.brookscole.com/geo**

The Virtual Cave

This web site contains a large number of images of every major type of cave feature and secondary mineral deposit from around the world. It also contains a link to a directory of U.S. caves so you can find out about caves you might want to visit on your next trip, as well as a link to the National Speleological Society.

U.S. Geological Survey Water Resources of the United States

This web site contains a wealth of information about streams, flooding, flood forecasting, flood warnings, and groundwater. At this web site there are links to water-use data sites, maps showing groundwater use in the United States, and information about groundwater including quality and problems with subsidence.

General Geyser Information

This web site is a collection of links to a wide variety of geyser sites. The home page is divided into headings with links under each heading. The headings include general information and photo sites; geyser reports, experiments, science fair projects, and reports (ideas for students/teachers); geyser research; museums, geyser exhibits; research labs; photographing geysers; and weird stuff.

United States Environmental Protection Agency Office of Ground Water & Drinking Water

This web site contains links to the following EPA sites: Source Water Protection Surface and Ground Water; Underground Injection Control; Drinking Water Standards Program; Public Drinking Water Supply Programs; Local Drinking Water Information; and Drinking Water and Health Basics.

 CD-ROM Exploration

Exploring your *Earth Systems Today* CD-ROM will add to your understanding of the material in this chapter.

Topic: Surficial Processes and Hydrosphere

Module: Groundwater

Explore activities in this module to see if you can discover the following for yourself:

Using this activity, examine the flow of water from a recharge area to an artesian well, a pumping or "standard" well, and a stream or river.

Using this activity, examine the relationship between precipitation and stream flow and also groundwater flow through the zone of aeration, capillary fringe, and zone of saturation.

Using this activity, study the development of a cone of depression. How do pumping rate and permeability affect development of a cone of depression?

Glaciers and Glaciation

Pense La Glacier in the Zanskar Valley, India.

During the Pleistocene Epoch, more popularly known as the Ice Age—a time from about 1.6 million to 10,000 years ago—glaciers were much more widespread than they are now. Since then, Earth has experienced a general warming trend periodically reversed by short, comparatively cool periods. One cool period lasting from the 1500s until the mid- to late-1800s is known as the *Little Ice Age*. During this time, small glaciers in mountain valleys expanded markedly, and at high latitudes, as in Iceland, sea ice persisted for much longer each year than it had previously. Furthermore, climatic changes during the Little Ice Age were responsible for shorter growing seasons, famines, and thousands of deaths in several countries.

The onset of the Little Ice Age is generally given as 1500, but the climatic conditions leading to this episode of cooler climates and expansion of glaciers actually began by about A.D. 1300. During the several preceding centuries, Europe experienced rather mild temperatures, and the North Atlantic was more storm-free than it is now. Explorers from Scandinavia dis-

covered and settled Iceland, and by A.D. 1200, about 80,000 people were residing there. They also sailed to Greenland and North America and established colonies in both areas. Unfortunately, as the climate changed, the North Atlantic became stormier, and sea ice was present farther south and persisted longer each year. Because of these poor sea conditions and political problems, all shipping ceased and the colonies in Greenland and North America eventually disappeared.

Glaciers are sensitive to climatic changes; they expand or contract depending on temperature changes and changes in the amount of precipitation as snow. During the Little Ice Age, glaciers in Europe and Iceland expanded and moved far down their valleys, reaching their greatest historic extent by the early 1800s (Figure 14.1). An area in Iceland that had formerly been ice-free was covered by a small ice cap, and in Canada, Alaska, and the western United States glaciers also expanded markedly. For instance, glaciers at Glacier Bay, Alaska, extended more than 100 km farther down their valleys than they do now.

OBJECTIVES

At the end of this chapter, you will have learned that

■ Moving bodies of ice on land known as glaciers cover about 10% of Earth's land surface, but they were much more widespread during the Pleistocene Epoch (Ice Age).

■ Glaciers constitute one reservoir in the hydrologic cycle.

■ In any area with a yearly net accumulation of snow, the snow is converted first to firn and eventually into glacial ice.

■ The concept of a glacial budget is important in considering the dynamics of any glacier.

■ Glaciers move by a combination of plastic flow and basal slip, but several factors determine the rate at which they move.

■ Glaciers effectively erode, transport, and deposit sediment, thus accounting for the origin of several distinctive landforms.

■ A current theory explaining the onset of ice ages relies on irregularities in Earth's rotation and orbit.

Figure 14.1 *During the Little Ice Age, many of the glaciers in Europe, such as this one in Switzerland, extended much farther down their valleys than they do at present. The Unterer Grindelwald was painted in 1826 by Samuel Birmann (1793–1847).*

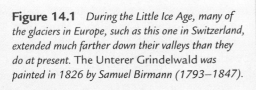

Advancing glaciers caused some problems in Europe where they covered roadways and pastures, destroyed some villages in Scandinavia, and threatened villages elsewhere. Overall, though, glaciers were the least of the problems faced by residents of northern Europe and Iceland. Cooler, wetter summers resulting in shorter growing seasons had a far greater impact on countries in the northern latitudes. Especially hard hit were Iceland and the Scandinavian countries, but at times several other parts of Europe including Great Britain, France, Germany, the Netherlands, and Belgium were affected as well.

Food shortages resulting from shorter growing seasons were responsible for a number of famines, especially in Northern Eu-

rope and Iceland. In fact, Iceland's population declined from its high of 80,000 in A.D. 1200 to about 40,000 by 1700. And between 1610 and 1870, on many occasions sea ice persisted near Iceland for three months each year, and each time sea ice persisted, poor growing seasons and food shortages followed.

Exactly when the Little Ice Age ended has not been resolved. Of course, the end of a long-term climatic event can never be determined exactly, only approximated. But some authorities say it ended as late as 1880, whereas others think that about 1850 is a more likely date. In either case, during the mid- to late-1800s, sea ice began retreating northward, glaciers started receding, and summer weather became more moderate.

Introduction

The Little Ice Age was just one of several climatic fluctuations that took place during the last few thousand years, but one that certainly had a dramatic impact on people living in the affected areas. Even regions far removed from the areas of glacier expansion and cooler, wetter summers experienced changes. Northern Europeans suffered during the Little Ice Age, but, in contrast, Italians enjoyed more rainfall and rather moderate temperatures. Unfortunately, our period of record keeping is too short to resolve the question of whether the last Ice Age and the Little Ice Age are truly events of the past or simply parts of long-term climatic events and likely to occur again. In any event, even though glaciers are much more restricted now, they are still capable of considerable geologic work.

Most people have some idea of what a glacier is and have heard of the Ice Age, or what geologists call the Pleistocene Epoch. By definition, a **glacier** is a mass of ice on land consisting of compacted and recrystallized snow that flows under its own weight. Accordingly, sea ice in the north polar region or the ice shelves adjacent to Antarctica are not glaciers, nor are drifting icebergs, even though they may have broken off from glaciers that flow into the sea. Snowfields in high mountains might persist for years, but these are not glaciers either because they are not actively flowing.

Presently, glaciers of one kind or another cover nearly 15 million km^2, or about one-tenth of Earth's land surface (Table 14.1). As a matter of fact, enough glacial ice is present to cover the United States and Canada to a depth of about 1.5 km! Small glaciers are common in the mountains of the western United States, especially Alaska, western Canada, the Andes in South America, the Alps of Europe, the Himalayas of Asia (see chapter opening photo), and in other high mountains. Even Mount Kilamanjaro in Africa is high enough to have glaciers, although it is close to the equator. In fact, Australia is the only continent lacking glaciers. Small glaciers in mountains are impressive and picturesque, but the truly vast glaciers are in Antarctica and Greenland. More than 95% of all glacial ice on this planet is in Antarctica (84.5%) and Greenland (12%), both of which are nearly covered by glaciers.

Table 14.1

Present-Day Ice-Covered Areas

Antarctica	12,653,000 km^2
Greenland	1,802,600
Northeast Canada	153,200
Central Asian ranges	124,500
Spitsbergen group	58,000
Other Arctic islands	54,000
Alaska	51,500
South American ranges	25,000
West Canadian ranges	24,900
Iceland	11,800
Scandinavia	5000
Alps	3600
Caucasus	2000
New Zealand	1000
USA (other than Alaska)	650
Others	about 800
	14,971,550

Total volume of present ice: 28 to 35 million km^3

SOURCE: C. Embleton and C. A. King, *Glacial Geomorphology* (New York: Halsted Press, 1975).

Why Should You Study Glaciers and Glaciation?

Why should you study glaciers and glaciation? Studying the possible causes of widespread past episodes of glaciation might help clarify some aspects of long-term climatic changes and possibly tell us something about the debate on global warming. Furthermore, present-day glaciers are sensitive even to short-term climatic changes, so they are closely monitored to see if they advance, remain stationary, or retreat.

At first glance glaciers appear static. Even a brief visit to a glacier might not dispel this idea, because although they move, they do so slowly. Nevertheless, just like other elements of the hydrosphere, such as running water, they are dynamic systems that continually adjust to changes. Thus, glaciers afford another opportunity to study the interactions among Earth's major systems. Glacial ice constitutes a huge reservoir of fresh water—about 75% of all fresh water is frozen in glaciers—and erosion by glaciers has yielded magnificent scenery in some of our national parks and monuments.

Glaciers—Part of the Hydrologic Cycle

One theme in this book is that Earth is a dynamic planet possessing several major systems that interact in complex ways (see Table 1.1). One of these systems, the hydrosphere, consists of all surface water in the oceans and on land, including water frozen in glaciers. Glaciers contain only 2.15% of all water on Earth, but fully 75% of all freshwater is found here. This small quantity of frozen water in glaciers, the dynamics of glaciers, and the geologic work done by glaciers are our main interests here.

The ultimate source of frozen water in glaciers, just like rainfall, is the ocean. It constitutes that part of the precipitation that falls over land as snow and is subsequently incorporated into these moving bodies of ice on land that we call glaciers. So, glaciers constitute one reservoir in the hydrologic cycle where water is stored for long periods, but even this water eventually returns to its original source, the oceans (see Figure 12.4). Many glaciers at high latitudes, as in Antarctica, Greenland, Alaska, and northern Canada, flow directly into the seas where they melt, or icebergs break off from them (a process known as *calving*) and drift out to sea where they eventually melt. At lower latitudes or areas more remote from the sea, glaciers flow from higher to lower elevations where melting takes place and the water yielded enters the groundwater system (another reservoir in the hydrologic cycle) or it returns to the seas by surface runoff. In some parts of the western United States and Canada, glaciers are important freshwater reservoirs that release water to streams during the dry season.

Melting is the most important process in returning glacial ice to the hydrologic cycle, but glaciers also lose water by *sublimation,* a process in which ice changes to water vapor without an intermediate liquid phase. Sublimation is easy to understand if you think of ice cubes stored in a container in a freezer. Because of sublimation the older ice cubes at the bottom of the container are much smaller than the more recently formed ones. The water vapor derived by sublimation from glaciers enters the atmosphere where it may condense and fall again as rain or snow, but in the long run all water in glaciers returns to the oceans.

How Do Glaciers Form and Move?

In Chapter 2 we briefly mentioned that ice is crystalline and possesses characteristic physical and chemical properties, and thus is a mineral. Accordingly, glacial ice must be a type of rock, but one that is easily deformed. Glacial ice forms in a very straightforward manner (Figure 14.2). In any area where more snow falls than melts during the warmer seasons, a net accumulation occurs. Freshly fallen snow has about 80% air-filled pore space and 20% solids, but it compacts as it accumulates, partly thaws, and refreezes, converting to a granular type of snow known as **firn.** As more snow accumulates, firn is buried and further compacted and recrystallizes until it becomes **glacial ice,** consisting of about 90% solids (Figure 14.2). As we men-

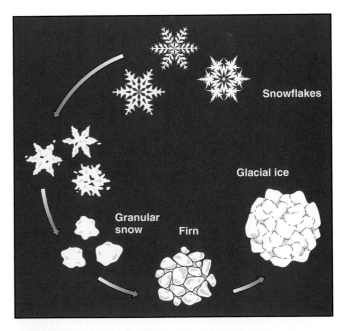

Figure 14.2 *The conversion of freshly fallen snow to firn and glacial ice.*

How Do Glaciers Form and Move? **401**

tioned in the Introduction, a *glacier* is a moving mass of re-crystallized snow on land, but just how is movement accomplished?

At this time it is useful to recall some of the terms from Chapter 10 on deformation. Remember that stress is force per unit area and strain is defined as a change in shape or volume or both of solids. When snow and ice reach a critical thickness of about 40 m, the stress on the ice at depth is great enough to induce **plastic flow,** a type of permanent deformation involving no fracturing. Glaciers move primarily by plastic flow, but they also slide over their underlying surface by **basal slip** (Figure 14.3). Basal slip is facilitated by the presence of water, which reduces frictional resistance between the underlying surface and a glacier. The total movement of a glacier, then, results from a combination of plastic flow and basal slip, although the former occurs continuously whereas the latter varies depending on the season. Indeed, if a glacier is solidly frozen to the surface below it moves only by plastic flow.

You now have some idea of how glaciers form in the first place, but what controls their distribution? As you probably suspect, temperature and the amount of snowfall are important factors. Of course temperature varies with elevation and latitude, so we would expect to find glaciers in high mountains or at high latitudes, if the areas receive enough snow. Many small glaciers are present in the Sierra Nevada of California, but only at elevations exceeding

3900 m. In fact, the high mountains in California, Oregon, and Washington all possess glaciers because in addition to their elevations they receive huge amounts of snow. Indeed, Mount Baker in Washington had almost 29 m of snow during the winter of 1998–1999 and average snowfalls of 10 m or more are common in many parts of these mountains. Glaciers are also present in the mountains along the Pacific Coast of Canada, which also receive considerable snow and in addition are farther north. Some of the higher peaks in the Rocky Mountains of both the United States and Canada also support glaciers.

What Kinds of Glaciers Are There?

All glaciers share some characteristics, but they also vary in several ways. Some are confined to mountain valleys or bowl-shaped depressions on mountainsides, and flow from higher to lower elevations. Others are of much greater thickness and extent; they flow outward from centers of accumulation and are completely unconfined by topography. Thus, we recognize two basic types of glaciers: *valley* and *continental*, and some variations on these basic types.

Valley Glaciers

A **valley glacier,** as its name implies, is confined to a mountain valley through which it flows from higher to lower elevations (Figure 14.4). We use the term *valley glacier* here, but *alpine glacier* and *mountain glacier* are synonyms. Many valley glaciers have smaller tributary glaciers entering them, just as rivers and streams have tributaries, thus forming a network of glaciers in a system of interconnected valleys. A valley glacier's shape is obviously con-

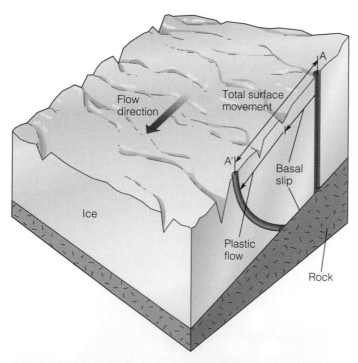

Figure 14.3 *Part of a glacier showing movement by a combination of plastic flow and basal slip. Plastic flow takes place as the ice is internally deformed, whereas basal slip involves the glacier sliding over the underlying surface. If a glacier is solidly frozen to its bed, it moves only by plastic flow. Notice that the top of the glacier moves farther in a given time than the bottom does.*

trolled by the shape of the valley it occupies, so these glaciers are long, narrow tongues of moving ice. Where a valley glacier flows from a valley onto a wider plain and spreads out, or where two or more valley glaciers coalesce at the base of a mountain range, they form a more extensive ice cover called a *piedmont glacier*.

Valley glaciers are invariably small compared to continental glaciers, but even so they may be several kilometers across, 200 km long, and hundreds of meters thick. For instance, the Bering Glacier in Alaska is about 200 km long, and the Saskatchewan Glacier in Canada is 555 m thick. Erosion and deposition by valley glaciers was responsible for much of the spectacular scenery in such places as Grand Teton National Park, Wyoming (see Chapter 10 Prologue); Glacier National Park, Montana; and Waterton, Banff, and Jasper National Parks in Canada.

Continental Glaciers

Continental glaciers, also known as *ice sheets,* cover at least 50,000 km^2 and are unconfined by topography (Figure 14.5). That is, their shape and movement are not controlled by the underlying landscape. Valley glaciers flow downhill within the confines of a valley, but continental glaciers flow outward in all directions from central areas of accumulation in response to variations in ice thickness.

Continental glaciers are currently present only in Greenland and Antarctica. In both areas, the ice is more

Figure 14.4 *A large valley glacier in Alaska. Notice the tributaries to the large glacier.*

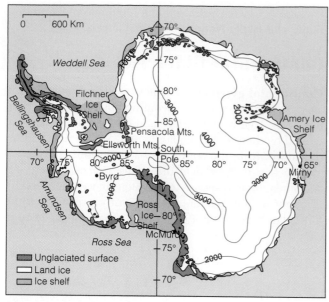

(a)

(b)

Figure 14.5 *(a) The West Antarctic and much larger East Antarctic ice sheet merge to form a nearly continuous ice cover averaging 2160 m thick and reaching a maximum thickness of 4000 m. (b) View of the margin of the ice sheet in Antarctica.*

50,000 km². The 6000 km² Penny Ice Cap on Baffin Island, Canada, is a good example. Some ice caps form when valley glaciers grow and overtop the divides and passes between adjacent valleys and coalesce to form a continuous ice cap. They also form on fairly flat terrain including some of the islands of the Canadian Arctic and Iceland.

Accumulation and Wastage— The Glacial Budget

Just as a savings account grows and shrinks as funds are deposited and withdrawn, glaciers expand and contract in response to accumulation and wastage. Their behavior can be described in terms of a **glacial budget,** which is essentially a balance sheet of accumulation and wastage. The upper part of a valley glacier is a **zone of accumulation** where additions exceed losses, and the glacier's surface is perennially covered by snow. In contrast, the same glacier at a lower elevation is in a **zone of wastage,** where losses from melting, sublimation, and calving of icebergs exceed the rate of accumulation (Figure 14.6).

At the end of winter, a glacier's surface is usually completely covered with the accumulated seasonal snowfall. During spring and summer, the snow begins to melt, first at lower elevations and then progressively higher up the glacier. The elevation to which snow recedes during a wastage season is called the *firn limit* (Figure 14.6). One can easily identify the zones of accumulation and wastage by noting the position of the firn limit.

Observations of a single glacier reveal that the position of the firn limit usually changes from year to year. If it does not change or shows only minor fluctuations, the glacier has a balanced budget; that is, additions in the zone of accumulation are exactly balanced by losses in the zone of wastage, and the distal end or *terminus* of the glacier remains stationary (Figure 14.6a). When the firn limit moves down the glacier, the glacier has a positive budget; its ad-

than 3000 m thick in their central areas, becomes thinner toward the margins, and covers all but the highest mountains (Figure 14.5b). The aerial extent of the continental glacier in Greenland is about 1,800,000 km², and in Antarctica the East and West Antarctic Glaciers merge to form a continuous ice sheet covering more than 12,650,000 km². During the Pleistocene Epoch, continental glaciers covered large parts of the Northern Hemisphere continents and account for many erosional and depositional landforms in Canada and the states from Washington to Maine.

Although valley and continental glaciers are easily differentiated by size and location, an intermediate variety called an *ice cap* is also recognized. Ice caps are similar to, but smaller than, continental glaciers, covering less than

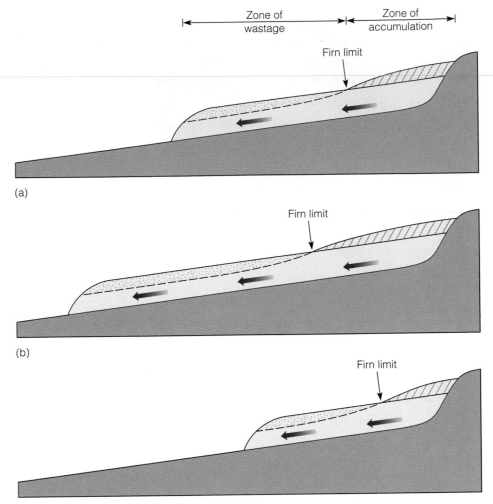

Figure 14.6 *Response of a hypothetical glacier to changes in its budget. (a) If losses in the zone of wastage, shown by stippling, equal additions in the zone of accumulation, shown by crosshatching, the terminus of the glacier remains stationary. (b) Gains exceed losses, and the glacier's terminus advances. (c) Losses exceed gains, and the glacier's terminus retreats, although the glacier continues to flow.*

ditions exceed its losses, and its terminus advances (Figure 14.6b). If the budget is negative, the glacier recedes and its terminus retreats up the glacial valley (Figure 14.6c). Even though a glacier's terminus may be receding, the glacial ice continues to move toward the terminus by plastic flow and basal slip. If a negative budget persists long enough, a glacier recedes and thins until it no longer flows and becomes a *stagnant glacier.*

Although we used a valley glacier as our example, the same budget considerations control the behavior of continental glaciers as well. For example, the entire Antarctic ice sheet is in the zone of accumulation, but it flows into the ocean where wastage occurs.

How Fast Do Glaciers Move?

In general, valley glaciers move more rapidly than continental glaciers, but the rates for both vary, ranging from centimeters to tens of meters per day. Valley glaciers mov-

ing down steep slopes flow more rapidly than glaciers of comparable size on gentle slopes, assuming that all other variables are the same. The main glacier in a valley glacier system contains a greater volume of ice and thus has a greater discharge and flow velocity than its tributaries (Figure 14.4). Temperature exerts a seasonal control on valley glaciers because although plastic flow remains rather constant year-round, basal slip is more important during warmer months when meltwater is more abundant.

Flow rates also vary within the ice itself. For example, flow velocity generally increases in the zone of accumulation until the firn limit is reached; from that point, the velocity becomes progressively slower toward the glacier's terminus. Valley glaciers are similar to rivers, in that the valley walls and floor cause frictional resistance to flow, so the ice in contact with the walls and floor moves more slowly than the ice some distance away (Figure 14.7).

Notice in Figure 14.7 that flow velocity increases upward until the top few tens of meters of ice are reached, but little or no additional increase occurs after that point.

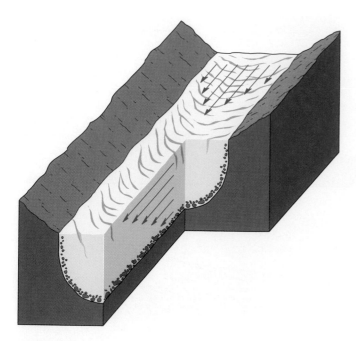

Figure 14.7 *Flow velocity in a valley glacier varies both horizontally and vertically. Velocity is greatest at the top-center of the glacier. Friction with the walls and floor of the glacial trough causes the flow to be slower adjacent to these boundaries. The length of the arrows in the figure is proportional to velocity.*

This upper ice constitutes the rigid part of the glacier that is moving as a consequence of basal slip and plastic flow below. The fact that this upper 40 m or so of ice behaves as a brittle solid is clearly demonstrated by large fractures called *crevasses* that develop when a valley glacier flows over a step in its valley floor where the slope increases or where it flows around a corner (Figure 14.8). In either case, the glacial ice is stretched (subjected to tension), and large crevasses develop, but they extend downward only to the zone of plastic flow. In some cases, a valley glacier descends over such a steep precipice that crevasses break up the ice into a jumble of blocks and spires, and an *ice fall* develops.

The flow rates of valley glaciers are complicated by *glacial surges*, which are bulges of ice that move through a glacier at a velocity several times faster than the normal flow. Although surges are best documented in valley glaciers, they take place in ice caps and continental glaciers as well. During a surge, a glacier's terminus may advance several kilometers during a year. In 1993, the terminus of the Bering Glacier in Alaska advanced 9 km in 17 months.

The causes of surges are not fully understood, but some of them occur following a period of unusually heavy precipitation in the zone of accumulation. Others develop when excessive amounts of snow and ice are dislodged

Figure 14.8 *Crevasses and an ice fall in a glacier in Alaska.*

from mountain peaks, and fall onto the upper parts of glaciers. In either case, rapid changes in the glacial budget occur.

One reason continental glaciers move comparatively slowly is that they exist at higher latitudes and are frozen to the underlying surface most of the time, which limits the amount of basal slip. Some basal slip does take place even beneath the Antarctic ice sheet, but most of its movement is by plastic flow. Nevertheless, some parts of continental glaciers manage to achieve extremely high flow rates. Near the margins of the Greenland ice sheet, the ice is forced between mountains in what are called *outlet glaciers.* In some of these outlets, flow velocities exceeding 100 m per day have been recorded.

Some areas of rapid flow known as ice streams in West Antarctica have flow rates considerably greater than in adjacent glacial ice. Drilling revealed a 5-m-thick layer of water-saturated sediment beneath these ice streams, which apparently acts to facilitate movement of the ice above. Some geologists think that geothermal heat from active volcanism melts the underside of the ice.

Erosion and Transport by Glaciers

Glaciers are moving solids that can erode and transport huge quantities of materials, especially unconsolidated sediment and soil. Important erosional processes associated with glaciers include bulldozing, plucking, and abrasion. Bulldozing, although not a formal geologic term, is fairly self-explanatory: a glacier simply shoves or pushes unconsolidated materials in its path. This effective process was aptly described in 1744 during the Little Ice Age by an observer in Norway:

> When at times [the glacier] pushes forward a great sound is heard, like that of an organ and it pushes in front of it unmeasurable masses of soil, grit and rocks bigger than any house could be, which it then crushes small like sand.*

Plucking, also called *quarrying,* results when glacial ice freezes in the cracks and crevices of a bedrock projection and eventually pulls it loose. One manifestation of plucking is a landform known as a *roche moutonnée,* which is French for "rock sheep." As shown in Figure 14.9, a glacier smooths the "upstream" side of an obstacle, such as a small hill, and plucks pieces of rock from the "downstream" side by repeatedly freezing and pulling away from the obstacle.

Bedrock over which sediment-laden glacial ice moves is effectively eroded by **abrasion,** and commonly develops a **glacial polish,** a smooth surface that glistens in reflected light (Figure 14.10a). Abrasion also yields *glacial striations,* consisting of rather straight scratches on rock surfaces (Figure 14.10b). Glacial striations are rarely more than a few millimeters deep, whereas *glacial grooves* are similar but much larger and deeper (Figure 14.10c). Abrasion also thoroughly pulverizes rocks so that they yield an aggregate of clay- and silt-sized particles having the consistency of flour, hence the name *rock flour.* Rock flour is so common in streams discharging from glaciers that the water generally has a milky appearance.

Continental glaciers can derive sediment from mountains projecting through them, and windblown dust settles on their surfaces. Otherwise, most of their sediment is obtained from the surface they move over and is transported in the lower part of the ice sheet. In contrast, valley glaciers carry sediment in all parts of the ice, but it is concentrated at the base and along the margins. Some of the marginal sediment is derived by abrasion and plucking, but much of it is supplied by mass wasting processes, as when soil, sediment, or rock falls or slides onto a glacier's surface.

Erosion by Valley Glaciers

Some of the world's most inspiring scenery results from erosion by valley glaciers. Many mountain ranges are scenic to begin with, but when modified by valley glaciers they take on a unique appearance of angular ridges and peaks in the midst of broad valleys (see Perspective 14.1). Several of the national parks and monuments in the western United States and in Canada owe their scenic appeal to erosion by valley glaciers. The erosional landforms resulting from valley glaciation are easily recognized and enable us to appreciate the tremendous erosive power of moving ice. (Figure 14.11).

U-Shaped Glacial Troughs

One of the most distinctive features of valley glaciation is a **U-shaped glacial trough** (Figure 14.11c). Mountain valleys eroded by running water are typically V-shaped in cross section; that is, they have valley walls that descend to a narrow valley bottom (Figure 14.11a). In contrast, valleys scoured by glaciers are deepened, widened, and straightened so that they possess very steep or vertical walls, but have broad, rather flat valley floors; thus, they exhibit a U-shaped profile (Figure 14.12a). Many glacial troughs contain triangular-shaped *truncated spurs,* which are cutoff or truncated ridges that extend into the preglacial valley (Figure 14.11c).

During the Pleistocene, when glaciers were more extensive, sea level was about 130 m lower than at present, so glaciers flowing into the sea eroded their valleys to

*Quoted in C. Officer and J. Page, *Tales of the Earth* (New York: Oxford University Press, 1993), p. 99.

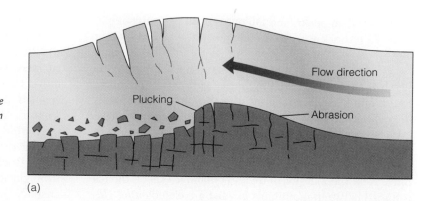

Figure 14.9 *(a) Origin of a roche moutonnée. As the ice moves over a hill, it smooths the "upstream" side by abrasion and shapes the "downstream" side by plucking. (b) A roche moutonnée.*

(a)

(b)

much greater depths than they do now. When the glaciers melted at the end of the Pleistocene, sea level rose, and the ocean filled the lower ends of the glacial troughs so that now they are long, steep-walled embayments called **fiords** (Figure 14.12b).

Lower sea level during the Pleistocene was not responsible for the formation of all fiords. Unlike running water, glaciers can erode a considerable distance below sea level. In fact, a glacier 500 m thick can stay in contact with the sea floor and effectively erode it to a depth of about 450 m before the buoyant effects of water cause the glacial ice to float! The depth of some fiords is quite impressive; some in Norway and southern Chile are about 1300 m deep.

Hanging Valleys

Waterfalls can form in several ways, but some of the world's highest and most spectacular are found in recently glaciated areas. For example, Bridalveil Falls in Yosemite National Park, California, plunges from a **hanging valley,** which is a tributary valley whose floor is at a higher level than that of the main valley (Figure 14.13). As Figure 14.11 shows, the large glacier in the main valley vigorously erodes, whereas the smaller glaciers in tributary valleys are less capable of erosion. When the glaciers disappear, the smaller tributary valleys remain as hanging valleys. Accordingly, streams flowing through hanging valleys plunge over vertical or steep precipices.

(a)

(b)

(c)

Figure 14.10 *When sediment-laden ice moves over rocks it abrades them and imparts a sheen known as glacial polish (a) as on this gneiss in Michigan. Glacial polish is also visible in (b) and so are straight scratches called glacial striations. The rock is basalt at Devil's Postpile National Monument, California. (c) These glacial grooves on Kelly's Island, Ohio, in Lake Erie are several meters deep.*

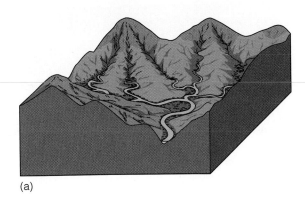

(a)

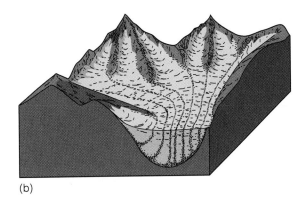

(b)

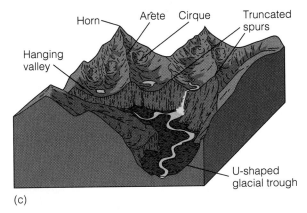

Horn　Arête　Cirque　Truncated spurs

Hanging valley

U-shaped glacial trough

(c)

Figure 14.11 *Erosional landforms produced by valley glaciers. (a) A mountain area before glaciation. (b) The same area during the maximum extent of valley glaciers. (c) After glaciation.*

Waterton Lakes and Glacier National Parks

Waterton Lakes National Park in Alberta, Canada, and Glacier National Park in Montana lie adjacent to one another and in 1932 were designated an international peace park, the first of its kind. Both parks have spectacular scenery, interesting wildlife such as mountain goats, bighorn sheep, and grizzly bears, and an impressive geologic history. The present-day landscapes resulted from deformation and uplift from Cretaceous to Eocene times, followed by deep erosion by streams and glaciers. Park visitors can see the results of the phenomenal forces at work during deformation by visiting sites where a large fault is visible, and glacial landforms such as U-shaped glacial troughs, arêtes, cirques, and horns are some of the finest in North America (Figure 1).

Most of the rocks exposed in Glacier National Park belong to the Late Proterozoic age Belt Supergroup,* whereas those in Waterton National Park are assigned to the Purcell Supergroup. The names differ north and south of the border, but the rocks are the same. These Belt-Purcell rocks are nearly 4000 m thick and were deposited between 1.45 billion and 850 million years ago. The rocks themselves are interesting, and some are quite attractive, especially thick limestone formations and red and green rocks consisting mostly of mud. In addition, many of the rocks contain a variety of sedimentary structures such as mud cracks, ripple marks, and cross-bedding that help geologists decipher their geologic history. Long after deposition of the Belt-Purcell rocks, mud and sand were deposited during the Cretaceous Period when a marine transgression took place that covered a large part of North America, including the area of the present-day parks.

The most impressive geologic structure in the parks is the Lewis Overthrust,** a fault along which Belt-Purcell rocks have moved at least 75 km eastward so that they now rest on much younger Cretaceous-aged rocks (Figure 2). If you take the trail from Marias Pass in Glacier National Park to get a closer look at the fault, you can see intense deformation of rocks lying below the fault. In any case, this large slab of ancient rocks has been deeply eroded by running water and glaciers, giving rise to the parks' present landscapes.

During the Pleistocene Epoch, glaciers formed and grew, overtopping the divides between valleys thus forming an ice cap that nearly covered the entire area. In fact, several episodes of Pleistocene glaciation took place, but the evidence for the most recent one is most obvious. These glaciers flowed outward in all directions, and on the east they merged with the continental glacier covering most of Canada and the northern states. Much of the parks' landscapes developed during these glacial episodes as valleys were deepened and widened, and cirques, arêtes, and horns developed (Figure 1).

*Supergroup is a geologic term for two or more groups that in turn are composed of two or more formations.

**An overthrust fault is simply a very-low angle thrust fault along which movement is usually measured in kilometers.

(a)

(b)

Figure 1 *The sharp angular peaks and ridges and rounded valleys are typical of areas eroded by valley glaciers such as (a) Glacier National Park, Montana, and (b) Waterton National Park, Alberta, Canada.*

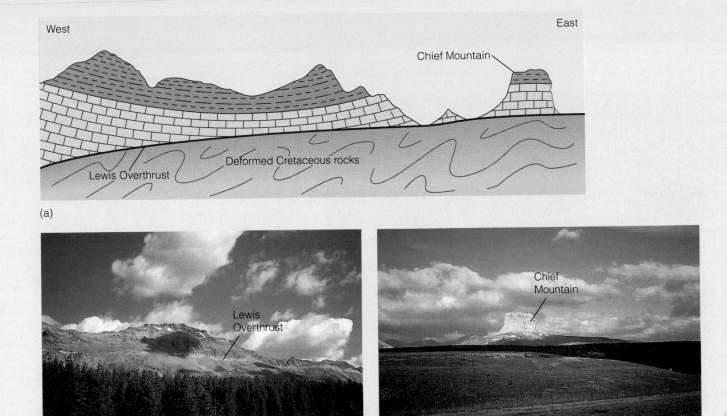

West   East

Chief Mountain

Deformed Cretaceous rocks

Lewis Overthrust

(a)

Lewis Overthrust

(b)

Chief Mountain

(c)

Figure 2 *(a) Diagrammatic view of the Lewis Overthrust in Glacier National Park, Montana. Rocks of Late Proterozoic age now rest upon deformed Cretaceous sedimentary rocks. (b) View from Marias Pass reveals the fault as a light-colored line on the mountainside. (c) Erosion has isolated Chief Mountain from the rest of the slab of overthrust rock.*

Today only a few dozen small glaciers remain active in the parks, but just like earlier ones they continue to erode, transport, and deposit sediment, only at a considerably reduced rate. In fact, many of the 150 or so glaciers present in Glacier National Park in 1850 are now gone or remain only as patches of stagnant ice. And even among the others it is difficult to determine exactly how many are active because they are so small and move so slowly, only a few meters per year. They did, however, expand markedly during the Little Ice Age, but have since retreated. For example, Grinnell Glacier covered only 0.88 km^2 in 1993 as opposed to 2.33 km^2 in 1850 (Figure 3), and during the same time Sperry Glacier was down to 0.87 km^2 from 3.76 km^2.

It seems that these small glaciers, which are very sensitive to climatic changes, are shrinking as a result of the 1°C increase in average summer temperatures in this region since 1900. According to one U.S. Geological Survey report, expected increased warming will eliminate the glaciers by 2030, and certainly by 2100 even if no additional warming takes place.

None of the active glaciers in either park can be reached by road, but several are visible from a distance. Nevertheless, Pleistocene glaciers and the remaining active ones were responsible for much of the striking scenery. Now that glaciers play only a minor role in the continuing evolution of the parks, weathering, mass wasting, and stream erosion are modifying the glacial landscape.

(continued)

Grinnell Glacier 1850–1981

1850

1968 1937

1981

Figure 3 *Grinnell Glacier in Glacier National Park, Montana. In 1850, at the end of the Little Ice Age, the glacier extended much farther and covered about 2.33 km². By 1981, its terminus had retreated to the position shown, and in 1993 it covered only about 0.88 km².*

Cirques, Arêtes, and Horns

Perhaps the most spectacular erosional landforms in areas of valley glaciation are at the upper ends of glacial troughs and along the divides separating adjacent glacial troughs. Valley glaciers form and move out from steep-walled, bowl-shaped depressions called **cirques** at the upper end of their troughs (Figure 14.11c). Cirques are typically steep-walled on three sides, but one side opens into a glacial trough (Figure 14.14a).

The details of cirque origin are not fully understood, but these depressions apparently form by erosion of a preexist-ing depression on a mountain side. As snow and ice accumulate in the depression, frost wedging and plucking enlarge it until it takes on the typical cirque shape. Abrasion, plucking, and several mass wasting processes cut deeper into mountain sides by headward erosion so cirques become wider and deeper. Thus, a combination of processes can erode a small mountainside depression into a large cirque; the largest one known is the Walcott Cirque in Antarctica, which is 16 km wide and 3 km deep. Many cirques have a lip or threshold, indicating that the glacial ice not only moves outward but rotates as well, scouring out a depression rimmed by rock.

(a)

(b)

Figure 14.12 *U-shaped glacial troughs. (a) This glacial trough is in the Bighorn Mountains on the Wyoming-Montana border. (b) Geirangerfjorden, a fiord in Norway.*

Figure 14.13 *Bridalveil Falls in Yosemite National Park, California, plunges about 190 m from a hanging valley. The valley in the foreground is a huge U-shaped glacial trough.*

(a)

(b)

Figure 14.14 *(a) This bowl-shaped depression on Mount Wheeler in Great Basin National Park, Nevada, is a cirque. Notice that it has steep walls on three sides but opens out into a glacial trough in the foreground. (b) Many cirques contain small lakes known as tarns such as these on Mount Whitney in California. (c) This knifelike ridge between glacial troughs in Alaska is an arête.*

Courtesy of John S. Shelton

(c)

These depressions commonly contain a small lake known as a *tarn* (Figure 14.14b).

The fact that cirques expand laterally and by headward erosion accounts for the origin of two other distinctive erosional features, *arêtes* and *horns*. **Arêtes**—narrow, serrated ridges—can form in two ways. In many cases, cirques form on opposite sides of a ridge, and headward erosion reduces the ridge until only a thin partition of rock remains (Figure 14.11c). The same effect results when erosion in two parallel glacial troughs reduces the intervening ridge to a thin spine of rock (Figure 14.14c).

The most majestic of all mountain peaks are **horns;** these steep-walled, pyramidal peaks are formed by headward erosion of cirques. In order for a horn to form, a mountain peak must have at least three cirques on its flanks, all of which erode headward (Figure 14.11c). Excellent examples of horns include Mount Assiniboine in the Canadian Rockies, the Grant Teton in Wyoming (see Chapter 10 Prologue), and the most famous of all, the Matterhorn in Switzerland (Figure 14.15).

Continental Glaciers and Erosional Landforms

Areas eroded by continental glaciers tend to be smooth and rounded because these glaciers bevel and abrade high areas that projected into the ice. Rather than yielding the sharp, angular landforms typical of valley glaciation, they produce a landscape of rather flat topography interrupted by rounded hills. These areas are characterized by deranged drainage (see Figure 12.24e), numerous lakes and swamps, low relief, extensive bedrock exposures, and little or no soil. They are generally referred to as *ice-scoured plains* (Figure 14.16).

In a large part of Canada, particularly the vast Canadian Shield region, continental glaciation has stripped off the soil and unconsolidated surface sediment, revealing extensive exposures of striated and polished bedrock. Similar though smaller bedrock exposures are also widespread in the northern United States from Maine through Minnesota.

Figure 14.15 *The Matterhorn in Switzerland is a well-known horn.*

Figure 14.16 *An ice-scoured plain in the Northwest Territories of Canada.*

Glacial Deposits

Both valley and continental glaciers effectively erode and transport, but eventually they deposit their sediment load as **glacial drift,** a general term for all deposits resulting from glacial activity. A vast sheet of Pleistocene glacial drift is present in the northern tier of the United States and adjacent parts of Canada. Smaller but similar deposits are also found where valley glaciers existed or remain active. The appearance of these deposits may not be as inspiring as are some landforms resulting from glacial erosion, but they are important as reservoirs of groundwater and in many areas they are exploited for their sand and gravel. As a matter of fact, glacial sand and gravel constitute a large part of the mineral extraction economies of several states and provinces.

All glacial drift has been carried far from its source area, but one conspicuous element of drift is rock fragments and boulders scattered around that were obviously not derived from the area in which they now rest. These stones, known as **erratics,** were eroded and transported from some distant source and then deposited (Figure 14.17). A good example is the popular decorative stone in Michigan known as puddingstone that consists of quartzite containing conspicuous pieces of red jasper that was eroded from surface exposures in Ontario, Canada.

Geologists generally define two types of glacial drift; till and stratified drift. **Till** consists of sediments deposited directly by glacial ice. These deposits are not sorted by particle size or density, and they exhibit no layering or stratification. The till of both valley and continental glaciers is similar, but that of continental glaciers is much more extensive and generally has been transported farther.

As opposed to till, **stratified drift** is layered (stratified) and invariably exhibits some degree of sorting. As a matter of fact, most of the deposits designated stratified drift are actually layers of sand and gravel or mixtures thereof that accumulated in braided stream channels. In Chapter 12 we mentioned that streams issuing from melting glaciers are commonly braided because they receive more sediment than they can effectively transport.

Landforms Composed of Till

Landforms composed of till include several types of *moraines* and elongated hills known as *drumlins.*

End Moraines

The terminus of either a valley or a continental glacier may become stabilized in one position for some period of time, perhaps a few years or even decades. Stabilization of the ice front does not mean that the glacier has ceased flowing, only that it has a balanced budget. When an ice front is stationary, the glacier continues to flow, and the sediment transported within or upon the ice is dumped as a pile of rubble at the glacier's terminus. These deposits are **end moraines,** which continue to grow as long as the ice front remains stabilized (Figure 14.18). End moraines of valley glaciers are commonly crescent-shaped ridges of till spanning the valley occupied by the

Figure 14.17 *A glacial erratic in Montana.*

glacier. Those of continental glaciers similarly parallel the ice front, but are much more extensive.

Following a period of stabilization, a glacier may advance or retreat, depending on changes in its budget. If it advances, the ice front overrides and modifies its former moraine. Should it have a negative budget, though, the ice front retreats toward the zone of accumulation. As the ice front recedes, till is deposited as it is liberated from the melting ice and forms a layer of **ground moraine** (Figure 14.18b). Ground moraine has an irregular, rolling topography, whereas end moraine consists of long ridgelike accumulations of sediment.

After a glacier has retreated for some time, its terminus may once again stabilize, and it deposits another end moraine. Because the ice front has receded, such moraines are called **recessional moraines** (Figure 14.18c). During the Pleistocene, continental glaciers in the mid-continent region extended as far south as southern Ohio, Indiana, and Illinois. Their outermost end moraines, marking the greatest extent of the glaciers, go by the special name **terminal moraine** (valley glaciers also deposit terminal moraines) (Figure 14.19). As the glaciers retreated from the positions where their terminal moraines were deposited, they temporarily ceased retreating numerous times and deposited dozens of recessional moraines.

Lateral and Medial Moraines

As we previously discussed, valley glaciers transport considerable sediment along their margins. Much of this sediment is abraided and plucked from the valley walls, but a significant amount falls or slides onto the glacier's surface by mass wasting processes. In any case, when a glacier melts, this sediment is deposited as long ridges of till called **lateral moraines** along the margin of the glacier (Figure 14.20).

Where two lateral moraines merge, as when a tributary glacier flows into a larger glacier, a **medial moraine** forms (Figure 14.20). Although medial moraines are identified by their position on a valley glacier, they are, in fact, formed from the coalescence of two lateral moraines. One can generally determine how many tributaries a valley glacier has by the number of its medial moraines.

Drumlins

In many areas where continental glaciers deposited till, the till has been reshaped into elongated hills known as **drumlins.** Some drumlins measure as much as 50 m high and 1 km long, but most are much smaller. From the side, a drumlin looks like an inverted spoon with the steep end on the side from which the glacial ice

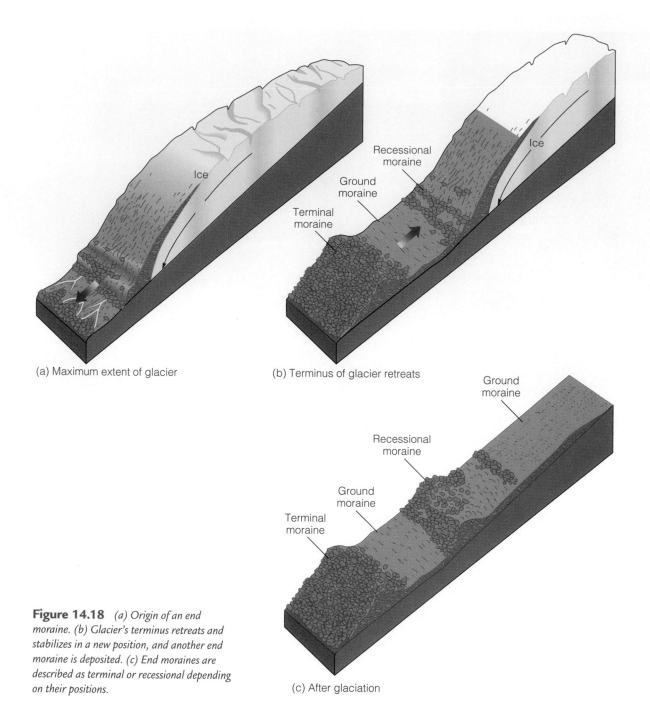

(a) Maximum extent of glacier

(b) Terminus of glacier retreats

Recessional
moraine

Ground
moraine

Terminal
moraine

Ice

Ground
moraine

Recessional
moraine

Ground
moraine

Terminal
moraine

(c) After glaciation

Figure 14.18 *(a) Origin of an end moraine. (b) Glacier's terminus retreats and stabilizes in a new position, and another end moraine is deposited. (c) End moraines are described as terminal or recessional depending on their positions.*

advanced, and the gently sloping end pointing in the direction of ice movement (Figure 14.21). Thus, drumlins are used to determine the direction of ice movement.

One hypothesis for the origin of drumlins holds that they form in the zone of plastic flow as glacial ice modifies till into streamlined hills. According to another hypothesis, drumlins form when huge floods of glacial meltwater modify deposits of till.

Drumlins rarely occur as single, isolated hills; instead they are usually found in *drumlin fields* in which hundreds

or thousands of drumlins are present. Drumlin fields are found in several states and Ontario, Canada, but perhaps the finest example is near Palmyra, New York.

Landforms Composed of Stratified Drift

Stratified drift is found in areas of both valley and continental glaciation, but as one would expect, it is more extensive in areas of continental glaciation.

End moraine

(a)

(b)

Figure 14.19 *(a) An end moraine deposited by a valley glacier. This particular end moraine is also a terminal moraine because it is the one most distant from the glacier's source. (b) Close-up of an end moraine. Notice that the deposit is not sorted by particle size, and it shows no layering or stratification.*

Outwash Plains and Valley Trains

Glaciers discharge meltwater laden with sediment most of the time, except perhaps during the coldest months. This meltwater forms a series of braided streams that radiate out from the front of continental glaciers over a wide region. So much sediment is supplied to these streams that much of it is deposited within the channels as sand and gravel bars. The vast blanket of sediments so formed is an **outwash plain** (Figure 14.22).

Valley glaciers discharge huge amounts of meltwater and, like continental glaciers, have braided streams extending from them. However, these streams are confined to the lower parts of glacial troughs, and their long, narrow deposits of stratified drift are known as **valley trains** (Figure 14.23a).

Outwash plains and valley trains commonly contain numerous circular to oval depressions, many of which contain small lakes. These depressions are *kettles* that form when a retreating ice sheet or valley glacier leaves a block of ice that is subsequently partly or wholly buried (Figure 14.22). When the ice block eventually melts, it leaves a depression; if the depression extends below the water table, it becomes the site of a small lake. Some outwash plains have so many kettles that they are called *pitted outwash plains*.

Kames and Eskers

Kames are conical hills as much as 50 m high composed of stratified drift (Figs. 14.22 and 14.23). Many kames form when a stream deposits sediment in a depression on a glacier's surface; as the ice melts, the deposit is lowered to the surface. They also form in cavities within or beneath stagnant ice.

Long sinuous ridges of stratified drift, many of which meander and have tributaries, are **eskers** (Figure 14.22 and 14.23). Some eskers are quite high, as much as 100 m, and can be traced for more than 100 km. Most eskers are in areas once covered by continental glaciers, but they are also associated with large valley glaciers. The sorting and stratification of the sediments within eskers clearly indicate deposition by running water. The properties of

(a)

(b)

Figure 14.20 *(a) Lateral and medial moraines on a glacier in Alaska. Notice that where the two tributary glaciers converge two lateral moraines merge to form a medial moraine. (b) The two parallel ridges extending from this mountain valley are lateral moraines.*

ancient eskers and observations of present-day glaciers indicate that they form in tunnels beneath stagnant ice (Figure 14.22).

Glacial Lake Deposits

Numerous lakes exist in areas of glaciation. Some formed as a result of glaciers scouring out depressions; others are found where a stream's drainage was blocked; and others are the result of water accumulating behind moraines or in kettles. Regardless of how they formed, glacial lakes, like all lakes, are areas of deposition. Sediment may be carried into them and deposited as small deltas, but of special interest are the fine-grained deposits. Mud deposits in glacial lakes are commonly finely laminated (having layers less than 1 cm thick) and consist of alternating light and dark layers. Each light-dark couplet is a *varve* (Figure 14.24), which represents an annual episode of deposition; the light layer forms during the spring and summer and consists of silt and clay; the dark layer forms

during the winter when the smallest particles of clay and organic matter settle from suspension as the lake freezes over. The number of varves indicates how many years a glacial lake existed.

Another distinctive feature of glacial lakes containing varved deposits is the presence of *dropstones* (Figure 14.24). These are pieces of gravel, some of boulder size, in otherwise very fine-grained deposits. Most of them were probably carried into the lakes by icebergs that eventually melted and released sediment contained in the ice.

What Causes Ice Ages?

How an individual glacier forms is well understood—if more snow falls than melts during the warm season, a net accumulation takes place, the snow gets deeper and deeper and at that depth is converted to glacial ice. And, as previously discussed, flow begins when the critical thickness of about 40 m is reached. So, we know how glaciers form,

Figure 14.21 *These streamlined hills are drumlins. Can you tell the direction of ice movement from their shape?*

something of their dynamics, and how they affect the land surface, but we have not addressed two questions: (1) What causes large-scale episodes of glaciation, and (2) why have there been so few episodes of widespread glaciation? Glaciers were not only much more widespread during the Pleistocene Epoch, but they also expanded and contracted several times. (See Chapter 24, "Cenozoic Earth and Life History," for more on the effects of Pleistocene glaciers.)

Only a few periods of glaciation are recognized in the geologic record, each separated from the others by long intervals of mild climate. Such long-term climatic changes probably result from slow geographic changes related to

plate tectonic activity. Moving plates can carry continents to high latitudes where glaciers can exist, provided that they receive enough precipitation as snow. Plate collisions, the subsequent uplift of vast areas far above sea level, and the changing atmospheric and oceanic circulation patterns caused by the changing shapes and positions of plates also contribute to long-term climatic change.

A theory explaining ice ages must address the fact that during the Pleistocene Ice Age (1.6 million to 10,000 years ago) several intervals of glacial expansion separated by warmer interglacial periods occurred. At least four major episodes of glaciation have been recognized in North

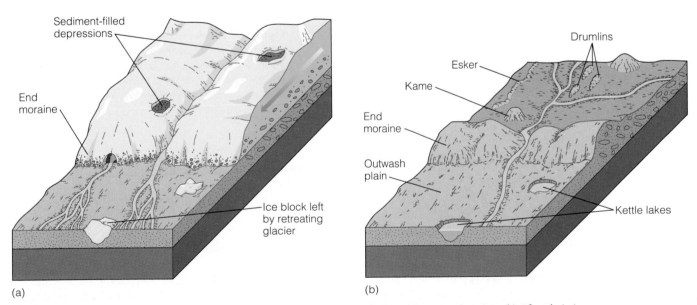

Figure 14.22 *Two stages in the origin of kettles, kames, eskers, and an outwash plain. (a) During glaciation. (b) After glaciation.*

(a)

(b)

(c)

Figure 14.23 *(a) Braided streams discharging from a valley glacier such as this one in the Yukon Territory, Canada, deposit a valley train of stratified drift. An outwash plain of a continental glacier is similar but much more extensive. (b) The sinuous ridge in this image is an esker. (c) This small, conical hill in Wisconsin is a kame.*

America, and six or seven glacial advances and retreats occurred in Europe. These intermediate-term climatic events take place on time scales of tens to hundreds of thousands of years. The cyclic nature of this most recent episode of glaciation has long been a problem in formulating a comprehensive theory of climatic change.

The Milankovitch Theory

A particularly interesting hypothesis for intermediate-term climatic events was put forth by the Yugoslavian as-

tronomer Milutin Milankovitch during the 1920s. He proposed that minor irregularities in Earth's rotation and orbit are sufficient to alter the amount of solar radiation that Earth receives at any given latitude and hence can affect climatic changes. Now called the **Milankovitch theory,** it was initially ignored, but has received renewed interest during the last 20 years.

Milankovitch attributed the onset of the Pleistocene Ice Age to variations in three parameters of Earth's orbit (Figure 14.25). The first of these is orbital eccentricity, which is the degree to which the orbit departs from a perfect circle.

Figure 14.24 *Glacial varves with a dropstone.*

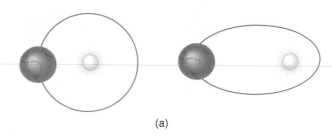

(a)

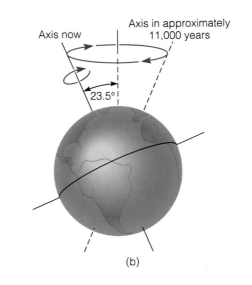

Axis now | Axis in approximately 11,000 years

23.5°

(b)

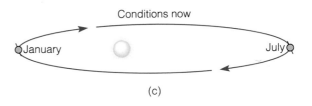

Conditions now

January | July

(c)

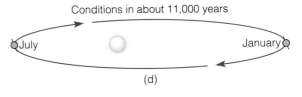

Conditions in about 11,000 years

July | January

(d)

Figure 14.25 *Minor irregularities in Earth's rotation and orbit may affect climatic changes. (a) Earth's orbit varies from nearly a circle (left) to an ellipse (right) and back again in about 100,000 years. (b) Earth moves around its orbit while spinning about its axis, which is tilted to the plane of its orbit around the Sun at 23.5° and points toward the North Star. Earth's axis of rotation slowly moves and traces out the path of a cone in space. (c) At present, Earth is closest to the Sun in January when the Northern Hemisphere experiences winter. (d) In about 11,000 years, as a result of precession, Earth will be closer to the Sun in July, when summer occurs in the Northern Hemisphere.*

Calculations indicate a roughly 100,000-year cycle between times of maximum eccentricity. This corresponds closely to 20 warm-cold climatic cycles that occurred during the Pleistocene. The second parameter is the angle between Earth's axis and a line perpendicular to the plane of its orbit around the Sun. This angle shifts about 1.5° from its current value of 23.5° during a 41,000-year cycle. The third parameter is the precession of the equinoxes, which causes the position of the equinoxes and solstices to shift slowly around Earth's elliptical orbit in a 23,000-year cycle.

Continuous changes in these three parameters cause the amount of solar heat received at any latitude to vary slightly over time. The total heat received by the planet, however, remains little changed. Milankovitch proposed, and now many scientists agree, that the interaction of

You are a geologist working in northern Canada and encounter deposits of poorly sorted sand and gravel showing no layering. In fact, the deposit contains boulders up to 2 m in diameter that obviously were not derived from the local bedrock. You conclude that this is glacial till, but a co-worker correctly points out that similar deposits also form by processes other than glaciation. What kinds of evidence might be present in and beneath the deposit that would convince this skeptic that your interpretation is correct?

these three parameters provides the triggering mechanism for the glacial-interglacial episodes of the Pleistocene.

Short-Term Climatic Events

Climatic events having durations of several centuries, such as the Little Ice Age, are too short to be accounted for by plate tectonics or Milankovitch cycles. Several hypotheses have been proposed, including variations in solar energy and volcanism.

Variations in solar energy could result from changes within the Sun itself or from anything that would reduce the amount of energy Earth receives from the Sun. The latter could result from the solar system passing through clouds of interstellar dust and gas or from substances in the atmosphere reflecting solar radiation back into space. Records kept over the past 75 years, however, indicate that during this time the amount of solar radiation has varied only slightly. Thus, although variations in solar energy may influence short-term climatic events, such a correlation has not been demonstrated.

During large volcanic eruptions, tremendous amounts of ash and gases are spewed into the atmosphere where they reflect incoming solar radiation and thus reduce atmospheric temperatures. Recall from Chapter 4 that small droplets of sulfur gases remain in the atmosphere for years and can have a significant effect on climate. Several large-scale volcanic events have been recorded, such as the 1815 eruption of Tambora, and are known to have had climatic effects. However, no relationship between periods of volcanic activity and periods of glaciation has yet been established.

Chapter Summary

1. Glaciers move by plastic flow and basal slip. They currently cover about 10% of the land surface and contain about 2.15% of all water on Earth.

2. Valley glaciers are confined to mountain valleys and flow from higher to lower elevations, whereas continental glaciers cover vast areas and flow outward in all directions from a zone of accumulation.

3. A glacier forms when winter snowfall in an area exceeds summer melt and therefore accumulates year after year. Snow is compacted and converted to glacial ice, and when the ice is about 40 m thick, pressure causes it to flow.

4. The behavior of a glacier depends on its budget, which is the relationship between accumulation and wastage. If a glacier possesses a balanced budget, its terminus remains stationary; a positive or negative budget results in advance or retreat of the terminus, respectively.

5. Glaciers move at varying rates depending on slope, discharge, and season. Valley glaciers tend to flow more rapidly than continental glaciers.

6. Glaciers are powerful agents of erosion and transport because they are solids in motion. They are particularly effective at eroding soil and unconsolidated sediment, and they can transport any size sediment supplied to them.

7. Continental glaciers transport most of their sediment in the lower part of the ice, whereas valley glaciers may carry sediment in all parts of the ice.

8. Erosion of mountains by valley glaciers yields several sharp, angular landforms including cirques, arêtes, and horns. U-shaped glacial troughs, fiords, and hanging valleys are also products of valley glaciation.

9. Continental glaciers abrade and bevel high areas, producing a smooth, rounded landscape known as an ice-scoured plain.

10. Depositional landforms include moraines, which are ridge-like accumulations of till. Several types of moraines are recognized, including terminal, recessional, lateral, and medial moraines.

11. Drumlins are composed of till that was apparently reshaped into streamlined hills by continental glaciers or floods.

12. Stratified drift in outwash and valley trains consists of sand and gravel deposited by meltwater streams issuing from glaciers. Ridges known as eskers and conical hills called kames are also composed of stratified drift.

13. Major glacial intervals separated by tens or hundreds of millions of years probably occur as a result of the changing positions of plates, which in turn cause changes in oceanic and atmospheric circulation patterns.

14. Currently, the Milankovitch theory is widely accepted as the explanation for glacial-interglacial intervals.

15. The reasons for short-term climatic changes, such as the Little Ice Age, are not understood. Two proposed causes are changes in the amount of solar energy received by Earth and volcanism.

Important Terms

abrasion
arête
basal slip
cirque
continental glacier
drumlin
end moraine
erratic
esker

fiord
firn
glacial budget
glacial drift
glacial ice
glacial polish
glacier
ground moraine
hanging valley

horn
kame
lateral moraine
medial moraine
Milankovitch theory
outwash plain
plastic flow
recessional moraine
stratified drift

terminal moraine
till
U-shaped glacial trough
valley glacier
valley train
zone of accumulation
zone of wastage

Review Questions

1. A pyramid-shaped peak known as a horn forms
   a. _____ by headward erosion of a group of cirques on a mountain peak; b. _____ when two or more valley glaciers merge to form a much larger glacier; c. _____ as till is modified by floods of glacial meltwater; d. _____ when a continental glacier freezes to its underlying surface and ceases moving; e. _____ when the terminus of a glacier recedes and deposits a recessional moraine.

2. The number of medial moraines on a valley glacier generally indicates the number of its
   a. _____ valley trains; b. _____ terminal moraines; c. _____ eskers; d. _____ drumlins; e. _____ tributary glaciers.

3. The two areas that presently have continental glaciers are _____ and _____
   a. _____ Glacier National Park/Waterton National Park; b. _____ Antarctica/Greenland; c. _____ Norway/Sweden; d. _____ Canada/Siberia; e. _____ North Pole/Mount Baker.

4. A glacially transported boulder now resting in an area far from its source is a(n) glacial _____
   a. _____ kame; b. _____ fiord; c. _____ erratic; d. _____ cirque; e._____ esker.

5. A cirque can be described as a(n)
   a. _____ deposit consisting of outwash and till; b. _____ erosional remnant of a mountain; c. _____ deep scratch caused by abrasion; d. _____ bowl-shaped depression at the upper end of a glacial trough; e. _____ type of lake in which dark- and light-colored layers are deposited.

6. Glaciers move mostly by
   a. _____ isostatic rebound; b. _____ plastic flow; c. _____ lateral compression; d. _____ basal slip; e. _____ surging.

7. Crevasses in glaciers extend down to
   a. _____ the bottom of the glacier; b. _____ variable depths depending on ice thickness; c. _____ about 300 m; d. _____ the area of accumulation; e. _____ the zone of plastic flow.

8. If a glacier has a balanced budget,
   a. _____ it stops moving; b. _____ its terminus will remain stationary; c. _____ its rate of wastage exceeds its rate of accumulation; d. _____ the glacier's length decreases; e. _____ crevasses no longer form.

9. A knifelike ridge separating adjacent cirques or glacial troughs is a(n)
   a. _____ horn; b. _____ pitted outwash plain; c. _____ arête; d. _____ medical moraine; e. _____ firn.

10. When freshly fallen snow compacts, partly melts, then re-freezes, it forms granular ice known as
    a. _____ till; b. _____ kame; c. _____ firn; d. _____ outwash; e. _____ drift.

11. The subdued landscape resulting from erosion by a continental glacier is a(n)
    a. _____ drumlin field; b. _____ valley train; c. _____ glacial striation; d. _____ ice-scoured plain; e. _____ zone of glacial abrasion.

12. During the Ice Age or _____ Epoch, glaciers covered about _____% of the land surface.
    a. _____ Cretaceous/75; b. _____ Proterozoic/10; c. _____ Mesozoic/15; d. _____ Pleistocene/30; e. _____ Paleozoic/50.

13. How do the erosional landforms of continental glaciers differ from those of valley glaciers? Give examples of landforms formed by each, and tell where you might see such examples.

14. How does glacial ice originate and why is it considered a rock?

15. Draw a diagram showing how a glacier can have four medial moraines on its surface.

16. Explain in terms of the glacial budget how a once active glacier becomes a stagnant one.

17. What are basal slip and plastic flow, what causes each, and how do they vary seasonally?

18. What is the firn limit on a glacier, and how does its position indicate whether a glacier has a negative, positive, or balanced budget?

19. Explain or diagram how a terminal moraine and a recessional moraine originate.

20. What kinds of evidence would you look for to demonstrate that an ice-free area was once covered by a continental glacier?

21. How does the Milankovitch theory explain the cyclic nature of glaciation during the Pleistocene Epoch?

22. In a roadside rock exposure, you observe a deposit of alternating light and dark laminated mud containing a few large boulders. Explain the sequence of events responsible for deposition.

23. A glacier has a cross-sectional area of 400,000 m^2 and a flow velocity of 2 m/day. How long would it take for a discharge of 1 km^3 to occur?

24. In North America, valley glaciers are common in Alaska and western Canada, and small ones are present in the mountains of the western United States; none, however, occur east of the Rocky Mountains. How can you explain this distribution of glaciers?

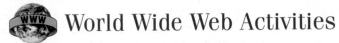

World Wide Web Activities

For these web site addresses, along with current updates and exercises, log on to **http://www.brookscole.com/geo**

Illinois State Museum

This site, maintained by the Illinois State Museum, contains information about programs, exhibits, collections, and a calendar of events at the museum.

Ice Ages and Glaciation

This site is maintained by the Department of Geological and Environmental Sciences at Hartwick College in New York. Take a virtual field trip through the Ice Age, and see maps and images of glacial features.

Wisconsin's Glacial Landscape

The Ice Age Park and Trail Foundation maintains this site, which has information about the Ice Age in Wisconsin along with maps and illustrations. When did the most recent continental glacier enter Wisconsin, how long was it present, and how much of the state did it cover? What is the Driftless Area of Wisconsin?

Glacial Lakes

The U.S. Geological Survey maintains this site, which is devoted largely to Glacial Lake Missoula. Click *Glacial Lake Missoula*, then click *Description: Glacial Lake Missoula and the Missoula Floods.* How many floods might have taken place? What caused the catastrophic floods of Glacial Lake Missoula?

Grand Teton National Park

The Grand Teton National Park, Wyoming, site is maintained by the National Park Service. At the park's home page click *geology*, and then scroll down and click *Journey through the Past: A Geology Tour.* Next go to the section on *glaciation*. What evidence indicates that glaciers were once much more widespread in the park than at present? What depositional and erosional glacial landforms can be observed in the park? When was the onset of widespread glaciation in this area, and when did the last extensive glaciers begin to melt?

The Work of Wind and Deserts

View looking along the ridge of a linear dune in the Rub'al-Khali in Saudi Arabia. The Rub'al-Khali, covering some 570,000 m², is the largest area of continuous sand in the world.

D uring the last few decades, deserts have been advancing across millions of acres of productive land, destroying rangelands, croplands, and even villages (Figure 15.1). Such expansion, estimated at 70,000 km^2 per year, has exacted a terrible toll in human suffering. Because of the relentless advance of deserts, hundreds of thousands of people have died of starvation or been forced to migrate as "environmental refugees" from their homelands to camps where the majority are severely malnourished. This expansion of deserts into formerly productive lands is **desertification.**

Most regions undergoing desertification lie along the margins of existing deserts. These margins have a delicately balanced ecosystem that serves as a buffer between the desert on one side and a more humid environment on the other. Their potential to adjust to increasing environmental pressures from natural causes or human activity is limited. Ordinarily, desert regions expand and contract gradually in response to natural processes such as climatic change, but much recent desertification has been greatly accelerated by human activities. In many areas, the natural vegetation has been cleared as crop cultivation has expanded into increasingly drier fringes to support the growing population. Because these areas are especially prone to droughts, crop failures are common occurrences, leaving the land bare and susceptible to increased wind and water erosion.

Because grasses constitute the dominant natural vegetation in most fringe areas, raising livestock is a common economic activity. Usually, these areas achieve a natural balance between vegetation and livestock as nomadic herders graze their animals on the available grasses. In many fringe areas, livestock numbers have been greatly increasing in recent years and now far exceed the land's capacity to

support them. As a result, the vegetation cover that protects the soil has diminished, causing the soil to crumble. This leads to further drying of the soil and accelerated soil erosion by wind and water.

Drilling water wells also contributes to desertification because human and livestock activity around a well site strips away the vegetation. With its vegetation gone, the topsoil blows away, and the resultant bare areas merge with the surrounding desert. In addition, the water used for irrigation from these wells sometimes contributes to desertification by increasing the salt content of the soil. As the water evaporates, a small amount of salt is deposited in the soil and is not flushed out as it would be in an area that receives more rain. Over time, the salt concentration becomes so high that plants can no longer grow. Desertification resulting from soil salinization is a major problem in North Africa, the Middle East, southwest Asia, and the western United States.

Collecting firewood for heating and cooking is another major cause of desertification, particularly in many less-developed countries where wood is the major fuel source. In the Sahel of Africa (a belt 300 to 1100 km wide that lies south of the Sahara), the expanding population has completely removed all trees and shrubs in the areas surrounding many towns and cities. Journeys of several days on foot to collect firewood are common there. Furthermore, the use of dried animal dung to supplement firewood has exacerbated desertification because important nutrients in the dung are not returned to the soil.

The Sahel averages between 10 and 60 cm of rainfall per year, 90% of which evaporates when it falls. Because drought is common in the Sahel, the region can support only a limited population of livestock and humans. Traditionally, herders and livestock existed in a natural balance with the vegetation,

At the end of this chapter, you will have learned that

■ Wind transports sediment and modifies the landscape through the processes of abrasion and deflation.

■ Dunes and loess are the result of deposition of material by wind.

■ Dunes form when wind flows over and around an obstruction.

■ There are four major dune types: barchan, longitudinal, transverse, and parabolic.

■ Loess is formed from windblown silt and clay and is derived from three main sources: deserts, Pleistocene glacial outwash deposits, and floodplains of rivers in semiarid regions.

■ The global pattern of air pressure belts and winds is responsible for Earth's atmospheric circulation patterns.

■ Deserts are dry; they receive less than 25 cm of rain per year, have high evaporation rates, typically have poorly developed soils, and are mostly or completely devoid of vegetation.

■ The majority of deserts are found in the dry climates of the low and middle latitudes.

■ Deserts have many distinctive landforms and these are produced by both wind and running water.

following the rains north during the rainy season and returning south to greener rangeland during the dry seasons. Some areas were alternately planted and left fallow to help regenerate the soil. During fallow periods, livestock fed off the stubble of the previous year's planting, and their dung helped fertilize the soil.

With the emergence of new nations and increased foreign aid to the Sahel during the 1950s and 1960s, nomads and their herds were restricted, and large areas of grazing land were converted to cash crops such as peanuts and cotton that have a short growing season. Expanding human and animal populations and more intensive agriculture increased the demands on the land. These factors, combined with the worst drought of the century (1968–1973), brought untold misery to the people of the Sahel. Without rains, the crops failed and the livestock denuded the land of what little vegetation remained. As a result, nearly 250,000 people and 3.5 million cattle died of starvation, and the adjacent Sahara expanded southward as much as 150 km.

The tragedy of the Sahel and prolonged droughts in other desert fringe areas serve to remind us of the delicate equilibrium of ecosystems in such regions. Once the fragile soil cover has been removed by erosion, it will take centuries for new soil to form (see Chapter 5).

Figure 15.1 *The Saharan community of El Gedida in western Egypt is slowly being overwhelmed by advancing sand.*

Introduction

Most people associate the work of wind with deserts. Wind is an effective geologic agent in desert regions, but it also plays an important role wherever loose sediment can be eroded, transported, and deposited, such as along shorelines or the plains (see Chapter 5). Therefore, we will examine the work of wind in general, as both an erosional and depositional geologic agent, and then discuss the distribution, characteristics, and landforms of deserts in particular.

Why Should You Study Deserts?

Why should you study deserts and why is it important to understand the processes under which they form, particularly because they tend to be inhospitable and generally uninhabited? One reason is that deserts cover large regions of Earth's surface. More than 40% of Australia is desert, and the Sahara occupies a vast part of northern Africa. While sparsely populated, desert regions are home to many peoples and cultures.

You have been asked to testify before a congressional committee charged with determining whether the National Science Foundation should continue to fund research devoted to the study of climate changes during the Cenozoic Era. Your specialty is the formation of deserts and desert landforms. What arguments would you make to convince the committee to continue funding research concerned with ancient climates?

Figure 15.2 *Most sand is moved near the ground surface by saltation. Sand grains are picked up by the wind and carried a short distance before falling back to the ground where they usually hit other grains, causing them to bounce and move in the direction of the wind.*

Furthermore, with the current debate about global warming, it is important to understand how desert processes operate and how global climate changes affect the various Earth systems and subsystems. By understanding how desertification operates, we can take steps to eliminate or reduce the destruction done, particularly in terms of human suffering. Understanding the underlying causes of climate change by examining ancient desert regions may provide insight into the possible duration and severity of present and future climatic changes.

More than 6000 years ago, the Sahara was a fertile savanna supporting a diverse fauna and flora, including humans. Then the climate changed and the area became a desert. How did this happen and will this region change back again in the future? These are some of the questions geoscientists hope to answer by studying deserts.

And finally, many of the agents and processes that have shaped deserts, do not appear to be limited to our planet. Many of the features found on Mars are apparently the result of the same wind-driven processes operating here on Earth (see Chapter 19).

How Does Wind Transport Sediment?

Wind is a turbulent fluid and therefore transports sediment in much the same way as running water. Although wind typically flows at a greater velocity than water, it has a lower density and, thus, can carry only clay- and silt-size particles as *suspended load*. Sand and larger particles are moved along the ground as *bed load*.

Bed Load

Sediments too large or heavy to be carried in suspension by water or wind are moved as bed load either by *saltation* or by rolling and sliding. As we discussed in Chapter 12, saltation is the process by which a portion of the bed load moves by intermittent bouncing along a streambed. Saltation also occurs on land. Wind starts sand grains rolling and lifts and carries some grains short distances before they fall back to the surface. As the descending sand grains hit the surface, they strike other grains causing them to bounce along by saltation (Figure 15.2). Wind-tunnel experiments show that once sand grains begin moving, they continue to move, even if the wind drops below the speed necessary to start them moving! This happens because once saltation begins, it sets off a chain reaction of collisions between sand grains that keeps the grains in constant motion.

Saltating sand usually moves near the surface, and even when winds are strong, grains are rarely lifted higher than about a meter. If the winds are very strong, these wind-whipped grains can cause extensive abrasion. A car's paint can be removed by sandblasting in a short time, and its windshield will become completely frosted and translucent from pitting.

Suspended Load

Silt- and clay-sized particles constitute most of a wind's suspended load. Even though these particles are much smaller and lighter than sand-sized particles, wind usually starts the latter moving first. The reason for this phenomenon is that a thin layer of motionless air lies next to the ground where the small silt and clay particles remain undisturbed. The larger sand grains, however, stick up into the turbulent air zone where they can be moved. Unless the stationary air layer is disrupted, the silt and clay particles remain on the ground providing a smooth surface. This phenomenon can be observed on a dirt road on a

Figure 15.3 *A dust storm in Death Valley, California.*

windy day. Unless a vehicle travels over the road, little dust is raised even though it is windy. When a vehicle moves over the road, it breaks the calm boundary layer of air and disturbs the smooth layer of dust, which is picked up by the wind and forms a dust cloud in the vehicle's wake.

In a similar manner, when a sediment layer is disturbed, silt- and clay-sized particles are easily picked up and carried in suspension by the wind, creating clouds of dust or even dust storms (Figure 15.3). Once these fine particles are lifted into the atmosphere, they may be carried thousands of kilometers from their source. For example, large quantities of fine dust from the southwestern United States were blown eastward and fell on New England during the Dust Bowl of the 1930s (see Perspective 5.2, Figure 1b).

How Does Wind Erode?

Even in arid regions, running water is still responsible for most erosional landforms, although stream channels are dry for most of the year. Another powerful agent, however, is wind, which produces many distinctive erosional features and is an extremely efficient sorting agent. Wind erodes material in two ways: abrasion and deflation.

Abrasion

Abrasion involves the impact of saltating sand grains on an object and is analogous to sandblasting. The effects of abrasion are usually minor because sand, the most common

WHAT WOULD YOU DO?

As an expert in desert processes, you have been assigned the job of teaching the first astronaut crew that will explore Mars all about deserts and their landforms. The reason is that many Martian features display evidence of having formed as a result of wind processes, and many landforms are the same as those found in deserts on Earth. Describe how you would teach the astronauts to recognize wind-formed features and where you would take the astronauts in the field to show them the types of landforms they may find on Mars.

agent of abrasion, is rarely carried more than 1 m above the surface. Rather than creating major erosional features, wind abrasion merely modifies existing features by etching, pitting, smoothing, or polishing. Nonetheless, wind abrasion can produce many strange-looking and bizarre-shaped features (Figure 15.4).

Ventifacts are a common product of wind abrasion; these are stones whose surfaces have been polished, pitted, grooved, or faceted by the wind (Figure 15.5). If the wind blows from different directions, or if the stone is moved, the ventifact will have multiple facets. Ventifacts are most common in deserts, yet they can also form wherever stones are exposed to saltating sand grains, as on beaches in humid regions and some outwash plains in New England.

Yardangs are larger features than ventifacts and also result from wind erosion (Figure 15.6). They are elongated and streamlined ridges that look like an overturned ship's hull. Yardangs are typically found grouped in clusters aligned parallel to the prevailing winds. They probably form by differential erosion in which depressions, parallel to the direction of wind, are carved out of a rock body, leaving sharp, elongated ridges. These ridges may then be further modified by wind abrasion into their characteristic shape. Although yardangs are fairly common desert features, interest in them was renewed when images radioed back from Mars showed that they are also widespread features on the Martian surface (see Chapter 20).

Deflation

Another important mechanism of wind erosion is **deflation,** which is the removal of loose surface sediment by the wind. Among the characteristic features of deflation in many arid and semiarid

Figure 15.4 *Wind abrasion has formed these structures by eroding the exposed limestone in Desierto Libico, Egypt.*

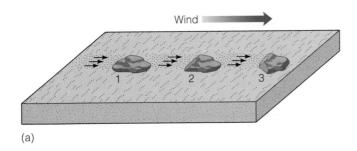

(a)

(b)

Figure 15.5 *(a) A ventifact forms when wind-borne particles (1) abrade the surface of a rock (2) forming a flat surface. If the rock is moved, (3) additional flat surfaces are formed. (b) Large ventifacts lying on desert pavement in Death Valley National Monument, California.*

How Does Wind Erode?

Figure 15.6 *Profile view of a streamlined yardang in the Roman playa deposits of the Kharga Depression, Egypt.*

regions are *deflation hollows* or *blowouts*. These shallow depressions of variable dimensions result from differential erosion of surface materials (Figure 15.7). Ranging in size from several kilometers in diameter and tens of meters deep to small depressions only a few meters wide and less than a meter deep, deflation hollows are common in the southern Great Plains region of the United States.

In many dry regions, the removal of sand-sized and smaller particles by wind leaves a surface of pebbles, cobbles, and boulders. As the wind removes the fine-grained material from the surface, the effects of gravity and occasional floodwaters rearrange the remaining coarse particles into a mosaic of close-fitting rocks called **desert pavement** (Figs. 15.5b and 15.8). Once a desert pavement forms, it protects the underlying material from further deflation.

Figure 15.7 *A deflation hollow in Death Valley, California.*

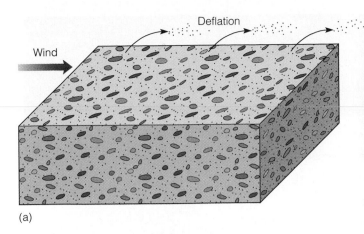

(a)

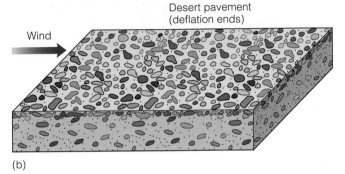

(b)

(c)

Figure 15.8 *Deflation and the origin of desert pavement. (a) Fine-grained material is removed by wind, (b) leaving a concentration of larger particles that form desert pavement. (c) Desert pavement in the Mojave Desert, California. Several ventifacts can be seen in the lower left of the photo.*

What Are the Different Types of Wind Deposits?

Although wind is of minor importance as an erosional agent, it is responsible for impressive deposits, which are primarily of two types. The first, dunes, occur in several distinctive types, all of which consist of sand-sized particles that are usually deposited near their source. The second is loess, which consists of layers of windblown silt and clay deposited over large areas downwind and commonly far from their source.

The Formation and Migration of Dunes

The most characteristic features associated with sand-covered regions are **dunes,** which are mounds or ridges of wind-deposited sand (Figure 15.9). Dunes form when wind flows over and around an obstruction, resulting in deposition of sand grains, which accumulate and build up a deposit of sand. As they grow, these sand deposits become self-generating in that they form ever-larger wind barriers that further reduce the wind's velocity, resulting in further sand deposition and growth of the dune.

Most dunes have an asymmetrical profile, with a gentle windward slope and a steeper downwind or leeward slope that is inclined in the direction of the prevailing wind (Figure 15.10a). Sand grains move up the gentle windward slope by saltation and accumulate on the leeward side forming an angle between 30° and 34° from the horizontal, which is the angle of repose of dry sand. When this angle is exceeded by accumulating sand, the slope collapses, and the sand slides down the leeward slope, coming to rest at its base. As sand moves from a dune's windward side and periodically slides down its leeward slope, the dune slowly migrates in the direction of the prevailing wind (Figure 15.10b). When preserved in the geologic record, dunes help geologists determine the prevailing direction of ancient winds (Figure 15.11).

Dune Types

Four major dune types are generally recognized (barchan, longitudinal, transverse, and parabolic), although intermediate forms between the major types also exist. The size, shape, and arrangement of dunes result from the interaction of such factors as sand supply, the direction and velocity of the prevailing wind, and the amount of vegetation. While dunes are usually found in deserts, they can also develop wherever there is an abundance of sand such as along the upper parts of many beaches.

Figure 15.9 *Large sand dunes in Death Valley, California. Well-developed ripple marks can be seen on the surface of the dunes. The prevailing wind direction is from left to right.*

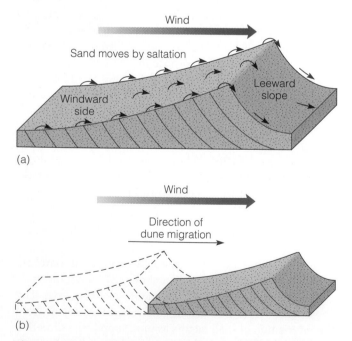

Wind

Sand moves by saltation

Leeward slope

Windward side

(a)

Wind

Direction of dune migration

(b)

Figure 15.10 *(a) Profile view of a sand dune. (b) Dunes migrate when sand moves up the windward side and slides down the leeward slope. Such movement of the sand grains produces a series of cross-beds that slope in the direction of wind movement.*

Figure 15.11 *Cross-bedding in this sandstone in Zion National Park, Utah, helps geologists determine the prevailing direction of the wind that formed these ancient sand dunes.*

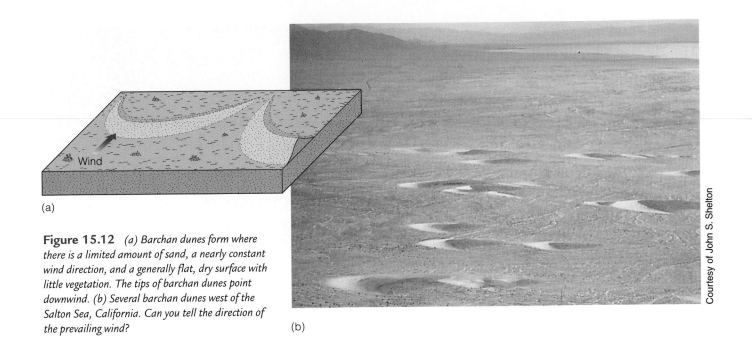

(a)

(b)

Figure 15.12 *(a) Barchan dunes form where there is a limited amount of sand, a nearly constant wind direction, and a generally flat, dry surface with little vegetation. The tips of barchan dunes point downwind. (b) Several barchan dunes west of the Salton Sea, California. Can you tell the direction of the prevailing wind?*

Courtesy of John S. Shelton

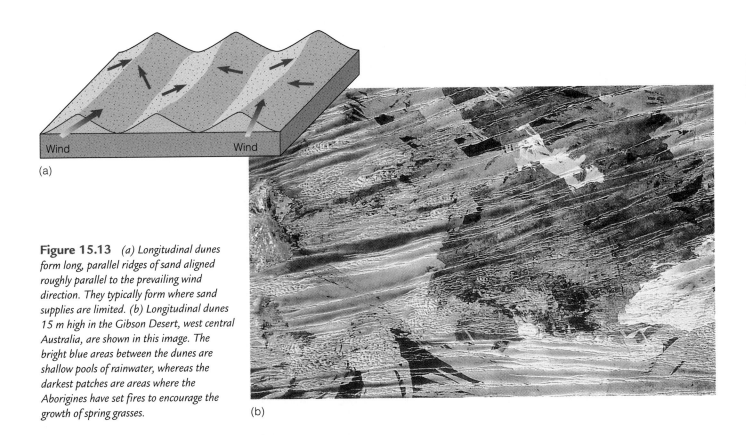

(a)

(b)

Figure 15.13 *(a) Longitudinal dunes form long, parallel ridges of sand aligned roughly parallel to the prevailing wind direction. They typically form where sand supplies are limited. (b) Longitudinal dunes 15 m high in the Gibson Desert, west central Australia, are shown in this image. The bright blue areas between the dunes are shallow pools of rainwater, whereas the darkest patches are areas where the Aborigines have set fires to encourage the growth of spring grasses.*

Barchan dunes are crescent-shaped dunes whose tips point downwind (Figure 15.12). They form in areas having a generally flat, dry surface with little vegetation, a limited supply of sand, and a nearly constant wind direction. Most barchans are small, with the largest reaching about 30 m high. Barchans are the most mobile of the major dune types, moving at rates that can exceed 10 m per year.

Longitudinal dunes (also called *seif dunes*) are long, parallel ridges of sand aligned generally parallel to the direction of the prevailing winds; they form where the sand supply is somewhat limited (Figure 15.13). Longitudinal

What Are the Different Types of Wind Deposits?

(a)

(b)

Figure 15.14 *(a) Transverse dunes form long ridges of sand that are perpendicular to the prevailing wind direction in areas of little or no vegetation and abundant sand. (b) Transverse dunes, Great Sand Dunes National Monument, Colorado. The prevailing wind direction is from lower left to upper right.*

Wind

dunes result when winds converge from slightly different directions to produce the prevailing wind. They range in size from about 3 m to more than 100 m high, and some stretch for more than 100 km. These dunes are especially well developed in central Australia, where they cover nearly one-fourth of the continent. They also cover extensive areas in Saudi Arabia, Egypt, and Iran.

Transverse dunes form long ridges perpendicular to the prevailing wind direction in areas where abundant sand is available and little or no vegetation exists (Figure 15.14). When viewed from the air, transverse dunes have a wavelike appearance and are therefore sometimes called *sand seas.* The crests of transverse dunes can be as high as 200 m, and the dunes may be as much as 3 km wide. Some transverse dunes develop a clearly distinguishable barchan form and may separate into individual barchan dunes along the edges of the dune field where there is less sand. Such intermediate-form dunes are known as *barchanoid dunes* (Figure 15.15).

Parabolic dunes are most common in coastal areas with abundant sand, strong onshore winds, and a partial cover of vegetation (Figure 15.16). Although parabolic dunes have a crescent shape like barchan dunes, their tips point upwind. Parabolic dunes form when the vegetation cover is broken and deflation produces a deflation hollow or blowout. As the wind transports the sand out of the depression, it builds up on the convex downwind dune crest. The central part of the dune is excavated by the wind, while vegetation holds the ends and sides fairly well in place.

Another type of dune commonly found in the deserts of North Africa and Saudi Arabia is the *star dune,* so named because of its resemblance to a multipointed star (Figure 15.17). Star dunes

Figure 15.15 *Barchanoid dunes at White Sands National Monument, New Mexico.*

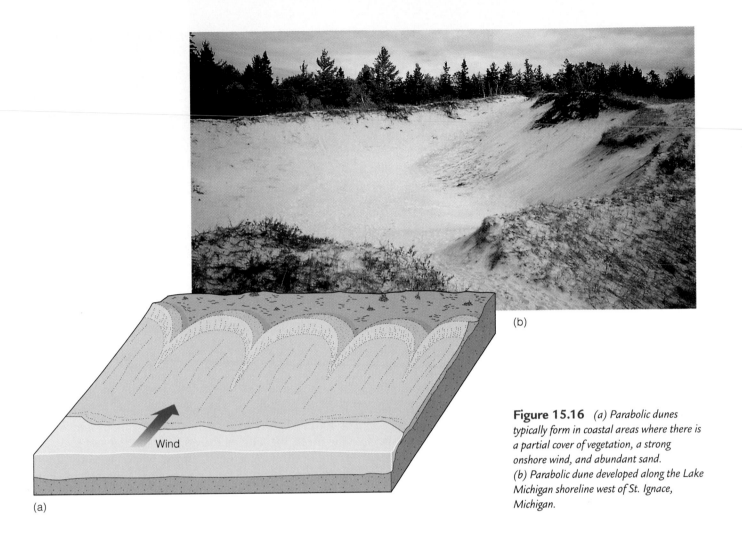

(b)

Figure 15.16 *(a) Parabolic dunes typically form in coastal areas where there is a partial cover of vegetation, a strong onshore wind, and abundant sand. (b) Parabolic dune developed along the Lake Michigan shoreline west of St. Ignace, Michigan.*

(a)

Wind

(a)

Figure 15.17 *(a) Star dunes are pyramidal hills of sand that develop where the wind direction is variable. (b) Ground-level view of star dunes in Namib-Naukluft Park, Namibia.*

(b)

What Are the Different Types of Wind Deposits? **439**

are among the tallest in the world, rising, in some cases, more than 100 m above the surrounding desert plain. They consist of pyramidal hills of sand, from which radiate several ridges of sand, and develop where the wind direction is variable. Star dunes can remain stationary for centuries at a time and have served as desert landmarks for many nomadic peoples.

Loess

Windblown silt and clay deposits composed of angular quartz grains, feldspar, micas, and calcite are known as **loess.** The distribution of loess shows that it is derived from three main sources: deserts, Pleistocene glacial outwash deposits, and the floodplains of rivers in semiarid regions. It must be stabilized by moisture and vegetation to accumulate. Consequently, loess is not found in deserts, even though they provide much of its material. Because of its unconsolidated nature, loess is easily eroded, and as a result, eroded loess areas are characterized by steep cliffs and rapid lateral and headward stream erosion (Figure 15.18).

Figure 15.18   *These steep banks along the Yukon River, Yukon Territory, Canada, are formed of loess.*

At present, loess deposits cover approximately 10% of Earth's land surface and 30% of the United States (Figure 15.19). The most extensive and thickest loess deposits occur in northeast China where accumulations greater than 30 m thick are common. The extensive deserts in central Asia are the source for this loess. Other important loess deposits are on the North European Plain from Bel-

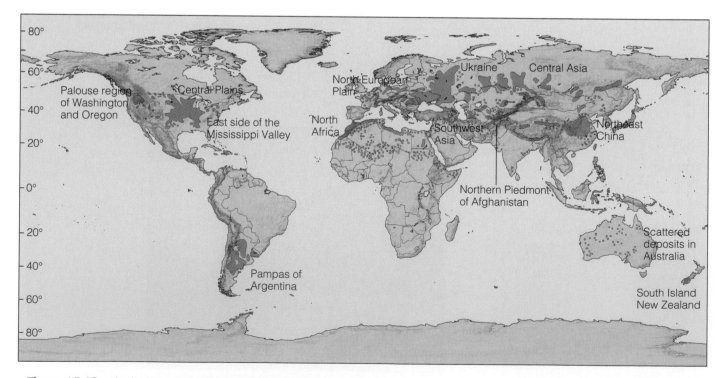

Figure 15.19   *The distribution of Earth's major loess-covered areas.*

gium eastward to Ukraine, in Central Asia, and the Pampas of Argentina. In the United States, they occur in the Great Plains, the Midwest, the Mississippi River Valley, and eastern Washington.

Loess-derived soils are some of the world's most fertile. It is therefore not surprising that the world's major grain-producing regions correspond to the distribution of large loess deposits such as the North European Plain, Ukraine, and the Great Plains of North America.

What Is the Distribution of Air Pressure Belts and Global Wind Patterns?

To understand the work of wind and the distribution of deserts, we need to consider the global pattern of air pressure belts and winds, which are responsible for Earth's atmospheric circulation patterns. Air pressure is the density of air exerted on its surroundings (that is, its weight). When air is heated, it expands and rises, reducing its mass for a given volume and causing a decrease in air pressure. Conversely, when air is cooled, it contracts and air pressure increases. Therefore, those areas of Earth's surface that receive the most solar radiation, such as the equatorial regions, have low air pressure, while the colder areas, such as the polar regions, have high air pressure.

Air flows from high-pressure zones to low-pressure zones. If Earth did not rotate, winds would move in a straight line from one zone to another. Because Earth rotates, however, winds are deflected to the right of their direction of motion (clockwise) in the Northern Hemisphere and to the left of their direction of motion (counterclockwise) in the Southern Hemisphere. Such a deflection of air between latitudinal zones resulting from Earth's rotation is known as the **Coriolis effect.** Therefore, the combination of latitudinal pressure differences and the Coriolis effect produces a worldwide pattern of east-west–oriented wind belts (Figure 15.20).

Earth's equatorial zone receives the most solar energy, which heats the surface air, causing it to rise. As the air rises, it cools and releases moisture that falls as rain in the equatorial region (Figure 15.20). The rising air is now much drier as it moves northward and southward toward each pole. By the time it reaches 20° to 30° north and south latitude, the air has become cooler and denser and begins to descend. Compression of the atmosphere warms the descending air mass and produces a warm, dry, high-pressure area, providing the perfect conditions for the formation of the low-latitude deserts of the Northern and Southern hemispheres (Figure 15.21).

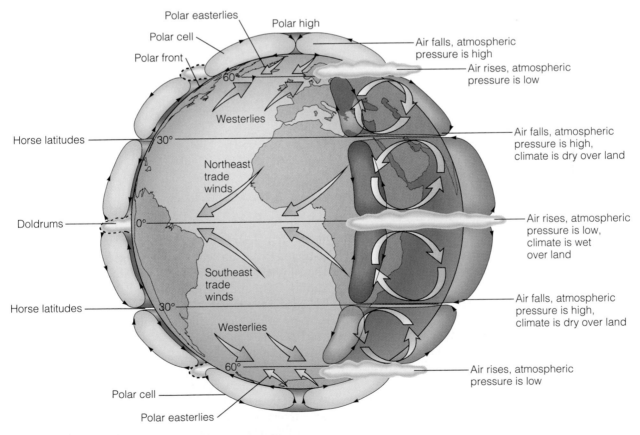

Figure 15.20 *The general circulation of Earth's atmosphere.*

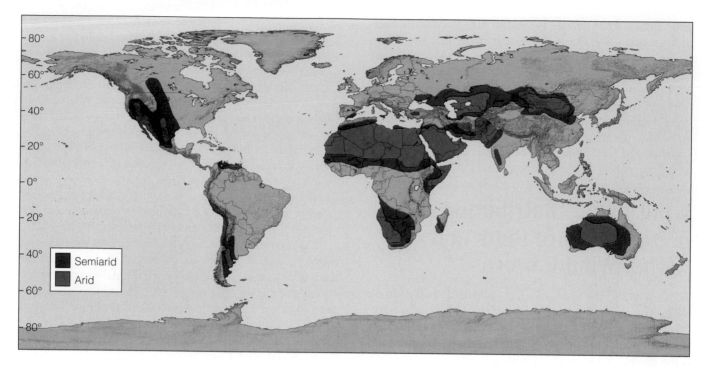

Figure 15.21 *The distribution of Earth's arid and semiarid regions.*

Legend:
- ■ Semiarid
- ■ Arid

Figure 15.22 *Many deserts in the middle and high latitudes are rainshadow deserts, so named because they form on the leeward side of mountain ranges. When moist marine air moving inland meets a mountain range, it is forced upward where it cools and forms clouds that produce rain. This rain falls on the windward side of the mountains. The air descending on the leeward side is much warmer and drier, producing a rainshadow desert.*

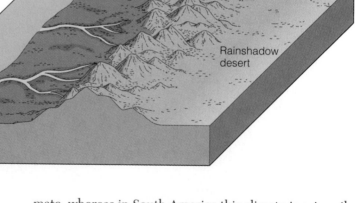

Moist marine air

Warm dry air

Rainshadow desert

Where Do Deserts Occur?

Dry climates occur in the low and middle latitudes where the potential loss of water by evaporation exceeds the yearly precipitation (Figure 15.21). Dry climates cover 30% of Earth's land surface and are subdivided into semiarid and arid regions. *Semiarid regions* receive more precipitation than arid regions, yet are moderately dry. Their soils are usually well developed and fertile and support a natural grass cover. *Arid regions,* generally described as **deserts,** are dry; they receive less than 25 cm of rain per year, have high evaporation rates, typically have poorly developed soils, and are mostly or completely devoid of vegetation.

The majority of the world's deserts are found in the dry climates of the low and middle latitudes (Figure 15.21). In North America, most of the southwestern United States and northern Mexico are characterized by this hot, dry cli-

mate, whereas in South America this climate is primarily restricted to the Atacama Desert of coastal Chile and Peru. The Sahara in northern Africa, the Arabian Desert in the Middle East, and the majority of Pakistan and western India form the largest essentially unbroken desert environment in the Northern Hemisphere. More than 40% of Australia is desert, and most of the rest of it is semiarid.

The remaining dry climates of the world are found in the middle and high latitudes, mostly within continental interiors in the Northern Hemisphere (Figure 15.21). Many of these areas are dry because of their remoteness from moist maritime air and the presence of mountain ranges that produce a **rainshadow desert** (Figure 15.22). When moist marine air moves inland and meets a mountain range, it is forced upward. As it rises, it cools, forming clouds and producing precipitation that falls on the windward side of the mountains. The air that descends on the leeward side of the mountain range is much warmer and drier, producing a rainshadow desert.

Three widely separated areas are included within the mid-latitude dry climate zone (Figure 15.21). The largest of these is the central part of Eurasia extending from just north of the Black Sea eastward to north-central China. The Gobi Desert in China is the largest desert in this region. The Great Basin area of North America is the second largest mid-latitude dry climate zone and results from the rainshadow produced by the Sierra Nevada. This region adjoins the southwestern deserts of the United States that formed as a result of the low-latitude subtropical high-pressure zone. The smallest of the mid-latitude dry climate areas is the Patagonian region of southern and western Argentina. Its dryness results from the rainshadow effect of the Andes. The remainder of the world's deserts are found in the cold but dry high latitudes, such as Antarctica.

What Are the Characteristics of Deserts?

To people who live in humid regions, deserts may seem stark and inhospitable. Instead of a landscape of rolling hills and gentle slopes with an almost continuous vegetation cover, deserts are dry, have little vegetation, and consist of

Figure 15.23 *Desert vegetation is typically sparse, widely spaced, and characterized by slow growth rates. The vegetation shown here in Organ Pipe National Monument, Arizona, includes saguaro and cholla cacti, paloverde trees, and jojoba bushes and is characteristic of the vegetation found in the Sonoran Desert of North America.*

nearly continuous rock exposures, desert pavement, or sand dunes. And yet despite the great contrast between deserts and more humid areas, the same geologic processes are at work, only operating under different climatic conditions.

Temperature, Precipitation, and Vegetation

The heat and dryness of deserts are well known. Many of the deserts of the low latitudes have average summer temperatures ranging between 32° and 38°C. It is not uncommon for some low-elevation inland deserts to record daytime highs of 46° to 50°C for weeks at a time. The highest temperature ever recorded was 58°C in El Azizia, Libya, on September 13, 1922.

During the winter months when the angle of the Sun is lower and there are fewer daylight hours, daytime temperatures average between 10° and 18°C. Winter nighttime lows can be quite cold, with frost and freezing temperatures common in the more poleward deserts. Winter daily temperature fluctuations in low-latitude deserts are among the greatest in the world, ranging between 18° and 35°C. Temperatures have been known to fluctuate from below 0°C to more than 38°C in a single day!

The dryness of the low-latitude deserts results primarily from the year-round dominance of the subtropical high-pressure belt, while the dryness of the mid-latitude deserts is due to their isolation from moist marine winds and the rainshadow effect created by mountain ranges. The dryness of both is further accentuated by their high temperatures.

Although deserts are defined as regions receiving, on average, less than 25 cm of rain per year, the amount of rain that falls each year is unpredictable and unreliable. It is not uncommon for an area to receive more than an entire year's average rainfall in one cloudburst and then to receive little rain for several years. Thus, yearly rainfall averages can be quite misleading.

Deserts display a wide variety of vegetation (Figure 15.23). Although the driest deserts, or those with large areas of shifting sand, are almost devoid of vegetation, most deserts support at least a sparse plant cover. Compared to humid areas, desert vegetation may appear monotonous. A closer examination though, reveals an amazing diversity of plants that have evolved the ability to live in the near-absence of water.

Desert plants are widely spaced, typically small, and slow-growing. Their stems and leaves are usually hard and waxy to minimize water loss by evaporation and to protect the plant from erosion. Most plants have a widespread shallow root system to absorb the dew that forms each morning in all but the driest deserts and to help anchor the plant in what little soil there may be. In extreme cases, many plants lie dormant during particularly dry years and spring to life after the first rain shower with a beautiful profusion of flowers.

Weathering and Soils

Mechanical weathering is dominant in desert regions. Daily temperature fluctuations and frost wedging are the primary forms of mechanical weathering (see Chapter 5). The breakdown of rocks by roots and from salt crystal growth is of minor importance. Some chemical weathering does occur, but its rate is greatly reduced by aridity and the scarcity of organic acids produced by the sparse vegetation. Most chemical weathering takes place during the winter months when more precipitation occurs, particularly in the mid-latitude deserts.

An interesting feature seen in many deserts is a thin, red, brown, or black shiny coating on the surface of many rocks. This coating, called *rock varnish,* is composed of iron and manganese oxides (Figure 15.24). Because many of the varnished rocks contain little or no iron and manganese oxides, the varnish is thought to result from either windblown iron and manganese dust that settles on the ground or from the precipitated waste of microorganisms.

Desert soils, if developed, are usually thin and patchy because the limited rainfall and the resultant scarcity of vegetation reduce the efficiency of chemical weathering and hence soil formation. Furthermore, the sparseness of the vegetative cover enhances wind and water erosion of what little soil actually forms.

Figure 15.24 *The shiny black coating on this rock exposed at Castle Valley, Utah, is rock varnish. It is composed of iron and manganese oxides.*

Mass Wasting, Streams, and Groundwater

When traveling through a desert, most people are impressed by such wind-formed features as moving sand, sand dunes, and sand and dust storms. They may also notice the dry washes and dry streambeds. Because of the lack of running water, most people would conclude that wind is the most important erosional geologic agent in deserts. They would be wrong. Running water, even though it occurs infrequently, causes most of the erosion in deserts. The dry conditions and sparse vegetation characteristic of deserts enhance water erosion. If you look closely, you will see the evidence of erosion and transportation by running water nearly everywhere except in areas covered by sand dunes.

Most of a desert's average annual rainfall of 25 cm or less comes in brief, heavy, localized cloudbursts. During these times, considerable erosion takes place because the ground cannot absorb all of the rainwater. With so little vegetation to hinder its flow, runoff is rapid, especially on moderately to steeply sloping surfaces, resulting in flash floods and sheetflows. Dry stream channels quickly fill with raging torrents of muddy water and mudflows, which carve out steepsided gullies and overflow their banks. During these times, a tremendous amount of sediment is rapidly transported and deposited far downstream.

While water is the major erosive agent in deserts today, recall that it was even more important during the Pleistocene Epoch when these regions were more humid (see Chapter 24). During that time, many of the major topographic features of deserts were forming. Today that topography is being modified by wind and infrequently flowing streams.

Most desert streams are poorly integrated and flow only intermittently. Many of them never reach the sea because the water table is usually far deeper than the channels of most streams, so they cannot draw upon groundwater to replace water lost to evaporation and absorption into the ground. This type of drainage in which a stream's load is deposited within the desert is called *internal drainage* and is common in most arid regions.

While the majority of deserts have internal drainage, some deserts have permanent through-flowing streams such as the Nile and Niger rivers in Africa, the Rio Grande and Colorado River in the southwestern United States, and the Indus River in Asia. These streams flow through desert regions because their headwaters are well outside the desert and water is plentiful enough to offset losses resulting from evaporation and infiltration. Demands for greater amounts of water for agriculture and domestic use from the Colorado River, however, are leading to increased salt concentrations in its lower reaches and causing political problems between the United States and Mexico.

The water table in most desert regions is below the stream channels and is only recharged for a short time after a rainfall. In deserts with through-flowing streams, the water table slopes away from the streams. The through-flowing streams help to recharge the groundwater supply and can support vegetation along their banks. Trees, which have high moisture requirements, are rare in deserts, but may occasionally occur along the banks of both ephemeral and permanent streams, where their roots can reach the higher water table.

Wind

Although running water does most of the erosional work in deserts, wind can also be an effective geologic agent capable of producing a variety of distinctive erosional and depositional features (Figure 15.4). It is effective in transporting and depositing unconsolidated sand, silt, and dust-sized particles. Contrary to popular belief, most deserts are not sand-covered wastelands, but rather consist of vast areas of rock exposures and desert pavement. Sand-covered regions, or sandy deserts, constitute less than 25% of the world's deserts. The sand in these areas has accumulated primarily by the action of wind.

What Types of Landforms Are Found in Deserts?

Because of differences in temperature, precipitation, and wind, as well as the underlying rocks and recent tectonic events, landforms in arid regions vary considerably. Although wind is an important geologic agent, many distinctive landforms are produced and modified by running water.

After an infrequent and particularly intense rainstorm, excess water not absorbed by the ground may accumulate in low areas and form *playa lakes* (Figure 15.25a). These lakes are temporary, lasting from a few hours to several months. Most of them are shallow and have rapidly shifting boundaries as water flows in or leaves by evaporation and seepage into the ground. The water is often very saline.

When a playa lake evaporates, the dry lake bed is called a **playa** or *salt pan* and is characterized by mudcracks and precipitated salt crystals (Figure 15.25b). Salts in some playas are thick enough to be mined commercially. For example, borates have been mined in Death Valley, California, for more than a hundred years.

Other common features of deserts, particularly in the Basin and Range region of the United States, are alluvial fans and bajadas. **Alluvial fans** form when sediment-laden streams flowing out from the generally straight, steep mountain fronts deposit their load on the relatively flat desert floor. Once beyond the mountain front where no valley walls contain streams, the sediment spreads out laterally, forming a gently sloping and poorly sorted fan-shaped sedimentary deposit (Figure 15.26). Alluvial fans are similar in origin and shape to deltas (see Chapter 12) but are formed entirely on land. Alluvial fans may coalesce to form a *bajada*, a broad alluvial apron typically with an undulating surface resulting from the overlap of adjacent fans (Figure 15.27).

Large alluvial fans and bajadas are frequently important sources of groundwater for domestic and agricultural use. Their outer portions are typically composed of fine-grained sediments suitable for cultivation, and their gentle slopes allow good drainage of water. Many alluvial fans and bajadas are also the sites of large

(a)

(b)

Figure 15.25 *(a) Playa lake formed after a rainstorm filled Croneis Dry Lake, Mojave Desert, California. (b) Racetrack Playa, Death Valley, California. The Inyo Mountains can be seen in the background.*

Figure 15.26 *Aerial view of an alluvial fan, Death Valley, California.*

Figure 15.27 *Coalescing alluvial fans forming a bajada at the base of the Black Mountains, Death Valley, California.*

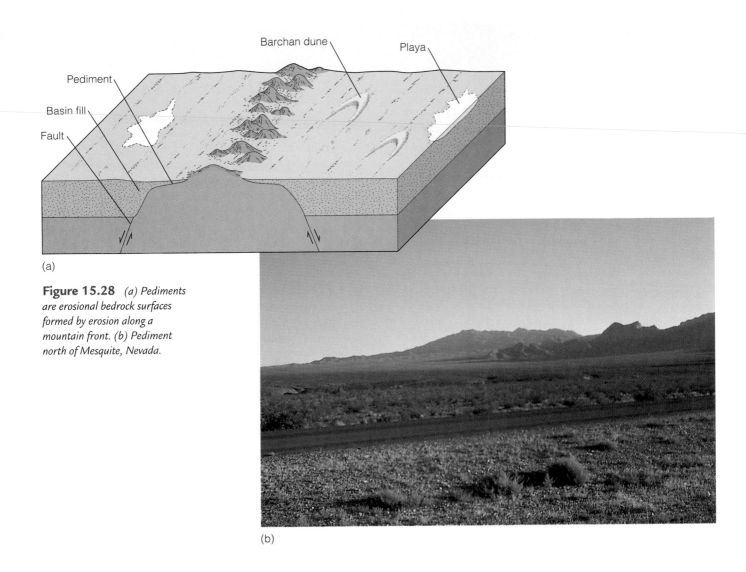

(a)

(b)

Figure 15.28 (a) Pediments are erosional bedrock surfaces formed by erosion along a mountain front. (b) Pediment north of Mesquite, Nevada.

towns and cities, such as San Bernardino, California; Salt Lake City, Utah; and Teheran, Iran.

Most mountains in desert regions, including those of the Basin and Range region, rise abruptly from gently sloping surfaces called pediments. **Pediments** are erosional bedrock surfaces of low relief that slope gently away from mountain bases (Figure 15.28). Most pediments are covered by a thin layer of debris or by alluvial fans or bajadas.

The origin of pediments has been the subject of much controversy. Most geologists agree that they are erosional features developed on bedrock in association with the erosion and retreat of a mountain front (Figure 15.28a). The disagreement concerns how the erosion occurred. While not all geologists would agree, it appears that pediments are produced by the combined activities of lateral erosion by streams, sheet flooding, and various weathering processes along the retreating mountain front. Thus, pediments grow at the expense of the mountains, and they will continue to expand as the mountains are eroded away or partially buried.

Rising conspicuously above the flat plains of many deserts are isolated steep-sided erosional remnants called **inselbergs,** a German word meaning "island mountain" (see Perspective 15.1). Inselbergs have survived for a longer period of time than other mountains because of their greater resistance to weathering.

Other easily recognized erosional remnants common to arid and semiarid regions are mesas and buttes (Figure 15.29). A **mesa** is a broad, flat-topped erosional remnant bounded on all sides by steep slopes. Continued weathering and stream erosion will form isolated pillarlike structures known as **buttes.** Buttes and mesas consist of relatively easily weathered sedimentary rocks capped by nearly horizontal, resistant rocks such as sandstone, limestone, or basalt. They form when the resistant rock layer is breached, allowing rapid erosion of the less resistant underlying sediment. One of the best-known areas of mesas and buttes in the United States is Monument Valley on the Arizona–Utah border.

The Geologic History of Uluru and Kata Tijuta

Rising majestically above the surrounding flat desert of central Australia is Uluru (Ayers Rock) and its neighbor to the northwest, Kata Tijuta (The Olgas) (Figure 1). Uluru and Kata Tijuta are the Aboriginal names for what most people know as Ayers Rock and The Olgas. Uluru was named Ayers Rock in 1873 by the explorer W. C. Gosse, after Sir Henry Ayers, who was at that time premier of South Australia. Kata Tijuta was named Mount Olga in 1872 by Ernest Giles, in honor of the King and Queen of Spain reigning then.

Archaeological evidence indicates there has been human activity in this region of Australia for at least the past 22,000 years, and probably even longer. Both Uluru and Kata Tijuta are an important part of Aboriginal culture and some sites are so sacred that only Aboriginal people can visit them. When the Australian government gave title to Uluru National Park back to the Aboriginal people in 1985, part of the agreement was to replace the European names of features and localities with Aboriginal (Anangu) ones.

The geologic history of Uluru and Kata Tijuta begins between about 900 million and 600 million years ago (Late Pro-

terozoic) when much of central Australia was beneath sea level and was part of the Amadeus Basin. This period of marine deposition was followed by the Petermann Ranges orogeny that took place during the Cambrian, resulting in a large mountain range. As this mountain range eroded, huge alluvial fans formed at its base. The sediments comprising these alluvial fans were eventually lithified into sedimentary rocks that would later become Uluru and Kata Tijuta. The sedimentary rocks of Uluru and Kata Tijuta are coarse-grained and poorly sorted arkoses (Uluru Arkose) and arkosic conglomerates (Mount Currie Conglomerate), indicative of alluvial fan deposition.

Sometime about 500 million years ago, the region was again covered by warm, shallow seas, and the alluvial fan deposits were buried by sandstones, mudstones, and limestones. Another orogeny, the Alice Springs orogeny, uplifted the area about 400 million years ago, resulting in much folding and faulting. Although the beds around Uluru were tilted almost vertically, the deformation in the Kata Tijuta region was not as severe. The area that is now Uluru National Park was raised above sea level during the Alice Springs orogeny and has remained above sea level ever since. Erosion of the area began to shape the uplifted rocks, giving form to what would eventually become Uluru and Kata Tijuta (Figure 2).

About 70 million years ago, a wide valley was eroded between Uluru and Kata Tijuta. The climate during this time was far different from the climate today. Instead of the arid conditions found presently, a rainy and more temperate climate prevailed, with year-round flowing streams and marshes a common feature of the landscape. Within the marshes, plant material accumulated, forming peat layers that later became converted to thin layers of low-grade coal.

By about 500,000 years ago, the climate became arid and has remained arid to the present day. Many of the sand dunes seen in Uluru Park today began forming about 30,000 years ago.

The spectacular and varied rock shapes of Uluru and Kata Tijuta are the result of millions of years of weathering and erosion by water, and to a minor extent, wind acting on the various fractures and joints caused by uplifting of the area. Differences in the composition of the arkoses and conglomerates that make up Uluru and Kata Tijuta has also played a role in the sculpting of these colorful and magnificent structures.

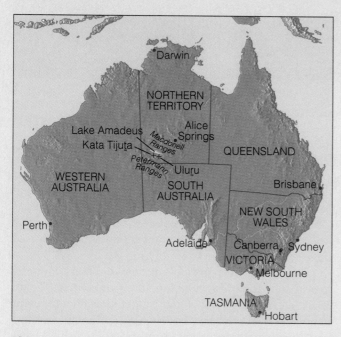

Figure 1 *Location of Uluru and Kata Tijuta in central Australia.*

(a)

(b)

Figure 2 *(a) Uluru. (b) Kata Tijuta.*

What Types of Landforms Are Found in Deserts? **449**

(a)

Figure 15.29 *(a) Spearhead Mesa, Monument Valley Navajo Tribal Park, Arizona/Utah. Other mesas and several buttes can be seen in the background, as well as a dune field in the foreground. (b) Left Mitten Butte and Right Mitten Butte in Monument Valley on the border of Arizona and Utah.*

(b)

Chapter Summary

1. Wind can transport sediment in suspension or as bed load, which involves saltation and surface creep.

2. Wind erodes material either by abrasion or deflation. Abrasion is a near-surface effect caused by the impact of saltating sand grains. Ventifacts are common wind-abraded features.

3. Deflation is the removal of loose surface material by wind. Deflation hollows resulting from differential erosion of surface material are common features of many deserts, as is desert pavement, which effectively protects the underlying surface from additional deflation.

4. The two major wind deposits are dunes and loess. Dunes are mounds or ridges of wind-deposited sand, whereas loess is wind-deposited silt and clay.

5. The four major dune types are barchan, longitudinal, transverse, and parabolic. The amount of sand available, the prevailing wind direction, the wind velocity, and the amount of vegetation present determine which type will form.

6. Loess is derived from deserts, Pleistocene glacial outwash deposits, and river floodplains in semiarid regions. Loess covers approximately 10% of Earth's land surface and weathers to a rich and productive soil.

7. Deserts are very dry (averaging less than 25 cm rain/year), have poorly developed soils, and are mostly or completely devoid of vegetation.

8. The winds of the major east-west–oriented air pressure belts resulting from rising and cooling air are deflected by the Coriolis effect. These belts help control the world's climate.

9. Dry climates are located in the low and middle latitudes where the potential loss of water by evaporation exceeds the yearly precipitation. Dry climates cover 30% of Earth's surface and are subdivided into semiarid and arid regions.

10. The majority of the world's deserts are in the low-latitude dry climate zone between 20° and 30° north and south latitudes. Their dry climate results from a high-pressure belt of descending dry air. The remaining deserts are in the middle latitudes where their distribution is related to the rainshadow effect, and in the dry polar regions.

11. Deserts are characterized by lack of precipitation and high evaporation rates. Furthermore, rainfall is unpredictable and, when it does occur, tends to be intense and of short duration. As a consequence of such aridity, desert vegetation and animals are scarce.

12. Mechanical weathering is the dominant form of weathering in deserts. The sparse precipitation and slow rates of chemical weathering result in poorly developed soils.

13. Running water is the dominant agent of erosion in deserts and was even more important during the Pleistocene Epoch when wetter climates resulted in humid conditions.

14. Wind is an erosional agent in deserts and is effective in transporting and depositing unconsolidated fine-grained sediments.

15. Important desert landforms include playas, which are dry lake beds; when temporarily filled with water, they form playa lakes. Alluvial fans are poorly sorted, fan-shaped sedimentary deposits that may coalesce to form bajadas.

16. Pediments are erosional bedrock surfaces of low relief gently sloping away from mountain bases. The origin of pediments is controversial, although most geologists think that they form by the combined activities of lateral erosion by streams, sheet flooding, and various weathering processes.

17. Inselbergs are isolated steep-sided erosional remnants that rise above the surrounding desert plains. Buttes and mesas are, respectively, pinnacle-like and flat-topped erosional remnants with steep sides.

Important Terms

abrasion	deflation	inselberg	pediment
alluvial fan	desert	loess	playa
barchan dune	desert pavement	longitudinal dune	rainshadow desert
butte	desertification	mesa	transverse dune
Coriolis effect	dune	parabolic dune	ventifact

Review Questions

1. The driest deserts in the world can be found between _____ and _____ latitude in both hemispheres.
   a. _____ 60°, 80°; b. _____ 40°, 60°; c. _____ 30°, 60°; d. _____ 20°, 30°; e. _____ 10°, 20°.

2. Bed load is primarily transported by which process?
   a. _____ saltation; b. _____ suspension; c. _____ precipitation; d. _____ abrasion; e. _____ deflation.

3. A crescent-shaped dune whose tips point downwind is a _____ dune.
   a. _____ star; b. _____ parabolic; c. _____ transverse; d. _____ barchan; e. _____ longitudinal.

4. The Coriolis effect causes wind to be deflected
   a. _____ to the right in the Northern Hemisphere and the left in the Southern Hemisphere; b. _____ to the left in the Northern Hemisphere and the right in the Southern Hemisphere; c. _____ only to the left for both hemispheres; d. _____ only to the right for both hemispheres; e. _____ not at all.

5. Air-borne particles abrading the surface of a rock produce these structures:
   a. _____ loess; b. _____ playas; c. _____ ventifacts; d. _____ dunes; e. _____ inselbergs.

6. Which of the following dunes are among the tallest in the world, consist of pyramidal hills of sand, and develop where the wind direction is variable?
   a. _____ barchan; b. _____ transverse; c. _____ parabolic; d. _____ star; e. _____ longitudinal.

7. Most of a wind's suspended load consists of which particle size?
   a. _____ sand; b. _____ silt; c. _____ clay; d. _____ answers (a) and (b); e. _____ answers (b) and (c).

8. In what area of the world are the thickest and most extensive loess deposits found?
   a. _____ China; b. _____ Ukraine; c. _____ Belgium; d. _____ Argentina; e. _____ United States.

9. Coalescing alluvial fans form.
   a. _____ buttes; b. _____ mesas; c. _____ playas; d. _____ inselbergs; e. _____ bajadas.

10. The major agent of erosion in deserts today is
    a. _____ wind; b. _____ running water; c. _____ abrasion; d. _____ glaciers; e. _____ none of these.

11. The primary cause of dryness in low-latitude deserts is
    a. _____ isolation from moist marine winds; b. _____ dominance of the subtropical high-pressure belt; c. _____ the Coriolis effect; d. _____ the rainshadow effect; e. _____ all of these.

12. Deserts
    a. _____ can be found in the low, middle, and high latitudes; b. _____ receive more than 25 cm of rain per year; c. _____ are mostly or completely devoid of vegetation; d. _____ answers (a) and (c); e. _____ answers (b) and (c).

13. Because much of the recent desertification has been greatly accelerated by human activities, can anything be done to slow the process?

14. Why are low latitude deserts so common?

15. Why are so many desert rock formations red in color?

16. Why is desert pavement important in a desert environment?

17. Is it possible to get the same types of sand dunes on Mars as on Earth? What does that tell us about the climate and geology of Mars?

18. How do dunes form and migrate? Why is dune migration a problem in some areas?

19. If deserts are dry regions in which mechanical weathering predominates, why are so many of their distinctive landforms the result of running water and not wind?

20. Considering what you now know about deserts, their location, how they form, and the various landforms associated with them, how can you use this information to determine where deserts may have existed in the past?

21. What is loess and why is it important?

22. How are temperature, precipitation, and vegetation interrelated in desert environments?

23. How do arid and semiarid regions differ from each other?

24. How are pediments formed?

25. As noted in the text, some large cities are built on alluvial fans. What are the advantages and disadvantages of such an arrangement?

 World Wide Web Activities

For these web site addresses, along with current updates and exercises, log on to **http://www.brookscole.com/geo**

USGS Deserts: Geology and Resources Online Edition

This web site is the online edition of the United States Geological Survey publication *Geology and Resources* by A. S. Walker. This publication discusses such topics as "What Is a Desert?" "Desert Features," "Mineral Resources in a Desert," and "Desertification" to name a few. In addition to the text, it has excellent diagrams and photographs.

Desertusa

This web site, devoted to information about deserts, contains a tremendous amount of information about the fauna, flora, culture, geology, physical environment, and other items of interest about deserts. You can also find out about the various deserts in the United States. This is an excellent site for anyone interested in deserts and visiting them.

U.S. Geological Survey National Park Service—Geology of Mojave National Preserve

This web site covers the geology and 2.7 billion year geologic history of Mojave National Preserve. This web site also has a field trip in which you can visit some of Mojave's most intriguing and interesting sites such as Kelso Dunes and Hole in the Wall.

 CD-ROM Exploration

Exploring your *Earth Systems Today* CD-ROM will add to your understanding of the material in this chapter.

Topic: Atmospheric Processes

Module: Weather and Climate

Explore activities in this module to see if you can discover the following for yourself:

How is Earth's weather directly influenced by the temperature of its oceans? What are El Niño and La Niña in the Pacific Ocean?

What key chemical compound in Earth's atmosphere limits the amount of ultraviolet light reaching the surface? What happens when that key compound is depleted over part of the Earth? Why does this depletion happen?

What are the relations between wind and pressure patterns, clouds and humidity, and precipitation and pressure fronts?

CHAPTER 16

Shorelines and Shoreline Processes

View of the Pacific Coast at Point Reyes National Seashore, California.

PROLOGUE

OBJECTIVES

At the end of this chapter, you will have learned that

■ Wind-generated waves and their associated nearshore currents effectively modify shorelines by erosion and deposition.

■ The gravitational attraction by the Moon and Sun and Earth's rotation are responsible for the rhythmic daily rise and fall of sea level known as tides.

■ Seacoasts and lakeshores are both modified by waves and nearshore currents, but seacoasts also experience tides that are insignificant in even the largest lakes.

■ The concept of a nearshore sediment budget considers equilibrium, losses, and gains in the amount of sediment in a coastal area.

■ Several distinctive erosional and depositional landforms such as wave-cut platforms, spits, and barrier islands are found along shorelines.

■ Shoreline management is complicated by rising sea level, and structures originally far from the shoreline are now threatened or have already been destroyed.

■ Several types of coasts are recognized based on criteria such as deposition and erosion and fluctuations in sea level.

Long, narrow islands lying a short distance offshore are found worldwide, but are particularly common along the United States' Atlantic coast from New York southward and along the Gulf coast of Texas. These *barrier islands* are the sites of many communities, beachfront homes, and recreation areas. Padre Island in Texas, for instance, is a favorite destination for many college students on spring break. Many barrier islands are present in the areas noted above, but here we are concerned only with Galveston Island, Texas, and the Outer Banks of North Carolina.

During hurricanes, strong winds drive large waves onshore, and as the storm moves over the ocean, low atmospheric pressure beneath the eye of the storm causes the ocean surface to bulge upward by as much as 0.5 m. When the eye reaches the shoreline, the bulge along with wind-driven waves piles up in a *storm surge* that might rise a few meters above normal high tide and inundate areas several kilometers inland. In fact, coastal flooding during these vast storms is responsible for much of the property damage and about 90% of all fatalities. And nowhere is more susceptible to flooding during hurricanes than barrier islands.

In 1900, Galveston, Texas, a city of about 38,000 on Galveston Island, was nearly destroyed when a hurricane struck and waves surged inland, eventually covering the entire island. Buildings and other structures closest to the shoreline were battered to pieces, and "great beams and railway ties were lifted by the [waves] and driven like battering rams into dwellings and business houses."* Finally, after the first four shoreline blocks were destroyed, the wreckage piled up high

*L. W. Bates, Jr., "Galveston—A City Built upon Sand," *Scientific American* 95 (1906): 64.

enough to form a protective barrier for the rest of the city. Unfortunately, between 6000 and 8000 of Galveston's residents were drowned and most of the city was reduced to ruins.

To protect the city from future flooding, a colossal two-part project was undertaken beginning in 1902. First, a massive seawall was constructed to protect the city from the direct impact of waves (Figure 16.1a). And next, because the highest part of the island was only 2.7 m above mean low tide, parts of the city were raised to the level of the top of the seawall (5 m above mean low tide). Buildings were elevated and supported on jacks while sand fill was pumped beneath them (Figure 16.1b). In addition, other structures such as streetcar lines, power poles, sewers, roadways, and sidewalks had to be elevated.

After seven years the project was completed, and so far the time and expense invested seems to have been justified. In 1961, for instance, hurricane Carla struck and flooded some of the city and some buildings were damaged by wind. But overall damage was minimal and there were no deaths.

The Outer Banks of North Carolina are simply several barrier islands extending about 240 km along the Atlantic coast. In most respects they are like barrier islands elsewhere, but with one notable exception—they lie much further offshore; a 48-km-wide lagoon separates them from the mainland (Figure 16.2a). Scenic appeal, historical interest, and recreational opportunities make the Outer Banks a favorite with tourists and residents of the region. Those interested in the history of aviation will recall that Orville and Wilbur Wright made their first flight near Kitty Hawk on December 17, 1903. And since 1526, more than 600 ships have been lost on the

(a)

(b)

Figure 16.1 *(a) Construction of this sea wall to protect Galveston, Texas, from storm waves began in 1902. Notice that the wall is curved to deflect waves upward. (b) Some of the nearly 3000 buildings that were raised and supported on stilts until sand fill could be pumped beneath them.*

dangerous shoals adjacent to the islands, particularly at Cape Lookout and Cape Hatteras. To reduce ship losses, a lighthouse funded by the federal government was erected at Cape Hatteras in 1802.

Most barrier islands, including those making up the Outer Banks, erode on their seaward sides during storms and waves carry sediment into their lagoons, so the entire barrier island complex migrates landward. In short, sediment is simply transported from the seaward sides of barrier islands to their landward sides. You can imagine the concern of homeowners on these islands as they see the shoreline march closer and closer to their dwellings, and, in many cases, destroy them.

As a matter of fact, shoreline erosion prompted the National Park Service to move the Cape Hatteras Lighthouse nearly 500 m inland and 760 m to the southwest (Figure 16.2b). When the current lighthouse was built in 1870, it was 457 m from the shoreline, but by the time it was moved in July 1999 it stood no more than 36.5 m inland! Given that the annual erosion rate is about 3 m per year, and that all previous attempts to stabilize the shoreline had failed, the Park Service found it prudent to act when it did. Shoreline erosion will continue, of course, but the Cape Hatteras Lighthouse, the world's tallest brick lighthouse, should be safe for several centuries.

(a)

(b)

Figure 16.2 *(a) This chain of barrier islands comprises the Outer Banks of North Carolina. (b) The Lighthouse on Cape Hatteras before it was moved in July 1999. The lighthouse was built 457 m from the shore in 1870, but by 1999 it was only 36.5 m inland, and all attempts to stabilize the shoreline had failed.*

Introduction

We can define the term **shoreline** as the land in contact with the sea, but it is useful to expand this definition somewhat by noting that it consists of the area between low tide and the highest level on land affected by storm waves. Accordingly, a shoreline is a long, rather narrow zone where marine processes, particularly waves, tides, and nearshore currents, continually modify existing shoreline features. In short, shorelines, like streams and glaciers, are dynamic areas where energy is expended, erosion takes place, and sediment is transported and deposited. Our main interest here is with ocean shorelines, but waves and nearshore currents are also quite effective in large lakes, where shorelines exhibit many of the features present along seashores. The most notable differences are that waves and nearshore currents are much more energetic along seashores and even the largest lakes lack appreciable tides.

The geologic processes operating along shorelines provide another example of interactions among Earth systems, in this case between the hydrosphere and solid Earth (see Table 1.1). In this chapter we concentrate on that narrow zone between land and sea or a lake, but we must be aware that a much larger element of the hydrosphere is involved. After all, waves are generated in the oceans and the gravitational attraction of the Moon and Sun affect all oceanic water, not just that along shorelines. It is, however, shorelines where much of the energy of waves and tides is expended.

The continents possess more than 400,000 km of shorelines. Some are rocky and steep, such as those in Maine, the Maritime Provinces of Canada, and along most of North America's west coast, whereas shorelines with broad sandy beaches are found in much of the eastern United States and along the Gulf coast. Whatever their type, all shorelines, including those of lakes, are dynamic areas of change where shoreline processes act on shore materials. On shorelines where energy levels are particularly high, erosion predominates, and the shoreline might retreat landward. Indeed, coastal erosion and landward migration of shorelines are problems in many areas. Yet some shorelines receive copious quantities of sediment and actually build outward. That is, they prograde seaward in much the same way that deltas do, although deltas tend to be lobe shaped, whereas shorelines are linear.

Living near a shoreline has both positive and negative aspects. Many of the world's major centers of commerce are at or near seashores, or on lakeshores or waterways where ships can navigate far inland as along the Mississippi and St. Lawrence Rivers. New York City, Los Angeles, and Chicago as well as Toronto and Vancouver in Canada are shoreline communities. And certainly the scenery and recreational opportunities of shorelines have great appeal for many people. Indeed, in 1998 about 3600 people in the United States moved to coastal areas every day. Communities such as Myrtle Beach, South Carolina; Fort Lauderdale, Florida; and many others depend heavily on tourists visiting and enjoying their shorelines.

However, in many parts of the world, including most of the United States and Canada, sea level is rising, and buildings that were once far inland are in peril and abandoned or already destroyed. Slumps and slides are common along steep shorelines eroded by waves, and narrow offshore barrier islands move landward because of erosion on their seaward sides and deposition on their landward sides. Hurricanes expend much of their energy on shorelines, resulting in extensive coastal flooding, numerous fatalities, and considerable property damage (see the Prologue).

Scientists from various disciplines have contributed to our understanding of shorelines and shoreline processes, making the subject truly interdisciplinary. The current debate concerning global warming has important implications for the world's shorelines. If we are entering a period of increasing atmospheric temperatures, this will cause accelerated melting of glaciers, resulting in a rise in sea level. Such a rise will dramatically affect the world's shorelines and have important political, economic, and social implications. It is therefore important we understand the dynamics affecting shorelines.

Why Should You Study Shorelines and Shoreline Processes?

Why should you study shorelines and shoreline processes? The hydrosphere is a vast system consisting of all water on Earth, the bulk of which is in the oceans where energy is transferred through the water to shorelines. Thus, understanding the geologic processes operating at shorelines is important to many people. Indeed, many of the world's major centers of commerce and much of the world's population is concentrated in a narrow band at or near shorelines.

Geologists, oceanographers, marine biologists, and coastal engineers, among others, are interested in the dynamic nature of shorelines. Elected officials and city planners of coastal communities must become familiar with shoreline processes so they can develop policies and zoning regulations that serve the public as well as protect the fragile shoreline environment. In short, the study of shorelines is not only interesting but also has many practical applications.

Shorelines and Shoreline Processes

In contrast to other geologic agents such as running water, wind, and glaciers that operate over vast areas, shoreline processes are restricted to a narrow zone at any particular time. But shorelines might migrate landward or seaward depending on changing sea level or uplift or subsidence of coastal regions. During a rise in sea level, for instance, the shoreline migrates landward, and as it does so, wave, tide, and nearshore-current activity shifts landward as well; and during times when sea level falls, just the opposite takes place. Recall from Chapter 5 that during marine transgressions and regressions, beach and nearshore sediments are deposited over vast regions.

In the marine realm, several biological, chemical, and physical processes are operating continuously. Organisms change the local chemistry of seawater and contribute their skeletons to nearshore sediments, and temperature and salinity changes and internal waves occur in the oceans. However, the processes most important for modifying shorelines are purely physical ones, especially waves, tides, and nearshore currents. We cannot totally discount some of these other processes though; offshore reefs, for instance, may protect a shoreline area from most of the energy of waves.

Tides

The surface of the oceans rises and falls twice daily in response to the gravitational attraction of the Moon and Sun. These regular fluctuations in the ocean's surface, or **tides,** result in most shoreline areas having two daily high tides and two low tides as sea level rises and falls anywhere from a few centimeters to more than 15 m (Figure 16.3). A complete tidal cycle includes a *flood tide* that progressively covers more and more of a nearshore area until high tide is reached, followed by *ebb tide* during which the nearshore area is once again exposed (Figure 16.3). These regular fluctuations in sea level constitute one largely untapped source of energy as do waves, ocean currents, and temperature differences in seawater (see Perspective 16.1).

Both the Moon and the Sun have sufficient gravitational attraction to exert tide-generating forces strong enough to deform the solid body of Earth, but they have a

(a)

(b)

Figure 16.3 *(a) Low tide and (b) high tide.*

Energy from the Oceans

The fact that several systems interact in complex ways has been stressed many times in previous chapters. Here too, we emphasize these interactions as we discuss the potential to extract energy from the oceans. Solar energy reaching Earth is responsible for differential heating of the atmosphere and thus air circulation as wind. Some of the energy of wind is transferred to the oceans where it causes waves and is partly responsible for oceanic currents, although Earth's rotation also plays a role in currents. Gravitational attraction between Earth and the Sun and Moon generate tides, and along with Earth's rotation results in most coastal areas experiencing a twice-daily rise and fall of sea level. In short, the oceans possess a tremendous reservoir of largely untapped energy.

If we can effectively harness the energy possessed by the oceans, an almost limitless, largely nonpolluting energy supply would be ensured. Unfortunately, ocean energy is diffuse, meaning that the amount of energy for a given volume of water is small and thus difficult to concentrate and use. Several ways of using ocean energy are being considered or under development, and one is currently in use, although it accounts for only a tiny proportion of all energy production. Of the several sources of ocean energy—temperature differences with depth, waves, currents, and tides—only the latter shows much promise for the near future.

Ocean water at depth might be as much as 25°C colder than surface water, which is the basis for ocean thermal energy conversion (OTEC). OTEC exploits this temperature difference to run turbines and generate electricity. The amount of energy available is enormous, but a number of practical problems must be solved before it can be used. For one thing, any potential site must be close to land and also have a sufficiently rapid change in depth to result in the required temperature difference. Furthermore, enormous quantities of warm and cold seawater would have to circulate through an electrical-generating plant, thus requiring that large surface areas be devoted to this purpose.

The concept of OTEC is more than a century old, but despite several decades of research no commercial OTEC plants are operating or even under construction, although small experimental ones have been tested in Hawaii and Japan.

Wind-generated ocean currents, such as the Gulf Stream that flows along the east coast of North America, also possess energy that might be tapped to generate electricity. Unlike streams that can be dammed to impound a reservoir, any electrical-generating facility exploiting oceanic currents would have to concentrate their diffuse energy and contend with any unpredictable changes in direction. In addition, whereas hydroelectric generating plants on land depend on water moving rapidly from a higher elevation to the turbines, the energy of

ocean currents comes from their flow velocity that is at most a few kilometers per hour.

The most obvious form of energy in the oceans is waves. Harnessing wave energy and converting it to electricity is not a new idea, and it is used on an extremely limited scale. Unfortunately, the energy possessed by a wave is distributed along its crest and is difficult to concentrate. Furthermore, any facility would have to be designed to withstand the effects of storms and saltwater corrosion. The Japanese have developed wave-energy devices to power lighthouses and buoys, and a facility capable of providing power to about 300 homes began operating in Scotland during September 2000.

Perhaps tidal power is the most promising form of ocean energy. In fact, it has been used for centuries in some coastal areas to run mills, but its use at present for electrical generation is limited. Most coastal areas experience a twice daily rise and fall of tides, but only a few areas are suitable for exploiting this energy source. One limitation is that the tidal range must be at least 5 m, and there must also be a coastal region where water can be stored following high tide.

Suitable sites for using tidal power are limited not only by tidal range, but also by location. Many areas along the U.S. Gulf Coast would certainly benefit from tidal power plants, but the tide range of generally less than 1 m precludes the possibility of development. Even an area with an appropriate tidal range in some remote region such as southern Chile or the Arctic islands of Canada offer little potential because of their great distances from population centers. Accordingly, in North America only a few areas show much potential for developing tidal energy: Cook Inlet, Alaska; the Bay of Fundy on the U.S.-Canadian border; and some areas along the New England coast, for instance.

The idea behind tidal power is simple, although putting it into practice is not easy. First, a dam with sluice gates to regulate water flow must be built across the entrance to a bay or estuary. When the water level has risen sufficiently high during flood tide, the sluice gates are closed. Water held on the landward side of the dam is then released and electricity is generated just as it is at a hydroelectric dam (see Figure 1, Perspective 12.1). Actually, a tidal power plant can operate during both flood and ebb tides (Figure 1).

The first tidal power-generating facility was constructed at the La Rance River estuary in France (Figure 2). This 240-megawatt (MW) plant began operation in 1966 and has been quite successful. In North America, a much smaller 20-MW tidal-power plant has been operating in the Bay of Fundy, Nova Scotia, where the tidal range, the greatest in the world, exceeds 16 m. The Bay of Fundy is part of the much larger Gulf of Maine where the United States and Canada have been con-

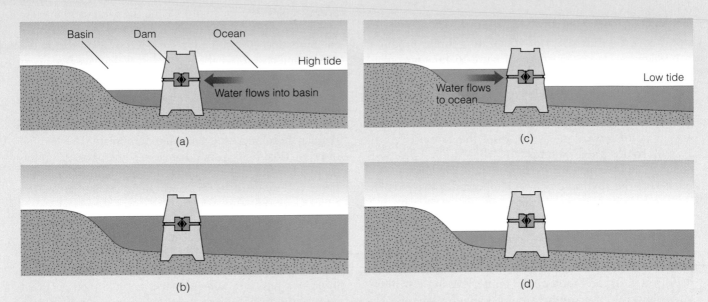

(a)

Basin   Dam   Ocean

High tide

Water flows into basin

(b)

(c)

Low tide

Water flows to ocean

(d)

Figure 1 *Rising and falling tides produce electricity by spinning turbines connected to generators just as at hydroelectric plants. (a) Water flows from ocean to basin during flood tide. (b) Basin full. (c) Water flows from basin to ocean during ebb tide. (d) Basin water is depleted, and the cycle begins once more.*

sidering a much larger tidal power-generating project for several decades.

Although tidal power shows some promise, it will not solve our energy needs even if developed to its fullest potential. Most analysts think that only 100 to 150 sites worldwide have sufficiently high tidal ranges and the appropriate

coastal configuration to exploit this energy resource. This, and the facts that construction costs are high and tidal energy systems can have disastrous effects on the ecology (biosphere) of estuaries, makes it unlikely that tidal energy will ever contribute more than a small percentage of all energy production.

Figure 2 *The La Rance River estuary tidal power-generating plant in western France. This 850-m-long dam was built where the maximum tidal range is 13.4 m. Electricity is generated during both flood and ebb tides.*

Shorelines and Shoreline Processes   **461**

much greater influence on the oceans. The Sun is 27 million times more massive than the Moon, but it is 390 times as far from Earth, and its tide-generating force is only 46% as strong as that of the Moon. Accordingly, the tides are dominated by the Moon, but the Sun plays an important role as well.

If we consider only the Moon acting on a spherical, water-covered Earth, its tide-generating forces produce two bulges on the ocean surface (Figure 16.4a). One bulge points toward the Moon because it is on the side of Earth where the Moon's gravitational attraction is greatest. The other bulge is on the opposite side of Earth, where the Moon's gravitational attraction is least. These two bulges always point toward and away from the Moon (Figure 16.4a), so as Earth rotates and the Moon's position changes, an observer at a particular shoreline location experiences the rhythmic rise and fall of tides twice daily, but the heights of two successive high tides may vary depending on the Moon's inclination with respect to the equator.

The Moon revolves around Earth every 28 days, so its position with respect to any latitude changes slightly each day. That is, as the Moon moves in its orbit and Earth rotates on its axis, it takes the Moon 50 minutes longer each day to return to the same position it was in the previous day. Thus, an observer would experience a high tide at 1:00 P.M. on one day, for example, and at 1:50 P.M. on the fol-lowing day. Tides are also complicated by the combined effects of the Moon and the Sun. Even though the Sun's tide-generating force is weaker than the Moon's, when the two are aligned every two weeks, their forces are added together and generate *spring tides* about 20% higher than average tides (Figure 16.4b). When the Moon and Sun are at right angles to one another, also at two-week intervals, the Sun's tide-generating force cancels some of that of the Moon, and *neap tides* about 20% lower than average occur (Figure 16.4c).

Tidal ranges are also affected by shoreline configuration. Broad, gently sloping continental shelves as in the Gulf of Mexico have low tidal ranges, whereas steep, irregular shorelines experience a much greater rise and fall of tides. Tidal ranges are greatest in some narrow, funnel-shaped bays and inlets. The Bay of Fundy in Nova Scotia has a tidal range of 16.5 m, and ranges greater than 10 m occur in several other areas.

Tides have an important impact on shorelines because the area of wave attack constantly shifts onshore and offshore as the tides rise and fall. Tidal currents themselves, however, have little modifying effect on shorelines, except in narrow passages where tidal current velocity is great enough to erode and transport sediment. Indeed, if it were not for strong tidal currents, some passageways would be blocked by sediments deposited by nearshore currents.

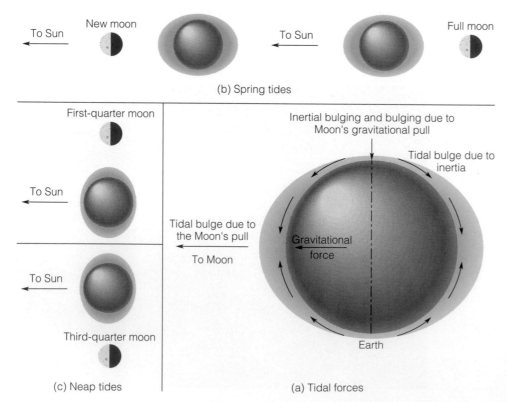

Figure 16.4 *(a) Tides are caused by the gravitational pull of the Moon and, to a lesser degree, the Sun. The Earth-Moon-Sun alignments at the time of the (b) spring and (c) neap tides are shown.*

Waves

Oscillations of a water surface, or simply **waves,** are seen on all bodies of water, but they are certainly best developed in the oceans and have their greatest impact on seashores. In fact, they are directly or indirectly responsible for most erosion, sediment transport, and deposition in coastal areas. A typical series of waves and the terms applied to them is illustrated in Figure 16.5. A **crest,** as you would expect, is the highest part of a wave, whereas the low area between crests is a **trough.** The distance from crest to crest (or trough to trough) is the *wavelength*, and the vertical distance from trough to crest is *wave height*. The speed at which a wave advances, generally called celerity (C), is calculated by

$$C = L/T$$

in which L is wave length, and T is wave period, that is, the time it takes for two successive wave crests, or troughs, to pass a given point.

The speed of wave advance (C) is actually a measure of the velocity of the wave form rather than a measure of the speed of the molecules of water in a wave. In fact, water waves are somewhat similar to waves moving across a grass-covered field; the grass moves forward and back as the wave passes, but the individual blades of grass remain in their original position. When waves move across a water surface, the water moves in circular orbits, but shows little or no net forward movement (Figure 16.5). Only the wave form moves forward, and as it does it transfers energy in the direction of wave movement.

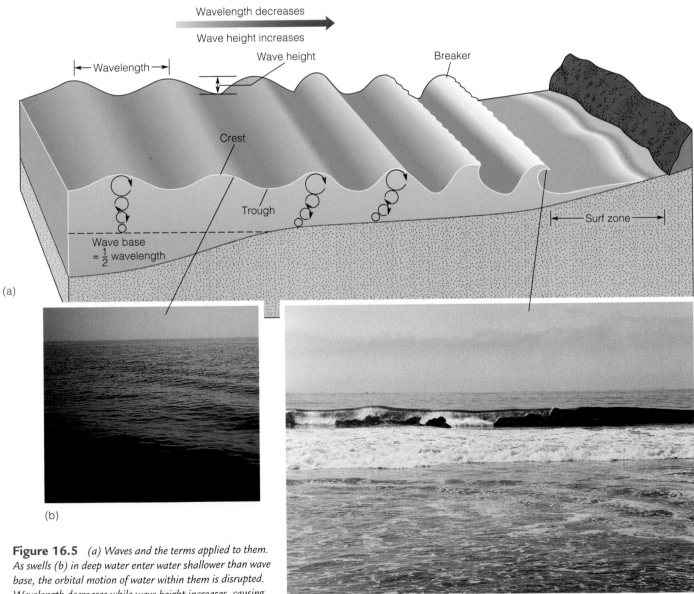

Figure 16.5 *(a) Waves and the terms applied to them. As swells (b) in deep water enter water shallower than wave base, the orbital motion of water within them is disrupted. Wavelength decreases while wave height increases, causing waves to oversteepen and plunge forward as breakers (c).*

Shorelines and Shoreline Processes　**463**

The diameters of the orbits followed by water in waves diminish rapidly with depth, and at a depth of about one-half wavelength (L/2), called **wave base,** they are essentially zero. Thus, at a depth exceeding wave base, the water and seafloor, or lake floor, are unaffected by surface waves (Figure 16.5). Wave base is an important consideration in some aspects of shoreline modification; it is explored more fully in later sections of this chapter.

Wave Generation

Everyone has seen waves but probably has little idea of how they form and what controls their size and shape. Actually, several processes generate waves. Landslides into the oceans and lakes displace water and generate waves, and so do faulting and volcanic eruptions. Some waves so-formed are huge and might be devastating to coastal areas, but most geologic work on shorelines is accomplished by wind-generated waves, especially storm waves. When wind blows over water, that is, one fluid (air) moves over another fluid (water), friction between the two results in energy transfer to the water, causing the water surface to oscillate.

In areas where waves are generated, as beneath a storm center at sea, sharp-crested, irregular waves called *seas* develop. Seas have various heights and lengths, and one wave cannot be easily distinguished from another. But as seas move out from their area of generation they are sorted into broad *swells* with rounded, long crests and all are about the same size.

As one would expect, the harder and longer the wind blows, the larger are the waves generated. Wind velocity and duration, however, are not the only factors controlling the size of waves. High-velocity wind blowing over a small pond will never generate large waves regardless of how long it blows. In fact, waves on ponds and most lakes are present only while the wind is blowing; once the wind stops, the water quickly smooths out. In contrast, the surface of the ocean is always in motion, and waves with heights of 34 m have been recorded during storms in the open sea.

The reason for the disparity between wave sizes on ponds and lakes and on the oceans is the **fetch,** which is the distance the wind blows over a continuous water surface. Fetch is limited by the available water surface, so on ponds and lakes it corresponds to their length or width, depending on wind direction. A wind blowing the length of Lake Superior, for instance, can generate large waves, but even larger ones develop in the oceans. To produce waves of greater length and height, more energy must be transferred from wind to water; hence large waves form beneath large storms at sea.

Shallow-Water Waves and Breakers

Swells moving out from the area of wave generation lose little energy as they travel long distances across the ocean.

In these deep-water swells, the water surface oscillates and water moves in circular orbits, but little net displacement of water takes place in the direction of wave travel (Figure 16.5). Of course wind blows some water from waves, thus forming white caps with foamy white crests, and surface currents transport water great distances, but deep-water waves themselves accomplish little actual water movement. But when these waves enter progressively shallow water, the wave shape changes and water is displaced in the direction of wave advance.

Broad, undulating deep-water waves are transformed into sharp-crested waves as they enter shallow water. This transformation begins at a water depth corresponding to wave base, that is, one-half wavelength. At this point the waves "feel" the seafloor, and the orbital motion of water within the waves is disrupted (Figure 16.5). As waves continue moving shoreward, the speed of wave advance and wavelength decrease but wave height increases. Thus as they enter shallow water, waves become oversteepened as the wave crest advances faster than the wave form, and eventually the crest plunges forward as a **breaker** (Figure 16.5). Breaking waves might be several times higher than their deep-water counterparts, and when they break they expend their kinetic energy on the shoreline.

The waves described above are the classic *plunging breakers* that crash onto shorelines with steep offshore slopes, such as those on the north shore of Oahu in the Hawaiian Islands (Figure 16.6a). In contrast, shorelines where the offshore slope is more gentle usually have *spilling breakers,* where the waves build up slowly and the wave's crest spills down the wave front (Figure 16.6b). Regardless of whether the breakers are plunging or spilling, the water rushes onto the shore and then returns seaward to become part of another breaking wave.

Nearshore Currents

The area extending seaward from the upper limit of the shoreline to just beyond the area of breaking waves is conveniently designated as the *nearshore zone.* Within the nearshore zone are the areas where waves break, the breaker zone, and a surf zone, where water from breaking waves rushes forward and then flows seaward as backwash. The nearshore zone's width varies depending on the length of approaching waves, because long waves break at a greater depth, and thus farther offshore, than do short waves. Incoming waves are responsible for two types of currents in the nearshore zone: *longshore currents* and *rip currents.*

Wave Refraction and Longshore Currents

Deep-water waves have long, continuous crests, but rarely are their crests parallel with the shoreline (Figure 16.7). In other words, they seldom approach a shoreline head-

(a)

(b)

Figure 16.6 *(a) Plunging breaker on the north shore of Oahu, Hawaii. (b) Spilling breaker.*

on, but rather at some angle. Thus, one part of a wave enters shallow water where it encounters wave base and begins breaking before other parts of the same wave. As a wave begins breaking, its velocity diminishes, but the part of the wave still in deep water races ahead until it too encounters wave base. The net effect of this oblique approach is that waves bend so that they more nearly parallel the shoreline, a phenomenon known as **wave refraction** (Figure 16.7).

Even though waves are refracted, they still usually strike the shoreline at some angle, causing the water between the breaker zone and the beach to flow parallel to the shoreline. These **longshore currents,** as they are called, are long and narrow and flow in the same general direction as the approaching waves. These currents are particularly important in transporting and depositing sediment in the nearshore zone.

Rip Currents

Waves carry water into the nearshore zone, so there must be a mechanism for mass transfer of water back out to sea. One way in which water moves seaward from the nearshore zone is in **rip currents,** narrow surface currents that flow out to sea through the breaker zone (Figure 16.8). Surfers commonly take advantage of rip currents for an easy ride out beyond the breaker zone, but these currents pose a danger to inexperienced swimmers. Some rip currents flow at several kilometers per hour, so if a swimmer is caught in one, it is useless to try to swim directly back to shore. Instead, because rip currents are narrow and usually nearly perpendicular to the shore, one can swim parallel to the shoreline for a short distance and then turn shoreward with no difficulty.

We can characterize rip currents as circulating cells fed by longshore currents. When waves approach a shoreline,

the amount of water builds up until the excess moves out to sea through the breaker zone. The rip currents are fed by nearshore currents that increase in velocity from midway between each rip current (Figure 16.8a).

Relief on the seafloor plays an important role in determining the location of rip currents. They commonly develop where wave heights are lower than in adjacent areas, and differences in wave height are controlled by variations in water depth. For instance, if waves move over a depression, the height of the waves over the depression tends to be less than in adjacent areas.

Deposition along Shorelines

Depositional features of shorelines include *beaches, spits, baymouth bars, tombolos,* and *barrier islands.* The characteristics of beaches are determined by wave energy and shoreline materials, and they are continually modified by waves and longshore currents. Spits, baymouth bars, and tombolos result from deposition by longshore currents, but the origin of barrier islands is not fully resolved. Rip currents play only a minor role in the configuration of shorelines, but they do transport fine-grained sediment seaward through the breaker zone.

Beaches

Beaches are the most familiar of all coastal landforms, attracting millions of visitors each year and providing the economic base for many communities. Depending on shoreline configuration and wave intensity, beaches may be discontinuous, existing only as *pocket beaches* in protected areas such as embayments, or they may be continuous for long distances (Figure 16.9).

Figure 16.7 *Wave refraction (wave crests are indicated by dashed lines). These waves are refracted as they enter shallow water and more nearly parallel the shoreline. The waves generate a longshore current that flows in the direction of wave approach, from upper left to lower right (arrow) in this example.*

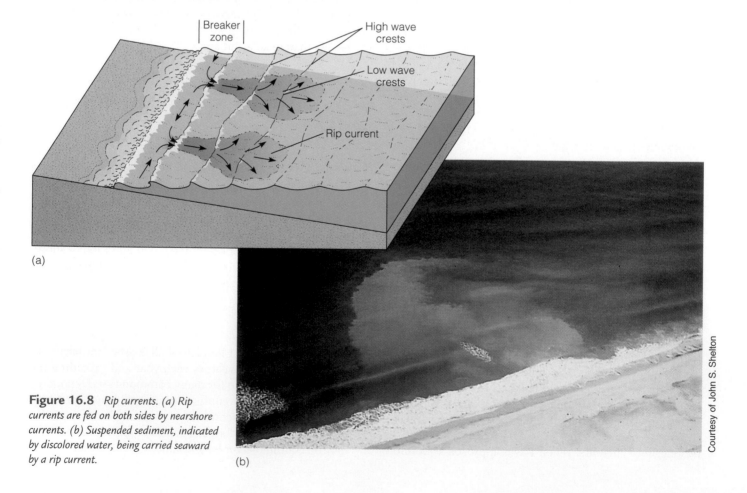

Breaker zone

High wave crests

Low wave crests

Rip current

(a)

(b)

Figure 16.8 *Rip currents. (a) Rip currents are fed on both sides by nearshore currents. (b) Suspended sediment, indicated by discolored water, being carried seaward by a rip current.*

Courtesy of John S. Shelton

By definition a **beach** is a deposit of unconsolidated sediment extending landward from low tide to a change in topography such as a line of sand dunes, a sea cliff, or the point where permanent vegetation begins. Typically, a beach has several component parts (Figure 16.10a) including a **backshore** that is usually dry, being covered by water only during storms or exceptionally high tides. The backshore consists of one or more **berms**, platforms composed of sediment deposited by waves; the berms are nearly horizontal or slope gently landward. The sloping area below a berm exposed to wave swash is the **beach face.** The beach face is part of the **foreshore,** an area covered by water during high tide but exposed during low tide.

Some of the sediment on beaches is derived from weathering and wave erosion of the shoreline, but most of it is transported to the coast by streams and redistributed along the shoreline by longshore currents (Figure 16.11a). As previously noted, waves usually strike beaches at some angle, causing the sand grains to move up the beach face at a similar angle; as the sand grains are carried seaward in the backwash, however, they move perpendicular to the long axis of the beach. Thus, individual sand grains move in a zigzag pattern in the direction of longshore currents. This movement is not restricted to the beach; it extends seaward to the outer edge of the breaker zone (Figure 16.11a).

In an attempt to widen a beach or prevent erosion, shoreline residents often build *groins,* structures that project seaward at right angles from the shoreline (Figure 16.11b). They interrupt the flow of longshore currents, causing sand deposition on their upcurrent sides, widening the beach at that location. However, erosion inevitably occurs on the downcurrent side of a groin.

Quartz is the most common mineral in most beach sands, but there are some notable exceptions. For exam-

(a)

(b)

Figure 16.9 *(a) Small pocket beaches along the California coast near San Francisco. (b) The Grand Strand of South Carolina, shown here at Myrtle Beach, is 100 km of nearly continuous beach.*

Figure 16.10 *(a) Cross section of a typical beach showing its component parts. (b) The backshore area of a pocket beach along the Pacific Coast. Notice that the berm ends at the rocks on the right and the beach face slopes steeply seaward.*

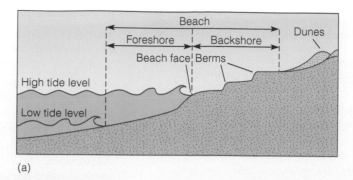

(a)

(b)

ple, the black sand beaches of Hawaii are composed of sand- and gravel-sized basalt rock fragments or small grains of volcanic glass, and some Florida beaches are composed of the fragmented calcium carbonate shells of marine organisms. In short, beaches are composed of whatever material is available; quartz is most abundant simply because it is available in most areas and is the most durable and stable of the common rock-forming minerals.

Seasonal Changes in Beaches

Wave energy is dissipated on beaches so the loose grains composing them are constantly moved by waves. But the overall configuration of a beach remains unchanged as long as equilibrium conditions persist. The beach profile consisting of a berm or berms and a beach face shown in Figure 16.10 can be thought of as a profile of equilibrium;

that is, all parts of the beach are adjusted to the prevailing conditions of wave intensity, nearshore currents, and materials composing the beach.

Tides and longshore currents affect the configuration of beaches to some degree, but by far the most important agent modifying their equilibrium profile is storm waves. In many areas, beach profiles change with the seasons; so we recognize *summer beaches* and *winter beaches*, each of which is adjusted to the conditions prevailing at those times (Figure 16.12). Summer beaches are generally covered with sand and have a wide berm, a gently sloping beach face, and a smooth offshore profile. Winter beaches on the other hand, tend to be coarser grained and steeper; they have a small berm or none at all, and their offshore profiles reveal sand bars paralleling the shoreline (Figure 16.12).

Seasonal changes in beach profiles are related to changing wave intensity. During the winter, energetic

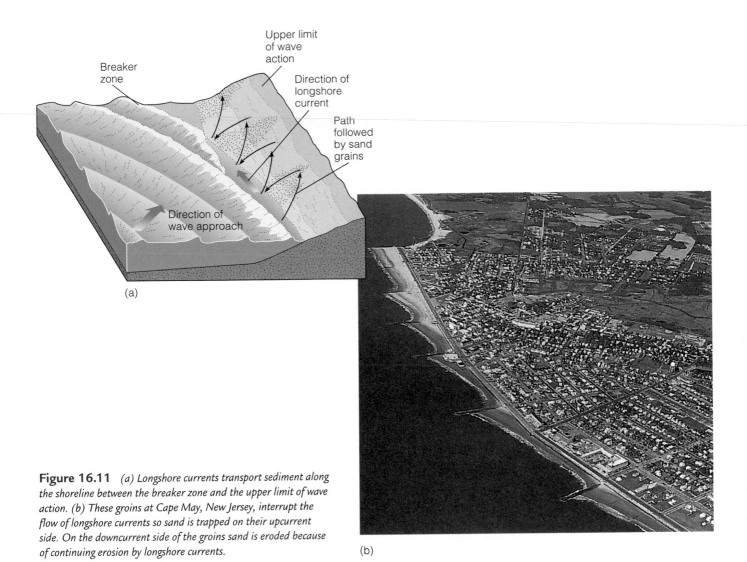

Figure 16.11 *(a) Longshore currents transport sediment along the shoreline between the breaker zone and the upper limit of wave action. (b) These groins at Cape May, New Jersey, interrupt the flow of longshore currents so sand is trapped on their upcurrent side. On the downcurrent side of the groins sand is eroded because of continuing erosion by longshore currents.*

storm waves erode the sand from beaches and transport it offshore where it is stored in sand bars (Figure 16.12). The same sand that was eroded from a beach during the winter returns the next summer when it is driven onshore by more gentle swells. The volume of sand in the system remains more or less constant; it simply moves farther offshore or onshore depending on the energy of waves.

The terms *winter* and *summer beaches,* although widely used, are somewhat misleading. A winter beach profile can develop at any time if there is a large storm, and likewise a summer beach profile can develop during a prolonged winter calm period.

Spits, Baymouth Bars, and Tombolos

Among the several deposition landforms of coasts, beaches are the most familiar, but spits, baymouth bars, and tombolos are also well represented in many areas. Actually a **spit**

is simply a fingerlike continuation of a beach forming a point, or "free end," projecting from the shoreline, commonly into a bay, and a **baymouth bar** is a spit that grew until it completely closed off a bay from the open sea (Figure 16.13). So both are variations of the same feature, and both are composed of sand, more rarely gravel, transported and deposited by longshore currents. Where these currents are weak, as in the deeper water at a bay's opening, sediment deposition builds these depositional landforms. Waves modify some spits so that their free ends are curved and go by the name *hook* or *recurved spit* (Figure 16.13a).

Another rarer type of spit is a **tombolo** that extends out from the shoreline to an island (Figure 16.14). A tombolo forms on the shoreward sides of an island as wave refraction around the island creates converging currents that turn seaward and deposit a sand bar. So, as opposed to spits and baymouth bars, which are usually nearly parallel with the shoreline, the long axes of tombolos are nearly at right angles to the shoreline.

(a)

(b)

(c)

Figure 16.12 (a) Seasonal changes in a beach profile. (b) and (c) San Gregorio State Beach, California. These photos were taken from nearly the same place but (c) was taken two years after (b). Much of the change from (b) to (c) is accounted for by beach erosion during 1997–1998 winter storms.

Spits, baymouth bars, and tombolos are most common along irregular seashores, but they can also be found in large lakes. Regardless of their setting, spits and baymouth bars especially constitute a continuing problem where bays must be kept open for pleasure boating and/or commercial shipping. Obviously a bay closed off by a sand bar is of little use for either endeavor, so they must be regularly dredged or protected from deposition by longshore currents. In some areas *jetties* are constructed that extend seaward (or lakeward) to interrupt the flow of longshore currents and thus protect the opening to a bay.

Barrier Islands

Long, narrow islands of sand lying a short distance offshore from the mainland are **barrier islands** (Figure 16.15). On their seaward sides they are smoothed by waves, but their landward margins are irregular because storm waves carry

WHAT WOULD YOU DO?

In 1945, the ore of mercury (the mineral cinnabar, HgS) was found in beach sands in Marin County, California. Suppose you made this discovery and wanted to exploit it. How do you think the cinnabar got into the beach sands, and how would you find its source? And once its source was identified, what considerations, geologic and otherwise, would be important in whether or not to open a mine?

sediment over the island and deposit it in a lagoon where it is little modified by further wave activity. The component parts of a barrier island include a beach, wind-blown sand dunes, and a marshy area on their landward sides.

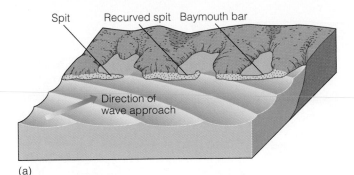

Spit   Recurved spit   Baymouth bar

Direction of
wave approach

(a)

Figure 16.13 *(a) Spits and baymouth bars form where longshore currents deposit sand in deeper water, as at the entrance to a bay. A spit (b) across a river's mouth and a baymouth bar (c).*

(b)

(c)

Everyone agrees that barrier islands form on gently sloping continental shelves where abundant sand is available, and where both wave energy and the tidal range are low. In fact, these are the reasons that so many are present along the United States' Atlantic and Gulf coasts. But even though it is well known where barrier islands typically form, the details of their origin are still unresolved. According to one model, they formed as spits that became detached from land, whereas another model holds that they formed as beach ridges that subsequently subsided (Figure 16.16).

Most barrier islands are migrating landward as a result of erosion on their seaward sides and deposition on their landward sides (see the Prologue). This is a natural part of barrier island evolution, and it takes place rather slowly. However, it takes place fast enough to cause many problems for island residents and communities.

The Nearshore Sediment Budget

We can think of the gains and losses of sediment in the nearshore zone in terms of a **nearshore sediment budget** (Figure 16.17). If a nearshore system has a bal-

anced budget, sediment is supplied to it as fast as it is removed, and the volume of sediment remains more or less constant, although sand may shift offshore and onshore with the changing seasons (Figure 16.12). A positive budget means gains exceed losses, whereas a negative budget results when losses exceed gains. If a negative budget prevails long enough, a nearshore system is depleted and beaches may disappear.

Erosion of sea cliffs provides some sediment to beaches, but in most areas probably no more than 5 to 10% of the total sediment comes from this source. There are exceptions, though; almost all the sediment on the beaches of Maine is derived from the erosion of shoreline rocks. Most sediment on typical beaches is transported to the shoreline by streams and then redistributed along the shoreline by longshore currents. Thus, longshore currents also play a role in the nearshore sediment budget because they continually move sediment into and away from beach systems.

The primary ways that a nearshore system loses sediment include offshore transport, wind, and deposition in submarine canyons. Offshore transport mostly involves fine-grained sediment carried seaward where it eventually settles in deeper water. Wind is an important process because it removes sand from beaches and blows it inland where it commonly piles up as sand dunes.

Figure 16.14 *(a) Wave refraction around an island and the origin of a tombolo. (b) A small tombolo. (c) Close-up view of the tombolo in (b).*

Tombolo

Wave crest

(a)

Tombolo

(b)

(c)

WHAT WOULD YOU DO?

While visiting a barrier island you notice some fine-grained, organic-rich deposits along the beach. In fact, so much organic matter is present that the sediments are black and have a foul odor. Their presence on the beach does not seem to make sense because you know that sediments like these were almost certainly deposited in a marsh on the landward (opposite) side of the island. How would you explain your observations?

If the heads of submarine canyons are nearshore, huge quantities of sand are funneled into them and deposited in deeper water. La Jolla and Scripps submarine canyons off the coast of southern California funnel off an estimated 2 million m³ of sand each year. In most areas, however, submarine canyons are too far offshore to interrupt the flow of sand in the nearshore zone.

It should be apparent from the preceding discussion that if a nearshore system is in equilibrium, its incoming supply of sediment exactly offsets its losses. Such a delicate balance tends to continue unless the system is somehow disrupted. One common change that affects this balance is the construction of dams across the streams

Texas

Corpus Christi

Gulf of Mexico

Padre Island

Laguna Madre

(a)

Figure 16.15 *(a) View from space of the barrier islands along the Gulf coast of Texas. Notice that a lagoon up to 20 km wide separates the long, narrow barrier islands from the mainland. (b) Aerial view of Padre Island, Texas. Laguna Madre is left of the island and the Gulf of Mexico is on the right.*

(b)

Spit

Tidal inlet   Barrier island

(a)

(b)

Beach ridge

(c)

Sea level rises

Barrier island   Lagoon

(d)

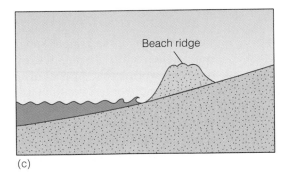

Figure 16.16 *Two models for the origin of barrier islands. (a) Longshore currents extend a spit along a coast. (b) During a storm, the spit is breached, forming a tidal inlet and a barrier island. (c) A beach ridge forms on land. (d) Sea level rises, partly submerging the beach ridge.*

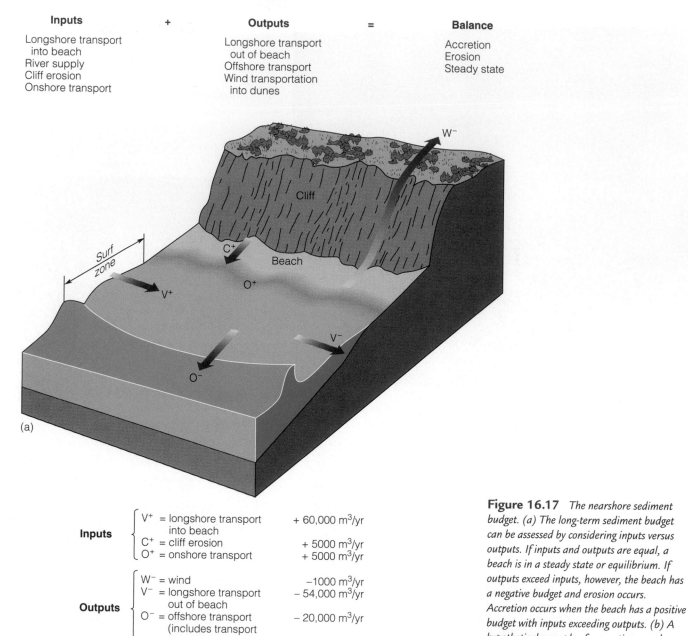

Inputs	+	Outputs	=	Balance
Longshore transport into beach		Longshore transport out of beach		Accretion
River supply		Offshore transport		Erosion
Cliff erosion		Wind transportation into dunes		Steady state
Onshore transport				

(a)

Inputs		
	V^+ = longshore transport into beach	+ 60,000 m³/yr
	C^+ = cliff erosion	+ 5000 m³/yr
	O^+ = onshore transport	+ 5000 m³/yr

Outputs		
	W^- = wind	−1000 m³/yr
	V^- = longshore transport out of beach	− 54,000 m³/yr
	O^- = offshore transport (includes transport to submarine canyons)	− 20,000 m³/yr

(b)

Balance	− 5000 m³/yr (net erosion)

Figure 16.17 *The nearshore sediment budget. (a) The long-term sediment budget can be assessed by considering inputs versus outputs. If inputs and outputs are equal, a beach is in a steady state or equilibrium. If outputs exceed inputs, however, the beach has a negative budget and erosion occurs. Accretion occurs when the beach has a positive budget with inputs exceeding outputs. (b) A hypothetical example of a negative nearshore sediment budget. In this example the beach is losing 5000 m³ a year to erosion.*

supplying sand. Once dams have been built, all sediment from the upper reaches of the drainage systems is trapped in reservoirs and thus cannot reach the shoreline.

How Are Shorelines Eroded?

Beaches are absent, poorly developed, or restricted to protected areas on seacoasts where erosion rather than deposition predominates (Figure 16.9a). Erosion also yields steep or vertical slopes known as *sea cliffs*. During storms, these cliffs are pounded by waves (hydraulic action) and worn by the impact of sand and gravel (abrasion), and more or less continuously eroded by corrosion involving the chemical breakdown of rocks by the solvent action of seawater. Tremendous energy from waves is concentrated on the bases of sea cliffs, and is most effective on those composed of sediments or highly fractured rocks. In any case, the net effect of these processes is erosion of the sea cliff and retreat of the cliff face landward.

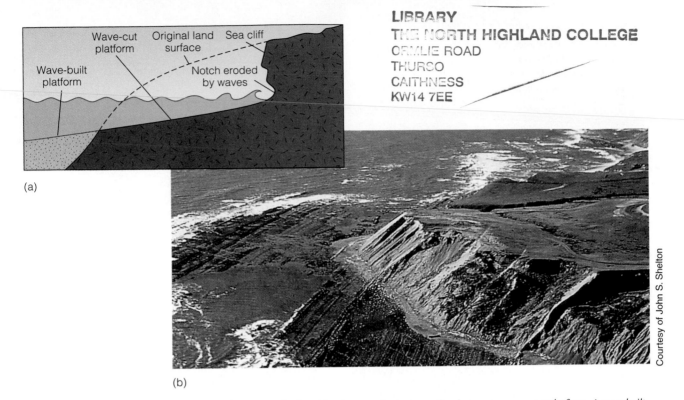

(a)

(b)

Courtesy of John S. Shelton

Figure 16.18 *(a) Wave erosion causes a sea cliff to migrate landward leaving a gently sloping surface known as a wave-cut platform. A wave-built platform originates by deposition at the seaward margin of the wave-cut platform. (b) Sea cliffs and a wave-cut platform.*

Wave-Cut Platforms and Associated Landforms

Wave intensity and the resistance of shoreline materials to erosion determine the rate at which a sea cliff retreats landward. A sea cliff of glacial drift on Cape Cod, Massachusetts, erodes as much as 30 m per century, and some parts of the White Cliffs of Dover in England retreat landward at more than 100 m per century. By comparison, sea cliffs of dense igneous or metamorphic rocks erode and retreat much more slowly.

Sea cliffs erode mostly as a result of hydraulic action and abrasion at their bases. As a sea cliff is undercut by erosion, the upper part is left unsupported and susceptible to mass wasting processes. Thus, sea cliffs retreat little by little, and as they do, they leave a beveled surface called a **wave-cut platform** that slopes gently seaward (Figure 16.18). Broad wave-cut platforms exist in many areas, but invariably the water over them is shallow because the abrasive planing action of waves is effective to a depth of only about 10 m. The sediment eroded from sea cliffs is transported seaward until it reaches deeper water at the edge of the wave-cut platform. There it is deposited and forms a *wave-built platform,* which is a seaward extension of the wave-cut platform (Figure 16.18). Wave-cut platforms now above sea level are known as **marine terraces** (Figure 16.19).

Figure 16.19 *This gently sloping surface along the Pacific Coast of California is a marine terrace. Notice the erosional remnants of the old shoreline rising above the terrace.*

Sea cliffs do not retreat uniformly because some of the materials of which they are composed are more resistant to erosion than others. **Headlands** are seaward-projecting parts of the shoreline that are eroded on both sides due to wave refraction (Figure 16.20). *Sea caves* may form on opposite sides of a headland and if these join, they form a *sea arch* (Figure 16.20a and b). Continued erosion generally causes the span of an arch to collapse, yielding isolated *sea*

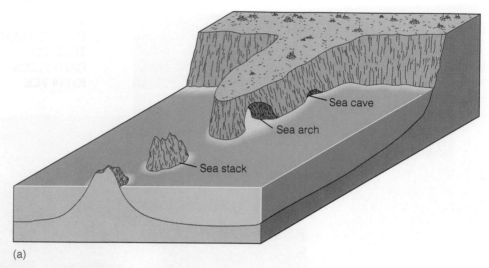

Figure 16.20 *(a) Erosion of a headland and the origin of sea caves, sea arches, and sea stacks. (b) This sea stack in Australia has an arch developed in it. (c) Sea stacks on California's central coast.*

(a)

(b)

(c)

stacks on wave-cut platforms (Figure 16.20c). In the long run, shoreline processes tend to straighten an initially irregular shoreline. They do so because wave refraction causes more wave energy to be expended on headlands and less on embayments. Thus, headlands become eroded, and some of the sediment yielded by erosion is deposited in the embayments.

How Are Coastal Areas Managed as Sea Level Rises?

Living near a shoreline is appealing, but it is not without risks. In many parts of the world, including most of the United States and much of Canada, sea level is rising, and buildings that were built some distance from the ocean are now being undermined and destroyed. Slumps and slides are common along rocky, steep shorelines; narrow offshore barrier islands migrate landward by erosion on their sea-

ward sides and deposition on their landward sides; and hurricanes expend much of their fury on coastal regions.

During the last century, sea level rose about 12 cm worldwide, and all indications are that it will continue to rise. The absolute rate of sea level rise in a particular shoreline region depends on two factors. The first is the volume of water in the ocean basins, which is increasing as a result of glacial ice melting and the thermal expansion of near-surface seawater. Many scientists think that sea level will continue to rise because of global warming resulting from concentrations of greenhouse gases in the atmosphere.

The second factor controlling sea level is the rate of uplift or subsidence of a coastal area. In some areas, uplift is occurring so fast that sea level is actually falling with respect to the land. In other areas sea level is rising while the coastal region is simultaneously subsiding, resulting in a net change in sea level of as much as 30 cm per century. Perhaps such a "slow" rate of sea level change seems insignificant; after all it amounts to only a few millimeters per year. But in gently sloping coastal areas, as in the eastern

United States from New Jersey southward, even a slight rise in sea level would eventually have widespread effects.

Many of the nearly 300 barrier islands along the east and Gulf coasts of the United States are migrating landward as sea level rises (Figure 16.21, see the Prologue). Landward migration of barrier islands would pose few problems if it were not for the numerous communities, resorts, and vacation homes located on them. Moreover, barrier islands are not the only threatened areas. For example, Louisiana's coastal wetlands, an important wildlife habitat and seafood-producing area, are currently being lost at a rate of about 90 km² per year. Much of this loss results from sediment compaction, but rising sea level exacerbates the problem.

Rising sea level also directly threatens many beaches upon which communities depend for revenue. The beach at Miami Beach, Florida, for instance, was disappearing at an alarming rate until the Army Corps of Engineers began replacing the eroded beach sand (Figure 16.22). The problem is even more serious in other countries. A rise in sea level of only 2 m would inundate large areas of the east and Gulf coasts, but would cover 20% of the entire country of Bangladesh. Other problems associated with sea level rise include increased coastal flooding during storms and saltwater incursions that may threaten groundwater supplies (see Chapter 13).

Armoring shorelines with *sea walls* (embankments of reinforced concrete or stone) (Figure 16.1a) and using *riprap* (piles of stones) (Figure 16.23a) protect beachfront structures, but both are initially expensive and during large storms are commonly damaged or destroyed. Sea walls do

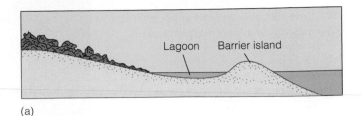

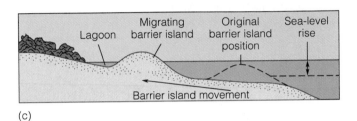

Figure 16.21 *Rising sea level and the landward migration of barrier islands. (a) Barrier island before landward migration in response to rising sea level. (b) Sea level rises and the barrier island migrates landward when storm waves wash sand from its seaward side and deposit it in the lagoon. (c) Over time, the entire complex migrates landward.*

(a)

(b)

Figure 16.22 *The beach at Miami Beach, Florida, (a) before and (b) after the U.S. Army Corps of Engineers' beach nourishment project.*

Figure 16.23 *(a) These large blocks of basalt were piled along this beach to prevent erosion and protect a luxury hotel just to the left out of the image. (b) Heavy equipment builds a berm on the seaward side of beach homes on the Isle of Palms, South Carolina, to protect them from waves. During the next spring tide, the berms disappear and must be rebuilt.*

(b)

(a)

afford some protection and are seen in many coastal areas along the oceans and large lakes, but some states, including North and South Carolina, Rhode Island, Oregon, and Maine, no longer allow their construction. And the futility of artificially maintaining beaches is aptly shown by efforts to protect homes on a South Carolina barrier island. After each spring tide, heavy equipment builds a sand berm to protect homes from the next spring tide (Figure 16.23b), only to see it disappear and then rebuild it in a never-ending cycle of erosion and expensive artificial replacement.

Because nothing can be done to prevent sea level from rising, engineers, scientists, planners, and political leaders must examine what can be done to prevent or minimize the effects of shoreline erosion. At present, only a few viable options exist. One is to put strict controls on coastal development. North Carolina, for example, permits large structures no closer to the shoreline than 60 times the annual erosion rate. Although a growing awareness of shoreline processes has resulted in similar legislation elsewhere, some states have virtually no restrictions on coastal development.

Regulating coastal development is commendable, but it has no impact on existing structures and coastal communities. A general retreat from the shoreline may be possible, but expensive, for individual dwellings and small communities, but it is impractical for large population centers. Such communities as Atlantic City, New Jersey; Miami Beach, Florida; and Galveston, Texas, have adopted one of two strategies to combat coastal erosion. One is to build protective barriers such as seawalls. Seawalls, such as the one at Galveston, Texas (see the Prologue), can be effective, but they are tremendously expensive to construct and maintain. More than $50 million has been spent to replenish the beach and build a seawall at Ocean City, Maryland, in just five years. Furthermore, barriers retard erosion only in the area directly behind them; Galveston Island west of its seawall has been eroded back about 45 m.

Another option, adopted by both Atlantic City, New Jersey, and Miami Beach, Florida, is to pump sand onto the beaches to replace that lost to erosion (Figure 16.22). This, too, is expensive as the sand must be replenished periodically because erosion is a continuing process. In many areas, groins are constructed to preserve beaches, but unless additional sand is artificially supplied to the beaches, longshore currents invariably erode sand from the downcurrent sides of the groins (Figure 16.11).

Types of Coasts

Coasts are difficult to classify because of variations in the factors controlling their development and variations in their composition and configuration. Rather than attempt to categorize all coasts, we shall simply note that two types of coasts have already been discussed, those dominated by deposition and those dominated by erosion, and shall look further at the changing relationships between coasts and sea level.

Depositional and Erosional Coasts

Depositional coasts, such as the U.S. Gulf Coast, are characterized by an abundance of detrital sediment and the presence of such depositional landforms as wide sandy beaches, deltas, and barrier islands. In contrast, erosional coasts are generally steep and irregular and typically lack well-developed beaches except in protected areas (Figure 16.9a). They are further characterized by sea cliffs, wave-cut platforms, and sea stacks. Many of the beaches along the west coast of North America fall into this category.

The following section examines coasts in terms of their changing relationships to sea level. But note that while some coasts, such as those in southern California, are described as emergent (uplifted), these same coasts may be erosional as well. In other words, coasts commonly possess features allowing them to be classified in more than one way.

Submergent and Emergent Coasts

If sea level rises with respect to the land or the land subsides, coastal regions are flooded and said to be **submergent** or *drowned* **coasts** (Figure 16.24). Much of the east coast of North America from Maine southward through South Carolina was flooded during the rise in sea level following the Pleistocene Epoch, so that it is extremely irregular. Recall that during the expansion of glaciers during the Pleistocene, sea level was as much as 130 m lower than at present, and that streams eroded their valleys more deeply and extended across continental shelves. When sea level rose, the lower ends of these valleys were drowned, forming *estuaries* such as Delaware and Chesapeake Bays (Figure 16.24). Estuaries are simply the seaward ends of river valleys where seawater and freshwater mix. The divides between adjacent drainage systems on submergent coasts project seaward as broad headlands or a line of islands.

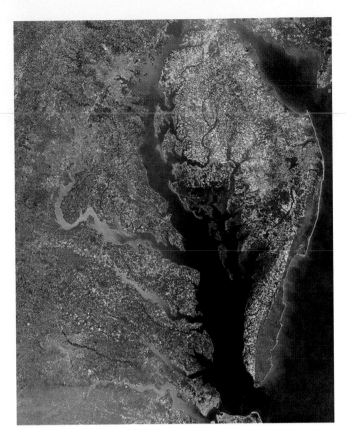

Figure 16.24 *Submergent coasts tend to be extremely irregular with estuaries such as Chesapeake Bay. It formed when the East Coast of the United States was flooded as sea level began rising at the end of the Pleistocene Epoch.*

Submerged coasts are also present at higher latitudes where Pleistocene glaciers flowed into the sea. When sea level rose, the lower ends of the glacial troughs were drowned, forming fiords (see Figure 14.12b).

Emergent coasts are found where the land has risen with respect to sea level (Figure 16.25). Emergence can take place when water is withdrawn from the oceans, as occurred during the Pleistocene expansion of glaciers. Presently, though, coasts are emerging as a result of isostasy or tectonism. In northeastern Canada and the Scandinavian

WHAT WOULD YOU DO?

You and other members of the city council of a beachfront community become alarmed when your pristine beach becomes noticeably narrower each year and tourist revenues fall off markedly. What would you do to identify the source of the problem, and what remedial action might you suggest?

Marine terrace

Figure 16.25 *Emergent coasts tend to be steep and straighter than submergent coasts. Notice the several sea stacks and the sea arch. Also, a marine terrace is visible in the distance.*

countries, for instance, the coasts are irregular because isostatic rebound is elevating formerly glaciated terrain from beneath the sea.

Coasts rising in response to tectonism, on the other hand, tend to be straighter because the seafloor topography being exposed as uplift is smooth. The west coasts of North and South America are rising as a consequence of plate tectonics. Distinctive features of these coasts are marine terraces (Figures 16.19 and 16.25), which are old wave-cut platforms now elevated above sea level. Uplift in such areas appears to be episodic rather than continuous, as indicated by the multiple levels of terraces in some places. In southern California, several terrace levels are present, each of which probably represents a period of tectonic stability followed by uplift. The highest terrace is now about 425 m above sea level.

Chapter Summary

1. Shorelines are continually modified by the energy of waves and longshore currents and, to a limited degree, by tidal currents.

2. The gravitational attraction of the Moon and Sun causes the ocean surface to rise and fall as tides twice daily in most shoreline areas.

3. Waves are oscillations on water surfaces that transmit energy in the direction of wave movement. Surface waves affect the water and seafloor only to wave base, which is equal to one-half the wavelength.

4. Little or no net forward motion of water occurs in waves in the open sea. When waves enter shallow water, they are transformed into waves in which water moves in the direction of wave advance.

5. Wind-generated waves, especially storm waves, are responsible for most geologic work on shorelines, but waves can also be generated by faulting, volcanic explosions, and rockfalls.

6. Breakers form where waves enter shallow water and the orbital motion of water particles is disrupted. The waves become oversteepened and plunge forward or spill onto the shoreline, thus expending their kinetic energy.

7. Waves approaching a shoreline at an angle generate a longshore current. These currents are capable of considerable erosion, transport, and deposition.

8. Rip currents are narrow surface currents that carry water from the nearshore zone seaward through the breaker zone.

9. Beaches, the most common shoreline depositional features, are continually modified by nearshore processes, and their profiles generally exhibit seasonal changes.

10. Spits, baymouth bars, and tombolos all form and grow as a result of longshore current transport and deposition.

11. Barrier islands are nearshore sediment deposits of uncertain origin. They parallel the mainland but are separated from it by a lagoon.

12. The volume of sediment in a nearshore system remains rather constant unless the system is somehow disrupted as when dams are built across streams supplying sand to the system.

13. Many shorelines are characterized by erosion rather than deposition. Such shorelines have sea cliffs and wave-cut platforms. Other features commonly present include sea caves, sea arches, and sea stacks. Depositional coasts are characterized by long sandy beaches, deltas, and barrier islands.

14. Submergent and emergent coasts are defined on the basis of their relationships to changes in sea level.

Important Terms

backshore	crest (wave)	nearshore sediment budget	trough (wave)
barrier island	emergent coast	rip current	wave
baymouth bar	fetch	shoreline	wave base
beach	foreshore	spit	wave-cut platform
beach face	headland	submergent coast	wave refraction
berm	longshore current	tide	
breaker	marine terrace	tombolo	

Review Questions

1. A barrier island
   a. _____ forms when a rocky, steep shoreline is eroded; b. _____ is a long, narrow, sandy island separated from the mainland by a lagoon; c. _____ has sea stacks and arches rising above it; d. _____ originates when shoreline erosion yields a surface sloping gently seaward; e. _____ is likely to be found along an emergent coast.

2. Longshore currents are generated when
   a. _____ isostatic rebound occurs more rapidly than sea level rises; b. _____ the orbital motion of water in waves stirs up seafloor sediment; c. _____ the amount of sand reaching a beach diminishes; d. _____ waves approach a shoreline at an angle; e. _____ sea level rises with respect to a continent.

3. A sand deposit extending into the mouth of a bay is a
   a. _____ headland; b. _____ sea stack; c. _____ spit; d. _____ wave-cut platform; e. _____ berm.

4. The continuous distance that wind blows over a water surface is known as the
   a. _____ fetch; b. _____ baymouth bar; c. _____ beach; d. _____ wave period; e. _____ tombolo.

5. Which one of the following is a depositional landform?
   a. _____ sea cave; b. _____ baymouth bar; c. _____ marine terrace; d. _____ wave base; e. _____ headland.

6. Wave refraction is the phenomenon of
   a. _____ excess water in the nearshore zone flowing out to sea through the breaker zone; b. _____ sediment deposition at the seaward margin of a wave-cut platform;

c. _____ waves oversteepening and plunging forward on beaches; d. _____ offshore sediment transport; e. _____ waves bending so they more nearly parallel a shoreline.

7. The time it takes for the crests (or troughs) of two successive waves to pass a given point is known as the
a. _____ wave period; b. _____ wave celerity; c. _____ wave form; d. _____ wave height; e. _____ wave fetch.

8. Although there are some exceptions, most beaches receive most of their sediment from
a. _____ coastal submergence; b. _____ erosion of shoreline rocks; c. _____ streams; d. _____ erosion of reefs; e. _____ breakers.

9. The part of the backshore zone consisting of a horizontal or gently landward-sloping surface is a
a. _____ berm; b. _____ beach face; c. _____ crest; d. _____ tide; e. _____ spit.

10. Erosional remnants of a shoreline rising above a wave-cut platform are
a. _____ embayments; b. _____ headlands; c. _____ tombolos; d. _____ sea stacks; e. _____ beach cusps.

11. Waves approaching a shoreline at an angle generate or cause
a. _____ flood tides; b. _____ longshore currents; c. _____ headland retreat; d. _____ deposition of sea stacks; e. _____ coastal emergence.

12. The solvent activity of salt water is _____ whereas the direct impact of water on a shoreline is _____.
a. _____ corrosion/hydraulic action; b. _____ wave oscillation/fetch; c. _____ terracing/wave reflection; d. _____ rip current/longshore drift; e. _____ salt wedging/shoreline progradation.

13. What are the characteristics of emergent and submergent coasts?

14. Why does an observer at a shoreline location experience two high and two low tides daily?

15. What is wave base, and how does it affect waves as they enter shallow water?

16. Explain how a longshore current is generated and how it transports and deposits sand.

17. Name and describe three coastal landforms resulting from shoreline deposition.

18. While driving along North America's west coast, you notice a rather flat surface dipping gently seaward with several isolated masses of rock rising above it. What is this landform, and how did it originate?

19. What are the positive and negative aspects of building groins along a shoreline?

20. How and why do summer and winter beaches differ?

21. What are the similarities and differences between spits and tombolos?

22. How do barrier islands migrate landward? What evidence indicates that some U.S. barrier islands are migrating? Give some specific examples.

23. Why are long, broad, sandy beaches more common in eastern North America than western North America?

24. A hypothetical nearshore area has a balanced sand budget, but a dam is constructed across the stream supplying most of the land-derived sand and a seawall is built to protect sea cliffs from erosion. What will likely happen in this area as a result of these projects?

 # World Wide Web Activities

For these web site addresses, along with current updates and exercises, log on to **http://www.brookscole.com/geo**

U.S. Geological Survey: The National Marine and Coastal Geology Program

This government site is dedicated to the coastal zone management program of the National Ocean Service. Investigate and see what the mission of the Coast Zone Management Program is.

The Natural History of Nova Scotia

The Nova Scotia Museum of Natural History maintains this site, which gives an overview of the geology, land and seascapes, plants, animals, and ecology of Nova Scotia. Under *Topics* see *Coastal Landforms* for images and discussions of various coastal features.

Park Geology Tour—Shoreline Geology

This is one of many sites maintained by the National Park Service. It has a menu listing many National Parks, National Monuments, and National Seashores and Lakeshores. Read about and see images of Acadia National Park, Maine; Cape Hatteras National Seashore, North Carolina; and Point Reyes National Seashore, California. How do the shoreline features in these three areas compare, and what processes are important in their geologic evolution?

U.S. Geological Survey Woods Hole Center, Coastal and Marine Geology

At this site, scroll down to *Earth History* and click *Chesapeake Bay Boldie*. What is a boldie, and how was one important in the geologic evolution of Chesapeake Bay?

CD-ROM Exploration

Exploring your *Earth Systems Today* CD-ROM will add to your understanding of the material in this chapter.

Topic: Surficial Processes and Hydrosphere

Module: Waves, Tides, and Currents

Explore activities in this module to see if you can discover the following for yourself:

Using activities in this module, explore the changes that occur in wind-generated waves as they move from deep to shallow water. Specifically, note how these waves behave in shallow water at the shoreline, and relate these to shoreline processes.

CHAPTER 17

Geologic Time: Concepts and Principles

OUTLINE

Three Sisters, in New South Wales, Australia. The Three Sisters are distinctive erosional remnants that formed over a long period of geologic time. They are a popular tourist attraction in Australia's Blue Mountains.

PROLOGUE

What is time? We seem obsessed with it, and organize our lives around it with the help of clocks, calendars, and appointment books. Yet most of us feel we don't have enough of it—we are always running "behind" or "out of time." According to biologists and psychologists, animals and children less than two years old exist in a "timeless present." Some scientists think that our early ancestors may also have lived in a state of timelessness with little or no perception of a past or future. According to Buddhist, Taoist, and Mayan beliefs, time is circular, and like a circle, all things are destined to return to where they once were. Thus, in these belief systems, there is no beginning or end, but rather a cyclicity to everything.

In some respects, time is defined by the methods used to measure it. Many prehistoric monuments are oriented to detect the summer solstice, and sundials were used to divide the day into measurable units. As civilization advanced, mechanical devices were invented to measure time, the earliest being the water clock, first used by the ancient Egyptians and further developed by the Greeks and Romans. The pendulum clock was invented in the 17th century and provided the most accurate timekeeping for the next two and one-half centuries.

Today the quartz watch is the most popular time-piece. Powered by a battery, a quartz crystal vibrates approximately 100,000 times per second. An integrated circuit counts these vibrations and converts them into a digital or dial reading on your watch face. An inexpensive quartz watch today is more accurate than the best mechanical watch, and precision-manufactured quartz clocks are accurate to within one second per 10 years.

Precise timekeeping is important in our technological world. Ships and aircraft plot their locations by satellite, relying on an extremely accurate time signal. Deep-space probes such as the *Voyagers* and the *Mars Pathfinder* and robotic rover *Sojourner* (see Chapter 19), require radio commands timed to billionths of a second, while physicists exploring the mo-

tion inside the nucleus of an atom deal in trillionths of a second as easily as we talk about minutes.

To achieve such accuracy, scientists use atomic clocks. First developed in the 1940s, these clocks rely on an atom's oscillating electrons, a rhythm so regular that they are accurate to within a few thousandths of a second per day. An atomic clock accurate to within one second per 3 million years was installed at the National Institute of Standards and Technology (NIST). Named the NIST-7, this clock uses lasers instead of a magnetic field to stimulate cesium 133 atoms, whose electrons oscillate at a predictable rate (9,192,631,770 vibrations per second).

Whereas physicists deal with incredibly short intervals of time, astronomers and geologists are concerned with geologic time measured in millions or billions of years. When astronomers look at a distant galaxy, they are seeing what it looked like billions of years ago. When geologists investigate rocks in the walls of the Grand Canyon, they are deciphering events that occurred over an interval of 2 billion years. Geologists can measure decay rates of such radioactive elements as uranium, thorium, and rubidium to determine how long ago an igneous rock formed. Furthermore, geologists know that Earth's rotational velocity has been slowing a few thousandths of a second per century as a result of the frictional effects of tides, ocean currents, and varying thicknesses of polar ice. Five hundred million years ago a day was only 20 hours long, and at the current rate of slowing, 200 million years from now a day will be 25 hours long.

Time is a fascinating topic that has been the subject of numerous essays and books. And although we can comprehend concepts such as milliseconds and understand how a quartz watch works, deep time, or geologic time, is still difficult for most people to comprehend. In fact, geology could not have advanced as a science until the foundation of geologic time was firmly established and accepted.

OBJECTIVES

At the end of this chapter, you will have learned that

■ The concept of geologic time and its measurement have changed through human history.

■ The principle of uniformitarianism is fundamental to geology.

■ Lord Kelvin seemingly invalidated the uniformitarian foundation of geology but his basic premise was false.

■ The fundamental principles of relative dating provide a means to interpret geologic history.

■ The three types of unconformities—disconformities, angular unconformities, and nonconformities—are erosional surfaces separating younger from older rocks and represent significant intervals of geologic time for which we have no record.

■ Time equivalency of rock units can be demonstrated by various correlation techniques.

■ Different absolute-dating methods are used to date geologic events in terms of years before the present.

■ The most accurate radiometric dates are obtained from igneous rocks.

■ The geologic time scale evolved primarily during the 19th century through the efforts of many people.

■ There are two fundamentally different kinds of stratigraphic units: those based on content and those related to geologic time.

Introduction

Time sets geology apart from most of the other sciences, and an appreciation of the immensity of geologic time is fundamental to understanding both the physical and biological history of our planet. The understanding and acceptance of the enormity of geologic time is one of the major contributions geology has made to the sciences.

Most people have difficulty comprehending geologic time because we tend to view time from the perspective of our own existence. Ancient history is what occurred among humans hundreds or even thousands of years ago, but when geologists talk of ancient geologic history, they are referring to events that happened millions or even billions of years ago!

Geologists use two different frames of reference when discussing geologic time. **Relative dating** involves placing geologic events in a sequential order as determined from their position in the geologic record. Relative dating will not tell us how long ago a particular event occurred, only that one event preceded another. A useful analogy for relative dating is a television guide without the times when programs are shown. In this example, you cannot tell what time a particular program will be shown, but by watching a few shows and checking the guide, you can determine whether you have missed the show or how many shows are scheduled before the one you want to see.

The various principles used to determine relative dates were discovered hundreds of years ago and since then have

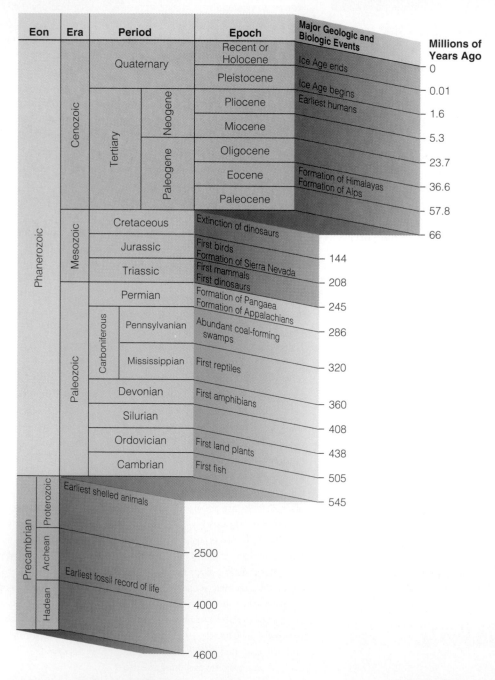

Figure 17.1 *The geologic time scale. Some of the major biological and geological events are indicated along the right-hand margin.*

been used to construct the *relative geologic time scale* (Figure 17.1). These principles are still widely used today.

Absolute dating results in specific dates for rock units or events expressed in years before the present. In our analogy of the television guide, the time when the programs were actually shown would be the absolute dates. In this way, you not only could determine whether you have missed a show (relative dating), but also would know how long it would be until a show you want to see would be shown (absolute dating).

Radiometric dating is the most common method of obtaining absolute ages. Such dates are calculated from the natural rates of decay of various radioactive elements present in trace amounts in some rocks. Not until the discovery of radioactivity near the end of the 19th century could absolute ages be accurately applied to the relative geologic time scale. Today the geologic time scale is really a dual scale: a relative scale based on rock sequences with radiometric dates expressed as years before the present (Figure 17.1).

Why Should You Study Geologic Time?

Besides fostering an appreciation for the immensity of geologic time, why is the study of geologic time important? One of the most important lessons to be learned in this chapter is how to reason and apply the fundamental geologic principles to solve geologic problems. Most students will not become professional geologists, but the logic used in applying the principles of relative dating to reconstruct the geologic history of an area are basic reasoning skills that can be transferred and used in almost any profession.

The study of geologic time is also important for a number of other reasons. The discovery of radioactivity at the end of the 19th century provided geologists with the means to determine the age of minerals and rocks. Advances and refinements in the various radiometric dating techniques during the 20th century have changed the way we view Earth in terms of when events occurred in the past and the rates of geologic change through time. The ability to accurately determine past climatic changes and their causes has important implications for the current debate on global warming and its effects on humans (see Perspective 17.1).

How Has the Concept of Geologic Time and Earth's Age Changed Throughout Human History?

The concept of geologic time and its measurement have changed through human history. Many early Christian scholars and clerics tried to establish the date of creation by analyzing historical records and the genealogies found in Scripture. Based on their analyses, they generally believed that Earth and all of its features were no more than about 6000 years old. The idea of a very young Earth provided the basis for most Western chronologies of Earth history prior to the 18th century.

During the 18th and 19th centuries, several attempts were made to determine Earth's age on the basis of scientific evidence rather than revelation. For example, the French zoologist Georges Louis de Buffon (1707–1788) assumed that Earth gradually cooled to its present condition from a molten beginning. To simulate this history, he melted iron balls of various diameters and allowed them to cool to the surrounding temperature. By extrapolating their cooling rate to a ball the size of Earth, he determined that Earth was at least 75,000 years old. While this age was much older than that derived from Scripture, it was still vastly younger than we now know the planet to be.

Other scholars were equally ingenious in attempting to calculate Earth's age. For example, if deposition rates could be determined for various sediments, geologists reasoned that they could calculate how long it would take to deposit any rock layer. They could then extrapolate how old Earth was from the total thickness of sedimentary rock in its crust. Rates of deposition vary, however, even for the same type of rock. Furthermore, it is impossible to estimate how much rock has been removed by erosion, or how much a rock sequence has been reduced by compaction. As a result of these variables, estimates ranged from less than 1 million years to more than 2 billion years.

Another attempt to determine Earth's age involved ocean salinity. Scholars assumed that Earth's ocean waters were originally fresh and that their present salinity was the result of dissolved salt being carried into the ocean basins by streams. Knowing the volume of ocean water and its salinity, John Joly, a 19th-century Irish geologist, measured the amount of salt currently in the world's streams. He then calculated that it would have taken at least 90 million years for the oceans to reach their present salinity level. This was still much younger than the now accepted age of 4.6 billion years for Earth, mainly because Joly had no way of calculating how much salt had been recycled or the amount of salt stored in continental salt deposits and seafloor clay deposits.

Besides trying to determine Earth's age, naturalists of the 18th and 19th centuries were also formulating some of the fundamental geologic principles that are still used in deciphering Earth history. From the evidence preserved in the geologic record, it was clear to them that Earth is very old and that geologic processes have operated over long periods of time.

Geologic Time and Climate Change

With all the debate concerning global warming and its possible implications, it is extremely important to be able to reconstruct past climatic regimes as accurately as possible. If we are to model how Earth's climate system has responded to changes in the past and use that information for simulations of future climatic scenarios, it is essential that we have as precise a geologic calendar as possible.

One way to study climatic changes is to examine lake sediment or ice cores that have organic matter in them. By taking closely spaced samples and dating the organic matter in the cores using the carbon 14 dating technique (see Radiocarbon and Tree-Ring Dating Methods section), a detailed chronology for each core examined can be constructed. Changes in isotope ratios, pollen, and plant and invertebrate fossil assemblages can then be accurately dated, and the time and duration of climate changes correlated over increasingly larger areas. Without a means of precise dating, there would be no way to accurately model past climatic changes with the precision needed to predict possible future climate changes on a human time-scale.

An interesting method that is becoming more common in reconstructing past climates is the analysis of stalagmites from caves. Stalagmites are icicle-shaped structures rising from a cave floor and formed of calcium carbonate precipitated from evaporating water (see Chapter 13). A stalagmite therefore records a layered history because each newly precipitated layer of calcium carbonate is younger than the previously precipitated layer. Thus a stalagmite's layers are oldest in the center at its base and get progressively younger outward. Using techniques developed during the past 10 years, geologists can achieve very precise absolute dates on individual layers of a stalagmite using high-precision ratios of uranium 234 to thorium 230. This technique lets geologists determine the age of materials much older than can be obtained by carbon 14 dating and is reliable back to about 500,000 years.

An interesting recent study of stalagmites from Crevice Cave in Missouri revealed a history of climatic and vegetation change in the midcontinent region of the United States during the interval between 75,000 and 25,000 years ago. By analyzing carbon 13 and oxygen 18 isotope profiles during this interval, geologists were able to deduce that average temperature fluctuations of 4°C correlated with major changes in vegetation. During the interval between 75,000 and 55,000 years ago, the climate oscillated between warm and cold, and vegetation varied from forest to savannah to prairie. Fifty-five thousand years ago the climate cooled and there was a sudden change from grasslands to forest, which persisted until 25,000 years ago. This corresponds to the time when global ice sheets began building and advancing.

High-precision uranium 234–thorium 230 dating techniques in stalagmite studies provides an accurate chronology allowing geologists to model climate systems of the past and perhaps determine what causes global climatic changes and their durations. Without these sophisticated dating techniques and others like them, geologists would not be able to make precise correlations and accurately reconstruct past environments and climates. By analyzing past environmental and climate changes and their durations, geologists hope they can use these data, sometime in the near future, to predict and possibly modify regional climatic changes.

Who Is James Hutton and Why Are His Contributions to Geology Important?

The Scottish geologist James Hutton (1726–1797) is considered by many to be the father of modern geology. His detailed studies and observations of rock exposures and present-day geological processes were instrumental in establishing the **principle of uniformitarianism** (see Chapter 1), the concept that the same processes have operated over vast amounts of time. Because Hutton relied on known processes to account for Earth history, he concluded that Earth must be very old and wrote that "we find no vestige of a beginning, and no prospect of an end."

Unfortunately, Hutton was not a particularly good writer, so his ideas were not widely disseminated or accepted. In 1830, Charles Lyell published a landmark book, *Principles of Geology,* in which he championed Hutton's concept of uniformitarianism. Instead of relying on catastrophic events to explain various Earth features, Lyell recognized that imperceptible changes brought about by present-day processes could, over long periods of time, have tremendous cumulative effects. Through his writings, Lyell firmly established uniformitarianism as the guiding principle of geology. Furthermore, the recognition of virtually limitless time was also necessary for, and instrumental in, the acceptance of Darwin's 1859 theory of evolution (see Chapter 18).

After finally establishing that present-day processes have operated over vast periods of time, geologists were nevertheless nearly forced to accept a young age for Earth when a highly respected English physicist, Lord Kelvin (1824–1907), claimed, in a paper written in 1866, to have destroyed the uniformitarian foundation of geology. Starting with the generally accepted belief that Earth was originally molten, Kelvin assumed that it has gradually been losing heat and that, by measuring this heat loss, he could determine its age.

Kelvin knew from deep mines in Europe that Earth's temperature increases with depth, and he reasoned that Earth is losing heat from its interior. By knowing the melting temperatures of rocks, the size of Earth, and the rate of heat loss, Kelvin was able to calculate the age at which Earth was entirely molten. From these calculations, he concluded that Earth could not be older than 400 million years or younger than 20 million years. This wide range in age reflected uncertainties over average temperature increases with depth and the various melting points of Earth's constituent materials.

After finally establishing that Earth was very old, and showing how present-day processes can be extended back over long periods of time to explain geological features, geologists were in a quandary. Either they had to accept Kelvin's dates and squeeze events into a shorter time frame or reject his calculations.

Kelvin's reasoning and calculations were sound but his basic premises were false, thereby invalidating his conclusions. Kelvin was unaware that Earth has an internal heat source, radioactivity, that has allowed it to maintain a fairly constant temperature through time.* His 40-year campaign for a young Earth ended with the discovery of radioactivity near the end of the nineteenth century. His calculations were therefore no longer valid, and his proof for a geologically young Earth collapsed. Moreover, although the discovery of radioactivity destroyed Kelvin's arguments, it provided geologists with a clock that could measure Earth's age and validate what geologists had been saying all along, namely, that Earth was indeed very old!

What Are Relative-Dating Methods and Why Are They Important?

Before the development of radiometric dating techniques, geologists had no reliable means of absolute dating and therefore depended solely on relative-dating methods. These methods allow events to be placed in sequential order and do not tell us how long ago an event took place.

*Actually, Earth's temperature has decreased through time because the original amount of radioactive materials has been decreasing and thus is not supplying as much heat. However, the temperature is decreasing at a rate considerably slower than would be required to lend any credence to Kelvin's calculations.

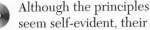 Although the principles of relative dating may now seem self-evident, their discovery was an important scientific achievement because they provided geologists with a means to interpret geologic history and develop a relative geologic time scale.

Six fundamental geologic principles are used in relative dating: *superposition, original horizontality, lateral continuity, cross-cutting relationships, inclusions,* and *fossil succession.*

Fundamental Principles of Relative Dating

The seventeenth century was an important time in the development of geology as a science because of the widely circulated writings of the Danish anatomist Nicolas Steno (1638–1686). Steno observed that when streams flood, they spread out across their floodplains and deposit layers of sediment that bury organisms dwelling on the floodplain. Subsequent floods produce new layers of sediments that are deposited or superposed over previous deposits. When lithified, these layers of sediment become sedimentary rock. Thus, in an undisturbed succession of sedimentary rock layers, the oldest layer is at the bottom and the youngest layer is at the top. This **principle of superposition** is the basis for relative age determinations of strata and their contained fossils (Figure 17.2).

Steno also observed that because sedimentary particles settle from water under the influence of gravity, sediment is deposited in essentially horizontal layers, thus illustrating the **principle of original horizontality** (Figure 17.2). Therefore, a sequence of sedimentary rock layers that is steeply inclined from the horizontal must have been tilted after deposition and lithification.

Steno's third principle, the **principle of lateral continuity,** states that sediment extends laterally in all directions until it thins and pinches out or terminates against the edge of the depositional basin (Figure 17.2).

James Hutton is credited with discovering the **principle of cross-cutting relationships.** Based on his detailed studies and observations of rock exposures in Scotland, Hutton recognized that an igneous intrusion or fault must be younger than the rocks it intrudes or displaces (Figure 17.3).

Although this principle illustrates that an intrusive igneous structure is younger than the rocks it intrudes, the association of sedimentary and igneous rocks may cause problems in relative dating. Buried lava flows and intrusive igneous bodies such as sills look very similar in a sequence of strata (Figure 17.4). A buried lava flow, however, is older than the rocks above it (principle of superposition), while a sill is younger than all the beds below it and younger than the bed immediately above it.

To resolve such relative age problems as these, geologists look to see if the sedimentary rocks in contact with

Figure 17.2 *The Grand Canyon of Arizona illustrates three of the six fundamental principles of relative dating. The sedimentary rocks of the Grand Canyon were originally deposited horizontally in a variety of marine and continental environments (principle of original horizontality). The oldest rocks are at the bottom of the canyon, and the youngest rocks are at the top, forming the rim (principle of superposition). The exposed rock layers extend laterally for some distance (principle of lateral continuity).*

(a)

(b)

Figure 17.3 *The principle of cross-cutting relationships. (a) A dark-colored dike has been intruded into older light-colored granite along the north shore of Lake Superior, Ontario, Canada. (b) A small fault displacing tilted beds along Templin Highway, Castaic, California.*

the igneous rocks show signs of baking or alteration by heat (see Chapter 6, Contact Metamorphism). A sedimentary rock showing such effects must be older than the igneous rock with which it is in contact. In Figure 17.4, for example, a sill produces a zone of baking immediately above and below it because it intruded into previously existing sedimentary rocks. A lava flow, in contrast, bakes only those rocks below it.

Another way to determine relative ages is by using the **principle of inclusions.** This principle holds that inclusions, or fragments of one rock contained within a layer of

another, are older than the rock layer itself. The batholith shown in Figure 17.5 contains sandstone inclusions, and the sandstone unit shows the effects of baking. Accordingly, we conclude that the sandstone is older than the batholith. In Figure 17.5b, however, the sandstone contains granite rock fragments, indicating that the batholith was the source rock for the inclusions and is therefore older than the sandstone.

Fossils have been known for centuries (see Chapter 18), yet their utility in relative dating and geologic mapping was not fully appreciated until the early 19th century. William

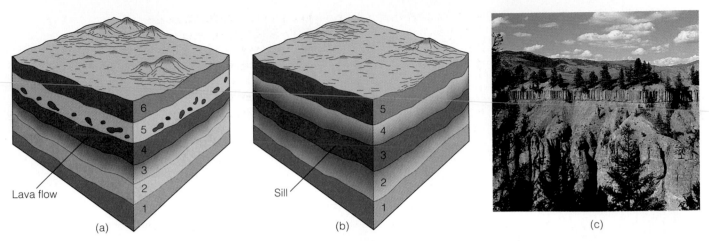

Figure 17.4 *Relative ages of lava flows, sills, and associated sedimentary rocks may be difficult to determine. (a) A buried lava flow in bed 4 baked the underlying bed, and bed 5 contains inclusions of the lava flow. The lava flow is younger than bed 3 and older than beds 5 and 6. (b) The rock units above and below the sill in bed 3 have been baked, indicating that the sill is younger than beds 2 and 4, but its age relative to bed 5 cannot be determined. (c) Buried lava flow, Yellowstone National Park, Wyoming. Lava flow displays columnar jointing.*

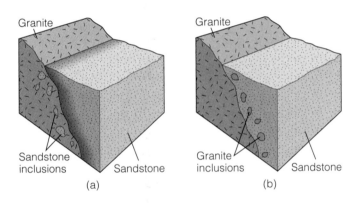

Figure 17.5 *The principle of inclusions. (a) The granite batholith is younger than the sandstone because the sandstone has been baked at its contact with the granite and the granite contains sandstone inclusions. (b) Granite inclusions in the sandstone indicate that the batholith was a source of the sandstone and therefore is older. (c) Outcrop in northern Wisconsin showing basalt inclusions (black) in granite (white). Accordingly, the basalt inclusions are older than the granite.*

Smith (1769–1839), an English civil engineer involved in surveying and building canals in southern England, independently recognized the principle of superposition by reasoning that the fossils at the bottom of a sequence of strata are older than those at the top of the sequence. This recognition served as the basis for the **principle of fossil succession** or the *principle of faunal and floral succession* as it is sometimes called (Figure 17.6).

According to this principle, fossil assemblages succeed one another through time in a regular and predictable order. The validity and successful use of this principle depends on three points: (1) Life has varied through time, (2) fossil assemblages are recognizably different from one another, and (3) the relative ages of the fossil assemblages can be determined. Observations of fossils in older versus younger strata clearly demonstrate that life-forms have

What Are Relative-Dating Methods and Why Are They Important?   **491**

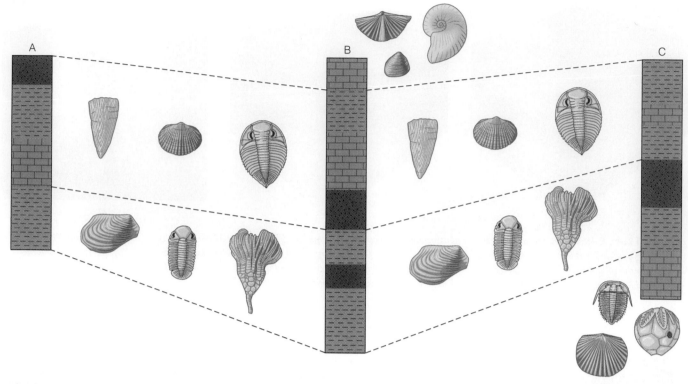

Figure 17.6 *This generalized diagram shows how geologists use the principle of fossil succession to identify strata of the same age in different areas. The rocks in the three sections encompassed by the dashed lines contain similar fossils and are thus the same age. Note that the youngest rocks in this region are in section B, whereas the oldest rocks are in section C.*

changed. Because this is true, fossil assemblages (point 2) are recognizably different. Furthermore, superposition can also be used to demonstrate the relative ages of the fossil assemblages.

Unconformities

Our discussion so far has been concerned with conformable sequences of strata, sequences in which no depositional breaks of any consequence occur. A sharp bedding plane (see Figure 5.25) separating strata may represent a depositional break of minutes, hours, years, or even tens of years, but it is inconsequential when considered in the context of geologic time.

Surfaces of discontinuity that encompass significant amounts of geologic time are **unconformities,** and any interval of geologic time not represented by strata in a particular area is a *hiatus* (Figure 17.7). Thus, an unconformity is a surface of nondeposition or erosion that separates younger strata from older rocks. As such, it represents a break in our record of geologic time. The famous 12-minute gap in the Watergate tapes of Richard Nixon's presidency is somewhat analogous. Just as we have no record of the conversations that were occurring during this period of time, we have no record of the events that occurred during a hiatus.

Three types of unconformities are recognized. A **disconformity** is a surface of erosion or nondeposition between younger and older beds that are parallel with one another (Figure 17.8). Unless a well-defined erosional surface separates the older from the younger parallel beds, the disconformity frequently resembles an ordinary bedding plane. Accordingly, many disconformities are difficult to recognize and must be identified on the basis of fossil assemblages.

An **angular unconformity** is an erosional surface on tilted or folded strata over which younger strata have been deposited (Figure 17.9). Both younger and older strata may dip, but if their dip angles are different (generally the older strata dip more steeply), an angular unconformity is present.

The angular unconformity illustrated in Figure 17.9b is probably the most famous in the world. It was here at Siccar Point, Scotland, that James Hutton realized that severe upheavals had tilted the lower rocks and formed mountains that were then worn away and covered by younger, flat-lying rocks. The erosional surface between the older tilted rocks and the younger flat-lying strata meant that there was a significant gap in the geologic record. Although Hutton did not use the term unconformity, he was the first to understand and explain the significance of such discontinuities in the geologic record.

The third type of unconformity is a **nonconformity.** Here an erosion surface cut into metamorphic or igneous

Figure 17.7 *Simplified diagram showing the development of an unconformity and a hiatus. (a) Deposition began 12 million years ago (MYA) and continued more or less uninterrupted until 4 MYA. (b) A 1-million-year episode of erosion occurred, and during that time strata representing 2 million years of geologic time were eroded. (c) A hiatus of 3 million years exists between the older strata and the strata that formed during a renewed episode of deposition that began 3 MYA. (d) The actual stratigraphic record. The unconformity is the surface separating the strata and represents a major break in our record of geologic time.*

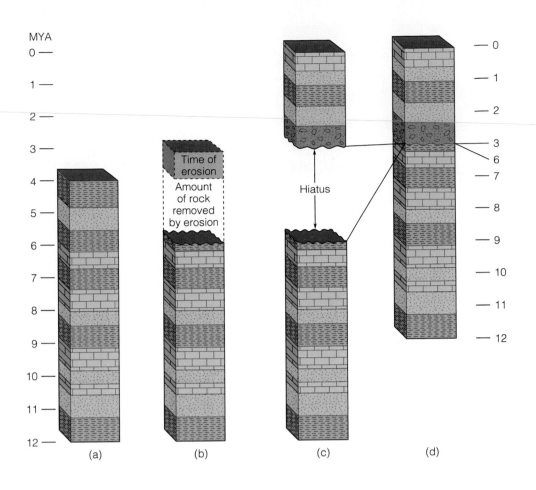

rocks is covered by sedimentary rocks (Figure 17.10). This type of unconformity closely resembles an intrusive igneous contact with sedimentary rocks. The principle of inclusions is helpful in determining whether the relationship between the underlying igneous rocks and the overlying sedimentary rocks is the result of an intrusion or erosion. In the case of an intrusion, the igneous rocks are younger, but in the case of erosion, the sedimentary rocks are younger. Being able to distinguish between a nonconformity and an intrusive contact is important because they represent different sequences of events.

Applying the Principles of Relative Dating

We can decipher the geologic history of the area represented by the block diagram in Figure 17.11 by applying the various relative-dating principles just discussed. The methods and logic used in this example are the same as those applied by nineteenth-century geologists in constructing the geologic time scale.

According to the principles of superposition and original horizontality, beds A–G were deposited horizontally; then they were either tilted, faulted (H), and eroded, or after deposition, they were faulted (H), tilted, and then

eroded (Figure 17.12a b, and c). Because the fault cuts beds A–G, it must be younger than the beds according to the principle of cross-cutting relationships.

Beds J–L were then deposited horizontally over this erosional surface producing an angular unconformity (I) (Figure 17.12d). Following deposition of these three beds, the entire sequence was intruded by a dike (M), which, according to the principle of cross-cutting relationships must be younger than all the rocks it intrudes (Figure 17.12e).

The entire area was then uplifted and eroded; next beds P and Q were deposited, producing a disconformity (N) between beds L and P and a nonconformity (O) between the igneous intrusion M and the sedimentary bed P (Figure 17.12f and g). We know that the relationship between igneous intrusion M and the overlying sedimentary bed P is a nonconformity because of the presence of inclusions of M in P (principle of inclusions).

At this point, there are several possibilities for constructing the geologic history of this area. According to the principle of cross-cutting relationships, dike R must be younger than bed Q because it intrudes into it. It could have intruded anytime *after* bed Q was deposited; however, we cannot determine whether R was formed right after Q, right after S, or after T was formed. For purposes of this history, we will say that it intruded after the deposition of bed Q (Figure 17.12g and h).

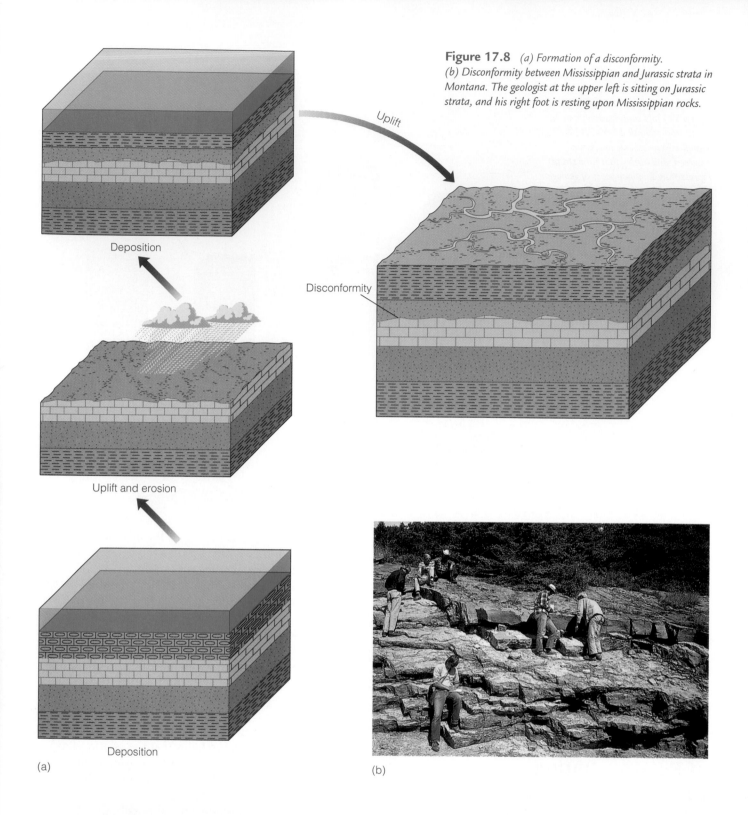

Figure 17.8 *(a) Formation of a disconformity.*
(b) Disconformity between Mississippian and Jurassic strata in Montana. The geologist at the upper left is sitting on Jurassic strata, and his right foot is resting upon Mississippian rocks.

Uplift

Deposition

Disconformity

Uplift and erosion

Deposition

(a)

(b)

Following the intrusion of dike R, lava S flowed over bed Q, followed by the deposition of bed T (Figure 17.12i and j). Although the lava flow (S) is not a sedimentary unit, the principle of superposition still applies because it flowed onto the surface, just as sediments are deposited on Earth's surface.

We have established a relative chronology for the rocks and events of this area by using the principles of relative dating. Remember, however, that we have no way of knowing how many years ago these events occurred unless we can obtain radiometric dates for the igneous rocks. With these dates we can establish the range of absolute ages between which the different sedimentary units were deposited and also determine how much time is represented by the unconformities.

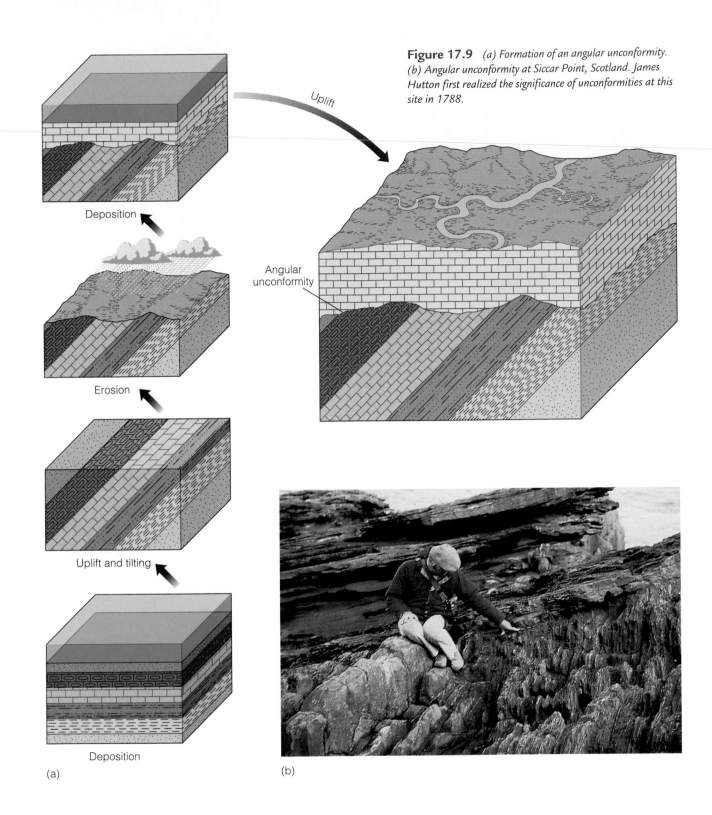

Figure 17.9 *(a) Formation of an angular unconformity. (b) Angular unconformity at Siccar Point, Scotland. James Hutton first realized the significance of unconformities at this site in 1788.*

Uplift

Angular unconformity

Deposition

Erosion

Uplift and tilting

Deposition

(a)

(b)

How Do Geologists Correlate Rock Units?

To decipher Earth history, geologists must demonstrate the time equivalency of rock units in different areas. This process is known as **correlation.**

If exposures are adequate, units may simply be traced laterally (principle of lateral continuity), even if occasional gaps exist (Figure 17.13). Other criteria used to correlate units are similarity of rock type, position in a sequence, and key beds. *Key beds* are units, such as coal beds or volcanic ash layers, that are sufficiently distinctive to allow identification of the same unit in different areas (Figure 17.13).

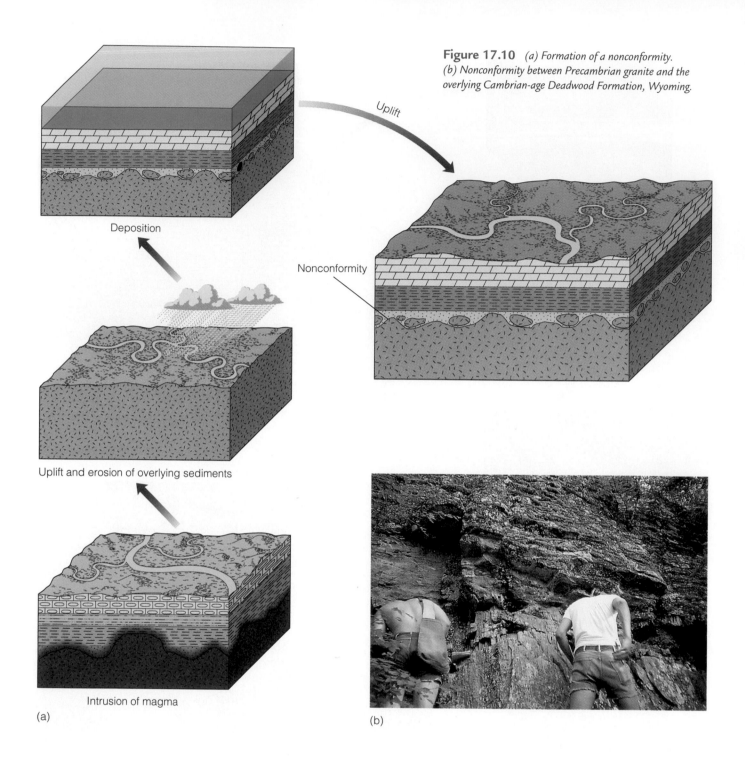

Uplift

Deposition

Nonconformity

Uplift and erosion of overlying sediments

Intrusion of magma

(a)

(b)

Generally, no single location in a region has a geologic record of all the events that occurred during its history; therefore geologists must correlate from one area to another in order to determine the complete geologic history of the region. An excellent example is provided by the history of the Colorado Plateau (Figure 17.14). A record of events occurring over approximately 2 billion years of Earth history is present in this region. Because of the forces of erosion, the entire record is not preserved at any single location. Within the walls of the Grand Canyon are

rocks of the Precambrian and Paleozoic Era, whereas Paleozoic and Mesozoic Era rocks are found in Zion National Park, and Mesozoic and Cenozoic Era rocks are exposed in Bryce Canyon (Figure 17.14). By correlating the uppermost rocks at one location with the lowermost equivalent rocks of another area, the history of the entire region can be deciphered.

Although geologists match up rocks on the basis of similar rock type and superposition, correlation of this type can only be done in a limited area where beds can be

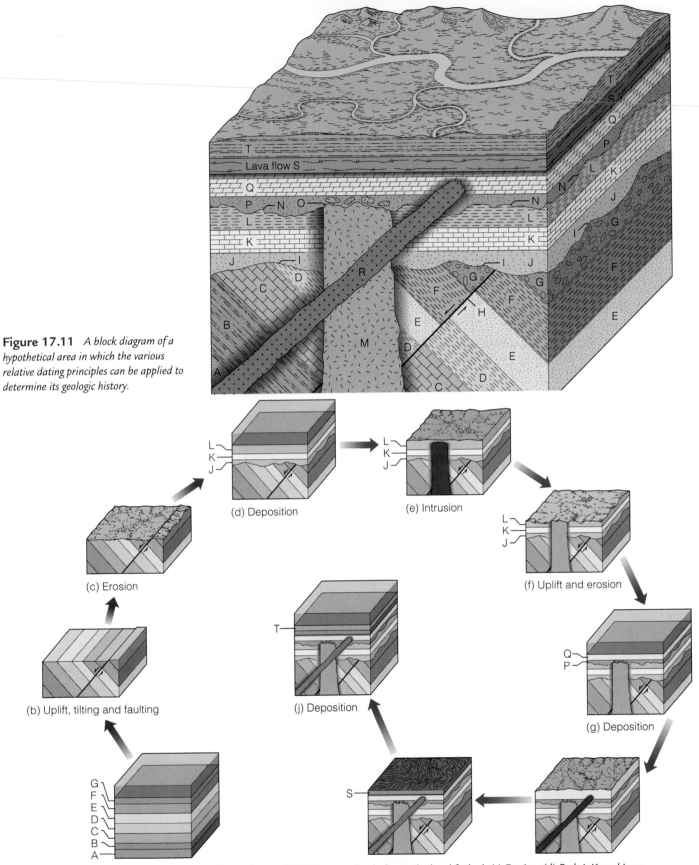

Figure 17.11 *A block diagram of a hypothetical area in which the various relative dating principles can be applied to determine its geologic history.*

Figure 17.12 *(a) Beds A, B, C, D, E, F, and G are deposited. (b) The preceding beds are tilted and faulted. (c) Erosion. (d) Beds J, K, and L are deposited, producing an angular unconformity. (e) The entire sequence is intruded by a dike. (f) The entire sequence is uplifted and eroded. (g) Beds P and Q are deposited, producing a disconformity (N) and a nonconformity (O). (h) Dike R intrudes. (i) Lava (S) flows over bed Q, baking it. (j) Bed T is deposited.*

traced from one site to another. To correlate rock units over a large area or to correlate age-equivalent units of different composition, fossils and the principle of fossil succession must be used.

Fossils are useful as time indicators because they are the remains of organisms that lived for a certain length of time during the geologic past. Fossils that are easily identified, are geographically widespread, and existed for a rather short geologic time are particularly useful. Such fossils are **guide fossils** or *index fossils* (Figure 17.15). The trilobite *Isotelus* and the clam *Inoceramus* meet all of these criteria and are therefore good guide fossils. In contrast, the brachiopod *Lingula* is easily identified and widespread, but its geologic range of Ordovician to Recent makes it of little use in correlation.

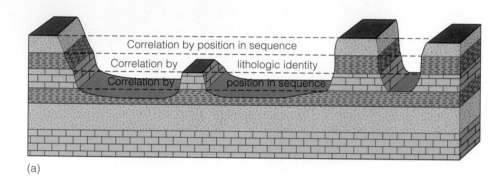

(a)

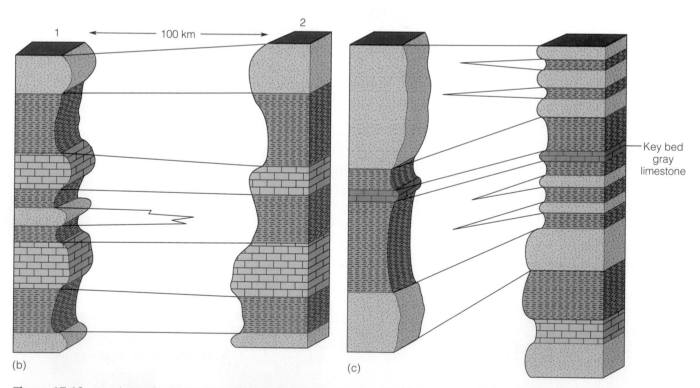

(b)

(c)

Key bed gray limestone

Figure 17.13 *Correlation of rock units. (a) In areas of adequate exposures, rock units can be traced laterally even if occasional gaps exist. (b) Correlation by similarities in rock type and position in a sequence. The sandstone in section 1 is assumed to intertongue or grade laterally into the shale at section 2. (c) Correlation using a key bed, a distinctive gray limestone.*

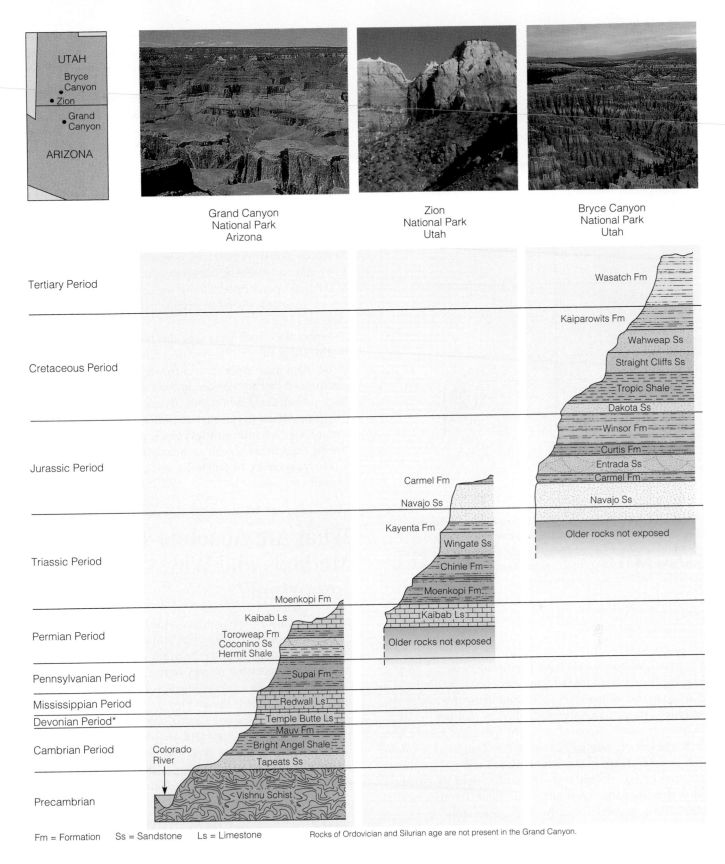

Grand Canyon
National Park
Arizona

Zion
National Park
Utah

Bryce Canyon
National Park
Utah

Tertiary Period

Cretaceous Period

Jurassic Period

Triassic Period

Permian Period

Pennsylvanian Period

Mississippian Period

Devonian Period*

Cambrian Period

Precambrian

Wasatch Fm

Kaiparowits Fm

Wahweap Ss

Straight Cliffs Ss

Tropic Shale

Dakota Ss

Winsor Fm

Curtis Fm

Entrada Ss

Carmel Fm

Carmel Fm

Navajo Ss

Navajo Ss

Older rocks not exposed

Kayenta Fm

Wingate Ss

Chinle Fm

Moenkopi Fm

Moenkopi Fm

Kaibab Ls

Kaibab Ls

Older rocks not exposed

Toroweap Fm
Coconino Ss
Hermit Shale

Supai Fm

Redwall Ls

Temple Butte Ls

Mauv Fm

Bright Angel Shale

Colorado
River

Tapeats Ss

Vishnu Schist

Fm = Formation   Ss = Sandstone   Ls = Limestone   Rocks of Ordovician and Silurian age are not present in the Grand Canyon.

Figure 17.14 *Correlation of rocks within the Colorado Plateau. By correlating the rocks from various locations, the history of the entire region can be deciphered.*

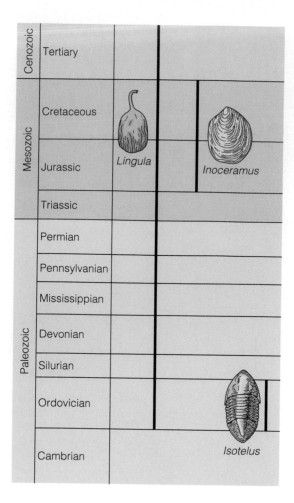

	Tertiary		
Cenozoic			
Mesozoic	Cretaceous		
	Jurassic	*Lingula*	*Inoceramus*
	Triassic		
Paleozoic	Permian		
	Pennsylvanian		
	Mississippian		
	Devonian		
	Silurian		
	Ordovician		*Isotelus*
	Cambrian		

Figure 17.15 *The geologic ranges of three marine invertebrates. The brachiopod* Lingula *is of little use in correlation because of its long geologic range. The trilobite* Isotelus *and the bivalve* Inoceramus *are good guide fossils because they are geographically widespread, are easily identified, and have short geologic ranges.*

Because most fossils have fairly long geologic ranges, geologists construct *assemblage range zones* to determine the age of the sedimentary rocks containing the fossils. Assemblage range zones are established by plotting the overlapping geologic ranges of different species of fossils. The first and last occurrences of two species are used to establish an assemblage zone's boundaries (Figure 17.16). Correlation of assemblage zones generally yields correlation lines that are considered time equivalent. In other words, the strata encompassed by the correlation lines are thought to be the same age.

Subsurface Correlation

In addition to the surface geology, geologists are also interested in subsurface geology, because it provides addi-

tional information about geologic features beneath Earth's surface. A variety of techniques and methods are used to acquire and interpret data about the subsurface geology of an area.

When drilling is being done for oil or natural gas, cores or rock chips called *well cuttings* are commonly recovered from the drill hole. These samples are studied under the microscope and reveal such important information as rock type, porosity (the amount of pore space), permeability (the ability to transmit fluids), and the presence of oil stains. In addition, the samples can also be processed for a variety of microfossils that aid in determining the geologic age of the rock and the environment of deposition.

Geophysical instruments may be lowered down the drill hole to record such rock properties as electrical resistivity and radioactivity, thus providing a record or *well log* of the rocks penetrated. Cores, well cuttings, and well logs are all extremely useful in making subsurface correlations (Figure 17.17).

Subsurface rock units may also be detected and traced by the study of seismic profiles. Energy pulses, such as those from explosions, travel through rocks at a velocity determined by rock density, and some of this energy is reflected from various horizons (contacts between contrasting layers) back to the surface, where it is recorded (see Figure 8.4). Seismic stratigraphy is particularly useful in tracing units in areas such as the continental shelves where it is very expensive to drill holes and other techniques have limited use.

What Are Absolute-Dating Methods and Why Are They Important?

Although most of the isotopes of the 92 naturally occurring elements are stable, some are radioactive and spontaneously decay to other more stable isotopes of elements, releasing energy in the process. The discovery, in 1903 by Pierre and Marie Curie, that radioactive decay produces heat meant that geologists finally had a mechanism for explaining Earth's internal heat that did not rely on residual cooling from a molten origin. Furthermore, geologists had a powerful tool to date geologic events accurately, and thus verify the long time periods postulated by Hutton, Lyell, and Darwin.

Atoms, Elements, and Isotopes

As we discussed in Chapter 2, all matter is made up of chemical elements, each of which is composed of extremely small particles called *atoms*. The nucleus of an atom is composed of *protons* and *neutrons* with *electrons* encircling it (see Figure 2.4). The number of protons de-

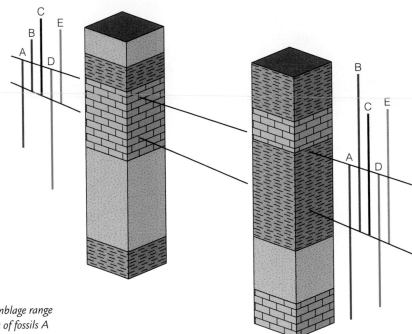

Figure 17.16 *Correlation of two sections by using assemblage range zones. These zones are established by the overlapping ranges of fossils A through E.*

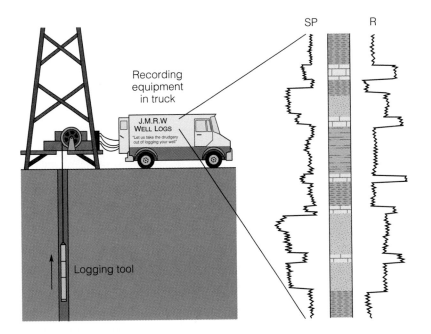

Figure 17.17 *A schematic diagram showing how well logs are made. As the logging tool is withdrawn from the drill hole, data are transmitted to the surface where they are recorded and printed as a well log. The curve labeled SP in this diagrammatic electric log is a plot of self-potential (electrical potential caused by different conductors in a solution that conducts electricity) with depth. The curve labeled R is a plot of electrical resistivity with depth. Electric logs yield information about the rock type and fluid content of subsurface formations. Electric logs are also used to correlate from well to well.*

fines an element's *atomic number* and helps determine its properties and characteristics. The combined number of protons and neutrons in an atom is its *atomic mass number*. However, not all atoms of the same element have the same number of neutrons in their nuclei. These variable forms of the same element are *isotopes*. Most isotopes are stable, but some are unstable and spontaneously decay to a more stable form. Geologists measure the decay rate of unstable isotopes to determine the absolute ages of rocks.

Radioactive Decay and Half-Lives

Radioactive decay is the process whereby an unstable atomic nucleus is spontaneously transformed into an atomic nucleus of a different element. Three types of radioactive decay are recognized, all of which result in a change of atomic structure (Figure 17.18). In **alpha decay,** two protons and two neutrons are emitted from the nucleus, resulting in a loss of two atomic numbers and four atomic mass numbers. In **beta**

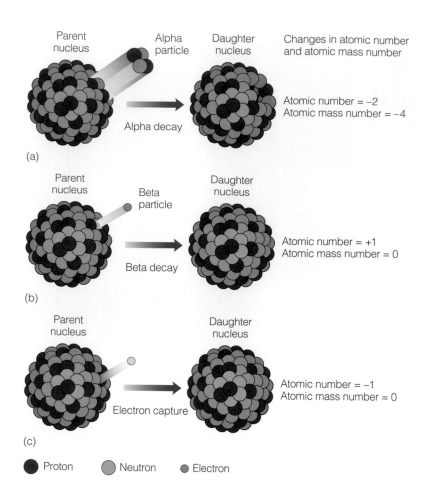

Figure 17.18 *Three types of radioactive decay. (a) Alpha decay, in which an unstable parent nucleus emits two protons and two neutrons. (b) Beta decay, in which an electron is emitted from the nucleus. (c) Electron capture, in which a proton captures an electron and is thereby converted to a neutron.*

Proton   Neutron   Electron

decay, a fast-moving electron is emitted from a neutron in the nucleus, changing that neutron to a proton and consequently increasing the atomic number by one, with no resultant atomic mass number change. **Electron capture** results when a proton captures an electron from an electron shell and thereby converts to a neutron, resulting in a loss of one atomic number and no change in the atomic mass number.

Some elements undergo only one decay step in the conversion from an unstable form to a stable form. For example, rubidium 87 decays to strontium 87 by a single beta emission, and potassium 40 decays to argon 40 by a single electron capture. Other radioactive elements undergo several decay steps. Uranium 235 decays to lead 207 by seven alpha and six beta steps, while uranium 238 decays to lead 206 by eight alpha and six beta steps (Figure 17.19).

When discussing decay rates, it is convenient to refer to them in terms of half-lives. The **half-life** of a radioactive element is the time it takes for one-half of the atoms of the original unstable *parent element* to decay to atoms of a new, more stable *daughter element*.

The half-life of a given radioactive element is constant and can be precisely measured. Half-lives of various radioactive elements range from less than a billionth of a second to 49 billion years.

Radioactive decay occurs at a geometric rate rather than a linear rate. Therefore, a graph of the decay rate produces a curve rather than a straight line (Figure 17.20). For example, an element with *1,000,000* parent atoms will have *500,000* parent atoms and 500,000 daughter atoms after one half-life. After two half-lives, it will have *250,000* parent atoms (one-half of the previous parent atoms, which is equivalent to one-fourth of the original parent atoms) and 750,000 daughter atoms. After three half-lives, it will have *125,000* parent atoms (one-half of the previous parent atoms or one-eighth of the original parent atoms) and 875,000 daughter atoms, and so on until the number of parent atoms remaining is so few that they cannot be accurately measured by present-day instruments.

By measuring the parent-daughter ratio and knowing the half-life of the parent (which has been determined in the laboratory), geologists can calculate the age of a sample contain-

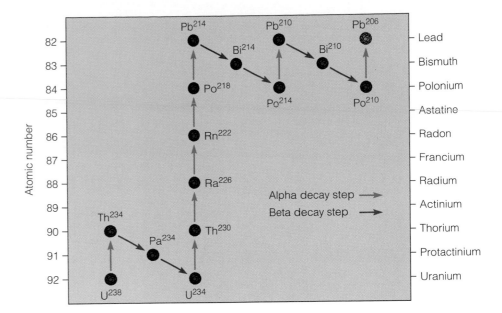

Figure 17.19 *Radioactive decay series for uranium 238 to lead 206. Radioactive uranium 238 decays to its stable end product, lead 206, by eight alpha and six beta decay steps. A number of different isotopes are produced as intermediate steps in the decay series.*

ing the radioactive element. The parent-daughter ratio is usually determined by a *mass spectrometer*, an instrument that measures the proportions of elements of different masses.

Sources of Uncertainty

The most accurate radiometric dates are obtained from igneous rocks. As a magma cools and begins to crystallize, radioactive parent atoms are separated from previously formed daughter atoms. Because they are the right size, some radioactive parent atoms are incorporated into the crystal structure of certain minerals. The stable daughter atoms, though, are a different size from the radioactive parent atoms and consequently cannot fit into the crystal structure of the same mineral as the parent atoms. Therefore a mineral crystallizing in a cooling magma will contain radioactive parent atoms but no stable daughter atoms (Figure 17.21). Thus, the time that is being measured is the time of crystallization of the mineral containing the radioactive atoms, and not the time of formation of the radioactive atoms.

Except in unusual circumstances, sedimentary rocks cannot be radiometrically dated, because one would be measuring the age of a particular mineral rather than the time that it was deposited as a sedimentary particle. One of the few instances in which radiometric dates can be obtained on sedimentary rocks is when the mineral glauconite is present. Glauconite is a greenish mineral containing radioactive potassium 40, which decays to argon 40 (Table 17.1). It forms in certain marine environments as a result of chemical reactions with clay minerals during the conversion from sediments to sedimentary rock. Thus, glauconite forms when the sedimentary rock forms, and a radiometric date indicates the time of the sedimentary rock's origin. Being a gas, however, the daughter product

argon can easily escape from a mineral. Therefore, any date obtained from glauconite, or any other mineral containing the potassium 40/argon 40 pair, must be considered a minimum age.

To obtain accurate radiometric dates, geologists must be sure that they are dealing with a *closed system*, meaning that neither parent nor daughter atoms have been added or removed from the system since crystallization and that the ratio between them results only from radioactive decay. Otherwise, an inaccurate date will result. If daughter atoms have leaked out of the mineral being analyzed, the calculated age will be too young; if parent atoms have been removed, the calculated age will be too great.

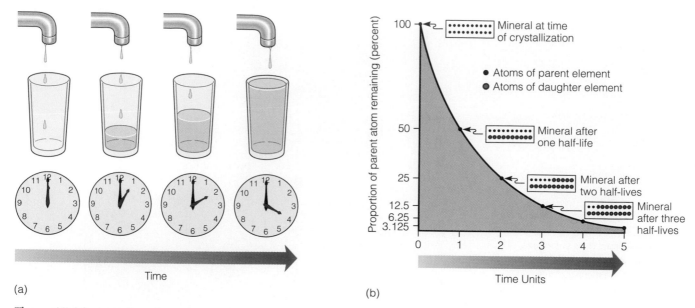

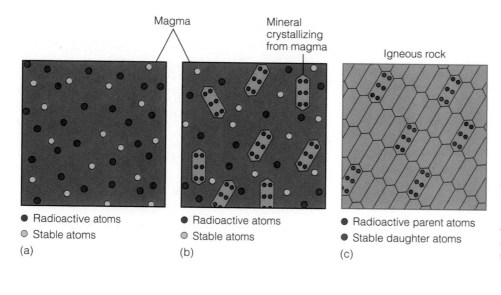

Figure 17.20 *(a) Uniform, linear change is characteristic of many familiar processes. In this example, water is being added to a glass at a constant rate. (b) Geometric radioactive decay curve, in which each time unit represents one half-life, and each half-life is the time it takes for one-half of the parent element to decay to the daughter element.*

Figure 17.21 *(a) A magma contains both radioactive and stable atoms. (b) As the magma cools and begins to crystallize, some radioactive atoms are incorporated into certain minerals because they are the right size and can fit into the crystal structure. Therefore, at the time of crystallization, the mineral will contain 100% radioactive parent atoms and 0% stable daughter atoms. (c) After one half-life, 50% of the radioactive parent atoms will have decayed to stable daughter atoms.*

Leakage may take place if the rock is heated or subjected to intense pressure as can sometimes occur during metamorphism. If this happens, some of the parent or daughter atoms may be driven from the mineral being analyzed, resulting in an inaccurate age determination. If the daughter product was completely removed, then one would be measuring the time since metamorphism (a useful measurement itself), and not the time since crystallization of the mineral (Figure 17.22). Because heat affects the parent-daughter ratio, metamorphic rocks are difficult to date accurately. Remember that while the parent-daughter ratio may be affected by heat, the decay rate of the parent element remains constant, regardless of any physical or chemical changes.

To obtain an accurate radiometric date, geologists must make sure that the sample is fresh and unweathered and that it has not been subjected to high temperatures or intense pressures after crystallization. Furthermore, it is sometimes possible to cross-check the date obtained by measuring the parent-daughter ratio of two different radioactive elements in the same mineral. For example, naturally occurring uranium consists of both uranium 235 and uranium 238 isotopes. Through various decay steps, uranium 235 decays to lead 207, whereas uranium 238 decays to lead 206 (Figure 17.19). If the minerals containing both uranium isotopes have remained closed systems, the ages obtained from each parent-daughter ratio should agree closely and therefore should indicate the time of crystal-

Table 17.1

Five of the Principal Long-Lived Radioactive Isotope Pairs Used in Radiometric Dating

Isotopes		Half-Life of Parent (Years)	Effective Dating Range (Years)	Minerals and Rocks That Can Be Dated	
Parent	**Daughter**				
Uranium 238	Lead 206	4.5 billion	10 million to 4.6 billion	Zircon	
				Uraninite	
Uranium 235	Lead 207	704 million			
Thorium 232	Lead 208	14 billion			
Rubidium 87	Strontium 87	48.8 billion	10 million to 4.6 billion	Muscovite Biotite Potassium feldspar Whole metamorphic or igneous rock	
Potassium 40	Argon 40	1.3 billion	100,000 to 4.6 billion	Glauconite Muscovite Biotite	Hornblende Whole volcanic rock

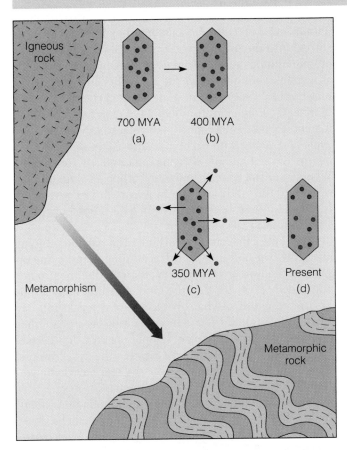

Figure 17.22 *The effect of metamorphism in driving out daughter atoms from a mineral that crystallized 700 million years ago (MYA). The mineral is shown immediately after crystallization (a), then at 400 million years (b), when some of the parent atoms had decayed to daughter atoms. Metamorphism at 350 MYA (c) drives the daughter atoms out of the mineral into the surrounding rock. (d) Assuming the rock has remained a closed chemical system throughout its history, dating the mineral today yields the time of metamorphism, while dating the whole rock provides the time of its crystallization, 700 MYA.*

lization of the magma. If the ages do not closely agree, other samples must be taken and ratios measured to see which, if either, date is correct.

Recent advances and the development of new techniques and instruments for measuring various isotope ratios have enabled geologists to analyze not only increasingly smaller samples, but to do so with a greater precision than ever before. Presently, the measurement error for many radiometric dates is typically less than 0.5% of the age, and in some cases is even better than 0.1%. Thus, for a rock 540 million years old (near the beginning of the Cambrian Period), the possible error could range from nearly 2.7 million years, to as low as less than 540,000 years.

Long-Lived Radioactive Isotope Pairs

Table 17.1 shows the five common, long-lived parent-daughter isotope pairs used in radiometric dating. Long-lived pairs have half-lives of millions or billions of years. All of these were present when Earth formed and are still present in measurable quantities. Other shorter-lived radioactive isotope pairs have decayed to the point that only small quantities near the limit of detection remain.

The most commonly used isotope pairs are the uranium-lead and thorium-lead series, which are used principally to date ancient igneous intrusives, lunar samples, and some meteorites. The rubidium-strontium pair is also used for very old samples and has been effective in dating the oldest rocks on Earth as well as meteorites. The potassium-argon method is typically used for dating fine-grained volcanic rocks from which individual crystals cannot be separated; hence the whole rock is analyzed. Because argon is a gas, great care must be taken to ensure that the sample has not been subjected to heat, which would allow argon to escape; such a sample would yield an age that is too young. Other long-lived radioactive isotope pairs exist, but they are rather rare and are used only in special situations.

Fission Track Dating

The emission of atomic particles resulting from the spontaneous decay of uranium within a mineral damages its crystal structure. The damage appears as microscopic linear tracks that are visible only after etching the mineral with hydrofluoric acid. The age of the sample is determined on the basis of the number of fission tracks present and the amount of uranium the sample contains: the older the sample, the greater the number of tracks (Figure 17.23).

Fission track dating is of particular interest to geologists because the technique can be used to date samples ranging from only a few hundred to hundreds of millions of years in age. It is most useful for dating samples from the time between about 40,000 and 1.5 million years ago, a period for which other dating techniques are not always particularly suitable. One of the problems in fission track dating occurs when the rocks have later been subjected to high temperatures. If this happens, the damaged crystal structures are repaired by annealing, and consequently the tracks disappear. In such instances, the calculated age will be younger than the actual age.

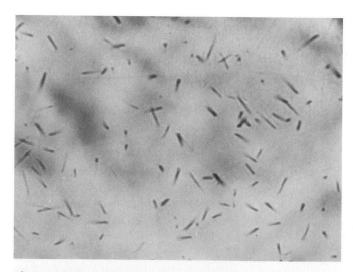

Figure 17.23 *Each fission track (about 16 μm in length) in this apatite crystal is the result of the radioactive decay of a uranium atom. The apatite crystal, which has been etched with hydrofluoric acid to make the fission tracks visible, comes from one of the dikes at Shiprock, New Mexico, and has a calculated age of 27 million years.*

Radiocarbon and Tree-Ring Dating Methods

Carbon is an important element in nature and is one of the basic elements found in all forms of life. It has three isotopes; two of these, carbon 12 and 13, are stable, whereas carbon 14 is radioactive (see Figure 2.5). Carbon 14 has a half-life of 5730 years plus or minus 30 years. The **carbon 14 dating technique** is based on the ratio of carbon 14 to carbon 12 and is generally used to date once-living material.

The short half-life of carbon 14 makes this dating technique practical only for specimens younger than about 70,000 years. Consequently, the carbon 14 dating method is especially useful in archeology and has greatly helped unravel the events of the latter portion of the Pleistocene Epoch.

Carbon 14 is constantly formed in the upper atmosphere when cosmic rays, which are high-energy particles (mostly protons), strike the atoms of upper-atmospheric gases, splitting their nuclei into protons and neutrons. When a neutron strikes the nucleus of a nitrogen atom (atomic number 7, atomic mass number 14), it may be absorbed into the nucleus and a proton emitted. Thus, the atomic number of the atom decreases by one, while the atomic mass number stays the same. Because the atomic number has changed, a new element, carbon 14 (atomic number 6, atomic mass number 14), is formed. The newly formed carbon 14 is rapidly assimilated into the carbon cycle and, along with carbon 12 and 13, is absorbed in a nearly constant ratio by all living organisms (Figure 17.24). When an organism dies, however, carbon 14 is not replenished, and the ratio of carbon 14 to carbon 12 decreases as carbon 14 decays back to nitrogen by a single beta decay step (Figure 17.24).

Currently, the ratio of carbon 14 to carbon 12 is remarkably constant in both the atmosphere and living organisms. There is good evidence, though, that the production of carbon 14, and thus the ratio of carbon 14 to carbon 12, has varied somewhat over the past several thousand years. This was determined by comparing ages established by carbon 14 dating of wood samples against those established by counting annual tree-rings in the same samples (Figure 17.25). As a result, carbon 14 ages have been corrected to reflect such variations in the past.

Tree-ring dating is another useful method for dating geologically recent events. The age of a tree can be determined by counting the growth-rings in the lower part of the stem. Each ring represents one year's growth, and the pattern of wide and narrow rings can be compared among trees to establish the exact year in which the rings were formed. The procedure of matching ring patterns from numerous trees and wood fragments in a given area is referred to as *cross-dating*. By correlating distinctive tree-ring sequences from living to nearby dead trees, a time scale can be constructed that extends back to about 14,000

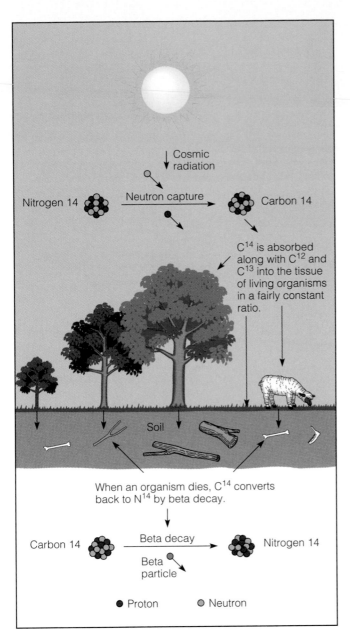

Figure 17.24 *The carbon cycle showing the formation, dispersal, and decay of carbon 14.*

Nitrogen 14

Cosmic radiation

Neutron capture

Carbon 14

C¹⁴ is absorbed along with C¹² and C¹³ into the tissue of living organisms in a fairly constant ratio.

Soil

When an organism dies, C¹⁴ converts back to N¹⁴ by beta decay.

Carbon 14

Beta decay

Beta particle

Nitrogen 14

● Proton ● Neutron

years ago (Figure 17.26). By matching ring patterns to the composite ring scale, wood samples whose ages are not known can be accurately dated.

The applicability of tree-ring dating is somewhat limited because it can only be used where continuous tree records are found. It is therefore most useful in arid regions, particularly the southwestern United States.

How Was the Geologic Time Scale Developed?

The geologic time scale is a hierarchical scale in which the 4.6-billion-year history of Earth is divided into time units of varying duration (Figure 17.1). It was not developed by any one individual, but rather evolved, primarily during the 19th century, through the efforts of many people. By applying relative dating methods to rock outcrops, geologists in England and western Europe defined the major geologic time units without the benefit of radiometric dating techniques. Using the principles of superposition and fossil succession, they could correlate various rock exposures and piece together a composite geologic section. This composite section is, in effect, a relative time scale because the rocks are arranged in their correct sequential order.

By the beginning of the 20th century, geologists had developed a relative geologic time scale, but did not yet have any absolute dates for the various time-unit boundaries. Following the discovery of radioactivity near the end of the last century, radiometric dates were added to the relative geologic time scale (Figure 17.1).

Because sedimentary rocks, with rare exceptions, cannot be radiometrically dated, geologists have had to rely on interbedded volcanic rocks and igneous intrusions to apply absolute dates to the boundaries of the various subdivisions of the geologic time scale (Figure 17.27). An ash fall or lava flow provides an excellent marker bed that is a time-equivalent surface, supplying a minimum age for the sedimentary rocks below and a maximum age for the rocks above. Ash falls are particularly useful because they may

Figure 17.25 *Discrepancies exist between carbon 14 dates and those obtained by counting annual tree rings. Back to about 600 B.C., carbon 14 dates are too old, and those from about 600 B.C. to about 5000 B.C. are too young. Consequently, corrections must be made to the carbon 14 dates for this time period.*

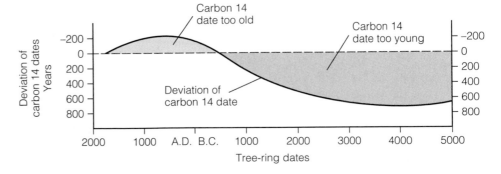

Carbon 14 date too old

Carbon 14 date too young

Deviation of carbon 14 date

Deviation of carbon 14 dates
Years

Tree-ring dates

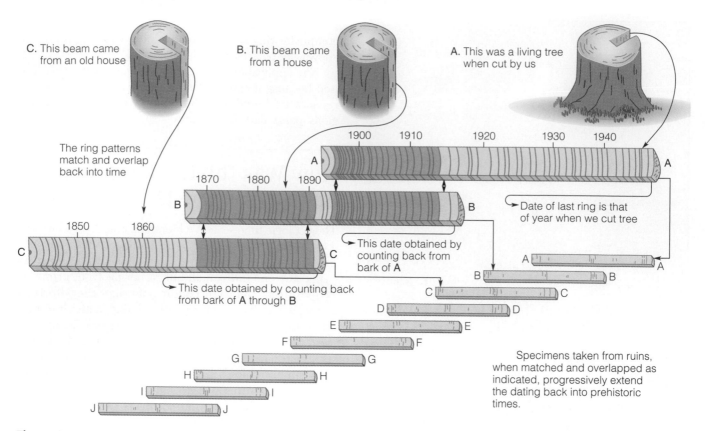

Figure 17.26 *In the cross-dating method, tree-ring patterns from different woods are matched against each other to establish a ring-width chronology backward in time.*

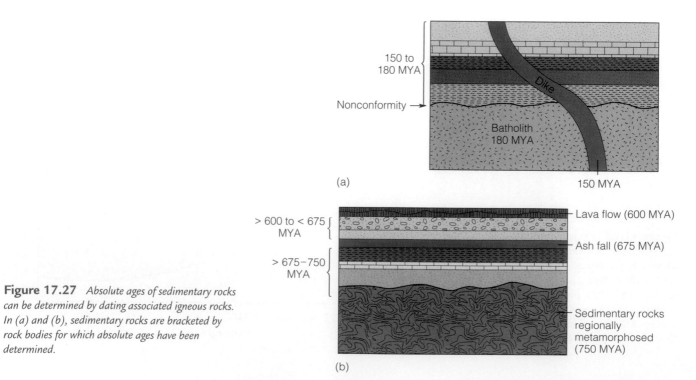

Figure 17.27 *Absolute ages of sedimentary rocks can be determined by dating associated igneous rocks. In (a) and (b), sedimentary rocks are bracketed by rock bodies for which absolute ages have been determined.*

fall over both marine and nonmarine sedimentary environments and can provide a connection between these different environments.

Thousands of absolute ages are now known for sedimentary rocks of known relative ages, and these absolute dates have been added to the relative time scale. In this way, geologists have been able to determine the absolute ages of the various geologic periods and to determine their durations (Figure 17.1).

Stratigraphy and Stratigraphic Terminology

The recognition of a relative time scale brought some order to stratigraphy (the study of the composition, origin, areal distribution, and age relationships of layered rocks). Problems remained, however, because many rock units are time transgressive, that is, a rock unit was deposited during one geologic period in a particular area and during another period elsewhere. In order to deal with both rocks and time, modern stratigraphic terminology includes two fundamentally different kinds of units: those defined by their content and those related to geologic time (Table 17.2).

Units defined by their content include **lithostratigraphic** and **biostratigraphic units.** Lithostratigraphic (*lith-* and *litho-* are prefixes meaning stone or stonelike) units are defined by physical attributes of the rocks, such as rock type, with no consideration of time of origin. The basic lithostratigraphic unit is the *formation*, which is a mappable rock unit with distinctive upper and lower boundaries (Figure 17.28). Formations may be lumped together into larger units called *groups* or *supergroups* or divided into smaller units known as *members* and *beds* (Table 17.2).

Biostratigraphic units are bodies of strata containing recognizably distinct fossils. Fossil content is the only criterion used to define them. Biostratigraphic unit bound-

Figure 17.28 *The Madison Group in Montana consists of two formations, the Lodgepole Formation and the overlying Mission Canyon Formation. The Mission Canyon Formation is the rock unit exposed on the skyline. The underlying Lodgepole Formation is the rock covered on the slopes below.*

aries do not necessarily correspond to lithostratigraphic boundaries (Figure 17.29). The fundamental biostratigraphic unit is the *biozone*. Several types of biozones are recognized, one of which, the *assemblage range zone*, was discussed in the section on correlation.

The units related to geologic time include **time-stratigraphic-units** (also known as chronstratigraphic units) and **time units** (Table 17.2). Time-stratigraphic units are units of rock that were deposited during a specific interval of time. The *system* is the fundamental time-stratigraphic unit. It is based on rocks in a particular area, the stratotype, and is recognized beyond the stratotype area primarily on the basis of fossil content.

Time units are simply units designating specific intervals of geologic time. The basic time unit is the *period*, but smaller units including epoch and age are also recognized. The time units period, epoch, and age correspond to the

Table 17.2

Classification of Stratigraphic Units

Units Defined by Content		Units Expressing or Related to Geologic Time	
Lithostratigraphic Units	**Biostratigraphic Units**	**Time-Stratigraphic Units**	**Time Units**
Supergroup	Biozones	Eonothem	Eon
Group		Erathem	Era
Formation		System	Period
Member		Series	Epoch
Bed		Stage	Age

time-stratigraphic units system, series, and stage, respectively (Table 17.2). For example, the Cambrian Period is defined as the time during which strata of the Cambrian System were deposited. Time units of higher rank than period also exist. Eras include several periods, while eons include two or more eras.

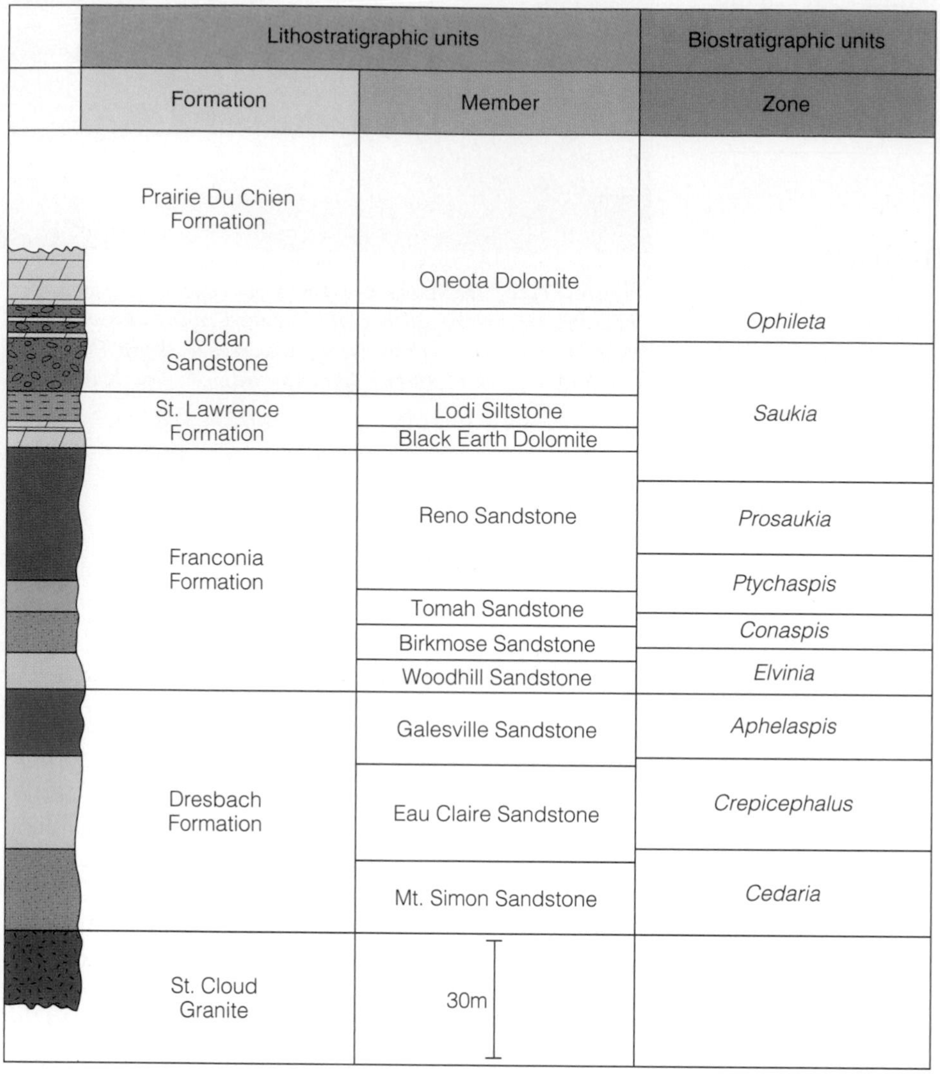

	Lithostratigraphic units		Biostratigraphic units
	Formation	Member	Zone
	Prairie Du Chien Formation		
		Oneota Dolomite	
			Ophileta
	Jordan Sandstone		
	St. Lawrence Formation	Lodi Siltstone	*Saukia*
		Black Earth Dolomite	
	Franconia Formation	Reno Sandstone	*Prosaukia*
			Ptychaspis
		Tomah Sandstone	*Conaspis*
		Birkmose Sandstone	
		Woodhill Sandstone	*Elvinia*
	Dresbach Formation	Galesville Sandstone	*Aphelaspis*
		Eau Claire Sandstone	*Crepicephalus*
		Mt. Simon Sandstone	*Cedaria*
	St. Cloud Granite	30m	

Figure 17.29 *Relationships of biostratigraphic units to lithostratigraphic units in southeastern Minnesota. Notice that the biozone boundaries do not necessarily correspond to lithostratigraphic boundaries.*

Chapter Summary

1. Relative dating involves placing geologic events in a sequential order as determined from their position in the geologic record. Absolute dating results in specific dates for events, expressed in years before the present.

2. During the 18th and 19th centuries, attempts were made to determine Earth's age based on scientific evidence rather than revelation. Although some attempts were quite ingenious, they yielded a variety of ages that now are known to be much too young.

3. James Hutton thought that present-day processes operating over long periods of time could explain all the geologic features of Earth. His observations were instrumental in establishing the principle of uniformitarianism.

4. Uniformitarianism, as articulated by Charles Lyell, soon became the guiding principle of geology. It holds that the laws of nature have been constant through time and that the same processes operating today have operated in the past, although not necessarily at the same rates.

5. Besides uniformitarianism, the principles of superposition, original horizontality, lateral continuity, cross-cutting relationships, inclusions, and fossil succession are basic for determining relative geologic ages and for interpreting Earth history.

6. Surfaces of discontinuity encompassing significant amounts of geologic time are common in the geologic record. Such surfaces are unconformities and result from times of nondeposition, erosion, or both.

7. Correlation is the practice of demonstrating equivalency of units in different areas. Time equivalence is most commonly demonstrated by correlating strata containing similar fossils.

8. Radioactivity was discovered during the late 19th century, and soon thereafter radiometric-dating techniques allowed geologists to determine absolute ages for geologic events.

9. Absolute dates for rock samples are usually obtained by determining how many half-lives of a radioactive parent element have elapsed since the sample originally crystallized. A half-life is the time it takes for one-half of the radioactive parent element to decay to a stable daughter element.

10. The most accurate radiometric dates are obtained from long-lived radioactive isotope pairs in igneous rocks. The most reliable dates are those obtained by using at least two different radioactive decay series in the same rock.

11. Carbon 14 dating can be used only for organic matter such as wood, bones, and shells and is effective back to about 70,000 years ago. Unlike the long-lived isotopic pairs, the carbon 14 dating technique determines age by the ratio of radioactive carbon 14 to stable carbon 12.

12. Through the efforts of many geologists applying the principles of relative dating, a relative geologic time scale was established.

13. Most absolute ages of sedimentary rocks and their contained fossils are obtained indirectly by dating associated metamorphic or igneous rocks.

14. Stratigraphic terminology includes two fundamentally different kinds of units: those based on content and those related to geologic time.

Important Terms

absolute dating
alpha decay
angular unconformity
beta decay
biostratigraphic unit
carbon 14 dating technique
correlation
disconformity

electron capture
fission track dating
guide fossil
half-life
lithostratigraphic unit
nonconformity
principle of cross-cutting
  relationships

principle of fossil
  succession
principle of inclusions
principle of lateral
  continuity
principle of original
  horizontality
principle of superposition

principle of uniformitarianism
radioactive decay
relative dating
time-stratigraphic unit
time unit
tree-ring dating
unconformity

Review Questions

1. Demonstrating time equivalency of rock units in different areas is called
   a. _____ absolute dating; b. _____ correlation; c. _____ relative dating; d. _____ unconformity; e. _____ none of these.

2. What step is essential to development of an angular unconformity?
   a. _____ initial deposition of layers; b. _____ uplift and tilting; c. _____ erosion; d. _____ deposition after erosion; e. _____ none of these.

3. In radiocarbon dating, what isotopic ratio decreases as carbon 14 decays back to nitrogen?
   a. _____ nitrogen 14 to carbon 14; b. _____ carbon 14 to carbon 12; c. _____ carbon 13 to carbon 12; d. _____ nitrogen 14 to carbon 12; e. _____ none of these.

4. Which of the following is not one of Steno's original principles?
   a. _____ lateral continuity; b. _____ fossil succession; c. _____ original horizontality; d. _____ superposition; e. _____ none of these.

5. Considering the half-life of potassium 40, which is 1.3 billion years, what fraction of the original potassium 40 can be expected within a given mineral crystal after 3.9 billion years?
   a. _____ 1/2; b. _____ 1/4; c. _____ 1/8; d. _____ 1/16; e. _____ 1/32.

6. Hutton's locality at Siccar Point, Scotland, is a good example of an unconformity called a(n)
   a. _____ disconformity; b. _____ angular unconformity; c. _____ nonconformity; d. _____ hiatus; e. _____ none of these.

7. If a feldspar grain within a sedimentary rock (e.g., a sandstone) is radiometrically dated, the date obtained will indicate when
   a. _____ the feldspar crystal formed; b. _____ the sedimentary rock formed; c. _____ the parent radioactive isotope formed; d. _____ the daughter radioactive isotope(s) formed; e. _____ none of these.

8. The geological principle demonstrated first by William Smith is today called
   a. _____ cross-cutting relationships; b. _____ superposition; c. _____ original horizontality; d. _____ fossil succession; e. _____ none of these.

9. Which of the following long-lived radioactive isotopes used in radiometric dating has a half-life less than 100,000 years?
   a. _____ uranium 238; b. _____ uranium 235; c. _____ thorium 232; d. _____ rubidium 87; e. _____ none of these.

10. In which type of radioactive decay is a neutron changed to a proton in the nucleus owing to emission of an electron?
    a. _____ alpha decay; b. _____ beta decay; c. _____ electron capture; d. _____ fission track; e. _____ none of these.

11. Placing geologic events in sequential or chronological order as determined by their position in the geologic record is
    a. _____ absolute dating; b. _____ relative dating; c. _____ historical dating; d. _____ uniformitarianism; e. _____ correlation.

12. How many half-lives are required to yield a mineral with 1250 atoms of U^{238} and 18,750 atoms of Pb^{206}?
    a. _____ 1; b. _____ 2; c. _____ 4; d. _____ 8; e. _____ 16.

13. What is the significance of an unconformity in correlation and relative dating? Define the types of unconformities and, in so doing, note their key features.

14. How does metamorphism affect the potential for accurate radiometric dating using any and all techniques discussed in this chapter? How would such radiometric dates be affected by metamorphism, and why?

15. In some places, where disconformities are particularly difficult to discern from a physical point of view, use of the principle of fossil succession helps us delineate such unconformities. How do you suppose using fossils could help us find such hard-to-see disconformities?

16. What is a mass spectrometer, and why would it be important in determining radiometric dates?

17. If a rock or mineral were radiometrically dated using two or more radioactive isotope pairs (e.g., uranium 238 to lead 206 and rubidium 87 to thorium 87) and the analysis for those isotope pairs yielded distinctly different results, what possible explanation could be offered to explain how this happened? How can one rock have two correct ages?

18. Describe the principle of uniformitarianism according to Hutton and Lyell. What is the significance of this principle?

19. Describe the uncertainties associated with trying to radiometrically date any sedimentary rock.

20. Explain the concept of an assemblage range zone and how that relates to index or guide fossils.

21. Briefly explain how well logs are made and their usefulness in the search for oil and gas. What is meant by SP and R on a well log?

22. Describe how the principle of inclusions would be important in recognizing a nonconformity.

23. Where did Lord Kelvin go wrong? Explain his rationale about age-of-Earth calculation and what discovery subsequently showed that Kelvin's calculations were, in fact, in error.

24. How do time-stratigraphic units differ from lithostratigraphic units?

World Wide Web Activities

For these web site addresses, along with current updates and exercises, log on to **http://www.brookscole.com/geo**

University of California Museum of Paleontology

Visit this site for an excellent description of any aspect of geologic time, paleontology, and evolution. Click on the *On-line Exhibits* to go to the *Paleontology without Walls* home page, which is an introduction to the UCMP Virtual Exhibits. Click on the *Geologic Time* site. It will take you to the *Geology and Geologic*

Time home page, which discusses geologic time. You can then click on one of the eras to learn more about what occurred during that time interval. Click on the *Cenozoic* site to learn more about how geologic time scales are constructed.

Radiocarbon Web-Info

This site is maintained by the radiocarbon labs of Waikato and Oxford universities. The site contains information about the basis of carbon 14 dating, applications, measurement methods, and other carbon 14 web sites.

 # CD-ROM Exploration

Exploring your *Earth Systems Today* CD-ROM will add to your understanding of the material in this chapter.

Topic: Earth's Processes

Module: Geologic Time

Explore activities in this module to see if you can discover the following for yourself:

Using the parent-daughter cross-plot graph, examine how the amount of daughter element changes as the amount of parent element changes. In addition, note how all samples of the same element have the same decay rate regardless of the element's age.

Evolution—The Theory and Its Supporting Evidence

An armadillo (top) and a restoration of an extinct armored mammal known as a glyptodont. The animals are not to scale. Armadillos measure about 0.7 m long, whereas some glyptodonts were gigantic, measuring more than 3 m long. The geographic distribution of armadillos and glyptodonts as well as their sharing several unusual characteristics were some of the evidence that convinced Charles Darwin that organisms had evolved.

PROLOGUE

On December 27, 1831, Charles Robert Darwin sailed from Devonport, England, aboard the H.M.S. *Beagle* as an unpaid naturalist. Nearly five years later the *Beagle* returned to England, and Darwin never ventured far from home again. Nevertheless, his 64,360-kilometer voyage was his most important experience, an experience that changed his view of nature and ultimately revolutionized all of science.

When Darwin departed from Devonport, he was a little-known recent graduate in theology from Christ's College. In fact, his father, Dr. Robert Darwin, had sent him to Christ's College as a last resort because Charles showed little aptitude for academics, except science, and he had already withdrawn from medical studies at Edinburgh. He completed his studies in theology at Christ's College but was rather indifferent to religion. He did, however, fully accept the biblical account of creation as historical fact, including the concept of *fixity of species*, meaning that existing species had been created in their present form. Darwin's belief in biblical creation initially endeared him to the *Beagle*'s captain, Robert Fitzroy, but his views changed during the voyage, and his relationship with Captain Fitzroy became strained.

When the voyage began, Darwin was nominally a clergyman with interests in botany, zoology, and geology. He suffered from prolonged bouts of seasickness but still managed to keep detailed notes on his observations and to collect, catalog, and dissect specimens. The *Beagle* made several lengthy stops in South America where Darwin explored rain forests, experienced an earthquake, and collected fossils. Some of the fossil mammals he collected differed from existing sloths, armadillos, and llamas but were clearly similar to them and implied that living species descended with change from ancient species. In short, he began to question the concept of fixity of species.

Darwin was particularly fascinated by the plants and animals of the Cape Verde Islands and the Galapagos Islands, which are comparable distances west of Africa and South America, respectively. These islands had their own unique plants and animals, yet these plants and animals most closely resembled those of the nearby continent. The Galapagos, for instance, are populated by 13 species of finches (Figure 18.1), but only 1 species exists in South America. The Galapagos finches are adapted to the different habitats occupied in South America by various species such as parrots, flycatchers, and toucans.

To account for his observations of organisms on these oceanic islands, Darwin reasoned that each had received colonists from the nearby continent and that "such colonists would be liable to modification—the principle of inheritance still betraying their original birthplace."* This statement reveals a change in Darwin's view of nature; he no longer accepted the idea of fixity of species but thought that change in organisms through time did take place. He thought, for example, that the finches on the Galapagos Islands were the modified descendants of an ancestral species that somehow reached the islands from South America. Once there, this species had differentiated into the various types he observed.

Darwin's views on the nature of the biological world were changing, as were his ideas about Earth history. During the voyage he had read Charles Lyell's *Principles of Geology* and came to accept the concept of uniformitarianism and Earth's great age. In other words, he began to view all of nature as dynamic rather than static.

* C. Darwin, *The Origin of Species* (New York: New American Library, 1958), p. 377.

OBJECTIVES

At the end of this chapter, you will have learned that

■ Although the idea of evolution is most commonly associated with Charles Darwin, the concept actually had a long development before he was even born.

■ The first widely accepted theory of evolution was proposed by Jean-Baptiste de Lamarck in 1809.

■ In 1859, Charles Darwin and Alfred Wallace simultaneously published their views on evolution and proposed a mechanism known as natural selection to account for evolution.

■ Experiments performed during the 1860s by an Austrian monk are the basis for our understanding of heredity, although more recent investigations have provided considerably more information.

■ Sexual reproduction and mutations provide most of the variation in populations.

■ Most species probably arise when a small population becomes isolated from its parent population.

■ Divergent evolution involves differentiation from a common ancestor, whereas convergent and parallel evolution entail development of similar structures.

■ Cladistics is a powerful tool for investigating relationships among living and fossil organisms.

■ Several predictive statements can be made from evolutionary theory, all of which are testable.

■ Fossils are much more abundant than most people realize, and they provide some of the evidence for ancestor-descendant relationships.

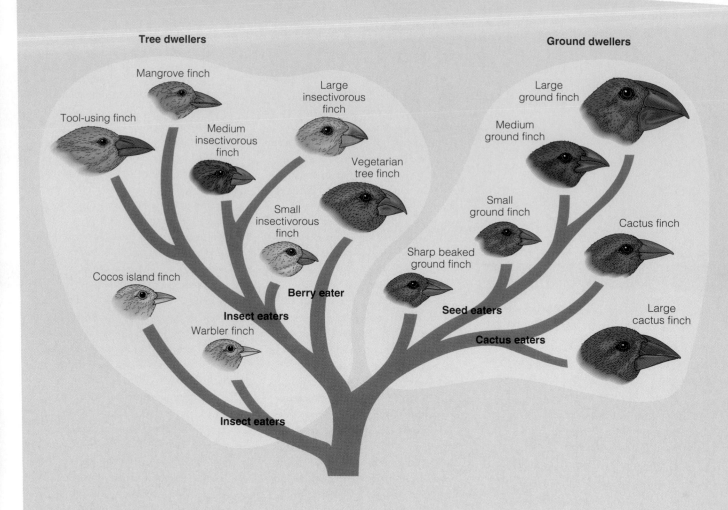

Figure 18.1 *Darwin's finches arranged to show evolutionary relationships. The six species on the right are ground dwellers, and the others are adapted for life in trees. Notice that the shapes of the beaks of finches in both groups vary depending on diet.*

Introduction

The term *evolution* comes from the Latin, meaning unrolled. Roman books were written on parchment and rolled on wooden rods, so they were unrolled or evolved, as they were read. In today's usage, evolution refers to change through time, and so defined it is a pervasive phenomenon; stars evolve, Earth has evolved and continues to do so, and languages and social systems evolve. In addition, studies of living organisms as well as evidence from the fossil record indicate that organisms have changed through time, and they too continue to evolve.

Our interest in this chapter is with the biological concept of evolution, which involves inheritable changes in organisms. Small-scale evolution is an observable phenomenon in many rapidly breeding species of bacteria, insects, and a few others. Large-scale evolutionary changes, such

as the origin of birds from reptiles, can be deduced from the fossil record as well as from studies of comparative anatomy, embryology, genetics, molecular biology, and biochemistry of living organisms. We noted in Chapter 1 that theories are naturalistic, testable scientific explanations for some phenomenon, and that they have a large body of supporting evidence. The central claim of the **theory of evolution** is that all present-day species are the modified descendants of life forms that lived during the past. Furthermore, this theory, as does any legitimate theory, allows scientists to make testable predictions (discussed in a later section).

Although much of the evidence supporting the theory of evolution comes from studies in biology, fossils are important in establishing the historical context for organisms. Some ancient Greek and Chinese scholars recognized fossils as the remains of organisms, but during most of historic time in the West they were variously attributed to spontaneous generation in rocks, growth of some kinds of mysterious seeds, or some kind of molding force. Some people even believed that they were objects placed in rocks by the Creator to test our faith or by Satan to sow seeds of doubt.

Some perceptive observers such as Leonardo de Vinci in 1508, Robert Hooke in 1665, and Nicolas Steno in 1667 recognized the true nature of fossils, but their views were largely ignored. But by the late 1700s, it had become clear that fossils were in fact the remains of plants and animals and that some fossils represented types of organisms now extinct. In addition to providing some of the evidence supporting the theory of evolution, fossils are useful in determining relative ages of rocks (see Chapter 17), and in determining environments of deposition of sedimentary rocks (see Chapter 5).

Why Should You Study Evolution—The Theory and Its Supporting Evidence?

Evolution, which involves changes in organisms through time, is fundamental to biology and paleontology. For instance, the theory explains why older and older rocks contain the remains of organisms increasingly different from those existing now. Thus it is a unifying theory that explains an otherwise encyclopedic collection of facts about living and fossil organisms, and it serves as the framework for discussions of life history in the following chapters.

Most people are not aware that almost everything in science, except observations, is based on some kind of theoretical underpinning, and the theory of evolution, like other theories, is a naturalistic, testable explanation supported by several kinds of evidence. It is, however, poorly understood, as is the theory of natural selection, one mechanism accounting for evolution. One hears objections to evolution and natural selection from outside the sciences that usually amount to misunderstandings of what they attempt to explain, distortions of evidence, and exaggerations of disagreements among scientists.

How Did the Idea of Evolution Develop?

Many people attribute the idea of evolution solely to Charles Darwin. He is indeed justly credited with elucidating the idea and providing a large body of supporting evidence, but the idea was seriously considered long before he was born. Even some ancient Greek philosophers speculated on interrelationships among organisms, and several theologians and philosophers during the Middle Ages discussed evolution in one form or another. Nevertheless, the prevailing belief well into the 1700s was that all important knowledge was contained in the works of Aristotle and the Bible, particularly in the first two chapters of Genesis.

Literally interpreted, Genesis was taken as the final word on the origin and diversity of life and much of Earth history. According to this view, all species of organisms were perfectly fashioned by God during the days of creation and had remained fixed and immutable ever since. To question any aspect of this interpretation was regarded as heresy. But the social and intellectual climate changed in 18th-century Europe, and the absolute authority of the Church in all matters declined. And ironically, the very naturalists who were determined to find physical evidence supporting Genesis as a factual, historical account were finding more and more evidence that could not be explained by a literal reading of Scripture.

A good example of these changing attitudes is the interpretation of sedimentary rocks that had previously been attributed to deposition during a single, worldwide flood. Naturalists examined these rocks and concluded that they showed clear evidence of deposition in environments like those existing now. They reasoned that this evidence truly reflected the conditions that prevailed when the rocks were deposited, for to infer otherwise implied a deception on the part of the Creator, a thesis they could not accept. These naturalists did not abandoned the Judeo-Christian tradition, but rather came to accept Genesis as a symbolic account containing important spiritual messages rather than a factual account of creation.

In this changing intellectual atmosphere, the concepts of uniformitarianism and Earth's great age were gradually accepted. The French zoologist Georges Cuvier demonstrated that many types of plants and animals had become extinct, and fossils clearly indicated that organisms differed through time. In view of the fossil evidence, as well as studies of living plants and animals, many naturalists became convinced that species were not fixed and immutable. Change from one species to another—that is, evolution—became an acceptable idea. What was lacking, though, was a theoretical framework to explain evolution.

Jean-Baptiste de Lamarck and His Ideas on Evolution

The first to propose a process whereby gradual evolution could take place was Erasmus Darwin (1731–1802), Charles Darwin's grandfather. But it was Jean-Baptiste de Lamarck (1744–1829), a French botanist-geologist, who was first taken seriously by his colleagues. In 1809, he proposed that evolution is an adaptive process whereby organisms acquire traits or characteristics during their lifetimes and pass these acquired traits on to their descendants.

Lamarck contributed much to our understanding of the natural world, but, unfortunately, he is best remembered for his mechanism for evolution. According to his theory of *inheritance of acquired characteristics* new features arise in organisms because of their needs and somehow these features are passed along to their descendants. In other words, characteristics acquired during the lifetime of an individual are inheritable.

Considering the data available at the time, Lamarck's explanation seemed logical and was accepted by many naturalists as a viable mechanism for evolution. In fact, his work on inheritance was not completely refuted until decades later when scientists discovered that the units of heredity known as *genes* cannot be altered by any effort on the part of an organism during its lifetime. In short, despite numerous attempts to demonstrate inheritance of acquired characteristics, including one in the former Soviet Union during this century, all have failed (see Perspective 18.1).

The Contributions of Charles Robert Darwin and Alfred Russel Wallace

In 1859, Charles Robert Darwin (1809–1882) (Figure 18.2) published *The Origin of Species* in which he outlined his ideas on evolution, amassed a huge amount of supporting evidence, and provided a mechanism to account for evolution. Most naturalists quickly recognized that evolution provided a unifying theory that explained an otherwise encyclopedic collection of biologic facts. While 1859 marks the beginning of modern evolutionary thought, Darwin had actually conceived his ideas more than 20 years earlier, but being aware of the furor they would cause, he was reluctant to publish them.

Darwin had concluded following his 1831–1836 voyage as naturalist aboard the H.M.S. *Beagle* that species are not fixed (see the Prologue). When he returned home in 1836 he had what he thought was good evidence that species had indeed changed through time, but he had no idea what might bring about change. But by 1838, his observations of the selection practiced by plant and animal breeders, and a chance reading of Thomas Malthus's essay on population, gave him the elements necessary to formulate his theories.

Animal and plant breeders routinely practice **artificial selection** when they select organisms for desirable traits, breed organisms with those traits, and thereby bring about a great amount of change. The fantastic variety of plants and animals resulting from this process made Darwin wonder if a process selecting among variant types in natural populations could also bring about change. He came to fully appreciate the power of selection when he read Malthus's book in which the author argued that far more animals are born than reach maturity, yet the numbers of adults of a species remain rather constant. Malthus reasoned that competition for resources resulted in a high infant mortality rate, thus limiting population size. Darwin proposed that a natural process was selecting only a few for survival, a process he called natural selection.

Natural Selection—Its Meaning and Significance

For 20 years Darwin kept his ideas on evolution and natural selection to himself, but in 1858, he received a letter from Alfred Russel Wallace (1823–1913), a young naturalist working in southern Asia. Wallace had also read Malthus's book and had come to exactly the same conclusion as Darwin. Darwin's and Wallace's ideas were presented simultaneously to the Linnaean Society in London. Their views on **natural selection** are summarized in the following four points:

1. Organisms in all populations possess heritable variations—size, speed, agility, visual acuity, digestive enzymes, color, and so forth.

Figure 18.2 *Charles Robert Darwin in 1850. Although Darwin had formulated his theory of natural selection by this time, he did not publish it until 1859.*

2. Some variations are more favorable than others; that is, some variant types have a competitive edge in acquiring resources and avoiding predators.

3. Not all young survive to reproductive maturity.

4. Those with favorable variations are *more likely* to survive and pass on their favorable variations.

Evolution by natural selection then is largely a matter of reproductive success, for only those that reproduce pass on their favorable variations. Of course, favorable variations do not guarantee survival for a particular individual, but in a population of perhaps thousands of individuals, those with favorable variations are more likely to survive and reproduce. But is evolution by natural selection substantially different from evolution by the inheritance of acquired characteristics? In other words, how does the Darwin-Wallace mechanism for evolution differ from Lamarck's?

Remember that Lamarck's mechanism for inheritance involves characteristics acquired during an organism's lifetime that are passed on. The Darwin-Wallace hypothesis, on the other hand, depends on favorable variations existing in a population, and it is this variation that natural selection works on.

A common misconception about natural selection is that among animals only the biggest, strongest, and fastest are likely to survive. It is true that size and strength are important when male bighorn sheep compete for mates, but remember that females also pass along their genes. Speed certainly is an advantage for some predators, but not all predators depend on speed. Badgers are not very fast but survive quite nicely, and some small cats depend on stealth and pouncing rather than speed to capture prey. In fact, in some animal populations, natural selection might favor the smallest, or most easily concealed, or those that adapt most readily to a new food source, or those that have the ability to detoxify some natural or human-made substance. Various microorganisms, some insects, and even rodents adapt so quickly that a constant race to develop more effective antibiotics, insecticides, and pesticides keeps chemists quite busy.

One further common point of confusion about evolution and natural selection needs clarification. Darwin and Wallace were attempting to explain how organisms changed through time—not how life originated. There are theories on the origin of life by natural processes, but their validity must be assessed independently of evolution by natural selection.

Mendel and the Birth of Genetics

Critics of natural selection were quick to point out that Darwin and Wallace could not account for the origin of variation or explain how it was maintained in populations. They reasoned that should a variant trait

WHAT WOULD YOU DO?

Suppose someone were to tell you that the geologic record was created with the appearance of a lengthy history. In other words, rocks that look as if they cooled from lava, or appear to be river deposits were simply created with that appearance. Furthermore, fossil shells and bones were also created with the rocks and are not really the remains of organisms. In short, their claim is that events inferred from the geologic record by scientists never really occurred. Actually, this idea was put forth during the 1850s, but was rejected by the public and scientists for philosophical as well as scientific reasons. Can you think of reasons this idea is less than appealing philosophically, and why it cannot be considered scientific?

arise, it would simply blend with other traits in the population and be lost. At the time, this was a valid criticism. Actually, information that could have given them the answer existed, but it remained in obscurity until 1900.

Mendel's Experiments

An Austrian monk may seem an unlikely candidate for the title "father of genetics," but during the 1860s Gregor Mendel was doing research that answered some of the inheritance problems that plagued Darwin and Wallace. Unfortunately, Mendel's work was published in an obscure journal and went largely unnoticed.

Mendel performed a series of controlled experiments with true-breeding strains of garden peas (strains that when self-fertilized always display the same trait, such as flower color). In one experiment, he transferred pollen from white-flowered plants to red-flowered plants, which yielded a second generation of all red-flowered plants. But when left to self-fertilize, these plants yielded a third generation with a ratio of red-flowered plants to white-flowered plants of slightly over 3 to 1 (Figure 18.3).

Mendel concluded that traits such as flower color are controlled by a pair of factors, or what we now call **genes.** He also concluded that genes controlling the same trait occur in alternate forms, or **alleles;** that one allele may be dominant over another; and that offspring receive one allele of each pair from each parent. When an organism produces sex cells—pollen and ovules in plants, and sperm and eggs in animals—only one allele for a trait is present in each sex cell (Figure 18.3). For example, if R represents the allele for red flowers, and r represents white, the offspring may inherit the combinations of alleles symbolized as RR,

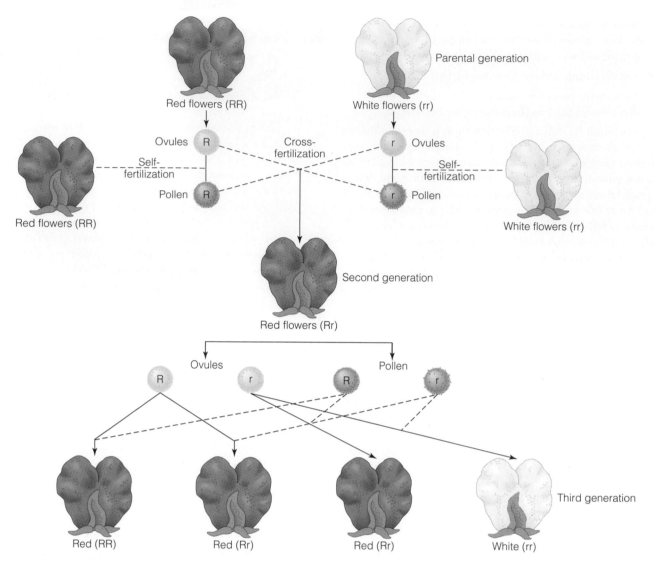

Figure 18.3 *In his experiments with flower color in garden peas, Mendel used true-breeding strains. Such plants, shown here as the parental generation, when self-fertilized always yield offspring with the same trait as the parent. However, if the parental generation is cross-fertilized, all plants in the second generation receive the combination of alleles indicated by Rr; these plants also have red flowers because R is dominant over r. The second generation of plants produces ovules and pollen with the alleles shown and, when left to self-fertilize, produces a third generation with a ratio of three plants with red flowers to one plant with white flowers.*

Rr, or rr. And because R is dominant over r, only those offspring with the rr combination will have white flowers.

The most important aspects of Mendel's work can be summarized as follows: the factors (genes) controlling traits do not blend during inheritance; and even though particular traits may not be expressed in each generation, they are not lost. Therefore, some variation in populations is accounted for by alternate expression of genes (alleles), and variation can be maintained.

Even though Mendelian genetics explains much about heredity, we now know the situation is much more complex. For example, our discussion has relied upon a single gene controlling a trait, but, in fact, most traits are controlled by many genes. Nevertheless, Mendel discovered the answers Darwin and Wallace needed, though they

went unnoticed until three independent researchers rediscovered them in 1900.

Genes and Chromosomes

The cells of all organisms contain threadlike **chromosomes,** which are complex, double-stranded, helical molecules of **deoxyribonucleic acid (DNA)** (Figure 18.4). Specific segments, or regions, of the DNA molecule are the basic hereditary units, the genes. The number of chromosomes is specific for a single species but varies among species. For example, fruit flies have 8 chromosomes, humans have 46, and horses have 64. However, chromosomes occur in pairs that carry genes controlling the same characteristics. Remember that the

The Tragic Lysenko Affair

In previous discussions we noted that theories are testable and that observations and/or experiments might lend credence to a theory or contradict it. Even well-supported theories are never proved beyond any question, and accumulating evidence can render a theory untenable, in which case it must be modified or abandoned. An excellent example of unquestioning acceptance of a discarded theory involved Trofim Denisovich Lysenko (1898–1972), who became president of the Soviet Academy of Agricultural Sciences in 1938.

Lysenko endorsed the theory of inheritance of acquired characteristics according to which plants and animals could be changed in desirable ways simply by exposing them to a new environment. For example, according to Lysenko, seeds exposed to dry conditions would acquire a resistance to drought, and this trait would be inherited by future generations. Lysenko accepted inheritance of acquired characteristics because of its apparent compatibility with Marxist-Leninist philosophy. As president of the academy, and with the endorsement of the Central Committee of the Soviet Union, Lysenko did not allow Soviet scientists to conduct any other research concerning inheritance mechanisms, and those who publicly disagreed with him lost their jobs or were even sent to labor camps.

Unfortunately for the Soviet people, inheritance of acquired characteristics had been discredited as a scientific theory more than 50 years before. The results of Lysenko's belief in the political correctness of this theory were widespread crop failures, starvation, and misery for millions of people. In fact, Lysenko's dismantling of genetic research was so complete that the former Soviet Union is still striving to catch up to the rest of the world in this field.

An interesting sidenote to this tragedy is that the only reputable Western biologist to support Lysenko was England's J. B. S. Haldane. Haldane, whose numerous contributions to genetics and evolutionary theory won him international recognition and respect, discovered Marxism during the 1930s and joined the Communist party in 1942. Unfortunately, his belief in the political correctness of the Marxist-Leninist doctrine appears to have blinded Haldane to the scientific absurdity of Lysenko's claims.

The lesson to be drawn from this example is that scientific research must be based on scientific realities. Science proceeds on the basis of the scientific method, not philosophical or political beliefs. Lysenko's ideas on inheritance were mandated as the correct way to proceed in agricultural research not because of their scientific merit but because they were deemed compatible with a belief system.

---

genes on chromosome pairs may be present in different forms, alleles.

In sexually reproducing organisms, the production of eggs and sperm results when parent cells undergo a type of cell division known as *meiosis.* Meiosis yields cells containing only one chromosome of each pair (Figure 18.5a), so eggs and sperm have only one-half the chromosome number of the parent cell.

During reproduction, a sperm fertilizes an egg (Figure 18.5b), producing a fertilized egg with the full set of chromosomes for that species. As Mendel deduced from his garden pea experiments, one-half of the genetic makeup of the fertilized egg comes from each parent. The fertilized egg, however, develops and grows by a cell division process called *mitosis* that does not reduce the chromosome number (Figure 18.5c).

WHAT WOULD YOU DO?

Suppose a powerful group in Congress were to mandate that all future genetic research had to conform to strict guidelines—namely, that plants and animals should be exposed to cold, heat, and aridity so that they would acquire characteristics that would allow agriculture in regions otherwise inhospitable. Furthermore, any other genetic research would not only not be funded but would be prohibited. Why would it be unwise to implement this program, and why is government-mandated science counterproductive?

The Modern View of Evolution

During the 1930s and 1940s, the ideas developed by geneticists, paleontologists, population biologists, and others were merged to form a **modern synthesis** or neo-Darwinian view of evolution. The chromosome theory of inheritance was incorporated into evolutionary thinking; mutations were seen as one source of variation in populations; Lamarck's concept of inheritance was completely rejected; and the importance of natural selection was reaffirmed. The modern synthesis also emphasized that evolution is a gradual process, a point that has been challenged.

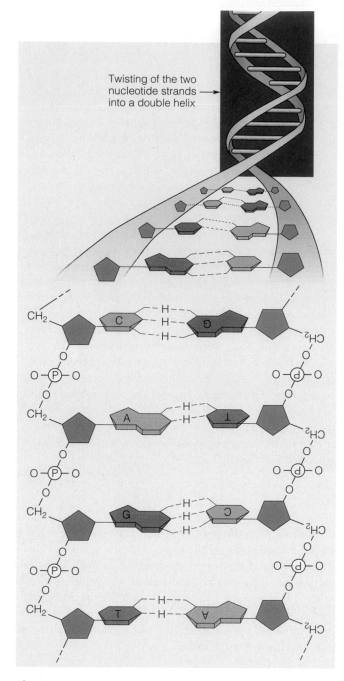

Twisting of the two nucleotide strands into a double helix →

Figure 18.4 *Chromosomes are double-stranded, helical molecules of deoxyribonucleic acid (DNA) shown here diagrammatically. Specific segments of chromosomes are genes. The two strands of the molecule are joined by hydrogen bonds (H).*

Sources of Variation

Evolution by natural selection works on variations in populations, most of which is accounted for by the reshuffling of alleles from generation to generation during sexual reproduction. Given that each of thousands of genes might have several alleles, and that offspring receive one-half of their genes from each parent, the potential for variation is

enormous. But even though this can account for considerable variation, it is variation already present in a population. Any new variations arise by **mutations,** involving some kind of change in the chromosomes or genes—that is, a change in the hereditary information.

Whether a *chromosomal mutation* (affecting a large segment of a chromosome) or a *point mutation* (a change in a particular gene), as long as it takes place in a sex cell it is inheritable. To fully understand mutations we must explore them further. For one thing, they are random with respect to fitness. This means that there is no predetermination of a mutation's effect; it may be beneficial, neutral, or harmful. However, the attributes of harmful versus beneficial can be considered only with respect to the environment.

If a species is well adapted, most mutations would not be particularly useful and perhaps be harmful. But what was a harmful mutation can become a useful one if the environment changes. For instance, some plants have developed a resistance for contaminated soils around mines. Plants of the same species from the normal environment do poorly or die in contaminated soils, while contaminant resistant plants do very poorly in the normal environment. Mutations for contaminant resistance probably occurred repeatedly in the population, but they were not beneficial until contaminated soils were present.

But how can a mutation be neutral? It would seem that any change would be either beneficial or harmful, but this is not the case. During protein synthesis in cells, information carried on chromosomes directs the formation of proteins by selecting the appropriate amino acids and arranging them into a specific sequence. However, some mutations have no effect on the type of protein synthesized. In other words, the same protein is synthesized before and after the mutation, and thus the mutation is neutral.

What causes mutations? Some are induced by *mutagens,* agents that bring about higher mutation rates. Exposure to some chemicals, ultraviolet radiation, X-rays, and extreme temperature changes might bring about mutations. But some mutations are spontaneous, taking place in the absence of any mutagen. They no doubt have some kind of physical and/or chemical cause, though.

Speciation and the Rate of Evolution

Speciation, the phenomenon of a new species arising from an ancestral species, is well documented, but the rate and ways in which it takes place vary. First, though, let us be clear on what we mean by **species,** a biological term for a population of similar individuals that in nature can interbreed and produce fertile offspring. Thus, a species is reproductively isolated from other species. This definition does not apply to organisms such as bacteria that reproduce asexually, but it is nevertheless useful for our discussion of plants and animals, the organisms with which we are most familiar.

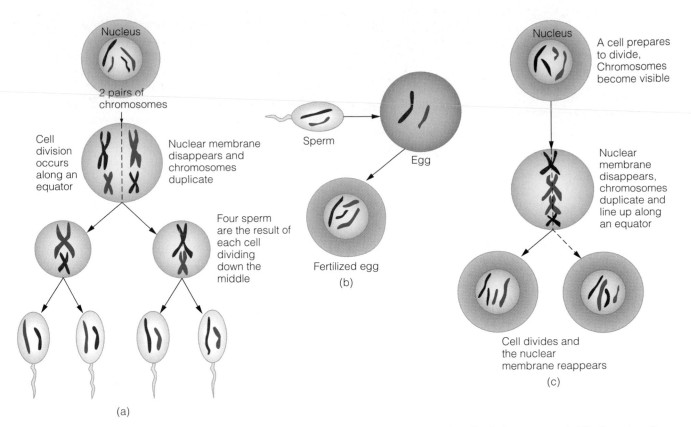

Figure 18.5 *(a) Meiosis is a type of cell division in which sex cells form; each cell contains one member of each chromosome pair. The formation of sperm cells is shown; eggs form in a similar manner, but only one of the four final cells is a functional egg. (b) When a sperm fertilizes an egg, the full complement of chromosomes is present in the fertilized egg. (c) Mitosis results in the complete duplication of a cell. In this simplified example, a cell with four chromosomes (two pairs) produces two cells, each with four chromosomes. Mitosis occurs in all body cells except sex cells. Once an egg is fertilized, the developing embryo grows by mitosis.*

Goats and sheep are easily distinguished by physical characteristics and they do not interbreed in nature. Thus, they must be separate species, yet in captivity they can produce fertile offspring. Lions and tigers can also interbreed in captivity, although they do not interbreed in nature and their offspring are sterile, so they too are separate species. Domestic horses that have gone wild can interbreed with zebras to yield a *zebroid* which is sterile, yet horses and zebras are separate species. It should be obvious from these examples that reproductive barriers are not complete in some species, indicating varying degrees of divergence from a common ancestral species.

The process of speciation involves a change in the genetic makeup of a population, which also might bring about changes in form and structure—lions and tigers resemble one another but also differ in several ways. According to the concept of **allopatric speciation,** species arise when a small part of a population becomes isolated from its parent population (Figure 18.6). Isolation might result from a marine transgression that effectively separates a once-interbreeding species, for instance. On the other hand, a few individuals might somehow get to a remote area and no longer exchange genes with the parent popu-

lation. Given these conditions and the fact that different selective pressures are likely, they might eventually give rise to a reproductively isolated species. Numerous examples of allopatric speciation have been well documented, especially in extremely remote areas such as oceanic islands (Figure 18.1).

Although widespread agreement exists on allopatric speciation, scientists disagree on how rapidly a new species might evolve. According to Darwin and reaffirmed by the modern synthesis, the gradual accumulation of minor changes eventually brings about the origin of a new species, a phenomenon called **phyletic gradualism** (Figure 18.7a). Another view, known as **punctuated equilibrium,** holds that little or no change takes place in a species during most of its existence, and then evolution occurs rapidly, giving rise to a new species in perhaps as little as a few thousands of years (Figure 18.7b).

Proponents of punctuated equilibrium argue that few examples of gradual transitions from one species to another are found in the fossil record. According to their view, transitional forms connecting ancestral and descendant species are unlikely to be preserved because species arise rapidly in small, geographically isolated populations.

The Modern View of Evolution   **523**

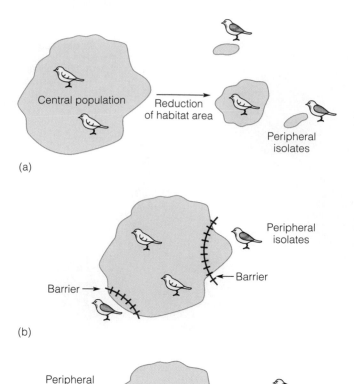

(a)

(b)

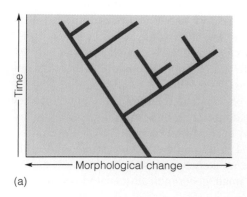

(c)

Figure 18.6 *Allopatric speciation. (a) Reduction of the area occupied by a species may leave small, isolated populations,* peripheral isolates, *at the periphery of the once more extensive range. In this example, members of both peripheral isolates have evolved into new species. (b) Barriers have formed across parts of a central population's range, thereby isolating small populations. (c) Out-migration and the origin of a peripheral isolate.*

And, as a matter of fact, many species do appear abruptly in the fossil record with no evidence of direct ancestors.

Critics, however, point out that neither Darwin nor those who formulated the modern synthesis insisted that all evolutionary change was gradual and continuous, a view shared by many present-day biologists and paleontologists. Indeed, they allowed for times during which evolutionary change in small populations could be quite rapid. Furthermore, deposition of sediments in most environments is not continuous; thus, the lack of gradual transitions in many cases is simply an artifact of the fossil record. And finally, despite the incomplete nature of the fossil record, a number of examples of gradual transitions from ancestral to descendant species are known.

Punctuated equilibrium is one way for evolution to take place, and apparently a rather common way. Nevertheless, one must remember that even according to this concept, a new species evolves in perhaps several thousands of years, which from the human perspective is still rather gradual.

Divergent, Convergent, and Parallel Evolution

The phenomenon of an ancestral species giving rise to diverse descendants adapted to various aspects of the environment is referred to as **divergent evolution.** At some time in the past, a population of finches, apparently from South America, reached the Galapagos Islands and diversified into a number of species (Figure 18.1). An even more impressive example involves the mammals whose diversification from a common ancestor during the Late Mesozoic gave rise to such varied animals as platypuses, armadillos, rodents, bats, primates, whales, and hoofed mammals such as rhinoceroses and deer (Figure 18.8).

Divergent evolution leads to descendants that differ markedly from their ancestors, whereas *convergent evolution* and *parallel evolution* are processes whereby similar

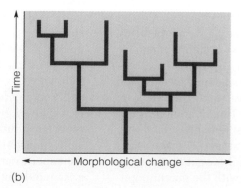

Figure 18.7 *Comparison of two models for differentiation of organisms from a common ancestor. (a) In the phyletic gradualism model, slow, continuous change takes place as one species evolves into another. (b) According to the punctuated equilibrium model, change occurs rapidly, and new species evolve rapidly. However, little or no change occurs in a species during most of its existence.*

(a)

(b)

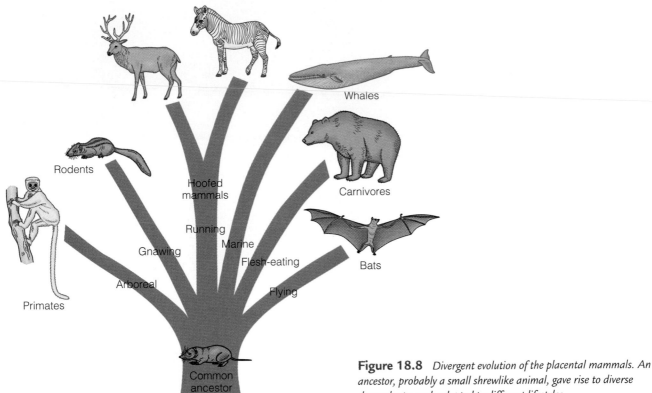

Figure 18.8 *Divergent evolution of the placental mammals. An ancestor, probably a small shrewlike animal, gave rise to diverse descendants, each adapted to different lifestyles.*

adaptations arise in different groups. Unfortunately, they differ in degree and are not always easy to distinguish. Nevertheless, both are common phenomena with **convergent evolution** involving the development of similar characteristics in distantly related organisms, whereas similar characteristics arising in closely related organisms is **parallel evolution.** In both cases, similar characteristics develop independently because the organisms in question adapt to similar environments. Perhaps the following examples will clarify the distinction between these two concepts.

During the Cenozoic Era, South America was an island continent, and its mammals evolved independently of those in North America. Nevertheless, a number of mammals on each continent adapted in similar ways and superficially resemble one another (Figure 18.9a). Of course, all are mammals, but their common ancestor existed in the remote past so this is an example of convergent evolution. Likewise, similarities between present-day marsupial (pouched) mammals of the Australian region and mammals elsewhere is another good example of convergence.

Parallel evolution is also a common phenomenon, but, as previously noted, it is not always easy to distinguish from convergent evolution. A good example, though, is the similar features seen in jerboas and kangaroo rats (Figure 18.9b). Both are closely related, but each independently developed similar features rather than inheriting them from a common ancestor.

Cladistics and Cladograms

Traditionally, evolutionary relationships have been depicted with *phylogenetic trees,* in which the horizontal axis represents anatomical differences and the vertical axis denotes time (Figure 18.10a). The patterns of ancestor-descendant relationships shown are based on a variety of characteristics, although the ones used are rarely specified. In contrast, a **cladogram** shows the relationships among members of a *clade,* a group of organisms including its most recent common ancestor (Figure 18.10b). Cladistics focus on derived as opposed to primitive characteristics. For instance, all land-dwelling vertebrate animals possess bone and paired limbs. Accordingly, these characteristics are primitive and of little use in establishing relationships among them. However, hair and mammary glands are derived characteristics, sometimes called *evolutionary novelties,* because only one subclade, the mammals, has them. If one considers only mammals, hair and mammary glands are primitive and of no further use, but live birth serves to distinguish most mammals from the egg-laying mammals.

Any association of organisms can be depicted in a cladogram, but the more shown, the more complex and difficult it is to construct. Let's use an example of only three animals—bats, dogs, and birds. Figure 18.11 shows three cladograms, each with a different interpretation of

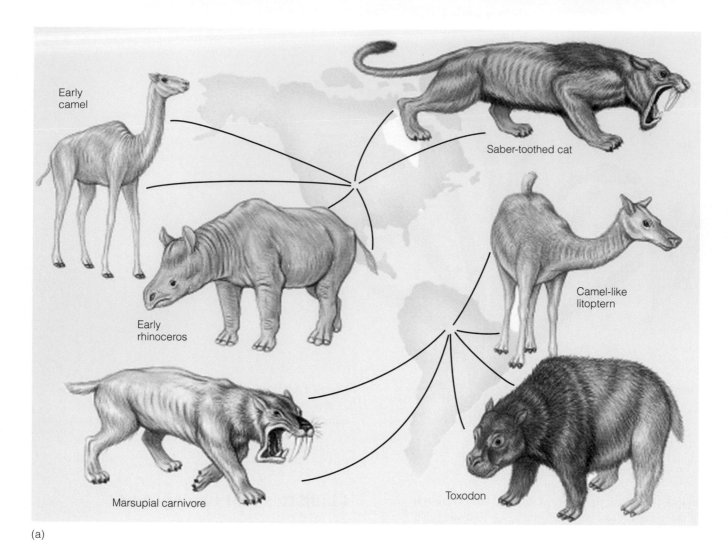

(a)

(b)

Figure 18.9 *(a) Convergent evolution. Before a land connection existed between North and South America, mammals on each continent evolved independently but adapted to similar environments. (b) Parallel evolution. Kangaroo rats and jerboas are closely related but have independently developed similar features.*

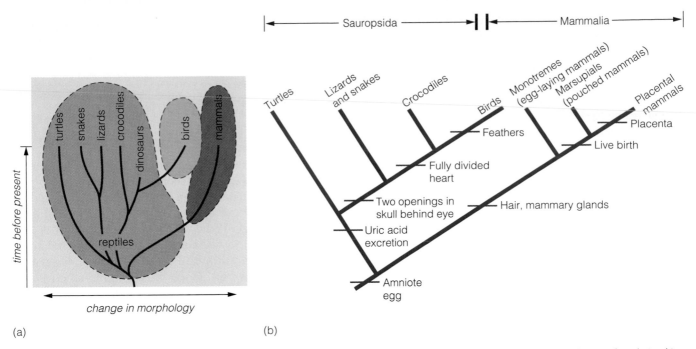

Figure 18.10 *(a) A phylogenetic tree showing the inferred relationships among reptiles, birds, and mammals. (b) Cladogram showing the relationships among living reptiles, birds, and mammals. Some of the characteristics used to differentiate subclades are indicated.*

the relationships among these animals. Bats and birds fly, so we might conclude that they are more closely related to one another than to dogs (Figure 18.11a). But if we concentrate on evolutionary novelties, such as hair and giving birth to live young, we conclude that bats and dogs are more closely related than either is to birds. Thus, the relationships among bats, dogs, and birds are best shown by the cladogram in Figure 18.11c.

Cladistics and cladograms work well for living organisms, but when applied to fossils care must be taken in determining what is a primitive versus a derived characteristic, especially in groups with poor fossil records. Even in the better-known evolutionary lineages, the earliest diversification is usually the most poorly represented by fossils. Furthermore, cladistic analysis depends solely on characteristics inherited from a common ancestor, so paleontologists must be especially careful of characteristics resulting from convergent evolution, such as wings in bats and birds. Nevertheless, cladistics is a powerful tool that has more clearly elucidated the relationships among many fossil lineages.

Extinctions

Judging from the fossil record, extinction is the rule in the history of life. Indeed, according to one estimate, more than 99% of all species that ever existed are now extinct.

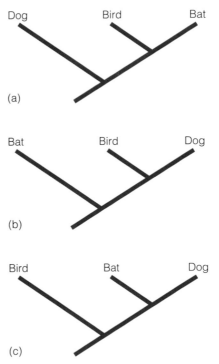

Figure 18.11 *Cladograms showing three hypotheses for the relationships among birds, bats, and dogs. Derived characteristics such as hair and giving birth to live young indicate that dogs and bats are most closely related, as shown in (c).*

The evolutionary descendants of some groups still exist, so the ancestral group, although extinct, has living relatives. Most paleontologists think birds are descendants of small, carnivorous dinosaurs. In other cases, however, a group simply dies out. A case in point is a group of large Cenozoic mammals known as titanotheres that died out during the Oligocene Epoch.

Extinction is a continual occurrence, but so is the origin of new species that usually quickly exploit the opportunities created when another species dies out. In the case of Mesozoic marine reptiles known as ichthyosaurs, however, their extinction was followed by a 30-million-year period during which no ecologically equivalent species existed. Not until well into the Cenozoic Era did the superficially similar dolphins and porpoises appear and fill this vacant niche.

The continual extinction in life history is generally referred to as *background extinction*, to clearly differentiate it from *mass extinctions*, during which extinction rates are greatly accelerated and Earth's biologic diversity is sharply reduced. Everyone knows about the extinction of dinosaurs at the end of the Mesozoic Era, which, in fact, was a great mass extinction. The one with the most drastic effects on organisms worldwide, though, took place at the end of the Paleozoic Era. These extinctions and some of lesser magnitude will be discussed in more detail in some of the following chapters.

What Kinds of Evidence Support Evolutionary Theory?

When Charles Darwin proposed his theory of evolution he cited supporting evidence such as classification, embryology, comparative anatomy, geographic distribution, and, to a limited extent, the fossil record. He had little knowledge of the mechanism of inheritance and both biochemistry and molecular biology were unknown during his time. Studies in these areas, coupled with a more complete and much better understood fossil record, have convinced scientists that the theory is as well supported by evidence as any other major theory. Of course, scientists still disagree on many details, but the central claim of the theory is well established and widely accepted.

But is the theory of evolution truly a scientific theory? That is, can testable predictive statements be made from

---

Table 18.1

Some Predictions from the Theory of Evolution

1. If present-day organisms descended with modification from species that lived during the past, ancient species must differ from those alive today.

2. If evolution by natural selection actually occurred, Earth must be very old, perhaps many millions of years.

3. If evolution has taken place, the oldest fossil-bearing rock layers should have remains of organisms very different from those existing now, and more recent rock layers should have fossils more similar to today's organisms.

4. If today's organisms descended with modification from ones in the past, there should be fossils showing characteristics connecting classes, orders, and so on.

5. If evolution is true, closely related species should be similar in details of their anatomy, biochemistry, genetics, and embryonic development, whereas distantly related species should show fewer similarities.

6. If the theory of evolution is correct, that is, living organisms descended from a common ancestor, that fact should be reflected in the system of classification.

7. If evolution actually took place, we would expect cave-dwelling plants and animals to be most similar to those immediately outside their respective caves rather than most similar to those in caves elsewhere.

8. If evolution actually took place, we would expect land-dwelling organisms on oceanic islands to most closely resemble those of nearby continents rather than those on other distant islands.

9. If evolution by natural selection has taken place, a mechanism should exist that accounts for the evolution of one species to another.

10. If evolution occurred, we would expect mammals to appear in the fossil record long after the appearance of the first fishes. Likewise, we would expect the reptiles to appear before the first mammals or birds.

Derived from several sources including Awbrey, F., and W. Thwaites, 1981, *Astec Lecture Notes,* San Diego State University, and Moore, J. A., 1993, *Science as a way of knowing: The foundations of modern biology,* Cambridge, Mass., Harvard University Press.

it? We have to be clear on what we mean by prediction. Evolutionary theory cannot predict the future—no one knows which existing species may become extinct, for example, or what the descendants, if any, of horses 10 million years from now will look like. Nevertheless, the theory does allow scientists to make a number of predictions about the present-day biological world as well as about many aspects of the fossil record (Table 18.1).

Perhaps some examples will help clarify the predictive nature of the theory of evolution. Foxes and wolves possess a number of anatomical similarities, but suppose their biochemistry, genetics, and embryological development were very different (see prediction 5 in Table 18.1). Obviously, our prediction would be less than satisfactory and we would be forced to at least modify our theory. If the theory also failed in several other areas—the first mammals appear in the fossil record before the first fishes, for instance—the theory would fail completely. Thus, the theory of evolution is truly scientific in that it can at least in principle be falsified, that is, proven wrong.

Classification—A Nested Pattern of Similarities

Carolus Linnaeus (1707–1778) proposed a classification in which organisms are given a two-part genus and species name; the coyote, for instance, is *Canis latrans*. Table 18.2 shows Linnaeus's classification scheme, which is a hierarchy of categories that becomes more inclusive as one proceeds up the list. The coyote (*Canis latrans*) and the wolf (*Canis lupus*) share numerous characteristics, so they are members of the same genus, whereas both share some but

Table 18.2

Expanded Linnaean Classification Scheme

The animal classified in this example is the coyote, *Canis latrans*.

Kingdom	Animalia
Phylum	Chordata
Subphylum	Vertebrata
Class	Mammalia
Order	Carnivora
Family	Canidae
Genus	*Canis*
Species	*latrans*

fewer characteristics with the red fox (*Vulpes fulva*) and all three are members of the family Canidae (Figure 18.12). All canids share some characteristics with cats, bears, and weasels and are grouped together in the order Carnivora, which is one of 18 living orders of the class Mammalia, all of whom are warm-blooded, possess hair, and have mammary glands.

Linnaeus clearly recognized shared characteristics among organisms, but his intent was simply to categorize species he thought were specially created and immutable. Following the publication of *The Origin of Species* in 1859, however, biologists quickly realized that shared characteristics constituted a strong argument for evolution. After all, if present-day organisms actually descended from ancient species we should expect a pattern of similarities between closely related species and fewer between more distantly related ones. In our example above, coyotes and wolves share many characteristics because they diverged from a common ancestor in the not too distant past, whereas the common ancestor of coyotes, wolves, and foxes existed in the more distant past (Figure 18.12).

Biological Evidence Supporting Evolution

If all existing organisms actually descended with modification from ancestors that lived during the past, fundamental similarities should exist among all life forms. As a matter of fact, all living things, be they bacteria, redwood trees, or whales, are composed mostly of carbon, nitrogen, hydrogen, and oxygen. Furthermore, their chromosomes consist of DNA, except bacteria which have RNA, and all cells synthesize proteins in essentially the same way.

Studies in biochemistry also provide evidence for evolutionary relationships. Blood proteins are similar among all mammals but also indicate that humans are most closely related to great apes, followed in order by Old World monkeys, New World monkeys, and lower primates such as lemurs. Biochemical tests support the idea that birds descended from reptiles, a conclusion supported by evidence in the fossil record.

The forelimbs of humans, whales, dogs, and birds are superficially dissimilar (Figure 18.13). Yet all are made up of the same bones; have basically the same arrangement of muscles, nerves, and blood vessels; all are similarly arranged with respect to other structures; and all have a similar pattern of embryonic development. These **homologous organs,** as they are called, are simply basic vertebrate forelimbs modified for different functions; that is, they indicate derivation from a common ancestor. There are, however, some similarities unrelated to evolutionary relationships. For instance, wings of insects and birds serve the same function but are quite dissimilar in both structure and development and are thus termed **analogous organs** (Figure 18.14).

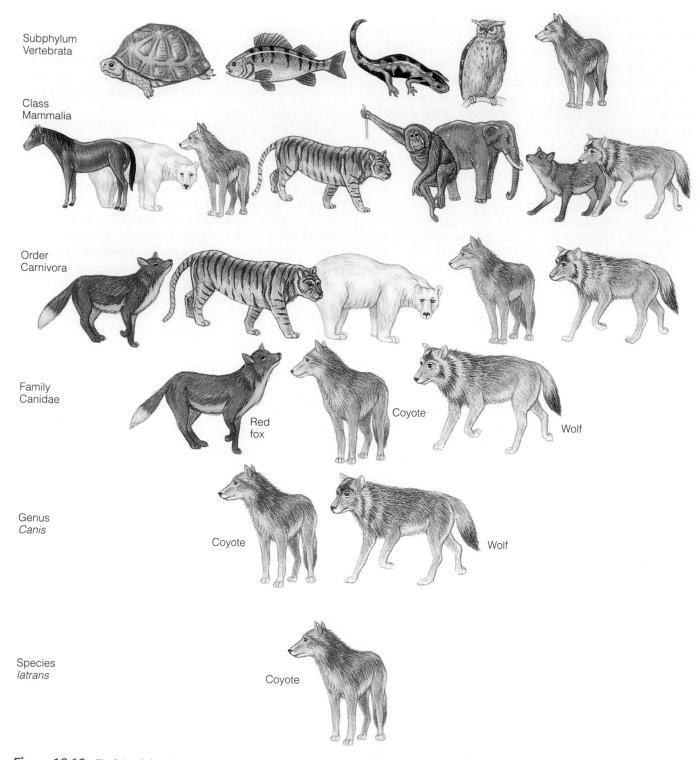

Subphylum
Vertebrata

Class
Mammalia

Order
Carnivora

Family
Canidae

Red
fox

Coyote

Wolf

Genus
Canis

Coyote

Wolf

Species
latrans

Coyote

Figure 18.12 *Traditional classification of organisms based on shared characteristics. All members of the subphylum Vertebrata, including fishes, amphibians, reptiles, birds, and mammals, have a segmented vertebral column. Among the vertebrates, though, only those warm-blooded animals having hair or fur and mammary glands are mammals. Eighteen living orders of mammals are recognized, including the order Carnivora shown here. All members of this order have teeth specialized for a diet of meat. The family Canidae includes only the doglike carnivores, and the genus Canis includes closely related species. The coyote, Canis latrans, stands alone as a species.*

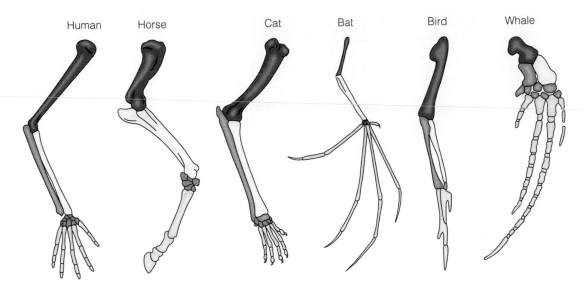

Figure 18.13 *Homologous organs such as the forelimbs of vertebrates may serve different functions but are composed of the same elements and undergo similar embryological development.*

Why do dogs have tiny, functionless toes on their back feet, or forefeet, or, in some cases, on all feet (Figure 18.15a)? These dewclaws are **vestigial structures**—that is, nonfunctional or partly functional remnants of structures that were functional in their ancestors. Ancestral dogs had five toes on each foot, all of which contacted the ground. But as they evolved they became toe-walkers with only four digits on the ground and the big toes and thumbs were lost or reduced to their present state. Almost all living species have some vestigial structures—extra functionless toes in pigs and deer, remnants of toes in horses (Fig-

ure 18.15b), whales with a pelvis but no rear limbs, wisdom teeth in humans, and a host of others. A Cretaceous-aged snake has stubby rear limb bones but no feet, and some living snakes retain a tiny remnant of a pelvis. Charles Darwin said "they [vestigial structures] have been partially retained by the power of inheritance, and relate to a former state of things."*

*The Origin of Species (New York: New American Library, 1958), pp. 419–420.

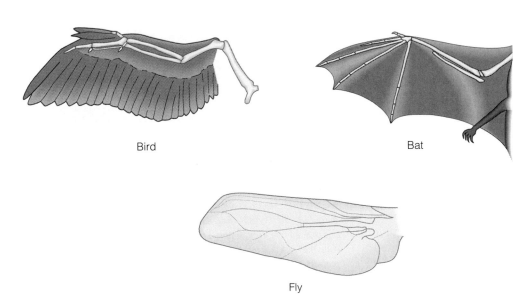

Bird

Bat

Fly

Figure 18.14 *The fly's wings serve the same function as wings of birds and bats but have a different embryological development. Organs that serve the same function are analogous, but the bird and bat wings are also homologous.*

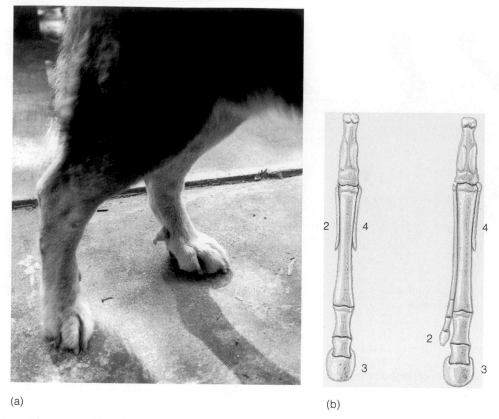

(a)   (b)

Figure 18.15 *Vestigial structures. (a) Notice the dewclaw, a vestige of the big toe, on the left back foot of this dog. (b) Normal condition of a horse's back foot (left); the only functional toe is the third, but side splits of toes 2 and 4 remain as vestiges. Occasionally, horses are born with one or both of these vestiges enlarged (right).*

Another type of evidence for evolution is observations of small-scale evolution in living organisms: the changing color of moths in England, for example (Figure 18.16). We have already mentioned one example of small-scale evolution—the adaptations of some plants to contaminated soils. As a matter of fact, small-scale changes take place rapidly enough that new insecticides and pesticides must be developed continually because insects and rodents develop resistance to existing ones. And development of antibiotic-resistant stains of bacteria is a continuing problem in medicine. Whether the variation in these populations previously existed or was established by mutations is irrelevant. In either case, some variant types lived and reproduced, bringing about a genetic change.

Fossils and Evolution

In Chapter 5, we briefly discussed **fossils,** the remains or traces of organisms preserved in rocks, and noted that they are useful for determining environments of deposition and are used for relative age determinations. The term **body**

Figure 18.16 *Small-scale evolution of the peppered moths of Great Britain. In pre-industrial England most of these moths were gray and blended well with the lichens on trees. Black varieties were known, but rare, because they stood out against the light background and were easily spotted by birds. With industrialization and pollution, the trees became soot-covered and dark, and the frequency of dark-colored moths increased while that of light-colored moths decreased. However, in rural areas unaffected by pollution, the light-dark frequency did not change.*

Table 18.3

Types of fossil preservation

Body fossils—unaltered remains	Original composition and structure retained
Freezing	Large Ice Age mammals in frozen ground (rare)
Mummification	Air drying and shriveling of soft tissues (rare)
Preservation in amber	Insects in hardened tree resin
Preservation in tar	Bones in asphalt-like substance at oil seeps (rare)
Body fossils—altered remains	Change in composition and/or structure of original
Permineralization	Addition of minerals to pores and cavities
Recrystallization	Change in crystal structure, e.g., aragonite recrystallized as calcite
Replacement	One chemical compound replaces another, e.g., pyrite replaces calcium carbonate; silicon dioxide replaces wood
Carbonization	Carbon film of leaf, insect, etc., when volatile elements lost
Trace fossils	Burrows, tracks, trails, nests, droppings (coprolites), or any other indication of organic activity
Molds and casts	Mold, a cavity having the shape of a bone or shell; cast, a mold filled by minerals or sediment

fossil applies to actual remains such as bones, teeth, and shells, whereas a **trace fossil** is an indication of organic activity, such as burrows and tracks. Fossilization takes place in several ways (Table 18.3 and Figure 18.17), and fossils in general are quite common although the chances of some organisms being preserved are much better than for others. In fact, the fossil record is strongly biased toward organisms with durable skeletons that live in areas where burial is likely. Accordingly, marine invertebrates such as clams and corals are better represented in the fossil record than bats, birds, worms, and jellyfish.

Fossil marine invertebrates far from the sea, even high in mountains, led early naturalists to conclude that the fossil-bearing rocks were deposited during a worldwide flood. In 1508, Leonardo de Vinci realized that the fossil distribution was not what one would expect from a rising flood, but the flood explanation persisted and John Woodward (1665–1728) proposed a testable hypothesis. According to him, the density of organic remains determined the order in which they settled from floodwaters, so, logically, it would seem that fossils in the oldest rocks should be denser than those in younger ones. Woodward's hypothesis was quickly rejected, because observations clearly did not support it; fossils of various densities are found throughout the fossil record.

The fossil record does show a sequence of different organisms, but not one based on density, size, shape, or habitat. Rather, the sequence consists of first appearances of more and more complex organisms through time (Figure 18.18). In fact, older and older rocks contain fossils of organisms increasingly different from those existing today (see Table 18.1). One-celled organisms appeared before multicelled ones, plants before animals, and invertebrates before vertebrates. Among vertebrates, fish appear first followed in succession by amphibians, reptiles, mammals, and birds.

Fossils are much more common than many people realize, so we might ask, "Where are the fossils showing the diversification of horses, rhinoceroses, and tapirs from a common ancestor; or the origin of birds from reptiles; or the evolution of whales from a land-dwelling ancestor?" It is true that the origin and initial diversification of a group is the most poorly represented in the fossil record, but in these cases, as well as many others, fossils of the kind we would expect are known.

Horses, rhinoceroses, and tapirs might seem an odd assortment of animals, but fossils and studies of living animals indicate that they, along with the extinct titanotheres and chalicotheres, share a common ancestor (Figure 18.19). If this statement is correct, then we can predict that as we trace these animals back in the fossil record, differentiating one from the other should become increasingly difficult. And, in fact, the earliest members of each are remarkably similar, differing mostly in size and details of their teeth. As their diversification proceeded, though, differences became more apparent.

Jurassic-aged fossils from Germany have anatomical characteristics much like those of small, carnivorous dinosaurs, including dinosaurlike teeth, a long tail, brain case, and hind limbs, yet they also clearly have feathers. These creatures, known as *Archaeopteryx*, have characteristics we would expect in a forerunner of birds. Whether *Archaeopteryx* was the actual ancestor of living birds or an early branch that led nowhere is debatable,

What Kinds of Evidence Support Evolutionary Theory?

(a)

(b)

(c)

(d)

(e)

(f)

Figure 18.17 *Fossils and fossilization. (a) Body fossils of Paleozoic horn corals. (b) Trace fossil. Notice the small amphibian footprint. (c) Fossilized sequoia stump at Florrisant Fossil Beds National Monument, Colorado. (d) Cast of a bivalve shell. (e) Fossilized feces (coprolite) of a carnivorous mammal. Specimen measures 5.5 cm long. (f) Insect preserved by carbonization.*

but that in no way diminishes the fact that it shows characteristics of both reptiles and birds. And during the last few years, dinosaurs with feathers have been found in China (see Chapter 23).

Until several years ago, the origin of whales was poorly understood. Fossil whales were common enough, but those linking fully aquatic animals with land-dwelling ancestors were largely absent. It turns out that this transition took place in a part of the world where the fossil record was poorly known. Now, however, a number of fossils are available that show how and when whales evolved (see Chapter 24).

WHAT WOULD YOU DO?

As a middle school teacher you are discussing fossils and fossilization in your science class. How would you explain to your students that by far most fossils are found in sedimentary rocks, especially in mudrocks, sandstone, and limestone. Why do you think this is so? Why are fossils largely absent from igneous and metamorphic rocks? Are there any exceptions, and if so, what are they? Also, how would you explain that even though most organisms are not preserved, fossils are nevertheless quite common?

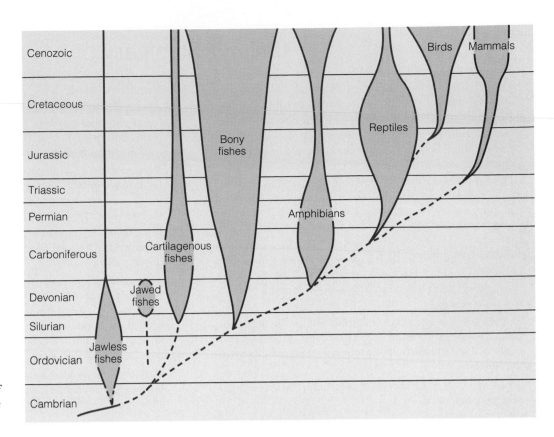

Figure 18.18 *The times of appearance of the major groups of vertebrate animals.*

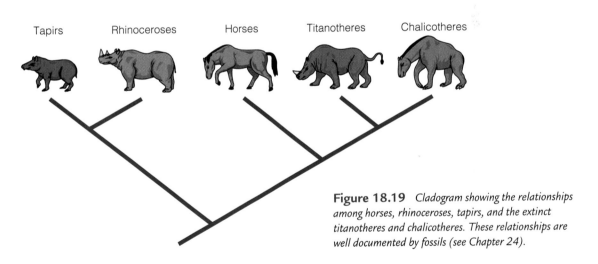

Figure 18.19 *Cladogram showing the relationships among horses, rhinoceroses, tapirs, and the extinct titanotheres and chalicotheres. These relationships are well documented by fossils (see Chapter 24).*

Of course, we will never have enough fossils to document the evolutionary history of all living creatures, simply because fossilization is a random process; remains of some organisms are more likely to be preserved than those of others, and the accumulation of sediments varies in both time and space. Nevertheless, more and more fossils are being found that more clearly illustrate ancestor-descendant relationships among many organisms. Fossils are important in studies of evolution, but you should always bear in mind that they constitute only one kind of evidence supporting the theory.

WHAT WOULD YOU DO?

An individual tells you evolution is "only a theory that has never been proven," and that "the fossil record shows a sequence of organisms that was determined by their density and habitat." Why is the first statement irrelevant to theories in general, and what kinds of observations could you cite to refute the second statement?

Chapter Summary

1. The first formal theory of evolution to be taken seriously was proposed by Jean-Baptiste de Lamarck. Inheritance of acquired characteristics was his mechanism for evolution.

2. In 1859 Charles Robert Darwin and Alfred Russel Wallace simultaneously published their views on evolution, and proposed natural selection as the mechanism for evolutionary change.

3. Darwin's observations of variation in natural populations and artificial selection, as well as his reading of Thomas Malthus's essay on population, helped him formulate the idea that natural processes select favorable variants for survival.

4. Gregor Mendel's breeding experiments with garden peas provided some of the answers regarding how variation is maintained and passed on. Mendel's work is the basis for modern genetics.

5. Genes are the hereditary determinants in all organisms. This genetic information is carried in the chromosomes of cells. Only the genes in the chromosomes of sex cells are inheritable.

6. Most variation is maintained in populations by sexual reproduction and mutations.

7. Evolution by natural selection is a two-step process. First, variation must be produced and maintained in interbreeding populations, and second, favorable variants must be selected for survival.

8. An important way in which new species evolve is by allopatric speciation. When a group is isolated from its parent population, gene flow is restricted or eliminated, and the isolated group is subjected to different selection pressures.

9. Divergent evolution involves an ancestral stock giving rise to diverse species. The development of similar adaptive types in different groups of organisms results from parallel and convergent evolution.

10. Most fossils are preserved in sedimentary rocks as unaltered remains, altered remains, molds and casts, or traces of organic activity.

11. Although fossils of some organisms are quite common, very few of the organisms that lived at any one time were actually preserved. Furthermore, the fossil record is biased toward those organisms with durable skeletons that lived in areas of active sedimentation.

12. The evidence for evolution includes the way organisms are classified, comparative anatomy and embryology, genetics, biochemical similarities, small-scale evolution, and the fossil record.

Important Terms

allele
allopatric speciation
analogous organ
artificial selection
body fossil
chromosome

cladogram
convergent evolution
divergent evolution
DNA (deoxyribonucleic acid)
fossil
gene

homologous organ
modern synthesis
mutation
natural selection
parallel evolution
phyletic gradualism

punctuated equilibrium
species
theory of evolution
trace fossil
vestigial structure

Review Questions

1. Biologists and paleontologists think that most new species arise by
   a. _____ mosaic evolution; b. _____ allopatric speciation; c. _____ differential reproduction; d. _____ homologous differentiation; e. _____ inheritance of vestigial structures.

2. According to the concept of punctuated equilibrium,
   a. _____ species change gradually and continually over millions of years; b. _____ most species that ever existed are now extinct; c. _____ the most common evolutionary trend is size increase; d. _____ new species arise rapidly, in perhaps only a few thousand years; e. _____ organs with the same basic structure and development are analogous.

3. Clams and brachiopods are very distantly related but possess similarities because of
   a. _____ convergent evolution; b. _____ adaptive divergence; c. _____ artificial selection; d. _____ parallel adaptation; e. _____ inheritance of the same alleles.

4. The theory of natural selection was proposed by Charles Darwin and
   a. _____ William Smith; b. _____ Erasmus Darwin; c. _____ Alfred Wallace; d. _____ Charles Lyell; e. _____ Jean-Baptiste de Lemarck.

5. The type of cell division by which sex cells such as eggs and sperm are produced is known as
   a. _____ meiosis; b. _____ divergence; c. _____ morphology; d. _____ speciation; e. _____ mosaic evolution.

6. In the classification scheme now used, an organism is referred by its _____ names.
   a. _____ family and phylum; b. _____ genus and species; c. _____ order and class; d. _____ subphylum and phylum; e. _____ kingdom and family.

7. Chromosomes are complex molecules of
   a. _____ deoxyribonucleic acid; b. _____ potassium, aluminum, and silicon; c. _____ carbon dioxide and helium; d. _____ calcium carbonate; e. _____ genes and homologous organs.

8. Divergent evolution involves the
   a. _____ interbreeding by unrelated species; b. _____ development of similar features in distantly related groups of plants; c. _____ inheritance of acquired characteristics; d. _____ diversification of a species into two or more descendant species; e. _____ replacement of one species by another following an extinction.
9. Organs such as bird and bat wings that have a similar development pattern and similar relationship to other organs are referred to as
   a. _____ artificial; b. _____ mosaic; c. _____ allopatric; d. _____ alleles; e. _____ homologous.
10. According to studies of living animals and fossils, horses are most closely related to which one of the following?
   a. _____ Pigs; b. _____ Birds; c. _____ Rhinoceroses; d. _____ Camels; e. _____ Reptiles.
11. Inheritance of acquired characteristics was the mechanism for evolution proposed by
   a. _____ Martin A. C. Hinton; b. _____ Jean-Baptiste de Lamarck; c. _____ Alfred Russel Wallace; d. _____ Gregor Mendel; e. _____ Philip Henry Gosse.
12. According to the idea of _____ , species remain unchanged during most of their time of existence, then evolve rapidly to produce a new species.
   a. _____ convergent evolution; b. _____ artificial selection; c. _____ adaptive divergence; d. _____ punctuated equilibrium; e. _____ allopatric speciation.
13. Explain how the fossil record provides evidence for evolution. Give an example.
14. How does sexual reproduction maintain variation in an interbreeding population?
15. Compare the concepts of punctuated equilibrium and phyletic gradualism. Is there any evidence for either? If so, what?
16. Discuss the concept of allopatric speciation and give an example of where it has occurred.
17. How did the experiments by Gregor Mendel help answer some of the criticisms of natural selection that plagued Darwin and Wallace?
18. What criteria are used to distinguish homologous organs from analogous organs?
19. Explain how the classification of organisms according to Linnaeus's scheme provides evidence for evolution.
20. Draw three cladograms showing possible relationships among sharks, whales, and bears. Which one most likely depicts the actual relationships among these animals, and what criteria did you use to make your decision?
21. Name some vestigial structures and explain why they constitute evidence supporting evolutionary theory.
22. A cladogram shows that among carnivorous mammals, cats, hyenas, and mongooses are more closely related to one another than to other carnivores. Yet, hyenas resemble dogs and mongooses look like weasels. What kinds of evidence from existing animals and fossils would lend credence to the idea that these three animals constitute a closely related group?
23. Evolution by natural selection works on variations in populations. Give examples of favorable, neutral, and harmful variations that might occur in a population of mammals. Would only favorable variations be selected for survival?
24. Give examples of convergent and parallel evolution. Why do these phenomena take place?

 # World Wide Web Activities

For these web site addresses, along with current updates and exercises, log on to **http://www.brookscole.com/geo**

University of California Museum of Paleontology

This huge site has excellent discussion and images on a variety of topics, including *On-Line Exhibits* with subheadings of Time Periods, Phylogeny (evolutionary history), Evolution, and Geology. In the Education and Public Outreach section you will find *A Collection of Classroom Activities* especially useful for students contemplating a career in teaching. Also, the *On-Line Exhibits* section has images and information on the life history, fossil record, and other aspects of numerous groups of organisms.

University of Miami Department of Biology

On the University of Miami Department of Biology page, click *courses* and then scroll down to *Biology 160—Evolution and Biodiversity.* Among the topics for this course with extensive notes are *Historical Perspective, Darwin and Natural Selection, The Species Concept,* and several others.

American Museum of Natural History: Understanding Cladistics

The American Museum of Natural History in New York City maintains this large site at which you can explore a variety of topics, including cladistics, evolution, and the fossil record. At this site see what cladograms are, how they are constructed, and why they are used.

CHAPTER 19

A History of the Universe, Solar System, and Planets

With a diameter of 3126 km, Jupiter's moon Europa (center) is the sixth largest moon in the solar system. Evidence from the Galileo mission indicates that Europa has an icy outer covering and perhaps liquid water below. Also shown in this image, but not to scale, are Jupiter (left center) and three of its other moons including Io (upper left) and Ganymede and Callisto (right).

At the end of this chapter, you will have learned that

■ The Big Bang theory for the origin of the universe is currently widely accepted.

■ The general characteristics of the solar system must be accounted for by any theory proposed to explain how it formed and evolved.

■ The three basic types of meteorites provide evidence for the composition and age of the solar system.

■ Several criteria such as density and composition are used to define terrestrial versus Jovian planets.

■ Earth formed as a gaseous sphere that became a planet with a core, mantle, and crust.

■ Scientists theorize that the Moon originated from material ejected from Earth following an asteroid impact.

Does life exist elsewhere in the universe or is it unique to Earth? This question has intrigued people for centuries. The discovery of extraterrestrial life would certainly create a major scientific and theological upheaval as the fact that we are not alone sinks into the collective human consciousness.

The search for life elsewhere in the solar system and beyond goes back at least as far as Galileo (1564–1642). His discovery that other planets existed fueled people's imagination about whether life may also exist in these places. In 1894, the wealthy amateur astronomer Percival Lowell built his own observatory in Flagstaff, Arizona, to further the study of Mars, a planet especially interesting to him. Inspired by earlier reports of dark lines on the Martian surface and his own observations, Lowell published a book in 1895 speculating that an advanced civilization had built canals (the dark crisscrossing lines some people observed) to bring water from the melting polar ice caps to the rest of the planet in response to a global drought.

Lowell's writings were instrumental in firing public interest in Mars and Martians and the possibility of extraterrestrial life forms. In 1898, H. G. Wells published his novel, *The War of the Worlds,* about an invasion of Earth by Martians. This theme was further amplified in a 1938 radio broadcast by Orson Welles in which Martians invaded New Jersey, and the 1953 movie titled *War of the Worlds.*

Based on what we know presently about Mars from telescope observations and the various space probes sent to study the planet, scientists are convinced that Mars does not harbor any higher forms of life—that is, complex organisms comparable to Earth's plants and animals. Yet the possibility remains that microorganisms may have lived on the Martian surface early in its history when water was present, or are living beneath its surface.

The human fascination with Mars is understandable, given that the so-called Red Planet has always had an air of mystery about it. Perhaps Mars does harbor living microorganisms or evidence of ancient microorganisms now long extinct. However, all attempts so far have failed to detect any organisms. Yet the search for life outside of Earth goes on, with another body in the solar system receiving serious attention.

Europa, one of Jupiter's moons, with a diameter of 3126 km is the sixth largest moon in the solar system (see chapter opening photo). Because of Jupiter's tremendous gravitational attraction and interactions with other Jovian moons, Europa experiences what is called *tidal flexing,* resulting in stretching and compression of the entire moon and especially its icy outer layer. It possesses few craters, indicating that it is a geologically active body.

Data from *Galileo* missions indicate that Europa has a vast ocean beneath its outer icy cover, an ocean perhaps 200 km deep. If that is correct, Europa has more liquid water than contained in all of Earth's oceans. In any event, scientists know that Europa is too small to retain any heat from its time of origin, and radioactive decay is insufficient to explain its geologic activity. Tidal heating is more probable, thus accounting for the liquid water. Because liquid water is likely present, scientists speculate that Europa's oceans may harbor some kinds of organisms, perhaps in environments similar to those adjacent to black smokers on Earth's seafloor. This question, however, will remain unanswered until a future mission reaches this small celestial body.

Do any other bodies in the solar system or elsewhere in the universe have the conditions necessary for life to evolve? Most scientists think that if extraterrestrial life exists, it will be carbon-based, because carbon is found in so many different and complex compounds, as well as being the fourth most abundant element in the universe. Carbon compounds are found in meteorites and comets, as well as in huge molecular clouds floating throughout the cosmos. In fact, some scientists think that life evolved from organic compounds or even living microbes that fell to Earth from space early in its history. Recent studies suggest that the core of some meteorites remains far below the sterilization temperature of 111°C, and thus organisms living inside such a meteorite could easily survive the plunge through Earth's atmosphere. Regardless of whether life originated elsewhere in the universe or on Earth, we do know that organisms were present on Earth at least 3.5 billion years ago.

If life is not unique to Earth, is it common elsewhere in the universe? Since the mid-1990s, more than 40 planet-bearing stars—that is, stars in which planets are circling them—have been discovered within the Milky Way Galaxy. What makes these discoveries so exciting is that, if these stars have planets circling them, the odds are greatly increased that other stars in the 100-billion-star Milky Way Galaxy have not only large, inhospitable planets circling them but also smaller, more life-friendly planets. Based on what we know about how life might have originated, we can say the probability is high that other planets in our galaxy also contain life.

Introduction

When considered in the context of the universe, Earth is a medium-sized planet circling a rather average star among billions in the Milky Way Galaxy, which is but one of billions of galaxies. A number of other planets have recently been discovered around other star systems and perhaps life exists elsewhere in the universe, but of all the known planets and moons in the solar system only Earth is known to have life. Viewed from the blackness of nearby space, Earth is a brilliant, bluish, shimmering, planet, wrapped in a veil of swirling white clouds. Vast oceans, seven continents, and numerous islands cover its surface, and organisms of one kind or another are present nearly everywhere. This unique planet, revolving around the Sun every 365.25 days, is a complex, evolving body. In short, it is a dynamic planet that has changed and continues to change in response to interactions among its major systems (see Chapter 1 opening photo).

Earth has not always looked the way it does today. Based on various kinds of evidence, many scientists think that it began as a homogeneous mass of rotating dust and gases that contracted, heated, and differentiated during its early history to form a medium-sized planet with a metallic core, a mantle composed of iron- and magnesium-rich rocks, and a thin crust (see Figure 1.8). Overlying this crust is an atmosphere now composed of 78% nitrogen and 21% oxygen.

As the third planet from the Sun, Earth seems to have formed at just the right distance from the Sun (150 million km) so that it is neither too hot nor too cold to support life as we know it. Furthermore, its size is just right to hold an atmosphere. If Earth were smaller, its gravity would be so weak that it could retain little, if any, atmosphere, and any surface water would vaporize.

Why Should You Study the History of the Universe, Solar System, and Planets?

Aside from the fundamental question of who we are and where did we come from, is there any good reason to study the origin and history of the universe, solar system, and planets? For one thing, humans are curious, and we like to know how we fit into the larger picture—in this case, the world beyond our planet. How did the universe begin? What has been its history? Is it infinite? What is its eventual fate? These are just some of the basic questions people have asked and wondered about since they first looked into the nighttime sky and saw the vastness of the universe beyond Earth. As a purely intellectual pursuit, people are curious about the planets, solar system, and universe. So that is certainly one good reason to study the history of these different entities. Another reason is to better understand the history of our own planet and how it formed. By learning more about the planets in our solar system, we are also learning about how our own planet formed and evolved. Another reason is that we might someday explore and colonize the planets, just as we have explored our Moon, so we need to learn as much as possible about our celestial neighbors.

We are living in a time when what we know about the world beyond Earth is rapidly changing and no textbook can be completely current in this area, particularly when it comes to information about the planets. New space probes are being launched at a rapid rate and sending back images and information about the planets, their moons, and asteroids at an astonishing rate. To keep up with this new information and be current in this field, it is important for students to check the web sites and peruse current issues of the various journals we've listed at the end of this chapter.

Origin of the Universe— Did It Begin with a Big Bang?

Most scientists think that the universe originated about 15 billion years ago in what is popularly called the **Big Bang.** In a region infinitely smaller than an atom, both time and space were set at zero. Therefore, there is no "before the Big Bang," only what occurred after it. The reason for this is that space and time are unalterably linked to form a space-time continuum demonstrated by Einstein's theory of relativity. Without space, there can be no time.

How do we know the Big Bang took place approximately 15 billion years ago? Why couldn't the universe have always existed as we know it today? Two fundamental phenomena indicate that the Big Bang occurred. The first is that the universe is expanding. When astronomers look beyond our own solar system, they observe that everywhere in the universe galaxies are apparently moving away from each other at tremendous speeds. By measuring this expansion rate, they can calculate how long ago the galaxies were all together at a single point. Second, there is a pervasive background radiation of 2.7° above absolute zero (absolute zero equals −273°C) everywhere in the universe. This background radiation is thought to be the faint afterglow of the Big Bang.

Early History of the Universe

At the time of the Big Bang, matter as we know it did not exist, and the universe consisted of pure energy.

Table 19.1

The Four Basic Forces of the Universe

Four forces appear to be responsible for all interactions of matter:

1. **Gravity** is the attraction of one body toward another.

2. **Electromagnetic force** combines electricity and magnetism into the same force and binds atoms into molecules. It also transmits radiation across the various spectra at wavelengths ranging from gamma rays (shortest) to radio waves (longest) through massless particles called *photons*.

3. **Strong nuclear force** binds protons and neutrons together in the nucleus of an atom.

4. **Weak nuclear force** is responsible for the breakdown of an atom's nucleus, producing radioactive decay.

Within the first second after the Big Bang, the four basic forces—gravity, electromagnetic force, strong nuclear force, and weak nuclear force (Table 19.1)—had all separated, and the universe experienced enormous expansion. Matter and antimatter collided and annihilated each other. Fortunately, there was a slight excess of matter left over that would become the universe. By the time the universe was three minutes old, temperatures were cool enough for protons and neutrons to fuse together to form the nuclei of hydrogen and helium atoms. Approximately 300,000 years later, electrons joined with the previously formed nuclei to make complete atoms of hydrogen and helium. At the same time, photons (the energetic particles of light) separated from matter, and light burst forth for the first time.

Changing Composition of the Universe

After about 200 million years, as the universe continued expanding and cooling, stars and galaxies began forming and the chemical makeup of the universe changed. Early in its history, the universe was 100% hydrogen and helium, whereas today it is 98% hydrogen and helium by weight.

Over the course of their history, stars undergo many nuclear reactions whereby lighter elements are converted into heavier elements by nuclear fusion in which atomic nuclei combine to form more massive nuclei. Such reactions, which convert hydrogen to helium, occur in the cores of all stars. The subsequent conversion of helium to heavier elements, such as carbon, depends on the mass of the star. When a star dies, often explosively, the heavier elements that were formed in its core are returned to interstellar space and are available for inclusion in new stars. When new stars form, they will have a small amount of these heavier elements, which may be converted to still heavier elements. In this way, the chemical composition of the galaxies, which are made up of billions of stars, is gradually enhanced in heavier elements.

The Solar System—Its Origin and Early Development

Having looked at the origin and brief history of the universe, we can now examine our own solar system, which is part of the Milky Way Galaxy, to learn how it formed. The solar system consists of the Sun, nine planets, 64 known moons, a tremendous number of asteroids—most of which orbit the Sun in a zone between Mars and Jupiter—and millions of comets and meteorites, as well as interplanetary dust and gases (Table 19.2).

Table 19.2

Characteristics of the Sun, Planets, and Moon

Object	Mean Distance to Sun (km × 10^6)	Orbital Period (days)	Rotational Period (days)	Tilt of Axis	Equatorial Diameter (km)	Mass (kg)	Mean Density (g/cm^3)	Number of Satellites
Sun	—	—	25.5	—	1,391,400	1.99 × 10^{30}	1.41	—
Terrestrial planets								
Mercury	57.9	88.0	58.7	28°	4880	3.33 × 10^{23}	5.43	0
Venus	108.2	224.7	243	3°	12,104	4.87 × 10^{24}	5.24	0
Earth	149.6	365.3	1	24°	12,760	5.97 × 10^{24}	5.52	1
Mars	227.9	687.0	1.03	24°	6787	6.42 × 10^{23}	3.96	2
Jovian planets								
Jupiter	778.3	4333	0.41	3°	142,796	1.90 × 10^{27}	1.33	16
Saturn	1428.3	10,759	0.43	27°	120,660	5.69 × 10^{26}	0.69	18
Uranus	2872.7	30,685	0.72	98°	51,200	8.69 × 10^{25}	1.27	18
Neptune	4498.1	60,188	0.67	30°	49,500	1.03 × 10^{26}	1.76	8
Pluto	5914.3	90,700	6.39	122°	2300	1.20 × 10^{22}	2.03	1
Moon	0.38 (from Earth)	27.3	27.32	7°	3476	7.35 × 10^{22}	3.34	—

Table 19.3

General Characteristics of the Solar System

1. Planetary orbits and rotation
   - Planetary and satellite orbits lie in a common plane.
   - Nearly all of the planetary and satellite orbital and spin motions are in the same direction.
   - The rotation axes of nearly all the planets and satellites are roughly perpendicular to the plane of the ecliptic.

2. Chemical and physical properties of the planets
   - The terrestrial planets are small, have a high density (4.0 to 5.5 g/cm^3), and are composed of rock and metallic elements.
   - The Jovian planets are large, have a low density (0.7 to 1.7 g/cm^3), and are composed of gases and frozen compounds.

3. The slow rotation of the Sun

4. Interplanetary material
   - The existence and location of the asteroid belt.
   - The distribution of interplanetary dust.

What Are the General Characteristics of the Solar System?

Recall from earlier discussions that a *theory* is an explanation for some phenomenon or related phenomena. As such, it must be testable in some fashion; that is, it must account for observations and experimental results. Thus, any theory formulated to explain the origin and evolution of the solar system must take into account its various general characteristics (Table 19.3).

One characteristic of the solar system is that all planets revolve around the Sun in the same direction, in nearly circular orbits, and in approximately the same plane (called the *plane of the ecliptic*), except for Pluto whose orbit is both highly elliptical and tilted 17° to the orbital plane of the rest of the planets (Figure 19.1).

All planets, except Venus and Uranus, and nearly all planetary moons, rotate counterclockwise when viewed from a point in space above Earth's North Pole. Furthermore, the axes of rotation of the planets, except for those of Uranus and Pluto, are nearly perpendicular to the plane of the ecliptic (Figure 19.1b).

The nine planets are divided into two groups based on their chemical and physical properties. The four inner planets—Mercury, Venus, Earth, and Mars—are all small and have high mean densities (Table 19.2), indicating that they are composed of rock and metallic elements. They are known as the **terrestrial planets** because they are similar to *terra*, which is Latin for "earth."

Figure 19.1 *Diagrammatic representation of the solar system showing (a) the relative sizes of the planets and (b) their orbits around the Sun.*

(a)

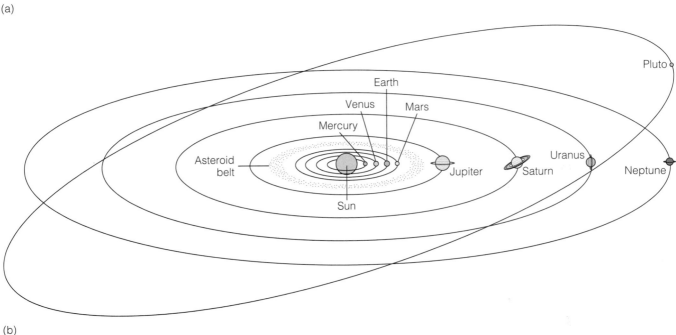

(b)

The next four planets—Jupiter, Saturn, Uranus, and Neptune—are known as **Jovian planets** because they all resemble Jupiter. The Jovian planets are large and have low mean densities, indicating they are composed of lightweight gases such as hydrogen and helium, as well as frozen compounds such as ammonia and methane. The outermost planet, Pluto, is small and has a low mean density of slightly more than 2.0 g/cm^3.

The slow rotation of the Sun is another feature that must be accounted for in any comprehensive theory of the origin of the solar system. If the solar system formed from the collapse of a rotating cloud of gas and dust as is currently accepted, the Sun, which was at the center of that cloud, should have a very rapid rate of rotation, instead of its leisurely 25-day rotation.

Finally, any theory of the origin of the solar system must accommodate the nature and distribution of the various interplanetary objects such as the asteroid belt, comets, and interplanetary gases and dust.

What Is the Current Theory for the Origin and Early History of the Solar System?

Various scientific theories of the origin of the solar system have been proposed, modified, and discarded since the French scientist and philosopher René Descartes first proposed, in 1644, that the solar system formed from a gigantic whirlpool within a universal fluid. Most theories have involved an origin from a primordial rotating cloud of gas and dust. Through the forces of gravity and rotation, this cloud then shrank and collapsed into a rotating disk. Detached rings within the disk condensed into planets, and the Sun condensed in the center of the disk.

The problem with most of these theories is that they failed to explain the slow rotation of the Sun, which according to the laws of physics should be rotating rapidly. This problem was finally solved with the discovery of the solar wind, which is an outflow of ionized gases from the Sun that interact with its magnetic field and slow down its rotation through a magnetic braking process.

As we discussed in Chapter 1, the currently accepted **solar nebula theory** (see Figure 1.7) for the origin of the solar system involves the condensation and collapse of interstellar material in a spiral arm of the Milky Way Galaxy. The collapse of this cloud of gases and small grains into a counterclockwise rotating disk caused about 90% of the material to be concentrated in the central part of the disk and resulted in the formation of an embryonic Sun, around which swirled a rotating cloud of material called a *solar nebula*. Within this solar nebula were localized eddies in which gas and solid particles condensed. Collisions between the various gases and particles resulted in accretion

into *planetesimals* (Figure 19.2). As the planetesimals collided and grew in size and mass they eventually became planets. According to estimates, a planet the size of Earth formed from planetesimals in as little as 1 million years.

The composition and evolutionary history of the planets are a consequence, in part, of their distance from the Sun. The terrestrial planets are composed of rock and metallic elements that condensed at the high temperatures of the inner nebula. The Jovian planets, all of which have small central rocky cores compared to their overall size, are composed mostly of hydrogen, helium, ammonia, and methane, which condense at low temperatures. Thus, the farther away from the Sun that condensation occurred, the lower the temperature, and hence the higher the percentage of *volatile elements*, which condense at low temperatures, relative to *refractory elements*, which condense at higher temperatures.

While the planets were accreting, material that had been pulled into the center of the nebula also condensed, collapsed, and was heated to several million degrees by gravitational compression. The result was the birth of a star, our Sun.

During the early accretionary phase of the solar system's history, collisions between various bodies were com-

Figure 19.2 *At the stage of development shown here, planetesimals have formed in the inner solar system, and large eddies of gas and dust remain at great distances from the protosun.*

mon, as indicated by the craters on many planets and moons. An unusually large collision involving Venus could explain why it rotates clockwise rather than counterclockwise, and a collision could also explain why Uranus and Pluto do not rotate nearly perpendicular to the plane of the ecliptic.

Asteroids probably formed as planetesimals in a localized eddy between what eventually became Mars and Jupiter in much the same way as other planetesimals formed the terrestrial planets. The tremendous gravitational field of Jupiter, however, prevented this material from ever accreting into a planet.

Comets, which are interplanetary bodies composed of loosely bound rocky and icy material, are thought to have condensed near the orbits of Uranus and Neptune. Each time the comets pass by Jupiter and Saturn, the gravitational effect of those planets increases their speed, forcing them further out into the solar system.

The solar nebula theory of the formation of the solar system thus accounts for most of the characteristics of the planets and their moons, the differences in composition between the terrestrial and Jovian planets, the slow rotation of the Sun, and the presence of the asteroid belt. Based on the available data, the solar nebula theory best explains the features of the solar system and provides a logical explanation for its evolutionary history.

Meteorites—Visitors from Outer Space

Meteorites are thought to be pieces of material that originated during the formation of the solar system 4.6 billion years ago. Early in the history of the solar system, a period of heavy meteorite bombardment occurred as the solar system cleared itself of the many pieces of material that had not yet accreted into planetary bodies or moons. Since then, meteorite activity has greatly diminished. Most meteorites that currently reach Earth are probably fragments resulting from collisions between asteroids.

Meteorites are classified into three broad groups based on their proportions of metals and silicate minerals (Figure 19.3a). About 93% of all meteorites are composed of iron and magnesium silicate minerals and thus are known as **stones** (Figure 19.3b). The many varieties of stones provide geologists with much information about the origin and history of the solar system.

Irons, the second group, accounting for about 6% of all meteorites, are composed primarily of a combination of iron and nickel alloys (Figure 19.3c). Their large crystal size and chemical composition indicate that they cooled slowly in large objects such as asteroids where the hot iron-nickel interior could be insulated from the cold of space. Collisions between such slowly cooling asteroids produced the iron meteorites that we find today.

Stony-irons, the third group, are composed of nearly equal amounts of iron and nickel and silicate minerals; they make up less than 1% of all meteorites (Figure 19.3d). Stony-irons are generally thought to represent fragments from the zone between the silicate and metallic portions of large differentiated asteroids.

Astronomers estimate that there are at least 750 asteroids whose orbits cross Earth's. Meteor Crater in Arizona formed when an asteroid hit Earth 25,000 to 50,000 years ago (Figure 19.4a). While asteroid-Earth collisions are rare, they do happen and could have devastating results if they occurred in a populated area (Figure 19.4b). Many scientists think that a collision with a meteorite about 10 km in diameter led to the extinctions of dinosaurs and several other groups of animals 66 million years ago (see Figure 23.32). We know that the ash and gases released into the atmosphere from volcanic eruptions have affected climates, and studies indicate that a collision with a large meteorite could produce enough dust to similarly affect global climate.

Planets and Moons

A tremendous amount of information about each planet in the solar system has been derived from Earth-based observations and measurements as well as from the numerous space probes launched during the past 30 years. Information about a planet's size, mass, density, composition, presence of a magnetic field, and atmospheric composition has allowed scientists to formulate hypotheses concerning the origin and history of the planets and their moons.

The Terrestrial Planets

All terrestrial planets appear to have experienced a similar early history during which volcanism and cratering from meteorite impacts were common. During the accretionary phase of formation and soon thereafter, each planet seems to have undergone differentiation as a result of heating by radioactive decay. The mass, density, and composition of the planets indicate that each formed a metallic core and a silicate mantle-crust during this phase. Images sent back by the various space probes also clearly show that volcanism and cratering by meteorites continued during the differentiation phase. Volcanic eruptions produced lava flows, and an atmosphere developed on each planet by **outgassing,** a process whereby light gases from the interior rise to the surface during volcanic eruptions.

Mercury

Mercury, the closest planet to the Sun, apparently has changed very little since it was heavily cratered during its early history (Figure 19.5a). Most of what we

Figure 19.3 *(a) Relative proportions of the three groups of meteorites. (b) Stony meterorite. (c) Iron meteorite. (d) Stony-iron meteorite.*

know about this small (4880 km diameter) planet comes from measurements and observations made during the fly-bys of *Mariner 10* in 1974 and 1975 (Table 19.2). Its high overall density of 5.4 g/cm^3 indicates that it has a large metallic core measuring 3600 km in diameter; the core accounts for 80% of Mercury's mass (Figure 19.5b). Furthermore, Mercury has a weak magnetic field (about 1% as strong as Earth's), indicating that the core is probably partially molten.

Images sent back by *Mariner 10* show a heavily cratered surface with the largest impact basins filled with what appear to be lava flows similar to the lava plains on Earth's moon. The lava plains are not deformed, however, indicating that there has been little or no tectonic activity. Another feature of Mercury's surface is a large number of scarps (cliffs usually produced by faulting or erosion) (Figure 19.5c). Some scientists think that these scarps formed when Mercury cooled and contracted.

Because Mercury is so small, its gravitational attraction is insufficient to retain atmospheric gases; any atmosphere that it may have held when it formed probably escaped into space quickly. Nevertheless, *Mariner 10* detected small quantities of hydrogen and helium, thought to have originated from the solar winds that stream by Mercury.

(a)

(b)

Figure 19.4 *(a) An Earth-asteroid impact about 25,000 to 50,000 years ago produced Meteor Crater, Arizona. The crater is 1.2 km wide and 180 m deep. (b) If a 48-km-diameter comet hit northern New Jersey, as depicted in this painting, everything seen here including the buildings of lower Manhattan in the foreground would be vaporized, and a plume of material would rise into the atmosphere and circulate around Earth.*

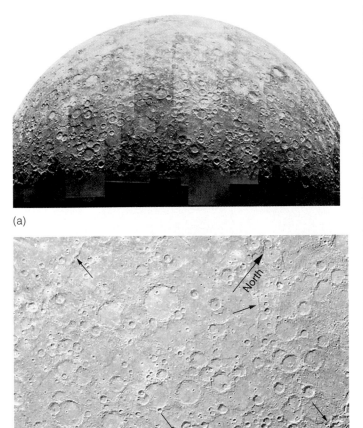

(a)

(c)

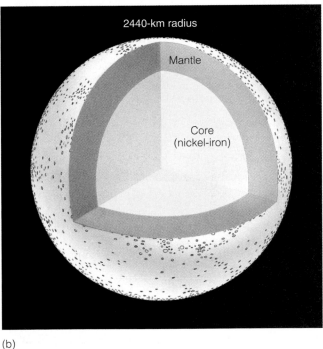

(b)

Figure 19.5 *(a) Mercury has a heavily cratered surface that has changed very little since its early history. (b) Internal structure of Mercury, showing its large solid core relative to its overall size. (c) Six scarps (indicated by arrows) can clearly be seen in this image. It is thought that these scarps might have formed when Mercury cooled and contracted early in its history.*

Venus

Of all the planets, Venus is the most similar in size and mass to Earth, but it differs in most other respects (Table 19.2). Venus is searingly hot with a surface temperature of 475°C and an oppressively thick atmosphere composed of 96% carbon dioxide and 3.5% nitrogen with traces of sulfur dioxide and sulfuric and hydrochloric acid (Figure 19.6a). From information obtained by the various space probes that have passed by, orbited Venus, and de-scended to its surface, we know that three distinct cloud layers composed of droplets of sulfuric acid envelop the planet. Furthermore, winds up to 360 km/hour occur at the top of the clouds, whereas the planet's surface remains calm.

Radar images from orbiting spacecraft and the Venusian surface reveal a wide variety of terrains (Figure 19.6b), some of which are unlike anything seen elsewhere in the solar system.

Even though no active volcanism has been observed on Venus, the presence of volcanoes, domes, extensive lava

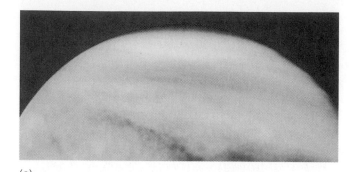

(a)

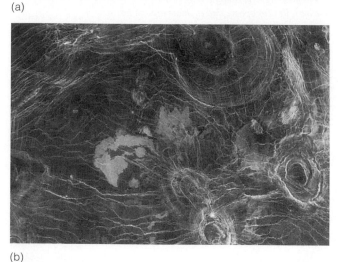

(b)

(c)

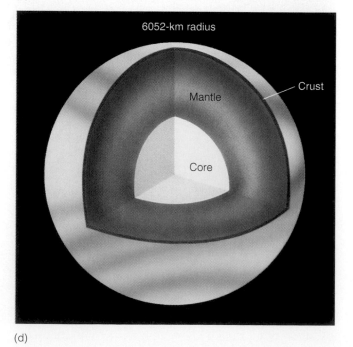

(d)

Figure 19.6 *(a) Venus has a searingly hot surface and is surrounded by an oppressively thick atmosphere composed largely of carbon dioxide. (b) This radar image of Venus made by the Magellan spacecraft reveals circular and oval volcanic features. A complex network of cracks and fractures extends outward from the volcanic features. Geologists think these features were created by blobs of magma rising from the interior of Venus, with dikes filling some of the cracks. (c) A nearly complete map of the northern hemisphere of Venus based on radar images beamed back to Earth from the Magellan space probe. (d) The internal structure of Venus.*

flows, folded mountain ranges, a complex network of faults, and what appear to be deep trenches comparable to the oceanic trenches on Earth indicate that internal and surface activity has occurred during the past (Figure 19.6c). There is, however, no evidence for active plate tectonics as on Earth.

Mars

Mars, the Red Planet, is differentiated, as are all the terrestrial planets, into a metallic core and a silicate mantle and crust (Figure 19.7a). The thin Martian atmosphere consists of 95% carbon dioxide, 2.7% nitrogen, 1.7% argon, and traces of other gases. Mars has distinct seasons during which its polar ice caps of frozen carbon dioxide expand and recede. The two small moons orbiting Mars are probably captured asteroids (see Perspective 19.1).

Perhaps the most striking aspect of Mars is its surface features, many of which have not yet been satisfactorily explained. Like the surfaces of Mercury and the Moon, the southern hemisphere is heavily cratered, attesting to a period of meteorite bombardment. *Hellas*, a crater with a diameter of 2000 km, is the largest known impact structure in the solar system and is found in the Martian southern hemisphere.

The northern hemisphere is much different, having large smooth plains, fewer craters, and evidence of extensive volcanism. The largest known volcano in the solar system, *Olympus Mons* has a basal diameter of 600 km, rises 27 km above the surrounding plains, and is topped by a huge circular crater 80 km in diameter. The large size of this volcano can probably be explained by the absence on Mars of moving plates, such as those found on Earth. Apparently, volcanism simply took place over a hot spot in this one area for a long period, building a gigantic volcano.

The northern hemisphere is also marked by huge canyons that are essentially parallel to the Martian equator. One of these canyons, *Valles Marineris*, is at least 4000 km long, 250 km wide, and 7 km deep and is the largest yet discovered in the solar system (Figure 19.7b). If it were present on Earth, it would stretch from San Francisco to New York! Geologists don't yet know how these vast canyons formed, although they postulate that the canyons may have started as large rift zones that were subsequently modified by running water and wind erosion. Such hypotheses are based on comparison to rift structures found on Earth and topographic features formed by geologic agents of erosion such as water and wind.

Because NASA has been sending spacecraft to Mars, much has been learned about the Red Planet. However, two recent failures have generated some criticism. In September 1999, the *Mars Climate Orbiter* was lost when it probably crashed into the planet, or burned up in the atmosphere, and in December 1999, the *Mars Polar Lander* fell silent. Despite the loss of these two spacecraft (at a

WHAT DO YOU THINK?

Considering our current knowledge of Mars, do you think it is possible for humans to colonize the planet effectively and become self-sufficient? What problems would our hypothetical colonists face, and how might these problems be solved with currently existing knowledge and technology?

cost of nearly $300 million) and criticism from some quarters, the Mars missions still enjoy considerable support from the public, probably in part because of the public's fascination with the planet (see the Prologue).

Tremendous wind storms have strongly influenced the surface of Mars and led to dramatic dune formations. Even more stunning than the dunes, however, are the outflow channels that appear to be the result of running water. Just as startling and exciting was the discovery in June 2000 of the possibility that water may recently have been seeping from craters in Mars's southern hemisphere. Images sent back by the *Mars Global Surveyor* show what appear to be fan-shaped depressions and gullies formed by running water near the rim of at least one crater (Figure 19.7c). Mars is currently too cold for surface water to exist, yet it might be possible that liquid water is present beneath its surface and that there was running water on the planet during the past.

The fresh-looking surfaces of its many volcanoes strongly suggest that Mars was tectonically active during the recent past and may still be. There is, however, no evidence that plate movement, such as occurs on Earth, has ever taken place.

The Jovian Planets

The Jovian planets are unlike any of the terrestrial planets in size or chemical composition (Table 19.2) and have had completely different evolutionary histories. Although they all apparently contain a small rocky core in relation to their overall size, the bulk of a Jovian planet is composed of volatile elements and compounds, such as hydrogen, helium, methane, and ammonia, that condense at low temperatures.

Jupiter

Jupiter is the largest of the Jovian planets (Table 19.2; Figure 19.8). With its moons, rings, and radiation belts, it is the most complex and varied planet in the solar system. Jupiter's density is only one-fourth that of

A persistent myth exists in which the British author Jonathan Swift (1667–1745) somehow knew that Mars had two small moons long before astronomers discovered them in 1877. Swift did in fact mention two moons close to Mars's surface in his work *Gulliver's Travels,* leading some present-day proponents of UFOs to claim that he learned this fact when he was visited by extraterrestrials. Furthermore, Swift wrote that the orbital periods of these two moons were 10 hours and 21.5 hours. But just how did these predictions come about, or were they predictions at all?

Actually, Swift was simply extrapolating on an idea, well known in his time, of the orderly progression in the number of moons possessed by the planets. Johannes Kepler (1571–1630), a famous German astronomer, noted that the closest planets to the Sun, Mercury, and Venus, had no moons, Earth had one, and Jupiter had four (actually it has 16, but only four were known then). Accordingly, Earth with one moon should be followed by Mars with two, then Jupiter with four, Saturn with eight, and so on, although the three outermost planets had not been discovered in Kepler's time.

Of course we now know that this proposed orderly progression in the numbers of moons is incorrect. Jupiter and Saturn have too many moons, Uranus probably has nearly the right number, but Neptune and Pluto have far too few (Table 19.2). Jonathan Swift was no doubt simply using this presumed knowledge in his reference to the moons of Mars, and just by coincidence happened to be right. But what about Swift's knowledge of the two moons' proximity to Mars's surface and their orbital periods? The fact that they had not been discovered made it apparent that if Mars had any moons at all they must be small and close to the planet. Furthermore, Swift was aware of Kepler's laws of planetary motion holding that if small moons did exist, they must have short orbital periods. The actual orbital periods are 7.7 hours and 30 hours, not very close to Swift's estimates if he really had some kind of special knowledge.

Kepler's idea and Swift's writings about Mars's moons are nothing more than interesting footnotes in the history of astronomy and literature, but the moons themselves are fascinating objects. Both are irregularly shaped, quite small, and probably represent captured asteroids rather than bodies that formed around Mars when the solar system originated (Figure 1). Their names, Phobos and Deimos, meaning Fear and Terror, respectively, were the names of the horses that pulled the chariot of Mars, the Greek god of war. Phobos measures only 28 km in its greatest dimension, and Deimos is no more than 12 km long. Both moons possess weak gravitational fields, so weak in fact that one could throw a baseball and put it into orbit around Deimos, unless it were thrown at more than 21 km/hr, in which case it would exceed escape velocity and never be seen again.

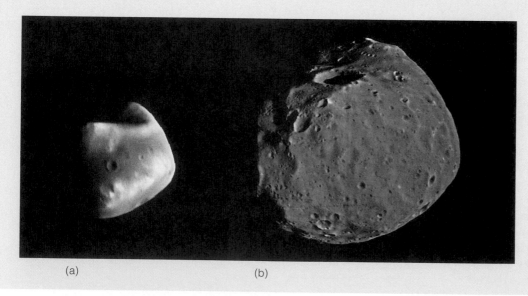

(a)　　　　(b)

Figure 1 *Both of Mars's moons are small and irregularly shaped. (a) Deimos is smoother than Phobos because it has a layer of dust covering smaller features. (b) Phobos at 28 km long is the larger of the two, whereas Deimos measures only 12 km in its greatest dimension.*

Earth, but it has 318 times the mass because it is so large (Table 19.2). It is an unusual planet in that it emits almost 2.5 times more energy than it receives from the Sun. One explanation is that most of the excess energy is left over from the time of its formation. When Jupiter formed, it heated up because of gravitational contraction (as did all the planets) and is still cooling. Jupiter's massive size insulates its interior, and hence it has cooled very slowly.

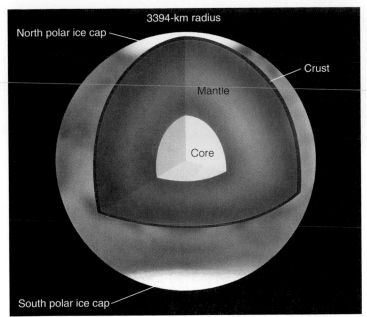

3394-km radius

North polar ice cap

Crust

Mantle

Core

South polar ice cap

(a)

(b)

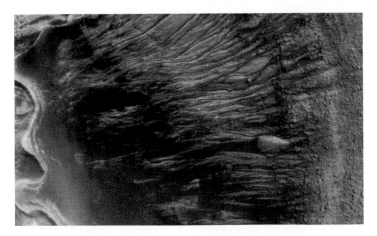

(c)

Figure 19.7 *(a) The internal structure of Mars. (b) A striking view of Mars is revealed in this mosaic of 102 Viking images. The largest canyon known in the solar system,* Valles Marineris, *can be clearly seen in the center of this image, while three of the planet's volcanoes are visible on the left side of the image. (c) The north wall of a small crater within Newton Crater on Mars shows numerous narrow gullies that were probably eroded by running water. The lobate and finger-like deposits at the base of the crater are further evidence that the material forming them was transported by running water.*

Figure 19.8 *A composite photograph, showing the Jovian planets and Earth to the same scale. The photographs of the Jovian planets were all obtained by the* Voyager *spacecraft.*

Jupiter has a relatively small central core of solid rocky material formed by differentiation. Above this core is a thick zone of liquid metallic hydrogen followed by a thicker layer of liquid hydrogen; above that is a thin layer of clouds (Figure 19.9). A strong magnetic field and an intense radiation belt surround Jupiter.

Jupiter has a dense, hot, and surprisingly dry atmosphere of hydrogen, helium methane, and ammonia, which some think are the same gases that composed Earth's first atmosphere. Jupiter's cloudy and violent atmosphere (with winds up to 650 km/hour) is divided into a series of different colored bands as well as a variety of spots (the Great Red Spot) and other features, all interacting in incredibly complex motions.

Revolving around Jupiter are 16 moons varying greatly in geologic activity (see chapter opening photo). A possi-

ble 17th moon was discovered in July 2000, but because of its small size, its existence is still to be confirmed. Also surrounding Jupiter is a thin, faint ring, a feature shared by all the Jovian planets.

Saturn

Saturn is slightly smaller than Jupiter, about one-third as massive, and about one-half as dense, but has a similar internal structure and atmosphere (Table 19.2). Like Jupiter, Saturn gives off more energy (2.2 times as much) than it gets from the Sun. Saturn's most conspicuous feature is its ring system, consisting of thousands of rippling, spiraling bands of countless particles (Figure 19.8).

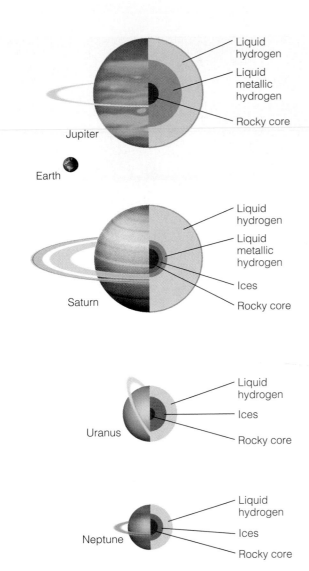

Figure 19.9 *Probable internal structure of the Jovian planets. Only Jupiter and Saturn are large enough that internal pressure is sufficient for liquid metallic hydrogen to exist. All Jovian planets have rings, but they are most obvious around Saturn.*

The composition of Saturn is similar to Jupiter's but consists of slightly more hydrogen and less helium. Saturn's core is not as dense as Jupiter's, and as in the case of Jupiter, a layer of liquid metallic hydrogen overlies the core, followed by a zone of liquid hydrogen and helium, and, lastly, a layer of clouds (Figure 19.9). Liquid metallic hydrogen can exist only at very high pressures, and because Saturn is smaller than Jupiter, such high pressures are found at greater depths in Saturn. Therefore, there is less of this conducting material than on Jupiter, and as a consequence, Saturn has a weaker magnetic field.

Even though the atmospheres of Saturn and Jupiter are similar, Saturn's atmosphere contains little ammonia because it is farther from the Sun and therefore colder. The cloud layer on Saturn is thicker than on Jupiter, but it lacks the contrast between the different bands. Unlike Jupiter, Saturn has seasons because its axis tilts 27°.

Saturn has 18 known moons, most of which are small with low densities. Titan with its nitrogen and methane atmosphere is the most distinctive and perhaps the most interesting moon in the solar system.

Uranus

Uranus is much smaller than Jupiter, but their densities are about the same (Table 19.2). It is the only planet that lies on its side (Figure 19.8); that is, its axis of rotation nearly parallels the plane of the ecliptic. Some scientists think that a collision with an Earth-sized body early in its history may have knocked Uranus on its side.

Data gathered by the flyby of *Voyager 2* indicate that Uranus has a water zone beneath its cloud cover. Because the planet's density is greater than if it were composed entirely of hydrogen and helium, it is thought that Uranus must have a dense, rocky core, and this core may be surrounded by a deep global ocean of liquid water (Figure 19.9).

The atmosphere of Uranus is similar to that of Jupiter and Saturn with hydrogen being the dominant gas, followed by helium and some methane. Uranus also has a banded atmosphere and a circulation pattern much like those of Jupiter and Saturn. Surrounding Uranus is a huge corkscrew-shaped magnetic field that stretches for millions of kilometers into space. Uranus has at least nine thin, faint rings and 18 small moons circling it (Figure 19.10).

Neptune

The flyby of *Voyager 2* in August 1989 provided the first detailed look at Neptune and showed it to be a dynamic, stormy planet (Figure 19.8). Its atmosphere is similar to those of the other Jovian planets, and it exhibits a pattern of zonal winds and giant storm systems comparable to those of Jupiter. Neptune's internal structure is similar to that of Uranus (Table 19.2) in that it has a rocky core surrounded by a semifrozen slush of water and liquid methane (Figure 19.9). Its thin atmosphere is composed of hydrogen and helium with some methane. Winds up to 2000 km/hour blow over the planet creating tremendous storms, the largest of which, the Great Dark Spot, is in Neptune's southern hemisphere. It is nearly as big as Earth and is similar to the Great Red Spot on Jupiter. One of the mysteries raised by *Voyager 2*'s discovery is where Neptune gets the energy to drive such a storm system.

Encircling Neptune are three faint rings and eight moons, the most interesting of which is Triton, Neptune's largest moon (Figure 19.11). Triton's mottled surface of delicate pinks, reds, and blues, consists primarily of water ice, with minor amounts of nitrogen and a methane frost. Geysers erupting carbon-rich material and frozen nitrogen particles some 8 km above its surface have been discovered making it only the place other than Earth and the Jovian moon Io with active volcanism.

Some areas of Triton are smooth while others have an irregular appearance indicating numerous episodes of

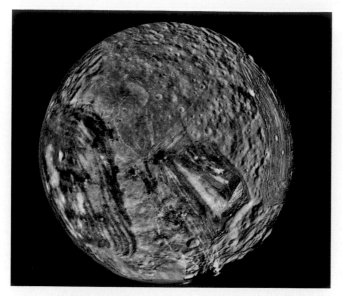

Figure 19.10 *Miranda, Uranus's inner moon, shows a spectacular surface that may reflect a catastrophic early history.*

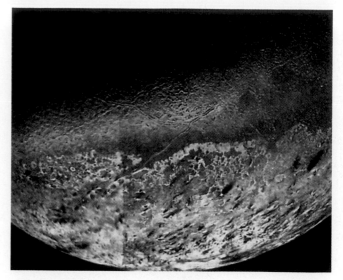

Figure 19.11 *Neptune's moon Triton is described by scientists as "a world unlike any other." In this composite of numerous high-resolution images taken by Voyager 2 during its August 1989 flyby, various features can be seen. The large south polar ice cap at the bottom consists mostly of frozen nitrogen that was deposited during the previous Tritonian winter and is slowly evaporating. The dark plumes in the lower right may be the result of volcanic activity. Smooth plains and fissures in the upper half are evidence of geologic activity in which the surface has been cracked and flooded by slushy ice that refroze.*

deformation. Heavily cratered areas bear witness to bombardment by meteorites or the collapse of its surface. Perhaps the most intriguing aspect of Triton is that it may have once been a planet—much like Pluto, which it resembles in size and possibly composition—that was captured by Neptune's gravitational field soon after the formation of the solar system.

Pluto

With a diameter of only 2300 km, Pluto is the smallest planet and, strictly speaking, it is not one of the Jovian planets (Table 19.2). Little is known about Pluto, but recent studies indicate it has a rocky core overlain by a mixture of methane gas and ice (Figure 19.12). It has a thin, two-layer atmosphere with a clear upper layer overlying a more opaque lower layer. Using the Hubble space telescope to take images of Pluto, researchers have found between seven and nine dark patches on Pluto's icy surface. These patches, each of which is hundreds of kilometers across, include a polar ice cap bisected by a dark strip.

Pluto differs from all the other planets in that it has a highly elliptical orbit that is tilted with respect to the plane of the ecliptic. It has one known moon, Charon, that is nearly half its size with a surface that appears to differ markedly from Pluto's.

Earth—Its Origin and Differentiation

As matter was accreting in the various turbulent eddies that swirled around the early Sun, enough material eventually gathered together in one eddy to form the planet Earth. Recall from Chapter 1 that Earth consists of concentric layers of different composition and densities (Figure 19.13). This differentiation into concentric layers is a fundamental characteristic of all the terrestrial planets, and presumably occurred early in their history.

Geologists know that Earth is 4.6 billion years old. The oldest known rocks, however, are 3.96-billion-year-old metamorphic rocks from Canada. Like younger crustal rocks, these rocks are composed of relatively light silicate minerals. It appears that a crust, a heavier silicate mantle, and an iron-nickel core were already present 3.96 billion years ago, or 640 million years after Earth formed.

Early Earth was probably cool, of generally uniform composition and density throughout, and composed mostly of silicate compounds, iron and magnesium oxides, and smaller amounts of all the other chemical elements (see Figure 1.8a). For the iron and nickel to concentrate in the core, Earth must have heated up enough for them to melt and sink through the surrounding lighter silicate minerals.

This initial heating could have occurred in three ways. Some heat was no doubt generated by the impact of meteorites; most of this heat was radiated back into space, but some was probably retained by the accreting planet. Heat was also generated within early Earth as gravitational compression reduced it to a smaller volume. Rock is a poor conductor of heat, so this heat accumulated within the evolving planet. The third cause of internal heating was the decay of radioactive elements such as uranium, thorium, and others. Even though these elements form only a

(a)

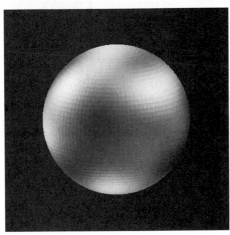

(b)

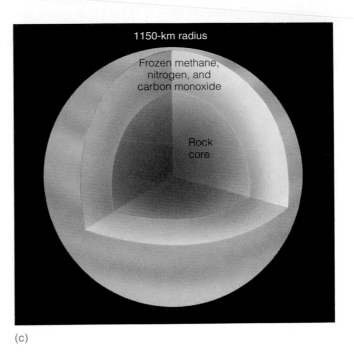

1150-km radius

Frozen methane, nitrogen, and carbon monoxide

Rock core

(c)

Figure 19.12 *(a–b) Even with the Hubble telescope, surface features on Pluto are barely discernible in these two images. Some of these areas might be darker, older terrain and lighter impact basins. The bright areas at the poles might be frosts. (c) The probable internal structure of Pluto.*

small portion of Earth, the heat generated during radioactive decay was absorbed by the surrounding rock.

The combination of meteorite impacts, gravitational compression, and heat from radioactive decay increased the temperature of early Earth enough to melt iron and nickel, which, being denser than silicate minerals, settled to Earth's center and formed the core (see Figure 1-8b). Simultaneously, the lighter silicates slowly flowed upward, beginning the differentiation of the mantle from the core.

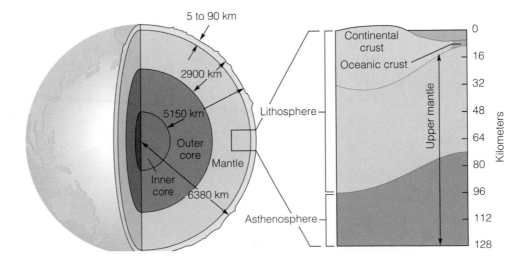

Figure 19.13 *Earth's internal structure consisting of a core, mantle, and crust. The expanded view shows details of the upper mantle and crust.*

Calculations indicate that with a uniform distribution of elements in an early solid Earth, enough heat could be generated to begin melting iron and nickel at depths between 400 and 850 km. Melting would have had to begin at shallow depths because the temperature at which melting begins increases with pressure; therefore the melting point of any material increases toward Earth's center.

The most significant event in Earth history was its differentiation into a layered planet, which led to the formation of a crust and eventually to continents. It also led to the outgassing of light volatile elements from the interior, which eventually resulted in the formation of the oceans and atmosphere.

The Earth-Moon System— How Did It Originate?

We probably know more about our Moon than any other celestial object except Earth (Figure 19.14). Nevertheless, even though the Moon has been studied for centuries through telescopes and has been sampled directly, many questions remain unanswered.

The Moon is one-fourth the diameter of Earth, has a low density (3.3 g/cm^3) relative to the terrestrial planets, and exhibits an unusual chemistry in that it is for the most part bone-dry, having been largely depleted of most volatile elements (Table 19.2). However, in 1995 a spacecraft cir-

cling the Moon in a vertical orbit discovered ice deposits in the south pole's Aitkin Basin. Although uncertain about where the water came from, scientists think that a collision with an icy meteor is the most likely source.

The Moon orbits Earth and rotates on its own axis at the same rate, so we always see the same side. Furthermore, the Earth-Moon system is unique among the terrestrial planets. Neither Mercury nor Venus has a moon, and the two small moons of Mars—Phobos and Deimos—are probably captured asteroids (see Perspective 19.1).

The surface of the Moon can be divided into two major parts: the low-lying dark colored plains, called *maria*, and the light-colored *highlands* (Figure 19.14). The highlands are the oldest parts of the Moon and are heavily cratered, providing striking evidence of the massive meteorite bombardment that occurred in the solar system more than four billion years ago.

Study of the several hundred kilograms of rocks returned by the *Apollo* missions indicates that three kinds of materials dominate the lunar surface: igneous rocks, breccias, and dust. Basalt, a common dark-colored igneous rock on Earth, is one of the several different types of igneous rocks on the Moon and makes up the greater part of the maria. The presence of igneous rocks that are essentially the same as those on Earth shows that magmas similar to those on Earth were generated on the Moon long ago.

The lunar surface is covered with a regolith estimated to be 3 to 4 m thick. This gray covering, which is composed of breccia, glass spherules, and small particles of dust, is thought to be the result of debris formed by meteorite impacts.

(a)

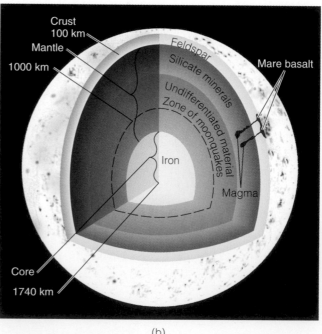

(b)

Figure 19.14 *(a) High-quality image of the Moon taken through a telescope on Earth. The light-colored lunar highlands are heavily cratered. The dark-colored areas are maria, which formed when lava flowed onto the surface. (b) The Moon's internal structure. Because seismic waves are not transmitted below 1000 km, it is likely that the innermost mantle is liquid.*

The interior structure of the Moon is quite different from that of Earth, indicating a different evolutionary history (Figure 19.14b). The highland crust is thick (65 to 100 km) and comprises about 12% of the Moon's volume. It was formed about 4.4 billion years ago, immediately following the Moon's accretion. The highlands are composed principally of the igneous rock anorthosite, which is made up of light-colored feldspar minerals that are responsible for their white appearance.

A thin covering (1 to 2 km thick) of basaltic lava fills the maria; lava covers about 17% of the lunar surface, mostly on the side facing Earth. These maria lavas came from partial melting of a thick underlying mantle of silicate composition. Moonquakes occur at a depth of about 1000 km, but below that depth seismic shear waves (S-waves) apparently are not transmitted. Because shear waves do not travel through liquid, their lack of transmission implies that the innermost mantle may be partially molten. There is increasing evidence that the Moon has a small (600 km to 1000 km diameter) metallic core.

The origin and earliest history of the Moon are still unclear, but the basic stages in its subsequent development are well understood. It formed some 4.6 billion years ago and shortly thereafter was partially or wholly melted, yielding silicate magma that cooled and crystallized to form the rock anorthosite. Because of the low density of the anorthite crystals and the lack of water in the silicate magma, the thick anorthosite highland crust formed. The remaining silicate magma cooled and crystallized to produce the zoned mantle, while the heavier metallic elements formed the small metallic core.

The formation of the lunar mantle was completed by about 4.4 to 4.3 billion years ago. The maria basalts, derived from partial melting of the upper mantle, were extruded during great lava floods between 3.8 and 3.2 billion years ago.

Numerous models have been proposed for the origin of the Moon, including capture from an independent orbit, formation with Earth as part of an integrated two-planet system, breaking off from Earth during accretion, and formation resulting from a collision between Earth and a large planetesimal (Figure 19.15). These various models are not mutually exclusive, and elements of some occur in others.

WHAT WOULD YOU DO?

Earth is a planet with considerable internal and surface activity, whereas its moon shows less activity, and bodies such as Mars's moon Phobos show hardly any. Why are these bodies so different in these aspects, considering that all are rocky bodies? Do you think these types of geologic activities on Earth will diminish and eventually cease? If so, why?

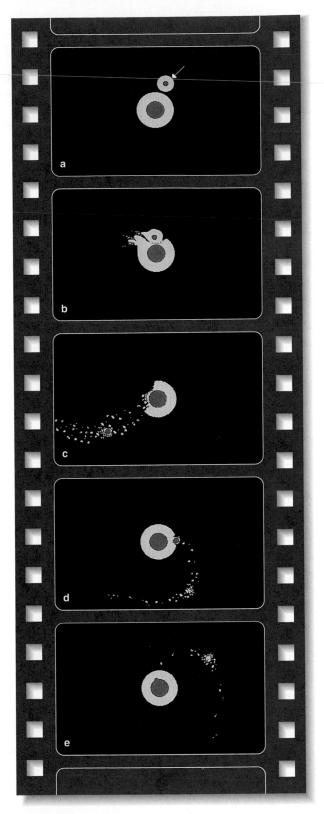

Figure 19.15 *According to one hypothesis, the Moon formed from the debris blasted into space during a collision between Earth and a Mars-sized planetesimal 4.4 to 4.6 billion years ago (a–c). The planetesimal's iron-rich core fell back to Earth (d), and the Moon began forming from the debris orbiting Earth (e).*

At this time, scientists cannot agree on a single model, as each has some inherent problems. However, the model that seems to account best for the Moon's particular composition and structure involves an impact by a large planetesimal with a young Earth.

In this model, a giant planetesimal, the size of Mars or larger, crashed into Earth about 4.6 to 4.4 billion years ago, causing the ejection of a large quantity of hot material that formed the Moon. The material that was ejected was mostly in the liquid and vapor phase and came primarily from the mantle of the colliding planetesimal. As it cooled, the various lunar layers crystallized out in the order we have discussed.

Chapter Summary

1. The universe began with a Big Bang approximately 15 billion years ago. Astronomers have deduced this age by observing that celestial objects are moving away from each other in what appears to be an ever expanding universe.

2. The universe has a background radiation of 2.7° above absolute zero, which is interpreted as the faint afterglow of the Big Bang.

3. About 4.6 billion years ago, the solar system formed from a rotating cloud of interstellar matter. As this cloud condensed, it eventually collapsed under the influence of gravity and flattened into a counterclockwise rotating disk. Within this rotating disk, the Sun, planets, and moons formed from the turbulent eddies of nebular gases and solids.

4. Meteorites provide vital information about the age and composition of the solar system. The three major groups are stones, irons, and stony-irons. Each has a different composition, reflecting a different origin.

5. Temperature as a function of distance from the Sun played a key role in the types of planets that evolved. The inner terrestrial planets are composed of rock and metallic elements that condense at high temperatures. The outer Jovian planets plus Pluto are composed mostly of hydrogen, helium, ammonia, and methane, all of which condense at lower temperatures.

6. All terrestrial planets are differentiated into a core, mantle, and crust, and all seem to have had a similar early history during which volcanism and cratering from meteorite impacts were common.

7. The Jovian planets differ from the terrestrial planets in size and chemical composition and followed completely different evolutionary histories. All Jovian planets have a small core compared to their overall size; they are mainly composed of volatile elements and compounds.

8. Earth formed from a swirling eddy of nebular material 4.6 billion years ago, and by at least 3.96 billion years ago, it had differentiated into its present-day structure. It accreted as a solid body and then underwent differentiation during a period of internal heating.

9. The Moon probably formed as a result of a Mars-sized planetesimal crashing into Earth 4.6 to 4.4 billion years ago and causing it to eject a large quantity of hot material. As the material cooled, the various lunar layers crystallized, forming a zoned body.

Important Terms

Big Bang
electromagnetic force
gravity
irons

Jovian planet
meteorite
outgassing

solar nebula theory
stones
stony-irons

strong nuclear force
terrestrial planet
weak nuclear force

Review Questions

1. What two observations lead scientists to conclude that the Big Bang occurred approximately 15 billion years ago?
   a. _____ a steady-state universe and 2.7° above absolute zero background radiation; b. _____ a steady-state universe and opaque background radiation; c. _____ an expanding universe and opaque background radiation; d. _____ an expanding universe and 2.7° above absolute zero background radiation; e. _____ a shrinking universe and opaque background radiation.

2. The composition of the universe has been changing since the Big Bang, yet 98% of it by weight still consists of the elements
   a. _____ hydrogen and carbon; b. _____ helium and carbon; c. _____ hydrogen and helium; d. _____ carbon and nitrogen; e. _____ hydrogen and nitrogen.

3. Which of the following is not one of the basic forces in the universe?
   a. _____ electromagnetic force; b. _____ gravity; c. _____ strong nuclear force; d. _____ quasar force; e. _____ none of these.

4. The current theory for the origin of the solar system is known as the
   a. _____ nebular theory; b. _____ encounter theory; c. _____ evolutionary theory; d. _____ planetesimal theory; e. _____ solar nebula theory.

5. The most abundant meteorites are
   a. _____ stones; b. _____ acondrites; c. _____ stony-irons; d. _____ irons; e. _____ carbonaceous chondrites.

6. The most widely accepted theory regarding the origin of the Moon involves

a. _____ capture from an independent orbit; b. _____ an independent origin from Earth; c. _____ breaking off from Earth during Earth's accretion; d. _____ formation resulting from a collision between Earth and a large planetesimal; e. _____ none of these.

7. The planets are separated into terrestrial and Jovian primarily on the basis of which property?
a. _____ size; b. _____ atmosphere; c. _____ density; d. _____ color; e. _____ none of these.

8. The atmosphere of Venus is
a. _____ thick and composed of carbon dioxide; b. _____ similar to Earth's; c. _____ nonexistent; d. _____ thin, like that of Mars; e. _____ none of these.

9. The only planet whose axis of rotation nearly parallels the plane of the ecliptic is
a. _____ Venus; b. _____ Saturn; c. _____ Uranus; d. _____ Neptune; e. _____ Pluto.

10. The largest planet in the solar system is
a. _____ Neptune; b. _____ Uranus; c. _____ Saturn; d. _____ Jupiter; e. _____ Mars.

11. Which planets give off more energy than they receive?
a. _____ Jupiter and Saturn; b. _____ Saturn and Neptune; c. _____ Neptune and Pluto; d. _____ Jupiter and Neptune; e. _____ Saturn and Pluto.

12. Which of the following events did all of the terrestrial planets experience early in their history?
a. _____ accretion; b. _____ differentiation; c. _____ volcanism; d. _____ meteorite impacting; e. _____ all of these.

13. Describe the two phenomena that indicate the Big Bang occurred.

14. Discuss the origin of the Earth-Moon system.

15. Describe the main groups of meteorites. Which is most common, and what do they tell about the origin of the solar system?

16. It is likely that the atmospheres of Jupiter and Saturn are very similar to that of early Earth. Why have these two planets' atmospheres remained essentially the same during the past 4.6 billion years, while those of the terrestrial planets have evolved?

17. What are the similarities and differences in the histories of the four terrestrial planets?

18. How do the Jovian planets differ from the terrestrial planets?

19. Why do you think that the largest volcano in the solar system is present on Mars rather than Earth?

20. What three sources of heat were important in the differentiation of the terrestrial planets?

21. How was the age of the universe determined?

22. What surface features on Mars indicate that it once had running water and that its surface is currently being modified by wind?

23. How does the solar nebula theory account for the general characteristics of the solar system? Are there any aspects of the solar system that it does not explain?

24. How and why do the atmospheres of Earth and Venus differ?

25. The Sun is thought to have been much dimmer during the early history of the solar system. Because of this, Venus probably had oceans early in its history, as did Mars. If the Sun was as bright during its early history as it is now, how might this have affected the evolution of the solar system, particularly the evolution of the terrestrial planets?

 # World Wide Web Activities

For these web site addresses, along with current updates and exercises, log on to **http://www.brookscole.com/geo**

Welcome to the Planets
This site, maintained by the California Institute of Technology with U.S. Government sponsorship under NASA Contract NAS7-1270, contains a collection of many of the best images of the planets from NASA's planetary exploration program. Each planet is listed as well as the various spacecrafts. In addition, it provides links to other sites around the world and a What's New section. Click on any of the planets for a planet profile, as well as spectacular images of the planet and specific regions of it and information about those regions.

NASA Home Page
This site, maintained by NASA, contains a tremendous amount of information about the organization of NASA, what it does, and what some of the projects are, as well as a Questions and Answers section. It also contains extensive photo archives of images of Earth from space.

1. Click on the *Today@NASA* site. This will take you to the latest breaking news about space as well as sites for information and images from the Hubble Space Telescope, the current Shuttle mission, and much more.

2. Click on the *Mission to Planet Earth* site. This page is dedicated to explaining the various ways Earth is changing and how humans are influencing those changes.

Views of the Solar System
This site is maintained by Calvin J. Hamilton and was created as an educational tour of the solar system. It contains information (more than 220 pages) and images (over 950 images and animations) about the planets, asteroids, comets, meteoroids and meteorites, and history of space exploration, as well as links to other astronomy web sites.

NASA Photo Archive
This site is the entry point to NASA's vast photo archives. Hubble Space telescope images, plus photographs and atlases are just a few of the things available.

Extrasolar Planets Catalog
A French site with references (mostly highly technical), but it also has a table of current data on all known or suspected extrasolar planets.

Precambrian Earth and Life History

Proterozoic sedimentary rocks exposed in the Rocky Mountains of Montana. Similar rocks are present in Wyoming, Idaho, and Alberta, Canada. As opposed to older Precambrian rocks, these have been little deformed or altered by metamorphism.

PROLOGUE

Imagine a barren, lifeless, waterless, hot planet with a poisonous atmosphere. Cosmic radiation is intense, meteorites and comets crash to the ground, and volcanoes erupt nearly continuously. The planet's crust of dark-colored igneous rock is thin and unstable. Storms form in the turbulent atmosphere, lightning flashes much of the time, but no rain falls. And because the atmosphere has no free oxygen, nothing burns, yet a continuous red glow comes from pools and streams of molten rock.

This might seem like a description from a science fiction novel, but it is probably a reasonably accurate account of what Earth was like shortly after it formed (Figure 20.1). "Probably" is the key word here, though, because no geologic record exists for the earliest chapter in Earth history, the interval from 4.6 to 3.96 billion years ago. So we can only speculate on the early Earth environment based on theories of how terrestrial planets form and our knowledge of other terrestrial planets.

When Earth formed (see Chapter 19), it had a tremendous reservoir of primordial heat, heat generated by colliding particles as Earth accreted, by gravitational compression, and by the decay of short-lived radioactive isotopes. In fact, many geologists are convinced that early Earth was so hot that it was partially or perhaps almost entirely molten. No one knows how hot its surface was, but we can be sure that it was too hot for liquid water to exist or for any organism to survive.

Assuming that we could somehow go back and visit this inhospitable early Earth, we would have to be protected from the high temperatures, and the atmosphere composed of water vapor, carbon dioxide, and probably methane and ammonia would be unbreathable. No ozone layer was present in the upper atmosphere so we would receive lethal doses of ultraviolet radiation, and of course the constant bombardment by meteorites and comets would be an additional danger. The view of the Moon would have been spectacular because it was closer to Earth, but its gravitational attraction would have caused massive Earth tides—no surface water was present yet, so we would not have to worry about ocean tides. And finally, we would experience a much shorter day because Earth rotated on its axis in as little as 10 hours.

Eventually, much of Earth's primordial heat dissipated into space and its surface cooled. As it cooled, water vapor condensed, rain fell, and surface waters accumulated. The incidence of meteorite and comet bombardment diminished considerably. By at least 3.96 billion years ago, some small areas of continental crust existed, and organisms made their appearance as much as 3.5 billion years ago. For many millions of years the only organisms were varieties of bacteria, but they were adding oxygen to the atmosphere and an ozone layer began to form.

Still, the land was barren, although undoubtedly some bacteria made a living on land. So even though the environment was becoming more like it is now, our hypothetical visitors would still have had difficulty breathing, enduring massive doses of ultraviolet radiation, and finding enough to eat. And finally, none of the landmasses would resemble today's continents, so if we could produce a globe of this early Earth, the geography would be quite unfamiliar.

OBJECTIVES

At the end of this chapter, you will have learned that

- Precambrian encompasses more than 88% of all geologic time, yet we know less about its history than we do about more recent intervals of geologic time.

- During the earliest part of the Precambrian, called the Hadian, Earth differentiated and crust began forming.

- The Archean Eon was a time during which several small continental nuclei began evolving, and the most common rocks are granite-gneiss complexes and greenstone belts.

- During the Proterozoic Eon an essentially modern style of plate tectonics developed, and crust that formed during the Archean amalgamated into a large craton we now call Laurentia.

- Proterozoic rocks consist of sandstone-carbonate-shale assemblages, red beds, banded iron formations, glacial deposits, and a variety of others.

- The Precambrian atmosphere evolved from one lacking free oxygen and an ozone layer but at the end of the Precambrian the atmosphere had much less free oxygen than it does now.

- Once Earth had cooled sufficiently, water vapor fell as precipitation, accounting for the origin of surface waters.

- During the Archean and Early Proterozoic only single-celled bacteria existed.

- During the Proterozoic, organisms that reproduced sexually evolved, and multicelled organisms appeared.

- Several natural resources are found in Precambrian rocks.

Figure 20.1 *Earth's surface as it is thought to have appeared shortly after it formed 4.6 billion years ago.*

Introduction

The term *Precambrian* is an informal one referring to both rocks and time. All rocks lying beneath strata of the Cambrian System are, by definition, Precambrian, and all geologic time from Earth's origin until the beginning of the Phanerozoic Eon is also called Precambrian. The Precambrian encompasses just over 88% of all geologic time, so if Earth's history were represented by a 24-hour day, slightly more than 21 hours of it would be Precambrian (Figure 20.2). Unfortunately for geologists, no rocks are known for the first 640 million years of Earth history. Rocks are present for the rest of Precambrian time, but because many have been complexly deformed and altered by metamorphism, deciphering their history is difficult, and few contain fossils of any use for making relative age determinations.

Geologists now subdivide Precambrian time into two eons; the *Archean Eon* (4.0 to 2.5 billion years ago), and the *Proterozoic Eon* (2.5 billion to 545 million years ago) (Table 20.1). The interval from Earth's origin 4.6 billion years ago to the onset of the Archean is informally designated as the *Hadean*. The Archean and Proterozoic are based solely on absolute ages, so they are geochronologic units rather than time-stratigraphic units. This departs from standard practice in which systems based on stratotypes are the basic time-stratigraphic units (see Chapter 17). Remember that geochronologic terms such as Cambrian Period are based on a body of rock known as time-stratigraphic units—the Cambrian System in this example. There are, however, no time-stratigraphic units for the Precambrian eons; thus, they are strictly geochronologic.

The terms defined above are used consistently in this book, but students should be aware that the U.S. Geological Survey (USGS) uses different designations for the Precambrian. Although their terms are also geochronologic, they are simply letter designations such as W, X, Y, and Z (Table 20.1). Some books and articles as well as USGS maps published since 1971 use these designations.

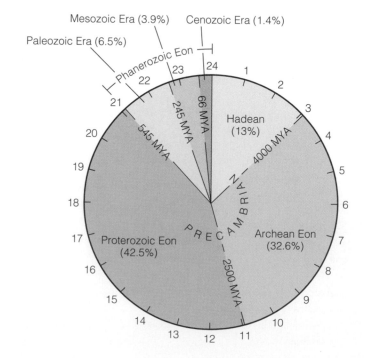

Figure 20.2 *Geologic time represented on a 24-hour clock. Precambrian time includes more than 21 hours on this clock, or more than 88% of all geologic time.*

Table 20.1

Two Classification Schemes for Precambrian Rocks and Time in the United States

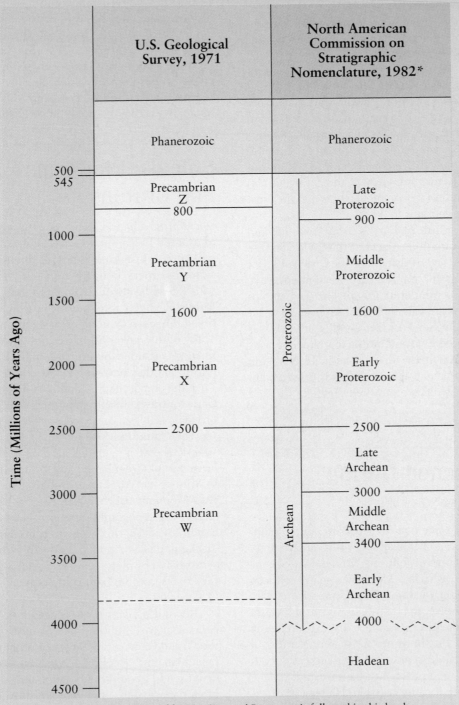

	U.S. Geological Survey, 1971	North American Commission on Stratigraphic Nomenclature, 1982*
	Phanerozoic	Phanerozoic
500 / 545	Precambrian Z	Late Proterozoic
	— 800 —	— 900 —
1000		
	Precambrian Y	Middle Proterozoic
1500	— 1600 —	— 1600 —
2000	Precambrian X	Early Proterozoic
2500	— 2500 —	— 2500 —
		Late Archean
3000		— 3000 —
	Precambrian W	Middle Archean
		— 3400 —
3500		Early Archean
4000		— 4000 —
		Hadean
4500		

Time (Millions of Years Ago) — vertical axis label

Proterozoic (Archean) — right scheme groupings

*This scheme, which was proposed by Harrison and Peterman, is followed in this book.

Why Should You Study Precambrian Earth and Life History?

Why should you study Precambrian Earth and life history? In Chapter 1, we introduced the concept of *systems* as a combination of related parts that operate in an organized fashion (see Figure 1.2). We gave examples of systems interactions in our discussions of several topics such as volcanism, plate tectonics, running water, and glaciation. Here our main interest is early Earth and life history, but once Earth formed, its various systems became operative although not all at the same time. For instance, Earth did not differentiate into a core and mantle until millions of years after it initially formed. Once it did differentiate, though, internal heat drove plate movements and the crust began evolving and continues to do so.

During its earliest history, Earth's surface was hot and dry, volcanism was ubiquitous, the atmosphere lacked free oxygen, and no ozone layer existed (see the Prologue). But gases derived from Earth's interior formed an atmosphere and surface waters accumulated. Organisms appeared as much as 3.5 billion years ago and had a profound impact on the evolving atmosphere. In fact, by this time Earth's systems were fully operative and interacting much as they do now. In short, Earth became an actively changing planet. Evidence of many of these changes is preserved in the geologic record, which we now investigate in this and the following chapters.

What Happened During the Hadean?

Geologists know that some continental crust existed more than 3.8 billion years ago because rocks this old are present in several areas including Minnesota, Greenland, and South Africa. Moreover, many of these rocks are metamorphic, meaning they formed from even older rocks. And of course, the 3.96-billion-year-old Acasta Gneiss in Canada indicates that some crust existed nearly 4.0 billion years ago! Furthermore, sedimentary rocks in Australia contain detrital zircons ($ZrSiO_4$) dated at 4.2 billion years, so older source rocks must have existed.

Geologists know that some very late Hadean crust existed, but they can only speculate on what existed previously. For example, little is known of this earliest crust's composition and structure, or whether it was worldwide or more restricted. Many think the earliest crust originated during an episode of partial melting and the origin of magma that rose to the surface to form a thin, unstable crust of ultramafic rock (<45% silica)—that is, crust with a comparatively low silica content. Rising basaltic magmas (45–52% silica) likely disrupted this early ultramafic crust that because of its density was consumed at subduction zones.

Following this earliest episode of crust formation, Earth's production of radiogenic heat diminished, and a second stage of crustal evolution began. Partial melting of earlier-formed basaltic crust yielded intermediate-composition magmas (52–65% silica) and perhaps even felsic magmas (>65% silica) that formed a number of volcanic island arcs. Plate movements accompanied by collisions between volcanic island arcs and emplacement of plutons resulted in the origin of several continental nuclei by the beginning of the Archean Eon.

Shields, Platforms, and Cratons

Recall from Chapter 6 that all continents have a vast area of exposed Precambrian rocks known as a **shield.** Extending outward from shields are **platforms,** consisting of Precambrian rocks beneath more recent rocks, especially widespread layers of sedimentary rock. A shield and platform collectively form a **craton,** which we can think of as the stable core or nucleus of a continent (Figure 20.3). Many of the rocks within a craton have been strongly deformed, intruded by plutons, and altered by metamorphism, but they have experienced little or no deformation since the end of the Precambrian. Their stability since that time contrasts sharply with their Precambrian history of orogenic activity.

The **Canadian Shield,** the exposed part of the North American craton, includes most of northeastern Canada; a large part of Greenland; parts of the Lake Superior region in Minnesota, Wisconsin, and Michigan; and the Adirondack Mountains of New York (Figure 20.3). It is a vast area of subdued topography, numerous lakes, and exposed Precambrian rocks, thinly covered in places by Pleistocene glacial deposits. Archean and Proterozoic rocks are well represented by plutonic rocks, lava flows, various sedimentary rocks, and metamorphic equivalents of all of these (Figure 20.4).

Beyond the Canadian Shield, exposures of Precambrian rocks are limited to areas of deep erosion, such as the Grand Canyon, and areas of orogeny such as the Appalachian and Rocky Mountains. Nevertheless, deep drilling and geophysical evidence indicate the Precambrian rocks underlie much of North America. These rocks form what is commonly called the *crystalline basement* of the continent, crystalline being the term that refers loosely to igneous and metamorphic rocks collectively.

The geologic history of the Canadian Shield is complicated and, given the complexity of many of the rocks, poorly understood. Nevertheless, geologists have delineated several smaller units within the shield, each of which

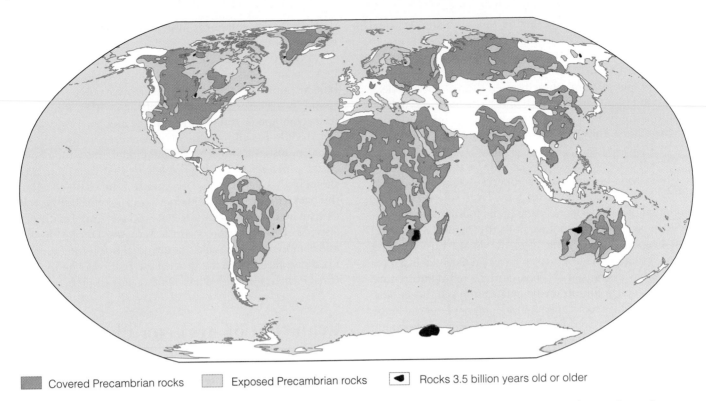

Figure 20.3 *Precambrian cratons of the world. The areas of exposed Precambrian rocks are the shields, while buried Precambrian rocks are the platforms. Shields and platforms collectively make up the cratons.*

is recognized by radiometric ages and trends of geologic structures. These smaller units, as well as others making up the platform, are the subunits that constitute the North American craton. Each smaller unit was likely an independent minicontinent that later assembled into a larger craton. The amalgamation of these units took place along deformation belts during the Early Proterozoic, a topic covered more fully later in this chapter.

Archean Earth History

By far the most common rocks of Archean age are complexes consisting of granite and gneiss, the latter being high-grade metamorphic rocks. Although subordinate in abundance, rock successions known as **greenstone belts** are reasonably common and tell some-

(a)

(b)

Figure 20.4 *(a) Archean gneiss at Georgian Bay, Ontario, Canada, is part of the Canadian Shield. (b) Beyond the Canadian Shield, Precambrian rocks are exposed in areas of uplift and deep erosion. This outcrop of the Archean-aged Brama Schist is in the deeper parts of the Grand Canyon in Arizona.*

thing of tectonic activity during the Archean. An ideal greenstone belt has three major rock units; the lower and middle units are mostly volcanic rocks, and the upper unit is sedimentary. Low-grade metamorphism and the origin of the mineral chlorite give the volcanic rocks a greenish color. Most greenstone belts have a synclinal structure and have been intruded by granitic plutons (Figure 20.5), they are commonly complexly folded and cut by thrust faults.

Much of the volcanism responsible for the volcanic units in greenstone belts must have been subaqueous because pillow lavas are common, but some large volcanic centers were built above sea level. Some of the most interesting volcanic rocks are ultramafic lava flows. In Chapter 3 we noted that eruptions producing these flows are rare in rocks younger than Archean, because Earth's radiogenic heat production has decreased and near-surface temperatures are not high enough for ultramafic magma to reach the surface.

Sedimentary rocks are a minor component in the lower parts of greenstone belts, but they become increasingly abundant toward the top (Figure 20.5). Associations of graywacke, a sandstone with abundant clay and volcanic rock fragments, and argillite, slightly metamorphosed mudrock, are particularly common. Small-scale graded bedding and cross bedding indicate that deposition by turbidity currents accounts for the graywacke-argillite successions. Some sedimentary rocks indicate deposition in delta, tidal flat, barrier island, and shallow marine shelf environments.

A variety of other sedimentary rocks are also present in Archean greenstone belts, including conglomerate, carbonates, and chert. Banded iron formations (BIFs) are also found in Archean greenstone belts, but BIFs are much more common in areas of Proterozoic rocks.

Models for the origin of greenstone belts rely on Archean plate movements. In one model greenstone belts develop in *back-arc marginal basins* that first open and then close. An early stage of extension takes place when the back-arc marginal basin opens, during which time volcanism and sedimentation take place, and finally an episode of compression occurs as it closes. During closure and compression the rocks are intruded by plutons and are metamorphosed, and the greenstone belt takes on its synclinal form as it is folded and faulted (Figure 20.5).

Evolution of Archean Cratons

As previously noted, several areas of continental crust existed more than 3.8 billion years ago, each representing a craton (Table 20.2). They must have been rather small, though, because rocks older than 3.0 billion years are of limited geographic extent, especially compared to those 2.5 to 3.0 billion years old, a time of rather rapid crustal

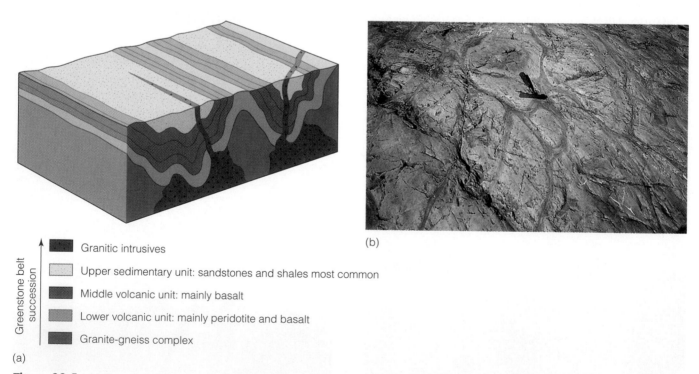

Greenstone belt succession

■ Granitic intrusives

▢ Upper sedimentary unit: sandstones and shales most common

■ Middle volcanic unit: mainly basalt

▨ Lower volcanic unit: mainly peridotite and basalt

■ Granite-gneiss complex

(a)

(b)

Figure 20.5 *(a) Two adjacent greenstone belts showing their synclinal structure. Older greenstone belts—those more than 2.8 billion years old—have an ultramafic lower unit succeeded upward by a basaltic unit as shown here. In younger greenstone belts, the succession is a basaltic lower unit overlain by an andesite-rhyolite unit. The upper unit in both older and younger greenstone belts consists of sedimentary rocks. (b) Pillow lavas in the Ispheming greenstone belt, Marquette, Michigan.*

Table 20.2

Chronologic Summary of Events Important in the Archean Development of Cratons (Ages in Thousands of Millions of Years)

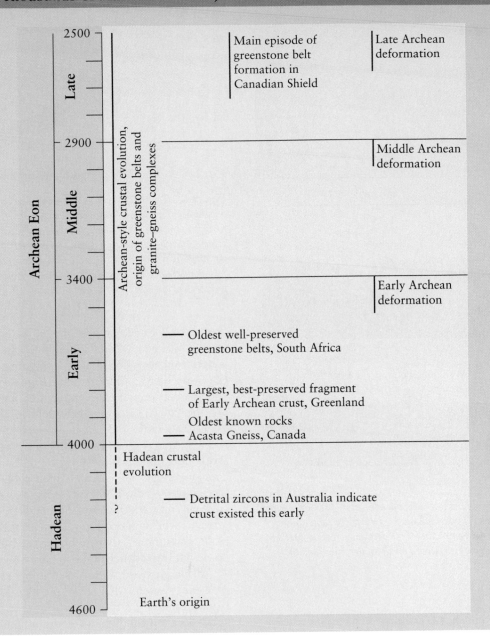

evolution. Developing Archean cratons probably evolved by a series of island-arc collisions accompanied by the origin or greenstone belts, plutonism, and deformation. The model in Figure 20.6 was proposed to account for the southern Superior craton of Canada, but it, or something very much like it, serves as a good model to account for the evolution of Archean cratons in general.

Events leading to the origin of the southern Superior craton (Figure 20.6) were part of a more extensive orogenic episode near the end of the Archean Eon. This period of deformation was responsible for the formation of the Superior and Slave cratons, as well as other parts of the Canadian Shield, and deformation of Archean rocks in Wyoming, Montana, and the Minnesota River Valley. By

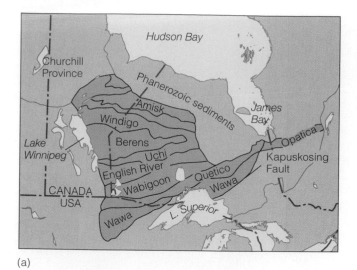

(a)

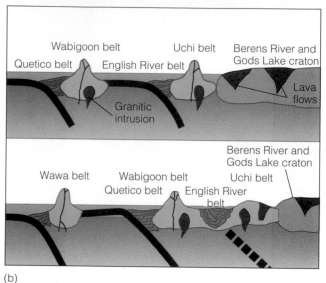

(b)

Figure 20.6 *Origin of the southern Superior craton. (a) Geologic map showing greenstone belts (green areas) and granite-gneiss subprovinces (tan areas). (b) Plate tectonic model for the development of the southern Superior craton. The figure represents a north-south section, and the upper diagram is an earlier stage of the lower diagram.*

the end of the Archean, 2.5 billion years ago, several cratons had formed that now constitute the older parts of the Canadian Shield. However, they were not amalgamated into a much larger cratonic unit until the Early Proterozoic.

Archean Plate Tectonics

Undoubtedly, the present regime of opening and closing oceans has been a primary agent in Earth evolution since at least the Early Proterozoic, and now geologists are convinced some kinds of plate movements were taking place during the Archean. We have already alluded to island-arc collisions and greenstone belt evolution in a plate tectonic

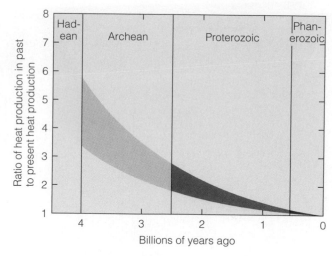

Figure 20.7 *Ratio of radiogenic heat production in the past to present heat production. The colored band encloses the ratios according to different models, all of which show an exponential decay of radioactive elements through time. Estimates of heat production for 4 billion years ago range from three to six times the present heat production, whereas heat production at the beginning of the Phanerozoic Eon was only slightly greater than it is now.*

framework. Furthermore, the rapid episode of crustal growth between 3.0 and 2.5 billion years ago probably resulted from accretion along convergent plate margins. On the other hand, ophiolites (see Figure 8.7), which mark younger convergent plate boundaries, are rare, but several are now known from the Late Archean.

Earth's radiogenic heat production has diminished through time (Figure 20.7). So during the Archean, when more heat was available, new crust was generated at spreading ridges more quickly, and seafloor spreading took place more rapidly. If this is correct, we would expect more volcanism at both divergent and convergent plate boundaries, as well as more rapid growth of continental crust along convergent plate boundaries. Both appear to be borne out by evidence in the geologic record.

Although plate tectonics was operating during the Archean, it seems that these plates behaved differently from those of the Proterozoic. For example, sedimentary sequences typical of passive continental margins are uncommon in the Archean but well represented in Proterozoic rocks. That is, broad continental shelves with adjacent slopes and rises were not common. Their scarcity during the Archean indicates that passive continental margins were either rare or poorly developed.

Proterozoic Earth History

The basic difference between Archean and Proterozoic Earth history is the style of crustal evolution. During the Archean, crust-forming processes generated greenstone belts and granite-gneiss complexes, which continued into the Proterozoic but at a considerably reduced rate. Most Archean rocks were altered by metamorphism, but many Proterozoic rocks have been little altered. And finally, widespread assemblages of sedimentary rocks deposited on passive continental margins are common in the Proterozoic but rare in the Archean.

In the preceding sections we noted that Archean cratons assembled through a series of island-arc and minicontinent collisions (Figure 20.6). These provided the nuclei around which Proterozoic continental crust accreted, thereby forming much larger cratons. One large landmass so formed, called **Laurentia,** consisted mostly of North America and Greenland, parts of northwestern Scotland, and perhaps parts of the Baltic shield of Scandinavia. We emphasize the geologic evolution of Laurentia in the following sections (Table 20.3).

Evolution of Proterozoic Crust

The first major episode of Proterozoic crustal evolution took place during the Early Proterozoic, between 2.0 and 1.8 billion years ago (BYA). Several major **orogens** developed—zones of deformed rocks, many of which were metamorphosed and intruded by plutons. So this was a time of continental accretion during which collisions between Archean-aged cratons formed a larger craton. Much of the North American craton had formed by 1.8 BYA (Figure 20.8a). For instance, the Slave and Rae cratons collided along the Thelon orogen, and rocks in the Trans-Hudson orogen provide a record of orogeny and suturing of the Superior, Hearne, and Wyoming cratons (Figure 20.8a).

Another notable Early Proterozoic event was the origin of the Wopmay orogen adjacent to the Slave craton in northwestern Canada. Rocks here record the oldest complete Wilson cycle—that is, opening and closing of an ocean basin. Furthermore, some rocks constitute a **sandstone-carbonate-shale assemblage,** a suite of rocks characteristic of passive continental margins that first became common and widespread during the Proterozoic.

Following the Early Proterozoic amalgamation of cratons, considerable accretion took place along Laurentia's southern margin. Between 1.8 and 1.6 BYA, accretion continued in what is now the southwestern and central United States as successively younger belts were sutured to the craton, forming the Yavapai and Mazatzal-Pecos orogens (Figure 20.8b). The net effect was the accretion of a belt of continental crust more than 1000 kilometers wide along the southern margin of Laurentia.

No major continental accretion took place between 1.6 and 1.3 BYA, but extensive igneous activity occurred that was unrelated to orogenic activity (Table 20.3; Figure 20.9). Laurentia did not increase in area, though, because magma was simply emplaced in or erupted onto previously existing continental crust. The rocks, mostly granitic plutons, and vast rhyolite and ash flows are deeply buried in most areas, but they are exposed in eastern Canada, Greenland, and the Baltic shield of Scandinavia. The origin of these Middle Proterozoic rocks is debated, but according to one hypothesis they resulted from large-scale upwelling of magma beneath a supercontinent.

Another important event in the evolution of Laurentia, the *Grenville orogeny* in the eastern United States and Canada, took place between 1.3 and 1.0 BYA (Figure 20.8c). Grenville rocks are found in Scandinavia, Greenland, and the Appalachian region of eastern North America (Figure 20.10). Some geologists think these Grenville rocks record an opening and then closing ocean basin, so perhaps they were deposited on a passive continental margin. But others are of the opinion that Grenville deformation took place in an intracratonic basin. Whatever the cause, it represents the final episode of Proterozoic continental accretion of Laurentia.

Contemporaneous with Grenville deformation was an episode of rifting in Laurentia resulting in the origin of the **Midcontinent rift** (Figure 20.8c). This huge feature has two branches extending south from the Lake Superior region. It cuts both Archean and Early Proterozoic rocks, but is buried by younger rocks except near Lake Superior where its rocks are well exposed.

Proterozoic Supercontinents

Thus far we have reviewed the Proterozoic evolution of Laurentia, a large landmass composed mostly of Greenland and North America (Figure 20.8). Studies in paleomagnetism and the continuation of the Grenville orogen into Scandinavia indicate that the Baltic shield may also have been part of or close to Laurentia. In fact, some geologists think that a collision between the Baltic and Laurentian cratons was responsible for Early Proterozoic deformation.

We certainly know that a supercontinent called Pangaea existed at the end of the Paleozoic Era. But what is not certain is the size, shape, and Precambrian associations of the elements that eventually combined to form Pangaea. According to one view, a large Pangaea-like supercontinent persisted from the Late Archean through the Proterozoic, but many geologists think uncertainties in the data make this view unwarranted. During the latest Proterozoic, however, large-scale deformation of all Southern Hemisphere continents may indicate that Gondwana was assembled then. In fact, evidence is accumulating that Laurentia, Australia, and Antarctica collided during the Late Proterozoic and formed a supercontinent known as *Rodinia*. It formed between 1.3 and 1.0 BYA, and then fragmented when rifting began 750 million years ago.

Table 20.3

Summary Chart Showing the Ages of Some of the Proterozoic Events Discussed in the Text

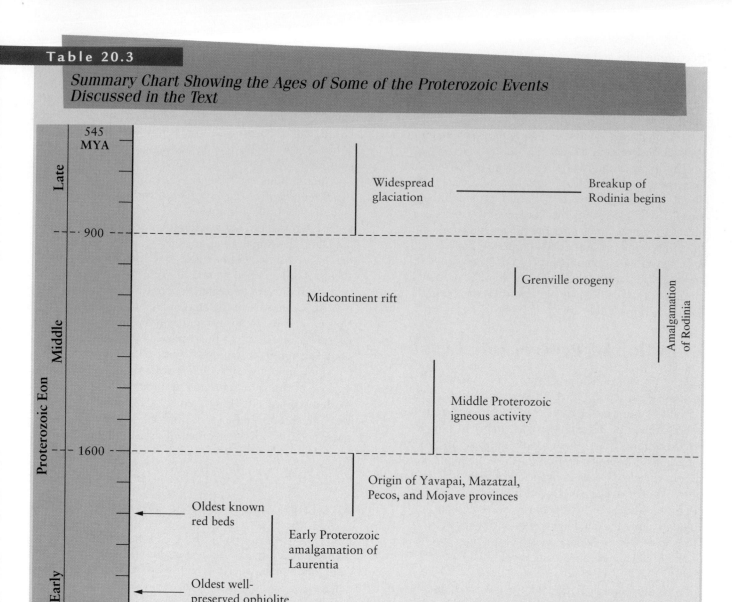

Proterozoic Rocks

Greenstone belts are most typical of the Archean, but continued to form during the Proterozoic although they differed in detail. A present-day regime of plate tectonics is indicated by the first well-preserved ophiolite, the 1.97-billion-year-old Jormua complex of Finland (Figure 20.11). Other rocks of interest include sedimentary rocks typical of passive continental margins, deposits indicating two major episodes of glaciation, banded iron formations, and red beds.

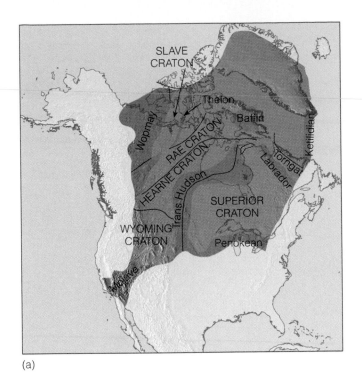

(a)

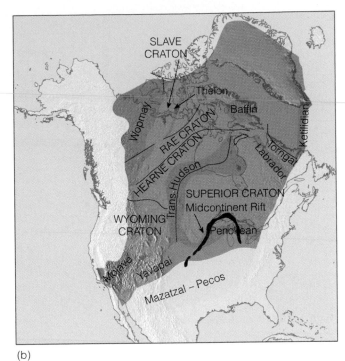

(b)

■	900 million – 1.2 billion
□	1.6 billion – 1.75 billion
■	1.75 billion – 1.8 billion
■	1.8 billion – 2.0 billion
■	2.5 billion – 3.0 billion

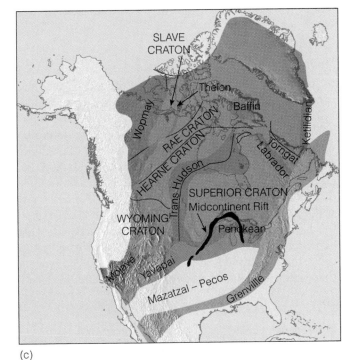

(c)

Figure 20.8 *Proterozoic evolution of the Laurentian craton.*
(a) During the Early Proterozoic, Archean cratons were sutured along
deformation belts called orogens. (b) Laurentia grew along its southern
margin by accretion of the Central Plains, Yavapai, and Mazatzal orogens.
(c) A final episode of Protoerozoic accretion occurred during the Grenville
orogeny.

Fully 60% of all known Proterozoic rocks consist of sandstone-carbonate-shale assemblages deposited on rifted continental margins and in intracratonic basins. Their widespread occurrence indicates that large stable cratons were present with adjacent depositional environments much like those of the present. Early Proterozoic sandstone-carbonate-shale assemblages are common in the Great Lakes region (Figure 20.12), and in the west-

ern United States and Canada they are found in three Middle to Late Proterozoic basins (Figure 20.13). Sedimentary structures such as wave-formed ripple marks and cross-bedding in the sandstones and stromatolite-bearing carbonate rocks indicate that most were deposited along continental margins, although some deposits probably accumulated in continental environments.

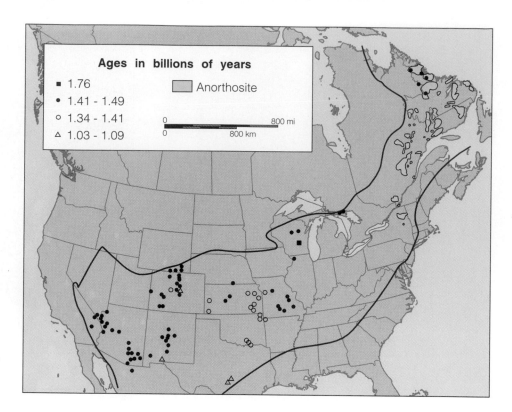

Ages in billions of years

- 1.76
- 1.41 – 1.49
- 1.34 – 1.41
- 1.03 – 1.09

☐ Anorthosite

0   800 mi

0   800 km

Figure 20.9 *Middle Proterozoic igneous rock complexes unrelated to orogenic activity are found within the heavy lines. Most of this activity took place between 1.34 and 1.49 billion years ago.*

North America probably had an extensive Early Proterozoic ice sheet centered southwest of Hudson Bay (Figure 20.14a). Deposits of about the same age and origin are also present in the United States, Australia, and South Africa, but the dates are not precise enough to determine whether there was a single widespread period of glaciation or a number of glacial events at different times in different areas. Another episode of widespread glaciation is indicated by Late Proterozoic glacial deposits on all continents except Antarctica (Table 20.3). Figure 20.14b shows the approximate distribution of these Late Proterozoic glaciers. But we emphasize that they are approximate, because the geographic extent of glacial ice is unknown, and these glaciers were not all present at the same time. Nev-

ertheless, these Late Proterozoic glaciers were so extensive that they were present even in near-equatorial areas.

In Chapter 5, we briefly discussed **banded iron formations (BIFs),** sedimentary rocks consisting of alternating thin layers of silica (chert) and iron oxide minerals such as hematite (Fe_2O_3) and magnetite (Fe_3O_4) (Figure 20.15a), although other iron minerals are present in some. Archean BIFs are small, lens-shaped bodies measuring a few meters across and a few meters thick. In contrast, Proterozoic BIFs are much thicker, cover vast areas, and appear to have been deposited in shallow marine environments.

Iron is a highly reactive element that in the presence of oxygen forms rustlike oxides that do not readily dissolve in water. If free oxygen is absent, though, iron goes into solution and can accumulate in the oceans. Given that the Archean atmosphere probably had little free oxygen, little iron was dissolved in seawater. But about 2.3 billion years ago there was an increase in the abundance of photosynthesizing bacteria, a corresponding increase in the amount of free oxygen in the atmosphere and oceans, and the precipitation of dissolved iron and silica to form BIFs.

One type of cement binding detrital particles together is iron oxide (see Chapter 5). Many sedimentary rocks in

Figure 20.10 *Outcrop near Alexandria, New York, of rocks metamorphosed during the Grenville orogeny. At this location an unconformity separates the Grenville rocks from overlying sedimentary rocks of Late Cambrian age.*

Lithology	Estimated thickness m	Interpretation		
Turbiditic graywacke and mudrocks	≫500	Deep-sea sediments		
Tuff, schist, chert, carbonate rock	< 200			
Pillow breccia pillow basalt massive lava	> 300	Submarine eruptions	Oceanic crust	
Locally no host rock, 100% mafic dikes, abundant interdikes gabbro and/or serpentinite	< 1000	Sheet dikes grading down into gabbro		
Gabbro	> 100			
Serpentinite	> 1000	Meta-morphosed peridotite	Upper mantle	

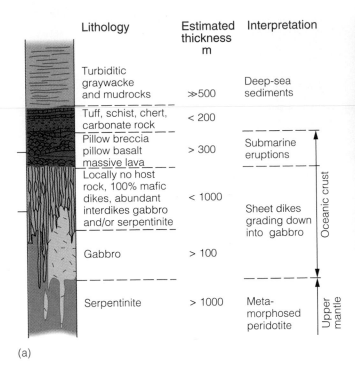

(a)

(b)

(c)

Figure 20.11 *(a) Reconstruction of the highly deformed Jormua mafic-ultramafic complex of Finland. These 1.97-billion-year-old rocks are considered to be the oldest known complete ophiolite sequence. (b) A metamorphosed basaltic pillow lava. The code plate is 12 cm long. (c) Metamorphosed gabbro between mafic dikes. The hammer shaft is 65 cm long.*

WHAT WOULD YOU DO?

Given that Precambrian rocks contain few fossils and those that are present are not particularly useful for relative age determinations, how would you go about demonstrating that Archean rocks are the same age as those elsewhere? Obviously the principle of fossil succession would be of little use in Precambrian rocks, but could you still use the principles of superposition, lateral continuity, original horizontality, and inclusions to decipher the geologic history in a single location? Explain.

the southwestern United States have small quantities of iron oxide cement and accordingly are reddish. These **red beds** of sandstone, siltstone, and shale do not appear in the geologic record until about 1.8 BYA (Figure 20.15b). Iron oxide cement forms under oxidizing conditions, which implies that Earth's atmosphere had some free oxygen by this time, although the amount was probably still only 1 or 2% of present levels.

Origin and Evolution of the Atmosphere and Oceans

In the Prologue we mentioned that during its earliest history Earth was "a barren, waterless, lifeless planet with a poisonous atmosphere." If we could somehow go back and visit this early Earth, we would witness numerous meteorite impacts and volcanic eruptions, and be subjected to intense ultraviolet radiation—quite an inhospitable environment for today's organisms. Now the atmosphere is rich in oxygen and contains important trace

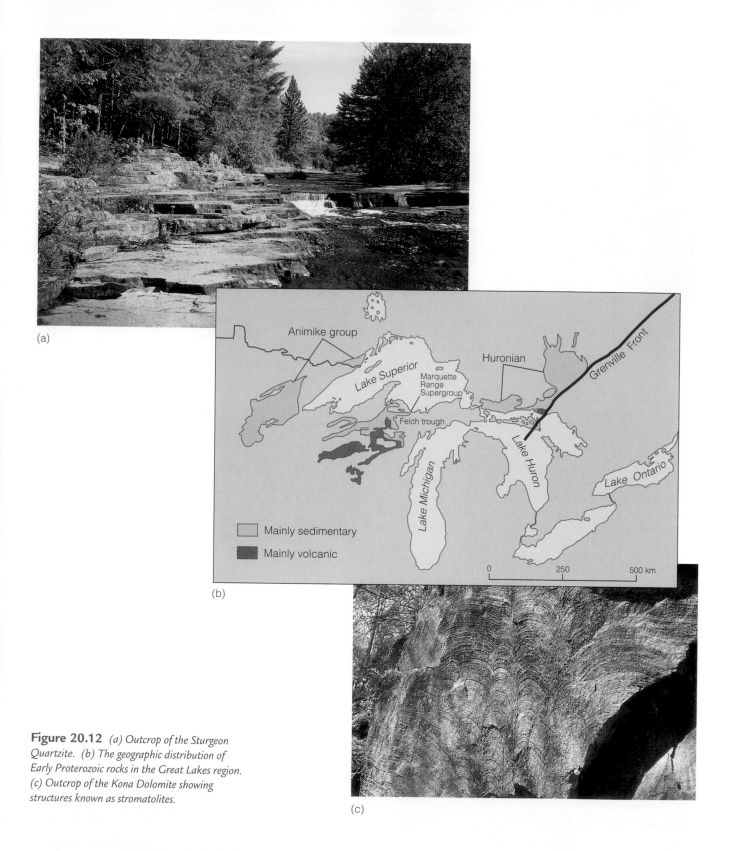

(a)

(b)

(c)

Figure 20.12 *(a) Outcrop of the Sturgeon Quartzite. (b) The geographic distribution of Early Proterozoic rocks in the Great Lakes region. (c) Outcrop of the Kona Dolomite showing structures known as stromatolites.*

amounts of carbon dioxide, water vapor, and other gases. Ozone in the upper atmosphere blocks most ultraviolet radiation, and now 71% of Earth's surface is water covered. The obvious question is, What brought about such remarkable changes?

Because hydrogen and helium are the most abundant elements in the universe they probably also formed Earth's earliest atmosphere. If so, Earth's gravitational attraction is too weak to retain gases with such low molecular weights and they would have escaped into space. And before Earth

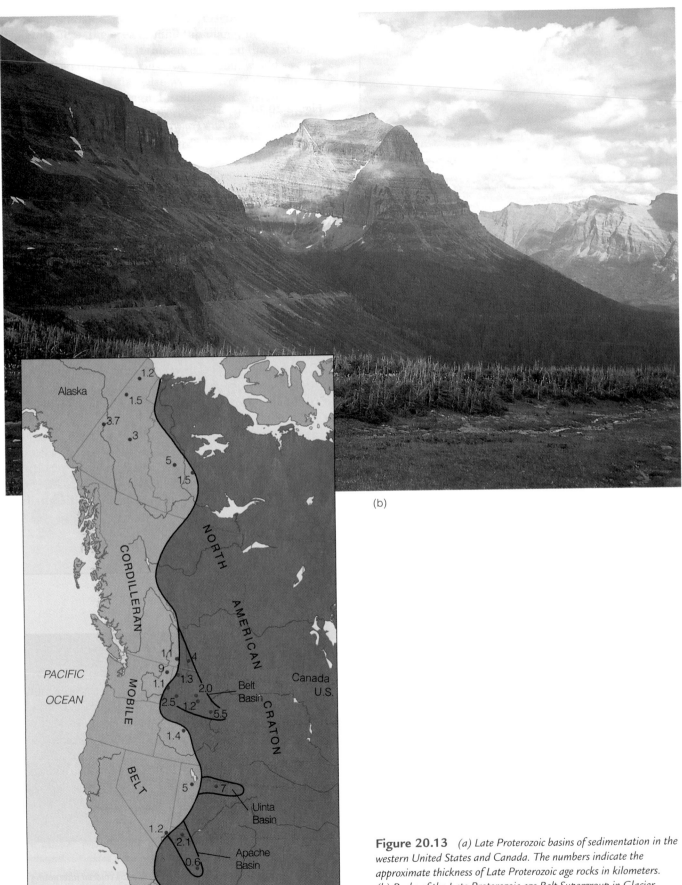

(a)

(b)

Figure 20.13 (a) Late Proterozoic basins of sedimentation in the western United States and Canada. The numbers indicate the approximate thickness of Late Proterozoic age rocks in kilometers. (b) Rocks of the Late Proterozoic age Belt Supergroup in Glacier National Park, Montana.

Archean Organisms

Prior to the mid-1950s, scientists assumed that the fossils so abundant in Cambrian rocks had a long earlier history, but they had little direct knowledge of Precambrian life. During the early 1900s, Charles Walcott described layered moundlike structures from the Early Proterozoic Gunflint Iron Formation of Canada that he proposed were constructed by algae. We now call these structures **stromatolites,** but not until 1954 did scientists demonstrate that these structures are actually the result of organic activity. In fact, stromatolites are still forming in a few areas where they originate by the entrapment of sediment on sticky mats of photosynthesizing cyanobacteria or what is better known as blue-green algae (Figure 20.20). Although widespread in Proterozoic rocks, they are now restricted to aquatic environments with especially high salinity where snails cannot live and graze on them.

Currently, the oldest known undisputed stromatolites are from 3.0-billion-year-old (BYO) rocks in South Africa, but other probable ones have been discovered in 3.3 to 3.5-BYO rocks near North Pole, Australia (Table 20.4). Even more ancient evidence of life comes from small carbon spheres in 3.8-BYO rocks in Greenland, but the evidence is not conclusive.

Cyanobacteria, the oldest known fossils, practice photosynthesis, a complex metabolic process that must have been preceded by a simpler process. So it seems reasonable that cyanobacteria were preceded by nonphotosynthesizing organisms for which we so far have no record. They must have resembled tiny bacteria, and because the atmosphere lacked free oxygen they must have been **anaerobic,** meaning they required no free oxygen. They also were likely **heterotrophic,** dependent on an external source of nutrients, rather than **autotrophic,** capable of manufacturing their own nutrients as in photosynthesis. And finally, they were **prokaryotic cells,** cells lacking a cell nucleus and other internal structures typical of more advanced *eukaryotic cells* (discussed in a later section).

We can characterize these very earliest organisms as single-celled, anaerobic, heterotrophic prokaryotes (Figure 20.21). Furthermore, they reproduced asexually. Their energy source was likely adenosine triphosphate (ATP), which can be synthesized from simple gases and phosphate, so it was no doubt available in the early Earth environment. These cells may have simply acquired their ATP directly from their surroundings, but this situation could not have persisted for long, because as more and more cells competed for the same resources, the supply should have diminished. Thus, a more sophisticated metabolic process developed, probably *fermentation,* an anaerobic process during which molecules such as sugars are split releasing carbon dioxide, alcohol, and energy. As a matter of fact, most living prokaryotes practice fermentation.

Of course the nature of the very earliest cells is informed speculation, but we can say that the most significant event in Archean life history was the development of the autotrophic process of photosynthesis. These more advanced cells were still anaerobic and prokaryotic, but as autotrophs they no longer depended on an external source of preformed organic molecules as a source of nutrients. The Archean fossils in Figure 20.21 belong to the kingdom Monera, which is represented today by bacteria and cyanobacteria.

Life of the Proterozoic

Our discussion of Archean organisms was rather brief because only bacteria existed during that time. In fact, the Early Proterozoic, just as the Archean, is characterized by a biota of single-celled, prokaryotic bacteria. No doubt thousands of varieties existed, but none of the more familiar organisms such as plants and animals were present. Before the appearance of cells capable of sexual reproduction, evolution was a comparatively slow process, accounting for the low organic diversity. But by the Middle Proterozoic, sexually reproducing cells appeared and the tempo of evolution picked up markedly thereafter.

(a)

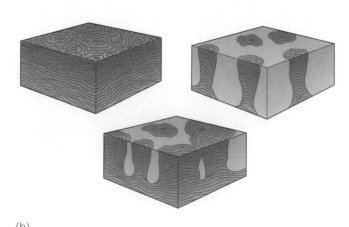
(b)

Figure 20.20 *(a) Present-day stromatolites, Shark Bay, Australia. (b) Different types of stromatolites include irregular mats, columns, and columns linked by mats.*

had a differentiated core, it lacked a magnetic field and a *magnetosphere,* the area around the planet in which the magnet field is confined. Its absence ensured that a strong *solar wind,* an outflow of ions from the Sun, would sweep away any gases that otherwise would have formed an atmosphere. Once a magnetosphere was established, though, gases from Earth's interior were released during volcanism and began to accumulate—a phenomenon known as **outgassing** (Figure 20.16).

Gases emitted by today's volcanoes are mostly water vapor, with lesser amounts of carbon dioxide, sulfur dioxide, carbon monoxide, nitrogen, hydrogen, and others. We have no reason to doubt that Archean volcanoes released the same gases, except more rapidly. Remember, Earth possessed more radiogenic heat at that time (Figure 20.7). These accumulating gases formed an atmosphere, but one notably deficient in free oxygen, that is, oxygen not combined with other elements. And with little or no free oxygen there could have been no ozone layer. Furthermore, as volcanic gases reacted chemically in the atmosphere they very likely yielded ammonia (NH_3) and methane (CH_4).

Detrital deposits containing pyrite (FeS_2) and uraninite (UO_2), both of which oxidize rapidly when oxygen is present, indicate that an oxygen-deficient atmosphere persisted throughout the Archean. Oxidized iron is quite common in Proterozoic rocks, however, so at least some free oxygen was present by that time. Two processes account for the introduction of free oxygen into the atmosphere.

One process that yielded free oxygen was **photochemical dissociation** of water vapor, in which water molecules

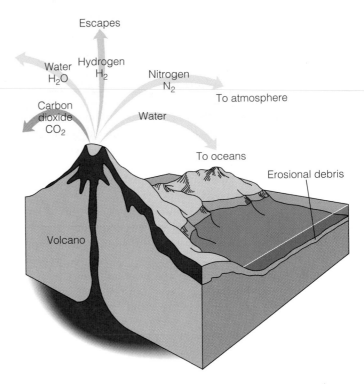

Figure 20.16 *Outgassing supplied gases to form an early atmosphere composed of the gases shown. Chemical reactions in the atmosphere also probably yielded methane (CH_4) and ammonia (NH_3).*

(H_2O) are broken up by ultraviolet radiation in the upper atmosphere (Figure 20.17). Perhaps 2% of present-day levels of free oxygen resulted from this process, but at that

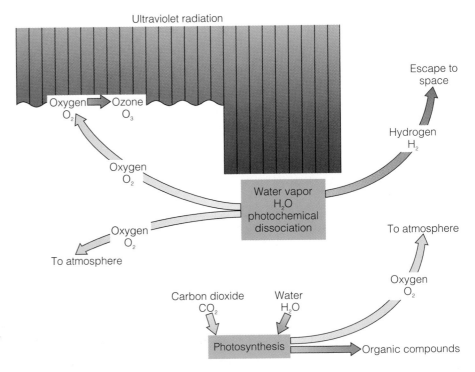

Figure 20.17 *Photochemical dissociation and photosynthesis added free oxygen to the atmosphere. Once free oxygen was present, an ozone layer formed in the upper atmosphere and blocked most incoming ultraviolet radiation.*

level ozone (O_3) forms, thus forming a barrier against incoming ultraviolet radiation. Accordingly, it is a self-limiting process and alone cannot account for all of the atmosphere's free oxygen.

The other process, **photosynthesis,** is much more important but could not take place until organisms were present. Photosynthesis involves organisms combining carbon dioxide and water into organic molecules needed for their survival, and releasing oxygen as a waste product (Figure 20.17). We know from fossils that organisms capable of photosynthesis existed as much as 3.5 billion years ago. But even with the oxygen-producing capacities of these organisms, the atmosphere at the end of the Archean may have had no more than 1% of its present oxygen level.

If the Archean atmosphere had little or no free oxygen, it was not strongly oxidizing as it is now. During the Proterozoic the free oxygen content increased from 1% to perhaps 10% of the present-day level, but probably not until well into the Paleozoic Era, about 400 million years ago, did the atmosphere attain its current oxygen concentration of 21%. Most of this free oxygen was released as a waste product by photosynthesizing bacteria, which became common about 2.3 BYA. The presence of banded iron formations and red beds (discussed earlier in this chapter) provide compelling evidence that Earth's Proterozoic atmosphere was an oxidizing one.

Water vapor, the most common gas emitted during eruptions, began to condense and fall as precipitation once Earth cooled sufficiently. And although the volume and extent of Earth's Archean oceans are unknown, we do know that large quantities of surface water were present, and certainly by Proterozoic time the oceans were no doubt nearly as extensive as at present. We can envision a hot Earth with numerous volcanoes and a rapid rate of surface water accumulation. Volcanism still takes place, so is the volume of oceanic waters still increasing? Probably so, but because Earth's radiogenic heat production has decreased (Figure 20.7) and volcanism is not as common, its rate of accumulation is quite limited in comparison with the existing volume of surface waters.

Life—Its Origin and Early History

Fossils are known from rocks as much as 3.5 billion years old, so organisms were present during a large part of the Archean, but compared to the present, this time was biologically impoverished. Today, Earth's biosphere includes millions of species of animals, plants, one-celled organisms, fungi, and bacteria, whereas only bacteria are known from the Archean. In Chapter 18, we considered the evolutionary processes whereby life diversified, but here our concern is with how life originated and its earliest history.

First we must be very clear on what is living versus nonliving. Everyone recognizes today's plants and animals as living, but just how do they differ from nonliving things, and is the distinction always clear? Minimally, an organism must reproduce and practice some kind of metabolism. Reproduction ensures the long-term survival of a species as it perpetuates itself, whereas metabolism ensures the short-term survival of an individual as a chemical system.

Using this reproduction-metabolism criterion, it would seem a simple matter to decide whether something is alive. Yet viruses, composed of a bit of DNA or RNA in a protein capsule, behave like living organisms when in an appropriate host cell, but outside a host cell they neither reproduce nor produce organic molecules. So, are they living or nonliving? Some biologists think they are nonliving; others think they represent another way of defining living. In either case, though, they illustrate that the distinction is not always clear.

Simple organic molecules known as microspheres form spontaneously and show greater organizational complexity than inorganic objects such as rocks. In fact, they grow and divide, but in ways that are more like random chemical processes. Consequently, they are not living, but they nevertheless show some characteristics of organisms. Furthermore, if life originated on Earth by some natural process, it must have passed through some kind of prebiotic stage, perhaps similar to microspheres.

The Origin of Life

As early as 1924, the Russian biochemist A. I. Oparin postulated that life originated from nonliving matter when Earth had little or no free oxygen and no ozone layer. Investigators agree that for life to originate, an energy source must have acted upon the appropriate chemical elements from which organic molecules could synthesize. The early atmosphere composed of carbon dioxide (CO_2), water vapor (H_2O), nitrogen (N_2), and likely ammonia (NH_3) and methane (CH_4) provided carbon, oxygen, hydrogen, and nitrogen, the primary elements making up all organisms. Lightning and ultraviolet radiation were two possible energy sources necessary for these elements to combine and form rather simple organic molecules known as **monomers,** such as amino acids.

Monomers are needed as basic building blocks of more complex organic molecules, but is it plausible they formed in the manner postulated? Experimental evidence indicates it is. During the 1950s, Stanley Miller circulated gases approximating Earth's early atmosphere through a closed glass vessel and subjected the mixture to an electric spark to simulate lightning. Within a few days, the mixture became turbid and an analysis showed that Miller had synthesized several amino acids typical of organisms (Figure 20.18). More recent experiments have successfully synthesized all 20 amino acids in organisms.

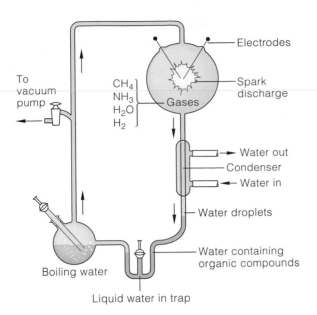

Figure 20.18 *Experimental apparatus used by Stanley Miller. Several amino acids characteristic of organisms were artifically synthesized during Miller's experiments.*

Making monomers in a test tube is one thing, but organisms are composed of more complex molecules called **polymers** such as nucleic acids (DNA and RNA) and proteins consisting of monomers linked together in a specific sequence. So how did this linking of monomers, or polymerization, take place? Researchers have successfully synthesized small molecules called *proteinoids*, consisting of more than 200 linked amino acids (Figure 20.19). In fact, when heated, dehydrated, and concentrated, amino acids spontaneously polymerize and form proteinoids.

We can call these artificially synthesized proteinoids *protobionts*, meaning they have characteristics between inorganic chemical compounds and living organisms. Suppose these protobionts came into existence in a manner similar to that outlined above. If this happened, they would have been diluted and would have ceased to exist if they had not developed some kind of outer covering. In other words, they had to be self-contained chemical systems as today's cells are. In the experiments just mentioned, proteinoids have spontaneously aggregated into microspheres (Figure 20.19) that have a cell-like outer covering and grow and divide somewhat like bacteria do.

Perhaps the first steps leading to life took place as monomers formed in great abundance and polymerized, but little is known about how a reproductive mechanism came about. Microspheres divide but, as noted earlier, do so in a nonbiologic fashion. In fact, for some time researchers were baffled because in present-day organisms either DNA or RNA is necessary for reproduction, but these nucleic acids cannot replicate without protein enzymes, yet protein enzymes cannot be made without nucleic acids. Or so it seemed until a few years ago, when researchers discovered that some small RNA molecules can in fact reproduce without the aid of protein enzymes. So, the first replicating system might have been an RNA molecule. Just how these molecules were synthesized under conditions that existed on early Earth has not been resolved.

A common theme among investigators is that life originated when organic molecules were synthesized from atmospheric gases. But even this has been questioned by those who think the same sequence of events—that is, formation of monomers, then polymers—took place adjacent to hydrothermal vent systems (black smokers) on the seafloor.

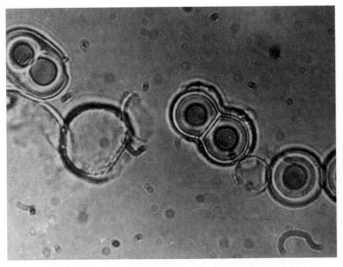

(a)

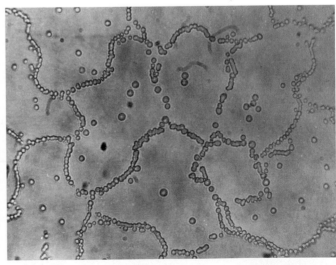

(b)

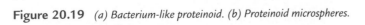

Figure 20.19 *(a) Bacterium-like proteinoid. (b) Proteinoid microspheres.*

Archean Organisms

Prior to the mid-1950s, scientists assumed that the fossils so abundant in Cambrian rocks had a long earlier history, but they had little direct knowledge of Precambrian life. During the early 1900s, Charles Walcott described layered moundlike structures from the Early Proterozoic Gunflint Iron Formation of Canada that he proposed were constructed by algae. We now call these structures **stromatolites,** but not until 1954 did scientists demonstrate that these structures are actually the result of organic activity. In fact, stromatolites are still forming in a few areas where they originate by the entrapment of sediment on sticky mats of photosynthesizing cyanobacteria or what is better known as blue-green algae (Figure 20.20). Although widespread in Proterozoic rocks, they are now restricted to aquatic environments with especially high salinity where snails cannot live and graze on them.

Currently, the oldest known undisputed stromatolites are from 3.0-billion-year-old (BYO) rocks in South Africa, but other probable ones have been discovered in 3.3 to 3.5-BYO rocks near North Pole, Australia (Table 20.4). Even more ancient evidence of life comes from small carbon spheres in 3.8-BYO rocks in Greenland, but the evidence is not conclusive.

Cyanobacteria, the oldest known fossils, practice photosynthesis, a complex metabolic process that must have been preceded by a simpler process. So it seems reasonable that cyanobacteria were preceded by nonphotosynthesizing organisms for which we so far have no record. They must have resembled tiny bacteria, and because the atmosphere lacked free oxygen they must have been **anaerobic,** meaning they required no free oxygen. They also were likely **heterotrophic,** dependent on an external source of nutrients, rather than **autotrophic,** capable of manufacturing their own nutrients as in photosynthesis. And finally, they were **prokaryotic cells,** cells lacking a cell nucleus and other internal structures typical of more advanced *eukaryotic cells* (discussed in a later section).

We can characterize these very earliest organisms as single-celled, anaerobic, heterotrophic prokaryotes (Figure 20.21). Furthermore, they reproduced asexually. Their energy source was likely adenosine triphosphate (ATP), which can be synthesized from simple gases and phosphate, so it was no doubt available in the early Earth environment. These cells may have simply acquired their ATP directly from their surroundings, but this situation could not have persisted for long, because as more and more cells competed for the same resources, the supply should have diminished. Thus, a more sophisticated metabolic process developed, probably *fermentation,* an anaerobic process during which molecules such as sugars are split releasing carbon dioxide, alcohol, and energy. As a matter of fact, most living prokaryotes practice fermentation.

Of course the nature of the very earliest cells is informed speculation, but we can say that the most significant event in Archean life history was the development of the autotrophic process of photosynthesis. These more advanced cells were still anaerobic and prokaryotic, but as autotrophs they no longer depended on an external source of preformed organic molecules as a source of nutrients. The Archean fossils in Figure 20.21 belong to the kingdom Monera, which is represented today by bacteria and cyanobacteria.

Life of the Proterozoic

Our discussion of Archean organisms was rather brief because only bacteria existed during that time. In fact, the Early Proterozoic, just as the Archean, is characterized by a biota of single-celled, prokaryotic bacteria. No doubt thousands of varieties existed, but none of the more familiar organisms such as plants and animals were present. Before the appearance of cells capable of sexual reproduction, evolution was a comparatively slow process, accounting for the low organic diversity. But by the Middle Proterozoic, sexually reproducing cells appeared and the tempo of evolution picked up markedly thereafter.

(a)

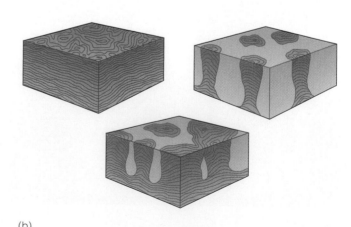

(b)

Figure 20.20 *(a) Present-day stromatolites, Shark Bay, Australia. (b) Different types of stromatolites include irregular mats, columns, and columns linked by mats.*

Table 20.4

Summary Chart for Proterozoic Events in Life History (dashed vertical lines indicate uncertainties in age ranges)

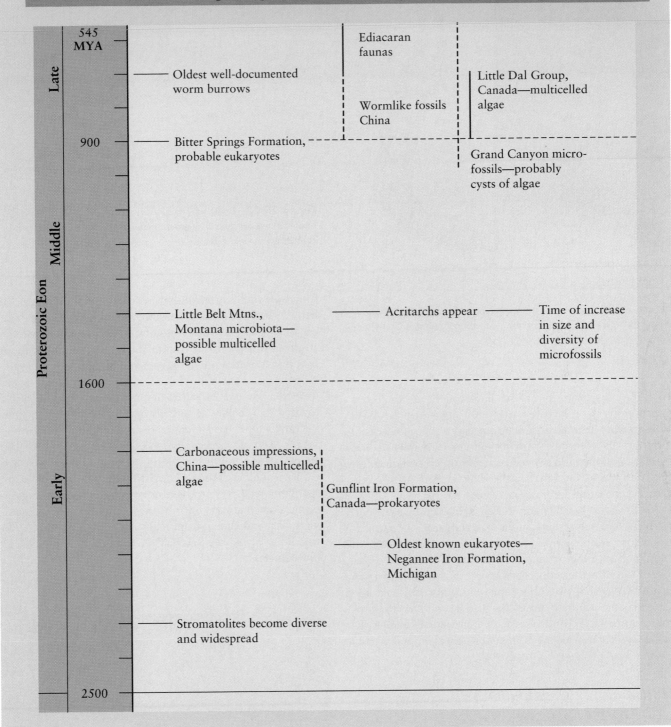

A New Type of Cell Appears

The origin of **eukaryotic cells** marks one of the most important events in life history (Table 20.4). These cells are much larger than prokaryotic cells; they have a membrane-bounded nucleus containing the genetic material, and most reproduce sexually (Figure 20.22). And most eukaryotes—that is, organisms made up of eukaryotic cells—are multicellular and aerobic, so they could not have existed until some free oxygen was present in the atmosphere.

No one doubts that eukaryotes were present by the Middle Proterozoic, and it is becoming increasingly clear that they existed even earlier (Figure 20.23). Rocks 1.4 to

Life—Its Origin and Early History

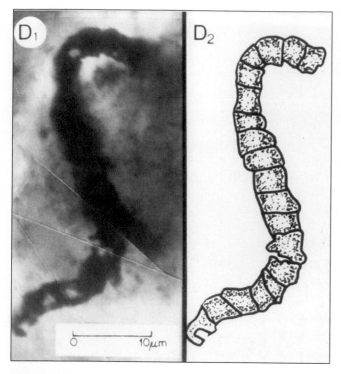

Figure 20.21 *Photomicrograph and schematic restoration of fossil prokaryote from the 3.3- to 3.5-billion-year-old Warrawoona Group, Western Australia.*

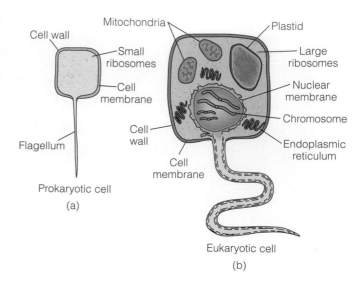

Figure 20.22 *Prokaryotic and eukaryotic cells. Note that eukaryotes have a cell nucleus containing genetic material and several organelles such as mitochondria and plastids.*

1.7 BYO in Montana and China have coiled, carbonaceous impressions of eukaryotes, probably algae, and 2.1-BYO rocks in Michigan have similar impressions. Some rocks in Australia have fossils showing evidence of mitosis and meiosis, cell division processes used only by eukaryotes. Theories of how eukaryotic cells evolved from prokaryotic cells are considered in Perspective 20.1.

Cells larger than 60 microns appear in abundance at least 1.4 BYA, and many show an increased degree of organizational complexity compared to that of prokaryotic cells. An internal membrane-bounded cell nucleus is present in some, for instance. Furthermore, microscopic, hollow fossils known as *acritarchs* that probably represent cysts of planktonic algae become common during the Middle and Late Proterozoic (Figure 20.24). And vase-shaped microfossils from rocks in the Grand Canyon have been tentatively identified as cysts of some kind of algae (Figure 20.24).

Multicellular Organisms

Multicellular organisms are not only composed of many cells, but also have cells specialized to perform specific functions such as reproduction and respiration. Unfortunately, the fossil record does not tell us how multicellular organisms arose from unicellular ancestors. However, studies of present-day organisms give some clues about how this transition might have taken place.

Suppose that a single-celled organism divided and formed a group of cells that did not disperse but remained together as a colony. The cells in some colonies may have become somewhat specialized similar to the situation in some living *colonial organisms* (Figure 20.25). Further specialization might have led to simple multicellular organisms such as sponges consisting of cells that carry out functions such as reproduction, respiration, and food gathering. Carbonaceous impressions of Proterozoic multicellular algae are known from many areas (Figure 20.23), but the oldest are in the 2.1-BYO Negaunee Iron Formation in Michigan.

Figure 20.23 *Carbonaceous impressions in Proterozoic rocks in the Little Belt Mountains, Montana. These may be impressions of multicellular algae, but this is uncertain.*

(a)   (b)   (c)

Figure 20.24 *(a) and (b) Acritarchs are probably the cysts of algae. These common Late Proterozoic microfossils are thought to represent eukaryotic organisms. (c) Vase-shaped microfossil, probably a cyst of some kind of algae.*

The Ediacaran Fauna

In 1947, the Australian geologist R. C. Sprigg discovered a unique assemblage of multicellular, soft-bodied animals preserved as molds and casts on the undersides of sandstone layers (Figure 20.26). Some investigators think that at least three present-day invertebrate phyla are represented: jellyfish and sea pens (phylum Cnidaria), segmented worms (phylum Annelida), and primitive members of the phylum Arthropoda. One wormlike Ediacaran fossil, *Spriggina,* has been cited as a possible ancestor of trilobites, and another, *Tribachidium,* may be a primitive echinoderm (Figure 20.26). On the other hand, some think Ediacara fossils represent an early evolutionary radiation quite distinct from the ancestry of existing invertebrates. These **Ediacaran faunas** existed between 670 and 570 million years ago, and are now known on all continents except South America and Antarctica. The animals were widespread, but their fossils are not very common because all of them lacked durable skeletons.

Other Proterozoic Animal Fossils

Although scarce, fossils or traces of organic activity are found in rocks older than those containing Ediacaran faunas. Worm burrows in rocks at least 700 million years old are known from several areas, and wormlike fossils associated with fossil algae were found in 700- to 900-million-year-old rocks in China (Figure 20.27). Perhaps these are worms, but their identity as well as the age of the rocks has been questioned. So for the present, we can consider these "fossils" as persuasive but not conclusive evidence for pre-Ediacaran animals.

By the latest Proterozoic, several animals with skeletons almost certainly existed. Evidence for this conclusion comes from denticles and minute scraps of shell-like material, as well as spicules, presumably from sponges. However, durable skeletons of silica and calcium carbonate did not appear in abundance until the beginning of the Paleozoic Era 545 million years ago. So even though multicelled animals were present by the Late Proterozoic, the only organisms of any kind on land were various types of bacteria and perhaps some worms.

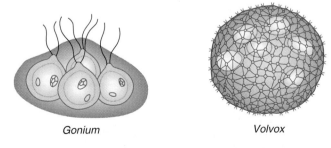

Gonium   *Volvox*

Figure 20.25 *Single-celled versus multicellular organisms. Although* Gonium *consists of as few as four cells, all cells are alike and can produce a new colony.* Volvox *has some cells specialized to perform different functions and has thus crossed the threshold that separates unicellular from multicellular organisms.*

What Kinds of Resources Are Found in Precambrian Rocks?

Precambrian rocks contain a variety of natural resources, but the one most commonly associated with those of Archean age is gold, whereas iron ore is the one that immediately comes to mind when discussing Proterozoic rocks. And of course some of the rocks themselves are resources used in aggregate for cement or for roadbeds, but rocks used for these purposes might be any age.

Origin of Eukaryotic Cells

A currently popular theory among evolutionary biologists holds that eukaryotic cells formed from several prokaryotic cells that had established a symbiotic relationship. Symbiosis, the living together of two or more dissimilar organisms, is quite common today. It may take the form of parasitism in which one organism lives at the expense of another, or the two may coexist with mutual benefit. For example, lichens, once considered to be plants, are actually symbiotic fungi and algae.

In a symbiotic relationship, each symbiont must be capable of metabolism and reproduction, but the degree of dependence in some symbiotic relationships is such that one symbiont cannot live independently. Many parasites, for example, cannot exist outside a host organism. This may have been the case with Proterozoic prokaryotes: two or more prokaryotes may have entered into a symbiotic relationship beneficial to both (Figure 1), and the symbionts became increasingly interdependent until the unit could exist only as a whole.

Supporting evidence for the symbiosis theory comes from the study of living eukaryotic cells. For example, eukaryotic cells contain internal structures called organelles that have their own genetic material. Although these organelles, such as plastids and mitochondria, cannot exist independently today, they apparently were once capable of reproduction as free-living organisms.

Another way of evaluating the symbiosis theory is to examine protein synthesis. Prokaryotic cells synthesize proteins, but can be thought of as a single system, whereas eukaryotes are a combination of protein-synthesizing systems; that is, some of the organelles within eukaryotes such as mitochondria and plastids are capable of protein synthesis. These organelles, with their own genetic material and protein-synthesizing capabilities, are thought to have been free-living bacteria that entered into a symbiotic relationship. With time, the interdependence of the various units grew until life was possible only as an integrated whole (Figure 1).

More recently, another theory for the origin of eukaryotic cells has been proposed. According to this theory, the sequence of events leading to the origin of eukaryotes began when early prokaryotes lost the ability to synthesize an essential component of cell walls. One solution to this problem was to develop an internal skeleton consisting of microtubules and microfilaments (Figure 2a). This cytoskeleton, as it is called, allowed for

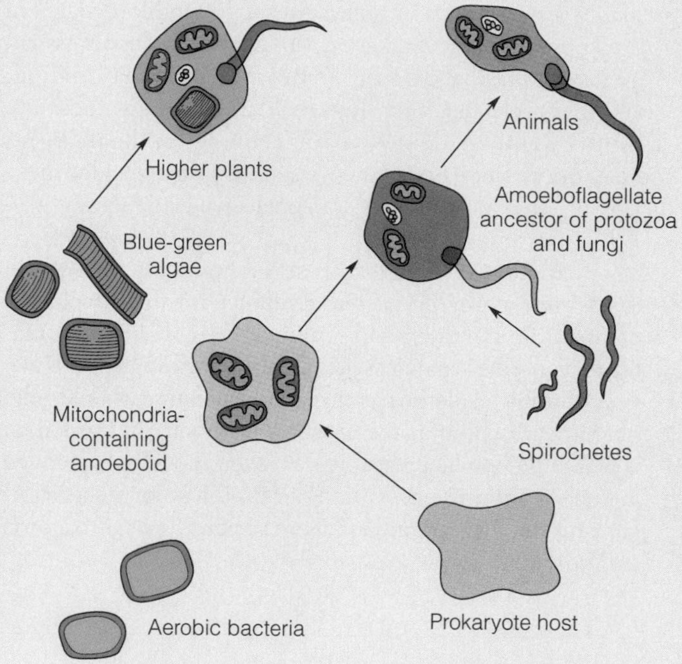

Figure 1 *Symbiosis theory for the origin of eukaryotic cells. An aerobic bacterium and a larger host of the kingdom Monera united to form a mitochondria-containing amoeboid. An amoeboflagellate was formed by a union of the amoeboid and a bacterium of the spirochete group; this amoeboflagellate was the direct ancestor of two kingdoms—Fungi and Animalia. Another kingdom, Plantae, originated when this amoeboflagellate formed a union with blue-green algae (cyanobacteria) that became plastids.*

Archean Mineral Resources

Archean and Proterozoic rocks near Johannesburg, South Africa, have yielded more than 50% of the world's gold since 1886, but a number of gold mining areas are also within the Superior craton in Canada. The second largest gold mine in the United States is in the Archean Homestake Formation at Lead, South Dakota.

A number of Archean massive sulfide deposits of zinc, copper, and nickel are known in Western Australia, Zimbabwe, and Canada. Similar deposits are currently forming adjacent to hydrothermal vents on the seafloor (see Per-

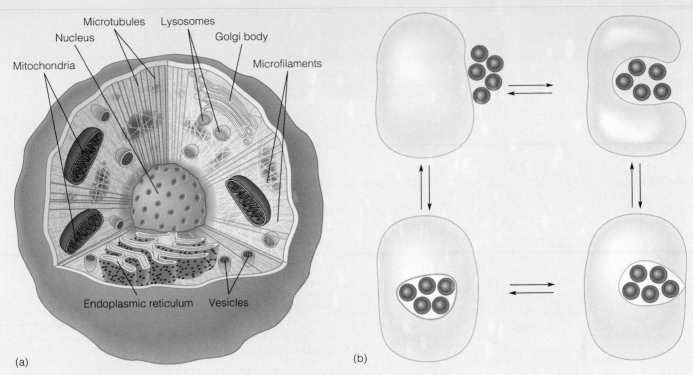

(a)

(b)

Figure 2 *(a) Diagrammatic view of a cell showing the cytoskeleton consisting of microtubules and microfilaments. (b) According to one theory for the origin of eukaryotic cells, cells acquired such structures as a nucleus, mitochondria, and plastids by infolding and enveloping materials.*

In figure (a): Microtubules, Lysosomes, Nucleus, Golgi body, Mitochondria, Microfilaments, Endoplasmic reticulum, Vesicles

an outer fluid cell membrane that infolded so that material coming into the cell could be enveloped. Such infolding is thought to have resulted in the origin of such intracellular structures as the cell nucleus (Figure 2b). According to this theory, the first eukaryotic cell was anaerobic; later it acquired a free-living aerobic organism by symbiosis, thus accounting for the origin of aerobic eukaryotes.

Although the fossil record does not record the acquisition of organelles or symbiosis, living eukaryotes can give some idea of what the first eukaryotes may have been like. The present-day giant amoeba *Pelomyxa,* which lives in the mud of

ponds, lacks mitochondria. Two types and hundreds of individual bacteria, however, have a symbiotic relationship with *Pelomyxa* and perform the same function as mitochondria.

Pelomyxa provides evidence for the symbiotic theory for the origin of eukaryotes. However, recent studies of *Giardia,* a single-celled eukaryote, seem to indicate that eukaryotes may have acquired their internal membrane-bounded organelles by infolding of the cell wall. Although *Giardia* is a eukaryote, it shares many characteristics with prokaryotes and is capable of acquiring materials from outside by infolding of the cell wall.

spective 8.1). About one-fourth of the world's chrome reserves are in Archean rocks, especially in Zimbabwe. These deposits appear to have formed when crystals settled and became concentrated in the lower parts of mafic and ultramafic sills and other plutons. The Stillwater Complex of Montana has low-grade chrome ores that were mined and stockpiled during both World Wars and the Korean War, but they have never been refined for their chrome. These

same rocks are also a potential source of platinum. Archean banded iron formations are mined in some areas, but Proterozoic ones are much more important.

Archean pegmatites are mostly granitic and of little economic importance. But some in Manitoba, Canada, and the Rhodesian Province in Africa contain valuable minerals. In addition to minerals of gem quality, a few Archean pegmatites are mined for lithium, beryllium, and rubidium.

What Kinds of Resources Are Found in Precambrian Rocks?

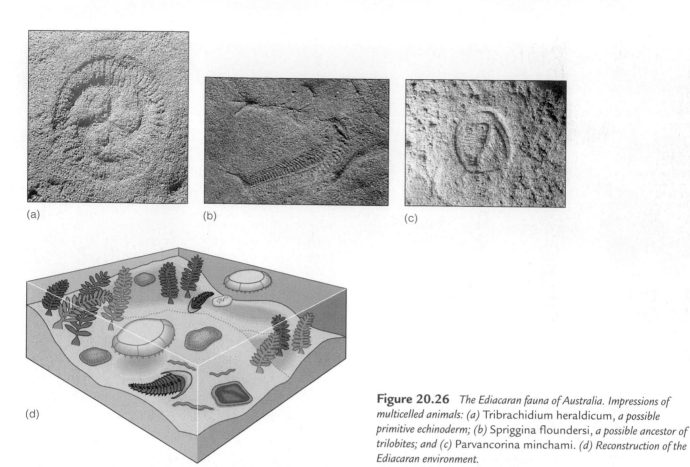

Figure 20.26 *The Ediacaran fauna of Australia. Impressions of multicelled animals: (a)* Tribrachidium heraldicum, *a possible primitive echinoderm; (b)* Spriggina floundersi, *a possible ancestor of trilobites; and (c)* Parvancorina minchami. *(d) Reconstruction of the Ediacaran environment.*

Figure 20.27 *Wormlike fossils from the Late Proterozoic of China.*

Proterozoic Mineral Resources

Banded iron formations (BIFs), the world's major source of iron ores, are present on all continents, and 92% of all BIFs were deposited during the Late Proterozoic (Figure 20.15a). Brazil, Australia, China, Ukraine, Sweden, South Africa, Canada, and the United States, in that order, are the largest producers of iron ore. And even though the United States is among the leaders, it still must import about 30% of the iron ore it uses, mostly from Canada and Venezuela. Most of the iron ore mined in North America comes from mines in the Lake Superior region and in eastern Canada.

The Sudbury mining district in Ontario, Canada, is an important area of nickel and platinum production. Nickel is essential for the manufacture of nickel alloys such as stainless steel and Monel metal (nickel plus copper), which are valued for their strength and resistance to corrosion and heat. The United States imports more than 50% of all the nickel it uses, most from the Sudbury mining district in Canada. Some platinum for jewelry, surgical instruments, and chemical and electrical equipment is also exported from Canada to the United States, but the largest exporter is South Africa. The United States also depends on South Africa for much of its chromite, the ore of chromium.

Economically recoverable oil and natural gas were discovered in Proterozoic rocks in China and Siberia, arousing interest in the Midcontinent Rift as a potential source of hydrocarbons. Some rocks within the rift are known to contain oil, but so far no oil or gas wells are operating.

Chapter Summary

1. Precambrian time is divided into two eons, the Archean and Proterozoic.

2. Each continent has an ancient, stable craton. The exposed part of a craton is a shield. The Canadian Shield of North America is made up of several units, which are delineated by age and structural trends.

3. Archean rocks are predominantly greenstone belts and granite-gneiss complexes. Greenstone belts have a synclinal structure and occur as linear bodies within much more extensive areas of granite and gneiss.

4. Archean cratons served as nuclei about which Proterozoic crust accreted. One large landmass that formed by this process is Laurentia. It consisted mostly of North America and Greenland.

5. The major events in the Proterozoic evolution of Laurentia were an Early Proterozoic episode of amalgamation of cratons, Middle Proterozoic igneous activity, and the Middle Proterozoic Grenville orogeny and Midcontinent rift.

6. Plate tectonics similar to that of the present did not begin until the Early Proterozoic, but many geologists think some type of plate tectonics took place during the Archean. Archean plates may have moved more rapidly, however, because Earth possessed more radiogenic heat.

7. Quartzite-carbonate-shale assemblages are known from the Late Archean but become common in Proterozoic rocks. These rock assemblages were deposited on passive continental margins and in intracratonic basins.

8. Widespread glaciation took place during the Early and Late Proterozoic.

9. The atmosphere and surface waters were derived from internally generated volcanic gases, a process known as outgassing. The atmosphere so formed was deficient in free oxygen, but it became richer in free oxygen during the Proterozoic.

10. During the period from 2.5 to 2.0 billion years ago, most of the world's iron ores were deposited as banded iron formations, and the first continental red beds were deposited about 1.8 billion years ago. The widespread occurrence of oxidized iron in sedimentary rocks indicates an oxidizing atmosphere.

11. Most models for the origin of life require a nonoxidizing atmosphere. Atmospheric gases contained the elements necessary for simple organic molecules. Ultraviolet radiation and lightning probably provided the energy for synthesizing organic molecules. Some investigators think RNA molecules may have been the first molecules capable of self-replication.

12. The Archean fossil record is poor. A few localities contain single-celled prokaryotic bacteria. Stromatolites formed by photosynthesizing bacteria may date from 3.5 billion years ago.

13. Eukaryotic cells probably appeared during the Early Proterozoic, but their presence in the Middle Proterozoic is unquestioned.

14. The oldest fossils of multicellular organisms are carbonaceous impressions, probably of algae, in rocks more than 1.5 billion years old.

15. The Late Proterozoic Ediacaran faunas include the oldest well-documented animal fossils other than burrows. Animals were widespread at this time, but all were soft-bodied, so fossils are not common.

Important Terms

anaerobic
autotrophic
banded iron formation (BIF)
Canadian Shield
craton
Ediacaran faunas
eukaryotic cell

greenstone belt
heterotrophic
Laurentia
Midcontinent rift
monomer
multicellular organism
orogen

outgassing
photochemical dissociation
photosynthesis
platform
polymer
prokaryotic cell
red bed

sandstone-carbonate-shale assemblage
shield
stromatolite

Review Questions

1. The uppermost unit in a typical greenstone belt is composed mostly of
   a. _____ sedimentary rocks; b. _____ ultramafic rocks; c. _____ andesite; d. _____ lava flows; e. _____ granite and gneiss.

2. The largest exposed area of the North American craton is the
   a. _____ Wyoming craton; b. _____ Grand Canyon; c. _____ Canadian Shield; d. _____ American platform; e. _____ Minnesota lowlands.

3. Earth's atmosphere was probably derived from volcanic gases, a process known as
   a. _____ dewatering; b. _____ photosynthesis; c. _____ autotrophism; d. _____ fermentation; e. _____ outgassing.

4. A polymer is an organic molecule
   a. _____ capable of metabolism and reproduction; b. _____ made up of amino acids linked together; c. _____ characterized as heterotrophic and anaerobic; d. _____ that binds together loose sediment to make stromatolites; e. _____ found mostly in the lava flows of greenstone belts.

5. Most of the free oxygen was probably added to Earth's atmosphere by the process known as
   a. _____ photosynthesis; b. _____ intracontinental rifting; c. _____ cratonization; d. _____ polymerization; e. _____ photochemical dissociation.

6. Some scientists think the first self-replicating system might have been a(n)
   a. _____ stromatolite; b. _____ RNA molcule; c. _____ bacterium; d. _____ ATP cell; e. _____ proteinoid.

7. Which one of the following rock associations is typical of passive continental margins?
   a. _____ sandstone-granite-basalt; b. _____ basalt-andesite-ash fall; c. _____ granite-tillite-banded iron formation; d. _____ stromatolite-prokaryote-lava; e. _____ sandstone-carbonate-shale.

8. Most of the evidence for the origin of eukaryotic cells from prokaryotic cells comes from
   a. _____ chemicals preserved in Archean rocks; b. _____ fossils in tillite; c. _____ studies of present-day organisms; d. _____ tracks and trails of trilobites; e. _____ Early Proterozoic carbonate rocks.

9. The Middle Proterozoic of Laurentia was a time of
   a. _____ the origin of most of Earth's greenstone belts; b. _____ igneous activity, rifting, and the Grenville orogeny; c. _____ appearance of the first stromatolites; d. _____ origin of animals with durable skeletons of chitin, calcium carbonate, and silicon dioxide; e. _____ widespread glaciation and meteorite impacts.

10. One indication of the presence of free oxygen in the Proterozoic atmosphere 1.8 billion years ago is
    a. _____ red beds; b. _____ animals much like those of the present; c. _____ extinction of stromatolites; d. _____ evolution of prokaryotes from protobionts; e. _____ deposition of passive margin sediments in the Lake Superior region.

11. Which one of the following statements is correct?
    a. _____ 92% of all banded iron formations formed during the Late Archean; b. _____ Eukaryotic cells have a membrane-bounded nucleus; c. _____ Proterozoic plate tectonic processes were considerably different than those of the present; d. _____ Photochemical dissociation accounts for the origin of multicellular organisms; e. _____ Glacial deposits indicate that the Proterozoic seas were chemically stratified.

12. The Archean Eon ended _____ billion years ago.
    a. _____ 545; b. _____ 1.3; c. _____ 2.5; d. _____ 3.8; e. _____ 3.96.

13. Why are Precambrian rocks so difficult to study compared with Phanerozoic rocks?

14. What is the evidence that some rocks were present before the beginning of the Archean Eon 4.0 billion years ago?

15. Why are ultramafic rocks so rare in rocks younger than Archean?

16. Summarize the experimental evidence that indicates both monomers and polymers could have formed by natural processes on early Earth.

17. Explain how photosynthesis and photochemical dissociation supplied oxygen to the Archean atmosphere.

18. How is it that geologists have few difficulties correlating Phanerozoic rocks of the same age, but find it hard or impossible to do the same with Archean rocks? Also, why are even radiometric dates of limited use in this endeavor?

19. What are stromatolites, and how do they form?

20. Explain how pre-Archean crust might have formed. Why is none of this ancient crust still present?

21. Summarize the evidence indicating that eukaryotic cells appeared as much as 2.1 billion years ago.

22. Explain how red beds, banded iron formations, and the earliest eukaryotic cells give us some idea of the composition of the Proterozoic atmosphere.

23. The Belt-Purcell Supergroup of the northwestern United States and adjacent parts of Canada is 4000 m thick and was deposited between 1.45 billion and 850 million years ago. Calculate the average rate of sediment accumulation in millimeters per year. Why is this figure unlikely to represent the real rate of sedimentation?

24. Discuss the endosymbiosis theory for the origin of eukaryotic cells.

World Wide Web Activities

For these web site addresses, along with current updates and exercises, log on to **http://www.brookscole.com/geo**

University of California Museum of Paleontology

The Museum of Paleontology at the University of California, Berkeley, maintains this huge site with information on a variety of topics of interest to geologists, biologists, and Earth scientists. Under *On-Line Exhibits* click *Time Periods* and then click *Hadean, Archean,* and *Proterozoic* and learn about Earth's oldest rocks and fossils, as well as how the atmosphere became oxygenated, and details and images of the Ediacaran fauna.

Travels With Geology—Grand Teton National Park

This is part of the Geology Gems site maintained by Winona State University, Winona, Minnesota. It has maps, images, and a discussion of the geologic history of the Teton Range of Wyoming.

What kinds of Archean and Proterozoic rocks are found in this range? When did the present-day range form and what processes are responsible for its present rugged topography?

Symbiotic Theory

This opens a site titled *Endosymbiotic Theory: Evolution from Simple Prokaryotes to Complex Eukaryotes,* which was designed, developed, and written by Joshua A. Bond, a student at Dekalb College/Georgia State University. It is a good two-page summary with figures about the endosymbiotic theory proposed by Lynn Margulis during the 1960s. What was Margulis's prediction about the organelles of eukaryotic cells?

Steel—2000 Million Years in the Making

This site presented by the University of South Australia Library has good images of banded iron formation rocks, as well as a brief discussion of mining and refining iron ore to produce iron and steel.

Paleozoic Earth History

Rainbow over the Grand Canyon, Grand Canyon National Park, Arizona.

PROLOGUE

"The Grand Canyon is the one great sight which every American should see," declared President Theodore Roosevelt. "We must do nothing to mar its grandeur." And so, in 1908, he named the Grand Canyon a national monument to protect it from exploitation. In 1919 the Grand Canyon National Monument was upgraded to a national park primarily because its scenery and the geology exposed in the canyon are unparalleled.

When people visit the Grand Canyon, many are astonished by the seemingly limitless time represented by the rocks exposed in the walls. For most people, staring down 1.5 km at the rocks in the canyon is their only exposure to the concept of geologic time.

Major John Wesley Powell was the first geologist to explore the Grand Canyon region. Major Powell, a Civil War veteran who lost his right arm in the battle of Shiloh, led a group of hardy explorers down the uncharted Colorado River through the Grand Canyon in 1869. Without any maps or other information, Powell and his group ran the many rapids of the Colorado River in fragile wooden boats, hastily recording what they saw. Powell wrote in his diary that "all about me are interesting geologic records. The book is open and I read as I run."

From this initial reconnaissance, Powell led a second expedition down the Colorado River in 1871. This second trip included a photographer, a surveyor, and three topographers. Members of this expedition made detailed topographic and geologic maps of the Grand Canyon area as well as the first photographic record of the region.

Probably no one has contributed as much to the understanding of the Grand Canyon as Major Powell. In recognition of his contributions, the Powell Memorial was erected on the South Rim of the Grand Canyon in 1969 to commemorate the hundredth anniversary of his first expedition.

When we stand on the rim and look down into the Grand Canyon, we are really looking far back in time, all the way back to the early history of our planet. More than one billion years of history are preserved in the rocks of the Grand Canyon, indicating episodes of mountain building as well as periods of transgressions and regressions of shallow seas.

The oldest rocks exposed in the Grand Canyon record two major mountain-building episodes during the Proterozoic Eon. The first episode, represented by the Vishnu and Brahma schists, records a time of uplift, deformation, and metamorphism. This mountain range was eroded to a rather subdued landscape and was followed by deposition of approximately 4000 m of sediments and lava flows of the Grand Canyon Supergroup.

These rocks and the underlying Vishnu and Brahma schists were uplifted and formed a second Proterozoic mountain range. This mountain range was also eroded to a nearly flat surface by the end of the Proterozoic Eon.

The first sea of the Paleozoic Era transgressed over the region during the Cambrian Period, depositing sandstones, siltstones, and limestones. A major unconformity separates the Cambrian rocks from the Mississippian limestones exposed as the cliff-forming Redwall Limestone. Another unconformity separates the Redwall Limestone from the overlying Permian Kaibab Limestone, which forms the rim of the Grand Canyon.

The Grand Canyon in all its grandeur is a most appropriate place to start our discussion of the Paleozoic history of North America.

OBJECTIVES

At the end of this chapter, you will have learned that

■ There were six major continents present at the beginning of the Paleozoic Era and through plate movement became assembled into one supercontinent, Pangaea, by the end of the Paleozoic Era.

■ The Paleozoic history of North America can be subdivided into six cratonic sequences, which represent major transgressive-regressive cycles.

■ During the transgressive portions of each cycle, the North American craton was partially to completely covered by shallow seas in which a variety of detrital and carbonate sediments were deposited resulting in widespread sandstone, reef, and coal deposits.

■ Mountain-building activity took place primarily along the eastern and southern margins (known as mobile belts) of the North American craton during the Paleozoic Era.

■ In addition to the large-scale plate interactions, microplate activity also played an important role in the formation of Pangaea.

■ Paleozoic-age rocks contain a variety of important mineral resources, including metallic and nonmetallic mineral deposits.

Introduction

Having reviewed the geologic history of the Archean and Proterozoic eons, we now turn our attention to the Phanerozoic Eon, comprising the remaining 11.8% of geologic time. At the beginning of the Phanerozoic, six major continental landmasses existed, four of which straddled the paleoequator. Plate movements during the Phanerozoic created a changing panorama of continents and ocean basins whose positions affected atmospheric and oceanic circulation patterns and created new environments for habitation by the rapidly evolving biota.

The Paleozoic history of most continents involves major mountain-building activity along their margins and numerous shallow-water marine transgressions and regressions over their interiors. These transgressions and regressions were caused by global changes in sea level probably related to plate activity and glaciation.

The following chapters present the geologic history of North America in terms of those major transgressions and regressions rather than a period-by-period chronology. While we focus on North American geologic history, we also place those events in a global context.

Why Should You Study the Paleozoic History of Earth?

Why should you study Earth's Paleozoic geologic history? One reason is to see how Earth's various systems, subsystems, and geologic processes have interacted to lay the groundwork for the distribution of continental landmasses, ocean basins, and the topography we have today.

Furthermore, by studying and understanding what happened during the Paleozoic Era, we can use that information to make reasonable assumptions about what might happen in the future and what effect that might have on humans and society. For example, an examination of the causes and duration of past global climate changes may help focus the current debate about global warming and the role and impact humans are having.

Lastly, the concentration and distribution of natural resources is related to geological processes and plate tectonics. Much of the world's coal is the result of cyclical changes in sea level in which huge amounts of plant material was buried during the Pennsylvanian Period in North America and elsewhere and later converted to coal by mountain-building activity near the end of the Paleozoic Era. Thick deposits of evaporites formed during the Paleozoic are not only mined for those minerals but also serve as reservoirs or caprocks for large petroleum and gas deposits.

As we know, Earth is a dynamic planet that is continuously changing. An understanding of its geologic history—in this case, the Paleozoic Era—allows us to better appreciate the complex interactions that are occurring today and what they hold for the future.

Continental Architecture: Cratons and Mobile Belts

During the Precambrian, continental accretion and orogenic activity led to the formation of sizable continents. At least three large continents existed during the Late Proterozoic, and some geologists think that these landmasses later collided to form a single Pangaea-like supercontinent. This supercontinent began breaking apart sometime during the latest Proterozoic, and by the beginning of the Paleozoic Era, six major continents were present. Each continent can be divided into two major components: a craton and one or more mobile belts.

Cratons are the relatively stable and immobile parts of continents and form the foundation upon which Phanerozoic sediments were deposited (Figure 21.1). Cratons typically consist of two parts: a shield and a platform. *Shields* are the exposed portion of the crystalline basement rocks of a continent and are composed of Precambrian metamorphic and igneous rocks (see Figures 6.2 and 20.3) that reveal a history of extensive orogenic activity during the Precambrian. During the Phanerozoic, however, shields were extremely stable and formed the foundation of the continents.

Extending outward from the shields are buried Precambrian rocks that constitute a *platform*, another part of the craton. Overlying the platform are flat-lying or gently dipping Phanerozoic detrital and chemical sedimentary rocks that were deposited in widespread shallow seas that transgressed and regressed over the craton. These seas, called **epeiric seas,** were a common feature of most Paleozoic cratonic histories. Changes in sea level, caused primarily by continental glaciation as well as by plate movement, were responsible for the advance and retreat of the seas.

Whereas most of the Paleozoic platform rocks are still essentially flat lying, in some places they were gently folded into regional arches, domes, and basins (Figure 21.1). In many cases some of these structures stood out as low islands during the Paleozoic Era and supplied sediments to the surrounding epeiric seas.

Mobile belts are elongated areas of mountain-building activity. They are located along the margins of continents where sediments are deposited in the relatively shallow waters of the continental shelf and the deeper waters at the base of the continental slope. During plate convergence along these margins, the sediments are deformed and intruded by magma, creating mountain ranges.

Four mobile belts formed around the margin of the North American craton during the Paleozoic; these were

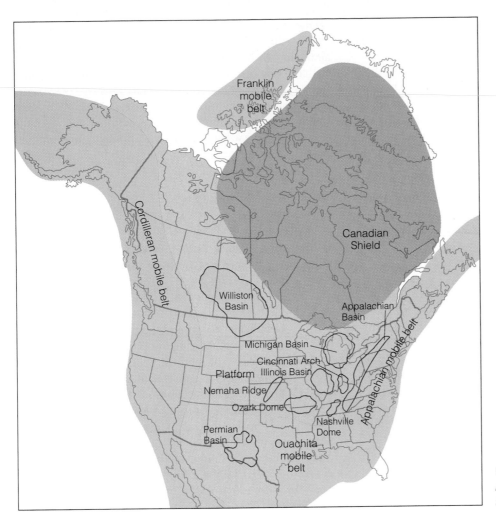

Figure 21.1 *The mobile belts and major cratonic structures of North America that formed during the Paleozoic Era.*

the **Franklin, Cordilleran, Ouachita,** and **Appalachian mobile belts** (Figure 21.1). Each was the site of mountain building in response to compressional forces along a convergent plate boundary and formed such mountain ranges as the Appalachians and Ouachitas.

Paleozoic Paleogeography

One of the results of plate tectonics is that Earth's geography is constantly changing. The present-day configuration of the continents and ocean basins is merely a snapshot in time. As the plates move about, the location of continents and ocean basins constantly changes. One of the goals of historical geology is to provide paleogeographic reconstructions of the world for the geologic past. By synthesizing all of the pertinent paleoclimatic, paleomagnetic, paleontologic, sedimentologic, stratigraphic, and tectonic data available, geologists can prepare paleogeographic maps of what the world looked like at a particular time in the geologic past.

The paleogeographic history of the Paleozoic Era, for example, is not as precisely known as the history of the Mesozoic and Cenozoic eras, in part because the magnetic anomaly patterns preserved in the oceanic crust were destroyed when much of the Paleozoic oceanic crust was subducted during the formation of Pangaea. Paleozoic paleogeographic reconstructions are therefore based primarily on structural relationships, climate-sensitive sediments such as red beds, evaporites, and coals, as well as the distribution of plants and animals.

Recall that six major continents were present at the beginning of the Paleozoic. Besides these large landmasses, geologists have also identified numerous small microcontinents and island arcs associated with various microplates that were present during the Paleozoic. We are primarily concerned, however, with the history of the six major continents and their relationship to each other. The six major Paleozoic continents are **Baltica** (Russia west of the Ural Mountains and the major part of northern Europe), **China** (a complex area consisting of at least three Paleozoic continents that were not widely separated and are here considered to include China, Indochina, and the Malay Peninsula), **Gondwana** (Africa, Antarctica, Australia, Florida, India, Madagascar, and parts of the Middle East and southern Europe), **Kazakhstania** (a triangular continent centered on Kazakhstan, but considered by some to be an extension of the Paleozoic Siberian continent), **Laurentia** (most of present North America, Greenland, northwestern Ireland, Scotland, and part of eastern Russia), and **Siberia** (Russia east of the Ural Mountains and Asia north of Kazakhstan and south of Mongolia). The paleogeographic reconstructions that follow (Figures 21.2, 21.3, and 21.4) are based on the methods used to determine and interpret the location, geographic features, and environmental conditions on the paleocontinents.

Early-Middle Paleozoic Global History

In contrast to today's global geography, the Cambrian world consisted of six major continents dispersed around the globe at low tropical latitudes (Figure 21.2a). Water circulated freely among ocean basins, and the polar regions were apparently ice-free. By the Late Cambrian, epeiric seas had covered large areas of Laurentia, Baltica, Siberia, Kazakhstania, and China, while major highlands were present in northeastern Gondwana, eastern Siberia, and central Kazakhstania.

During the Ordovician and Silurian periods, plate movement played a major role in the changing global geography (Figure 21.2b and c). Gondwana moved southward during the Ordovician and began to cross the South Pole as indicated by Upper Ordovician tillites found today in the Sahara Desert. In contrast to the passive continental margin Laurentia exhibited during the Cambrian, an active convergent plate boundary formed along its eastern margin during the Ordovician as indicated by the Late Ordovician *Taconic orogeny* that occurred in New England. During the Silurian, Baltica moved northwestward relative to Laurentia and collided with it to form the larger continent of **Laurasia.** This collision, which closed the northern Iapetus Ocean, is marked by the *Caledonian orogeny.* Following this orogeny, the southern part of the Iapetus Ocean still remained open between Laurentia and Gondwana (Figure 21.2c). Siberia and Kazakhstania moved from a southern equatorial position during the Cambrian to north temperate latitudes by the end of the Silurian Period.

During the Devonian, as the southern Iapetus Ocean narrowed between Laurasia and Gondwana, mountain building continued along the eastern margin of Laurasia with the *Acadian orogeny* (Figure 21.3a). The erosion of the resulting highlands provided vast amounts of reddish fluvial sediments that covered large areas of northern Europe (Old Red Sandstone) and eastern North America (the Catskill Delta). Other Devonian tectonic events, probably

related to the collision of Laurentia and Baltica, include the Cordilleran *Antler orogeny,* the *Ellesmere orogeny* along the northern margin of Laurentia (which may reflect the collision of Laurentia with Siberia), and the change from a passive continental margin to an active convergent plate boundary in the Uralian mobile belt of eastern Baltica. The distribution of reefs, evaporites, and red beds, as well as the existence of similar floras throughout the world, suggests a rather uniform global climate during the Devonian Period.

Late Paleozoic Global History

During the Carboniferous Period, southern Gondwana moved over the South Pole, resulting in extensive continental glaciation (Figures 21.3b and 21.4a). The advance and retreat of these glaciers produced global changes in sea level that affected sedimentation patterns on the cratons. As Gondwana continued moving northward, it first collided with Laurasia during the Early Carboniferous and continued suturing with it during the rest of the Carboniferous (Figures 21.3b and 21.4a). Because Gondwana rotated clockwise relative to Laurasia, deformation generally progressed in a northeast-to-southwest direction along the Hercynian, Appalachian, and Ouachita mobile belts of the two continents. The final phase of collision between Gondwana and Laurasia is indicated by the Ouachita Mountains of Oklahoma, which were formed by thrusting during the Late Carboniferous and Early Permian.

Elsewhere, Siberia collided with Kazakhstania and moved toward the Uralian margin of Laurasia (Baltica), colliding with it during the Early Permian. It has been suggested that the northwestern margin of China collided with the southwestern margin of Siberia during the Late Carboniferous. By the end of the Carboniferous, the various continental land-masses were fairly close together as Pangaea began taking shape.

The Carboniferous coal basins of eastern North America, western Europe, and the Donets Basin of Ukraine all lay in the equatorial zone, where rainfall was high and temperatures were consistently warm. The absence of strong seasonal growth rings in fossil plants from these coal basins is indicative of such a climate. The fossil plants found in the coals of Siberia and China, however, show well-developed growth rings, signifying seasonal growth with abundant rainfall and distinct seasons such as occur in the temperate zones (latitudes 40 degrees to 60 degrees north).

Glacial conditions and the movement of large continental ice sheets in the high southern latitudes are indicated by widespread tillites and glacial striations in southern Gondwana (see Figure 9.6). These ice sheets spread toward the equator and, at their maximum growth, extended well into the middle temperate latitudes.

The assembly of Pangaea was essentially concluded during the Permian with the completion of many of the

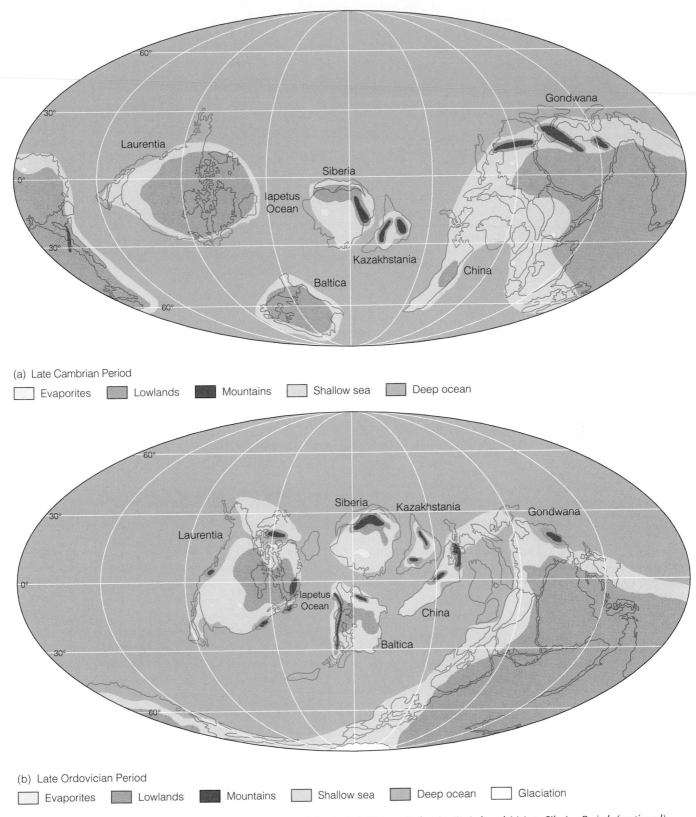

(a) Late Cambrian Period

Evaporites   Lowlands   Mountains   Shallow sea   Deep ocean

(b) Late Ordovician Period

Evaporites   Lowlands   Mountains   Shallow sea   Deep ocean   Glaciation

Figure 21.2 *Paleogeography of the world for the (a) Late Cambrian Period, (b) Late Ordovician Period, and (c) Late Silurian Period. (continued)*

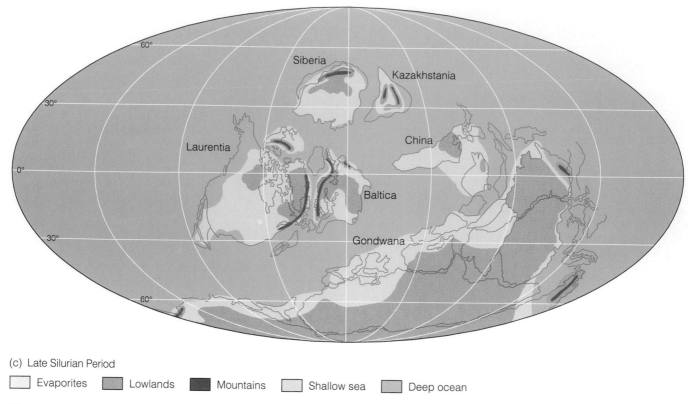

(c) Late Silurian Period

☐ Evaporites ▨ Lowlands ■ Mountains ☐ Shallow sea ▨ Deep ocean

Figure 21.2 *Paleogeography of the world for the (c) Late Silurian Period.*

continental collisions that began during the Carboniferous (Figure 21.4b). Although geologists generally agree on the configuration and location of the western half of the supercontinent, there is no consensus on the number or configuration of the various terranes and continental blocks that composed the eastern half of Pangaea. Regardless of the exact configuration of the eastern portion, geologists know that the supercontinent was surrounded by various subduction zones and moved steadily northward during the Permian. Furthermore, an enormous single ocean, **Panthalassa,** surrounded Pangaea and spanned Earth from pole to pole. Waters of this ocean probably circulated more freely than at present, resulting in more equable water temperatures.

The formation of a single large landmass had climatic consequences for the terrestrial environment as well. Terrestrial Permian sediments indicate that arid and semi-arid conditions were widespread over Pangaea. The mountain ranges produced by the *Hercynian, Alleghenian,* and *Ouachita orogenies* were high enough to create rainshadows that blocked the moist, subtropical, easterly winds—much as the southern Andes Mountains do in western South America today. This produced dry conditions in North America and Europe, as evident from the

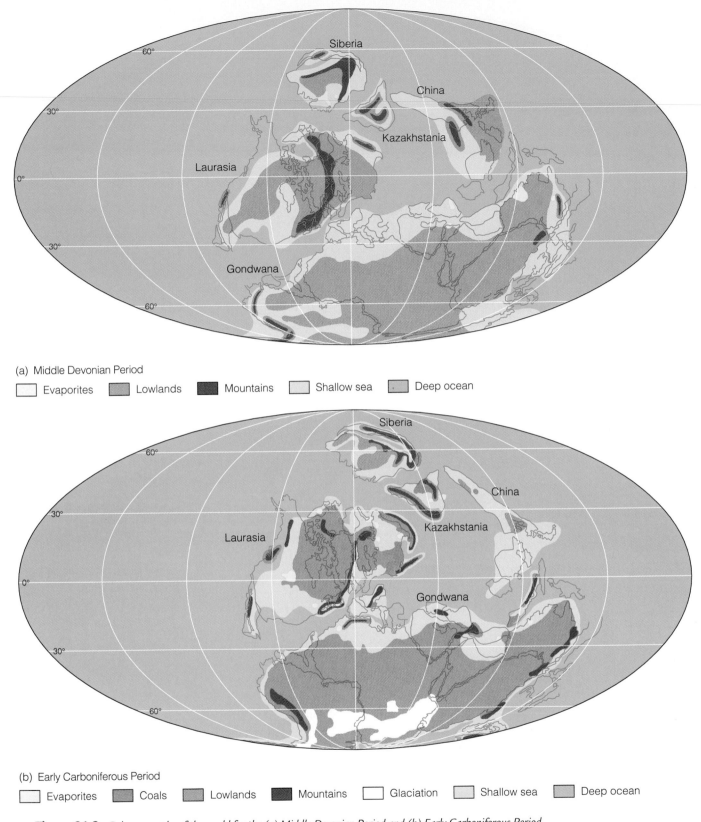

(a) Middle Devonian Period

☐ Evaporites ▨ Lowlands ■ Mountains ☐ Shallow sea ▨ Deep ocean

(b) Early Carboniferous Period

☐ Evaporites ▨ Coals ▨ Lowlands ■ Mountains ☐ Glaciation ☐ Shallow sea ▨ Deep ocean

Figure 21.3 *Paleogeography of the world for the (a) Middle Devonian Period and (b) Early Carboniferous Period.*

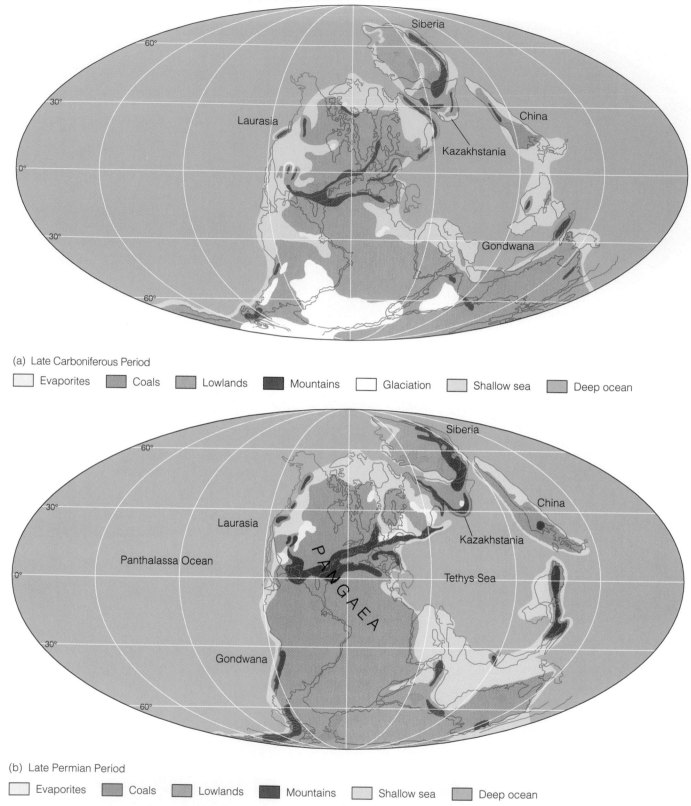

(a) Late Carboniferous Period

☐ Evaporites ▨ Coals ▨ Lowlands ■ Mountains ☐ Glaciation ☐ Shallow sea ▨ Deep ocean

(b) Late Permian Period

☐ Evaporites ▨ Coals ▨ Lowlands ■ Mountains ☐ Shallow sea ▨ Deep ocean

Figure 21.4 *Paleogeography of the world for the (a) Late Carboniferous Period and (b) Late Permian Period.*

extensive Permian evaporites found in western North America, central Europe, and parts of Russia. Permian coals, indicative of abundant rainfall, were mostly limited to the northern temperate belts (latitude 40 degrees to 60 degrees north), while the last remnants of the Carboniferous ice sheets retreated to the mountainous regions of eastern Australia.

Paleozoic Evolution of North America

It is convenient to divide the history of the North American craton into two parts: the first dealing with the relatively stable continental interior over which epeiric seas transgressed and regressed, and the other with the mobile belts where mountain building occurred.

In 1963 the American geologist Laurence L. Sloss proposed that the sedimentary-rock record of North America could be subdivided into six cratonic sequences. A **cratonic sequence** is a large-scale (greater than supergroup) lithostratigraphic unit representing a major transgressive-regressive cycle bounded by cratonwide unconformities (Figure 21.5). The transgressive phase, which is usually covered by younger sediments, commonly is well preserved, whereas the regressive phase of each sequence is marked by an unconformity. Where rocks of the appropriate age are preserved, each of the six unconformities can be shown to extend across the various sedimentary basins of the North American craton and into the mobile belts along the cratonic margin.

Geologists have also recognized major unconformity bounded sequences in cratonic areas outside North America. Such global transgressive and regressive cycles are caused by sea level changes and are thought to result from major tectonic and glacial events.

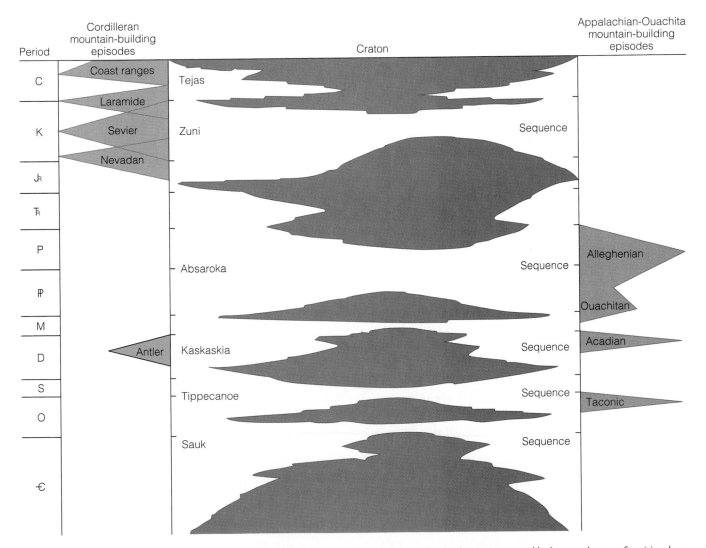

Figure 21.5 *Cratonic sequences of North America. The white areas represent sequences of rocks that are separated by large-scale unconformities shown as brown areas. The major Cordilleran orogenies are shown on the left side of the figure, and the major Appalachian orogenies are shown on the right side.*

The realization that rock units can be divided into cratonic sequences and that these sequences can be further subdivided and correlated provides the foundation for an important concept in geology that allows high-resolution analysis of time and facies relationships within sedimentary rocks. **Sequence stratigraphy** is the study of rock relationships within a time-stratigraphic framework of related facies bounded by erosional or nondepositional surfaces. The basic unit of sequence stratigraphy is the *sequence*, which is a succession of rocks bounded by unconformities and their equivalent conformable strata. Sequence boundaries form as a result of a relative drop in sea level. Sequence stratigraphy is an important tool in geology because it allows geologists to subdivide sedimentary rocks into related units that are bounded by time-stratigraphically significant boundaries. Geologists use sequence stratigraphy for high-resolution correlation and mapping as well as interpreting and predicting depositional environments.

The Sauk Sequence

Rocks of the **Sauk sequence** (Late Proterozoic–Early Ordovician) record the first major transgression onto the North American craton (Figure 21.5). During the Late Proterozoic and Early Cambrian, deposition of marine sediments was limited to the passive shelf areas of the Appalachian and Cordilleran borders of the craton. The craton itself was above sea level and experiencing extensive weathering and erosion. Because North America was located in a tropical climate at this time and there is no evidence of any terrestrial vegetation, weathering and erosion of the exposed Precambrian basement rocks must have proceeded rapidly. During the Middle Cambrian, the transgressive phase of the Sauk began with epeiric seas encroaching over the craton (see Perspective 21.1). By the Late Cambrian, the Sauk Sea had covered most of North America, leaving only a portion of the Canadian Shield and a few large islands above sea level (Figure 21.6). These islands, collectively named the **Transcontinental Arch,** extended from New Mexico to Minnesota and the Lake Superior region.

The sediments deposited on both the craton and along the shelf area of the craton margin show abundant evidence of shallow-water deposition. The only difference between the shelf and craton deposits is that the shelf deposits are thicker. In both areas, the sands are generally clean and well sorted, and commonly contain ripple marks and small-scale cross-bedding. Many of the carbonates are bioclastic (composed of fragments of organic remains), contain stromatolites, or have oolitic (small, spherical calcium carbonate grains) textures. Such sedimentary structures and textures indicate shallow-water deposition.

The Tippecanoe Sequence

As the Sauk Sea regressed from the craton during the Early Ordovician, it revealed a landscape of low relief. The rocks exposed were predominantly limestones and dolostones that experienced deep and extensive erosion because North America was still located in a tropical environment (Figure 21.7). The resulting cratonwide unconformity marks the boundary between the Sauk and Tippecanoe sequences.

Like the Sauk sequence, deposition of the **Tippecanoe sequence** (Middle Ordovician–Early Devonian) began with a major transgression onto the craton. This transgressing sea deposited clean, well-sorted quartz sands over most of the craton. The best known of the Tippecanoe basal sandstones is the St. Peter Sandstone, an almost pure quartz sandstone used in manufacturing glass. It occurs throughout much of the midcontinent and resulted from numerous cycles of weathering and erosion of Proterozoic and Cambrian sandstones deposited during the Sauk transgression (Figure 21.8).

The Tippecanoe basal sandstones were followed by widespread carbonate deposition (Figure 21.7). The limestones were generally the result of deposition by calcium carbonate–secreting organisms such as corals, brachiopods, stromatoporoids, and bryozoans. Besides the limestones, there were also many dolostones. Most of the dolostones formed as a result of magnesium replacing calcium in calcite, thus converting the limestones into dolostones.

In the eastern portion of the craton, the carbonates grade laterally into shales. These shales mark the farthest extent of detrital sediments derived from weathering and erosion of the Taconic highlands, a tectonic event we will discuss later.

Tippecanoe Reefs and Evaporites

Organic reefs are limestone structures constructed by living organisms, some of which contribute skeletal materials to the reef framework (Figure 21.9). Today, corals and calcareous algae are the most prominent reef builders, but in the geologic past other organisms played a major role. Regardless of the organisms dominating reef communities, reefs appear to have occupied the same ecological niche in the geologic past that they do today. Because of the ecological requirements of reef-building organisms, reefs today are confined to a narrow latitudinal belt between 30 degrees north and south of the equator. Corals, the major reef-building organisms today, require warm, clear, shallow water of normal salinity for optimal growth.

The size and shape of a reef are largely the result of the interaction between the reef-building organisms, the

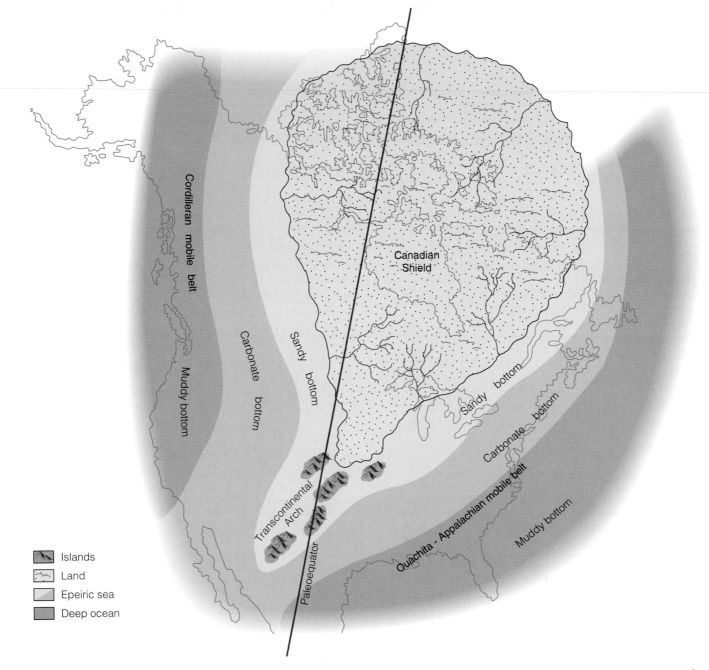

Figure 21.6 *Paleogeography of North America during the Cambrian Period. Note the position of the Cambrian paleoequator. During this time North America straddled the equator as indicated in Figure 21.2a.*

bottom topography, wind and wave action, and subsidence of the seafloor. Reefs also alter the area around them by forming barriers to water circulation or wave action.

Reefs have been common features since the Cambrian and have been built by a variety of organisms. The first skeletal builders of reeflike structures were archaeocyathids. These conical-shaped organisms lived during the Cambrian and had double, perforated, calcareous shell walls. Archaeocyathids built small mounds that have been

found on all continents except South America (see Figure 22.5). Beginning in the Middle Ordovician, stromatoporoid-coral reefs became common in the low latitudes, and similar reefs remained so throughout the rest of the Phanerozoic Eon. The burst of reef building seen in the Late Ordovician through Devonian probably occurred in response to evolutionary changes triggered by the appearance of extensive carbonate seafloors and platforms beyond the influence of detrital sediments.

Pictured Rocks National Lakeshore

Exposed along the south shore of Lake Superior between Au Sable Point and Munising in Michigan's Upper Peninsula is the beautiful and imposing wavecut sandstone called Pictured Rocks cliffs (Figure 1). The rocks exposed in this area, part of which is designated a national lakeshore, comprise the Upper Cambrian Munising Formation, which is divided into two members: the lower Chapel Rock Sandstone and the upper Miner's Castle Sandstone (Figure 1). The Munising Formation unconformably overlies the Upper Proterozoic Jacobsville Sandstone and is unconformably overlain by the Middle Ordovician Au Train Formation. The reddish brown, coarse-grained Jacobsville Sandstone was deposited in streams and lakes over an irregular erosion surface (Figure 1). Following deposition, the Jacobsville was slightly uplifted and tilted.

By the Late Cambrian, the transgressing Sauk Sea reached the Michigan area, and the Chapel Rock Sandstone was deposited. The principal source area for this unit was the Northern Michigan highlands, an area that corresponds to the present Upper Peninsula. Following deposition of the Chapel Rock Sandstone, the Sauk Sea retreated from the area.

During a second transgression of the Sauk Sea in this area, the Miner's Castle Sandstone was deposited. This second transgression covered most of the Upper Peninsula of Michigan and drowned the highlands that were the source for the older Chapel Rock Sandstone.

The source area for the Miner's Castle Sandstone was the Precambrian Canadian Shield area to the north and northeast. The Miner's Castle Sandstone contains rounder, better sorted, and more abundant quartz grains than the Chapel Rock Sandstone, indicating a different source area. A major unconformity separates the Miner's Castle Sandstone from the overlying Middle Ordovician Au Train Formation.

One of the most prominent features of Pictured Rocks National Lakeshore is Miner's Castle, a wavecut projection along the shoreline (Figure 1). The lower sandstone unit at water level is the Chapel Rock Sandstone, while the rest of the feature is composed of the Miner's Castle Sandstone. The two turrets of the castle formed as sea stacks during a time following the Pleistocene when the water level of Lake Superior was much higher.

Figure 1 *Location of Pictured Rocks National Lakeshore and stratigraphy of rocks exposed in this area. The photograph of the Munising Formation shows the projection of the shoreline called Miner's Castle. The two turrets at the top of Miner's Castle formed by wave action when the water level of Lake Superior was higher. The contact between the Miner's Castle and Chapel Rock sandstone members is located just above lake level.*

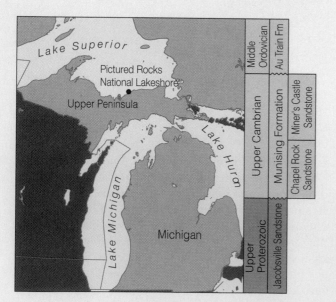

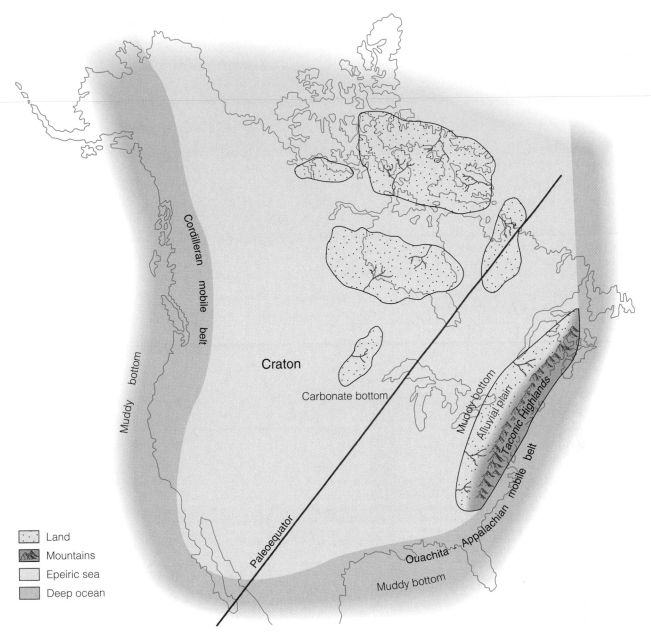

Figure 21.7 *Paleogeography of North America during the Ordovician Period. Note that the position of the equator has changed, indicating North America was rotating in a counterclockwise direction.*

The Middle Silurian rocks (Tippecanoe sequence) of the present-day Great Lakes region are world famous for their reef and evaporite deposits and have been extensively studied (Figure 21.10). The most famous structure in the region, the Michigan Basin, is a broad, circular basin surrounded by large barrier reefs. No doubt these reefs contributed to increasingly restricted circulation and the precipitation of Upper Silurian evaporites within the basin (Figure 21.11).

Within the rapidly subsiding interior of the basin, other types of reefs are found. *Pinnacle reefs* are tall, spindly structures up to 100 m high. They reflect the rapid upward growth needed to maintain themselves near sea level dur-

ing subsidence of the basin (Figure 21.11). Besides the pinnacle reefs, bedded carbonates and thick sequences of salt and anhydrite are also found in the Michigan Basin.

As the Tippecanoe Sea gradually regressed from the craton during the Late Silurian, precipitation of evaporite minerals occurred in the Appalachian, Ohio, and Michigan basins. In the Michigan Basin alone, approximately 1500 m of sediments were deposited, nearly half of which are halite and anhydrite. How did such thick sequences of evaporites accumulate? One possibility is that when sea level dropped the tops of the barrier reefs were as high as or above sea level, thus preventing the influx of new seawater into the

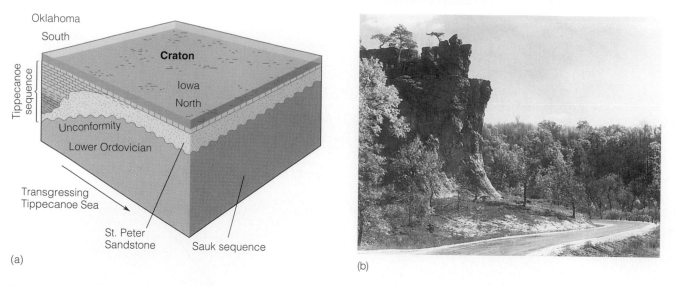

Figure 21.8 *(a) The transgression of the Tippecanoe Sea resulted in the deposition of the St. Peter Sandstone (Middle Ordovician) over a large area of the craton. (b) Wisconsin Department of Natural Resources.*

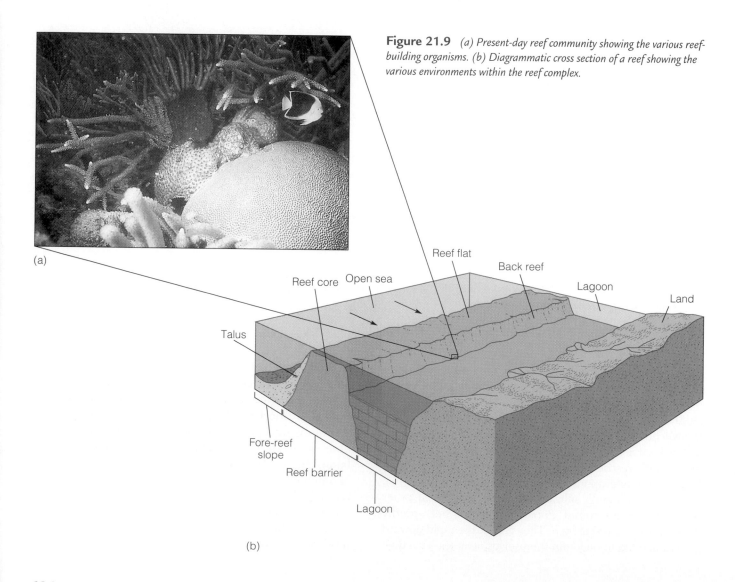

Figure 21.9 *(a) Present-day reef community showing the various reef-building organisms. (b) Diagrammatic cross section of a reef showing the various environments within the reef complex.*

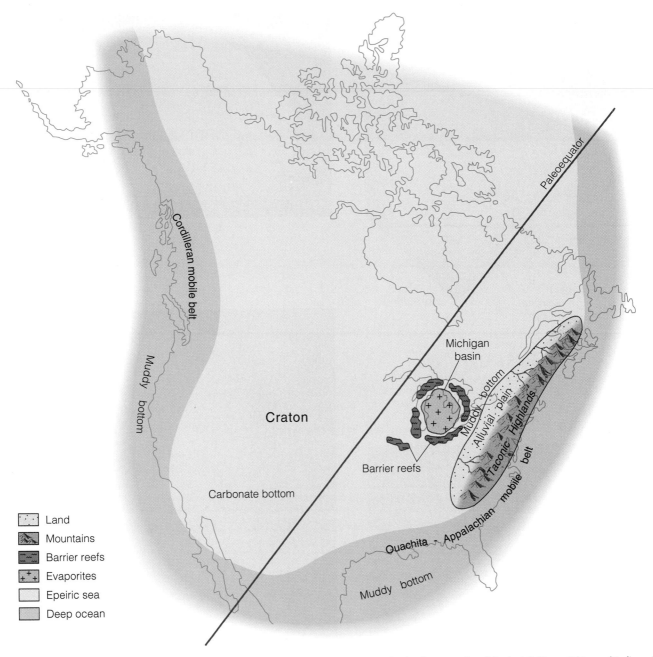

Figure 21.10 *Paleogeography of North America during the Silurian Period. Note the development of reefs in the Michigan, Ohio, and Indiana-Illinois-Kentucky areas.*

basin. Evaporation of the basinal seawater would result in the precipitation of salts. A second possibility is that the reefs grew upward so close to sea level that they formed a sill or barrier that eliminated interior circulation (Figure 21.12).

The End of the Tippecanoe Sequence

By the Early Devonian, the regressing Tippecanoe Sea had retreated to the craton margin exposing an extensive low-land topography. During this regression, marine deposition was initially restricted to a few interconnected cratonic basins and, finally by the end of the Tippecanoe, to only the mobile belts surrounding the craton.

As the Tippecanoe Sea regressed during the Early Devonian, the craton experienced mild deformation resulting in the formation of many domes, arches, and basins. These structures were mostly eroded during the time the craton was exposed so that they were eventually covered by deposits from the encroaching Kaskaskia Sea.

(a)

Meters
0
100

Laminar
stromatoporoid
Barrier
reef   Anhydrite   Halite

Evaporite
Carbonate

Pinnacle
reef

Stromatoporoid
Barrier reef

Stromatolites
Algal
Coral algal
Crinoidal
Laminar
stromatoporoid

Niagara Fm.

Clinton Fm.

(d)

(b)

(c)

Figure 21.11 *(a) Generalized cross section of the northern Michigan Basin during the Silurian Period. (b) Stromatoporoid barrier-reef facies. (c) Evaporite facies. (d) Carbonate facies.*

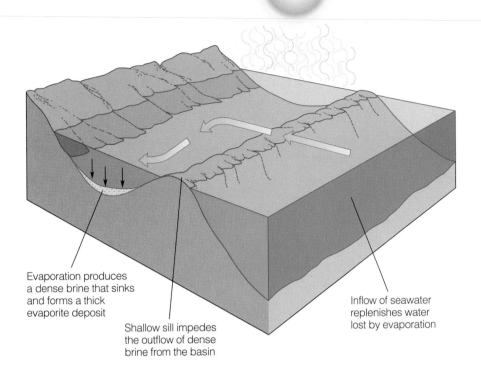

Evaporation produces a dense brine that sinks and forms a thick evaporite deposit

Shallow sill impedes the outflow of dense brine from the basin

Inflow of seawater replenishes water lost by evaporation

Figure 21.12 *Silled basin model for evaporite sedimentation by direct precipitation from seawater. Vertical scale is greatly exaggerated.*

The Kaskaskia Sequence

The boundary between the Tippecanoe sequence and the overlying **Kaskaskia sequence** (Middle Devonian–Middle Mississippian) is marked by a major unconformity. As the Kaskaskia Sea transgressed over the low relief landscape of the craton, the majority of the basal beds deposited consisted of clean, well-sorted, quartz sandstones.

The source areas for the basal Kaskaskia sandstones were primarily the eroding highlands of the Appalachian mobile belt area (Figure 21.13), exhumed Cambrian and Ordovician sandstones cropping out along the flanks of the Ozark Dome, and exposures of the Canadian Shield in the Wisconsin area. The lack of similar sands in the Silurian carbonate beds below the Tippecanoe-Kaskaskia unconformity indicates that the source areas of the basal Kaskaskia detrital rocks were submerged when the Tippecanoe sequence was deposited. Stratigraphic studies show that these source areas were uplifted and the Tippecanoe carbonates removed by erosion prior to the Kaskaskia transgression. Kaskaskian basal rocks elsewhere on the craton consist of carbonates that are frequently difficult to differentiate from the underlying Tippecanoe carbonates unless they are fossiliferous.

Except for widespread Upper Devonian and Lower Mississippian black shales, the majority of Kaskaskian rocks are carbonates, including reefs, and associated evaporite deposits. In many other parts of the world, such as southern England, Belgium, central Europe, Australia, and Russia, the Middle and early Late Devonian epochs were times of major reef building.

Reef Development in Western Canada

The Middle and Late Devonian reefs of western Canada contain large reserves of petroleum and have been widely studied from outcrops and in the subsurface (Figure 21.14). These reefs began forming as the Kaskaskia Sea transgressed southward into western Canada. By the end of the Middle Devonian, they had coalesced into a large barrier-reef system that restricted the flow of oceanic water into the back-reef platform, thus creating conditions for evaporite precipitation. In the back reef area, up to 300 m of evaporites were precipitated in much the same way as in the Michigan Basin during the Silurian (Figure 21.11). More than half of the world's potash, which is used in fertilizers, comes from these Devonian evaporites. By the middle of the Late Devonian, reef growth stopped in the western Canada region, although nonreef carbonate deposition continued.

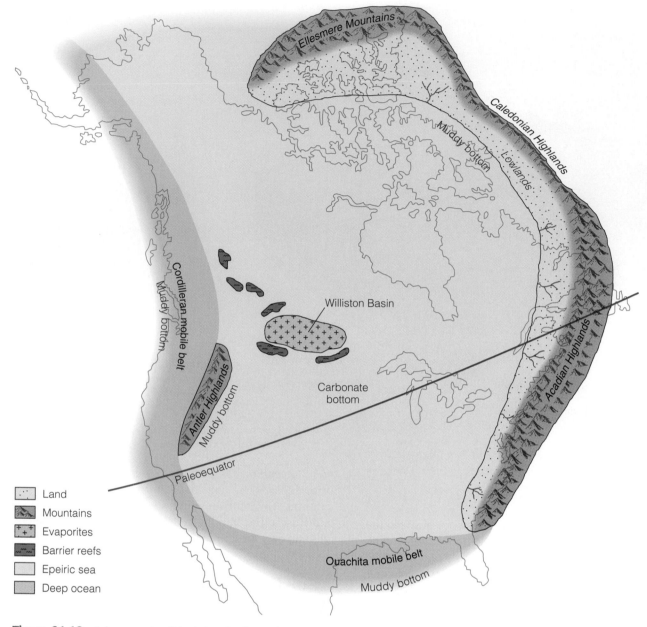

Figure 21.13 *Paleogeography of North America during the Devonian Period.*

Black Shales

In North America, many areas of carbonate-evaporite deposition gave way to a greater proportion of shales and coarser detrital rocks beginning in the Middle Devonian and continuing into the Late Devonian. This change to detrital deposition resulted from the formation of new source areas brought on by the mountain-building activity associated with the Acadian orogeny in North America (Figure 21.13).

As the Devonian Period ended, a conspicuous change in sedimentation occurred over the craton with the appearance of widespread black shales. In the eastern United

States, these black shales are commonly called the Chattanooga Shale, but are known by a variety of local names elsewhere (for example, New Albany Shale and Antrim Shale). Although these black shales are best developed from the cratonic margins along the Appalachian mobile belt to the Mississippi Valley, correlative units can also be found in many western states and in western Canada (Figure 21.15).

The Upper Devonian–Lower Mississippian black shales of North America are typically noncalcareous, thinly bedded, and usually less than 10 m thick. Fossils are usually rare, but some Upper Devonian black shales do contain rich conodont faunas. Because most black shales lack

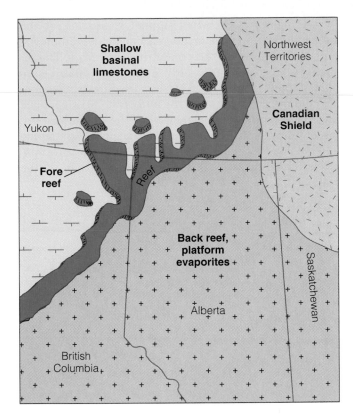

Figure 21.14 *Reconstruction of the extensive Devonian Reef complex of western Canada. These extensive reefs controlled the regional facies of the Devonian epeiric seas.*

body fossils, they are difficult to date and to correlate. In places where they can be dated, usually by conodonts (microscopic animals), acritarchs (microscopic algae), or plant spores, the lower beds are Late Devonian, and the upper beds are Early Mississippian in age.

Although the origin of these extensive black shales is still being debated, the essential features required to produce them include undisturbed anaerobic bottom water, a reduced supply of coarser detrital sediment, and high organic productivity in the overlying oxygenated waters. High productivity in the surface waters leads to a shower of organic material, which decomposes on the undisturbed seafloor and depletes the dissolved oxygen at the sediment-water interface.

The wide extent of such apparently shallow-water black shales in North America remains puzzling. Nonetheless, these shales are rich in uranium and are an important source rock of oil and gas in the Appalachian region.

The Late Kaskaskia—A Return to Extensive Carbonate Deposition

Following deposition of the widespread Upper Devonian–Lower Mississippian black shales, carbonate sedimentation on the craton dominated the remainder of the Mississippian Period (Figure 21.16). During this time, a variety of carbonate sediments were deposited in the epeiric sea as indicated

(a) (b)

Figure 21.15 *(a) The extent of the Upper Devonian to Lower Mississippian Chattanooga Shale and its equivalent units in North America. (b) Upper Devonian New Albany Shale, Button Mold Knob Quarry, Kentucky.*

by the extensive deposits of crinoidal limestones (rich in crinoid fragments), oolitic limestones, and various other limestones and dolostones (Figure 21.17). These Mississippian carbonates display cross-bedding, ripple marks, and well-sorted fossil fragments, all of which are indicative of a shallow-water environment. Analogous features can be observed on the present-day Bahama Banks. In addition, numerous small organic reefs occurred throughout the craton during the Mississippian. These were all much smaller than the large barrier-reef complexes that dominated the earlier Paleozoic seas.

During the Late Mississippian regression of the Kaskaskia Sea from the craton, carbonate deposition was replaced by vast quantities of detrital sediments. The resulting sandstones, particularly in the Illinois Basin, have been studied in great detail because they are excellent petroleum reservoirs. Prior to the end of the Mississippian, the Kaskaskia Sea had retreated to the craton margin, once again exposing the craton to widespread weathering and erosion that resulted in a cratonwide unconformity when the Absaroka Sea began transgressing back over the craton.

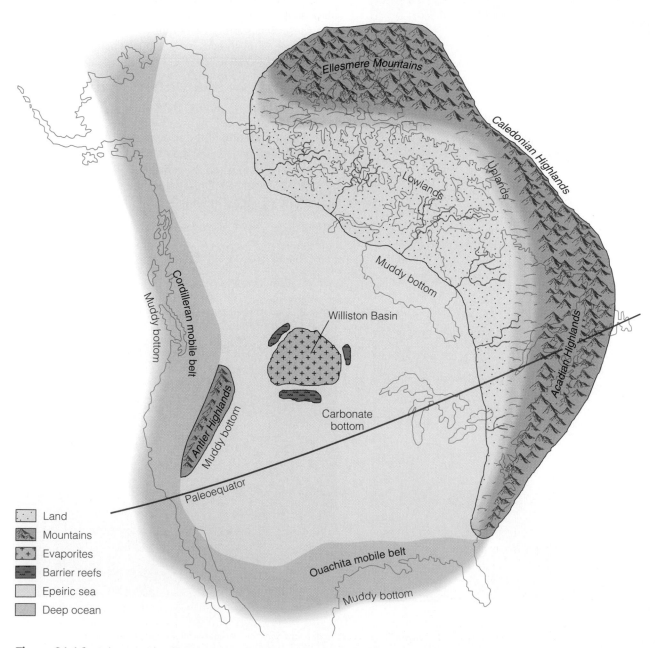

Figure 21.16 *Paleogeography of North America during the Mississippian Period.*

Figure 21.17 *Mississippian limestones exposed near Bowling Green, Kentucky.*

The Absaroka Sequence

The **Absaroka sequence** includes rocks deposited during the latest Mississippian through Early Jurassic. In this chapter, however, we will be concerned only with the Paleozoic rocks of the Absaroka sequence. The extensive unconformity separating the Kaskaskia and Absaroka sequences essentially divides the strata into the North American Mississippian and Pennsylvanian systems. These two systems are equivalent to the European Lower and Upper Carboniferous systems, respectively. The rocks of the Absaroka sequence are not only different from those of the Kaskaskia sequence, but they are also the result of quite different tectonic regimes.

The lowermost sediments of the Absaroka sequence are confined to the margins of the craton. These deposits are generally thickest in the east and southeast, near the emerging highlands of the Appalachian and Ouachita mobile belts, and thin westward onto the craton. The lithologies also reveal lateral changes from nonmarine detrital rocks and coals in the east, through transitional marine-nonmarine beds, to largely marine detrital rocks and limestones farther west (Figure 21.18).

What Are Cyclothems and Why Are They Important?

One of the characteristic features of Pennsylvanian rocks is their cyclical pattern of alternating marine and nonmarine strata. Such rhythmically repetitive sedimentary sequences are known as **cyclothems.** They result from repeated alternations of marine and nonmarine environments, usually in areas of low relief. Though seemingly simple, cyclothems reflect a delicate interplay between nonmarine deltaic and shallow-marine interdeltaic and shelf environments.

For purposes of illustration, we can look at a typical coal-bearing cyclothem from the Illinois Basin (Figure 21.19). Such a cyclothem contains nonmarine units, capped by a coal and overlain by marine units. Figure 21.19 shows the depositional environments that produced the cyclothem. The initial units represent deltaic and fluvial deposits. Above them is an underclay that frequently contains root casts from the plants and trees that comprise the overlying coal. The coal bed results from accumulations of plant material and is overlain by marine units of alternating limestones and shales, usually with an abundant marine invertebrate fauna. The marine cycle ends with an erosion surface. A new cyclothem begins with a nonmarine deltaic sandstone. All the beds illustrated in the idealized cyclothem are not always preserved because of abrupt changes from marine to nonmarine conditions or removal of some units by erosion.

Cyclothems represent transgressive and regressive sequences with an erosional surface separating one cyclothem from another. Thus, an idealized cyclothem passes upward from fluvial-deltaic deposits, through coals, to detrital shallow-water marine sediments, and finally to limestones typical of an open marine environment.

Such regularity and cyclicity in sedimentation over a large area requires an explanation. The hypothesis currently favored by most geologists is a rise and fall of sea level related to advances and retreats of Gondwanan continental glaciers. When the Gondwanan ice sheets advanced, sea level dropped, and when they melted, sea level rose. Late Paleozoic cyclothem activity on all the cratons closely corresponds to Gondwanan glacial-interglacial cycles.

Cratonic Uplift—The Ancestral Rockies

Recall that cratons are stable areas, and when they do experience deformation, it is usually mild. The Pennsylvanian Period, however, was a time of unusually severe cratonic deformation, resulting in uplifts of sufficient magnitude to expose Precambrian basement rocks. In addition to newly formed highlands and basins, many previously formed arches and domes, such as the Cincinnati Arch, Nashville Dome, and Ozark Dome, were also reactivated (Figure 21.1).

During the Pennsylvanian, the area of greatest deformation occurred in the southwestern part of the North American craton where a series of fault-bounded uplifted blocks formed the **Ancestral Rockies** (Figure 21.20a). These mountain ranges had diverse geologic histories and were not all elevated at the same time. Uplift of these mountains, some of which were elevated more than 2 km along near-vertical faults, resulted in the erosion of the overlying Paleozoic sediments and exposure of the Precambrian igneous and metamorphic basement rocks (Figure 21.20b). As the mountains eroded, tremendous

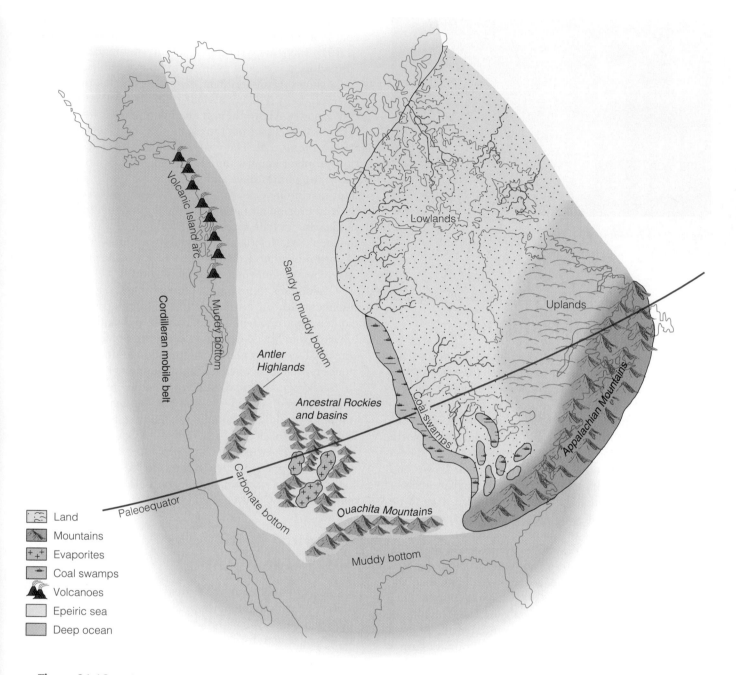

Figure 21.18 *Paleogeography of North America during the Pennsylvanian Period.*

quantities of coarse, red arkosic sand and conglomerate were deposited in the surrounding basins. These sediments are preserved in many areas, including the rocks of the Garden of the Gods near Colorado Springs (Figure 21.20c) and at the Red Rocks Amphitheatre near Morrison, Colorado.

Intracratonic mountain ranges are unusual, and their cause has long been debated. It is now thought that the collision of Gondwana with Laurasia (Figure 21.4a) produced great stresses in the southwestern region of the North American craton. These crustal stresses were re-

lieved by faulting that resulted in uplift of cratonic blocks and downwarp of adjacent basins, forming a series of ranges and basins.

The Late Absaroka—More Evaporite Deposits and Reefs

While the various intracratonic basins were filling with sediment during the Late Pennsylvanian, the Absaroka Sea slowly began retreating from the craton. During the Early

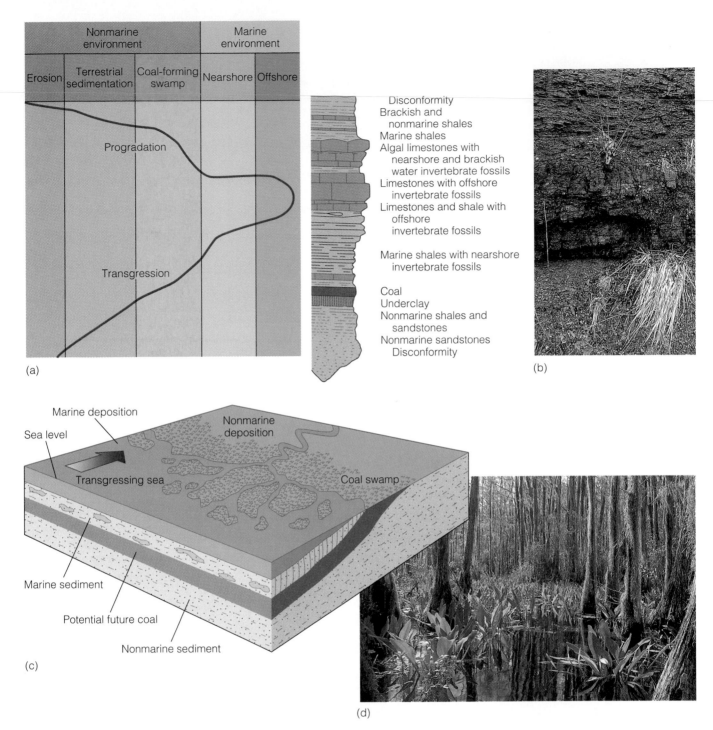

Figure 21.19 *(a) Columnar section of a complete cyclothem. (b) Pennsylvanian coal bed, West Virginia. (c) Reconstruction of the environment of a Pennsylvanian coal-forming swamp. (d) The Okefenokee Swamp, Georgia, is a modern example of a coal-forming environment, similar to those occurring during the Pennsylvanian Period.*

Permian, the Absaroka Sea occupied a narrow region from Nebraska through West Texas (Figure 21.21). By the Middle Permian, the sea had retreated to West Texas and southern New Mexico. The thick evaporite deposits in Kansas and Oklahoma provide evidence of the restricted nature of the Absaroka Sea during the Early and Middle

Permian and its southwestward retreat from the central craton.

During the Middle and Late Permian, the Absaroka Sea was restricted to West Texas and southern New Mexico, forming an interrelated complex of lagoonal, reef, and open-shelf environments (Figure 21.22). Three basins

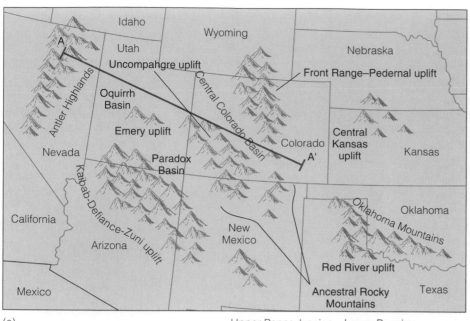

(a)

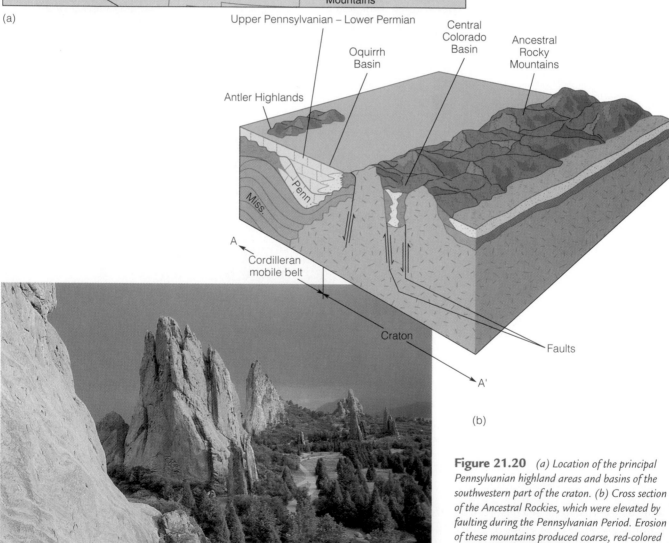

(b)

(c)

Figure 21.20 (a) Location of the principal Pennsylvanian highland areas and basins of the southwestern part of the craton. (b) Cross section of the Ancestral Rockies, which were elevated by faulting during the Pennsylvanian Period. Erosion of these mountains produced coarse, red-colored sediments that were deposited in the adjacent basins. (c) Garden of the Gods, storm sky view from Near Hidden Inn, Colorado Springs, Colorado.

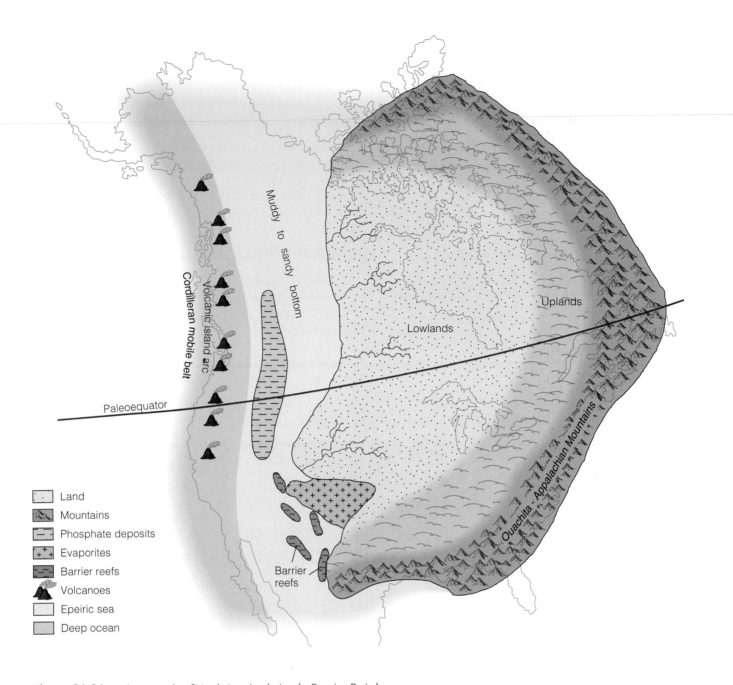

Figure 21.21 *Paleogeography of North America during the Permian Period.*

Land

Mountains

Phosphate deposits

Evaporites

Barrier reefs

Volcanoes

Epeiric sea

Deep ocean

separated by two submerged platforms formed in this area during the Permian. Massive reefs grew around the basin margins (Figure 21.23), while limestones, evaporites, and red beds were deposited in the lagoonal areas behind the reefs. As the barrier reefs grew and the passageways between the basins became more restricted, Late Permian evaporites gradually filled the individual basins.

Spectacular deposits representing the geologic history of this region can be seen today in the Guadalupe

Mountains of Texas and New Mexico where the Capitan Limestone forms the caprock of these mountains (Figure 21.24). These reefs have been extensively studied because of the tremendous oil production that comes from this region.

By the end of the Permian Period, the Absaroka Sea had retreated from the craton exposing continental red beds that had been deposited over most of the southwestern and eastern region.

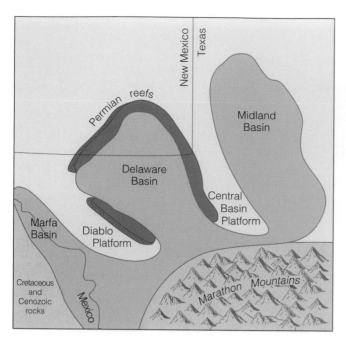

Figure 21.22 *Location of the West Texas Permian basins and surrounding reefs.*

History of the Paleozoic Mobile Belts

Having examined the Paleozoic history of the craton, we now turn our attention to the orogenic activity in the mobile belts. The mountain building that occurred during this time had a profound influence on the climate and sedimentary history of the craton. In addition, it was part of the global tectonic regime that sutured the continents together, forming Pangaea by the end of the Paleozoic Era.

Appalachian Mobile Belt

Throughout Sauk time (Late Proterozoic–Early Ordovician), the Appalachian region was a broad, passive, continental margin. Sedimentation was closely balanced by subsidence as thick, shallow marine sands were succeeded by extensive carbonate deposits. During this time, the **Iapetus Ocean** was widening as a result of movement along a divergent plate boundary (Figure 21.25a).

Figure 21.23 *A reconstruction of the Middle Permian Capitan Limestone reef environment. Shown are brachiopods, corals, bryozoans, and large glass sponges.*

Figure 21.24 *The prominent Capitan Limestone forms the caprock of the Guadalupe Mountains. The Capitan Limestone is rich in fossil corals and associated reef organisms.*

Taconic Orogeny

Beginning with the subduction of the Iapetus plate beneath Laurentia (an oceanic-continental convergent plate boundary), the Appalachian mobile belt was born (Figure 21.25b). The resulting **Taconic orogeny,** named after the present-day Taconic Mountains of eastern New York, central Massachusetts, and Vermont, was the first of several orogenies to affect the Appalachian region.

The Appalachian mobile belt can be divided into two depositional environments. The first is the extensive, shallow-water carbonate platform that formed the broad eastern continental shelf and stretched from Newfoundland to Alabama (Figure 21.25a). It formed during the Sauk Sea transgression onto the craton when carbonates were deposited in a vast, shallow sea.

Carbonate deposition ceased along the east coast during the Middle Ordovician and was replaced by deep-water deposits characterized by thinly bedded black shales, graded beds, coarse sandstones, graywackes, and associated volcanics. This suite of sediments marks the onset of mountain building, in this case, the Taconic orogeny. The subduction of the Iapetus plate beneath Laurentia resulted in volcanism and downwarping of the car-

bonate platform, forming an area where sediments accumulated (Figure 21.25b).

The final piece of evidence for the Taconic orogeny is the development of a large **clastic wedge,** an extensive accumulation of mostly detrital sediments deposited adjacent to an uplifted area. These deposits are thickest and coarsest nearest the highland area and become thinner and finer grained away from the source area, eventually grading into the carbonate cratonic facies (Figure 21.26). The clastic wedge resulting from the erosion of the Taconic Highlands is referred to as the **Queenston Delta.** Careful mapping and correlation of these deposits indicate that more than 600,000 km^3 of rock were eroded from the Taconic Highlands. Based on this figure, geologists estimate the Taconic Highlands were at least 4000 m high.

Caledonian Orogeny

The Caledonian mobile belt extends along the western border of Baltica and includes the present-day countries of Scotland, Ireland, and Norway (Figure 21.2c). During the Middle Ordovician, subduction along the boundary between the Iapetus plate and Baltica (Europe) began, forming a mirror image of the convergent plate boundary off the east coast of Laurentia (North America).

The culmination of the **Caledonian orogeny** occurred during the Late Silurian and Early Devonian with the formation of a mountain range along the margin of Baltica. Red-colored sediments deposited along the front of the Caledonian highlands formed a large clastic wedge known as the *Old Red Sandstone.*

Acadian Orogeny

The third Paleozoic orogeny to affect Laurentia and Baltica began during the Late Silurian and concluded at the end of the Devonian Period. The **Acadian orogeny** affected the Appalachian mobile belt from Newfoundland to Pennsylvania as sedimentary rocks were folded and thrust against the craton.

As with the preceding Taconic and Caledonian orogenies, the Acadian orogeny occurred along an oceanic-continental convergent plate boundary. As the northern Iapetus Ocean continued to close during the Devonian, the plate carrying Baltica finally collided with Laurentia, forming a continental-continental convergent plate boundary along the zone of collision (Figure 21.3a).

Weathering and erosion of the Acadian Highlands produced the **Catskill Delta,** a thick clastic wedge named for the Catskill Mountains in northern New York where it is well exposed. The Catskill Delta, composed of red, coarse conglomerates, sandstones, and shales, contains nearly three times as much sediment as the Queenston Delta.

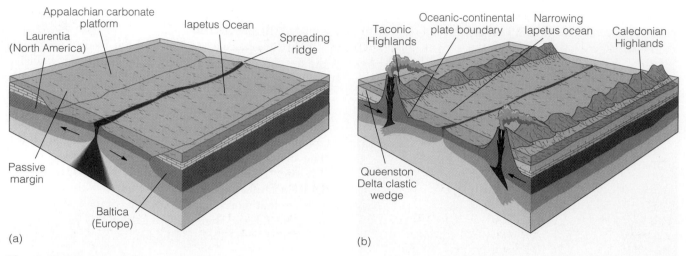

(a)

(b)

Figure 21.25 *Evolution of the Appalachian mobile belt from the Late Proterozoic to the Late Ordovician. (a) During the Late Proterozoic to the Early Ordovician, the Iapetus Ocean was opening along a divergent plate boundary. Both the east coast of Laurentia and the west coast of Baltica were passive continental margins where large carbonate platforms existed. (b) Beginning in the Middle Ordovician, the passive margins of Laurentia and Baltica became oceanic-continental plate boundaries resulting in orogenic activity.*

The Devonian rocks of New York are among the best studied on the continent. A cross section of the Devonian strata clearly reflects an eastern source (Acadian Highlands) for the Catskill facies (Figure 21.27). These clastic rocks can be traced from eastern Pennsylvania, where the coarse clastics are approximately 3 km thick, to Ohio, where the deltaic facies are only about 100 m thick and consist of cratonic shales and carbonates.

The red beds of the Catskill Delta derive their color from the hematite found in the sediments. Plant fossils and oxidation of the hematite indicate that the beds were deposited in a continental environment. Toward the west,

the red beds grade laterally into gray sandstones and shales containing fossil tree trunks, which indicate a swamp or marsh environment.

The Old Red Sandstone

The red beds of the Catskill Delta have a European counterpart in the Devonian Old Red Sandstone of the British Isles (Figure 21.27). The Old Red Sandstone was a Devonian clastic wedge that grew eastward from the Caledonian Highlands onto the Baltica craton. The Old Red Sandstone, just like its North American Catskill counter-

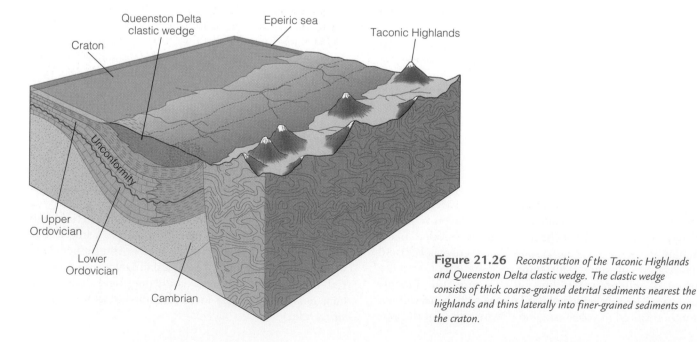

Figure 21.26 *Reconstruction of the Taconic Highlands and Queenston Delta clastic wedge. The clastic wedge consists of thick coarse-grained detrital sediments nearest the highlands and thins laterally into finer-grained sediments on the craton.*

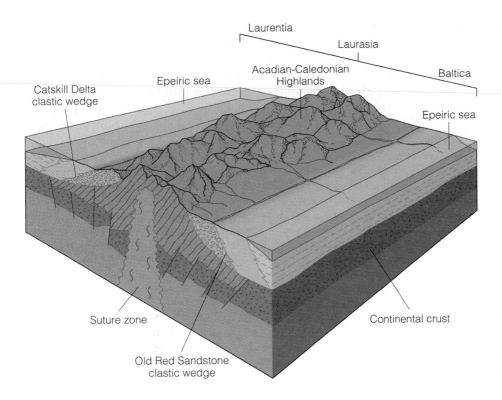

Laurentia
Laurasia
Baltica
Acadian-Caledonian
Highlands
Epeiric sea
Catskill Delta
clastic wedge
Epeiric sea
Suture zone
Old Red Sandstone
clastic wedge
Continental crust

Figure 21.27 *Cross section showing the area of collision between Laurentia and Baltica. Note the bilateral symmetry of the Catskill Delta clastic wedge and the Old Red Sandstone and their relationship to the Acadian and Caledonian Highlands.*

part, contains numerous fossils of freshwater fish, early amphibians, and land plants.

By the end of the Devonian Period, Baltica and Laurentia were sutured together, forming Laurasia (Figure 21.27). The red beds of the Catskill Delta can be traced north, through Canada and Greenland, to the Old Red Sandstone of the British Isles and into northern Europe. These beds were deposited in similar environments along the flanks of developing mountain chains formed by tectonic forces at convergent plate boundaries.

The Taconic, Caledonian, and Acadian orogenies were all part of the same major orogenic event related to the closing of the Iapetus Ocean (Figures 21.25 and 21.27). This event began with paired oceanic-continental convergent plate boundaries during the Taconic and Caledonian orogenies and culminated along a continental-continental convergent plate boundary during the Acadian orogeny as Laurentia and Baltica became sutured. Following this, the Hercynian-Alleghenian orogeny began, followed by orogenic activity in the Ouachita mobile belt.

Hercynian-Alleghenian Orogeny

The Hercynian mobile belt of southern Europe and the Appalachian and Ouachita mobile belts of North America mark the zone along which Europe (part of Laurasia) collided with Gondwana (Figure 21.3). While Gondwana and southern Laurasia collided during the Pennsylvanian and Permian periods in the area of the Ouachita mobile belt, eastern Laurasia (Europe and southeast-

ern North America) joined together with Gondwana (Africa) as part of the **Hercynian-Alleghenian orogeny** (Figure 21.4).

Initial contact between eastern Laurasia and Gondwana began during the Mississippian Period along the Hercynian mobile belt. The greatest deformation occurred during the Pennsylvanian and Permian periods and is referred to as the *Hercynian orogeny*. The central and southern parts of the Appalachian mobile belt (from New York to Alabama) were folded and thrust toward the craton as eastern Laurasia and Gondwana were sutured. This event in North America is referred to as the *Alleghenian orogeny*.

These three Late Paleozoic Orogenies (Hercynian, Alleghenian, and Ouachita) represent the final joining of Laurasia and Gondwana into the supercontinent Pangaea during the Permian.

Cordilleran Mobile Belt

During the Late Proterozoic and Early Paleozoic, the Cordilleran area was a passive continental margin along which extensive continental shelf sediments were deposited. Thick sections of marine sediments graded laterally into thin cratonic units as the Sauk Sea transgressed onto the craton. Beginning in the Middle Paleozoic, an island arc, formed off the western margin of the craton. A collision between this eastward-moving island arc and the western border of the craton took place during the Late Devonian and Early Mississippian resulting in a highland area.

This orogenic event, the **Antler orogeny,** was caused by subduction and resulted in the closing of the narrow ocean basin that separated the island arc from the craton (Figure 21.28). Erosion of the resulting Antler Highlands produced large quantities of sediment that were deposited to the east in the epeiric sea covering the craton and to the west in the deep sea. The Antler orogeny was the first in a series of orogenic events to affect the Cordilleran mobile belt. During the Mesozoic and Cenozoic, this area was the site of major tectonic activity caused by oceanic-continental convergence and accretion of various terranes.

Ouachita Mobile Belt

The Ouachita mobile belt extends for approximately 2100 km from the subsurface of Mississippi to the Marathon region of Texas. Approximately 80% of the former mobile belt is buried beneath a Mesozoic and Cenozoic sedimentary cover. The two major exposed areas in this region are the Ouachita Mountains of Oklahoma and Arkansas and the Marathon Mountains of Texas. Based on extensive study of the subsurface geology and the Ouachita and Marathon mountains, geologists have learned that this region had a complex geologic history (Figure 21.29).

During the Late Proterozoic to Early Mississippian, shallow-water detrital and carbonate sediments were slowly deposited on a broad continental shelf, while in the deeper-water portion of the adjoining mobile belt, bedded cherts and shales were also slowly accumulating (Figure 21.29a). Beginning in the Mississippian Period, the rate of sedimentation increased dramatically as the region changed from a passive continental margin to an active convergent plate boundary (Figure 21.29b). Rapid deposition of sediments continued into the Pennsylvanian with the formation of a clastic wedge that thickened to the south. As much as 16,000 m of Mississippian- and Penn-

sylvanian-age rocks crop out in the Ouachita Mountains attesting to the rapid rate of sedimentation during this time. The formation of a clastic wedge marks the beginning of uplift of the area and formation of a mountain range during the **Ouachita orogeny.**

Thrusting of sediments continued throughout the Pennsylvanian and Early Permian as a result of the compressive forces generated along the zone of subduction as Gondwana collided with Laurasia (Figure 21.29c). The collision of Gondwana and Laurasia is marked by the formation of a large mountain range, most of which was eroded during the Mesozoic Era. Only the rejuvenated Ouachita and Marathon mountains remain of this once lofty mountain range.

The Ouachita deformation was part of the general worldwide tectonic activity that occurred when Gondwana united with Laurasia. The Hercynian, Appalachian, and Ouachita mobile belts were continuous, and marked the southern boundary of Laurasia (Figure 21.4). The tectonic activity that resulted in the uplift in the Ouachita mobile belt was very complex and involved not only the collision of Laurasia and Gondwana but also several microplates between the continents that eventually became part of Central America. The compressive forces impinging on the Ouachita mobile belt also affected the craton by causing broad uplift of the southwestern part of North America.

What Role Did Microplates Play in the Formation of Pangaea?

We have presented the geologic history of the mobile belts, bordering the Paleozoic continents in terms of subduction along convergent plate boundaries. It is becoming increasingly clear, however, that accretion along the continental margins is more complicated than the somewhat simple, large-scale plate interactions that we have described. Geologists now recognize that numerous terranes or microplates existed during the Paleozoic and were involved in the orogenic events that occurred during that time.

A careful examination of the Paleozoic global paleogeographic maps (Figures 21.2, 21.3, and 21.4) shows numerous microplates, and their location and role during the formation of Pangaea must be taken into account. For example, the small continent of Avalonia is composed of some coastal parts of New England, southern New Brunswick, much of Nova Scotia, the Avalon Peninsula of eastern Newfoundland, southeastern Ireland, Wales, England, and parts of Belgium and northern France. This microplate existed as a separate continent during the Ordovician and collided with Baltica

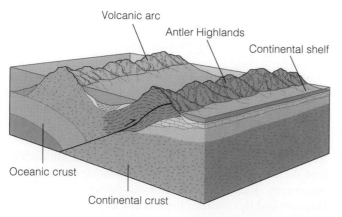

Figure 21.28 *Reconstruction of the Cordilleran mobile belt during the Early Mississippian, showing the effects of the Antler orogeny.*

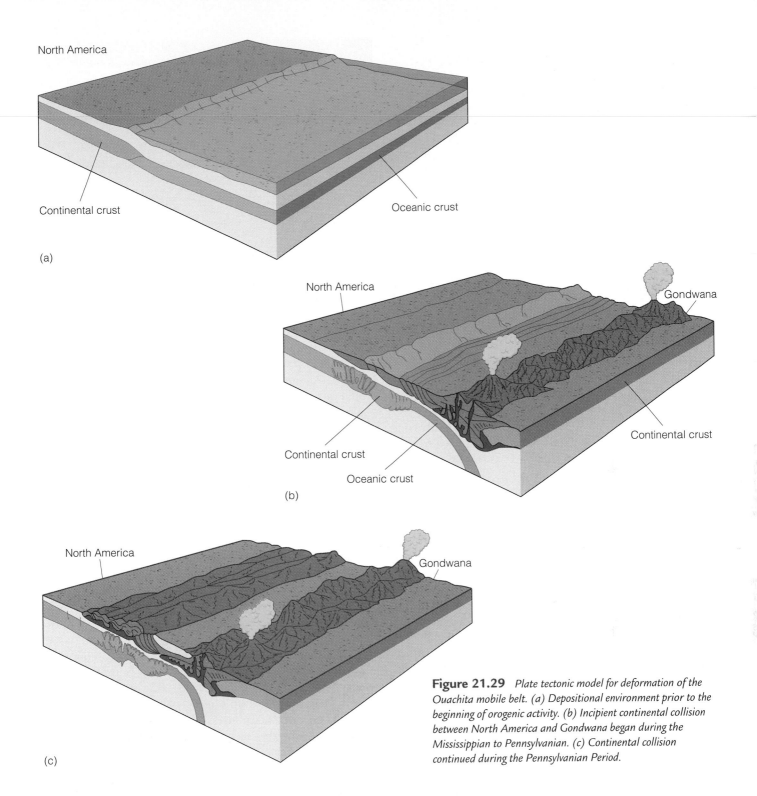

North America

Continental crust

Oceanic crust

(a)

North America

Gondwana

Continental crust

Continental crust

Oceanic crust

(b)

North America

Gondwana

(c)

Figure 21.29 *Plate tectonic model for deformation of the Ouachita mobile belt. (a) Depositional environment prior to the beginning of orogenic activity. (b) Incipient continental collision between North America and Gondwana began during the Mississippian to Pennsylvanian. (c) Continental collision continued during the Pennsylvanian Period.*

during the Silurian and Laurentia during the Devonian (Figures 21.2 and 21.3).

Florida and parts of the eastern seaboard of North America make up the Piedmont microplate that was part of the larger Gondwana continent. This microplate became sutured to Laurasia during the Pennsylvanian Period. Numerous microplates occupied the region between Gondwana and Laurasia that eventually became part of

Central America during the Pennsylvanian collision between these continents.

Thus, while the basic history of the formation of Pangaea during the Paleozoic remains the same, geologists now realize that microplates also played an important role. Furthermore, the recognition of terranes within mobile belts helps explain some previously anomalous geologic situations.

What Role Did Microplates Play in the Formation of Pangaea?

Paleozoic Mineral Resources

Paleozoic-age rocks contain a variety of important mineral resources, including energy resources, metallic and nonmetallic mineral deposits, and sand and gravel for construction. Important sources of industrial or silica sand are the Upper Cambrian Jordan Sandstone of Minnesota and Wisconsin, the Middle Ordovician St. Peter Sandstone, the Lower Silurian Tuscarora Sandstone in Pennsylvania and Virginia, the Devonian Ridgeley Formation in West Virginia, Maryland, and Pennsylvania, and the Devonian Sylvania Sandstone in Michigan.

Silica sand has a variety of uses, including the manufacture of glass, refractory bricks for blast furnaces, and molds for casting iron, aluminum, and copper alloys. Some silica sands, called hydraulic fracturing sands, are pumped into wells to fracture oil- or gas-bearing rocks and provide permeable passageways for the oil or gas to migrate to the well.

Thick deposits of Silurian evaporites, mostly rock salt (NaCl) and rock gypsum ($CaSO_4 \cdot H_2O$) altered to rock anhydrite ($CaSO_4$), underlie parts of Michigan, Ohio, New York, and adjacent areas in Ontario, Canada. These rocks are important sources of various salts. In addition, barrier and pinnacle reefs in carbonate rocks associated with these evaporites are reservoirs for oil and gas in Michigan and Ohio.

The Zechstein evaporites of Europe extend from Great Britain across the North Sea and into Denmark, the Netherlands, Germany, and eastern Poland and Lithuania. In addition to the evaporites themselves, Zechstein deposits form the caprock for the large reservoirs of the gas fields of the Netherlands and part of the North Sea region.

Other important evaporite mineral resources include those of the Permian Delaware Basin of West Texas and New Mexico, and Devonian evaporites in the Elk Point basin of Canada. In Michigan, gypsum is mined and used in the construction of wallboard. Upper Paleozoic limestones from many areas in North America are used in the manufacture of cement. Limestone is also mined and used in blast furnaces when steel is produced.

Metallic mineral resources including tin, copper, gold, and silver are known from Late Paleozoic-age rocks, especially those that have been deformed during mountain building. The host rocks for deposits of lead and zinc in southeast Missouri are Cambrian dolostones, although some Ordovician rocks contain these metals as well. These deposits have been mined since 1720 but have been largely depleted. Now most lead and zinc mined in Missouri come from Mississippian-age sedimentary rocks.

The Silurian Clinton Formation crops out from Alabama north to New York, and equivalent rocks are found in Newfoundland. This formation has been mined for iron in many places. In the United States, the richest ores and most extensive mining occurred near Birmingham, Alabama, but only a small amount of ore is currently produced in that area.

Petroleum and natural gas are recovered in commercial quantities from rocks ranging in age from Devonian through Permian. For example, Devonian-age rocks in the Michigan Basin, Illinois Basin, and the Williston Basin of Montana, South Dakota, and adjacent parts of Alberta, Canada, have yielded considerable amounts of hydrocarbons. Permian reefs and other strata in the western United States, particularly Texas, have also been important producers.

Although Permian-age coal beds are known from several areas including Asia, Africa, and Australia, much of the coal in North America and Europe comes from Pennsylvanian (Late Carboniferous) deposits. Large areas in the Appalachian region and the midwestern United States are underlain by vast coal deposits (Figure 21.30). These coal deposits formed from the lush vegetation that flourished in Pennsylvanian coal swamps (see Chapter 22).

Much of this coal is characterized as bituminous coal, which contains about 80% carbon. It is a dense, black coal that has been so thoroughly altered that plant remains can be seen only rarely. Bituminous coal is used to make *coke*, a hard, gray substance made up of the fused ash of bituminous coal. Coke is used to fire blast furnaces during the production of steel.

Some of the Pennsylvanian coal from North America is *anthracite*, a metamorphic type of coal containing up to 98% carbon. Most anthracite is in the Appalachian region (Figure 21.30). It is an especially desirable type of coal because it burns with a smokeless flame and it yields more heat per unit volume than other types of coal. Unfortunately, it is the least common type of coal, so much of the coal used in the United States is bituminous.

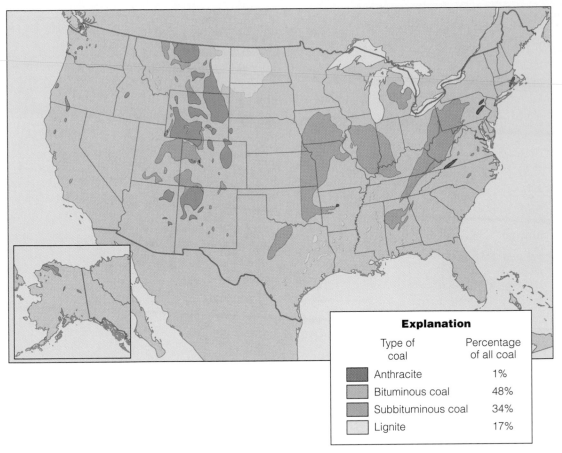

Explanation

Type of coal	Percentage of all coal
Anthracite	1%
Bituminous coal	48%
Subbituminous coal	34%
Lignite	17%

Figure 21.30 *Distribution of coal deposits in the United States. The age of the coals in the midwestern states and the Appalachian region are mostly Pennsylvanian, whereas those in the west are mostly Cretaceous and Tertiary.*

Chapter Summary

Tables 21.1 and 21.2 provide a summary of the geologic history of the North American craton and mobile belts as well as global events and sea-level changes during the Paleozoic Era.

1. Six major continents existed at the beginning of the Paleozoic Era; four of them were located near the paleoequator.

2. During the Early Paleozoic (Cambrian-Silurian), Laurentia was moving northward and Gondwana moved to a south polar location, as indicated by glacial deposits.

3. During the Late Paleozoic, Baltica and Laurentia collided, forming Laurasia. Siberia and Kazakhstania collided and finally were sutured to Laurasia. Gondwana moved over the South Pole and experienced several glacial-interglacial periods, resulting in global changes in sea level and transgressions and regressions along the low-lying craton margins.

4. Laurasia and Gondwana underwent a series of collisions beginning in the Carboniferous. During the Permian, the formation of Pangaea was completed. Surrounding the supercontinent was a global ocean, Panthalassa.

5. Most continents consist of two major components: a relatively stable craton over which epeiric seas transgressed and regressed, surrounded by mobile belts in which mountain building took place.

6. The geologic history of North America can be divided into cratonic sequences that reflect cratonwide transgressions and regressions.

7. The Sauk Sea was the first major transgression onto the craton. At its maximum, it covered the craton except for parts of the Canadian Shield and the Transcontinental Arch, a series of large, northeast-southwest trending islands.

8. The Tippecanoe sequence began with deposition of an extensive sandstone over the exposed and eroded Sauk landscape. During Tippecanoe time, extensive carbonate deposition took place. In addition, large barrier reefs enclosed basins, resulting in evaporite deposition within these basins.

9. The basal beds of the Kaskaskia sequence that were deposited on the exposed Tippecanoe surface consisted of either sandstones, derived from the eroding Taconic Highlands, or carbonate rocks.

10. Most of the Kaskaskia sequence is dominated by carbonates and associated evaporites. The Devonian Period was a time of major reef building in western Canada, southern England, Belgium, Australia, and Russia.

Table 21.1

Summary of Early Paleozoic Geologic Events

Geologic Period	Sequence	Relative Changes in Sea Level		Cordilleran Mobile Belt
		Rising	Falling	
Silurian	Tippecanoe			
Ordovician				
	Sauk	Present sea level		
Cambrian				

Age (Millions of Years): 408, 438, 505, 545

Craton	Ouachita Mobile Belt	Appalachian Mobile Belt	Major Events Outside North America
		Acadian orogeny	
Extensive barrier reefs and evaporites common.			Caledonian orogeny
Queenston Delta clastic wedge.		Taconic orogeny	Continental glaciation in Southern Hemisphere.
Transgression of Tippecanoe Sea.			
Regression exposing large areas to erosion.			
Canadian Shield and Transcontinental Arch only areas above sea level.			
Transgression of Sauk Sea.			

Table 21.2

Summary of Late Paleozoic Geologic Events

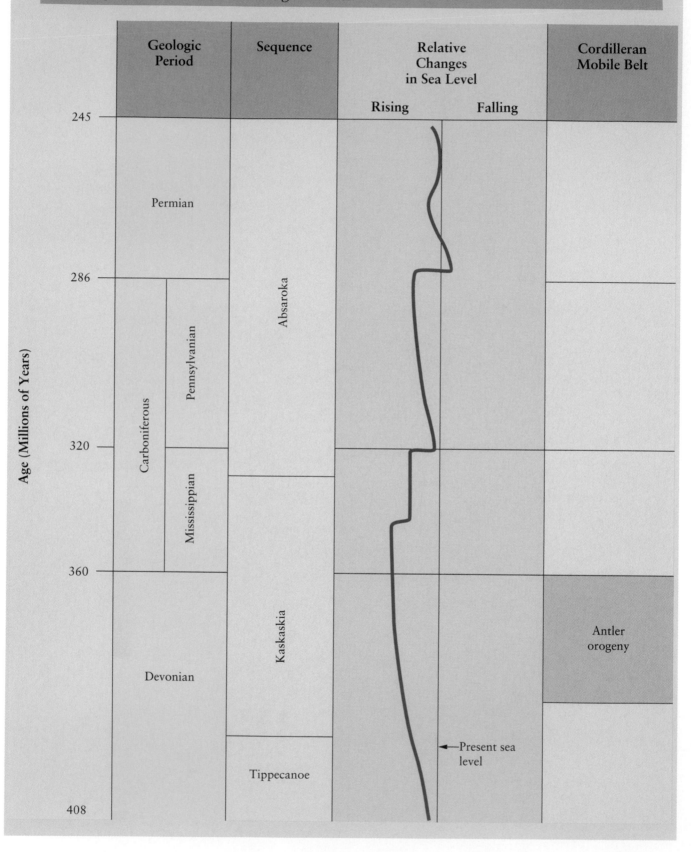

Craton	Ouachita Mobile Belt	Appalachian Mobile Belt	Major Events Outside North America
Deserts, evaporites, and continental red beds in southwestern United States. Extensive reefs in Texas area.		Formation of Pangaea.	
		Allegheny orogeny	Hercynian orogeny
Coal swamps common. Formation of Ancestral Rockies.	Ouachita orogeny		Continental glaciation in Southern Hemisphere.
Transgression of Absaroka Sea. Widespread black shales and limestones.			
Widespread black shales. Catskill Delta clastic wedge. Extensive barrier-reef formation in Western Canada. Transgression of Kaskaskia Sea.		Acadian orogeny	Old Red Sandstone clastic wedge in British Isles.
			Caledonian orogeny

11. A persistent and widespread black shale, the Chattanooga Shale and its equivalents, was deposited over a large area of the craton during the Late Devonian and Early Mississippian.

12. The Mississippian Period was dominated by carbonate deposition.

13. Transgressions and regressions over the low-lying craton resulted in cyclothems and the formation of coals during the Pennsylvanian Period.

14. Cratonic mountain building occurred during the Pennsylvanian Period, and thick nonmarine detrital rocks and evaporites were deposited in the intervening basins.

15. By the Early Permian, the Absaroka Sea occupied a narrow zone of the south-central craton. Here, several large reefs and associated evaporites developed. By the end of the Permian Period, the Absaroka Sea had retreated from the craton.

16. The eastern edge of North America was a stable carbonate platform during Sauk time. During Tippecanoe time an oceanic-continental convergent plate boundary formed, resulting in the Taconic orogeny, the first of several orogenies to affect the Appalachian mobile belt. The newly formed Taconic Highlands shed sediments into the western epeiric sea, producing the Queenston Delta, a large clastic wedge.

17. The Caledonian, Acadian, Hercynian, and Alleghenian orogenies were all part of the global tectonic activity resulting from the assembly of Pangaea.

18. The Cordilleran mobile belt was the site of the Antler orogeny, a minor Devonian orogeny during which deep-water sediments were thrust eastward over shallow-water sediments.

19. Mountain building occurred in the Ouachita mobile belt during the Pennsylvanian and Early Permian. This tectonic activity was partly responsible for the cratonic uplift that took place in the southwest, producing the Ancestral Rockies.

20. During the Paleozoic Era, numerous microplates existed and played an important role in the formation of Pangaea.

21. Paleozoic age rocks contain a variety of mineral resources including building stone, limestone for cement, silica sand, evaporites, petroleum, coal, ores of iron, lead, and zinc, and other metallic deposits.

Important Terms

Absaroka sequence	clastic wedge	Iapetus Ocean	Panthalassa Ocean
Acadian orogeny	Cordilleran mobile belt	Kaskaskia sequence	Queenston Delta
Ancestral Rockies	craton	Kazakhstania	Sauk sequence
Antler orogeny	cratonic sequence	Laurasia	sequence stratigraphy
Appalachian mobile belt	cyclothem	Laurentia	Siberia
Baltica	epeiric sea	mobile belt	Taconic orogeny
Caledonian orogeny	Franklin mobile belt	organic reef	Tippecanoe sequence
Catskill Delta	Gondwana	Ouachita mobile belt	Transcontinental Arch
China	Hercynian-Alleghenian orogeny	Ouachita orogeny	

Review Questions

1. A major transgressive-regressive cycle bounded by craton-wide unconformities is (a)n
   a. _____ biostratigraphic unit; b. _____ cratonic sequence; c. _____ orogeny; d. _____ shallow sea; e. _____ cyclothem.

2. Cyclothems are a characteristic feature of which Paleozoic cratonic sequence?
   a. _____ Sauk; b. _____ Tippecanoe; c. _____ Kaskaskia; d. _____ Absaroka; e. _____ Zuni.

3. Weathering of which highlands produced the Catskill Delta clastic wedge?
   a. _____ Transcontinental Arch; b. _____ Acadian; c. _____ Taconic; d. _____ Sevier; e. _____ Nevadan.

4. The first Paleozoic orogeny to occur in the Cordilleran mobile belt was the
   a. _____ Acadian; b. _____ Alleghenian; c. _____ Antler; d. _____ Caledonian; e. _____ Cordilleran.

5. The Taconic orogeny resulted from what type of plate movement?
   a. _____ oceanic-oceanic convergent; b. _____ oceanic-continental convergent; c. _____ continental-continental convergent; d. _____ divergent; e. _____ transform.

6. Which formation is a source of Early Paleozoic-age iron ore?
   a. _____ St. Peter; b. _____ Tuscarora; c. _____ Jordan; d. _____ Oriskany; e. _____ Clinton.

7. During which sequence was the eastern margin of Laurentia a passive plate margin?
   a. _____ Sauk; b. _____ Tippecanoe; c. _____ Kaskaskia; d. _____ Absaroka; e. _____ Zuni.

8. An elongated area marking the site of mountain building is a(n)
   a. _____ craton; b. _____ platform; c. _____ shield; d. _____ epeiric sea; e. _____ mobile belt.

9. Rhythmically repetitive sedimentary sequences are
   a. _____ tillites; b. _____ cyclothems; c. _____ orogenies; d. _____ reefs; e. _____ evaporites.

10. Uplift in the southwestern part of the craton during the Late Absaroka resulted in which mountainous region?
    a. _____ Ancestral Rockies; b. _____ Antler Highlands; c. _____ Appalachians; d. _____ Marathon; e. _____ none of these.

11. The European counterpart to the Devonian Catskill Delta clastic wedge is the
    a. _____ Phosphoria Formation; b. _____ Oriskany Sandstone; c. _____ Capitan Limestone; d. _____ Old Red Sandstone; e. _____ none of these.

12. Which orogeny was not part of the closing of the Iapetus Ocean?
    a. _____ Acadian; b. _____ Alleghenian; c. _____ Antler;
    d. _____ Caledonian; e. _____ Taconic.
13. The Paleozoic ocean separating Laurentia from Siberia and Baltica was the
    a. _____ Panthalassa; b. _____ Tethys; c. _____ Caledonian;
    d. _____ Iapetus; e. _____ Kaskaskia.
14. Discuss how plate movement during the Paleozoic Era affected worldwide weather patterns.
15. Provide a geologic history for the Iapetus Ocean during the Paleozoic Era.
16. Compare the Taconic, Caledonian, and Acadian orogenies in terms of the tectonic forces that caused them and the sedimentary features that resulted.
17. What are some of the methods geologists can use to determine the locations of continents during the Paleozoic Era?
18. Where was the Transcontinental Arch? What evidence indicates that it existed?
19. Discuss how evaporites of the Michigan Basin may have formed during the Silurian Period.
20. What were the major differences between the Appalachian, Ouachita, and Cordilleran mobile belts during the Paleozoic?

21. How did the formation of Pangaea and Panthalassa affect the world's climate at the end of the Paleozoic Era?
22. According to estimates made from mapping and correlation, the Queenston Delta contains more than 600,000 km³ of rock eroded from the Taconic Highlands. Based on this figure, geologists estimate the Taconic Highlands were at least 4000 m high. They also estimate that the Catskill Delta contains three times as much sediment as the Queenston Delta. From what you know about the geographic distribution of the Taconic Highlands and the Acadian Highlands, can you estimate how high the Acadian Highlands might have been?
23. Discuss how sequence stratigraphy can be used to make global correlations and why it is so useful in reconstructing past events.
24. Discuss the role plate tectonics plays in the formation and distribution of Paleozoic-age mineral resources.
25. Based on the discussion of Milankovitch cycles and their role in causing glacial-interglacial cycles (see Chapter 14), could these cycles be partly responsible for the transgressive-regressive cycles that resulted in cyclothems during the Pennsylvanian Period?

 World Wide Web Activities

For these web site addresses, along with current updates and exercises, log on to **http://www.brookscole.com/geo**

Stratigraphy of the Black Hills
This site contains information about the geology of the Black Hills, South Dakota. You can view a nice relief image of the Black Hills, as well as a geologic map of South Dakota. In addition, a stratigraphic column of the Black Hills is provided, with links to the various formations of the Black Hills. Under the Cambrian to Jurassic stratigraphic column, click on any of the Paleozoic formations to learn more about that particular formation.

Union College Geology Department
This site is maintained by the Union College Geology Department, which is located in Schenectady, New York. This particular site provides information about the sediments that were deposited during the Middle and Late Ordovician Period in the present location of the Mohawk River Valley of New York. It also

contains sites showing the paleogeography of the world and New York State during the Ordovician Period.

Global Earth History
This site, maintained at the University of Northern Arizona, contains a series of plate tectonic reconstructions to show what Earth looked like at various times in the geologic past. Included are paleogeographic maps showing global, North American, and European reconstructions from the Cambrian to the present, as well as tectonic, sedimentation, and paleogeographic maps of the North Atlantic region.

North American Orogenies
This site is part of the *Global Earth History* site, and shows simple restored cross sections of the major orogenies taking place in North America from the Proterozoic through Mesozoic. Click on the various orogenies that occurred during the Paleozoic Era.

Paleozoic Life History

Diorama of the environment and biota of the Phyllopod bed of the Burgess Shale, British Columbia, Canada. In the background is a vertical wall of a submarine escarpment with algae growing on it. The large cylindrical ribbed organisms on the muddy bottom in the foreground are sponges.

PROLOGUE

OBJECTIVES

At the end of this chapter you will have learned that

■ Animals with skeletons appeared abruptly at the beginning of the Paleozoic Era and experienced a short period of rapid evolutionary diversification.

■ During the Paleozoic Era, the invertebrates experienced times of diversification followed by extinction, culminating in the greatest mass extinction in Earth's history at the end of the Permian Period.

■ Vertebrates first evolved during the Cambrian Period and the fish diversified rapidly during the Paleozoic.

■ Amphibians first appear in the fossil record during the Devonian, having made the transition from water to land and became extremely abundant during the Pennsylvanian Period.

■ The evolution of the amniote egg allowed reptiles to colonize all parts of the land beginning in the Late Mississippian.

■ The pelycosaurs or finback reptiles were the dominant reptile group during the Permian and were the ancestors to the therapsids or mammal-like reptiles.

■ The earliest land plants are known from the Ordovician Period while the oldest known vascular land plants first appear in the Middle Silurian.

■ Seedless vascular plants were very abundant during the Pennsylvanian Period.

■ With the onset of arid conditions during the Permian Period, the gymnosperms became the dominant element of the world's flora.

On August 30 and 31, 1909, near the end of the summer field season, Charles D. Walcott, geologist and head of the Smithsonian Institution, was searching for fossils along a trail on Burgess Ridge between Mount Field and Mount Wapta, near Field, British Columbia, Canada. On the west slope of this ridge, he discovered the first soft-bodied fossils from the Burgess Shale, a discovery of immense importance in deciphering the early history of life. During the following week, Walcott and his collecting party split open numerous blocks of shale, many of which yielded the impressions of a number of soft-bodied organisms beautifully preserved on bedding planes. Walcott returned to the site the following summer and located the shale stratum that was the source of his fossil-bearing rocks in the steep slope above the trail. He quarried the site and shipped back thousands of fossil specimens to the United States National Museum of Natural History, where he later cataloged and studied them.

The importance of Walcott's discovery is not that it was another collection of well-preserved Cambrian fossils, but rather that it allowed geologists a rare glimpse into a world previously almost unknown—that of the soft-bodied animals that lived some 530 million years ago. The beautifully preserved fossils from the Burgess Shale present a much more complete picture of a Middle Cambrian community than deposits containing only fossils of the hard parts of organisms. Specifically, the Burgess Shale contains species of trilobites, sponges, brachiopods, mollusks, and echinoderms, all of which have hard parts and are characteristic of Cambrian faunas throughout the world. But in addition to the diverse skeletonized fauna, a large and varied fossil assemblage of soft-bodied animals is also present. In fact, 60% of the total fossil assemblage is composed of soft-bodied animals, which usually are not preserved. In all, more than 100 genera of animals, at least 60 of which were soft-bodied and preserved as impressions, have been recovered from the Burgess Shale. This proportion of soft-bodied animals to those with hard parts is comparable to the makeup of present-day marine communities.

What conditions led to the remarkable preservation of the Burgess Shale fauna? The site of deposition of the Burgess Shale was located at the base of a steep submarine escarpment. The animals whose exquisitely preserved fossil remains are found in the Burgess Shale lived in and on mud banks that formed along the top of this escarpment. Periodically, this unstable area would slump and slide down the escarpment as a turbidity current. At the base, the mud and animals carried with it were deposited in a deep-water anaerobic environment devoid of life. In such an environment, bacterial degradation did not destroy the buried animals; and they were compressed by the weight of the overlying sediments and eventually preserved as carbonaceous impressions.

Introduction

In this chapter, we examine the history of Paleozoic life as a series of interconnected biologic and geologic events in which the theories and processes of evolution and plate tectonics played a major role. The opening and closing of ocean basins, transgressions and regressions of epeiric seas, and the changing positions of the continents had a profound effect on the evolution of the marine and terrestrial communities.

A time of tremendous biologic change began with the appearance of skeletonized animals near the Precambrian-Cambrian boundary. Following this event, marine invertebrates began a period of adaptive radiation and evolution during which the Paleozoic marine invertebrate community greatly diversified. Indeed, the history of the Paleozoic marine invertebrate community was one of diversifications and extinctions.

Vertebrates also evolved during the Paleozoic. The earliest fossil records of vertebrates are of fish, which evolved during the Cambrian. One group of fish was ancestral to the first land animals, the amphibians, which evolved during the Devonian. Reptiles evolved from a group of amphibians during the Mississippian Period and were the dominant vertebrate animals on land by the end of the Paleozoic Era.

Plants preceded animals onto the land. Both plants and animals were confronted with the same basic problems in making the transition from water to land. The method of reproduction proved to be the major barrier to expansion into new environments for both groups. With the evolution of the seed in plants and the amniote egg in animals, this limitation was removed, and both groups were able to move into all terrestrial environments.

The end of the Paleozoic Era was a time of major extinctions. The marine invertebrate community was greatly decimated, and many amphibians and reptiles on land also became extinct.

Why Should You Study Paleozoic Life?

Why should you study the Paleozoic life history of earth? Perhaps the easiest answer is actually another question that most humans ask at some time in their life and that is, "Where did we come from?" It is a question as old as human consciousness, and one that a study of the fossil record can help answer. In this chapter we examine the panorama of skeletonized life from its explosive radiation at the beginning of the Cambrian Period to the greatest mass extinction the world has ever known at the close of the Paleozoic Era. In between we will examine the Paleozoic evolutionary history of fish, amphibians, and reptiles, as well as the world's flora.

The fossil record provides us with clear evidence of evolution's major tenet that all existing organisms are the evolutionary descendants of life forms that lived during the past. This in itself is reason enough to study the history of life. But there are other equally valid reasons. For instance, we can examine the history of life, and in this case life during the Paleozoic, as a system whose parts consist of a series of interconnected biologic and geologic events. The underlying processes of evolution and plate tectonics are the forces that drove this system. Furthermore, examining and understanding the underlying causes of the greatest mass extinction in the past can provide us with insights into changes taking place today in the biosphere.

What Was the Cambrian Explosion?

At the beginning of the Paleozoic Era, animals with skeletons appeared rather abruptly in the fossil record. In fact, their appearance is described as an explosive development of new types of animals and is referred to as the "Cambrian explosion" by most scientists. This sudden and rapid appearance of new animals in the fossil record is rapid, however, only in the context of geologic time, having taken place over millions of years during the Early Cambrian Period.

This seemingly sudden appearance of animals in the fossil record is not a recent discovery. Early geologists observed that the remains of skeletonized animals appeared rather abruptly in the fossil record. Charles Darwin addressed this problem in *On the Origin of Species* and observed that without a convincing explanation, such an event was difficult to reconcile with his newly expounded evolutionary theory.

The sudden appearance of shelled animals during the Early Cambrian contrasts sharply with the biota living during the preceding Proterozoic Eon. Up until the evolution of the Ediacaran fauna, Earth was populated primarily by single-celled organisms. Recall from Chapter 20 that the Ediacaran fauna, which is found on all continents except Antarctica, consists primarily of multicelled soft-bodied organisms. Microscopic calcareous tubes, presumably housing wormlike suspension feeding organisms, have also been found at some localities. In addition, trails and burrows, which represent the activities of worms and other sluglike animals, are also found associated with Ediacaran faunas throughout the world. The trails and burrows are similar to those made by present-day soft-bodied organisms.

Until recently, it appeared that there was a fairly long period of time between the extinction of the Ediacaran fauna and the first Cambrian fossils. That gap has been considerably narrowed in recent years with the discovery of new Proterozoic fossiliferous localities. Now, Proterozoic fossil assemblages continue right to the base of the Cambrian. Furthermore, recent work from Namibia indicates that Ediacaran-like fossils are even present above the first occurrence of Cambrian index fossils.

Nonetheless, the cause of the sudden appearance of so many different animal phyla during the Early Cambrian is still a hotly debated topic. Newly developed molecular techniques that allow evolutionary biologists to compare the similarity of molecular sequences of the same gene from different species is being applied to the phylogeny of many organisms. In addition, new fossil sites and detailed stratigraphic studies are shedding light on the early history and ancestry of the various invertebrate phyla.

It appears likely that the Cambrian explosion probably had its roots firmly planted in the Proterozoic. However, the mechanism that triggered this event is still unknown and was likely a combination of factors, both biological and geological. For example, geological evidence indicates that Earth was glaciated one or more times during the Proterozoic, followed by global warming during the Cambrian. These global environmental changes may have stimulated evolution and contributed to the Cambrian explosion. Whatever the ultimate cause of the Cambrian explosion, the appearance of a skeletonized fauna and the rapid diversification of that fauna during the Early Cambrian was a major event in Earth's history.

The Emergence of a Shelly Fauna

The earliest organisms with hard parts are Proterozoic calcareous tubes found associated with Ediacaran faunas from several locations throughout the world. These are fol-

lowed by other microscopic skeletonized fossils from the Early Cambrian (Figure 22.1) and the appearance of large skeletonized animals during the Cambrian explosion. Along with the question of why animals appeared so suddenly in the fossil record is the equally intriguing one of why they initially acquired skeletons and what selective advantage this provided. A variety of explanations have been proposed, but as yet, none are completely satisfactory or universally accepted.

The formation of an exoskeleton confers many advantages on an organism: (1) It provides protection against ultraviolet radiation, allowing animals to move into shallower waters; (2) it helps prevent drying out in an intertidal environment; (3) it provides protection against predators. Recent evidence of actual fossils of predators and specimens of damaged prey, as well as antipredatory adaptations in some animals, indicates that the impact of predation during the Cambrian was great. With predators playing an important role in the Cambrian marine ecosystem, any mechanism or feature that protected an animal would certainly be advantageous and confer an adaptive advantage to the organism. (4) A supporting skeleton, whether an exo- or endoskeleton, allows animals to increase their size and provides attachment sites for muscles.

There is currently no clear answer to why marine organisms evolved mineralized skeletons during the Cambrian explosion and shortly thereafter. They undoubtedly

(a)

(b)

(c)

Figure 22.1 *Three small Early Cambrian shelly fossils. (a) A conical sclerite (a piece of the armor covering) of* Lapworthella *from Australia. (b)* Archaeooides, *an enigmatic spherical fossil from the Mackenzie Mountains, Northwest Territories, Canada. (c) The tube of an anabaritid from the Mackenzie Mountains, Northwest Territories, Canada.*

evolved because of a variety of biologic and environmental factors. Whatever the reason, the acquisition of a mineralized skeleton was a major evolutionary innovation allowing invertebrates to successfully occupy a wide variety of marine habitats.

Paleozoic Invertebrate Marine Life

Rather than focusing on the history of each invertebrate phylum (Table 22.1), we will survey the evolution of the Paleozoic marine invertebrate communities through time, concentrating on the major features and changes that took place. To do that, we need to briefly examine the nature and structure of living marine communities so that we can make a reasonable interpretation of the fossil record.

The Present Marine Ecosystem

In analyzing the present-day marine ecosystem, we must look at where organisms live, how they get around, as well as how they feed (Figure 22.2). Organisms that live in the water column above the sea floor are called *pelagic*. They are divided into two main groups: the floaters, or **plankton,** and the swimmers, or **nekton.**

Plankton are mostly passive and go where the current carries them. Plant plankton such as diatoms, dinoflagellates, and various algae, are called *phytoplankton* and are mostly microscopic. Animal plankton are called *zooplankton* and are also mostly microscopic. Examples of zooplankton include foraminifera, radiolarians, and jellyfish. The nekton are swimmers and are mainly vertebrates such as fish; the invertebrate nekton include cephalopods.

Organisms that live on or in the seafloor make up the **benthos.** They are characterized as *epifauna* (animals) or *epiflora* (plants), for those that live on the seafloor, or as *infauna,* which are animals living in and moving through the sediments. The benthos can be further divided into those organisms that stay in one place, called *sessile,* and those that move around on or in the seafloor, called *mobile.*

The feeding strategies of organisms are also important in terms of their relationships with other organisms in the marine ecosystem. There are basically four feeding groups: **suspension-feeding** animals remove or consume microscopic plants and animals as well as dissolved nutrients from the water; **herbivores** are plant eaters; **carnivore-scavengers** are meat eaters; and **sediment-deposit feeders** ingest sediment and extract the nutrients from it.

We can define an organism's place in the marine ecosystem by where it lives and how it eats. For example, a brachiopod is a benthonic, epifaunal suspension feeder, whereas a cephalopod is a nektonic carnivore.

Table 22.1

The Major Invertebrate Groups and Their Stratigraphic Ranges

Phylum Protozoa	Cambrian-Recent	**Phylum Mollusca**	Cambrian-Recent
Class Sarcodina	Cambrian-Recent	Class Monoplacophora	Cambrian-Recent
Order Foraminifera	Cambrian-Recent	Class Gastropoda	Cambrian-Recent
Order Radiolaria	Cambrian-Recent	Class Bivalvia	Cambrian-Recent
Phylum Porifera	Cambrian-Recent	Class Cephalopoda	Cambrian-Recent
Class Demospongea	Cambrian-Recent	**Phylum Annelida**	Precambrian-Recent
Order Stromatoporoida	Cambrian-Oligocene	**Phylum Arthropoda**	Cambrian-Recent
Phylum Archaeocyatha	Cambrian	Class Trilobita	Cambrian-Permian
Phylum Cnidaria	Cambrian-Recent	Class Crustacea	Cambrian-Recent
Class Anthozoa	Ordovician-Recent	Class Insecta	Silurian-Recent
Order Tabulata	Ordovician-Permian	**Phylum Echinodermata**	Cambrian-Recent
Order Rugosa	Ordovician-Permian	Class Blastoidea	Ordovician-Permian
Order Scleractinia	Triassic-Recent	Class Crinoidea	Cambrian-Recent
Phylum Bryozoa	Ordovician-Recent	Class Echinoidea	Ordovician-Recent
Phylum Brachiopoda	Cambrian-Recent	Class Asteroidea	Ordovician-Recent
Class Inarticulata	Cambrian-Recent	**Phylum Hemichordata**	Cambrian-Recent
Class Articulata	Cambrian-Recent	Class Graptolithina	Cambrian-Mississippian

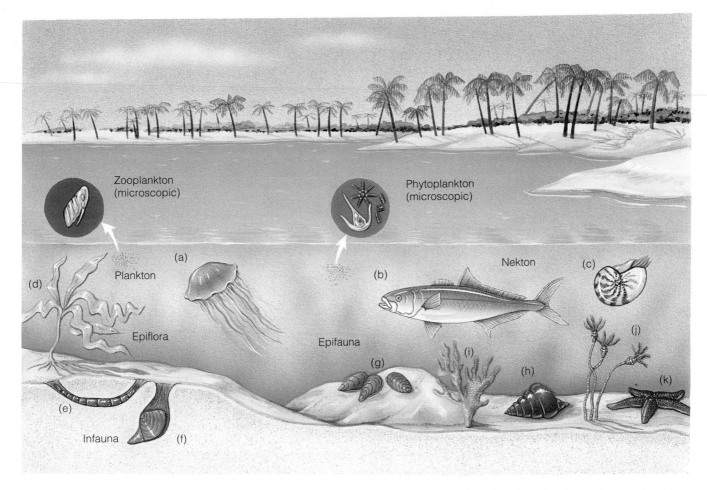

Figure 22.2 *Where and how animals and plants live in the marine ecosystem. Plankton: (a) jellyfish. Nekton: (b) fish and (c) cephalopod. Benthos: (d) through (k). Sessile epiflora: (d) seaweed. Sessile epifauna: (g) bivalve,(i) coral, and (j) crinoid. Mobile epifauna: (k) starfish and (h) gastropod. Infauna: (e) worm and (f) bivalve. Suspension feeders: (g) bivalve, (i) coral, and (j) crinoid. Herbivores: (h) gastropod. Carnivores-scavengers: (k) starfish. Sediment-deposit feeders: (e) worm.*

An ecosystem includes several *trophic levels,* which are tiers of food production and consumption within a feeding hierarchy. The feeding hierarchy and hence energy flow in an ecosystem comprise a food web of complex interrelationships among the producers, consumers, and decomposers (Figure 22.3). The **primary producers,** or *autotrophs,* are those organisms that manufacture their own food. Virtually all marine primary producers are phytoplankton. Feeding on the primary producers are the primary consumers, which are mostly suspension feeders. Secondary consumers feed on the primary consumers, and thus are predators, while tertiary consumers, which are also predators, feed on the secondary consumers. Besides the producers and consumers, there are also transformers and decomposers. These are bacteria that break down the dead organisms that have not been consumed into organic compounds that are then recycled.

When we look at the marine realm today, we see a complex organization of organisms interrelated by trophic interactions and affected by changes in the physical environment. When one part of the system changes, the whole structure changes, sometimes almost insignificantly, other times catastrophically.

As we examine the evolution of the Paleozoic marine ecosystem, keep in mind how geologic and evolutionary changes can have a significant impact on its composition and structure. For example, the major transgressions onto the craton opened up vast areas of shallow seas that could be inhabited. The movement of continents affected oceanic circulation patterns as well as causing environmental changes.

Cambrian Marine Community

The Cambrian Period was a time during which many new body plans evolved and animals moved into new niches. As might be expected, the Cambrian witnessed a higher percentage of such experiments than any other period of geologic history.

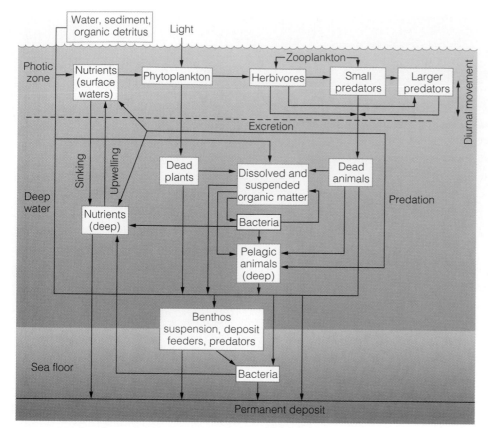

Figure 22.3 *Marine food web showing the relationships among the producers, consumers, and decomposers.*

Although almost all the major invertebrate phyla evolved during the Cambrian Period, many were represented by only a few species. While trace fossils are common, and echinoderms diverse, trilobites, brachiopods, and archaeocyathids comprised the majority of Cambrian skeletonized life (Figure 22.4).

Trilobites were by far the most conspicuous element of the Cambrian marine invertebrate community and made up about half of the total fauna. Trilobites were benthonic mobile sediment-deposit feeders that crawled or swam along the seafloor.

Cambrian *brachiopods* were mostly primitive types and were benthonic sessile suspension feeders. Brachiopods really became abundant during the Ordovician Period.

The third major group of Cambrian organisms were *archaeocyathids* (Figure 22.5). These organisms were ben-

Figure 22.4 *Reconstruction of a Cambrian marine community. Floating jellyfish, swimming arthropods, benthonic sponges, and scavenging trilobites are shown.*

Figure 22.5 *Restoration of a Cambrian reeflike structure built by archaeocyathids.*

thonic sessile suspension feeders that constructed reeflike structures. The rest of the Cambrian fauna consisted of representatives of the other major phyla, including many organisms that were short-lived evolutionary experiments.

The Burgess Shale Biota

No discussion of Cambrian life would be complete without mentioning one of the best examples of a preserved soft-bodied fauna and flora, the Burgess Shale biota. As the Sauk Sea transgressed from the Cordilleran shelf onto the western edge of the craton, Early Cambrian sands were covered by Middle Cambrian black muds that allowed a diverse soft-bodied benthonic community to be preserved. As we discussed in the Prologue, these fossils were discovered in 1909 by Charles D. Walcott near Field, British Columbia. They represent one of the most significant fossil finds of the last century because they consist of impressions of soft-bodied animals and plants (Figure 22.6), which are rarely preserved in the fossil record.

In recent years, the reconstruction, classification, and interpretation of many of the Burgess Shale fossils have undergone a major change that has led to new theories and explanations of the Cambrian explosion of life. Recall that during the Late Proterozoic multicellular organisms evolved, and shortly thereafter animals with hard parts made their first appearance. These were followed by an explosion of invertebrate phyla during the Cambrian, some of which are now extinct. These Cambrian phyla represent the root stock and basic body plans from which all present-day invertebrates evolved. The question that paleontologists are still debating is how many phyla arose during the Cambrian, and at the center of that debate are the Burgess Shale fossils. For years, most paleontologists placed the bulk of the Burgess Shale organisms into existing phyla, with only a few assigned to phyla that are now extinct. Thus, the phyla of the Cambrian world were viewed as being essentially the same in number as the phyla of the present-day world, but with fewer species in each phylum. According to this view, the history of life has been simply a gradual increase in the diversity of species within each phylum through time. The number of basic body plans has therefore remained more or less constant since the initial radiation of multicellular organisms.

This view, however, has been challenged by other paleontologists who think that the initial explosion of varied life-forms in the Cambrian was promptly followed by a short period of experimentation and then extinction of many phyla. The richness and diversity of modern life-forms are the result of repeated variations of the basic body plans that survived the Cambrian extinctions. In other words, life was much more diverse in terms of phyla during the Cambrian than it is today. The reason members of the Burgess Shale biota look so strange to us is that no living organisms possess their basic body plan, and therefore many of them have been placed into new phyla.

Discoveries of new Cambrian fossils at localities such as Sirius Passet, Greenland, and Yunnan, China, have resulted in reassignment of some of the Burgess Shale specimens back into extant phyla. If these reassignments to known phyla prove to be correct, then no massive extinction followed the Cambrian explosion, and life has gradually increased in diversity through time. Currently, there is no clear answer to this debate, and the outcome will probably be decided as more fossil discoveries are made.

Ordovician Marine Community

A major transgression that began during the Middle Ordovician (Tippecanoe sequence) resulted in the most widespread inundation of the craton. This vast epeiric sea, which experienced a uniformly warm climate during this time, opened numerous new marine habitats that were soon filled by a variety of organisms.

Not only did sedimentation patterns change dramatically from the Cambrian to the Ordovician, but the fauna underwent equally striking changes. Whereas the Cambrian invertebrate community was dominated by trilobites, brachiopods, and archaeocyathids, the Ordovician was characterized by the continued diversification of brachiopods and by the adaptive radiation of many other animal phyla, (such as bryozoans and corals), with a consequent dramatic increase in the diversity of the total shelly fauna (Figure 22.7).

During the Cambrian, archaeocyathids were the main builders of reeflike structures, but bryozoans, stromatoporoids, and tabulate and rugose corals assumed that role beginning in the Middle Ordovician. Many of these reefs were small patch reefs similar in size to those of the Cambrian but of a different composition, whereas others were quite large. As with present-day reefs, Ordovician reefs exhibited a high diversity of organisms and were dominated by suspension feeders.

The end of the Ordovician was a time of mass extinctions in the marine realm. More than 100 families of

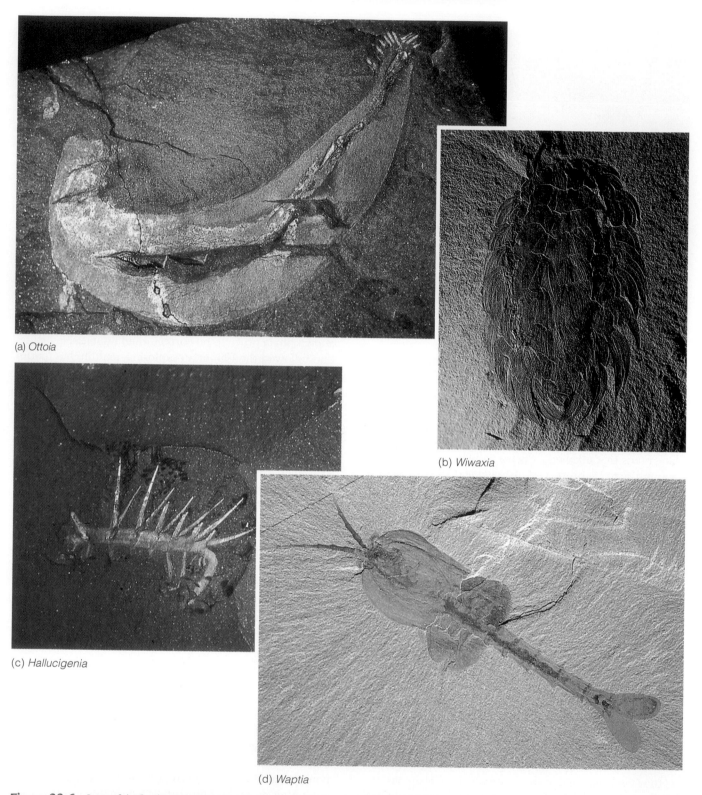

(a) *Ottoia*

(b) *Wiwaxia*

(c) *Hallucigenia*

(d) *Waptia*

Figure 22.6 *Some of the fossil animals preserved in the Burgess Shale. (a) Ottoia, a carnivorous worm. (b) Wiwaxia, a scaly armored sluglike creature whose affinities remain controversial. (c) Hallucigenia, a velvet worm. (d) Waptia, an anthropod.*

marine invertebrates became extinct, and in North America alone, approximately one-half of the brachiopods and bryozoans died out. What caused such an event? Many geologists think that these extinctions were the result of the extensive glaciation that occurred in Gondwana at the end of the Ordovician Period (see Chapter 21).

Figure 22.7 *Recreation of a Middle Ordovician seafloor fauna. Cephalopods, crinoids, colonial corals, graptolites, trilobites, and brachiopods are shown.*

Silurian and Devonian Marine Communities

The mass extinction at the end of the Ordovician was followed by rediversification and recovery of many of the decimated groups. Brachiopods, bryozoans, gastropods, bivalves, corals, crinoids, and graptolites were just some of the groups that rediversified beginning during the Silurian.

As we discussed in Chapter 21, the Silurian and Devonian were times of major reef building. While most of the Silurian radiations of invertebrates represented repopulating of niches, organic reef-builders diversified in new ways, building massive reefs larger than any produced during the Cambrian or Ordovician. This repopulation was probably due in part to renewed transgressions over the craton, and although a major drop in sea level occurred at the end of the Silurian, the Middle Paleozoic sea level was generally high (see Table 21.1).

The Silurian and Devonian reefs were dominated by tabulate and colonial rugose corals and stromatoporoids (Figure 22.8). While the fauna of these Silurian and Devonian reefs was somewhat different from that of earlier reefs and reeflike structures, the general composition and structure are the same as in present-day reefs.

The Silurian and Devonian periods were also the time when *eurypterids* (arthropods with scorpionlike bodies and impressive pincers) were abundant, especially in brackish and freshwater habitats (Figure 22.9). Ammonoids, a subclass of the cephalopods, evolved from nautiloids during the Early Devonian and rapidly diversified. With their distinctive suture patterns, short stratigraphic ranges, and widespread distribution, ammonoids are excellent guide fossils for the Devonian through Cretaceous periods (Figure 22.8).

Another mass extinction occurred near the end of the Devonian and resulted in a worldwide near-total collapse of the massive reef communities. On land, however, the seedless vascular plants were seemingly unaffected, although the diversity of freshwater fish was greatly reduced.

The demise of the Middle Paleozoic reef communities serves to highlight the geographic aspects of the Late Devonian mass extinction. The tropical groups were most severely affected; in contrast, the polar communities were seemingly little affected. Apparently, an episode of global cooling was largely responsible for the extinctions near the end of the Devonian. During such a cooling, the disappearance of tropical conditions would have had a severe effect on reef and other warm-water organisms. Cool-water species, on the other hand, could have simply migrated toward the equator. While cooling temperatures certainly played an important role in the Late Devonian extinctions, the closing of the Iapetus Ocean and the orogenic events of this time undoubtedly also played a role by reducing the area of shallow shelf environments where many marine invertebrates lived.

Carboniferous and Permian Marine Communities

The Carboniferous invertebrate marine community responded to the Late Devonian extinctions in much the same way the Silurian invertebrate marine community responded to the Late Ordovician extinctions—that is, by renewed adaptive radiation and rediversification. The brachiopods and ammonoids quickly recovered and again assumed important ecologic roles, while other groups, such as the lacy bryozoans and crinoids, reached their greatest diversity during the Carboniferous. With the decline of

Figure 22.8 *Reconstruction of a Middle Devonian reef from the Great Lakes area. Shown are corals, ammonoids, trilobites, and brachiopods.*

Figure 22.9 *Restoration of a Silurian brackish-marine bottom scene near Buffalo, New York. Shown are algae, eurypterids, worms, and shrimp.*

the stromatoporoids and the tabulate and rugose corals, large organic reefs like those existing earlier in the Paleozoic virtually disappeared and were replaced by small patch reefs. These reefs were dominated by crinoids, blastoids, lacy bryozoans, brachiopods, and calcareous algae and flourished during the Late Paleozoic (Figure 22.10).

The Permian invertebrate marine faunas resembled those of the Carboniferous, but were not as widely distributed because of the restricted size of the shallow seas on the cratons and the reduced shelf space along the continental margins (see Figure 21.21). The spiny and odd-shaped productids dominated the brachiopod assemblage and constituted an important part of the reef complexes that formed in the Texas region during the Permian (Figure 22.11). The fusulinids, (spindle-shaped foraminifera)

which first evolved during the Late Mississippian and greatly diversified during the Pennsylvanian, experienced a further diversification during the Permian. Because of their abundance, diversity, and worldwide occurrence, fusulinids are important guide fossils for the Pennsylvanian and Permian. Bryozoans, sponges, and some types of calcareous algae also were common elements of the Permian invertebrate fauna.

The Permian Marine Invertebrate Mass Extinction

The greatest recorded mass extinction to affect Earth occurred at the end of the Permian Period (Figure 22.12).

Figure 22.10 *Marine life during the Mississippian based on an Upper Mississippian fossil site at Crawfordville, Indiana. Invertebrate animals shown include crinoids, blastoids, lacy bryozoans, and small corals.*

Before the Permian ended, roughly 50% of all marine invertebrate families and about 90% of all marine invertebrate species became extinct. Fusulinids, rugose and tabulate corals, several bryozoan and brachiopod orders as well as trilobites and blastoids did not survive the end of the Permian. All of these groups had been very successful during the Paleozoic Era. In addition, more than 65% of all amphibians and reptiles, as well as nearly 33% of insects on land also became extinct.

What caused such a crisis for both the marine as well as land-dwelling organisms? Various hypotheses have been proposed, but no completely satisfactory answer has yet been found. One hypothesis that can be ruled out as causing the extinctions is a meteorite impact. No evidence, either direct or indirect, of a meteorite impact has been discovered; furthermore, it appears that the Permian mass extinctions took place over an 8-million-year interval at the end of the Permian Period.

Some hypotheses put forth to explain the extinctions include (1) a reduction of shelf space due to the suturing of the continents to form Pangaea, (2) a global drop in sea level due to glacial conditions, (3) a reduction of shelf

Figure 22.11 *A Permian patch-reef community from the Glass Mountains of West Texas. Shown are algae, productid brachiopods, cephalopods, and corals.*

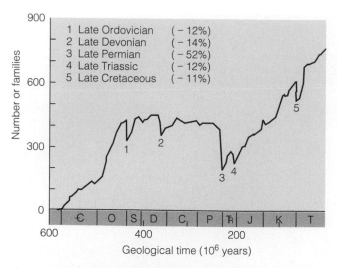

1 Late Ordovician (−12%)
2 Late Devonian (−14%)
3 Late Permian (−52%)
4 Late Triassic (−12%)
5 Late Cretaceous (−11%)

Figure 22.12 *Phanerozoic diversity for marine invertebrate and vertebrate families. Note the three episodes of Paleozoic mass extinctions, with the greatest occurring at the end of the Permian Period.*

the atmosphere and contributing to increased climatic instability and ecologic collapse. At the end of the Permian, a rise in sea level flooded and destroyed the nearshore terrestrial habitats, thus causing the extinction of many terrestrial plants and animals.

In addition to the above scenario, scientists also suggest that a global turnover of deep-ocean waters would have increased the amount of carbon dioxide in the surface waters as well as releasing large amounts of carbon dioxide into the atmosphere. The end result would be the same: global warming and disruption of the marine and terrestrial ecosystem.

Regardless of the ultimate cause of the Permian mass extinctions, the fact is that Earth's biota was dramatically changed. The resulting Triassic marine faunas were of low diversity, but the surviving species tended to be abundant and widely distributed around the world. This fauna provided the root stock from which the Mesozoic marine fauna evolved and repopulated the world's oceans (see Chapter 23).

space due to a large marine regression, and climatic changes. The first two of these can be eliminated because most of the collisions of the continents had already taken place by the end of the Permian and the large-scale formation of glaciers took place during the Pennsylvanian Period. In addition, current evidence indicates a time of sudden warming at the end of the Permian.

Currently, many scientists think that a large-scale marine regression coupled with climatic changes in the form of global warming—due to an increase in carbon dioxide levels—may have been responsible for the mass extinctions recorded in the fossil record. In this scenario, a widespread lowering of sea level occurred near the end of the Permian, which greatly reduced the amount of shallow shelf space for marine organisms and exposed the shelf to erosion. Oxidation of the organic matter trapped in the sediments ensued, which reduced atmospheric oxygen levels as well as releasing large quantities of carbon dioxide into the atmosphere, resulting in increased global warming. During this time, widespread volcanic eruptions also took place, further releasing additional carbon dioxide into

Vertebrate Evolution

A chordate (Phylum Chordata) is an animal that has, at least during part of its life cycle, a notochord, a dorsal hollow nerve cord, and gill slits (Figure 22.13). **Vertebrates,** which are animals with backbones, are simply a subphylum of chordates.

The ancestors and early members of the phylum Chordata were soft-bodied organisms that left few fossils (Figure 22.14). Consequently, we know little about the early evolutionary history of the chordates or vertebrates. Surprisingly, a close relationship exists between echinoderms and chordates. They may even have shared a common ancestor, because the development of the embryo is the same in both groups and differs completely from other invertebrates (Figure 22.15). Furthermore, the biochemistry of muscle activity and blood proteins, and the larval stages are similar in both echinoderms and chordates.

The evolutionary pathway to vertebrates thus appears to have taken place much earlier and more rapidly than

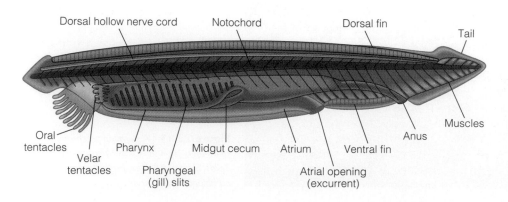

Figure 22.13 *The structure of the lancelet Amphioxus illustrates the three characteristics of a chordate: a notochord, a dorsal hollow nerve cord, and gill slits.*

Figure 22.14 *Found in 525-million-year-old rocks in Yunnan province, China,* Yunnanozoon lividum, *a 5 cm-long animal is the oldest known chordate.*

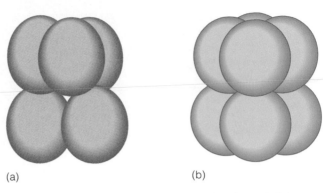

(a)   (b)

Figure 22.15 *(a) Arrangement of cells resulting from spiral cleavage. In this arrangement, cells in successive rows are nested between each other. Spiral cleavage is characteristic of all invertebrates except echinoderms. (b) Arrangement of cells resulting from radial cleavage is characteristic of chordates and echinoderms. In this configuration, cells are directly above each other.*

many scientists have long thought. Based on fossil evidence and recent advances in molecular biology, one scenario suggests that vertebrates evolved shortly after an ancestral chordate, probably resembling *Yunnanozoon,* acquired a second set of genes. According to this hypothesis, a random mutation produced a duplicate set of genes, allowing the ancestral vertebrate animal to evolve entirely new body structures, which proved to be evolutionarily advantageous. Not all scientists accept this hypothesis and the evolution of vertebrates is still hotly debated.

Fish

The most primitive vertebrates are fish, and the oldest fish remains are found in the Upper Cambrian Deadwood Formation in northeastern Wyoming. Here phosphatic scales and plates of *Anatolepis,* a primitive member of the class Agnatha (jawless fish) have been recovered from marine sediments. All known Cambrian and Ordovician fossil fish have been found in shallow, nearshore marine deposits, while the earliest nonmarine fish remains have been found in Silurian strata. This does not prove that fish originated in the oceans, but it does lend strong support to the idea.

As a group, fish range from the Late Cambrian to the present (Figure 22.16). The oldest and most primitive of the class Agnatha are the **ostracoderms,** whose name means "bony skin" (Table 22.2). These are armored jawless fish that first evolved during the Late Cambrian, reached their zenith during the Silurian and Devonian, and then became extinct.

The majority of ostracoderms lived on the sea bottom. *Hemicyclaspis* is a good example of a bottom-dwelling ostracoderm (Figure 22.17). Vertical scales allowed *Hemicy-*

claspis to wiggle sideways, propelling itself along the seafloor, while the eyes on the top of its head allowed it to see such predators as cephalopods and jawed fish approaching from above. While moving along the sea bottom, it probably sucked up small bits of food and sediments through its jawless mouth.

The evolution of jaws was a major advance among primitive vertebrates. While their jawless ancestors could only feed on detritus, jawed fish could chew food and become active predators, thus opening many new ecological niches. The vertebrate jaw is an excellent example of evolutionary opportunism. Various studies suggest that the jaw originally evolved from the first three gill arches of jawless fish. Because the gills are soft, they are supported by gill arches composed of bone or cartilage. The evolution of the jaw may thus have been related to respiration rather than feeding (Figure 22.18). By evolving joints in the forward gill arches, jawless fish could open their mouths wider. Every time a fish opened and closed its mouth, it would pump more water past the gills, thereby increasing the oxygen intake. The modification from rigid to hinged forward gill arches enabled fish to increase both their food consumption and oxygen intake, and the evolution of the jaw as a feeding structure rapidly followed.

The fossil remains of the first jawed fish are found in Silurian rocks and belong to the **acanthodians,** a group of enigmatic fish characterized by large spines, scales covering much of the body, jaws, teeth, and reduced body armor (Figure 22.17c). Many scientists think the acanthodians included the probable ancestors of the present-day bony and cartilaginous fish groups.

The other jawed fish that evolved during the Late Silurian was the **placoderms,** whose name means "plate-skinned" (Table 22.2). Placoderms were heavily armored

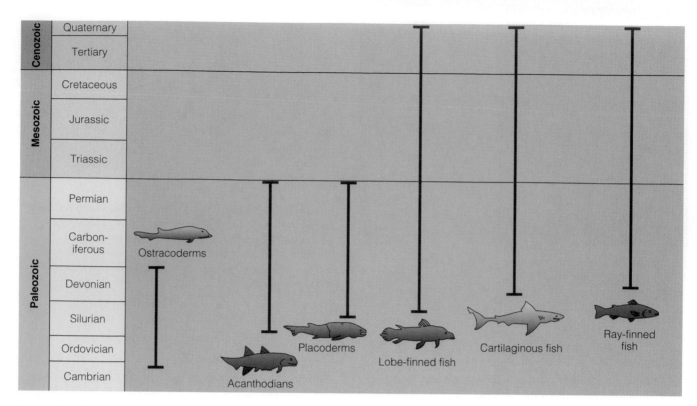

Figure 22.16 *Geologic ranges of the major fish groups.*

Table 22.2

Brief Classification of Fish Showing the Groups Referred to in the Text

Classification	Geologic Range	Living Example
Class Agnatha (jawless fish)	Late Cambrian-Recent	Lamprey, hagfish
Early members of the class are called ostracoderms		No living ostracoderms
Class Acanthodii (the first fish with jaws)	Early Silurian-Permian	None
Class Placodermii (armored jawed fish)	Late Silurian-Permian	None
Class Chondrichthyes (cartilagenous fish)	Devonian-Recent	Sharks, rays, skates
Class Osteichthyes (bony fish)	Devonian-Recent	Tuna, perch, bass, pike, catfish, trout, salmon, lungfish, *Latimeria*
Subclass Actinopterygii (ray-finned fish)	Devonian-Recent	Tuna, perch, bass, pike, catfish, trout, salmon
Subclass Sarcopterygii (lobe-finned fish)	Devonian-Recent	Lungfish, *Latimeria*
Order Dipnoi	Devonian-Recent	Lungfish
Order Crossopterygii	Devonian-Recent	*Latimeria*
Suborder Rhipidistia	Devonian-Permian	None

Figure 22.17 *Recreation of a Devonian seafloor showing (a) an ostracoderm* (Hemicyclapis), *(b) a placoderm* (Bothriolepis), *(c) an acanthodian* (Parexus), *and (d) a ray-finned fish* (Cheirolepis).

jawed fish that lived in both freshwater and the ocean. The placoderms exhibited considerable variety, including small bottom dwellers (Figure 22.17b) as well as large major predators such as *Dunkleosteus*, a Late Devonian fish that lived in the mid-continental North American epeiric seas (Figure 22.19a). It was by far the largest fish of the time, attaining a length of more than 12 m. It had a heavily armored head and shoulder region, a huge jaw lined with razor-sharp bony teeth, and a flexible tail, all features consistent with its status as a ferocious predator.

Besides the abundant acanthodians, placoderms, and ostracoderms, other fish groups, such as the cartilaginous and bony fish, also evolved during the Devonian Period. It is small wonder that the Devonian is informally called the "Age of Fish," because all major fish groups were present during this time period.

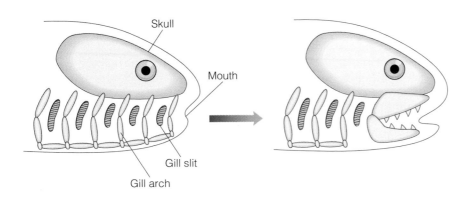

Figure 22.18 *The evolution of the vertebrate jaw is thought to have occurred from the modification of the first two or three anterior gill arches. This theory is based on the comparative anatomy of living vertebrates.*

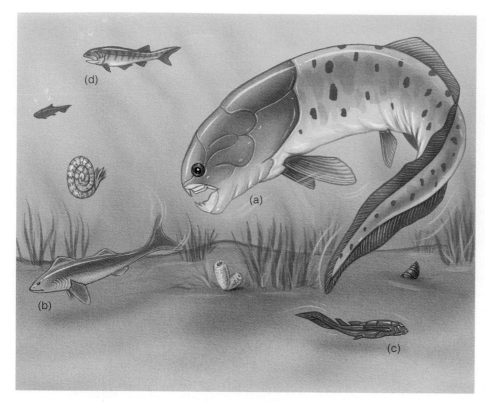

Figure 22.19 *A Late Devonian marine scene from the midcontinent of North America. (a) The giant placoderm* Dunkleosteus *(length more than 12 m) is pursuing (b) the shark* Cladoselache *(length up to 1.2 m). Also shown are (c) the bottom-dwelling placoderm* Bothriolepsis *and (d) the swimming ray-finned fish* Cheirolepsis, *both of which attained a length of 40–50 cm.*

Cartilaginous fish, class Chrondrichthyes (Table 22.2), represented today by sharks, rays, and skates, first evolved during the Middle Devonian, and by the Late Devonian, primitive marine sharks such as *Cladoselache* were quite abundant (Figure 22.19b). Cartilaginous fishes have never been as numerous nor as diverse as their cousins, the bony fishes, but they were, and still are, important members of the marine vertebrate fauna.

Along with the cartilaginous fish, the **bony fish,** class Osteichthyes (Table 22.2), also first evolved during the Devonian. Because bony fish are the most varied and numerous of all the fishes, and because the amphibians evolved from them, their evolutionary history is particularly important. There are two groups of bony fish: the common *ray-finned fish* (Figure 22.19d) and the less familiar *lobe-finned fish* (Table 22.2).

The term ray-finned refers to the way the fins are supported by thin bones that spread away from the body (Figure 22.20a). From a modest freshwater beginning during the Devonian, ray-finned fish, which include most of the familiar fish such as trout, bass, perch, salmon, and tuna, rapidly diversified to dominate the Mesozoic and Cenozoic seas.

Present-day lobe-finned fish are characterized by muscular fins. The fins do not have radiating bones, but rather

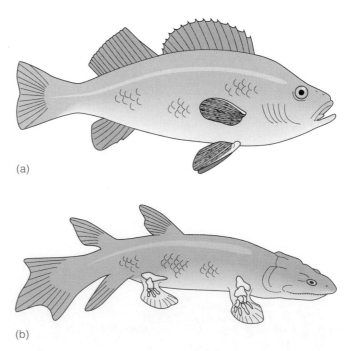

Figure 22.20 *Arrangement of fin bones for (a) a typical ray-finned fish and (b) a lobe-finned fish. The muscles extend into the fin of the lobe-finned fish, allowing greater flexibility of movement than for the ray-finned fish.*

articulating bones with the fin attached to the body by a fleshy shaft (Figure 22.20b). Two major groups of lobe-finned fish are recognized: *lung fish* and *crossopterygians* (Table 22.2).

Lung fish were fairly abundant during the Devonian, but today only three freshwater genera exist, one each in South America, Africa, and Australia. Their present-day distribution presumably reflects the Mesozoic breakup of Gondwana.

The **crossopterygians** are an important group of lobe-finned fish because amphibians evolved from them. During the Devonian, two separate branches of crossopterygians evolved. One led to the amphibians, while the other invaded the sea. This latter group, the *coelacanths*, were thought to have become extinct at the end of the Cretaceous. In 1938, however, fishermen caught a coelacanth in the deep waters off Madagascar, and since then several dozen more have been caught.

The group of crossopterygians that is ancestral to amphibians are *rhipidistians* (Table 22.2). These fish, attaining lengths of over 2 m, were the dominant freshwater predators of the Late Paleozoic. *Eusthenopteron*, a good example of a rhipidistian crossopterygian, had an elongate body that enabled it to move swiftly in the water, as well as paired muscular fins that could be used for locomotion on land (Figure 22.21). The structural similarity between crossopterygian fish and the earliest amphibians is striking and one of the better documented transitions from one major group to another (Figure 22.22).

Amphibians—Vertebrates Invade the Land

Although amphibians were the first vertebrates to live on land, they were not the first land-living organisms. Land plants, which probably evolved from green algae, first evolved during the Ordovician. Furthermore, insects, millipedes, spiders, and even snails invaded the

Figure 22.21 Eusthenopteron, *a member of the rhipidistian crossopterygians. The crossopterygians are the group from which the amphibians are thought to have evolved.* Eusthenopteron *had an elongate body and paired fins that could be used for moving about on land.*

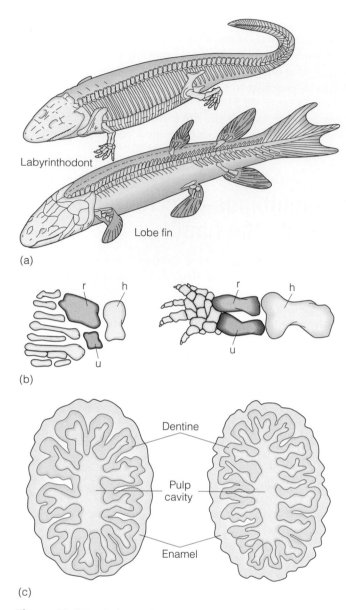

(a)

(b)

(c)

Figure 22.22 *Similarities between the crossopterygian lobe-finned fish and the labyrinthodont amphibians. (a) Skeletal similarity. (b) Comparison of the limb bones of a crossopterygian (left) and amphibian (right); color identifies the bones (u = ulna, shown in blue, r = radius, mauve, h = humerus, gold) that the two groups have in common. (c) Comparison of tooth cross sections shows the complex and distinctive structure found in both the crossopterygians (left) and amphibians (right).*

land before amphibians. Fossil evidence indicates that such land-dwelling arthropods as scorpions and flightless insects had evolved by at least the Devonian.

The transition from water to land required that several barriers be surmounted. The most critical for animals were desiccation, reproduction, the effects of gravity, and the extraction of oxygen from the atmosphere by lungs rather than from water by gills. These problems were partly solved by the crossopterygians; they already had a back-

bone and limbs that could be used for walking and lungs that could extract oxygen (Figure 22.22).

The oldest amphibian fossils are found in the Upper Devonian Old Red Sandstone of eastern Greenland. These amphibians had streamlined bodies, long tails, and fins. In addition, they had four legs, a strong backbone, a rib cage, and pelvic and pectoral girdles, all of which were structural adaptations for walking on land (Figure 22.23). The earliest amphibians appear to have had many characteristics that were inherited from the crossopterygians with little modification.

The Late Paleozoic amphibians did not at all resemble the familiar frogs, toads, newts, and salamanders that make up the modern amphibian fauna. Rather they displayed a broad spectrum of sizes, shapes, and modes of life (Figure 22.24). One group of amphibians were the **labyrinthodonts,** so named for the labyrinthine wrinkling and folding of the chewing surface of their teeth (Figure 22.22). Most labyrinthodonts were large animals, as much as 2 m in length. These typically sluggish creatures lived in swamps and streams, eating fish, vegetation, insects, and other small amphibians (Figure 22.24).

Labyrinthodonts were abundant during the Carboniferous when swampy conditions were widespread, but soon declined in abundance during the Permian, perhaps in response to changing climatic conditions. Only a few species survived into the Triassic.

Figure 22.23 *A Late Devonian landscape in the eastern part of Greenland. Shown is* Ichthyostega, *an amphibian that grew to a length of about 1 m. The flora of the time was diverse, consisting of a variety of small and large seedless vascular plants.*

Figure 22.24 *Reconstruction of a Carboniferous coal swamp. The varied amphibian fauna of the time is shown, including the large labyrinthodont amphibian* Eryops *(foreground), the larval* Branchiosaurus *(center), and the serpentlike* Dolichosoma *(background).*

Evolution of the Reptiles—The Land Is Conquered

Amphibians were limited in colonizing the land because they had to return to water to lay their fishlike gelatinous eggs. The evolution of the **amniote egg** (Figure 22.25) freed reptiles from this constraint. In such an egg, the developing embryo is surrounded by a liquid-filled sac called the *amnion* and provided with both a yolk, or food sac, and an allantois, or waste sac. In this way the emerging reptile is in essence a miniature adult, bypassing the need for a larval stage in the water. The evolution of the amniote egg allowed vertebrates to colonize all parts of the land because they no longer had to return to the water as part of their reproductive cycle.

Many of the differences between amphibians and reptiles are physiological and are not preserved in the fossil record. Nevertheless, amphibians and reptiles differ sufficiently in skull structure, jawbones, ear location, and limb and vertebral construction to suggest that reptiles evolved from labyrinthodont ancestors by the Late Mississippian.

Some of the oldest known reptiles are from the Lower Pennsylvania Joggins Formation in Nova Scotia, Canada. Here, remains of *Hylonomus* are found in the sediments filling in tree trunks. These earliest reptiles were small and agile and fed largely on grubs and insects. They are loosely grouped together as **protorothyrids,** whose members include the earliest reptiles (Figure 22.26). During the Permian Period, reptiles diversified and began displacing many amphibians. The success of the reptiles is due partly to their advanced method of reproduction and their more advanced jaws and teeth as well as to their ability to move rapidly on land.

The **pelycosaurs,** or finback reptiles, evolved from the protorothyrids during the Pennsylvanian and were the dominant reptile group by the Early Permian. They

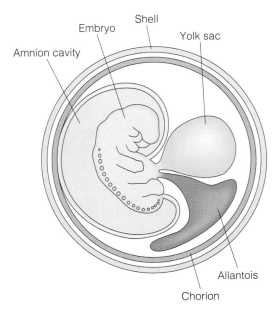

Figure 22.25 *In an amniote egg, the embryo is surrounded by a liquid sac (amnion cavity) and provided with a food source (yolk sac) and waste sac (allantois). The evolution of the amniote egg freed reptiles from having to return to the water for reproduction and allowed them to inhabit all parts of the land.*

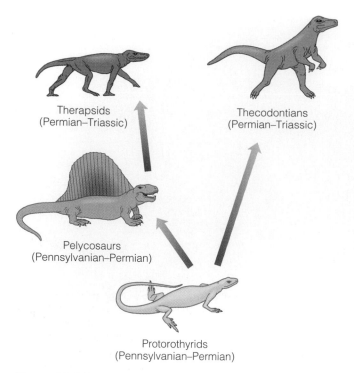

Therapsids
(Permian–Triassic)

Thecodontians
(Permian–Triassic)

Pelycosaurs
(Pennsylvanian–Permian)

Protorothyrids
(Pennsylvanian–Permian)

Figure 22.26 *Evolutionary relationship among the Paleozoic reptiles.*

evolved into a diverse assemblage of herbivores, exemplified by *Edaphosaurus*, and carnivores such as *Dimetrodon* (Figure 22.27). An interesting feature of the pelycosaurs is their sail. It was formed by vertebral spines that, in life, were covered with skin. The sail has been variously explained as a type of sexual display, a means of protection, and a display to look more ferocious. The current consensus seems to be that the sail served as some type of thermoregulatory device, raising the reptile's temperature by catching the sun's rays or cooling it by facing the wind. Because pelycosaurs are considered to be the group from which therapsids evolved, it is interesting that they may have had some sort of body-temperature control.

The pelycosaurs became extinct during the Permian and were succeeded by the **therapsids,** mammal-like reptiles that evolved from the carnivorous pelycosaur lineage and rapidly diversified into herbivorous and carnivorous lineages (Figure 22.28). Therapsids were small to medium-size animals displaying the beginnings of many mammalian features: fewer bones in the skull due to fusion of many of the small skull bones; enlargement of the lower jawbone; differentiation of the teeth for various functions such as nipping, tearing, and chewing food; and a more vertical position of the legs for greater flexibility, as opposed to the way the legs sprawled out to the side in primitive reptiles.

Furthermore, some paleontologists think therapsids were *endothermic,* or warm-blooded, enabling them to maintain a constant internal body temperature. This characteristic would have allowed them to expand into a variety of habitats, and indeed the Permian rocks in which their fossil remains are found have a wide latitudinal distribution.

As the Paleozoic Era came to an end, the therapsids constituted about 90% of the known reptile genera and occupied a wide range of ecological niches. The mass extinctions that decimated the marine fauna at the close of the Paleozoic had an equally great effect on the terrestrial population. By the end of the Permian, about 90% of all marine invertebrate species were extinct compared with more than two-thirds of all amphibians and reptiles. Plants, on the other hand, apparently did not experience as great a turnover as animals.

Figure 22.27 *Most pelycosaurs, or finback reptiles have a characteristic sail on their back. One hypothesis explains the sail as a type of thermoregulatory device. Other hypotheses are that it was a type of sexual display or a device to make the reptile look more intimidating. Shown here are (left) the carnivore* Dimetrodon *and (right) the herbivore* Edaphosaurus.

Figure 22.28 *A Late Permian scene in southern Africa showing various therapsids including* Dicynodon *(left foreground) and* Moschops *(right). Many paleontologists think therapsids were endothermic and may have had a covering of fur as shown here.*

Plant Evolution

When plants made the transition from water to land, they had to solve most of the same problems that animals did: desiccation, support, and the effects of gravity. Plants did so by evolving a variety of structural adaptations that were fundamental to the subsequent radiations and diversification that occurred during the Silurian, Devonian, and later periods (Table 22.3). Most experts agree that the ancestors of land plants first evolved in a marine environment, then moved into a freshwater environment and finally onto land. In this way, the differences in osmotic pressures between salt and freshwater were overcome while the plant was still in the water.

The higher land plants are composed of two major groups, the nonvascular and vascular plants. Most land plants are **vascular,** meaning they have a tissue system of specialized cells for the movement of water and nutrients. The **nonvascular** plants, such as bryophytes (liverworts, hornworts, and mosses) and fungi, do not have these specialized cells and are typically small and usually live in low, moist areas.

The earliest land plants from the Middle to Late Ordovician were probably small and bryophyte-like in their overall organization (but not necessarily related to bryophytes). The evolution of vascular tissue in plants was an important step as it allowed for the transport of food and water.

Discoveries of probable vascular plant megafossils and characteristic spores indicate to many paleontologists that the evolution of vascular plants occurred well before the Middle Silurian. Sheets of cuticlelike cells—that is, the cells that cover the surface of present-day land plants—and tetrahedral clusters that closely resemble the spore tetrahedrals of primitive land plants have been reported from Middle to Upper Ordovician rocks from western Libya and elsewhere.

The ancestor of terrestrial vascular plants was probably some type of green alga. While no fossil record of the transition from green algae to terrestrial vascular plants exists, comparison of their physiology reveals a strong link. Primitive **seedless vascular plants** such as ferns resemble green algae in their pigmentation, important metabolic enzymes, and type of reproductive cycle. Furthermore, the green algae are one of the few plant groups to have made the transition from salt water to freshwater. The evolution of terrestrial vascular plants from an aquatic, probable green algal ancestry was accompanied by various modifications that allowed them to occupy this new and harsh environment.

Silurian and Devonian Floras

The earliest known vascular land plants are small Y-shaped stems assigned to the genus *Cooksonia* from the Middle Silurian of Wales and Ireland. Together with Upper Silurian and Lower Devonian species from Scotland, New York State, and the Czech Republic, these earliest plants were small, simple, leafless stalks with a spore-producing structure at the tip (Figure 22.29); they are known as seedless vascular plants because they did not produce seeds. They also did not have a true root system. A *rhizome,* the underground part of the stem, transferred water from the soil to the plant and anchored the plant to the ground. The sedimentary rocks in which these plant fossils are found indicate that they lived in low, wet, marshy, freshwater environments.

An interesting parallel can be seen between seedless vascular plants and amphibians. When they made the transition from water to land, they had to overcome the problems such a transition involved. Both groups, while successful, nevertheless required a source of water in order to reproduce. In the case of amphibians, their gelatinous egg had to remain moist, while the seedless vascular plants required water for the sperm to travel through to reach the egg.

Table 22.3

Major Events in the Evolution of Land Plants. The Devonian Period was a time of rapid evolution for the land plants. Major events were the appearance of leaves, heterospory, secondary growth, and the emergence of seeds.

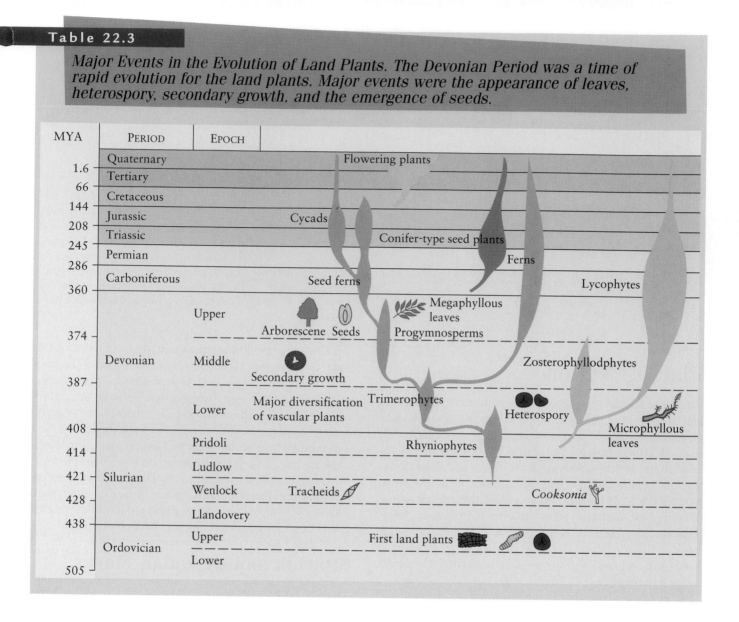

From this simple beginning, the seedless vascular plants evolved many of the major structural features characteristic of modern plants such as leaves, roots, and secondary growth. These features did not all evolve simultaneously but rather at different times, a pattern known as *mosaic evolution*. This diversification and adaptive radiation took place during the Late Silurian and Early Devonian and resulted in a tremendous increase in diversity. During the Devonian, the number of plant genera remained about the same, yet the composition of the flora changed. Whereas the Early Devonian landscape was dominated by relatively small, low-growing, bog-dwelling types of plants, the Late Devonian witnessed forests of large tree-size plants up to 10 m tall.

In addition to the diverse seedless vascular plant flora of the Late Devonian, another significant floral event took place. The evolution of the seed at this time liberated land plants from their dependence on moist conditions and allowed them to spread over all parts of the land.

Seedless vascular plants require moisture for successful fertilization because the sperm must travel to the egg on the surface of the gamete-bearing plant (gametophyte) to produce a successful spore-generating plant (sporophyte). Without moisture, the sperm would dry out before reaching the egg (Figure 22.30a). In the seed method of reproduction, the spores are not released to the environment as they are in the seedless vascular plants, but are retained on the spore-bearing plant, where they grow into the male and female forms of the gamete-bearing generation. In the case of the **gymnosperms,** or flowerless seed plants, these are male and female cones (Figure 22.30b). The male cone produces pollen, which contains the sperm and has a waxy coating to prevent desiccation, while the egg, or embryonic seed, is contained in the female cone. After fertilization, the seed then develops into a mature, cone-bearing plant. In this way the need for a moist environment for the gametophyte generation is solved. The significance of this development is that seed plants, like reptiles, were no longer restricted to wet

Figure 22.29 *The earliest known fertile land plant was Cooksonia, seen in this fossil from the Upper Silurian of South Wales. Cooksonia consisted of upright, branched stems terminating in sporangia (spore-producing structures). It also had a resistant cuticle and produced spores typical of a vascular plant. These plants probably lived in moist environments such as mud flats. This specimen is 1.49 cm long.*

areas, but were free to migrate into previously unoccupied dry environments. While the seedless vascular plants dominated the flora of the Carboniferous coal-forming swamps, the gymnosperms made up an important element of the Late Paleozoic flora, particularly in the nonswampy areas.

Late Carboniferous and Permian Floras

As discussed earlier, the rocks of the Pennsylvania Period (Late Carboniferous) are the major source of the world's coal. Coal results from the alteration of plant remains accumulating in low, swampy areas. The geologic and geographic conditions of the Pennsylvanian were ideal for the growth of seedless vascular plants, and consequently these coal swamps had a very diverse, flora (Figure 22.31) (see Perspective 22.1).

It is evident from the fossil record that whereas the Early Carboniferous flora was similar to its Late Devonian

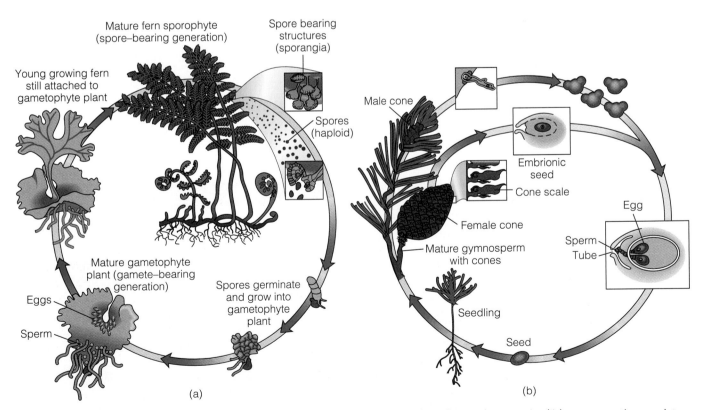

Figure 22.30 *(a) Generalized life history of a seedless vascular plant. The mature sporophyte plant produces spores, which upon generation grow into small gametophyte plants that produce sperm and eggs. The fertilized eggs grow into the spore-producing mature plant, and the sporophyte-gametophyte life cycle begins again. (b) Generalized life history of a gymnosperm plant. The mature plant bears both male cones that produce sperm-bearing pollen grains and female cones that contain embryonic seeds. Pollen grains are transported to the female cones by the wind. Fertilization occurs when the sperm moves through a moist tube growing from the pollen grain and unites with the embryonic seed which then grows into a cone-bearing mature plant.*

Figure 22.31 *Reconstruction of a Pennsylvanian coal swamp with its characteristic vegetation. The amphibian is* Eogyrinus.

counterpart, a great deal of evolutionary experimentation was occurring that would lead to the highly successful Late Paleozoic flora of the coal swamps and adjacent habitats. Among the seedless vascular plants, the *lycopsids* and *sphenopsids* were the most important coal-forming groups of the Pennsylvanian Period.

The lycopsids were the dominant element of the coal swamps, achieving heights up to 30 m in such genera as *Lepidodendron* and *Sigillaria*. The Pennsylvanian lycopsid trees are interesting because they lacked branches except at their top. The leaves were elongate and similar to the individual palm leaf of today. As the trees grew, the leaves were replaced from the top, leaving prominent and characteristic rows or spirals of scars on the trunk. Today, the lycopsids are represented by small temperate-forest ground pines.

The sphenopsids, the other important coal-forming plant group, are characterized by being jointed and having horizontal underground stem-bearing roots. Many of these

The Mazon Creek Fauna and Flora

The unusual state of preservation of the Pennsylvanian-age Mazon Creek biota of northeastern Illinois provides us with significant insights about the soft-part anatomy of organisms rarely preserved in the fossil record. The biota is divided into marine and nonmarine components and contains the only known or oldest fossil representatives of several major animal groups.

The Mazon Creek fossils occur in spheroidal- to elliptical-shaped iron-carbonate concretions ranging from 1 to 30 cm long (Figure 1). Rapid burial and the formation of concretions around the organisms were primarily responsible for the excellent preservation of the fossils. Not only are the hard parts of organisms preserved, but also impressions and carbonaceous films of the soft-bodied animals and plants.

The environment in which these plants and animals lived was a large delta where sluggish southward-flowing rivers emptied into a subtropical epeiric sea that covered most of the present state of Illinois. Two major habitats are represented: a swampy forested lowland of the subaerial delta and the shallow-marine environment of the actively prograding delta.

A diverse marine fauna lived in the warm, shallow waters of the delta front and included cnidarians, mollusks, echinoderms, arthropods, worms, and fish. From this fauna have come the only known fossils of several animal groups, including the lamprey, whose gills and liver can be discerned in the impressions.

More than 350 plant species lived in the swampy lowlands surrounding the delta. Almost all the plants were seedless vascular plants, typical of the kinds that comprised the Pennsylvanian coal-forming swamps of North America. Also found in the swampy lowlands were numerous insects, including millipedes and centipedes as well as spiders and other animals such as scorpions and amphibians. In the ponds, lakes, and rivers were many fish, shrimp, and ostracodes.

Local farmers and townspeople began collecting the Mazon Creek biota in the mid-1800s. Their efforts were soon followed by scientific studies by such famous geologists and paleontologists as J. D. Dana, L. Lesquereux, and E. D. Cope. Study of the

Figure 1 *Seedless vascular plant fossil in iron-carbonate concretion from Pennsylvanian-age deposits, Mazon Creek locality. Illinois.*

Mazon Creek fossils has been uneven through the years, with much of it concentrating on the descriptions of the many plant and animal species recovered. Today, research is still largely concerned with describing the complete assemblage, but researchers are also interested in determining the phylogenetic relationships of the organisms and the paleoecologic significance of the assemblage.

plants, such as *Calamites*, average 5 to 6 m tall. Living sphenopsids include the horsetail (*Equisetum*) and scouring rushes (Figure 22.32). Small seedless vascular plants and seed ferns formed a thick undergrowth or ground cover beneath these treelike plants.

Not all plants were restricted to the coal-forming swamps. Among those plants occupying higher and drier ground were some of the *cordaites*, a group of tall gymnosperm trees that grew up to 50 m and probably formed vast forests (Figure 22.33). Another important nonswamp dweller was *Glossopteris*, the famous plant so abundant in Gondwana (see Fig-

ure 9.1), whose distribution is cited as critical evidence that the continents have moved through time.

The floras that were abundant during the Pennsylvanian persisted into the Permian, but due to climatic and geologic changes resulting from tectonic events they declined in abundance and importance. By the end of the Permian, the cordaites became extinct, while the lycopsids and sphenopsids were reduced to mostly small, creeping forms. Those gymnosperms, with lifestyles more suited to the warmer and drier Permian climates, diversified and came to dominate the Permian, Triassic, and Jurassic landscapes.

Figure 22.32 *Living sphenopsids include the horsetail Equisetum.*

Figure 22.33 *A cordaite forest from the Late Carboniferous. Cordaites were a group of gymnosperm trees that grew up to 50 m tall.*

Chapter Summary

Table 22.4 summarizes the major evolutionary and geologic events of the Paleozoic Era and shows their relationships to each other.

1. Multicelled organisms presumably had a long Precambrian history during which they lacked hard parts. Invertebrates with hard parts suddenly appeared during the Early Cambrian explosion. Skeletons provided such advantages as protection against predators and support for muscles, enabling organisms to grow large and increase locomotor efficiency. Hard parts probably evolved as a result of various geologic and biologic factors rather than a single cause.

2. Marine organisms are classified as plankton if they are floaters, nekton if they swim, and benthos if they live on or in the seafloor.

3. Marine organisms are divided into four basic feeding groups: suspension feeders, which consume microscopic plants and animals as well as dissolved nutrients from water; herbivores, which are plant eaters; carnivore-scavengers, which are meat eaters; and sediment-deposit feeders, which ingest sediment and extract nutrients from it.

4. The marine ecosystem consists of various trophic levels of food production and consumption. At the base are primary producers, upon which all other organisms are dependent. Feeding on the primary producers are the primary consumers, which in turn can be fed upon by higher levels of consumers. The decomposers are bacteria that break down the complex organic compounds of dead organisms and recycle them within the ecosystem.

5. The Cambrian invertebrate community was dominated by three major groups: trilobites, brachiopods, and archaeocyathids. Little specialization existed among the invertebrates, and most phyla were represented by only a few species.

6. The Ordovician marine invertebrate community marked the beginning of the dominance by the shelly fauna and the start of large-scale reef building. The end of the Ordovician Period was a time of major extinctions for many invertebrate phyla.

7. The Silurian and Devonian periods were times of diverse faunas dominated by reef-building animals, while the Carboniferous and Permian periods saw a great decline in invertebrate diversity.

8. Chordates are characterized by a notochord, dorsal hollow nerve cord, and gill slits. The earliest chordates were soft-bodied organisms that were rarely fossilized. Vertebrates are a subphylum of the chordates.

9. Fish are the earliest known vertebrates, with their first fossil occurrence in Upper Cambrian rocks. They have had a long and varied history, including jawless and jawed armored forms (ostracoderms and placoderms), cartilaginous forms, and bony forms. Crossopterygians, a group of lobe-finned fish, gave rise to the amphibians.

10. The link between crossopterygians and the earliest amphibians is convincing and includes a close similarity of bone and tooth structures. The transition from fish to amphibians occurred during the Devonian. During the Carboniferous, the labyrinthodont amphibians were the dominant terrestrial vertebrate animals.

11. The earliest fossil record of reptiles is from the Late Mississippian. The evolution of an amniote egg was the critical factor in the reptiles' ability to colonize all parts of the land.

12. Pelycosaurs were the dominant reptile group during the Early Permian, whereas therapsids dominated the landscape for the rest of the Permian Period.

13. Plants had to overcome the same basic problems as animals, namely desiccation, reproduction, and gravity in making the transition from water to land.

14. The earliest fossil record of land plants is from Middle to Upper Ordovician rocks. These plants were probably small and bryophyte-like in their overall organization.

15. The evolution of vascular tissue was an important event in plant evolution as it allowed food and water to be transported throughout the plant and provided the plant with additional support.

16. The ancestor of terrestrial vascular plants was probably some type of green alga based on such similarities as pigmentation, metabolic enzymes, and the same type of reproductive cycle.

17. The earliest seedless vascular plants were small, leafless stalks with spore-producing structures on their tips. From this simple beginning, plants evolved many of the major structural features characteristic of today's plants.

18. By the end of the Devonian Period, forests with tree-sized plants up to 10 m had evolved. The Late Devonian also witnessed the evolution of the flowerless seed plants (gymnosperms) whose reproductive style freed them from having to stay near water.

19. The Carboniferous Period was a time of vast coal swamps, where conditions were ideal for the seedless vascular plants. With the onset of more arid conditions during the Permian, the gymnosperms became the dominant element of the world's flora.

20. A major extinction occurred at the end of the Paleozoic Era, affecting the invertebrates as well as the vertebrates. Its cause is still being debated.

Table 22.4

Major Evolutionary and Geologic Events of the Paleozoic Era

Age (Millions of Years)	Geologic Period		Invertebrates	Vertebrates
245 —	Permian		Largest mass extinction event to affect the invertebrates.	Acanthodians, placoderms, and pelycosaurs become extinct. Therapsids and pelycosaurs the most abundant reptiles.
286 —	Carboniferous	Pennsylvanian	Fusulinids diversify.	Amphibians abundant and diverse.
320 —		Mississippian	Crinoids, lacy bryozans, and blastoids become abundant. Renewed adaptive radiation following extinctions of many reef-builders.	Reptiles evolve.
360 —	Devonian		Extinctions of many reef-building invertebrates near end of Devonian. Reef building continues. Eurypterids abundant.	Amphibians evolve. All major groups of fish present—Age of Fish.
408 —	Silurian		Major reef building. Diversity of invertebrates remains high.	Ostracoderms common. Acanthodians, the first jawed fish, evolve.
438 —	Ordovician		Extinctions of a variety of marine invertebrates near end of Ordovician. Major adaptive radiation of all invertebrate groups. Suspension feeders dominant.	Ostracoderms diversify.
505 —	Cambrian		Many trilobites become extinct near end of Cambrian. Trilobites, brachiopods, and archaeocyathids are most abundant.	Earliest vertebrates—jawless fish called ostracoderms.
545				

Plants	Major Geologic Events
Gymnosperms diverse and abundant.	Formation of Pangaea. Alleghenian orogeny. Hercynian orogeny.
Coal swamps with flora of seedless vascular plants and gymnosperms.	Coal-forming swamps common. Formation of Ancestral Rockies. Continental glaciation in Gondwana.
Gymnosperms appear (may have evolved during Late Devonian).	Ouachita orogeny.
First seeds evolve. Seedless vascular plants diversify.	Widespread deposition of black shale. Antler orogeny. Acadian orogeny.
Early land plants—seedless vascular plants.	Caledonian orogeny. Extensive barrier reefs and evaporites.
Plants move to land?	Continental glaciation in Gondwana. Taconic orogeny.
	First Phanerozoic transgression (Sauk) onto North American craton.

Important Terms

acanthodian	crossopterygian	ostracoderm	sediment-deposit feeder
amniote egg	gymnosperm	pelycosaur	seedless vascular plant
benthos	herbivore	placoderm	suspension feeder
bony fish	labyrinthodont	plankton	therapsid
carnivore-scavenger	nekton	primary producer	vascular
cartilaginous fish	nonvascular	protorothyrids	vertebrate
chordate			

Review Questions

1. Which of the following three groups of invertebrates comprised the majority of Cambrian skeletonized life?
   a. _____ trilobites, echinoderms, corals; b. _____ trilobites, brachiopods, corals; c. _____ echinoderms, corals, bryozoans; d. _____ trilobites, archaeocyathids, brachiopods; e. _____ brachiopods, archaeocyathids, corals.

2. The greatest recorded mass extinction to affect Earth occurred at the end of which period?
   a. _____ Cambrian; b. _____ Ordovician; c. _____ Devonian; d. _____ Permian; e. _____ Cretaceous.

3. Which of the following fish groups was the first to evolve jaws?
   a. _____ ostracoderms; b. _____ placoderms; c. _____ acanthodians; d. _____ bony; e. _____ lobe-finned.

4. Amphibians evolved from which of the following groups?
   a. _____ reptiles; b. _____ placoderms; c. _____ lungfish; d. _____ archaeocyathids; e. _____ lobe-finned fish.

5. The most significant evolutionary change that allowed reptiles to colonize all of the land was the evolution of
   a. _____ a scaly covering; b. _____ limbs and a backbone capable of supporting the animals on land; c. _____ tear ducts; d. _____ an egg that contained a food and waste sac and surrounded the embryo in a fluid-filled sac; e. _____ the middle ear bones.

6. Which algal group was the probable ancestor to vascular plants?
   a. _____ green; b. _____ blue-green; c. _____ brown; d. _____ red; e. _____ yellow.

7. The fossils of the Burgess Shale are significant because they provide a rare glimpse of
   a. _____ the first shelled animals; b. _____ the soft-part anatomy of extinct groups; c. _____ soft-bodied animals; d. _____ answers (a) and (b); e. _____ answers (b) and (c).

8. Which of the following must an organism possess during at least part of its life cycle to be classified as a chordate?
   a. _____ vertebrae, dorsal hollow nerve cord, gill slits; b. _____ notochord, ventral solid nerve cord, lungs; c. _____ notochord, dorsal hollow nerve cord, gill slits; d. _____ vertebrae, dorsal hollow nerve cord, lungs; e. _____ notochord, dorsal solid nerve cord, lungs.

9. Based on similarity of embryo development, which invertebrate phylum is most closely allied with the chordates?
   a. _____ Arthropoda; b. _____ Annelida; c. _____ Porifera; d. _____ Echinodermata; e. _____ Mollusca.

10. Which reptile group gave rise to the mammals?
    a. _____ labyrinthodonts; b. _____ protothyrids; c. _____ therapsids; d. _____ pelycosaurs; e. _____ acanthodians.

11. The first plant group that did not require a wet area for part of its life cycle was the
    a. _____ seedless vascular; b. _____ gymnosperms; c. _____ naked seed bearing; d. _____ angiosperms; e. _____ flowering.

12. What type of invertebrates dominated the Ordovician invertebrate community?
    a. _____ epifaunal benthonic sessile suspension feeders; b. _____ infaunal benthonic sessile suspension feeders; c. _____ epifaunal benthonic mobile suspension feeders; d. _____ infaunal nektonic carnivores; e. _____ epifloral planktonic primary producers.

13. In which period were amphibians and seedless vascular plants most abundant?
    a. _____ Silurian; b. _____ Devonian; c. _____ Mississippian; d. _____ Pennsylvanian; e. _____ Permian.

14. Discuss how changing geologic conditions affected the evolution of life during the Paleozoic Era.

15. Describe the problems that had to be overcome before organisms could inhabit and completely colonize the land.

16. What are the major differences between the seedless vascular plants and the gymnosperms and why are these differences significant in terms of exploiting the terrestrial environment?

17. Discuss some of the possible causes for the Permian mass extinction.

18. Discuss how the incompleteness of the fossil record may play a role in such theories as the Cambrian explosion of life.

19. If the Cambrian explosion of life was partly the result of the filling of unoccupied niches, why don't we see such rapid evolution following mass extinctions such as occurred at the end of the Permian and Cretaceous periods?

20. Why is it likely that fish evolved in the seas and then migrated to fresh water environments? Could fish have evolved in fresh water and then migrated to the seas?

21. Outline the evolutionary history of fish.

22. Discuss the significance and advantages of the pelycosaur sail.

23. Draw a marine food web that shows the relationships among the producers, consumers, and decomposers.

24. Discuss the significance of the appearance of the first shelled animals and possible causes for the acquisition of a mineralized exoskeleton.

25. What are the major differences between the Cambrian marine community and the Ordovician marine community?

World Wide Web Activities

For these web site addresses, along with current updates and exercises, log on to **http://www.brookscole.com/geo**

University of California Museum of Paleontology

This is an excellent site to visit for any aspect of geologic time, paleontology, and evolution. At this site you can learn about the life history, ecology, and fossil record of any animal or plant group, as well as the history of life for any period or era.

Illinois State Museum Mazon Creek Fossils

This site contains some of the more interesting and dramatic types of fossils recovered from the Francis Creek Shale in Illinois. Included is a section on the importance of the Mazon Creek fossils.

Kevin's Trilobite Home Page

This site contains information, photos, and line drawings of trilobites. In addition to the Table of Contents, it has links to other paleontology sites, a trilobite literature section, trilobite collectors and specialists, trilobite classification, in other words, just about anything you would want to know about trilobites. It is maintained by Kevin Brett at the University of Alberta, Alberta, Canada.

CHAPTER 23

Mesozoic Earth and Life History

Restoration of the Cretaceous dinosaur Giganotosaurus carolinii *from South America. Its length of nearly 13 m and weight of about 7.3 metric tons makes it one of the largest carnivorous dinosaurs. Giganotosaurs lived about 25 million years before the similar and better-known dinosaur* Tyrannosaurus rex.

Approximately 150 to 210 million years after the emplacement of massive plutons created the Sierra Nevada (Nevadan orogeny), gold was discovered at Sutter's Mill on the South Fork of the American River at Coloma, California. On January 24, 1848, James Marshall, a carpenter building a sawmill for John Sutter, found bits of the glittering metal in the mill's tailrace. Soon, settlements throughout the state were completely abandoned as word of the chance for instant riches spread throughout California. Within a year after the news of the gold discovery reached the East Coast, the Sutter's Mill area was swarming with more than 80,000 prospectors, all hoping to make their fortune (Figure 23.1). In all, at least 250,000 gold seekers prospected the Sutter's Mill area, and though most were Americans, they came from all over the world, even as far away as China. Most of them thought the gold was simply waiting to be taken.

Of course, no one gave any thought to the consequences of so many people converging on the Sutter's Mill area, all intent on making easy money. In reality, only a small percentage of prospectors ever hit it big or were even moderately successful. The rest barely eked out a living until they eventually abandoned their dream and went home.

Although some prospectors dug $30,000 worth of gold dust a week out of a single claim and gold was found practically on the surface of the ground, most of this easy gold was recovered very early during the gold rush. Most prospectors made only a living wage working their claims. Nevertheless, during the five years from 1848

OBJECTIVES

At the end of this chapter, you will have learned that

■ The Mesozoic four-stage breakup of Pangaea profoundly affected geologic and biologic events.

■ North America was above sea level during much of the Mesozoic, except for two times when it was partly covered by seaways.

■ Western North America was affected by four interrelated orogenies that took place at an oceanic-continental plate boundary.

■ Marine invertebrates that survived the Paleozoic extinctions diversified and repopulated the seas.

■ Land-plant communities changed markedly when flowering plants evolved during the Cretaceous.

■ Reptile diversification, which began during the Mississippian, continued throughout the Mesozoic with the origin and eventual extinction of dinosaurs, flying reptiles, and marine reptiles.

■ Birds evolved from a reptile stock, probably a small carnivorous dinosaur, whereas mammals descended from mammal-like reptiles.

■ The proximity of continents and mild Mesozoic climates allowed many plants and animals to spread over extensive geographic areas.

■ Extinctions at the end of the Mesozoic resulted in the demise of dinosaurs and some of their close relatives as well as several varieties of marine invertebrates.

Figure 23.1 *By 1852, mining operations were well under way on the American River near Sacramento.*

to 1853 that constituted the gold rush proper, more than $200 million in gold was extracted.

Would-be prospectors could follow three basic routes to the gold fields. The most popular was the overland journey by wagon train from the East to California—a route fraught with peril, including the threat of starvation, disease, and the crossing of the Sierra Nevada. Those who took this route and survived finally arrived at Sutter's Mill and dispersed from there to prospect. The Panama route to California began with a voyage by ship to Chagres on the Isthmus of Panama. Here the potential prospectors and their goods were transported to Panama City by boat and mules and then traveled by ship to California. For the most part, the prospectors who took this route to California were young, vigorous men without wives and children. They faced the usual hardships of steamy jungles, disease, and overcrowded ships, many of which were unseaworthy. The third route, by ship around the southern tip of South America, was the longest, but it avoided the overcrowding encountered on the Panama route and the perils of the overland journey. Those who went by ship arrived at San Francisco, where many decided to stay after seeing that there were opportunities to make money in this new town.

The earliest prospectors came from the West Coast and were, for the most part, honest and hardworking. Those who arrived from elsewhere in 1849 and later were also mostly honest and hardworking, but, unfortunately, some were thieves, outlaws, and murderers. Life in the mining camps was extremely hard and expensive. Frequently, the shopowners and traders made more money than the prospectors did. Gambling halls and saloons sprang up all over because the men had little to do with their time except gamble and drink. Family life was virtually nonexistent as women comprised only 2% of the population.

The gold these prospectors sought was mostly in the form of placer deposits. Weathering of gold-bearing igneous rocks and mechanical separation of minerals by density during stream transport forms placer deposits. Many prospectors searched for the mother lode, but all of the gold recovered during the gold rush came from placers. A common method of mining these deposits is by panning. A prospector dips a shallow pan into a streambed, swirls the material around, and pours off the lighter material. The gold, being about six times heavier than most sand grains and rock chips, concentrates on the bottom of the pan and can then be picked out.

Introduction

The Mesozoic Era (245 to 66 million years ago) is divided into three periods beginning with the Triassic, followed by the Jurassic, and finally the Cretaceous. The stratotypes for the systems from which these periods derive their names are in the Hercynian Mountains of Germany (Triassic), the Jura Mountains of Switzerland (Jurassic), and the Paris Basin of France (Cretaceous).

The dawn of the Mesozoic ushered in a new era in Earth history. The major geologic event was the breakup of Pangaea, which affected oceanic and climatic circulation patterns and influenced the evolution of the terrestrial biotas. Because most of the Mesozoic geologic history of North America involves the continental margins, we focus our attention on the eastern, Gulf, and western coastal regions. The transgressions and regressions of the final two epeiric seas and their depositional sequences (Absaroka and Zuni) will be incorporated into the geologic history of each of these regions.

Why Should You Study Mesozoic Geologic and Life History?

Why should you study Mesozoic geologic and life history? The Mesozoic was an important time in Earth history during which Pangaea began breaking up, resulting in major geologic and biologic changes. Just a few important Mesozoic events include plate movements accounting for the origin of the Atlantic Ocean basin and the Rocky Mountains; accumulation of vast salt deposits that eventually formed salt domes adjacent to which oil and natural gas were trapped; and the emplacement of huge batholiths accounting for the origin of some valuable mineral resources. Furthermore, the Mesozoic Era is known as the "Age of Reptiles," a phrase emphasizing the dominance of reptiles, and particularly dinosaurs, among the land-dwelling vertebrates. Birds, mammals, and flowering plants also first evolved and diversified during this time. The end of the Mesozoic witnessed another mass extinction, the cause(s) of which is still being debated.

The Breakup of Pangaea

Just as the formation of Pangaea influenced geologic and biologic events during the Paleozoic, the breakup of this supercontinent profoundly affected geologic and biologic events during the Mesozoic. The movement of continents affected the global climatic and oceanic regimes as well as the climates of the individual continents. Populations became isolated or were brought into contact with other populations, leading to evolutionary changes in the biota. So great was the effect of this breakup on the world that it forms the central theme of this chapter.

Geologic, paleontologic, and paleomagnetic data indicate that the breakup of Pangaea took place in four general stages. The first stage involved rifting between Laurasia and Gondwana during the Late Triassic. By the end of the Triassic, the expanding Atlantic Ocean separated North America from Africa (Figure 23.2a). This change was followed by the rifting of North America from South America sometime during the Late Triassic and Early Jurassic.

Separation of the continents allowed water from the Tethys Sea to flow into the expanding central Atlantic Ocean, while Pacific Ocean waters flowed into the newly formed Gulf of Mexico, which at that time was little more than a restricted bay (Figure 23.3). During that time, these areas were in the low tropical latitudes where high temperatures and high rates of evaporation were ideal for the formation of thick evaporite deposits.

The second stage in Pangaea's breakup involved rifting and movement of the various Gondwana continents during the Late Triassic and Jurassic periods. As early as the Late Triassic, Antarctica and Australia, which remained sutured together, separated from South America and Africa, while India split away from all four Gondwana continents and began moving northward (Figures 23.2a and b).

The third stage of breakup began during the Late Jurassic, when South America and Africa began separating (Figure 23.2b). The rifting and subsequent separation of these two continents formed a narrow basin where thick evaporite deposits accumulated from the evaporation of southern ocean waters (Figure 23.3). During this stage, the eastern end of the Tethys Sea began closing as a result of the clockwise rotation of Laurasia and the northward movement of Africa. This narrow Late Jurassic and Cretaceous seaway between Africa and Europe was the forerunner of the present Mediterranean Sea.

By the end of the Cretaceous, Australia and Antarctica had separated, India had nearly reached the equator, South America and Africa were widely separated, and the eastern side of what is now Greenland had begun separating from Europe (Figure 23.2c).

The final stage in Pangaea's breakup occurred during the Cenozoic. During this time, Australia continued moving northward, and Greenland completely separated from Europe and rifted from North America to form a separate landmass.

The Effects of the Breakup of Pangaea on Global Climates and Ocean Circulation Patterns

By the end of the Permian Period, Pangaea extended from pole to pole, covered about one-fourth of Earth's surface, and was surrounded by Panthalassa, a global ocean that encompassed about 300 degrees of longitude. Such a configuration exerted tremendous influence on the world's climate and resulted in generally arid conditions over large parts of Pangaea's interior.

The world's climates result from the complex interaction between wind and ocean currents and the location and topography of the continents. In general, dry climates occur on large landmasses in areas remote from sources of moisture and where barriers to moist air, such as mountain ranges, exist. Wet climates occur near large bodies of water or where winds can carry moist air over land.

Past climatic conditions can be inferred from the distribution of climate-sensitive deposits. Evaporites are deposited where evaporation exceeds precipitation. While desert dunes and red beds may form locally in humid regions, they are characteristic of arid regions. Coal forms in both warm and cool humid climates. Vegetation that is eventually converted into coal requires at least a good seasonal water supply; thus, coal deposits are indicative of humid conditions.

Widespread Triassic evaporites, red beds, and desert dunes in the low and middle latitudes of North and South America, Europe, and Africa indicate dry climates in those regions, while coal deposits are found mainly in the high latitudes, indicating humid conditions. These high-latitude coals are analogous to today's Scottish peat bogs or Canadian muskeg. The lands bordering the Tethys Sea were probably dominated by seasonal monsoon rains resulting from the warm moist winds and warm oceanic currents impinging against the east-facing coast of Pangaea.

The temperature gradient between the tropics and the poles also affects oceanic and atmospheric circulation. The greater the temperature difference between the tropics and the poles, the steeper the temperature gradient, and the faster the circulation of the oceans and atmosphere. Oceans absorb about 90% of the solar radiation they receive, while continents absorb only about 50%, even less if they are snow covered. The rest of the solar radiation is reflected back into space. Therefore, areas dominated by seas are warmer than those dominated by continents. By knowing the distribution of continents and ocean basins, geologists can calculate the average annual temperature for any region on Earth, as well as determining a temperature gradient.

The breakup of Pangaea during the Late Triassic caused the global temperature gradient to increase because the Northern Hemisphere continents moved further northward, displacing higher-latitude ocean waters. Due to the steeper global temperature gradient caused by a decrease in temperature in the high latitudes and the changing positions of the continents, oceanic and atmospheric circulation patterns greatly accelerated during the Mesozoic (Figure 23.4). Though the temperature gradient and seasonality on land were increasing during the Jurassic and Cretaceous, the middle- and higher-latitude oceans were still quite warm, because warm waters from the Tethys Sea were circulating to the higher latitudes. The result was a relatively equable worldwide climate through the end of the Cretaceous.

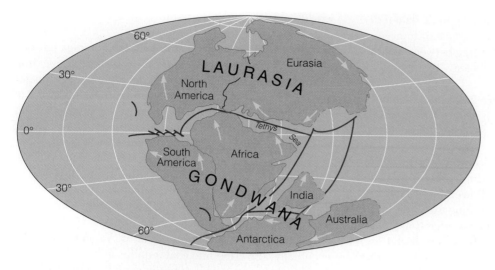

(a) Triassic Period (245–208 MYA)

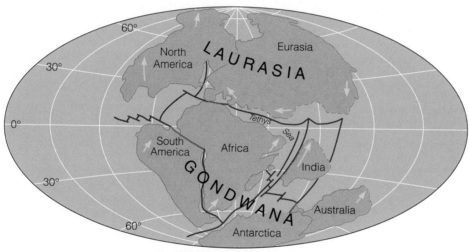

(b) Jurassic Period (208–144 MYA)

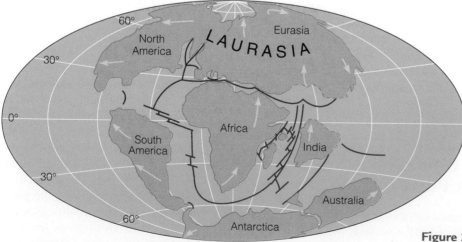

(c) Cretaceous Period (144–66 MYA)

Figure 23.2 *Paleogeography of the world during the Mesozoic. Yellow arrows show the direction of movement for the continents. (a) The Triassic Period, (b) the Jurassic Period, and (c) the Cretaceous Period.*

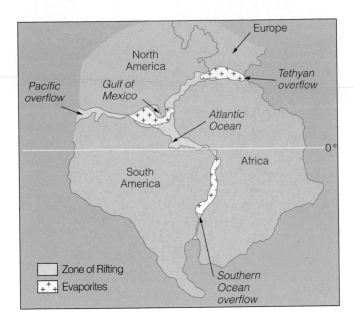

Figure 23.3 *Evaporites accumulated in shallow basins as Pangaea broke apart during the Early Mesozoic. Water from the Tethys Sea flowed into the central Atlantic Ocean, while water from the Pacific Ocean flowed into the newly formed Gulf of Mexico. Marine water from the south flowed into the southern Atlantic Ocean.*

Mesozoic History of North America

The beginning of the Mesozoic Era was essentially the same in terms of tectonism and sedimentation as the preceding Permian Period in North America. Terrestrial sedimentation continued over much of the craton, while block faulting and igneous activity began in the Appalachian region as North America and Africa began separating (Figure 23.5). The newly forming Gulf of Mexico experienced extensive evaporite deposition during the Late Triassic and Jurassic as North America separated from South America (Figure 23.6).

A global rise in sea level during the Cretaceous resulted in worldwide transgressions onto the continents (Figure 23.7). These transgressions were caused by higher heat flow along the oceanic ridges due to increased rifting and the consequent expansion of oceanic crust. By the Middle Cretaceous, sea level probably was as high as at any time since the Ordovician, and approximately one-third of the present land area was inundated by epeiric seas.

Marine deposition was continuous over much of the North American Cordillera. A volcanic island arc system that formed off the western edge of the craton during the Permian was sutured to North America sometime later during the Permian or Triassic. This event is referred to as

the *Sonoma orogeny* and will be discussed later in this chapter. During the Jurassic, the entire Cordilleran area was involved in a series of major mountain-building episodes that resulted in the formation of the Sierra Nevada, the Rocky Mountains, and other lesser mountain ranges. While each orogenic episode has its own name, the entire mountain-building event is simply called the *Cordilleran orogeny* (also discussed later in this chapter). With this simplified overview of the Mesozoic history of North America in mind, we will now examine the specific regions of the continent.

Continental Interior

Recall that the history of the North American craton can be divided into unconformity-bound sequences reflecting advances and retreats of epeiric seas over the craton (see Figure 21.5). While these transgressions and regressions played a major role in the Paleozoic geologic history of the continent, they were not as important during the Mesozoic. Most of the continental interior during the Mesozoic was well above sea level and did not experience epeiric sea inundation. Consequently, the two Mesozoic cratonic sequences, the *Absaroka sequence* (Late Mississippian to Early Jurassic) and the *Zuni sequence* (Early Jurassic to Early Paleocene) (see Figure 21.5) are incorporated as part of the history of the three continental margin regions of North America.

Eastern Coastal Region

During the Early and Middle Triassic, coarse detrital sediments derived from the erosion of the recently uplifted Appalachians (Alleghenian orogeny) filled the various intermontane basins and spread over the surrounding areas. As erosion continued during the Mesozoic, this once lofty mountain system was reduced to a low-lying plain. During the Late Triassic, the first stage in the breakup of Pangaea began with North America separating from Africa. Fault-block basins developed in response to upwelling magma beneath Pangaea in a zone stretching from present-day Nova Scotia to North Carolina (Figure 23.8). Erosion of the adjacent fault-block mountains filled these basins with great quantities (up to 6000 m) of poorly sorted red-colored nonmarine detrital sediments known as the *Newark Group.* Reptiles roamed along the margins of the various lakes and streams that formed in these basins, leaving their footprints and trackways in the soft sediments. Although the Newark rocks contain numerous dinosaur footprints, they are almost completely devoid of dinosaur bones! The Newark Group is mostly Late Triassic in age, but in some areas deposition began in the Early Jurassic.

Concurrent with sedimentation in the fault-block basins were extensive lava flows that blanketed the basin floors as well as intrusions of numerous dikes and sills. The

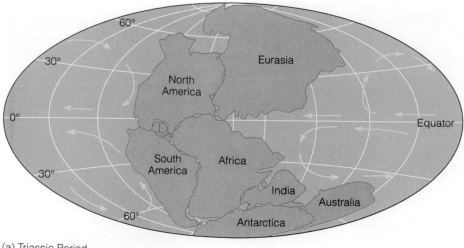

(a) Triassic Period

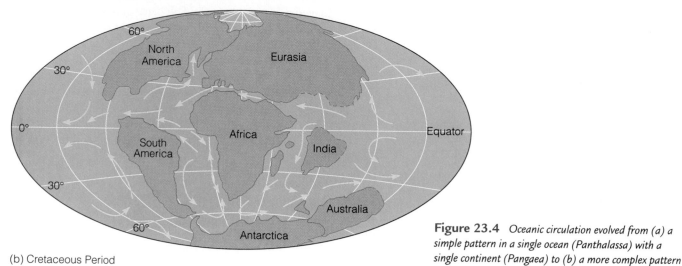

(b) Cretaceous Period

Figure 23.4 *Oceanic circulation evolved from (a) a simple pattern in a single ocean (Panthalassa) with a single continent (Pangaea) to (b) a more complex pattern in the newly formed oceans of the Cretaceous Period.*

most famous intrusion is the prominent Palisades sill along the Hudson River in the New York–New Jersey area (Figure 23.8d).

As the Atlantic Ocean grew, rifting ceased along the eastern margin of North America, and this once active plate margin became a passive, trailing continental margin. The fault-block mountains that were produced by this rifting continued eroding during the Jurassic and Early Cretaceous until all that was left was a large low-relief area. The sediments produced by this erosion contributed to the growing eastern continental shelf. During the Cretaceous Period, the Appalachian region was reelevated and once again shed sediments onto the continental shelf, forming a gently dipping, seaward-thickening wedge of rocks up to 3000 m thick. These rocks are currently exposed in a belt extending from Long Island, New York, to Georgia.

Gulf Coastal Region

The Gulf Coastal region was above sea level until the Late Triassic (Figure 23.5). As North America separated from South America during the Late Triassic, the Gulf of Mexico began to form (Figure 23.6). With oceanic waters flowing into this newly formed, shallow, restricted basin, conditions were ideal for evaporite formation. More than 1000 m of evaporites were precipitated at this time, and most geologists think that these Jurassic evaporites are the source for the Tertiary salt domes found today in the Gulf of Mexico and southern Louisiana.

By the Late Jurassic, circulation in the Gulf of Mexico was less restricted, and evaporite deposition ended. Normal marine conditions returned to the area with alternating transgressing and regressing seas. The resulting sediments were covered and deeply buried by thousands of meters of Cretaceous and Cenozoic sediments.

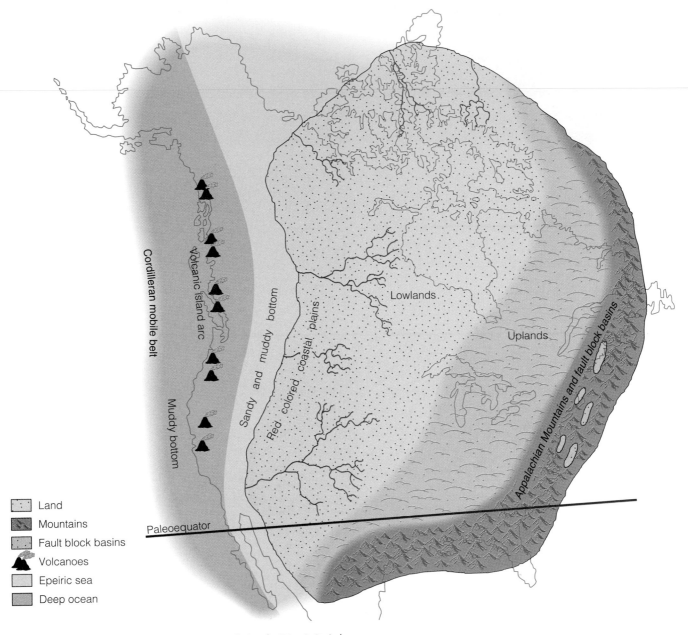

Figure 23.5 *Paleogeography of North America during the Triassic Period.*

During the Cretaceous, the Gulf Coastal region, like the rest of the continental margin, was inundated by northward-transgressing seas (Figure 23.7). As a result, nearshore sandstones are overlain by finer sediments characteristic of deeper waters. Following an extensive regression at the end of the Early Cretaceous, a major transgression began during which a wide seaway extended from the Arctic Ocean to the Gulf of Mexico (Figure 23.7). Sediments that were deposited in the Gulf Coastal region during the Cretaceous formed a seaward-thickening wedge.

Reefs were also widespread in the Gulf Coastal region during the Cretaceous. Bivalves called *rudists* were the main constituent of many of these reefs. Because of their high porosity and permeability, rudistoid reefs make excellent petroleum reservoirs.

Western Region

Mesozoic Tectonics

The Mesozoic geologic history of the North American Cordilleran mobile belt is very complex, involving the eastward subduction of the oceanic Pacific plate under the continental North American plate. Activity along this oceanic-continental convergent plate boundary resulted in an eastward movement of deformation. This

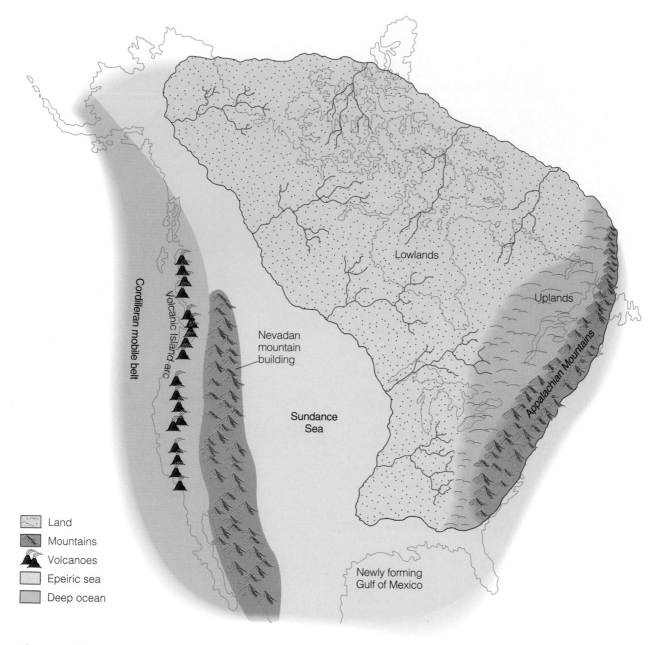

Figure 23.6 *Paleogeography of North America during the Jurassic Period.*

orogenic activity progressively affected the trench and continental slope, the continental shelf, and the cratonic margin, causing a thickening of the continental crust.

Except for the Late Devonian-Early Mississippian Antler orogeny (see Figure 21.28), the Cordilleran region of North America experienced little tectonism during the Paleozoic. However, during the Permian, an island arc and ocean basin formed off the western North American craton (Figure 23.5). This was followed by subduction of an oceanic plate beneath the island arc and the thrusting of oceanic and island arc rocks eastward against the craton margin (Figure 23.9). This event, known as the **Sonoma orogeny,** occurred at or near the Permian-Triassic boundary.

Following the Late Paleozoic-Early Mesozoic destruction of the volcanic island arc during the Sonoma orogeny, the western margin of North America became an oceanic-continental convergent plate boundary. During the Late Triassic, a steeply dipping subduction zone developed along the western margin of North America in response to the westward movement of North America over the Pacific plate. This newly created oceanic-continental plate boundary controlled Cordilleran tectonics for the rest of the Mesozoic and for most of the Cenozoic Era; this subduction zone marks the beginning of the modern circum-Pacific orogenic system.

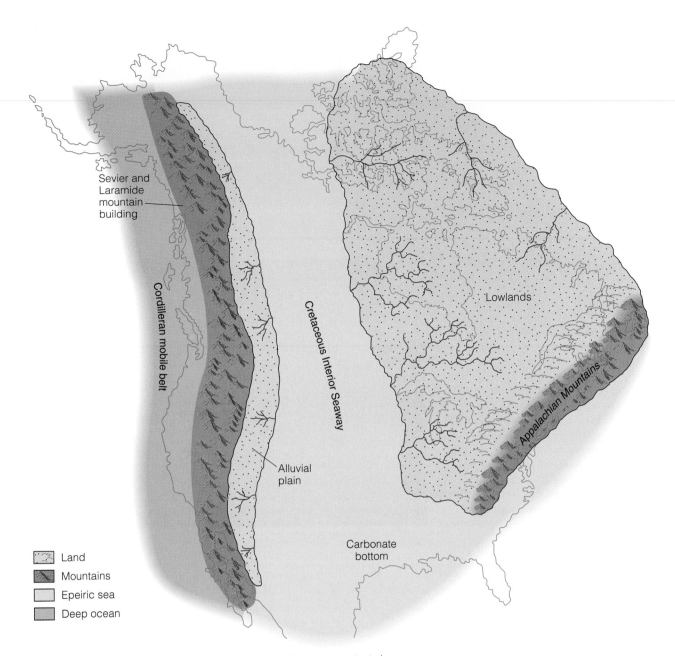

Figure 23.7 *Paleogeography of North America during the Cretaceous Period.*

Two subduction zones dipping in opposite directions from each other formed off the west coast of North America during the Middle and early Late Jurassic (Figure 23.10). The more westerly subduction zone was eliminated by the westward-moving North American plate, which overrode the oceanic Pacific plate.

The *Franciscan Complex*, which is up to 7000 m thick, is an unusual rock unit consisting of a chaotic mixture of rocks that accumulated during the Late Jurassic and Cretaceous. The various rock types such as graywacke, volcanic breccia, siltstone, black shale, chert, pillow basalt, and blueshist metamorphic rocks, suggest that continental shelf, slope, and deep-sea environments were brought together in

a submarine trench when North America overrode the subducting Pacific plate (Figure 23.10) (see Perspective 10.2).

East of the Franciscan complex and currently separated from it by a major thrust fault is the *Great Valley Group.* It consists of more than 16,000 m of Cretaceous conglomerates, sandstones, siltstones, and shales. These sediments were deposited on the continental shelf and slope at the same time the Francisan deposits were accumulating in the submarine trench (see Figure 6.22).

The general term **Cordilleran orogeny** is applied to the mountain-building activity that began during the Jurassic and continued into the Cenozoic (Figure 23.11). The Cordilleran orogeny consisted of a series of individual

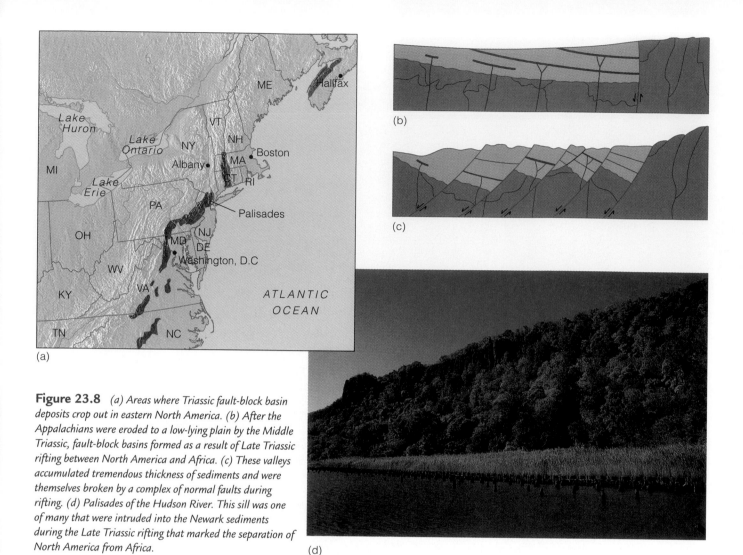

(a)

(b)

(c)

(d)

Figure 23.8 (a) Areas where Triassic fault-block basin deposits crop out in eastern North America. (b) After the Appalachians were eroded to a low-lying plain by the Middle Triassic, fault-block basins formed as a result of Late Triassic rifting between North America and Africa. (c) These valleys accumulated tremendous thickness of sediments and were themselves broken by a complex of normal faults during rifting. (d) Palisades of the Hudson River. This sill was one of many that were intruded into the Newark sediments during the Late Triassic rifting that marked the separation of North America from Africa.

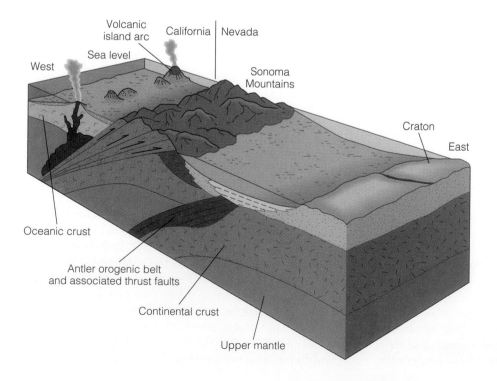

Figure 23.9 Tectonic activity that culminated in the Permian-Triassic Sonoma orogeny in western North America. The Sonoma orogeny was the result of a collision between the southwestern margin of North America and an island arc system.

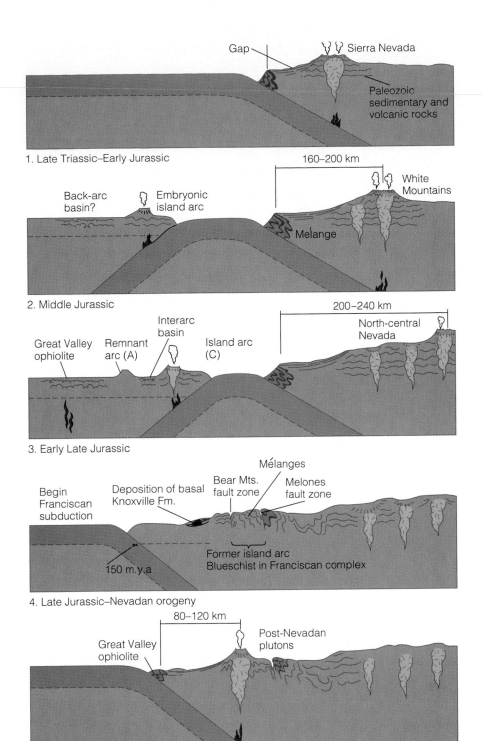

1. Late Triassic–Early Jurassic

2. Middle Jurassic

3. Early Late Jurassic

4. Late Jurassic–Nevadan orogeny

5. Latest Jurassic

Figure 23.10 *Interpretation of the tectonic evolution of the Sierra Nevada during the Mesozoic Era.*

mountain-building events that occurred in different regions at different times. Most of this Cordilleran orogenic activity is related to the continued westward movement of the North American plate.

The first phase of the Cordilleran orogeny, the **Nevadan orogeny** (Figure 23.11), began during the Late Jurassic and continued into the Cretaceous as large volumes of granitic magma were generated at depth beneath

the western edge of North America. These granitic masses ascended as huge batholiths that are now recognized as the Sierra Nevada, Southern California, Idaho, and Coast Range batholiths (Figure 23.12).

By the Late Cretaceous, most of the volcanic and plutonic activity had migrated eastward into Nevada and Idaho. This migration was probably caused by a change from high-angle to low-angle subduction, resulting in the

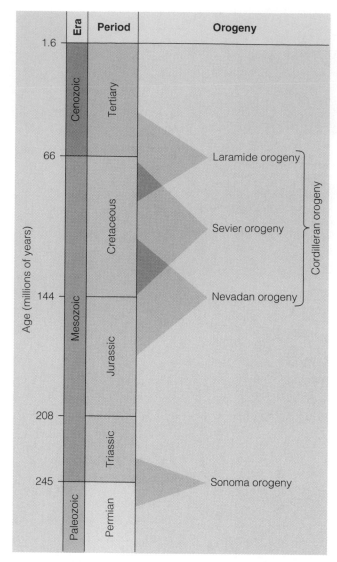

Figure 23.11 *Mesozoic orogenies occurring in the Cordilleran mobile belt.*

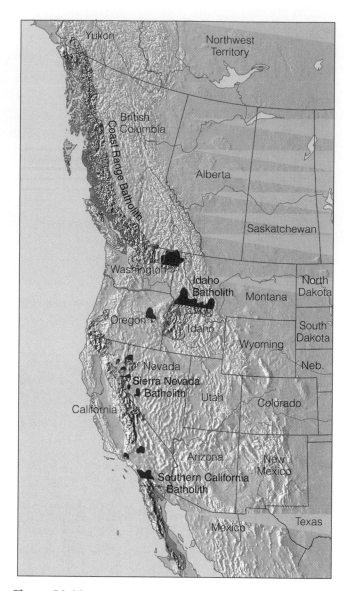

Figure 23.12 *Location of Jurassic and Cretaceous batholiths in western North America.*

subducting oceanic plate reaching its melting depth farther east (Figure 23.13). Thrusting occurred progressively further east so that by the Late Cretaceous, it extended all the way to the Idaho-Washington border.

The second phase of the Cordilleran orogeny, the **Sevier orogeny,** was mostly a Cretaceous event (Figure 23.11). Subduction of the Pacific plate beneath the North American plate continued during this time, resulting in numerous overlapping, low-angle thrust faults in which blocks of older strata were thrust eastward on top of younger strata (Figure 23.14). This deformation produced generally north-south-trending mountain ranges that stretch from Montana to western Canada.

During the Late Cretaceous to Early Cenozoic, the final pulse of the Cordilleran orogeny occurred (Figure 23.11). The **Laramide orogeny** developed east of the Sevier orogenic belt in the present-day Rocky Mountain areas of New Mexico, Colorado, and Wyoming. Most of the features of the present-day Rocky Mountains resulted from the Cenozoic phase of the Laramide orogeny, and for that reason, it will be discussed in Chapter 24.

Mesozoic Sedimentation

Concurrent with the tectonism in the Cordilleran mobile belt, Early Triassic sedimentation on the western continental shelf consisted of shallow-water marine sandstones, shales, and limestones. During the Middle and Late Triassic, the western shallow seas regressed further west, exposing large areas of former seafloor to erosion. Marginal marine and nonmarine Triassic rocks, particularly red beds, contribute to the spectacular and colorful scenery of the region.

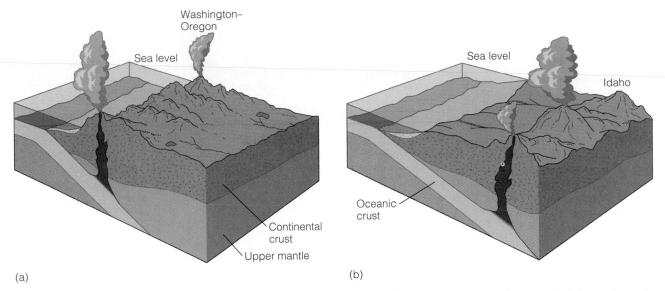

Figure 23.13 *A possible cause for the eastward migration of igneous activity in the Cordilleran region during the Cretaceous Period was a change from (a) high-angle to (b) low-angle subduction. As the subducting plate moved downward at a lower angle, the depth of melting moved farther to the east.*

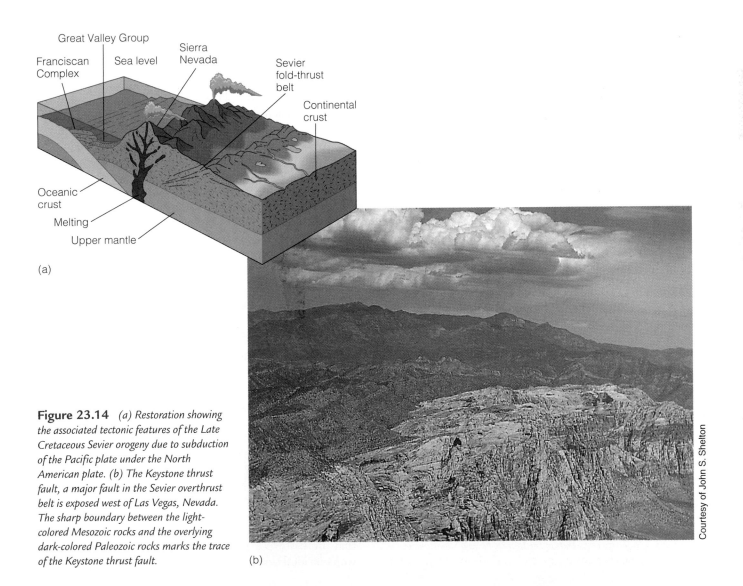

Figure 23.14 *(a) Restoration showing the associated tectonic features of the Late Cretaceous Sevier orogeny due to subduction of the Pacific plate under the North American plate. (b) The Keystone thrust fault, a major fault in the Sevier overthrust belt is exposed west of Las Vegas, Nevada. The sharp boundary between the light-colored Mesozoic rocks and the overlying dark-colored Paleozoic rocks marks the trace of the Keystone thrust fault.*

Mesozoic History of North America **675**

Figure 23.15 *Petrified Forest National Park, Arizona. All of the logs here are* Araucarioxylon, *which is the most abundant tree in the park. The petrified logs have been weathered from the Chinle Formation and are mostly in the position in which they were buried some 200 million years ago.*

These rocks represent a variety of continental depositional environments. The Upper Triassic *Chinle Formation,* for example, is widely exposed throughout the Colorado Plateau region and is probably most famous for its petrified wood spectacularly exposed in Petrified Forest National Park, Arizona (Figure 23.15). This formation, as well as other Triassic formations in the Southwest, also contains the fossilized remains and tracks of various amphibians and reptiles.

Early Jurassic deposits in a large part of the western region consist mostly of clean, cross-bedded sandstones indicative of windblown deposits. The thickest and most prominent of these is the *Navajo Sandstone,* a widespread cross-bedded sandstone that accumulated in a coastal dune environment along the southwestern margin of the craton. The sandstone's most distinguishing feature is its large-scale cross-beds, some of which are more than 25 m high (Figure 23.16).

Marine conditions returned to the region during the Middle Jurassic when a wide seaway called the **Sundance Sea,** twice flooded the interior of western North America (Figure 23.6). The resulting deposits, the *Sundance Formation,* were produced from erosion of tectonic highlands to the west that paralleled the shoreline. These highlands resulted from intrusive igneous activity and associated volcanism that began during the Triassic.

During the Late Jurassic, a mountain chain formed in Nevada, Utah, and Idaho as a result of the deformation produced by the Nevadan orogeny (Figure 23.17a). As the mountain chain grew and shed sediments eastward, the Sundance Sea began retreating northward. A large part of the area formerly occupied by the Sundance Sea was then covered by multicolored sandstones, mudstones, shales, and occasional lenses of conglomerates that comprise the world-famous *Morrison Formation* (Figure 23.17b). The Morrison Formation contains the world's richest assemblage of Jurassic dinosaur remains. Although most of the dinosaur skeletons are broken up, as many as 50 individuals have been found together in a small area.

Shortly before the end of the Early Cretaceous, Arctic waters spread southward over the craton, forming a large inland sea in the Cordilleran foreland basin area. Mid-Cretaceous transgressions also occurred on other continents, and all were part of the global mid-Cretaceous rise in sea level that resulted from accelerated seafloor spreading as Pangaea continued to fragment.

By the beginning of the Late Cretaceous, this incursion joined the northward-transgressing waters from the Gulf area to create an enormous **Cretaceous Interior Seaway** that occupied the area east of the Sevier orogenic belt. Extending from the Gulf of Mexico to the Arctic Ocean, and more than 1500 km wide at its maximum extent, this seaway effectively divided North America into two large landmasses until just before the end of the Late Cretaceous (Figure 23.7).

Cretaceous deposits less than 100 m thick indicate that the eastern margin of the Cretaceous Interior Seaway subsided slowly and received little sediment from the

Figure 23.16 *Large cross-beds of the Jurassic Navajo Sandstone in Zion National Park, Utah.*

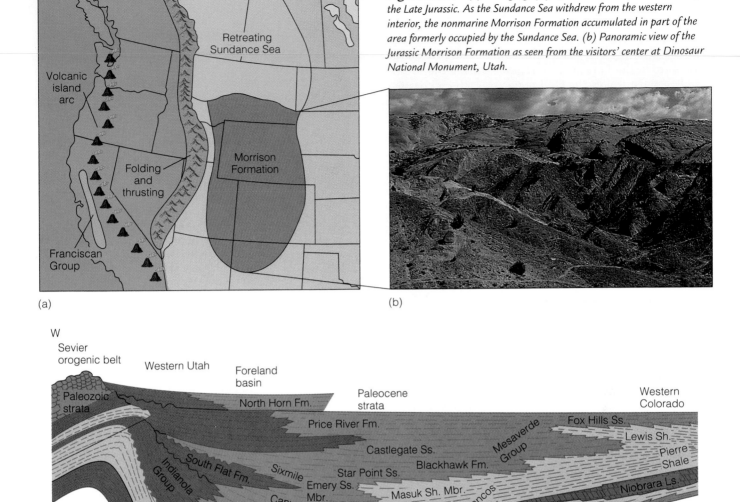

Figure 23.17 *(a) Paleogeography of western North America during the Late Jurassic. As the Sundance Sea withdrew from the western interior, the nonmarine Morrison Formation accumulated in part of the area formerly occupied by the Sundance Sea. (b) Panoramic view of the Jurassic Morrison Formation as seen from the visitors' center at Dinosaur National Monument, Utah.*

(a)

(b)

Figure 23.18 *This restored west-east cross section of Cretaceous facies of the western Cretaceous Interior Seaway shows their relationship to the Sevier orogenic belt.*

emergent, low-relief craton to the east. The western shoreline, however, shifted back and forth, primarily in response to fluctuations in the supply of sediment from the Cordilleran Sevier orogenic belt to the west. The facies relationships show lateral changes from conglomerate and coarse sandstone adjacent to the mountain belt through finer sandstones, siltstones, shales, and even limestones and chalks in the east (Figure 23.18). During times of particularly active mountain building, these coarse clastic wedges of gravel and sand prograded even further east.

As the Mesozoic Era ended, the Cretaceous Interior Seaway withdrew from the craton. During this regression, marine waters retreated to the north and south, and marginal marine and continental deposition formed widespread coal-bearing deposits on the coastal plain.

Mesozoic Mineral Resources

Although much of the coal in North America is Pennsylvanian or Tertiary in age, important Mesozoic coals occur in the Rocky Mountain states. These are mostly lignite and bituminous coals, but some local anthracites are present as well. Particularly widespread in western North America are coals of Cretaceous age. Mesozoic coals are also known from Alberta and British Columbia, Canada, as well as from Australia, Russia, and China.

Large concentrations of petroleum occur in many areas of the world, but more than 50% of all proven reserves are in the Persian Gulf region. During the Mesozoic Era, what is now the Gulf region was a broad passive continental margin extending eastward from Africa. This continental margin lay near the equator where countless microorganisms lived in the surface waters, particularly during the Cretaceous Period when most of the petroleum formed. The remains of these organisms accumulated with the bottom sediments and were buried, beginning the complex processes of petroleum generation.

Similar conditions existed in what is now the Gulf Coast region of the United States and Central America. Here petroleum and natural gas also formed on a broad shelf over which transgressions and regressions occurred. In this region, the hydrocarbons are largely in reservoir rocks that were deposited as distributary channels on deltas and as barrier-island and beach sands. Some of these hydrocarbons are associated with structures formed adjacent to rising salt domes. The salt, called the *Louann Salt,* initially formed in a long, narrow sea when North America separated from Europe and North Africa during the fragmentation of Pangaea (Figure 23.3).

The richest uranium ores in the United States are widespread in Mesozoic rocks of the Colorado Plateau area of Colorado and adjoining parts of Wyoming, Utah, Arizona, and New Mexico.

As noted in Chapter 20, Proterozoic banded iron formations are the main sources of iron ores. There are, however, some important exceptions. For example, the Jurassic-age "Minette" iron ores of Western Europe, composed of oolitic limonite and hematite, are important ores in France, Germany, Belgium, and Luxembourg. In Great Britain, low-grade iron ores of Jurassic age consist of oolitic siderite, which is an iron carbonate. And in Spain, Cretaceous rocks are the host rocks for iron minerals.

South Africa, the world's leading producer of gem-quality diamonds and among the leaders in industrial diamond production, mines these minerals from conical igneous intrusions called kimberlite pipes. Kimberlite pipes are composed of dark gray or blue igneous rock known as kimberlite. Diamonds, which form at great depth where pressure and temperature are high, are brought to the surface during the explosive volcanism that forms kimberlite pipes. Although kimberlite pipes have formed throughout geologic time, the most intense episode of such activity in South Africa and adjacent countries was during the Cretaceous Period. Emplacement of Triassic and Jurassic diamond-bearing kimberlites also occurred in Siberia.

In the Prologue we noted that the mother lode or source for the placer deposits mined during the California gold rush is in Jurassic-age intrusive rocks of the Sierra Nevada. Gold placers are also known in Cretaceous-age conglomerates of the Klamath Mountains of California and Oregon.

Porphyry copper was originally named for copper deposits in the western United States mined from porphyritic granodiorite, but the term now applies to large, low-grade copper deposits disseminated in a variety of rocks. These porphyry copper deposits are an excellent example of the relationship between convergent plate boundaries and the distribution, concentration, and exploitation of valuable metallic ores. Magma generated by partial melting of a subducting plate rises toward the surface, and as it cools, it precipitates and concentrates various metallic ores. The world's largest copper deposits were formed during the Mesozoic and Tertiary in a belt along the western margins of North and South America (see Figure 9.30).

Life of the Mesozoic Era

Most people find the Mesozoic Era the most interesting episode in life history, because flying reptiles, marine reptiles, and dinosaurs lived during that time. All of these animals have been popularized in novels, children's books, TV specials, and several movies including *Jurassic Park* (1993), *The Lost World* (1997), and *Dinosaur* (2000). Part of their popularity is probably related to the greatly increased knowledge scientists have acquired about Mesozoic land animals, especially during the last few decades. In fact, new discoveries are made so routinely that it is difficult to keep up with the current literature and interpretations of fossil dinosaurs and several other Mesozoic animals.

The Mesozoic Era is commonly referred to as the "Age of Reptiles," indicating that among land-dwelling vertebrate animals, reptiles were the most diverse and abundant. Certainly the Mesozoic diversification of reptiles is an important event in life history, but other equally although not as well-known events also took place. For instance, mammals made their appearance during the Triassic, having evolved from mammal-like reptiles, whereas birds evolved from reptiles, probably small carnivorous dinosaurs, by the Jurassic Period.

Vast changes took place in plant communities too, when the first flowering plants (angiosperms) evolved and soon became the most diverse and common plants on land. And even though the major land-plant groups from the Paleozoic persisted and still exist, they now constitute less than 10% of all land plants. Marine invertebrates also made a notable resurgence following the mass extinctions at the end of the Permian.

The breakup of Pangaea, begun during the Triassic, continued throughout the Mesozoic and Cenozoic eras, eventually producing the present distribution of land and sea. Nevertheless, the proximity of continents and mild Mesozoic climates made it possible for plants and animals to occupy extensive geographic areas. But as the fragmentation of Pangaea continued, some continents, especially Australia and South America, became isolated; as their faunas evolved independently, they became increasingly different from those of the other continents.

The mass extinctions at the end of the Paleozoic were the most severe to affect life on Earth, but once again widespread extinctions took place at the end of the Mesozoic Era. Because dinosaurs and their relatives were victims of the Mesozoic extinctions, these events have received much more publicity than Paleozoic extinctions. So once again organic diversity declined markedly, but, as before, it recovered during the following interval of geologic time.

Marine Invertebrates

Following the wave of extinctions at the end of the Paleozoic, the Mesozoic was a time when marine invertebrates repopulated the seas. The Early Triassic invertebrate marine fauna was not diverse but by the Late Triassic the seas were once again richly populated with invertebrates. The mollusks became increasingly diverse and abundant throughout the Mesozoic. The brachiopods, however, never completely recovered from their near extinction at the end of the Paleozoic. In areas of warm, relatively clear, shallow marine waters, corals again proliferated but these corals were of a new and familiar type, the *scleractinians*. Echinoids, which were rare during the Paleozoic, greatly diversified during the Mesozoic.

One of the major differences between Paleozoic and Mesozoic marine invertebrate communities was the increased abundance and diversity of burrowing organisms. With few exceptions, Paleozoic burrowers were soft-bodied animals such as worms. The bivalves and echinoids, which were epifaunal elements during the Paleozoic, evolved various means of entering infaunal habits. This trend toward an infaunal existence may reflect an adaptive response to increasing predation from the rapidly evolving fish and cephalopods.

Beginning with the primary producers, we will now examine some of the major Mesozoic marine plant and invertebrate groups. *Coccolithophores* are an important group of living phytoplankton (Figure 23.19a). They first evolved during the Jurassic and diversified tremendously during the Cretaceous. *Diatoms* evolved during the Cretaceous, but were more important as primary producers during the Cenozoic. Diatoms construct their shells of silica and are presently most abundant in cooler waters (Figure 23.19b). *Dinoflagellates* were common during the Mesozoic and today are the major primary producers in warm water (Figure 23.19c).

Mollusks were the major invertebrate phylum of the Mesozoic. The mollusks include six classes, only three of which—the gastropods, bivalves, and cephalopods—are significant members of the marine invertebrate fauna. The gastropods increased in abundance and diversity during the Mesozoic, becoming most abundant during the Cretaceous, when carnivorous forms appeared. It was the bivalves and cephalopods, however, that dominated the invertebrate community of the Mesozoic.

Mesozoic bivalves diversified to inhabit many epifaunal and infaunal niches. Oysters and clams became particularly diverse and abundant epifaunal suspension feeders and, despite a reduction at the end of the Cretaceous, continued to be important throughout the Cenozoic to the present (Figure 23.20a). Reef-forming rudists were a significant group of Mesozoic bivalves. Bivalves also expanded

Figure 23.19 (a) A Miocene coccolith from the Gulf of Mexico (left); a Pliocene-Miocene coccolith from the Gulf of Mexico (right). (b) Upper Miocene diatons from Java (left and right). (c) Eocene dinoflagellates from Alabama (left) and the Gulf of Mexico (right).

(a)

(b)

Figure 23.20 *(a) Bivalves, represented here by two Cretaceous forms, were diverse and quite common during the Mesozoic. (b) Cephalopods such as the Late Cretaceous ammonoids* Baculites *(foreground) and* Helioceros *were present throughout the Mesozoic, but were most abundant during the Jurassic and Cretaceous.*

into the infaunal niche during the Mesozoic. By burrowing into the sediment, they escaped predation from cephalopods and fish.

Cephalopods were one of the most important Mesozoic invertebrate groups. Their rapid evolution and nek-

tonic lifestyle make them excellent guide fossils (Figure 23.20b). The Ammonoidea, cephalopods with wrinkled sutures, are divided into three groups: the goniatites, ceratites, and ammonites. The ammonites, which are characterized by extremely complex suture patterns, were present during all three Mesozoic periods but were most prolific during the Jurassic and Cretaceous. While most ammonites were coiled, some attaining diameters of 2 m, others were uncoiled and led a near benthonic existence.

Although the ammonites became extinct at or near the end of the Cretaceous, two other groups of cephalopods survived into the Cenozoic—the nautiloids and the belemnoids, a group of squidlike cephalopods that were highly successful during the Jurassic and Cretaceous. The stalked echinoderms were minor members of the Mesozoic marine invertebrate community. However, the echinoids, which were exclusively epifaunal during the Paleozoic, branched out into the infaunal habitat and became very diverse and abundant. Bryozoans, although rare in Triassic strata, diversified and expanded during the Jurassic and Cretaceous.

As is true today, where shallow marine waters were warm and clear, coral reefs proliferated. Mesozoic corals belong to the order Scleractinia. Whether scleractinian corals evolved from the rugose order or from an as yet unknown soft-bodied group that left no fossil record is still unresolved.

Last, the foraminifera underwent an explosive radiation during the Jurassic and Cretaceous that continued to the present. The planktonic forms in particular underwent rapid diversification, but most genera became extinct at the end of the Cretaceous.

In general terms, we can think of the Mesozoic as a time of increasing complexity of the marine invertebrate community. At the beginning of the Triassic, diversity was low and food chains were short. Near the end of the Cretaceous the marine invertebrate community was highly complex with interrelated food chains.

Plants—Primary Producers on Land

Just as during the Late Paleozoic, Triassic and Jurassic land-plant communities consisted of seedless vascular plants and various gymnosperms. Among the gymnosperms, though, the large seed ferns became extinct by the end of the Triassic, but *ginkgos* remained abundant throughout the Mesozoic Era, and *conifers* continued to diversify. A new type of gymnosperm, the *cycads*, evolved during the Triassic. Cycads superficially resemble palms, and several varieties still exist in tropical and subtropical areas.

The long dominance of seedless plants and gymnosperms ended during the Early Cretaceous when many were replaced by **angiosperms,** or flowering plants (Figure 23.21). Angiosperms probably evolved from specialized gymnosperms. Indeed, recent studies have identified

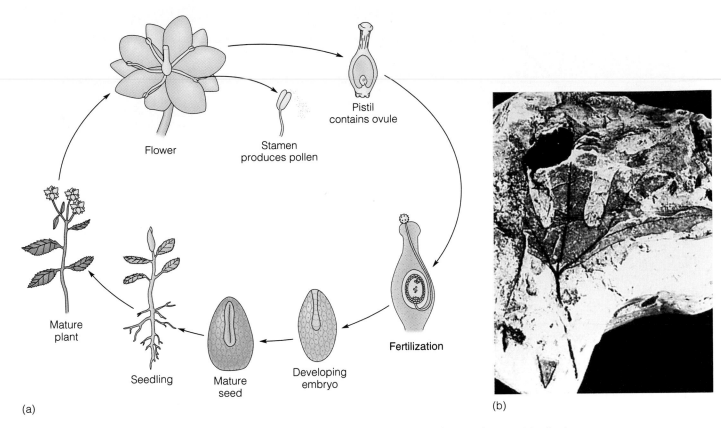

(a)   (b)

Figure 23.21 *(a) The reproductive cycle in angiosperms. (b) Early Cretaceous angiosperm from Cecil County, Maryland.*

both fossil and living gymnosperms that show close relationships to angiosperms. And the oldest known fossil flowers are similar to those of present-day magnolias, which are among the more primitive living angiosperms. In any case, since they first evolved angiosperms have adapted to nearly every terrestrial habitat. Several factors account for their phenomenal success, but chief among them is their method of reproduction (Figure 23.21). Two developments were particularly important: the evolution of flowers, which attract animal pollinators, especially insects; and the evolution of enclosed seeds.

Seedless vascular plants and gymnosperms are important and still flourish in many environments; in fact, many botanists regard ferns and conifers as emerging groups. Nevertheless, a measure of angiosperms' success is that today with 250,000 to 300,000 species they account for more than 90% of all land plant species, and they occupy some habitats in which other land plants do poorly or cannot exist.

The Diversification of Reptiles

Reptile diversification began during the Mississippian Period with the evolution of the protorothyrids, apparently the first animals to lay amniotic eggs (see Chapter 22). From this basic stock of so-called *stem reptiles,* all other reptiles as well as birds and mammals evolved

(Figure 23.22). Recall from Chapter 22 that pelycosaurs were the dominant land vertebrates of the Pennsylvania and Permian. Here we continue our story of reptile diversification with a group called *archosaurs.*

Archosaurs and the Origin of Dinosaurs

The reptiles known as **archosaurs** (*archo* meaning "ruling," and *sauros* meaning "lizard") include crocodiles, pterosaurs (flying reptiles), dinosaurs, and birds. Including such diverse animals in a single group implies that they share a common ancestor, and indeed they possess several characteristics that unite them. For instance, all have teeth set in individual sockets, except today's birds, but even the earliest birds had this feature. We now turn to a discussion of dinosaurs, but will have more to say about pterosaurs and birds later in this chapter.

All dinosaurs possess a number of shared characteristics, yet differ enough for us to recognize two distinct orders, the **Saurischia** and **Ornithischia**. Each order is characterized by a distinctive pelvic structure: saurischian dinosaurs have a lizardlike pelvis and are thus called lizard-hipped dinosaurs; ornithischians have a birdlike pelvis and are called bird-hipped dinosaurs (Figure 23.23). For decades, paleontologists thought each order evolved independently during the Late Triassic, but it is now clear that

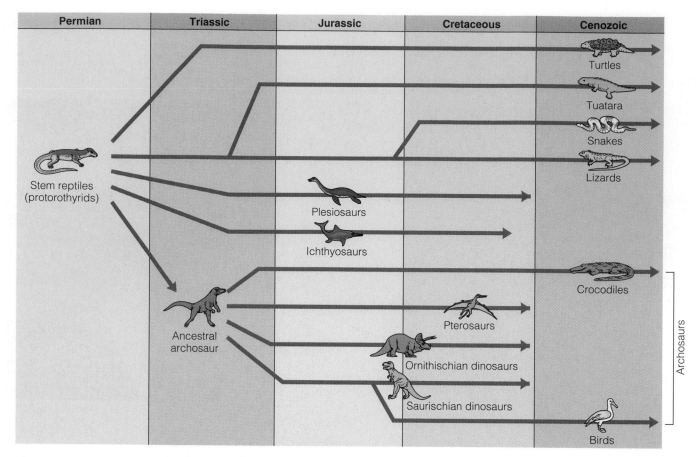

Figure 23.22 *Relationships among fossil and living reptiles and birds.*

they had a common ancestor much like archosaurs known from Middle Triassic rocks in Argentina. These dinosaur ancestors were small (less than 1 m long), long-legged carnivores that walked and ran on their hind limbs, so they were **bipedal,** as opposed to **quadrupedal** animals that move on all four limbs.

Dinosaurs

The term *dinosaur* was proposed by Sir Richard Owen in 1842 to mean "fearfully great lizard," although now "fearfully" has come to mean "terrible," thus the characterization of dinosaurs as "terrible lizards." But of course they were not terrible, or at least no more terrible than animals living today, and they were definitely not lizards. Nevertheless, dinosaurs more than any other kind of animal have inspired awe and thoroughly captured the public imagination. Unfortunately, their popularization in many cartoons, books, and movies has commonly been inaccurate and has contributed to many misunderstandings. For instance, many people think that all dinosaurs were large, and because they are extinct they must have been poorly adapted. It is true that many were large—indeed, the largest animals ever to live on land. But not all were large. In fact, dinosaurs

varied from giants weighing several tens of metric tons to those no larger than a chicken. And to consider them poorly adapted is to ignore the fact that dinosaurs were extremely diverse and widespread for more than 140 million years!

Although various media are now portraying dinosaurs as more active animals, the misconception that they were lethargic beasts persists. Evidence now available indicates that some were quite active and perhaps even warm-blooded. It also appears that some species cared for their young long after hatching, a behavioral characteristic most often associated with birds and mammals. And while many questions remain unanswered about dinosaurs, their fossils and the rocks containing them are revealing more and more about their evolutionary relationships and behavior (see Perspective 23.1).

We have already noted that scientists recognize two distinct orders of dinosaurs, the Saurischia and Ornithischia (Figure 23.23, Table 23.1). Among the saurischians, two suborders are defined: theropods and sauropods. All *theropods* were bipedal carnivores that ranged in size from tiny *Compsognathus* (Figure 23.24a) to comparative giants such as *Tyrannosaurus* and similar but even larger species from Africa and Argentina (see chapter opening photo). The movie *Jurassic Park* and its sequel *The Lost*

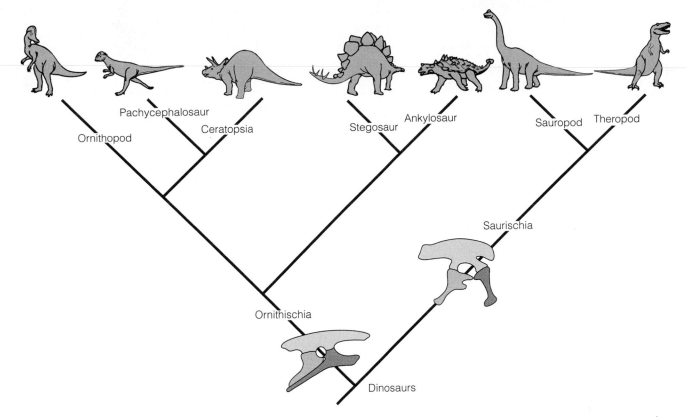

Figure 23.23 *Cladogram showing dinosaur relationships. Pelvises of ornithischian and saurischian dinosaurs are shown for comparison. Among the seven subgroups of dinosaurs theropods were carnivorous and all others were herbivores.*

World popularized some of the smaller theropods such as *Velociraptor* with a large sickle-shaped claw on each hind foot. *Velociraptor* and its relative *Deinonychus* likely used these claws in a slashing type of attack (Figure 23.24b). Some remarkable discoveries beginning in 1996 by Chinese paleontologists have yielded species of small theropods with feathers (Figure 23.24c). The significance of these fossils is discussed in later sections.

Included among the sauropods are the giant, quadrupedal herbivores such as *Apatosaurus*, *Diplodicus*, and *Brachiosaurus*, the largest land animals of any kind (Table 23.1). According to one estimate, *Brachiosaurus*, a giant even by sauropod standards, weighed more than 75 metric tons! Partial remains discovered in Colorado, New Mexico, and elsewhere indicate that even larger sauropods existed, but just how large they were is unresolved. The sauropods were preceded in the fossil record by the smaller *prosauropods*.

As an indication of the great diversity of ornithischians, five distinct suborders are recognized: ornithopods, pachycephalosaurs, ankylosaurs, stegosaurs, and ceratopsians (Figure 23.25, Table 23.1). All ornithischians were herbivores but some were bipeds whereas others were quadrupeds. In fact, ornithopods such as the well-known duck-billed dinosaurs were primarily bipeds but their well-developed fore-

limbs allowed them to walk on all fours as well. Also included among the ornithischians were the heavily armored ankylosaurs, the stegosaurs with plates on their backs as well as spikes for defense on their tails, the horned ceratopsians, and the peculiar dome-headed pachycephalosaurs (Table 23.1).

Warm-Blooded Dinosaurs?

Could dinosaurs maintain a rather constant body temperature despite the outside temperature? In other words, were they *endotherms* (warm-blooded) like today's mammals and birds, or were they *ectotherms* (cold-blooded) as are all of today's reptiles? Almost everyone now agrees that some compelling evidence exists for dinosaur endothermy, but opinion is still divided among (1) those holding that all dinosaurs were endotherms; (2) those who think only some were endotherms; and (3) those proposing that dinosaur metabolism, and thus their ability to regulate body temperature, changed as they matured.

Bones of endotherms typically have numerous passageways that, when the animals are alive, contain blood vessels, but considerably fewer passageways are present in bones of ectotherms. Proponents of dinosaur endothermy note that dinosaur bones are more similar to

Dinosaur Behavior

Fossil bones tell us much about how dinosaurs moved about and their teeth indicate whether they were herbivores or carnivores, but the fossil record reveals little about behavior. Even coprolites (fossilized feces) tell something of their diet, but as for their social interactions, mating rituals, and various aspects of their daily lives, we can only provide some informed speculation. For example, it is reasonable to assume that carnivorous dinosaurs behaved much like today's carnivores. That is, they went for the easy kill—the young, old, or disabled—and avoided large dangerous prey, or if large themselves they simply chased other predators away from their kill. Nevertheless, carnivorous dinosaurs are still depicted in movies as ferocious, aggressive beasts, and titanic battles between *Tyrannosaurus* and some kind of large dinosaur with armor, spikes, or clubs is common fare.

No one doubts that many carnivorous dinosaurs actively hunted, but today just about any carnivore will dine on carrion if available and we have every reason to think that dinosaurs did likewise. But were carnivorous dinosaurs pack hunters? The evidence is sketchy, but some such as *Deinonychus* (Figure 23.24b) and *Velociraptor* of *Jurassic Park* fame as well as diminutive *Coelophysis* probably did hunt in packs.

Undoubtedly, some dinosaurs were solitary animals, but the available evidence indicates that many herbivores were gregarious, congregating in vast herds. Two kinds of evidence support this conclusion. One consists of fossil trackways indicating that many dinosaurs of the same type moved together. The other kind of evidence is so-called bone beds in which only one species of dinosaur is found. These concentrations of fossils indicate that large numbers of animals perished quickly as the result of some kind of catastrophe, such as drowning while crossing a river or suffocation by volcanic ash. Several sauropods were trapped in mud at what is now Howe Quarry, Wyoming, and several thousand duckbilled dinosaurs died en masse during the Late Cretaceous in Montana. Bone beds of ceratopsian dinosaurs in Alberta, Canada, indicate that these dinosaurs also congregated in herds.

Several dinosaur features might have served some kind of behavioral function. The crests of ornithopods are the most obvious, but horns, frills, and spikes in ceratopsians might also have functioned in species recognition or displays to attract mates. And no doubt grunting, bellowing, and visual displays were used in defense and territorial disputes. Paleontologists think that stegosaur plates were some kind of device to absorb and dissipate heat, but these too might have been used for display and/or species recognition. Color is important in many living reptiles and birds and may have been so in dinosaurs, too; but unfortunately, even if skin or feathers are preserved, color does not fossilize.

Nesting sites discovered in Montana clearly demonstrate that at least three types of Late Cretaceous dinosaurs not only nested in colonies but also used the same nesting sites repeatedly, much as some birds do today (Figure 1). One of these dinosaurs, named *Maiasaura* (good mother dinosaur), laid eggs in 2-m-diameter nests spaced about 7 m apart, or about the length of an adult. Some nests contain juveniles up to 1 m long, which is much greater than their length at the time of hatching. So the young must have remained in the nest area for some time during which adults fed and protected them.

those of living endotherms. Yet crocodiles and turtles have this so-called endothermic bone, but they are ectotherms, and some small mammals have bone more typical of ectotherms. Perhaps bone structure is related more to body size and growth patterns than to endothermy, so this evidence is obviously not conclusive for dinosaur endothermy.

Endotherms must eat more than comparable sized ectotherms because their metabolic rates are so much higher. Consequently, endothermic predators require large prey populations and thus constitute a much smaller proportion of the total animal population than their prey, usually only a few percent. In contrast, the proportion of ectothermic predators to prey might be as high as 50%. Where data are sufficient to allow an estimate, dinosaur predators made up 3% to 5% of the total population. Nevertheless, uncertainties in the data make this less than a convincing argument for many paleontologists.

A large brain in comparison to body size is not necessary for endothermy, but endothermy does seem to be a prerequisite for having a large brain, because a complex nervous system needs a rather constant body temperature. Some dinosaurs did have large brains compared to body size, especially the small- and medium-sized carnivores, but many did not. So brain size for some dinosaurs might be a convincing argument, but even more compelling evidence for theropod endothermy comes from their probable relationship to birds, and the recent discoveries in China of theropods with feathers or a feather-like outer covering (Figure 23.24c). Today, only endotherms have hair, fur, or feathers for insulation.

(a)

(b)

Figure 1 *(a) Nest of the theropod dinosaur* Troodon, *from Egg Mountain in northern Montana. (b) Scene from the Late Cretaceous of northern Montana. A female* Maiasaura *leads her young to a feeding area.*

Nests and eggs of the ostrichlike dinosaur *Oviraptor* are well known in Late Cretaceous-age rocks of Mongolia. It has not been determined whether they nested in colonies, but one remarkable discovery in 1993 shows a 2.4-m-long adult *Oviraptor* that might have been incubating about 15 eggs. The dinosaur's forelimbs encircle the eggs and the back legs are folded underneath the animal's body, but this posture may have simply been to protect the eggs from a predator.

In 1998 American and Argentine scientists reported the first eggs and embryos of sauropod dinosaurs, probably sauropods known as titanosaurs. A Late Cretaceous nesting ground with thousands of eggs, some with unhatched embryos, was discovered in the badlands of Patagonia in Argentina. Apparently, adult titanosaurs gathered by the thousands and nested on floodplains where some of the nests and eggs were buried during floods. Certainly these dinosaurs nested in colonies, but it is not known whether they cared for their young after they hatched.

Some scientists point out that certain duck-billed dinosaurs grew and reached maturity much more quickly than would be expected for ectotherms and conclude that they must have been warm-blooded. Furthermore, a recently prepared fossil ornithopod discovered in 1993 has a preserved four-chambered heart much like that of living mammals and birds. Three-dimensional imaging of this structure, now on display at the North Carolina Museum of Natural Sciences, has convinced many scientists that this animal was an endotherm.

Good arguments for endothermy exist for several types of dinosaurs, although the large sauropods were probably not endothermic but nevertheless were capable of maintaining a rather constant body temperature. Large animals heat up and cool down more slowly than smaller ones because they have a small surface area compared to their volume. With their comparatively smaller surface area for heat loss, sauropods probably retained heat more effectively than their smaller relatives.

In general, a fairly good case can be made for endothermy in many theropods and in some ornithopods. Nevertheless, disagreement still exists and for some dinosaurs the question is still open.

Flying Reptiles

Paleozoic insects were the first animals to achieve flight, but the first among vertebrates were the pterosaurs, or flying reptiles, which were common in the skies from the Late Triassic until their extinction at the end of the Cretaceous (Figure 23.26). Adaptations for flight include a wing membrane supported by an elongated fourth

Table 23.1

Summary Chart for the Orders and Suborders of Dinosaurs. Lengths and weights approximate from several sources.

Order	Suborder	Familiar Genera	Comments
Saurischia	Theropoda	*Allosaurus, Coelophysis, Compsognathus, Deinonychus, Tyrannosaurus*, Velociraptor*	Bipedal carnivores. Late Triassic to end of Cretaceous. Size from 0.6 to 15 m long, 2 or 3 kg to 7.3 metric tons. Some smaller genera may have hunted in packs.
	Sauropoda	*Apatosaurus, Brachiosaurus, Camarasaurus, Diplodocus, Titanosaurus*	Giant quadrupedal herbivores. Late Triassic to Cretaceous, but most common during Jurassic. Size up to 27 m long, 75 metric tons.** Trackways indicate sauropods lived in herds. Preceded in fossil record by the smaller prosauropods.
Ornithischia	Ornithopoda	*Anatosaurus, Camptosaurus, Hypsilophodon, Iguanodon, Parasaurolophus*	Some ornithopods, such as *Anatosaurus,* had a flattened bill-like mouth and are called duck-billed dinosaurs. Size from a few meters long up to 13 m and 3.6 metric tons. Especially diverse and common during the Cretaceous. Primarily bipedal herbivores, but could also walk on all fours.
	Pachycepha-losauria	*Stegoceras*	*Stegoceras* only 2 m long and 55 kg, but larger species known. Thick bones of skull cap might have been for butting contests for dominance and mates. Bipedal herbivores of Cretaceous.
	Ankylosauria	*Ankylosaurus*	*Ankylosaurus* more than 7 m long and about 2.5 metric tons. Heavily armored with bony plates on top of head, back, and sides. Quadrupedal herbivore.
	Stegosauria	*Stegosaurus*	A variety of stegosaurs are known, but *Stegosaurus* with bony plates on its back and a spiked tail is best known. Plates probably were for absorbing and dissipating heat. Quadrupedal herbivores that were most common during the Jurassic. *Stegosaurus* 9 m long, 1.8 metric tons.
	Ceratopsia	*Triceratops*	Numerous genera known. Some early ones bipedal, but later large animals were quadrupedal herbivores. Much variation in size; *Triceratops* to 7.6 m long and 5.4 metric tons, with large bony frill over top of neck, three horns on skull, and beaklike mouth. Especially common during the Cretaceous.

*Until recently *Tyrannosaurs* at 4.5 metric tons was the largest known theropod, but now similar, larger animals are known from Argentina and Africa.

**Partial remains indicate even larger brachiosaurs existed, perhaps measuring 30 m long and weighing 135 metric tons.

finger, light hollow bones, and development of those parts of the brain associated with muscular coordination and sight. Because at least one pterosaur species had a coat of hair or hairlike feathers, possibly it, and perhaps all pterosaurs, were endotherms.

Pterosaurs are generally depicted in movies as large creatures, but some were no bigger than today's sparrows, robins, and crows. There were, however, a few species with wingspans of several meters, and one Cretaceous pterosaur found in Texas had a wingspan of at least 12 m! Nevertheless, even the very largest species probably weighed no more than a few tens of kilograms.

Experiments and studies of fossils indicate that the bones of large pterosaurs such as *Pteranodon* (Figure 23.26b) were too weak for sustained wing flapping. These comparatively large animals probably took advantage of thermal updrafts to stay airborne, mostly by soaring but oc-

casionally flapping their wings for maneuvering. In contrast, smaller pterosaurs probably stayed aloft by vigorously flapping their wings just as present-day small birds do.

Marine Reptiles

The most familiar of the Mesozoic marine reptiles are the rather porpoise-like *ichthyosaurs* (Figure 23.27a). Most of these fully aquatic animals were about 3 m long, but one species reached about 12 m. All ichthyosaurs had a streamlined body, a powerful tail for propulsion, and flipperlike forelimbs for maneuvering. Their numerous sharp teeth indicate that they were carnivores, and preserved stomach contents reveal a diet of fish, cephalopods, and other marine organisms. Ichthyosaurs were so completely aquatic that is doubtful they could come onto land, so females probably retained eggs within their bodies and

(a)

(c)

(b)

Figure 23.24 *Theropod dinosaurs. (a)* Compsognathus *weighed only 2 or 3 kg. Bones found within its ribcage indicate it ate lizards. (b) Lifelike restoration of* Deinonychus *in its probable attack posture. (c) Restoration of the Early Cretaceous feathered dinosaur* Caudipteryx *from China.*

gave birth to live young. A few fossils with small ichthyosaurs within the appropriate part of the body cavity support this interpretation.

Another well-known group of Mesozoic marine reptiles, the *plesiosaurs,* belonged to one of two groups: short-

necked and long-necked (Figure 23.27b). Most were modest sized animals 3.6 to 6 m long, but one species found in Antarctica measures 15 m. Short-necked plesiosaurs might have been bottom feeders, but their long-necked cousins probably used their necks in a snakelike fashion to capture

(a)

Figure 23.25 *Representatives of two of the five suborders of ornithischians. (a) The ankylosaur* Euoplocephalus. *Notice the heavy armor and bony club on the tail. (b) Skeleton of the ceratopsian* Triceratops *in the Natural History Museum, London, England. This rhinoceros-size dinosaur was very common during the Late Cretaceous in western North America.*

(b)

(a)

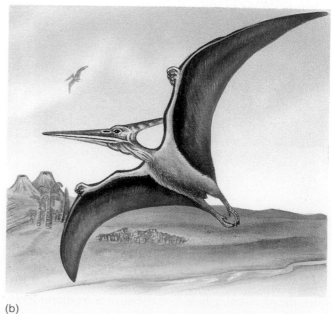
(b)

Figure 23.26 *(a) This long-tailed pterosaur from the Jurassic of Europe had a wingspan of about 0.6 m.* Pteranodon *(b) was a Cretaceous pterosaur with a wingspan of more than 6 m.*

(a)

(c)

(b)

Figure 23.27 *Mesozoic marine reptiles. (a) Ichthyosaurs. (b) A long-necked plesiosaur. (c) Tylosaurus, a Late Cretaceous mosasaur, measured up to 9 m long.*

fish with their numerous sharp teeth. These animals probably came ashore to lay their eggs.

Mosasaurs were Late Cretaceous marine lizards related to the present-day Komodo dragon or monitor lizard. Some species measured no more than 2.5 m long, but a few such as *Tylosaurus* were giants, measuring up to 9 m long (Figure 23.27c). Mosasaur limbs resemble paddles and were probably used mostly for maneuvering whereas the long tail provided propulsion. All were predators, and preserved stomach contents indicate that they ate fish, birds, smaller mosasaurs, and a variety of invertebrates including ammonoids.

From Reptiles to Birds

Several fossils with feather impressions have been discovered in the Jurassic Solnhofen Limestone of Germany, but in almost every other known physical feature these fossils are most similar to small theropods. The birdlike creature known as *Archaeopteryx* retained dinosaur-like teeth, tail, brain size, and hind limb structure, but it also had feathers and a wishbone, characteristics typical of birds (Figure 23.28).

Most paleontologists now think that some kind of small theropod was the ancestor of birds. Even the wishbone, consisting of fused clavicles, so typical of birds, is found in a number of theropods, and recent discoveries of theropods in China with some kind of feathery covering is further

Figure 23.28 Archaeopteryx, *a Jurassic-age animal from Germany, has feathers and a wishbone and is therefore classified as a bird. In almost all other anatomical features, though, it more closely resembles theropod dinosaurs.*

evidence of this relationship (Figure 23.24c). However, another candidate for bird ancestry is a small lizardlike reptile known as *Longisquama* that was discovered during the 1960s in Kyrgystan. *Longisquama* is probably an archosaur, and it too appears to have feathers, but some paleontologists think these structures are actually scales. In any event, the bird-reptile relationship is firmly established, but disagreement exists on the exact bird ancestor.

Two more Mesozoic fossils shed more light on bird evolution. One specimen, from China, is slightly younger than *Archaeopteryx* and possesses both primitive and advanced features. For instance, it retains abdominal ribs similar to those of *Archaeopteryx* and theropods, but it has a reduced tail more typical of present-day birds. Another Mesozoic bird from Spain is also a mix of primitive and advanced characteristics, but it appears to lack abdominal ribs. One of the oldest known toothless birds comes from the Late Jurassic or Early Cretaceous of Asia.

Archaeopteryx's fossil record is not good enough to resolve whether it is actually the ancestor of today's birds or an animal that died out without leaving descendants. Of course, that in no way diminishes its having both reptile and bird characteristics. However, fossils of two crow-sized individuals known as *Protoavis* are claimed by some to be an even earlier bird than *Archaeopteryx*. These Late Triassic fossils have hollow bones and the breastbone structure of birds, but because no feature impressions were found, many paleontologists think they are specimens of small theropods.

Origin and Earth History of Mammals

Therapsids, or the advanced mammal-like reptiles, were briefly described in Chapter 22. They diversified into numerous species of herbivores and carnivores,

Table 23.2

Summary Chart Showing Some Characteristics and How They Changed During the Transition from Reptiles to Mammals

	Typical Reptile	Cynodont	Mammal
Lower Jaw	Dentary and several other bones	Dentary enlarged, other bones reduced	Dentary bone only, except in earliest mammals
Jaw–Skull Joint	Articular–quadrate	Articular–quadrate; some advanced cynodonts had both the reptile jaw-skull joint and the mammal jaw-skull joint	Dentary–squamosal
Middle-Ear Bones	Stapes	Stapes	Stapes, incus, malleus
Secondary Palate	Absent	Partly developed	Well developed
Teeth	No differentiation	Some differentiation	Fully differentiated into incisors, canines, and chewing teeth
Tooth Replacement	Teeth replaced continuously	Only two sets of teeth in some advanced cynodonts	Two sets of teeth
Occlusion (chewing teeth meet surface to surface to allow grinding)	No occlusion	Occlusion in some advanced cynodonts	Occlusion
Endothermic vs. Ectothermic	Ectothermic	Probably endothermic	Endothermic
Body Covering	Scales	One fossil shows it had skin similar to that of mammals	Skin with hair or fur

and during the Permian were the most diverse and numerous terrestrial vertebrates. One particular group of carnivorous therapsids called *cynodonts* was the most mammal-like of all and by the Late Triassic gave rise to mammals.

The transition from **cynodonts** to mammals is well documented by fossils and is so gradational that classification of some fossils as either reptile or mammal is difficult. We can easily recognize living mammals as warm-blooded animals with hair or fur that have mammary glands and, except for the platypus and spiny anteater, give birth to live young.

Obviously these criteria for recognizing living mammals are inadequate for classifying fossils. For them, we must use skeletal structure only. Several skeletal modifications characterize the transition from mammal-like reptiles to mammals, but distinctions between the two groups are based largely on details of the middle ear, the lower jaw, and the teeth (Table 23.2).

Reptiles have only one small bone in the middle ear—the stapes—while mammals have three—the incus, the malleus, and the stapes. Also, the lower jaw of a mammal is composed of a single bone called the *dentary*, but a reptile's jaw is composed of several bones (Figure 23.29). In addition, a reptile's jaw is hinged to the skull at a contact between the articular and quadrate bones, while in mammals the dentary contacts the squamosal bone of the skull (Figure 23.29).

During the transition from cynodonts to mammals, the quadrate and articular bones that had formed the joint between the jaw and skull in reptiles were modified into the incus and malleus of the mammalian middle ear (Figure 23.29). Fossils clearly document the progressive enlargement of the dentary until it became the only element in the mammalian jaw. Likewise, a progressive change from the reptile to mammal jaw joint is documented by fossil evidence. In fact, some of the most advanced cynodonts were truly transitional because they had a compound jaw joint consisting of (1) the articular and quadrate bones typical of reptiles and (2) the dentary and squamosal bones as in mammals (Table 23.2).

In Chapter 18 we noted that the study of embryos provides evidence for evolution. Opossum embryos clearly show that the middle-ear bones of mammals were originally part of the jaw. In fact, even when opossums are born, the middle-ear elements are still attached to the dentary (Figure 23.29), but as they develop further, these elements migrate to the middle ear, and a typical mammal jaw joint develops.

Several other aspects of cynodonts also indicate that they were ancestors of mammals. Their teeth were somewhat differentiated into distinct types in order to perform specific functions. In mammals the teeth are fully differentiated into incisors, canines, and chewing teeth, but typical

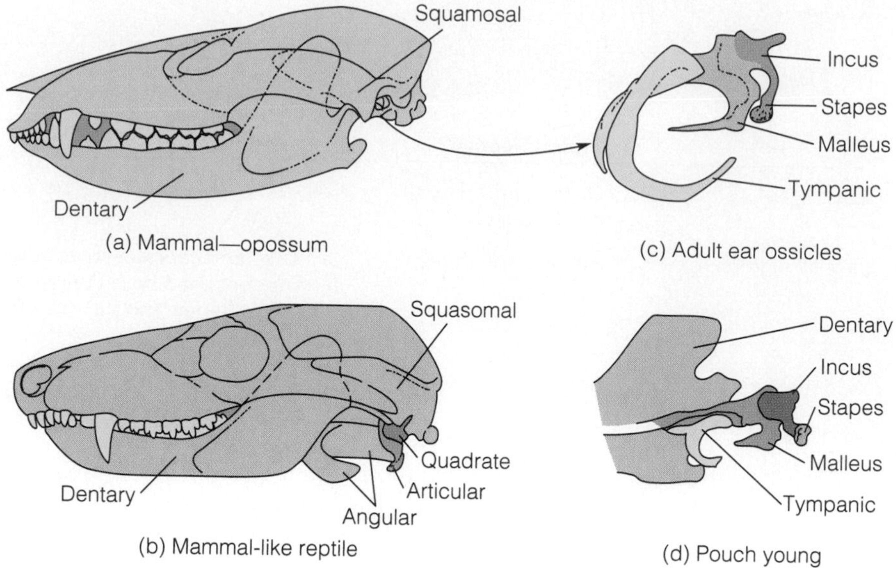

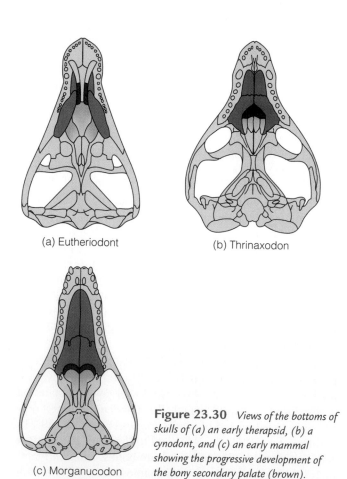

Figure 23.29 *(a) The skull of an opossum showing the typical mammalian dentary-squamosal jaw joint. (b) The skull of a cynodont shows the articular-quadrate jaw joint of reptiles. (c) Enlarged view of an adult opossum's middle ear bones. (d) View of the inside of a young opossum's jaw showing that the elements of the middle ear are attached to the dentary during early development. This is the same arrangement of bones that is found in the adults of the ancestral mammals.*

(a) Mammal—opossum

(c) Adult ear ossicles

(b) Mammal-like reptile

(d) Pouch young

reptiles do not have differentiated teeth. In addition, mammals have only two sets of teeth during their lifetimes—a set of baby teeth and the permanent adult teeth. Typical reptiles have teeth replaced continuously throughout their lives. The notable exception is seen in some cynodonts who in mammalian fashion had only two sets of teeth. Another important feature of mammal teeth is occlusion; that is, the chewing teeth meet surface to surface to allow grinding. Thus, mammals chew their food, but typical reptiles, amphibians, and fish do not. However, tooth occlusion is known in some advanced cynodonts (Table 23.2).

Another mammalian feature, the secondary palate, was partially developed in advanced cynodonts (Figure 23.30). This secondary palate, a bony shelf separating the nasal passages from the mouth cavity, is an adaptation for eating and breathing at the same time, a necessary requirement for endotherms with their high demands for oxygen.

Even though mammals evolved during the Late Triassic, their diversity remained low during the rest of the Mesozoic. The first mammals retained several reptilian characteristics, but had mammalian features as well. For example, the Triassic triconodonts had the fully differentiated teeth typical of mammals, but they had both the reptile and mammal types of jaw joints. In short, some mammalian features evolved more rapidly than others, thereby accounting for animals with characteristics of both reptiles and mammals.

The early mammals diverged into two distinct branches. One branch includes the triconodonts (Figure 23.31a) and their probable evolutionary descendants, the **monotremes,** or egg-laying mammals, which includes the platypus and spiny anteater of the Australian region. The second evolutionary branch included the **marsupial** (pouched) **mammals,** and the **placental mammals** and their ancestors, the

eupantotheres. All living mammals except monotremes have ancestries that can be traced back through this branch.

Eupantotheres were shrew-sized animals with a poor fossil record, but details of their teeth indicate they were

(a) Eutheriodont

(b) Thrinaxodon

(c) Morganucodon

Figure 23.30 *Views of the bottoms of skulls of (a) an early therapsid, (b) a cynodont, and (c) an early mammal showing the progressive development of the bony secondary palate (brown).*

(a)

(b)

Figure 23.31 *Mesozoic mammals. (a) One of the earliest mammals, the triconodont* Triconodon. *(b) The first known placental mammals were members of the Order Insectivora such as those in this scene from the Late Cretaceous. None of these animals measured more than a few centimeters long.*

ancestral to both marsupial and placental mammals. The divergence of marsupials and placentals from a common ancestor probably occurred during the Early Cretaceous, but undoubtedly both were present by the Late Cretaceous. The earliest known placental mammals were members of the order *Insectivora* (Figure 23.31b), an order represented today by shrews, moles, and hedgehogs.

Mesozoic Climates and Paleogeography

Fragmentation of the supercontinent Pangaea began by the Late Triassic and continues to the present, but during much of the Mesozoic, close connections existed between the various landmasses. The proximity of these landmasses, however, is not sufficient to explain Mesozoic biogeographic distributions, because climates are also effective barriers to wide dispersal. During much of the Mesozoic, though, climates were more equable and lacked the strong north and south zonation characteristic of the present. In short, Mesozoic plants and animals had greater opportunities to occupy much more extensive geographic ranges.

Pangaea persisted as a single unit through most of the Triassic (Figure 23.2a), and the Triassic climate was warm-temperate to tropical, although some areas, such as the present southwestern United States, were arid. Mild temperatures extended 50 degrees north and south of the equator, and even the polar regions may have been temperate. The fauna had a truly worldwide distribution. Some dinosaurs had continuous ranges across Laurasia and Gondwana, the peculiar gliding lizards were in New

Jersey and England, and reptiles known as phytosaurs lived in North America, Europe, and Madagascar.

By the Late Jurassic, Laurasia had become partly fragmented by the opening North Atlantic, but a connection still existed (Figure 23.2b). The South Atlantic had begun to open so that a long, narrow sea separated the southern parts of Africa and South America. Otherwise the southern continents were still close together.

The mild Triassic climate persisted into the Jurassic. Ferns, whose living relatives are now restricted to the tropics of southeast Asia, are known from areas as far as 63° south latitude and 75° north latitude. Dinosaurs roamed widely across Laurasia and Gondwana. For example, the giant sauropod *Brachiosaurus* is known from western North America and eastern Africa. Stegosaurs and some families of carnivorous dinosaurs lived throughout Laurasia and in Africa.

By the Late Cretaceous, the North Atlantic had opened further, and Africa and South America were completely separated (Figure 23.2c). South America remained an island continent until late in the Cenozoic. Its fauna, evolving in isolation, became increasingly different from faunas of the other continents. Marsupial mammals reached Australia from South America via Antarctica, but the South American connection was eventually severed. Placentals, other than bats and a few rodents, never reached Australia, thus explaining why marsupials continue to dominate the continent's fauna even today.

Cretaceous climates were more strongly zoned by latitude, but they remained warm and equable until the close of that period. Climates then became more seasonal and cooler, a trend that persisted into the Cenozoic. Dinosaur and mammal fossils demonstrate that interchange was still possible, especially between the various components of Laurasia.

Mass Extinctions—A Crisis in the History of Life

The greatest mass extinction took place at the end of the Paleozoic Era (see Chapter 22), but the one at the close of the Mesozoic has attracted more attention because among its casualties were dinosaurs, flying reptiles, marine reptiles, and several kinds of marine invertebrates. Among the latter were ammonites, which had been so abundant through the Mesozoic, rudistid bivalves, and some planktonic organisms.

Hypotheses proposed to explain Mesozoic extinctions are numerous, but most have been dismissed as improbable, untestable, or inconsistent with the available data. A proposal that has become popular since 1980 is based on a discovery at the Cretaceous–Tertiary boundary in Italy—a 2.5-cm-thick clay layer with a remarkably high concentration of the platinum group element iridium (Figure 23.32). High iridium concentrations have now been identified at many other Cretaceous–Tertiary boundary sites.

The significance of this *iridium anomaly* is that iridium is rare in crustal rocks but is found in much higher concentrations in some meteorites. Accordingly, some investigators propose a meteorite impact to explain the iridium anomaly, and further postulate that the impact of a meteorite perhaps 10 km in diameter set in motion a chain of events leading to extinctions. Some Cretaceous–Tertiary boundary sites also contain soot and shock-metamorphosed quartz grains, both of which are cited as additional evidence of an impact.

According to the impact hypothesis, about 60 times the mass of the meteorite was blasted from the crust high into the atmosphere, and the heat generated at impact started raging forest fires that added more particulate matter to the atmosphere. Sunlight was blocked for several months, causing a temporary cessation of photosynthesis, food chains collapsed, and extinctions followed. Furthermore, with sunlight greatly diminished, Earth's surface temperatures were drastically reduced, adding to the biologic stress. Another proposed consequence of an impact is that sulfuric acid (H_2SO_4) and nitric acid (HNO_3) resulted from vaporized rock and atmospheric gases. Both would have contributed to strongly acid rain that might have had devastating effects on vegetation and marine organisms.

The iridium anomaly is real, but its origin and significance are debateable. We know very little about the distribution of iridium in crustal rocks or how it may be distributed and concentrated. Some geologists suggest that the iridium was derived from within Earth by volcanism, but this idea is not conclusively supported by evidence.

Some now claim that a probable impact site centered on the town of Chicxulub on the Yucatán Peninsula of Mexico has been found. The structure lies beneath layers of sedimentary rock and measures 180 km in diameter. Furthermore, it appears to be the right age. Evidence supporting the conclusion that the Chicxulub structure is an impact crater includes shocked quartz, what appear to be the deposits of huge waves, and tektites, which are small pieces of rock that were melted during the proposed impact and hurled into the atmosphere.

Even if a meteorite did hit Earth, did it lead to these extinctions? If so, both terrestrial and marine extinctions must have occurred at the same time. To date, strict time equivalence between terrestrial and marine extinctions has not been demonstrated. The selective nature of the extinctions is also a problem. In the terrestrial realm, large animals were the most drastically affected, but not all dinosaurs were large, and crocodiles, close relatives of dinosaurs, were unaffected. Some paleontologists think that dinosaurs, some marine invertebrates, and many plants were already on the decline and headed for extinction before the end of the Cretaceous. A meteorite impact may have simply hastened the process.

Figure 23.32 *View of the Cretaceous–Tertiary boundary in the Raton Basin, Colorado. The boundary, the thin white clay layer, is at the level of the knee of R. Farley Fleming.*

In the final analysis, Mesozoic extinctions have not been explained to everyone's satisfaction. Most geologists now concede that a large meteorite impact occurred, but we also know that vast outpourings of lava were taking place in what is now India. Perhaps these brought about detrimental atmospheric changes. Furthermore, the vast shallow seas that covered large parts of the continents had mostly withdrawn by the end of the Cretaceous, and the mild equable Mesozoic climates became harsher and more seasonal by the end of that era. But the fact remains that these extinctions were very selective, and no single explanation accounts for all aspects of this crisis in life history.

Chapter Summary

Tables 23.3 and 23.4 provide summaries of Mesozoic geologic and biologic events, respectively.

1. The breakup of Pangaea can be divided into four stages.

   a. The first stage involved the separation of North America from Africa during the Late Triassic, followed by the separation of North America from South America.

   b. The second stage involved the separation of Antarctica, India, and Australia from South America and Africa during the Jurassic. During this stage, India broke away from the still-united Antarctica and Australia landmass.

   c. During the third stage, which began in the Late Jurassic, South America separated from Africa, while Europe and Africa began converging.

   d. In the last stage, during the Cenozoic Era, Greenland separated from North America and Europe.

2. The breakup of Pangaea influenced global climate and ocean circulation patterns. While the temperature gradient from the tropics to the poles gradually increased during the Mesozoic, overall global temperatures remained equable.

3. Except for incursions along the continental margins and two major transgressions (the Sundance Sea and the Cretaceous Interior Seaway), the North American craton was above sea level during the Mesozoic Era.

4. The Eastern Coastal Plain was the initial site of the separation of North America from Africa that began during the Late Triassic. During the Cretaceous Period, it was inundated by marine transgressions.

5. The Gulf Coastal region was the site of major evaporite accumulation during the Jurassic as North America rifted from South America. During the Cretaceous, it was inundated by a transgressing sea, which, at its maximum, connected with a sea transgressing from the north to create the Cretaceous Interior Seaway.

6. Mesozoic rocks of the western region of North America were deposited in a variety of continental and marine environments. One of the main controls of sediment distribution patterns was tectonism.

7. Western North America was affected by four interrelated orogenies: the Sonoma, Nevadan, Sevier, and Laramide. Each involved igneous intrusions as well as eastward thrust faulting and folding.

8. The cause of the Sonoma, Nevadan, Sevier, and Laramide orogenies was the changing angle of subduction of the oceanic Pacific plate under the continental North American plate. The timing, rate, and to some degree the direction of plate movement was related to seafloor spreading and the opening of the Atlantic Ocean.

9. Mesozoic rocks contain a variety of resources, including coal, petroleum, uranium, gold, and copper.

10. Among the marine invertebrates, survivors of the Permian extinction diversified and gave rise to increasingly complex Mesozoic marine invertebrate communities.

11. Triassic and Jurassic land-plant communities were composed of seedless plants and gymnosperms. Angiosperms, or flowering plants, evolved during the Early Cretaceous, diversified rapidly, and soon became the dominant land plants.

12. Dinosaurs evolved during the Late Triassic, but were most abundant and diverse during the Jurassic and Cretaceous. Based on pelvic structure, two distinct orders of dinosaurs are recognized—Saurischia (lizard-hipped) and Ornithischia (bird-hipped).

13. Pterosaurs were the first flying vertebrate animals. Small pterosaurs were probably active, wing-flapping fliers, while large ones may have depended on thermal updrafts and soaring to stay aloft. At least one pterosaur species had hair or feathers, so it was likely endothermic.

14. The fish-eating, porpoise-like ichthyosaurs were thoroughly adapted to an aquatic life. Female ichthyosaurs probably retained eggs within their bodies and gave birth to live young. Plesiosaurs were heavy-bodied marine reptiles that probably came ashore to lay eggs.

15. Birds probably evolved from small carnivorous dinosaurs. The oldest known bird, *Archaeopteryx*, appeared during the Jurassic, but few other Mesozoic birds are known. *Protoavis* in Triassic rocks may represent a bird older than *Archaeopteryx*.

16. The earliest mammals evolved during the Late Triassic, but they are difficult to distinguish from advanced cynodonts. Details of the teeth, the middle ear, and lower jaw are used to distinguish the two.

17. Several types of Mesozoic mammals existed, but all were small, and their diversity was low. A group of Mesozoic mammals known as eupantotheres gave rise to both marsupials and placentals during the Cretaceous.

18. Because the continents were close together during much of the Mesozoic and climates were mild even at high latitudes, animals and plants dispersed widely.

19. Mesozoic mass extinctions account for the disappearance of dinosaurs, several other groups of reptiles, and a number of marine invertebrates. One hypothesis holds that the extinctions were caused by the impact of a large meteorite. Many paleontologists think a meteorite impact occurred, but that volcanism and climate changes also contributed to extinctions.

Table 23.3

Summary of Mesozoic Geologic Events

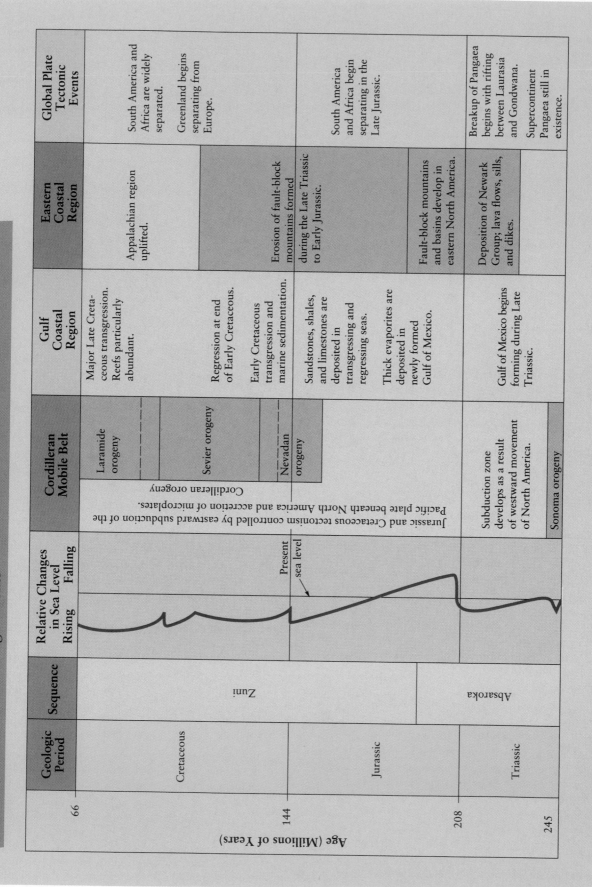

Age (Millions of Years)	Geologic Period	Sequence	Relative Changes in Sea Level (Rising / Falling)	Cordilleran Mobile Belt	Gulf Coastal Region	Eastern Coastal Region	Global Plate Tectonic Events
66	Cretaceous	Zuni	Present sea level	Laramide orogeny · Sevier orogeny · Nevadan orogeny (Cordilleran orogeny) — Jurassic and Cretaceous tectonism controlled by eastward subduction of the Pacific plate beneath North America and accretion of microplates.	Major Late Cretaceous transgression. Reefs particularly abundant. — Regression at end of Early Cretaceous. — Early Cretaceous transgression and marine sedimentation.	Appalachian region uplifted. — Erosion of fault-block mountains formed during the Late Triassic to Early Jurassic.	South America and Africa are widely separated. Greenland begins separating from Europe.
144	Jurassic			Subduction zone develops as a result of westward movement of North America.	Sandstones, shales, and limestones are deposited in transgressing and regressing seas. — Thick evaporites are deposited in newly formed Gulf of Mexico.	Fault-block mountains and basins develop in eastern North America.	South America and Africa begin separating in the Late Jurassic.
208	Triassic	Absaroka		Sonoma orogeny	Gulf of Mexico begins forming during Late Triassic.	Deposition of Newark Group; lava flows, sills, and dikes.	Breakup of Pangaea begins with rifting between Laurasia and Gondwana. Supercontinent Pangaea still in existence.
245							

Table 23.4

Summary of Mesozoic Biological Events

Age (Millions of Years)	Geologic Period	Invertebrates	Vertebrates	Plants	Climate	Plate Tectonics
66	Cretaceous	Extinction of ammonites, rudists, and most planktonic foraminifera at end of Creataceous. Continued diversification of ammonites and belemnoids. Rudists become major reef-builders.	Extinctions of dinosaurs, flying reptiles, marine reptiles, and some marine invertebrates. Placental and marsupial mammals diverge.	Angiosperms evolve and diversify rapidly. Seedless plants and gymnosperms still common but less varied and abundant.	North-south zonation of climates more marked, but remains equable. Climate becomes more seasonal and cooler at end of Cretaceous.	Further fragmentation of Pangaea. South America and Africa have separated. Australia separated from South America but remains connected to Antarctica. North Atlantic continues to open.
144	Jurassic	Ammonites and belemnoid cephalopods increase in diversity. Scleractinian coral reefs common. Appearance of rudist bivalves.	GREATEST DIVERSITY OF DINOSAURS First birds (may have evolved in Late Triassic). Time of giant sauropod dinosaurs.	Seedless vascular plants and gymnosperms only.	Much like Triassic. Ferns with living relatives restricted to tropics live at high latitudes, indicating mild climates.	Fragmentation of Pangaea continues, but close connections exist among all continents.
208	Triassic	The seas are repopulated by invertebrates that survived the Permian extinction event. Bivalves and echinoids expand into the infaunal niche.	Mammals evolve from cynodonts. Cynodonts become extinct. Ancestral archosaur gives rise to dinosaurs. Flying reptiles and marine reptiles evolve.	Land flora of seedless vascular plants and gymnosperms as in Late Paleozoic.	Warm-temperate to tropical. Mild temperatures extend to high latitudes; polar regions may have been temperate. Local areas of aridity.	Fragmentation of Pangaea begins in Late Triassic.
245						

Important Terms

angiosperm
archosaur
bipedal
Cordilleran orogeny
Cretaceous Interior Seaway

cynodont
eupantothere
Laramide orogeny
marsupial mammal
monotreme

Nevadan orogeny
Ornithischia
placental mammal
quadrupedal
Saurischia

Sevier orogeny
Sonoma orogeny
Sundance Sea
therapsid

Review Questions

1. Evidence for the breakup of Pangaea includes
   a. _____ sills; b. _____ rift valleys; c. _____ great quantities of poorly sorted nonmarine detrital sediments; d. _____ dikes; e. _____ all of the previous.

2. The breakup of Pangaea began with initial Triassic rifting between which two continental landmasses?
   a. _____ India and Australia; b. _____ North America and Eurasia; c. _____ Antarctica and India; d. _____ Laurasia and Gondwana; e. _____ South America and Africa.

3. The orogeny responsible for the present-day Rocky Mountains is the
   a. _____ Antler; b. _____ Sevier; c. _____ Sonoma; d. _____ Nevadan; e. _____ Laramide.

4. The time of greatest post-Paleozoic inundation of the craton occurred during which geologic period?
   a. _____ Triassic; b. _____ Jurassic; c. _____ Cretaceous; d. _____ Paleogene; e. _____ Neogene.

5. The Jurassic formation or complex famous for dinosaur fossils is the
   a. _____ Navajo; b. _____ Chinle; c. _____ Sundance; d. _____ Morrison; e. _____ Franciscan.

6. A possible cause for the eastward migration of igneous activity in the Cordilleran region during the Cretaceous was a change from
   a. _____ oceanic-oceanic convergence to oceanic-continental convergence; b. _____ high-angle to low-angle subduction; c. _____ divergent to convergent plate margin activity; d. _____ divergent plate margin activity to subduction; e. _____ subduction to divergent plate margin activity.

7. A typical mammal's jaw-skull joint is between which of these pairs of bones?
   a. _____ stapes-quadrate; b. _____ dentary-squamosal; c. _____ articular-malleus; d. _____ ethemoid-jugular; e. _____ occipital-incus.

8. All carnivorous dinosaurs belong to the group known as
   a. _____ theropods; b. _____ ankylosaurs; c. _____ mosasaurs; d. _____ ammonites; e. _____ gymnosperms.

9. Which one of the following pairs of animals were marine reptiles?
   a. _____ angiosperm-ginkgo; b. _____ ichthyosaur-plesiosaur; c. _____ cephalopod-rudist; d. _____ cynodont-marsupial; e. _____ monotreme-endotherm.

10. An important group of marine phytoplankton, the _____, first evolved during the Jurassic and remain numerous today.
    a. _____ oysters; b. _____ burrowing worms; c. _____ coccolithophores; d. _____ belemnoids; e. _____ teleosts.

11. Dinosaurs, crocodiles, pterosaurs, and birds are collectively known as
    a. _____ placentals; b. _____ archosaurs; c. _____ mosasaurs; d. _____ invertebrates; e. _____ stem reptiles.

12. An important Mesozoic event in the history of land plants was the
    a. _____ origin of ferns and horsetail rushes; b. _____ extinction of cycads and conifers; c. _____ first appearance and diversification of angiosperms; d. _____ dominance of seedless vascular plants; e. _____ prevalence of ginkgos and ceratopsians.

13. Discuss the changes that took place in the jaw and middle ear when cynodonts gave rise to mammals.

14. How does the breakup of Pangaea help us better understand the Mesozoic distribution of plants and animals?

15. Briefly summarize the evidence for and against endothermy in dinosaurs.

16. How do the Mesozoic marine invertebrate communities differ from those of the Paleozoic?

17. Mesozoic fossil cephalopods, especially ammonites, and foraminifera are very good guide fossils. Explain why.

18. Briefly outline the major changes in the composition of Mesozoic marine invertebrate communities.

19. Discuss the similarities and differences between the orogenic activity that occurred in the Appalachian mobile belt during the Paleozoic Era and that which occurred in the Cordilleran mobile belt during the Mesozoic Era.

20. How did the Mesozoic rifting that took place on the East Coast of North America affect the tectonics in the Cordilleran mobile belt?

21. The breakup of Pangaea influenced the distribution of continental landmasses, ocean basins, and oceanic and atmospheric circulation patterns, which in turn affected the distribution of natural resources, landforms, and the evolution of the world's biota. Reconstruct a hypothetical history of the world for a different breakup of Pangaea—one in which the continents separate in a different order or rift apart in a different configuration. How would such a scenario affect the distribution of natural resources? Would the distribution of coal and petroleum reserves be the same? How might evolution be affected? Would human history be different?

22. Discuss the modifications that occurred for flight in pterosaurs and for an aquatic life in ichthyosaurs.

23. How did land plant communities of the Triassic and Jurassic differ from those of the Cretaceous?

24. Briefly summarize the evidence for and against the hypothesis that a meteorite impact caused Mesozoic mass extinctions.

World Wide Web Activities

For these web site addresses, along with current updates and exercises, log on to **http://www.brookscole.com/geo**

Global Earth History

This site, maintained at the University of Northern Arizona, contains a series of plate tectonic reconstructions to show what Earth looked like at various times in the geologic past. Included are paleogeographic maps showing global, North American, and European reconstructions from the Cambrian to the present, as well as tectonic, sedimentation, and paleogeographic maps of the North Atlantic region.

North American Orogenies

This site is part of the *Global Earth History* site and shows simple, restored cross sections of the major orogenies taking place in North America from the Proterozoic through Mesozoic. Click on the various orogenies that occurred during the Mesozoic.

Morrison Research Initiative

The Morrison Research Initiative is a multidisciplinary study of the Upper Jurassic Morrison Formation, which is funded by the National Park Service and the USGS.

Dinosaur Provincial Park

This site regarding Dinosaur Provincial Park, Alberta, Canada, is brief but has good descriptions of the fossils and the park, along with several photographs. What was the climate like when the dinosaur fossils were preserved in this area? Also, how many types of dinosaurs have been found here and what other types of animal fossils have been recovered?

Yale Peabody Museum

A site maintained by the Peabody Museum of Natural History at Yale University. On the main page click *China's Feathered Dinosaurs* and then click *Online Exhibits* and see the images of feathered dinosaurs and find out about their discovery, age, and relationships. Click on the names *Sinosauropteryx, Protarchaeopteryx,* and *Caudipteryx* to see images and discussions of the genera of feathered dinosaurs or those with protofeathers.

Return to the main page and click *The Age of Reptiles*, then click each of the Mesozoic periods, *Triassic, Jurassic,* and *Cretaceous,* for discussions of dinosaurs from these times.

Other Sites

Many web sites have information on dinosaurs and other Mesozoic animals and plants. A search using "dinosaurs" will yield numerous results, but one can find reliable information and good images at the following sites, to name a few: American Museum of Natural History (New York), Field Museum of Natural History (Chicago), Museum of the Rockies (Bozeman, Montana), Los Angeles County Museum (Los Angeles), Natural History Museum (London), Carnegie Museum of Natural History (Pittsburgh), Canadian Museum of Nature (Ottawa, Canada), Denver Museum of Natural History (Denver), University of Michigan Museum of Paleontology (Ann Arbor), Florida Museum of Natural History (Gainesville), and University of Kansas Natural History Museum (Lawrence).

CHAPTER 24

Cenozoic Earth and Life History

Well-preserved leaves of Metasequoia *(dawn red-wood) from the Oligocene to Miocene John Day Formation of Oregon.*

Only 1.4% of all geologic time is represented by the Cenozoic Era, yet rocks of this age are the most commonly encountered in many areas, especially in western North America. In contrast, the east was dominated by erosion during much of this time and accordingly few Cenozoic-age rocks are present. The notable exceptions are the Cenozoic sedimentary rocks of the Atlantic and Gulf Coastal Plains, and rocks exposed in parts of Florida and in the Calvert Cliffs of Maryland. And Pleistocene glaciers deposited gravel, sand, and mud, in many parts of Canada and the northern tier of eastern states.

In the west, Cenozoic tectonism yielded a number of basins that became the sites of sediment deposition, and a vast sheet of sediment was deposited on the Great Plains east of the present-day Rocky Mountains. Nearly all of these rocks were deposited in continental depositional environments—the rocks in Bryce Canyon National Park, Utah (see Chapter 5 Prologue), are a good example. Farther west, in California, tectonism was responsible for several basins that subsided below sea level where Cenozoic marine sedimentary rocks were deposited; many of these are prolific oil and natural gas producers.

Cenozoic sedimentary rocks in Badlands National Park, South Dakota, as well as sedimentary and volcanic rocks in John Day Fossil Beds National Monument, Oregon, are scenic and contain excellent records of land-dwelling mammals and plants. In Badlands National Park, differential weathering and erosion of some of the rocks has yielded closely spaced small gullies and deep ravines separated by sharp angular slopes, ridges, and pinnacles, whereas other rocks form smooth rounded hills (Figure 24.1).

OBJECTIVES

After reading this chapter, you will have learned that

■ The Mesozoic breakup of Pangaea continued during the Cenozoic accounting for orogenies in two major belts.

■ The North American Cordillera experienced deformation, the origin of mountains, volcanism, uplift, and deep erosion.

■ North America's interior lowlands were occupied briefly by an epeiric sea.

■ Thick deposits accumulated along the Gulf and Atlantic Coastal plains.

■ Cenozoic uplift and erosion account for the present expression of the Appalachian Mountains.

■ Pleistocene glaciers covered large areas on the Northern Hemisphere continents.

■ Cenozoic rocks contain several resources such as oil and gold.

■ Marine invertebrates such as foraminifera and mollusks were abundant during the Cenozoic.

■ Mammals diversified and eventually gave rise to today's familiar mammals.

■ A variety of large mammals and birds existed during the Pleistocene.

■ Primates were present throughout the Cenozoic, but human evolution took place during the Pliocene and Pleistocene.

Figure 24.1 *Cenozoic sedimentary rocks in Badlands National Park, South Dakota. Weathering and erosion of the Oligocene-age Brule Formation (on the skyline) yielded typical badlands topography, whereas smooth, rounded slopes developed on the underlying Chadron Formation (Eocene to Oligocene). The gently dipping yellowish rocks in the gully belong to the Cretaceous-age Pierre Shale. The contact between the Pierre Shale and Chadron Formation is an angular unconformity.*

(a)

(b)

(c)

Figure 24.2 *(a) (b) Two exposures of Cenozoic sedimentary rocks in John Day Fossil Beds, National Monument, Oregon. (c) Fossils from the Eocene Clarno Formation indicate the climate was subtropical. Lush forests were occupied by many mammals including (1) titanotheres, (2) a carnivore, (3) ancient horses, (4) tapirs, and (5) an early rhinoceros.*

These Eocene and Oligocene rocks, all of which were deposited in river channels, on floodplains, and in lakes, contain numerous mammal fossils some of which the Park Service has left but protected for viewing by park visitors.

The colors and erosional patterns of 6- to 54-million-year-old rocks in John Day Fossil Beds National Monument in Oregon provide some truly inspiring scenery (Figure 24.2a,b). Furthermore, some of the rocks clearly illustrate that the categories we so easily define in terms of the rock cycle (see Figure 1.13) are not always so readily distinguished. Many of the rocks here are *volcanoclastic*, meaning they contain large quantities of volcanic rock debris and/or pyroclastic materials, especially ash. Added to this, few contain sedimentary structures, a combination that resulted in their being ignored for many years by geologists interested in sedimentary rocks. Similarly, geologists who study igneous rocks and processes also ignored them because they thought the rocks were sedimentary.

Many of the rocks are truly sedimentary (Figure 24.2b), but undisputed igneous rocks such as lava flows are also found in the monument (see Figure 4.6). All have been well studied, some contain numerous fossil mammals and plants, and the many ash beds present make the rocks easily dated by absolute dating techniques. Today, the region is semiarid, but rocks and fossils of Eocene age clearly show it was an area of semitropical rainforest of palms, avocados, and ferns occupied by ancient horses, rhinoceroses, and carnivorous mammals, as well as the now extinct oreodonts and titanotheres (Figure 24.2c). Younger rocks and fossils provide evidence of changing conditions until those similar to the present existed.

Equally interesting and scenic Cenozoic sedimentary and igneous rocks are present in many other parts of western North America. They provide part of the record of events that took place on this continent during this most recent interval of geologic time.

Introduction

In Chapter 20 we noted that Precambrian time encompasses more than 88% of all geologic time, or more than 21 hours in our analogy of geologic time represented by a 24-hour clock (see Figure 20.2). At 66 million years long, the Cenozoic Era is rather brief, accounting for only about 20 minutes of our 24-hour day. Nevertheless, 66 million years is an extremely long time by any measure, certainly long enough for significant evolution of both Earth and its biota. And Cenozoic rocks provide a more easily interpreted record of geologic history because they are accessible at or near the surface and generally have been less altered by deformation and metamorphism than older rocks.

Traditionally geologists have divided the Cenozoic Era into two periods—the *Tertiary* (66 to 1.6 million years ago) and the *Quaternary* (1.6 million years ago to the present). Each period is further divided into shorter time intervals known as *epochs* (see Figure 17.1). These designations for Cenozoic time are widely used, but another scheme is becoming popular. In this system Quaternary is retained, but Tertiary has been replaced by the *Paleogene Period* (66 to 24 million years ago) and the *Neogene Period* (24 to 1.6 million years ago) (see Figure 17.1). We follow the traditional usage of Tertiary and Quaternary in this book.

Why Should You Study Cenozoic Geologic and Life History?

Why should you study Cenozoic geologic and life history? For one thing, the present distribution of land and sea as well as Earth's current topography resulted from processes operating during the Cenozoic. The study of the climatic effects responsible for Pleistocene (Ice Age) glaciation might provide some answers in the debate about global warming. Mammals evolved into the familiar forms we know today, thus the Cenozoic is known as the "Age of Mammals," and flowering plants continued to diversify and dominate the world's flora. From our own perspective, it was during the Late Cenozoic that the first humans evolved.

Cenozoic Plate Tectonics and Orogeny—An Overview

The Late Triassic fragmentation of the supercontinent Pangaea (see Figure 23.2a) began an episode of plate motions that continues even now. As a result, Cenozoic orogenic activity has been largely concentrated in two major zones or belts, the *Alpine-Himalayan belt* and the *circum-Pacific belt* (see Figure 10.22). The Alpine-Himalayan belt includes the mountainous regions of southern Europe and north Africa and extends eastward through the Middle East and India and into southeast Asia, whereas the circum-Pacific belt, as its name implies, nearly encircles the Pacific Ocean basin.

Within the Alpine-Himalayan orogenic belt, the *Alpine orogeny* began during the Mesozoic, but major deformation also took place from the Eocene to Late Miocene as the African and Arabian plates moved northward against Eurasia. Deformation resulting from plate convergence formed the Pyrenees Mountains between Spain and France, the Alps of mainland Europe, the Apennines of Italy, and the Atlas Mountains of North Africa (see Figure 10.22). Active volcanoes in Italy and seismic activity in much of southern Europe and the Middle East indicate that this orogenic belt remains geologically active.

Farther east in the Alpine-Himalayan orogenic belt, the Himalayan orogen resulted from the collision of India with Asia (see Figure 10.25). The exact time of this collision is uncertain, but sometime during the Eocene India's northward drift rate decreased abruptly, indicating the probable time of collision. In any event, a *collision orogen* resulted as two continental plates became sutured, accounting for the location of the present-day Himalayas far inland rather than at a continental margin.

Plate subduction in the circum-Pacific orogenic belt occurred throughout the Cenozoic, giving rise to orogens in the Aleutians, the Philippines, Japan, and along the west coasts of North, Central, and South America. For example, the Andes Mountains in western South America formed as a result of convergence of the Nazca and South American plates (see Figure 10.24). Spreading at the East Pacific Rise and subduction of the Cocos and Nazca plates beneath Central and South America, respectively, account for continuing seismic, volcanic, and orogenic activity in these regions.

Evolution of North America During the Tertiary Period

Many of Earth's features are much older than 66 million years, but the present geographic distribution of continents and oceans, the topographic expression of continents, and many landforms resulted from processes operating during the Cenozoic. For instance, the Appalachian Mountains began their evolution during the Precambrian, but their present expression is largely the product of Cenozoic uplift and erosion. Likewise, the uplift of the Sierra Nevada of California and Nevada and the origin of the Himalayas of Asia took place during the Cenozoic, although in both cases mountain building was preceded by events going back many hundreds of millions of years. And many of the distinctive landforms such as badlands, glacial valleys, deep canyons, lava flows, and volcanoes in our national parks developed during the past few thousands to several millions of years. In short, many of Earth's distinctive aspects formed very recently in the context of geologic time and in many areas they continue to evolve.

The North American Cordillera

The **North American Cordillera** is a complex mountainous segment of the circum-Pacific orogenic belt extending from Alaska, through Canada and the United States, and into central Mexico (Figure 24.3). It was more or less continuously deformed during the Late Jurassic to Early Tertiary as the Nevadan, Sevier, and Laramide orogenies progressively affected areas from west to east (see Figure 23.11). But the final episode of deformation, the Late Cre-

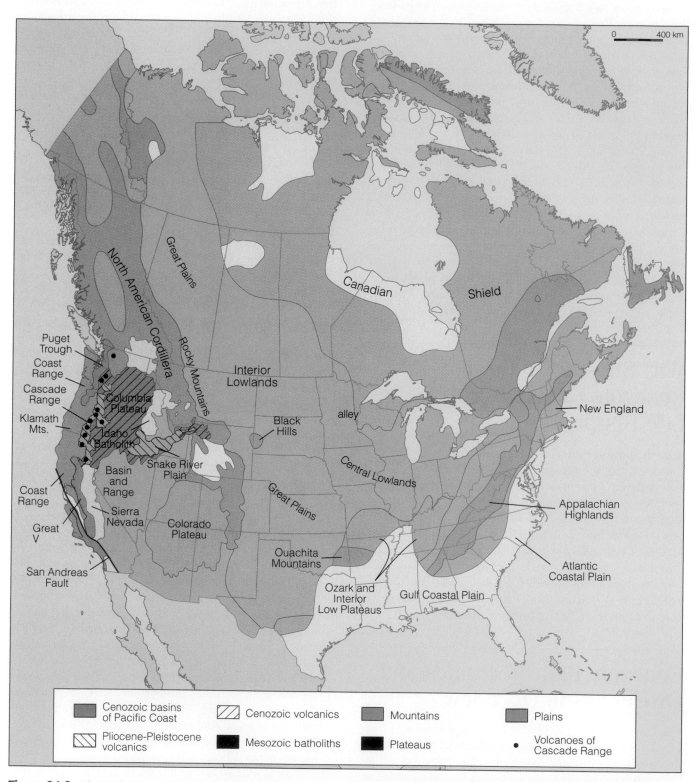

Figure 24.3 *The North American Cordillera, which is a segment of the circum-Pacific organic belt, and the major provinces of North America discussed.*

taceous to Eocene **Laramide orogeny,** differed from the previous orogenies. It took place much further inland than is typical; deformation was mostly in the form of vertical uplifts as opposed to compression-induced folding and faulting, and little volcanism and intrusion of plutons took place during deformation.

Orogenies resulting from oceanic-continental convergence usually take place very near the continental plate margin (see Figure 10.24). To account for the Laramide orogeny far from a plate boundary, geologists have proposed a model involving shallow subduction and cessation of igneous activity. During the Late Cretaceous, a subduction zone existed along the western margin of North America where the Farralon plate was subducted at about a 50° angle, and igneous activity took place 150 to 200 km inland from the trench (Figure 24.4). By Early Tertiary time,

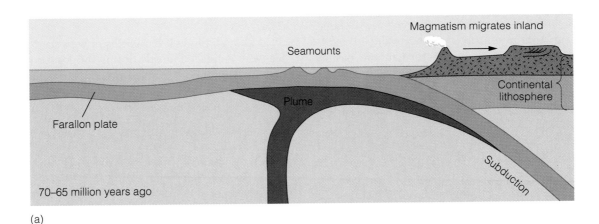

(a)

(b)

(c)

Figure 24.4 *The Laramide orogeny resulted when the Farallon plate was subducted beneath North America during the Late Cretaceous to Eocene. (a) As North America moved westward over the Farallon plate, beneath which was the deflected head of a mantle plume, the angle of subduction decreased and magmatism shifted inland. (b) With nearly horizontal subduction, magmatism ceased and the continental crest was deformed mostly by vertical forces. (c) Disruption of the oceanic plate by the mantle plume marked the onset of renewed volcanism.*

Evolution of North America During the Tertiary Period

the westward moving North American plate had overridden part of the Farallon plate beneath which was the deflected head of a mantle plume. The lithosphere immediately above the plume was buoyed up, thus accounting for the change from steep to shallow subduction (Figure 24.4). As a result, igneous activity migrated farther inland and eventually ceased because the descending Farallon plate no longer penetrated to the mantle.

Another consequence of the decreased angle of subduction was a change in the style of deformation. The fold-thrust tectonism of the Sevier orogeny gave way to large scale buckling and fracturing, which produced fault-bounded uplifts with intervening intermontane basins. These basins were the depositional sites for Early Tertiary sediments eroded from the uplifts (Figure 24.5).

The Laramide orogen is centered in the middle and southern Rockies in Wyoming and Colorado, but deformation also took place far to the north and south. In Montana and Alberta, Canada, large slabs of strata were transported to the east along overthrust faults (see Figure 2, Perspective 14.1). A major fold-thrust belt also developed in northern Mexico, and Laramide structural trends can be traced south into the area of Mexico City. Cessation of Laramide deformation apparently coincided with an increasing angle of descent of the Farallon plate (Figure 24.4).

Although the vast batholiths of western North America were emplaced mostly during the Mesozoic, intrusive activity continued into the Tertiary. Numerous small plutons were emplaced, including copper and molybdenum-

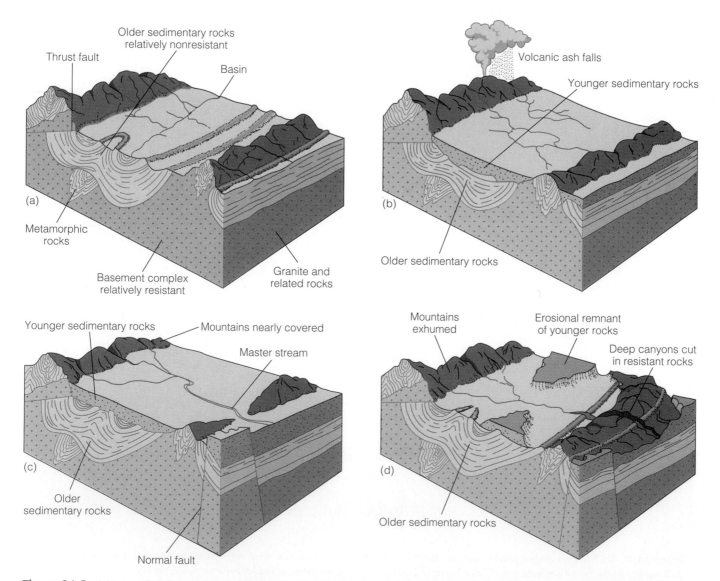

Figure 24.5 (a) Laramide deformation yielded fault-bounded uplifted blocks with intervening intermontane basins. (b) and (c) Sediment eroded from the uplifted blocks nearly filled the basins. (d) Renewed erosion removed some of the basin sediments, and streams cut deep canyons into the uplifted blocks.

(a)

(b)

(c)

Figure 24.6 *Cenozoic volcanism in the North American Cordillera. (a) San Francisco Peaks, Arizona, a cluster of volcanoes that erupted from 2.8 million to about 200,000 years ago. (b) Basalt lava flows of the Snake River Plain exposed at Twin Falls, Idaho. (c) Cenozoic lava flows exposed near Chico, California.*

bearing stocks in Utah, Nevada, Arizona, and New Mexico. In addition, Devil's Tower, Wyoming (see Figure 3, Perspective 3.2), and a number of other volcanic pipes were emplaced during the Cenozoic.

The Cordillera was also the site of considerable Cenozoic volcanism although it varied in eruptive style and location (Figure 24.6). A large area in Washington and adjacent parts of Oregon and Idaho is underlain by about 200,000 km³ of Miocene basalt lava flows known as the

Columbia River basalts (see Figure 4.17). The flows issued from long fissures and are now well exposed in the deep gorges cut by the Snake and Columbia rivers. The Snake River Plain in Idaho (Figure 24.6b) is actually a depression filled mostly by Pliocene and younger basalt lava flows. Bordering the Snake River Plain on the northeast is the Yellowstone Plateau of Wyoming, an area of Late Pliocene and Quaternary rhyolitic and some basaltic volcanism. The volcanoes of the Cascade Range were built by andesitic volcanism during the Pliocene, Pleistocene, and Recent epochs (see Perspective 4.2).

A large area in the Cordillera known as the **Basin and Range Province** (Figure 24.3) has experienced crustal extension and the development of north-south oriented normal faults since the Late Miocene. Differential movement on these faults has produced a basin-and-range structure, consisting of mountain ranges separated by broad valleys (Figure 24.7).

The Colorado Plateau (Figure 24.3) is a vast area of deep canyons, broad mesas, volcanic mountains, and brilliantly colored rocks (Figure 24.8). During the Early Tertiary this area, which was formerly near sea level, was uplifted and deformed into broad anticlines and arches and basins. A number of large normal faults also cut the area, but overall deformation was far less intense than it was elsewhere in the Cordillera. Late Tertiary uplift elevated the region to the 1200 to 1800 m elevations seen today. As uplift proceeded, deposition ceased and erosion of the deep canyons began.

The present plate tectonic elements of the Pacific Coast developed as a result of the westward drift of the North American plate, the partial consumption of the Farallon plate, and the collision of North America with the Pacific-Farallon ridge (Figure 24.9). Most of the Farallon plate was consumed at a subduction zone along the west coast of North America, and now only two small remnants exist—the Juan de Fuca and Cocos plates. Westward drift of the North American plate also resulted in its collision

with the Pacific-Farallon ridge and the origin of the Queen Charlotte and San Andreas transform faults along the west coasts of British Columbia, Canada, and California, respectively (Figure 24.9).

Much of the Cordilleran deformation and volcanism occurred during the Tertiary Period, but these activities continued through the Quaternary to the present. For example, many of California's coastal oil- and gas-producing fold structures formed during the Pleistocene. Pleistocene tectonism also affected the Sierra Nevada in California, the Teton Range of Wyoming, and parts of the central and northern Rocky Mountains. Continued subduction of the Juan de Fuca and Cocos plates accounts for Pleistocene and Recent seismic activity and volcanism in the Pacific Northwest and Mexico, respectively.

The Interior Lowlands

During the Cretaceous, most of the western part of the Interior Lowlands (Figure 24.3) was covered by the Zuni epeiric sea. By Early Tertiary time, the Zuni Sea had withdrawn from most of North America, but a sizable remnant arm of the sea was still present in North Dakota during the Paleocene Epoch. Sediments derived from Laramide highlands to the west and southwest were transported to this remnant sea where they were deposited in transitional and marine environments. Elsewhere within the western Interior Lowlands, sediment was transported eastward from the Cordillera and deposited in continental environ-

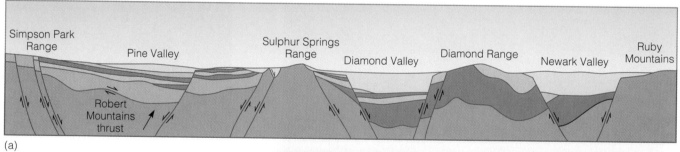

(a)

(b)

Figure 24.7 *(a) Cross section of part of the Basin and Range Province in Nevada. The ranges and valleys are bounded by normal faults. (b) View of the western margin of the Basin and Range Province. The Sierra Nevada are bounded on the east by normal faults and rise 3000 m above the basins to the east.*

(a)

(b)

(c)

Figure 24.8 *Rocks of the Colorado Plateau. (a) The Grand Canyon, Arizona. (b) A volcanic neck in northern Arizona. (c) Valley of the Gods, Utah. The rocks in (a) and (c) are sedimentary.*

ments. It formed large, eastward-thinning wedges that now underlie the Great Plains (Figure 24.10).

Igneous activity was not widespread in the Interior Lowlands, but in some local areas it was significant. In northeastern New Mexico, for example, Late Tertiary extrusive volcanism produced volcanoes and numerous lava flows. A number of small intrusive bodies were emplaced in Colorado, Montana, South Dakota, and Wyoming.

Eastward, beyond the Great Plains section of the Interior Lowlands, Cenozoic deposits, other than those of Pleistocene glaciers, are uncommon. Much of the Interior Lowlands was subjected to erosion during the Tertiary. Of course, the eroded material had to be deposited somewhere, and that was on the Gulf Coastal Plain (Figure 24.3).

The Gulf Coastal Plain

Following the final withdrawal of the Cretaceous Zuni Sea, the Cenozoic **Tejas epeiric sea** made a brief appearance on the continent. But even at its maximum extent, it was

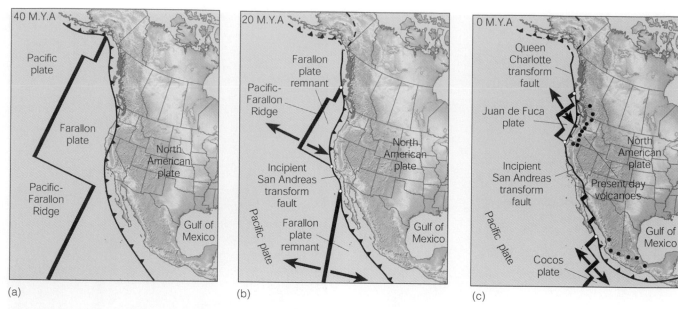

Figure 24.9 *(a), (b), and (c) Three stages in the westward drift of North America, its collision with the Pacific-Farallon Ridge, and the origin of the Queen Charlotte and San Andreas faults. As the North American plate overrode the ridge, its margin became bounded by transform faults rather than a subduction zone.*

largely restricted to the Atlantic and Gulf Coastal plains. In fact, its greatest incursion onto North America was in the area of the Mississippi Valley, where it extended as far north as southern Illinois.

Sedimentary facies development on the Gulf Coastal Plain was controlled largely by a regression of the Cenozoic Tejas epeiric sea. Its regression, however, was periodically reversed by minor transgressions; eight transgressive-regressive episodes are recorded in Gulf Coastal Plain sedimentary rocks.

The Gulf Coast sedimentation pattern was established during the Jurassic and persisted through the Cenozoic. Sediments were derived from the eastern Cordillera, western Appalachians, and Interior Lowlands and were trans-

Figure 24.10 *Cenozoic sedimentary rocks of the Great Plains at Scott's Bluff, Nebraska.*

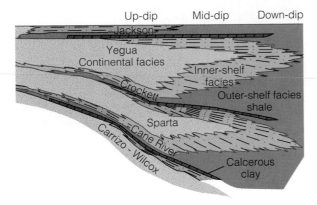

Figure 24.11 *Cenozoic deposition on the Gulf Coastal Plain. This cross section of the Eocene Claiborne Group shows facies changes and the seaward thickening of the deposits.*

ported toward the Gulf of Mexico. In general, the sediments form seaward-thickening wedges that grade from terrestrial facies in the north to progressively more off-shore marine facies in the south (Figure 24.11).

Much of the Gulf Coastal Plain was dominated by detrital sediment deposition during the Cenozoic. In the Florida section of the coastal plain and the Gulf Coast of Mexico, though, thick sequences of carbonate rocks were deposited. A carbonate platform was established in Florida during the Cretaceous, and shallow-water carbonate deposition continued through the Early Tertiary. Carbonate deposition continues in Florida Bay, the Florida Keys, and the Great Bahama Bank (Figure 24.12).

Eastern North America

The eastern seaboard has been a passive continental margin since Late Triassic rifting separated North America from North Africa and Europe. Some seismic activity still occurs there, (see Figure 7.10) but overall the region lacks the geologic activity characteristic of active continental margins.

The present distinctive topography of the Appalachian Mountains is the product of Cenozoic uplift and erosion. By the end of the Mesozoic, the Appalachian Mountains had been eroded to a plain (Figure 24.13). Cenozoic uplift rejuvenated the streams, which responded by renewed downcutting. As the streams eroded downward, they were superposed on resistant strata and cut large canyons across these rocks. For example, the distinctive topography of the Valley and Ridge Province is the product of Cenozoic erosion and preexisting geologic structures. It consists of northeast-southwest trending ridges of resistant upturned strata and intervening valleys eroded into less resistant strata (Figure 24.13).

The Atlantic continental margin includes the Atlantic Coastal Plain (Figure 24.3) and extends seaward across the continental shelf, slope, and rise (Figure 24.14). It possesses a number of Mesozoic and Cenozoic sedimentary basins that formed as a result of rifting. Deposition in these basins began during the Jurassic, and even though sediments of this age are known only from a few deep wells, they are presumed to underlie the entire margin of the continent. The distribution of Cretaceous and Cenozoic sediments is better known because both are present in the Atlantic Coastal Plain, and both have been penetrated by wells on the continental shelf.

In general, the sedimentary rocks of the Atlantic Coastal Plain are part of a seaward-thickening wedge that dips gently seaward. In some places, such as off the coast of New Jersey, these deposits are up to 14 km thick. The best-studied seaward-thickening wedge of sedimentary rocks is in the Baltimore Canyon Trough, an area that also exhibits the structures typical of a passive continental margin (Figure 24.14).

Pleistocene Glaciation

In 1837, the Swiss naturalist Louis Agassiz argued that large boulders (erratics), polished and striated bedrock, U-shaped valleys, and deposits of sand and

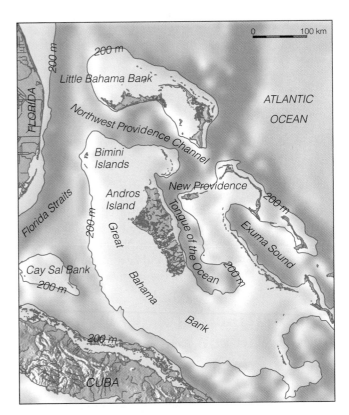

Figure 24.12 *The Great Bahama Bank and areas adjacent to Florida are underlain by thick Cenozoic carbonate rocks, and carbonate deposition continues in this region.*

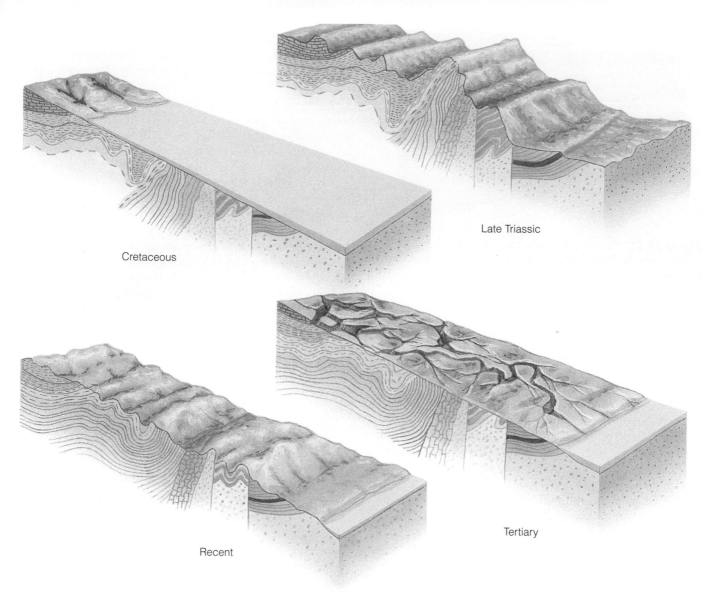

Cretaceous

Late Triassic

Tertiary

Recent

Figure 24.13 *The origin of the present topography of the Appalachian Mountains. Erosion in response to Cenozoic uplift accounts for this topography.*

gravel found in parts of Europe resulted from huge glaciers moving over the land. Although the idea initially met with considerable resistance, scientists finally came to realize that Agassiz was correct and accepted the idea that an Ice Age had taken place in the recent geologic past.

The Distribution and Extent of Pleistocene Glaciers

We know today that the Pleistocene, or what is commonly called the Ice Age, began 1.6 million years ago and ended about 10,000 years ago. During this time, several intervals of widespread continental glaciation took place, especially on the Northern Hemisphere continents, each separated by warmer interglacial periods. In addition, valley glaciers were more common at lower elevations and latitudes, and many extended much farther than their shrunken remnants do today (see Figure 3, Perspective 14.1). Unfortunately, scientists do not know whether we are still in an interglacial period or entering another cooler glacial interval.

As one would expect, the climatic effects responsible for Pleistocene glaciers were worldwide. Nevertheless, Earth was not as frigid as portrayed in movies and cartoons, nor was the onset of the climatic conditions leading to glaciation very rapid. Indeed, evidence from several types of investigations indicates that the climate gradually cooled from the Eocene through the Pleistocene. Furthermore, evidence from oxygen isotope data (the ratio of O^{18} to O^{16}) from deep-sea cores shows that 20 major warm-cold cycles have occurred during the last 2 million years.

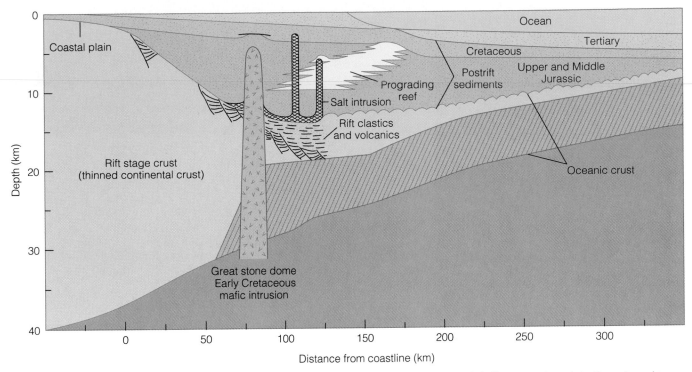

Figure 24.14 *The continental margin in eastern North America. The coastal plain and the continental shelf are covered mostly by Cenozoic sandstones and shales. Beneath these sediments are Cretaceous-age and probably Jurassic-age sedimentary rocks.*

The Effects of Glaciation

Glaciation has had many direct and indirect effects. Movement of glaciers over Earth's surface has produced distinctive landscapes in much of Canada, the northern tier of states, and in the mountains of the West (see Chapter 14). Sea level has risen and fallen with the formation and melting of glaciers, and these changes in turn have affected the margins of continents. Glaciers have also altered the world's climate, causing cooler and wetter conditions in some areas that are arid to semiarid today. In addition to the usual evidence of glacial activity (Figure 24.15), one of the largest floods in history was caused by the collapse of an ice dam in eastern Washington.

More than 70 million km³ of snow and ice blanketed the continents during the maximum glacial coverage of the Pleistocene. The storage of ocean waters in glaciers lowered sea level 130 m and exposed large areas of the present-day continental shelves, which were soon covered by vegetation.

Lowering of sea level also affected the base level of rivers and streams. When sea level dropped, streams eroded downward as they sought to adjust to a lower base level.

From such glacial features as terminal moraines, erratics, and drumlins (see Chapter 14), it seems that at their greatest extent Pleistocene glaciers covered about three times as much of Earth's surface as they do now

(Figure 24.15a). That is, they covered more than 40 million km² (see Table 14.1), and like the vast ice sheets now present in Greenland and Antarctica they were probably 3 km thick. Geologists have identified four major Pleistocene glacial episodes that took place in North America—the *Nebraskan, Kansan, Illinoian,* and *Wisconsinan* (Figure 18.15b), each named for the states in which the most southerly glacial deposits are well exposed. The three interglacial stages are named for localities of well-exposed soils and other deposits (Figure 24.15b). In Europe, six or seven major glacial advances and retreats are recognized.

Recent studies show that there were an as yet undetermined number of pre-Illinoian glacial advances and retreats in North America, and that the glacial history of this continent is more complex than previously thought. In view of this evidence, the traditional four-part subdivision of the North American Pleistocene will have to be modified. Stream channels in coastal areas were extended and deepened along the emergent continental shelves. When sea level rose with the melting of the glaciers, the lower ends of stream valleys along the east coast of North America were flooded and are now important harbors, while just off the west coast, they form impressive submarine canyons. Great amounts of sediment eroded by the glaciers were transported by streams to the sea and thus contributed to the growth of submarine fans along the base of the continental slope.

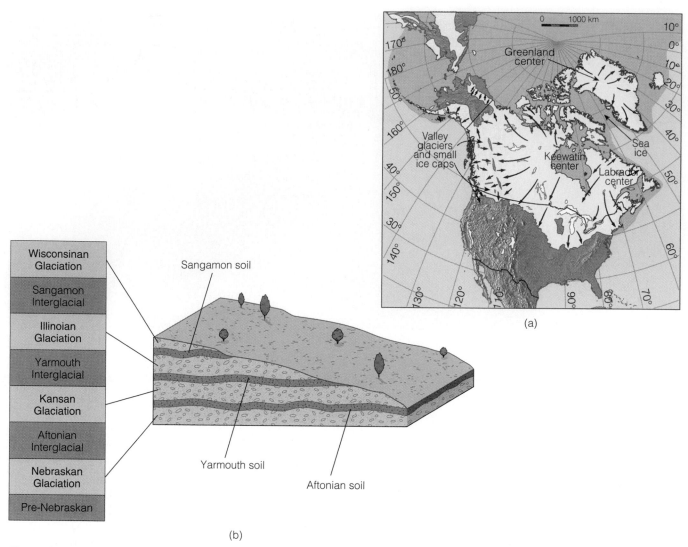

Figure 24.15 *(a) Centers of ice accumulation and maximum extent of Pleistocene glaciers in North America. (b) Standard terminology for Pleistocene glacial and interglacial stages in North America.*

We noted earlier that as the Pleistocene ice sheets formed and increased in size, the weight of the ice caused the crust to slowly subside deeper into the mantle. In some places, Earth's surface was depressed as much as 300 m below the preglacial elevations. As the ice sheets retreated by melting, the downwarped areas gradually rebounded to their former positions.

During the Wisconsinan glacial stage many large lakes existed in what are now dry basins in the southwestern United States. These lakes formed as a result of greater precipitation and overall cooler temperatures (especially during the summer), which lowered the evaporation rate. At the same time, increased precipitation and runoff helped maintain high water levels. Lakes that formed during those times are *pluvial lakes,* and they correspond to the expansion of glaciers elsewhere. The largest of these lakes was Lake Bonneville, which attained a maximum size

of 50,000 sq km² and a depth of at least 335 m (Figure 24.16). The vast salt deposits of the Bonneville Salt Flats west of Salt Lake City, Utah, formed as parts of this ancient lake dried up; Great Salt Lake is simply the remnant of this once vast lake. Another large pluvial lake (Lake Manly) existed in Death Valley, California, which is now the hottest, driest place in North America.

In contrast to pluvial lakes, which form far from glaciers, *proglacial lakes* are formed by the meltwater accumulating along the margins of glaciers. In fact, in many proglacial lakes one shoreline is the ice front itself, while the other shorelines consist of moraines. Lake Agassiz was a large proglacial lake covering about 250,000 km² of North Dakota, Manitoba, Saskatchewan, and Ontario. It persisted until the glacial ice along its northern margin melted, at which time the lake was able to drain northward into Hudson Bay.

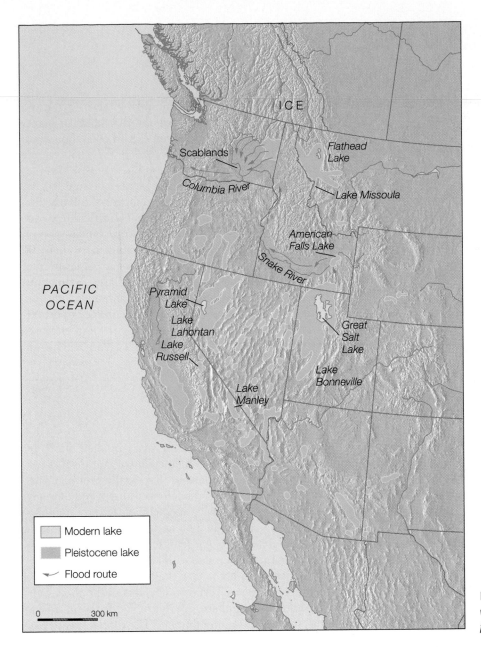

Figure 24.16 *Pleistocene lakes in the western United States. Lake Missoula was a proglacial lake, but the others were pluvial lakes.*

Cenozoic Mineral Resources

The United States is the third largest producer of petroleum, accounting for about 17% of the world's total. Much of this production comes from Tertiary reservoirs of the Gulf Coastal Plain and the adjacent continental shelf. On the Gulf Coastal Plain, most of the petroleum is in structural traps related to salt domes and other structures. Several Tertiary basins in southern California are also important areas of petroleum production. The Green River Formation of Wyoming, Utah, and Colorado has huge reserves of oil shale and evaporites.

In 1989, more than half of Florida's total mineral resources came from Upper Miocene phosphate-bearing rock in the Central Florida Phosphate District (see Figure 2, Perspective 8.2). Phosphorus from phosphate rock has a variety of uses in metallurgy, preserved foods, ceramics, and matches.

Diatomite is a sedimentary rock composed of the microscopic shells of diatoms, which are single-celled marine and freshwater plants that secrete skeletons of silica (SiO_2). This rock, also called diatomaceous earth, is used chiefly in gas purification and to filter a number of liquids such as molasses, fruit juices, water, and sewage. The United States is the leader in diatomite production, mostly from Tertiary deposits in California, Oregon, and Washington.

Huge deposits of low-grade lignite and subbituminous coals in the Northern Great Plains are becoming increasingly important resources. These coal deposits are Late

According to one estimate, present-day technology is capable of yielding 80 billion barrels of oil from oil shale in the Green River Formation of Wyoming, Colorado, and Utah, yet none is currently being produced. Given that the United States' domestic oil production is about 8 million barrels per day and we must import about half of what we consume, what contribution would the Green River Formation make, assuming that its entire potential could be realized? Do you see any problems with your projections? What kinds of economic, technological, and environmental problems might be encountered?

Cretaceous to Early Tertiary in age and are most extensive in the Williston and Powder River basins of Montana, Wyoming, and North and South Dakota. In addition to having a low sulfur content, some of these coals are in beds 30 to 60 m thick!

Gold production from the Pacific Coast, particularly California, comes mostly from Tertiary and Quaternary gravels. The gold is found in placer deposits, which formed as concentrations of minerals separated from weathered debris by fluvial processes.

A variety of other Tertiary mineral deposits are important. For example, the United States must import almost all manganese used in the manufacture of steel. The largest manganese deposits are in Lower Tertiary rocks in Russia. One Tertiary molybdenum deposit in Colorado accounts for much of the world production of this element. Tertiary sand and gravel, as well as evaporites, building stone, and clay deposits, are quarried from areas around the world.

Sand and gravel deposits resulting from glacial activity are a valuable Quaternary resource in many formerly glaciated areas. Most Pleistocene sand and gravel deposits originated as floodplain gravels, outwash sediment, or esker deposits. The bulk of the sand and gravel in the United States and Canada is used in construction and as roadbase and fill for highway and railway construction.

The periodic evaporation of pluvial lakes in the Death Valley region of California during the Pleistocene led to the concentration of many evaporite minerals such as borax. During the 1880s, borax was transported from Death Valley by the famous 20-mule team wagon trains.

Another Quaternary resource is peat, a vast potential energy resource that has been developed in Canada and Ireland. Peatlands formed from plant assemblages as the result of particular climate conditions.

Life History of the Tertiary Period

Earth's flora and fauna continued to evolve during the Cenozoic Era as more and more familiar kinds of plants and animals made their appearance. And even though we emphasize the evolution of mammals in this chapter, one should be aware of other important life events. Flowering plants continued to dominate land plant communities, the present-day groups of birds evolved during the Early Tertiary, and some marine invertebrates continued to diversify, eventually giving rise to today's marine fauna.

Mammals coexisted with dinosaurs for more than 100 million years, yet their Mesozoic fossil record indicates that they were not abundant, diverse, or very large. Extinctions at the end of the Mesozoic eliminated dinosaurs and some of their relatives, thereby creating the adaptive opportunities that mammals quickly exploited. The "Age of Mammals," as the Cenozoic Era is commonly called, had begun.

Marine Invertebrates and Phytoplankton

The Cenozoic marine ecosystem was populated mostly by those plants, animals, and single-celled organisms that survived the Mesozoic extinctions. Especially prolific Cenozoic invertebrate groups were foraminifera, radiolarians, corals, bryozoans, mollusks, and echinoids. The marine invertebrate community in general became more provincial during the Cenozoic because of changing ocean currents and temperature difference with latitude.

Only a few species in each major group of phytoplankton survived into the Tertiary. The coccolithophores, diatoms, and dinoflagellates all recovered from their Late Cretaceous reduction in numbers. The diatoms were particularly abundant during the Miocene, probably because of increased volcanism during this time. Volcanic ash provided increased dissolved silica in seawater and was used by the diatoms to construct their skeletons.

The foraminifera were a major component of the Cenozoic marine invertebrate community. Though dominated by relatively small forms (Figure 24.17), it included some exceptionally large forms that lived in the warm waters of the Cenozoic Tethys Sea. Shells of these larger foraminifera accumulated to form thick limestones, some of which were used by the ancient Egyptians to construct the Sphinx and the Pyramids of Gizeh.

The corals, having relinquished their reef-building role to the rudists during the mid-Cretaceous, again became the dominant reef-builders during the Cenozoic. Other suspension feeders such as bryozoans and crinoids were also abundant and successful during the Cenozoic. Perhaps the least important of the Cenozoic marine inverte-

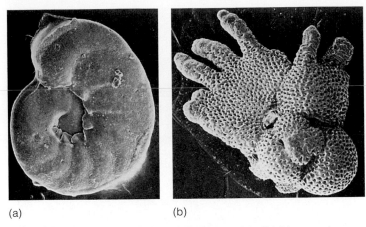

(a)   (b)

Figure 24.17 *Foraminifera of the Cenozoic Era. (a)* Cibicides americanus, *a benthonic form from the Early Miocene of California. (b) A planktonic form,* Globigerinoides fistulosus, *from the Pleistocene, South Pacific Ocean.*

brates were the brachiopods, with fewer than 60 genera surviving today.

Just as during the Mesozoic, bivalves and gastropods were two of the major groups of marine invertebrates, and they had a markedly modern appearance. Following the extinction of the ammonites and belemnites at the end of the Cretaceous, the Cenozoic cephalopod fauna consisted of nautiloids and cephalopods lacking shells, such as squids and octopuses.

The echinoids continued their expansion in the infaunal habitat and were particularly prolific during the Tertiary. New forms such as sand dollars evolved during this time from biscuit-shaped ancestors (Figure 24.18).

Diversification of Mammals

Among living mammals **monotremes,** such as the platypus, lay eggs, whereas **marsupials** and **placentals** give birth to live young. Marsupial mammals are born in an immature, almost embryonic condition, and then undergo further development in their mother's pouch. In placental mammals, on the other hand, the amnion of the amniote egg (see Figure 22.25) has fused with the walls of the uterus, forming a *placenta.* Nutrients and oxygen carried from the mother to the embryo through the placenta permit the young to develop much more fully before birth.

A measure of the success of placental mammals is that more than 90% of all mammals, fossil and living, are placental. In contrast, judging from the fossil record, monotremes have never been very common, and the only living ones are platypuses and spiny anteaters of the Australian region. Marsupials have been more successful, at least in terms of number of species and geographic distribution, but even they have been largely restricted to South America and the Australian region.

Although mammals first appeared during the Triassic, a major adaptive radiation began during the Paleocene and continued throughout the Cenozoic. Several of the groups of Paleocene mammals are considered *archaic,* meaning that they were holdovers from the Mesozoic Era or they did not give rise to any of today's mammals (Figure 24.19). But also among these mammals were the first rodents, rabbits, primates, carnivores, and hoofed mammals. However, even these had not yet become clearly differentiated from their ancestors, and the differences between herbivores and carnivores were slight. And most were small—large mammals were not present until the Late Paleocene, and the first giant terrestrial mammals did not appear until the Eocene (Figure 24.20).

Diversification continued during the Eocene when several more types of mammals appeared, but if we could go back and visit this time, we would probably not recognize many of these animals. Some would be vaguely familiar, but the ancestors of horses, camels, rhinoceroses, and elephants would bear little resemblance to their living descendants. By Oligocene time, all the basic groups of existing mammals—that is, the orders—were present, but diversification continued as more familiar families and genera appeared. Miocene and Pliocene mammals were mostly mammals that we could readily identify, although a few unusual types still existed (Figure 24.21).

Figure 24.18 *Echinoids were particularly abundant during the Tertiary, and new infaunal forms such as this sand dollar evolved from their Mesozoic biscuit-shaped ancestors.*

Figure 24.19 *The archaic mammals of the Paleocene Epoch included such animals as (1)* Protictis, *an early carnivore; (2) insectivores; (3)* Ptilodus; *and (4)* Pantolambda, *which stood about 1 m tall.*

Tertiary Mammals

Mammals arose from mammal-like reptiles known as cynodonts during the Late Triassic, so two-thirds of their evolution was during the Mesozoic Era (see Chapter 23). However, following the Mesozoic extinctions, mammals began an adaptive radiation and soon became the most abundant land-dwelling vertebrates. Now, more than 4000 species exist, ranging from tiny shrews to giants such as elephants and whales.

When people consider mammals, they think mostly of larger species such as elephants, horses, deer, dogs, and cats, but they fail to realize that most mammals are small, weighing less than 1 kg. In fact, with few exceptions, rodents, insectivores, rabbits, and bats fall into this category and they constitute fully 75% of all mammal species. These animals adapted to the microhabitats unavailable to larger mammals, or in the case of bats, became the only flying mammals. With this in mind we now turn our attention to some of the larger ones, especially hoofed mammals, carnivores, elephants, and whales.

Hoofed Mammals

Ungulate is a general term referring to several types of mammals but especially the orders Artiodactyla and Perissodactyla. About 170 living species of antelope, camels, giraffes, deer, goats, peccaries, pigs, and several others are even-toed hoofed mammals, or **artiodactyls,** the most common living ungulates. In marked contrast, only 16 species of horses, rhinoceroses, and tapirs are **perissodactyls,** or odd-toed hoofed mammals. As even- and odd-toed imply, artiodactyls have two or four toes while perissodactyls have one or three.

All ungulates are herbivores but some are **grazers,** meaning they feed on grasses, and others are **browsers,** feeding on the tender shoots, twigs, and leaves of trees and bushes. When grasses grow through soil, they pick up tiny pieces of sand that are quite abrasive to teeth, so the grazing ungulates developed high-crowned chewing teeth resistant to abrasion (Figure 24.22a). Browsers, on the other hand, never developed these kinds of chewing teeth.

Some ungulates are quite small and depend on concealment to avoid predators; others, such as rhinoceroses,

Figure 24.20 *The uintatheres were Eocene rhinoceros-sized mammals with three pairs of bony protuberances on the skull and saberlike upper canine teeth.*

are so large that size alone is enough to discourage predators, at least for adults. But many of the more modest-size ungulates are speedy runners. Adaptations for running include elongation of some of the limb bones as well as reduction in the number of bony limb elements, especially toes. Accordingly, the limbs of speedy ungulates are long and slender (Figure 24.22b).

Rabbit-size ancestral artiodactyls of the Early Eocene differed little from their ancestors, but gave rise to numerous families, several of which are now extinct (Figure 24.23). Small four-toed camels, for instance, appeared early in this diversification and were common in North America well into the Pleistocene. In fact, most of their evolution took place on this continent and only during the Pliocene did they migrate to South America and Asia where they now exist.

Certainly the most common existing artiodactyls are the *bovids* represented by cattle, goats, bison, and many species of antelope. They first appeared in the Miocene fossil record and continued to diversify throughout the rest of the Cenozoic. Most bovid evolution took place in Europe and northern Asia, but these creatures have since migrated to southern Asia and Africa where they are most common today.

The perissodactyls—horses, rhinoceroses, tapirs—and the extinct titanotheres and chalicotheres are united by several shared characteristics (see Figure 18.19). Furthermore, the fossil record shows they all evolved from a common ancestor during the Eocene. Their diversity increased through the Oligocene, but since then they have declined markedly and now constitute less than 10% of the world's hoofed mammal fauna.

Horses have a particularly good fossil record which shows that present-day *Equus* evolved from a tiny Eocene ancestor (Figure 24.24). Most of their evolution took place in North America. Fossils show several trends such as increased size, lengthening of the limbs, and development of high-crowned chewing teeth as horses became speedy grazing animals (Table 24.1). There was, however, another branch of horses leading to three-toed browsers that became extinct during the Pleistocene.

Other Mammals—Carnivores, Elephants, and Whales

During the Paleocene, the *miacids*, small carnivorous mammals with short heavy limbs, made their appearance. These small creatures were ancestors to all later members of the order Carnivora, which includes, among others, today's dogs, cats, hyenas, bears, weasels, and seals. All have well-developed canine teeth for slashing and tearing and most also developed a pair of large shearing teeth (Figure 24.25a). Some of the better known fossil carnivores are the saber-toothed cats, or what are more commonly called saber-toothed tigers (Figure 24.25b).

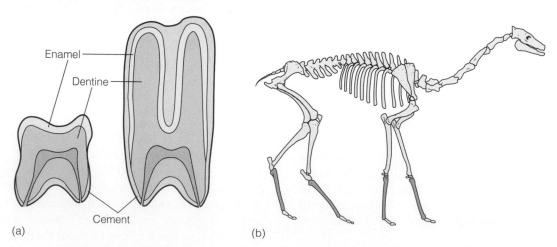

Figure 24.21 *Mural showing Pliocene mammals of the western North America grasslands. The animals shown include (1) Amebeledon, a shovel-tusked mastodon; (2) Teleoceras, a short-legged rhinoceros; (3) Cranioceras, a horned, hoofed mammal; (4) a rodent; (5) a rabbit; (6) Merycodus, an extinct pronghorn; (7) Synthetoceras, a hoofed mammal with a horn on its snout; and (8) Pliohippus, a one-toed grazing horse.*

Enamel

Dentine

Cement

(a)   (b)

Figure 24.22 *(a) Comparison of low-crowned and high-crowned chewing teeth. Grazing hoofed mammals developed high-crowned teeth as an adaptation for eating abrasive grasses. The cusps of high-crowned teeth are elevated into tall, slender pillars, and the teeth are covered with enamel and cement, both of which are hard substances. (b) Oxydactylus, a Miocene camel, shows the trend of limb elongation seen in many hoofed mammals. The bones between the wrist and toes and the ankle and toes became longer, thereby increasing the length of the limbs.*

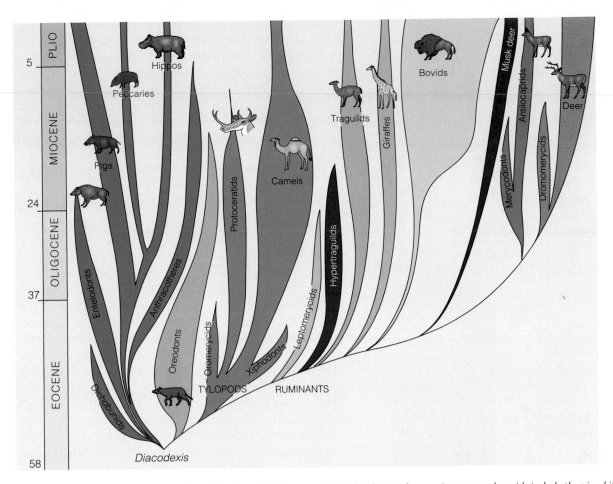

Figure 24.23 *History of the artiodactyls. Early in their history, artiodactyls split into three major groups: the suids include the pigs, hippopotamuses, and extinct giant hogs; the tylopoda are represented by the camels; and the ruminants consist of the cud-chewing animals.*

Table 24.1

Overall Trends in the Cenozoic Evolution of the Present-Day Horse Equus. *A number of horse genera existed during the Cenozoic that evolved differently. For instance, some horses were browsers rather than grazers and never developed high-crowned chewing teeth, and retained three toes.*

Trend

1. Size increase.
2. Legs and feet become longer, an adaptation for running.
3. Lateral toes reduced to vestiges. Only toe three remains functional in *Equus*.
4. Straightening and stiffening of the back.
5. Incisor teeth become wider.
6. Molarization of premolars yielded a continuous row of teeth for grinding vegetation.
7. The chewing teeth, molars and premolars, become high-crowned and cement covered for grinding abrasive grasses.
8. Chewing surfaces of premolars and molars become more complex. Also an adaptation for grinding abrasive grasses.
9. Front part of skull and lower jaw become deeper to accommodate high-crowned premolars and molars.
10. Face in front of eye becomes longer to accommodate high-crowned teeth.
11. Larger, more complex brain.

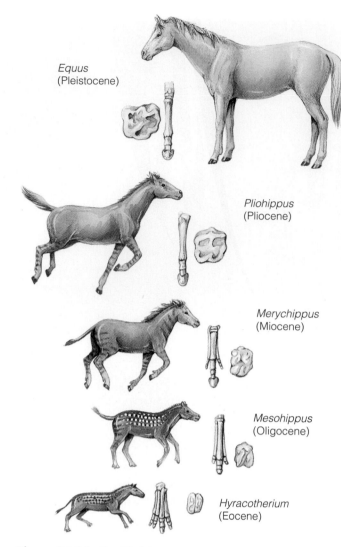

Equus
(Pleistocene)

Pliohippus
(Pliocene)

Merychippus
(Miocene)

Mesohippus
(Oligocene)

Hyracotherium
(Eocene)

Figure 24.24 *Simplified diagram showing some of the trends from the earliest known horse to the one-toed grazing horses of the present. Trends shown include size increase, reduction in the number of toes and lengthening of the legs, and development of high-crowned teeth with complex chewing surfaces. Notice that* Merychippus *had three toes whereas* Pliohippus *had only one. Another evolutionary lineage of horses, not shown here, led to the now extinct three-toed browsers.*

Elephants (order Proboscidea), the largest land mammals, evolved from pig-sized ancestors during the Eocene. And by Oligocene time, they clearly showed the trend toward large size and the development of a long snout (proboscis) and large tusks. Mastodons with teeth adapted for browsing were present by the Miocene, and during the Pliocene the present-day elephants and mammoths diverged from a common ancestor. During most of the Cenozoic, elephants were widespread on the northern continents, but now only two species exist in southern Asia and Africa.

We briefly mentioned whales in Chapter 18 in Fossils and Evolution, noting that until quite recently little was known about their transition from land-dwelling ancestor to fully aquatic whales. And while a number of questions remain unanswered, the fossils now available indicate that whales appeared during the Early Eocene and by the Late Eocene had become diverse and widespread. Eocene whales still possessed vestigial rear limbs, their teeth resembled those of their land-dwelling ancestors, their nostrils (blowhole) were not on top of the head, and they were proportioned quite differently from living whales (Figure 24.26). By Oligocene time, both groups of living whales—the toothed whales and the baleen whales—had evolved.

Pleistocene Faunas

One of the most remarkable aspects of the Pleistocene mammalian fauna is that so many large species existed. In North America, for example, there were mastodons and mammoths, giant bison, huge ground sloths, giant camels, and beavers nearly 2 m tall at the shoulder (Figure 24.27). Kangaroos standing 3 m tall, wombats the size of rhinoceroses, leopard-sized marsupial lions, and large platypuses characterized the Pleistocene fauna of Australia. In Europe and parts of Asia lived cave bears, elephants, and the giant deer commonly called the Irish elk with an antler spread of 3.35 m (see Perspective 24.1). The evolutionary trend toward large body size was perhaps an adaptation to the cooler temperatures of the Pleistocene. Large animals have proportionately less surface area compared to their volume and thus retain heat more effectively than do smaller animals.

In addition to mammals, some other Pleistocene vertebrate animals were of impressive proportions. The giant moas of New Zealand and the elephant birds of Madagascar were very large, and Australia had giant birds standing 3 m tall and weighing nearly 500 kg and a lizard 6.4 m long and weighing 585 kg. The tar pits of Rancho La Brea in southern California contain the remains of at least 200 kinds of animals. Many of these are fossils of dire wolves, saber-toothed cats, and other mammals, but some are the remains of birds, especially birds of prey, and a giant vulture with a wingspan of 3.6 m.

Primate Evolution

Primates are difficult to characterize as an order because they lack the strong specializations found in most other mammalian orders. We can, however, point to several trends in their evolution that help define primates and are related to their *arboreal,* or tree-dwelling, ancestry. These include changes in the skeleton and mode of locomotion; an increase in brain size; a shift toward smaller, fewer, and less specialized teeth; and the evolution of stereoscopic vision and a grasping hand with opposable thumb. Not all of these trends took place in every primate group, nor did they evolve at the same rate in each group.

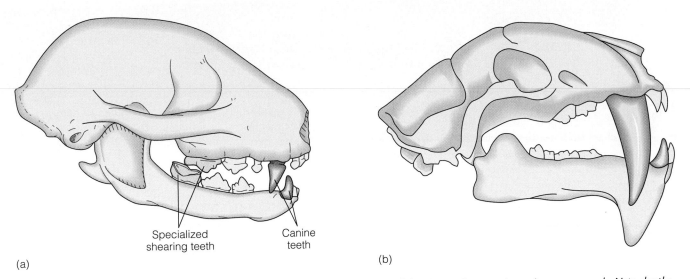

Specialized
shearing teeth

Canine
teeth

(a)

(b)

Figure 24.25 *(a) This present-day cat skull and jaw show the specialized sharp-crested shearing teeth present in carnivorous mammals. Note also the enlarged canines. (b) A number of Cenozoic saber-toothed cats had huge canine teeth. This one is the Oligocene saber-tooth* Euismilus.

In fact, some primates have retained certain primitive features, whereas others show all or most of these trends.

The primate order is divided into two suborders (Table 24.2). The *prosimians*, or lower primates, include the lemurs, lorises, tarsiers, and tree shrews; they are the oldest primate lineage, with a fossil record extending back to the Paleocene. Sometime during the Late Eocene, the *anthropoids*, or higher primates that include monkeys, apes,

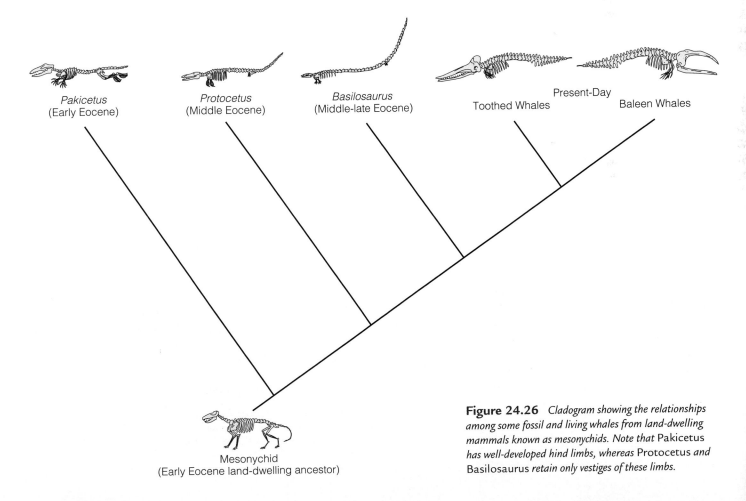

Pakicetus
(Early Eocene)

Protocetus
(Middle Eocene)

Basilosaurus
(Middle-late Eocene)

Present-Day

Toothed Whales

Baleen Whales

Mesonychid
(Early Eocene land-dwelling ancestor)

Figure 24.26 *Cladogram showing the relationships among some fossil and living whales from land-dwelling mammals known as mesonychids. Note that* Pakicetus *has well-developed hind limbs, whereas* Protocetus *and* Basilosaurus *retain only vestiges of these limbs.*

The Giant Irish "Elk" and Other Mammals with Horns or Antlers

Most living artiodactyls are ruminants—that is, cud-chewing herbivores with complex three- or four-chambered stomachs. Included in this group are all antelope, bison, cattle, goats, camels, and giraffes, most of which have some kinds of horns or antlers. Horns such as those in cattle have a bony core surrounded by a permanent sheath, whereas antlers are bony outgrowths of the skull that are shed annually as done by deer, elk, and moose. Some of these animals use their horns and antlers in defense, but they are probably more important as displays for attracting mates and for intraspecific tests of strength to establish dominance. Whatever their use, horns and antlers come in a fantastic variety of shapes and sizes in today's ruminants, and even larger and more unusual ones are found in a number of extinct animals.

Some of the more unusual antlers occurred in such extinct animals as *Synthetoceros*, which belonged to a group of ruminants known as protoceratids, and in the extinct deer *Eucladoceros* and *Hoplitomeryx* (Figure 1). The latter genus, in addition to having a five-horned skull, possessed saberlike canine teeth. Some small present-day deer in Asia use enlarged canine teeth to strip bark from trees, so perhaps *Hoplitomeryx* used its

teeth in a similar fashion. Some extinct cattle had horns measuring more than 4 m across, but probably the most spectacular antlers belonged to the so-called Irish elk that existed during the Late Pleistocene.

Megaloceros, commonly called the Irish elk, was neither an elk nor was it restricted to Ireland, although many fossils have been found there. In fact, it was a fallow deer, with large males weighing more than 700 kg. That is, they were the size of moose, but they had an antler spread of up to 3.5 m (Figure 2)! Their antlers, present only in males, probably weighed 34 to 40 kg. These truly spectacular animals ranged over Europe and much of Asia. *Megaloceros* seems to have been a fast-running, open-woodlands animal that declined markedly beginning about 12,000 years ago and finally became extinct. Some, however, might have survived in parts of Europe until about 500 B.C.

As is typical of hoofed mammals, *Megaloceros* was a herbivore, but its teeth indicate that it had a varied diet. A diet of shoots, twigs, and leaves probably sustained the animals through part of the year, but incisors similar to those of today's true elk indicate that grazing was also important, perhaps at different times of the year. The overall appearance of

(a) *Synthetoceros*   (b) *Eucladoceros*   (c) *Hoplitomeryx*

Figure 1 *Some of the antlered, hoofed mammals of the Cenozoic. (a)* Synthetoceros *from the Late Miocene and Pliocene of North America had two antlers on its skull and a Y-shaped one on its snout. (b)* Eucladoceros *had an antler spread of only 1.7 m, but it had 12 tines on each antler. It lived in Europe during the Pleistocene. (c)* Hoplitomeryx *from the Late Miocene of Italy had five bony outgrowths from its skull and enlarged canine teeth.*

Figure 2 *Restoration of the giant deer* Megalocereos giganteus, *commonly called the Irish elk. It lived in Europe and Asia during the Pleistocene. Large males had an antler spread of about 3.35 m.*

Megaloceros is well known from numerous fossils and from cave paintings.

As one would expect from the name "Irish elk," fossils of this animal are common in Ireland. Indeed, they are, especially in bog deposits. Probably the largest concentrations of such fossils come from Ballybetagh Bog near Dublin, where the remains of more than 100 males have been found. One explanation for this unusual number of fossils, all of antlered males, is that the animals became mired in mud and drowned because of their heavy antlers.

The fact that all were males might seem to support this hypothesis, but several other observations indicate that it is almost certainly incorrect. For instance, if this hypothesis is correct, one would expect males with relatively large antlers, but they are actually smaller than some 150 other specimens from other areas. In addition, if large animals were trapped and died in mud deposits, one would expect the deposits to be disrupted

by thrashing animals in their attempts to escape, and the bones of at least the legs should be found penetrating the mud deposits. Neither of these is found, thus making the miring-drowning hypothesis an unlikely explanation.

Another more likely hypothesis is based on observations of the fossils and of present-day deer that visit the bog. Today, mostly male deer visit the bog, and many of the fossil bones are broken and show signs of scavenging—indicating that the animals probably died near the lakeshore where they were trampled and gnawed. Also, the deaths seem to have been seasonal, probably mostly in winter, and many of the fossils are from animals that were in poor physical condition. One might also expect juveniles to be common in the deposits, but this is not the case. However, this is not too surprising because juveniles with smaller and less durable bones have a lower potential for preservation and are typically underrepresented in most fossil accumulations.

Figure 24.27 *Among the diverse Pliocene and Pleistocene mammals of Florida were 6-m-long giant sloths, and armored mammals known as glyptodonts that weighed more than 2 metric tons. Horses, camels, elephants, carnivores, and a variety of rodents were also common.*

and humans, evolved from a prosimian lineage. By the Oligocene, the anthropoids were a well-established group with both Old World monkeys (Africa, Asia) and New World monkeys (Central and South America) having evolved during this epoch. The *hominoids,* the group containing apes and humans diverged from Old World monkeys sometime before the Miocene, but exactly when is still being debated. It is generally accepted, however, that hominoids evolved in Africa.

Hominids

The **hominids** (family Hominidae), the primate family that includes present-day humans and their extinct ancestors (Table 24.2), have a fossil record extending back 4.4 million years. Several features distinguish them from other hominoids. Hominids are bipedal; that is, they have an upright posture, which is indicated by several modifications in their skeleton (Figure 24.28a and b). In addition, they show a trend toward a large and internally reorganized brain (Figure 24.28c–e). Other features include a reduced face and reduced canine teeth, omnivorous feeding, increased manual dexterity, and the use of sophisticated tools.

Many anthropologists think that these hominid features evolved in response to major climatic changes that began during the Miocene and continued into the Pliocene. During this time vast savannas replaced the African tropical rain forests where the lower primates and Old World monkeys had been so abundant. As the savannas and grasslands continued to expand, the hominids made the transition from true forest dwelling to life in an environment of mixed forests and grasslands.

The oldest known hominid is *Ardipithecus ramidus* (Figure 24.29). Discovered by Tim White and his Ethiopian colleagues at Aramis, Ethiopia, this nearly complete skeleton has been dated at 4.4 million years. As its bones are studied in more detail, its place in the hominid lineage will become clearer.

Australopithecines

Australopithecine is a collective term for all members of the genus *Australopithecus*. Currently five species are recognized: *A. anamensis, A. afarensis, A. africanus, A. robustus,* and *A. boisei.* Many paleontologists accept the evolutionary scheme in which *A. anamensis,* the oldest known australopithecine, is ancestral to *A. afarensis,* who in turn is ancestral to *A. africanus* and the genus *Homo,* as well as the side branch of australopithecines represented by *A. robustus* and *A. boisei.*

Table 24.2

Classification of the Primates

Order Primates: Lemurs, lorises, tarsiers, monkeys, apes, humans

Suborder Prosimii: Lemurs, lorises, tarsiers (lower primates)

Suborder Anthropoidea: monkeys, apes, humans (higher primates)

Superfamily Hominoidea: apes, humans

Family Hominidae: humans

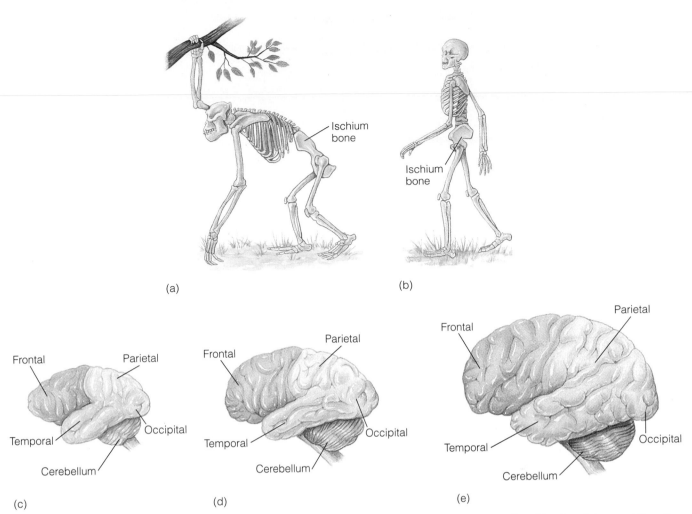

(a)

(b)

Ischium bone

Ischium bone

Frontal

Parietal

Temporal

Occipital

Cerebellum

(c)

Frontal

Parietal

Temporal

Occipital

Cerebellum

(d)

Frontal

Parietal

Temporal

Occipital

Cerebellum

(e)

Figure 24.28 *Comparison between quadrupedal and bipedal locomotion in gorillas and humans. (a) In gorillas the ischium bone is long, and the entire pelvis is tilted toward the horizontal. (b) In humans the ischium bone is much shorter, and the pelvis is vertical. (c through e) An increase in brain size and organization is apparent in comparing the brains of (c) a New World monkey, (d) a great ape, and (e) a present-day human.*

The oldest known australopithicine is *Australopithecus anamensis*. Discovered at Kanapoi, a site near Lake Turkana, Kenya, by Meave Leakey of the National Museums of Kenya and her colleagues, this 4.2-million-year-old bipedal species has many features in common with its younger relative, *Australopithecus afarensis*, yet is more primitive in other characteristics, such as its teeth and skull.

Australopithecus afarensis (Figure 24.30), which lived 3.9–3.0 million years ago, was fully bipedal and exhibited great variability in size and weight. It had a brain size of 380–450 cubic centimeters (cc), which is greater than the 300–400 cc of a chimpanzee, but much smaller than that of present-day humans (1350 cc average). The skull of *A. afarensis* retained many apelike features, including massive brow ridges and a forward-jutting jaw, but its teeth were intermediate between those of apes and humans. The heavily enameled molars were probably an adaptation to chewing fruits, seeds, and roots (Figure 24.30).

A. afarensis was succeeded by *Australopithecus africanus*, which lived 3.0–2.3 million years ago. The differences between the two species are relatively minor, although *A. africanus* was slightly larger, had a flatter face, and a somewhat larger brain.

Both *A. afarensis* and *A. africanus* differ markedly from the so-called robust species *A. robustus* (2.0–1.2 million years ago) and *A. boisei* (2.6–1.0 million years ago), neither of which had any evolutionary descendants.

Most scientists accept the idea that the robust australopithecines form a separate lineage from the other australopithecines that went extinct 1 million years ago.

The Human Lineage

The earliest member of our own genus **Homo** is *Homo habilis*, which lived 2.5–1.6 million years ago. *H. habilis* evolved from the *A. afarensis* and *A. africanus* lineage and coexisted with *A. africanus* for about 200,000 years

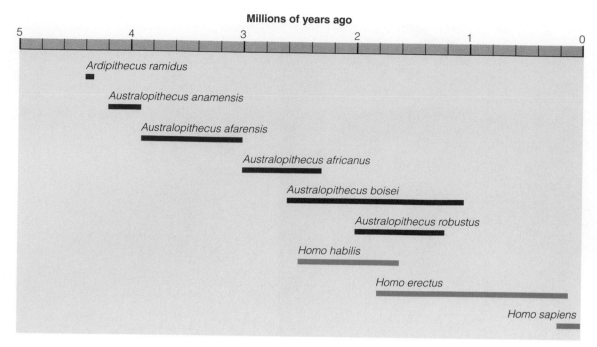

Millions of years ago

Ardipithecus ramidus

Australopithecus anamensis

Australopithecus afarensis

Australopithecus africanus

Australopithecus boisei

Australopithecus robustus

Homo habilis

Homo erectus

Homo sapiens

Figure 24.29 *The geologic age ranges for the commonly accepted species of hominids.*

(Figure 24.30). *H. habilis* had a larger brain (700 cc average) than its australopithicine ancestors, but smaller teeth. The evolutionary transition from *H. habilis* to *H. erectus* appears to have occurred in a short time, between 1.8 and 1.6 million years ago.

In contrast to the australopithecines and *H. habilis,* which are unknown outside Africa, *Homo erectus* was a widely distributed species, having migrated from Africa during the Pleistocene (Figure 24.31). Although *H. erectus* developed regional variations in form, the species differed from modern humans in several ways. Its brain size of 800–1300 cc, though much larger than that of *H. habilis,* was still less than the average for *Homo sapiens* (1350 cc). The skull of *H. erectus* was thick-walled, its face was massive, it had prominent brow ridges, and its teeth were slightly larger than those of present-day humans. *H. erectus* were comparable in size to modern humans, standing between 1.6 and 1.8 m tall and weighing between 53 and 63 kg. The archaeological record indicates that *H. erectus* made tools, and probably used fire and lived in caves, an advantage for those living in more northerly climates.

Currently, a heated debate surrounds the transition from *Homo erectus* to our own species, *Homo sapiens.* Paleoanthropologists are split into two camps. On the one side are those who support the "out of Africa" view, in which our ancestors, *Homo erectus,* migrated from Africa perhaps as recently as 100,000 years ago and populated Europe and Asia, driving the earlier hominid population to extinction.

The alternative explanation, the "multiregional" hypothesis, maintains that early modern humans did not have an isolated origin in Africa but rather established separate populations throughout Europe and Asia. Occasional contact and interbreeding between these populations enabled our species to maintain its overall cohesiveness while still preserving the regional differences in people we see today. At this time, not enough evidence points to which theory, if either, is correct and the emergence of *H. sapiens* from ancestral *H. erectus* is still not resolved.

Perhaps the most famous of all fossil humans are the **Neanderthals** who inhabited Europe and the Near East from about 200,000 to 30,000 years ago (Figure 24.32). The most notable difference between Neanderthals and present-day humans is in the skull. Neanderthal skulls were long and low

WHAT WOULD YOU DO?

Because of the recent controversy concerning the teaching of evolution in the public schools, your local school board has asked you to make a 30-minute presentation on the evolutionary history of humans and how the fossil record of humans and their ancestors is evidence that evolution is a valid scientific theory. With only 30 minutes to make your case, what evidence in the fossil record would you emphasize and how would you go about convincing the school board that humans have indeed evolved from earlier hominids?

Figure 24.30 *Recreation of a Pliocene landscape showing members of* Australopithecus afarensis *gathering and eating various fruits and seeds.*

with heavy brow ridges, a projecting mouth, and a weak, receding chin. Their brain was slightly larger on average than our own, and somewhat differently shaped. The Neanderthal body was somewhat more massive and heavily muscled than ours, with rather short limbs, much like those of other cold-adapted people of today.

Based on specimens from more than 100 sites, we now know that Neanderthals were not much different from us, only more robust. Neanderthals were the first humans to move into truly cold climates. Their remains are found chiefly in caves and hut like structures, which also contain a variety of specialized stone tools and weapons.

About 30,000 years ago, humans closely resembling modern Europeans moved into the region inhabited by Neanderthals and completely replaced them. **Cro-Magnons,** the name given to the successors of Neanderthals in France, lived from about 35,000 to 10,000 years ago; during this period the development of art and technology far exceeded anything the world had seen before. Using paints made from manganese and iron oxides, Cro-Magnon people painted hundreds of scenes on the ceilings and walls of caves in France and Spain.

With the appearance of Cro-Magnons, human evolution has become almost entirely cultural rather than biological. Since the evolution of the Neanderthals about 200,000 years ago, humans have gone from a stone culture to a technology that has allowed us to visit other planets with space probes and land men on the Moon. It remains to be seen how we will use this technology in the future and whether we will continue as a species, evolve

Figure 24.31 *Recreation of a Pleistocene setting in Europe in which members of* Homo erectus *are using fire and stone tools.*

into another species, or become extinct as many groups have before us.

Pleistocene Extinctions

Near the end of the Pleistocene many terrestrial mammals became extinct. Although this extinction was modest com-

pared to previous ones, it was unusual in that it affected mostly mammals weighing more than 40 kg. Furthermore, its effects were much greater in North and South America and Australia than in the Old World, where comparatively few extinctions occurred.

Two hypotheses for this extinction are currently being debated. The first holds that the large mammals became

Figure 24.32
Archaeological evidence indicates that Neanderthals lived in caves and participated in ritual burials as depicted in this painting of a burial ceremony that occurred approximately 60,000 years ago at Shanidar Cave, Iraq.

extinct because they could not adapt to the rapid climatic changes at the end of the Pleistocene. According to the second hypothesis, the mammals were killed off by human hunters, a hypothesis known as *prehistoric overkill.*

Researchers favoring a climatic cause point to the rapid changes in climate and vegetation that occurred during the Late Pleistocene. While rapid changes in climate with their accompanying effects on vegetation can certainly result in changes in animal populations, the hypothesis encounters several problems. First, why didn't the large mammals migrate to more suitable habitats as the climate and vegetation changed? After all, many other animal species did. The second argument against this hypothesis is that previous changes in climate during the Pleistocene were not marked by episodes of mass extinctions.

Paul Martin of the University of Arizona at Tucson, the leading proponent of the prehistoric overkill hypothesis, argues that the mass extinctions in North and South America and Australia coincided closely with the arrival of humans in each area. According to Martin, hunters had a tremendous impact on the faunas of North and South America about 11,000 years ago because the animals had no previous experience with humans. The same thing happened much earlier in Australia soon after people arrived about 40,000 years ago. No large-scale extinctions occurred in Africa and most of Europe because animals in those regions had long been familiar with humans.

Several arguments can also be made against the prehistoric overkill hypothesis. One problem is the difficulty of imagining how a few hunters could have decimated so many species of large mammals, even if they hunted mostly females and young animals. A second problem is the comparison to the habits of present-day hunters, who concentrate on smaller, abundant, and less dangerous animals. And finally, few human artifacts are found among the remains of extinct animals in North and South America, and there is usually little evidence that the animals were hunted.

The reason for the extinctions of large mammals at the end of the Pleistocene is still unresolved and probably will be for some time. It may turn out that the extinction resulted from a combination of many different circumstances.

Chapter Summary

1. Cenozoic orogenic activity was concentrated in two major belts—the Alpine-Himalayan and the circum-Pacific orogenic belts. Each belt is composed of smaller units called orogens.

2. The North American Cordillera is a complex mountainous region extending from Alaska into Mexico. Its Cenozoic evolution included deformation during the Laramide orogeny, extensional tectonics that formed the basin-and-range structures, intrusive and extrusive volcanism, and uplift and erosion.

3. Subduction of the Farallon plate beneath North America resulted in the vertical uplifts of the Laramide orogeny. The Laramide orogen is centered in the middle and southern Rockies, but deformation occurred from Alaska to Mexico.

4. The westward drift of North America resulted in its collision with the Pacific-Farallon ridge. Subduction ceased, and the continental margin became bounded by major transform faults, except where the Juan de Fuca plate continues to collide with North America.

5. Sediments eroded from Laramide uplifts were deposited in intermontane basins, on the Great Plains, and in a remnant of the Cretaceous epeiric sea in North Dakota. A seaward-thickening wedge of sediments pierced by salt domes on the Gulf Coastal Plain contains large quantities of oil and natural gas.

6. Cenozoic uplift and erosion were responsible for the present topography of the Appalachian Mountains. Much of the sediment eroded from the Appalachians was deposited on the Atlantic Coastal Plain.

7. During the Pleistocene Epoch, glaciers covered about 30% of the land surface. About 20 warm-cold Pleistocene climatic cycles are recognized from paleontologic and oxygen isotope data derived from deep-sea cores.

8. Several intervals of widespread glaciation, separated by interglacial periods, occurred in North America. The other Northern Hemisphere continents were also affected by widespread Pleistocene glaciation.

9. Marine invertebrate groups that survived the Mesozoic extinctions continued to expand and diversify during the Cenozoic.

10. The Paleocene mammalian fauna was composed of Mesozoic holdovers and a number of new orders. This time of diversification among mammals also saw several orders become extinct. Most living mammalian orders were present by the Eocene.

11. Placental mammals owe much of their success to their method of reproduction. Shrewlike placental mammals that appeared during the Cretaceous were the ancestral stock for the placental adaptive radiation of the Cenozoic.

12. The perissodactyls and artiodactyls evolved during the Eocene. Ungulate adaptations include modifications of the teeth for grinding up vegetation and limb modifications for speed.

13. The evolutionary history of horses is particularly well documented by fossils. The earliest horses and present-day horses differ considerably, but a continuous series of intermediate fossils shows that they are related.

14. The primates evolved during the Paleocene. Several trends help characterize primates and differentiate them from other mammalian orders, including a change in overall skeletal structure and mode of locomotion, an increase in brain size, stereoscopic vision, and evolution of a grasping hand with opposable thumb.

15. The primates are divided into two suborders: the prosimians, which are the oldest primate lineage and include lemurs, lorises, tarsiers, and tree shrews; and the anthropoids, which include monkeys, apes, and humans and their ancestors.

16. The earliest hominids evolved 4.4 million years ago, and the human lineage began about 2.5 million years ago in Africa with the evolution of *Homo habilis*. *Homo erectus* evolved from *H. habilis* about 1.8 million years ago and was the first hominid to migrate out of Africa.

17. The transition from *Homo erectus* to *Homo sapiens* is still unresolved. There currently is not enough evidence to determine which of the two competing hypotheses, the "out of Africa" or the "multiregional" hypothesis is correct.

18. Neanderthals inhabited Europe and the Near East between 200,000 and 30,000 years ago, and were not much different from us, only more robust. Cro-Magnons were the successors of the Neanderthals and lived from about 35,000 to 10,000 years ago.

19. Modern humans succeeded the Cro-Magnons about 10,000 years ago and have spread throughout the world.

Important Terms

artiodactyl	grazer	monotreme	primate
australopithecine	hominid	Neanderthal	Tejas epeiric sea
Basin and Range Province	*Homo*	North American Cordillera	ungulate
browser	Laramide orogeny	perissodactyl	
Cro-Magnon	marsupial mammal	placental mammal	

Review Questions

1. Uplift and erosion were the main processes responsible for the present topographic expression of the
   a. _____ Cascade Range; b. _____ Appalachian Mountains;
   c. _____ Gulf Coastal Plain; d. _____ Juan de Fuca plate;
   e. _____ East-African Rift system.

2. The classic model for orogenies at oceanic-continental plate boundaries was modified for the Laramide orogeny because it is
   a. _____ an example of mountains made up exclusively of extrusive igneous rocks; b. _____ the oldest known orogeny to occur in North America; c. _____ farther inland than most other orogenies; d. _____ the result of subduction of continental lithosphere beneath oceanic lithosphere;
   e. _____ restricted to the northern part of Canada.

3. A vast area of overlapping lava flows in the northwestern United States is known as the
   a. _____ Andes Mountains; b. _____ Basin and Range Province; c. _____ Colorado Plateau; d. _____ Tejas epeiric sea; e. _____ Columbia River basalts.

4. Large-scale block-faulting took place during the Cenozoic in the
   a. _____ Basin and Range Province; b. _____ Appalachian Mountains; c. _____ Atlantic Coastal Plain; d. _____ Florida-Bahamas carbonate platform; e. _____ Farallon plate.

5. As a result of ongoing subduction of the Juan de Fuca plate beneath North America, volcanism continues in the
   a. _____ Canadian Shield; b. _____ Great Plains; c. _____ Cascade Range; d. _____ interior lowlands; e. _____ Great Lakes region.

6. The youngest Pleistocene glacial stage recognized in North America is the
   a. _____ Missoluian; b. _____ Kentuckian; c. _____ Wisconsinan; d. _____ Booneville; e. _____ Labradoran.

7. Hoofed mammals with teeth adapted to a diet of leaves are known as
   a. _____ browsers; b. _____ monotremes; c. _____ bryozoans; d. _____ primates; e. _____ dinoflagellates.

8. The only existing egg-laying mammals are
   a. _____ mosasaurs; b. _____ pterosaurs; c. _____ proboscideans; d. _____ monotremes; e. _____ insectivores.

9. Tertiary bryozoans were particularly abundant and successful
   a. _____ carnivorous mammals; b. _____ flightless birds; c. _____ suspension feeders; d. _____ marine reptiles; e. _____ dinoflagellates.

10. Large body size of Pleistocene mammals might have been an adaptation for
    a. _____ cooler temperatures; b. _____ increased predation; c. _____ higher elevations; d. _____ longer summers; e. _____ dwelling on grass-covered plains.

11. During the Tertiary Period, the _____ Sea was the last of the epeiric seas to invade North America.
    a. _____ Sauk; b. _____ Laramide; c. _____ Ungulate; d. _____ Illinoian; e. _____ Tejas.

12. Which of the following features distinguish hominids from other hominoids?
    a. _____ bipedalism; b. _____ a large and internally reorganized brain; c. _____ a reduced face and reduced canine teeth; d. _____ use of sophisticated tools; e. _____ all of the previous.

13. The oldest primate lineage is the
    a. _____ anthropoids; b. _____ prosimians; c. _____ hominids; d. _____ pelycosaurs; e. _____ insectivores.

14. Which were the first hominids to migrate out of Africa and from which we evolved?
    a. _____ *Australopithecus robustus*; b. _____ *Ardipithecus ramidus*; c. _____ *Homo habilis*; d. _____ *Homo erectus*; e. _____ *Homo sapiens*.

15. Explain how the Laramide orogeny differs from most orogenies at convergent plate boundaries.

16. What kinds of evidence indicate that widespread glaciers were present during the Pleistocene?
17. How have the teeth of some ungulates been modified for grazing?
18. What are the three basic groups of mammals and how do they differ from one another? Also, discuss the success of each in terms of numbers of species and geographic distribution.
19. During the Paleozoic, eastern North America experienced considerable deformation, whereas during the Mesozoic and Cenozoic, most deformation took place in the western part of the continent. What accounts for this changing pattern of deformation?
20. How does the fossil record demonstrate that animals as different as horses and rhinoceroses evolved from a common ancestor? (Hint—also see Chapter 18.)

21. Briefly outline the geologic history of the Appalachian Mountains.
22. What kinds of sediments accumulated in the Interior Lowlands and on the Gulf Coastal Plain, and what was the source of these sediments?
23. Summarize the evidence for and against the hypothesis called prehistoric overkill.
24. Briefly discuss the two hypotheses regarding the transition from *Homo erectus* to *Homo sapiens*.
25. What factors do you think will influence the future course of human evolution, and can we as a species control the direction evolution takes?

 # World Wide Web Activities

For these web site addresses, along with current updates and exercises, log on to **http://www.brookscole.com/geo**

Origins of Humankind

This site is a comprehensive Internet resource concerning human origins. Because this field of research is so broad and changing so rapidly, rather than refer to specific sites to visit here, we suggest that you click on any of the sites listed on the home page to learn more about controversial theories and to find links to other sites concerned with human evolution.

Royal Tyrrell Museum

This site contains considerable information on ancient organisms. It is supported by the Royal Tyrrell Museum Cooperating Society in Drumheller, Alberta, Canada, "a nonprofit society funded to support the scientific, educational, recreational and public operations of the Royal Tyrrell Museum of Paleontology." The home page contains icons for Directions to the Museum, Admissions, Virtual Tours, Programmes, and Educational Programming. Click on the *Age of Mammals* icon. This site has a summary of Cenozoic life, with emphasis on mammals. A variety of restorations of extinct mammals are shown along with explanations of how they lived and what descendants they have, if any. Check out the extinct mammals *Uintatherium* and *Dinictis*. Where did these mammals live and what did they eat?

The Age of Mammals: Life in the Cenozoic Era

This National Park Service site is an excellent place to find links to any of the Tertiary Period epochs, as well as the Pleistocene. It also has links to other sites at which Cenozoic fossils are found. These include *Agate Fossil Beds, Badlands, Florissant Fossil Beds, Fossil Butte,* and *Hagerman Fossil Beds.*

The Cenozoic Era

This site is part of the "Continents, Oceans, and Life–A New View of . . . The Third Planet," which is maintained by the Milwaukee Public Museum. Take the tour through *Grasslands of Western Nebraska, The Glacier's Edge in Wisconsin 12,000 Years Ago* and others to see what the environment and life was like during the Cenozoic Era.

UC Berkeley Museum of Paleontology

The University of California at Berkeley maintains this huge site with information on a variety of topics. On the main page click *Time Periods,* which brings up a geologic time scale. On the time scale click *Cenozoic Era* and see why the Cenozoic is sometimes called the Age of Mammals. What are the two main subdivisions of the Cenozoic? Click *Stratigraphy* at the bottom of the page. Read about and explain how the Tertiary and Quaternary Periods were named. Also, click *Localities.* What is the age of the fossils in the La Brea Tar Pits? How and when was the Monterey Formation deposited?

Illinois State Museum

This site, maintained by the Illinois State Museum, contains information about programs, exhibits, collections, and a calendar of events at the museum. Click on the *Exhibits* icon, then click on the *Ice Ages.* Here you can learn about the ice ages. What are ice ages? When did they occur? What causes ice ages? Click on the *Exhibits* icon, then click on *The Midwestern U.S. 16,000 Years Ago.* This site shows what the Midwest looked like 16,000 years ago. Check on the *wander through the exhibit* and find out more about the environment, plants, and animals of the midwestern United States during the Ice Age.

Periodic Table of the Elements

47	— Atomic Number
Ag	— Symbol of Element
silver	— Name of Element
107.9	— Atomic Mass Number (rounded to four significant figures)

Representative Elements	Transition Elements	Inner-Transition Elements	Noble Gases

Period	(1)* I A	(2) II A	(3) III B	(4) IV B	(5) V B	(6) VI B	(7) VII B	(8)	(9) VIII B
1	1 **H** hydrogen 1.008								
2	3 **Li** lithium 6.941	4 **Be** beryllium 9.012							
3	11 **Na** sodium 22.99	12 **Mg** magnesium 24.31							
4	19 **K** potassium 39.10	20 **Ca** calcium 40.08	21 **Sc** scandium 44.96	22 **Ti** titanium 47.90	23 **V** vanadium 50.94	24 **Cr** chromium 52.00	25 **Mn** manganese 54.94	26 **Fe** iron 55.85	27 **Co** cobalt 58.93
5	37 **Rb** rubidium 85.47	38 **Sr** strontium 87.62	39 **Y** yttrium 88.91	40 **Zr** zirconium 91.22	41 **Nb** niobium 92.91	42 **Mo** molybdenum 95.94	43 **Tc** technetium 98.91	44 **Ru** ruthenium 101.1	45 **Rh** rhodium 102.9
6	55 **Cs** cesium 132.9	56 **Ba** barium 137.3	57 **La** lanthanum 138.9	72 **Hf** hafnium 178.5	73 **Ta** tantalum 180.9	74 **W** tungsten 183.9	75 **Re** rhenium 186.2	76 **Os** osmium 190.2	77 **Ir** iridium 192.2
7	87 **Fr** francium (223)	88 **Ra** radium 226.0	89 **Ac** actinium (227)	104 **Rf** rutherfordium (261)	105 **Db** dubnium (262)	106 **Sg** seaborgium (263)	107 **Bh** bohrium (262)	108 **Hs** hassium (265)	109 **Mt** meitnerium (266)

Lanthanides	58 **Ce** cerium 140.1	59 **Pr** praseodymium 140.9	60 **Nd** neodymium 144.2	61 **Pm** promethium (147)	62 **Sm** samarium 150.4
Actinides	90 **Th** thorium 232.0	91 **Pa** protactinium 231.0	92 **U** uranium 238.0	93 **Np** neptunium 237.0	94 **Pu** plutonium (244)

( ) Indicates mass number of isotope with longest known half-life.

* Number in ( ) heading each column represents the group designation recommended by the American Chemical Society Committee on Nomenclature.

			(13) III A	(14) IV A	(15) V A	(16) VI A	(17) VII A	(18) Noble Gases
								2 **He** helium 4.003
			5 **B** boron 10.81	6 **C** carbon 12.01	7 **N** nitrogen 14.01	8 **O** oxygen 16.00	9 **F** fluorine 19.00	10 **Ne** neon 20.18
(10)	(11) I B	(12) II B	13 **Al** aluminum 26.98	14 **Si** silicon 28.09	15 **P** phosphorus 30.97	16 **S** sulfur 32.06	17 **Cl** chlorine 35.45	18 **Ar** argon 39.95
28 **Ni** nickel 58.71	29 **Cu** copper 63.55	30 **Zn** zinc 65.37	31 **Ga** gallium 69.72	32 **Ge** germanium 72.59	33 **As** arsenic 74.92	34 **Se** selenium 78.96	35 **Br** bromine 79.90	36 **Kr** krypton 83.80
46 **Pd** palladium 106.4	47 **Ag** silver 107.9	48 **Cd** cadmium 112.4	49 **In** indium 114.8	50 **Sn** tin 118.7	51 **Sb** antimony 121.8	52 **Te** tellurium 127.6	53 **I** iodine 126.9	54 **Xe** xenon 131.3
78 **Pt** platinum 195.1	79 **Au** gold 197.0	80 **Hg** mercury 200.6	81 **Tl** thallium 204.4	82 **Pb** lead 207.2	83 **Bi** bismuth 209.0	84 **Po** polonium (210)	85 **At** astatine (210)	86 **Rn** radon (222)
110 **Uun** ununnilium (269)	111 **Uuu** unununium (272)	112 **Uub** ununbium (277)	113	114 **Uuq** ununquadium (289)	115	116 **Uuh** ununhexium (289)	117	118 **Uuo** ununoctium (293)

63 **Eu** europium 152.0	64 **Gd** gadolinium 157.3	65 **Tb** terbium 158.9	66 **Dy** dysprosium 162.5	67 **Ho** holmium 164.9	68 **Er** erbium 167.3	69 **Tm** thulium 168.9	70 **Yb** ytterbium 173.0	71 **Lu** lutetium 175.0

95 **Am** americium (243)	96 **Cm** curium (247)	97 **Bk** berkelium (247)	98 **Cf** californium (251)	99 **Es** einsteinium (254)	100 **Fm** fermium (257)	101 **Md** mendelevium (258)	102 **No** nobelium (255)	103 **Lr** lawrencium (256)

English-Metric Conversion Chart

English Unit	Conversion Factor	Metric Unit	Conversion Factor	English Unit
Length				
Inches (in)	2.54	Centimeters (cm)	0.39	Inches (in)
Feet (ft)	0.305	Meters (m)	3.28	Feet (ft)
Miles (mi)	1.61	Kilometers (km)	0.62	Miles (mi)
Area				
Square inches (in^2)	6.45	Square centimeters (cm^2)	0.16	Square inches (in^2)
Square feet (ft^2)	0.093	Square meters (m^2)	10.8	Square feet (ft^2)
Square miles (mi^2)	2.59	Square kilometers (km^2)	0.39	Square miles (mi^2)
Volume				
Cubic inches (in^3)	16.4	Cubic centimeters (cm^3)	0.061	Cubic inches (in^3)
Cubic feet (ft^3)	0.028	Cubic meters (m^3)	35.3	Cubic feet (ft^3)
Cubic miles (mi^3)	4.17	Cubic kilometers (km^3)	0.24	Cubic miles (mi^3)
Weight				
Ounces (oz)	28.3	Grams (g)	0.035	Ounces (oz)
Pounds (lb)	0.45	Kilograms (kg)	2.20	Pounds (lb)
Short tons (st)	0.91	Metric tons (t)	1.10	Short tons (st)
Temperature				
Degrees Fahrenheit (°F)	$-32° \times 0.56$	Degrees centigrade (Celsius)(°C)	$\times 1.80 + 32°$	Degrees Fahrenheit (°F)

Examples:

10 inches = 25.4 centimeters; 10 centimeters = 3.9 inches

100 square feet = 9.3 square meters; 100 square meters = 1080 square feet

50°F = 10.1°C; 50°C = 122°F

Mineral Identification Tables

Metallic Luster

Mineral	Chemical Composition	Color	Hardness Specific Gravity	Other Features	Comments
Chalcopyrite	$CuFeS_2$	Brassy yellow	3.5–4 / 4.1–4.3	Usually massive; greenish-black streak; iridescent tarnish	Most common copper mineral. Important source of copper. Mostly in hydrothermal rocks.
Galena	PbS	Lead gray	2.5 / 7.6	Cubic crystals; 3 cleavages at right angles	The ore of lead. Mostly in hydrothermal rocks.
Graphite	C	Black	1–2 / 2.09–2.33	Greasy feel; writes on paper; 1 direction of cleavage	Used for pencil "leads" and as a dry lubricant. Mostly in metamorphic rocks.
Hematite	Fe_2O_3	Red brown	6 / 4.8–5.3	Usually granular or massive; reddish-brown streak	Most important ore of iron. An accessory mineral in many rocks.
Magnetite	Fe_3O_4	Black	5.5–6.5 / 5.2	Strong magnetism	An important ore of iron. An accessory mineral in many rocks.
Pyrite	FeS_2	Brassy yellow	6.5 / 5.0	Cubic and octahedral crystals	The most common sulfide mineral. Found in some igneous and hydrothermal rocks, and in sedimentary rocks associated with coal.

Nonmetallic Luster

Mineral	Chemical Composition	Color	Hardness Specific Gravity	Other Features	Comments
Anhydrite	$CaSO_4$	White, gray	3.5 / 2.9–3.0	Crystals with 2 cleavages; usually in granular masses	Found in limestones, evaporite deposits, and the cap rocks of salt domes. Used as a soil conditioner.
Apatite	$Ca_5(PO_4)_3F$	Blue, green, brown, yellow, white	5 / 3.1–3.2	6-sided crystals; in massive or granular masses	An accessory mineral in many rocks. The main constituent of bone and dentine. A source of phosphorous for fertilizer.

Nonmetallic Luster

Mineral	Chemical Composition	Color	Hardness Specific Gravity	Other Features	Comments
Augite	$Ca(Mg,Fe,Al)(Al,Si)_2O_6$	Black, dark green	6 3.25–3.55	Short 8-sided crystals; 2 cleavages; cleavages nearly at right angles	The most common pyroxene mineral. Found mostly in mafic igneous rocks.
Barite	$BaSO_4$	Colorless, white, gray	3 4.5	Tabular crystals; high specific gravity for a nonmetallic mineral	Commonly found with ores of a variety of metals, and in limestones and hot spring deposits. A source of barium.
Biotite (Mica)	$K(Mg,Fe)_3AlSi_3O_{10}(OH)_2$	Black, brown	2.5 2.9–3.4	1 cleavage direction; cleaves into thin sheets	In felsic and mafic igneous rocks, in metamorphic rocks, and in some sedimentary rocks.
Calcite	$CaCO_3$	Colorless, white	3 2.71	3 cleavages at oblique angles; cleaves into rhombs; reacts with dilute hydrochloric acid	The most common carbonate mineral. Main component of limestone and marble. Also common in hydrothermal rocks.
Cassiterite	SnO_2	Brown to black	6.5 7.0	High specific gravity for a nonmetallic mineral	The main ore of tin. Most is concentrated in alluvial deposits because of its high specific gravity.
Chlorite	$(Mg,Fe)_3(Si,Al)_4O_{10}$ $(Mg,Fe)_3(OH)_6$	Green	2 2.6–3.4	1 cleavage; occurs in scaly masses	Common in low-grade metamorphic rocks such as slate.
Corundum	Al_2O_3	Gray, blue, pink, brown	9 4.0	6-sided crystals and great hardness are distinctive	An accessory mineral in some igneous and metamorphic rocks. Used as a gemstone and for abrasives.
Dolomite	$CaMg(CO_3)_2$	White, yellow, gray, pink	3.5–4 2.85	Cleavage as in calcite; reacts with dilute hydrochloric acid when powdered	The main constituent of dolostone. Also found associated with calcite in some limestones and marble.
Fluorite	CaF_2	Colorless, purple, green, brown	4 3.18	4 cleavage directions; cubic and octahedral crystals	Occurs mostly in hydrothermal rocks and in some limestones and dolostones. Used in the manufacture of steel and the preparation of hydrofluoric acid.
Garnet	$Fe_3Al_2(SiO_4)_3$	Dark red	7–7.5 4.32	12-sided crystals common; uneven fracture	Found mostly in gneiss and schist. Used as a semiprecious gemstone and for abrasives.

Mineral	Chemical Composition	Color	Hardness Specific Gravity	Other Features	Comments
Gypsum	$CaSO_4 \cdot 2H_2O$	Colorless, white	2 2.32	Elongate crystals; fibrous and earthy masses	The most common sulfate mineral. Found mostly in evaporite deposits. Used to manufacture plaster of Paris and cements.
Halite	$NaCl$	Colorless, white	3–4 2.2	3 cleavages at right angles; cleaves into cubes; cubic crystals; salty taste	Occurs in evaporite deposits. Used as a source of chlorine and in the manufacture of hydrochloric acid, many sodium compounds, and food seasoning.
Hornblende	$NaCa_2(Mg,Fe,Al)_5$ $(Si,Al)_8O_{22}(OH)_2$	Green, black	6 3.0–3.4	Elongate, 6-sided crystals; 2 cleavages intersecting at 56° and 124°	A common rock-forming amphibole mineral in igneous and metamorphic rocks.
Illite	$(Ca,Na,K)(Al,Fe^{+3},$ $Fe^{+2},Mg)_2$ $(Si,Al)_4O_{10}(OH)_2$	White, light gray, buff	1–2 2.6–2.9	Earthy masses; particles too small to observe properties	A clay mineral common in soils and clay-rich sedimentary rocks.
Kaolinite	$Al_2Si_4O_{10}(OH)_8$	White	2 2.6	Massive; earthy odor; particles too small to observe properties	A common clay mineral formed by chemical weathering of aluminum-rich silicates. The main ingredient of kaolin clay used for the manufacture of ceramics.
Muscovite (Mica)	$KAl_2Si_3O_{10}(OH)_2$	Colorless	2–2.5 2.7–2.9	1 direction of cleavage; cleaves into thin sheets	Common in felsic igneous rocks, metamorphic rocks, and some sedimentary rocks. Used as an insulator in electrical appliances.
Olivine	$(Fe,Mg)_2SiO_4$	Olive green	6.5 3.3–3.6	Small mineral grains in granular masses; conchoidal fracture	Common in mafic igneous rocks.
Plagioclase feldspars	Varies from $CaAl_2Si_2O_8$ to $NaAlSi_3O_8$	White, gray, brown	6 2.56	2 cleavages at right angles	Common in igneous rocks and a variety of metamorphic rocks. Also in some arkoses.
Microcline	$KAlSi_3O_8$	White, pink, green			Common in felsic igneous rocks, some metamorphic rocks, and arkoses. Used in the manufacture of porcelain.
Orthoclase	$KAlSi_3O_8$	White, pink	6 2.56	2 cleavages at right angles	

Potassium feldspar

Nonmetallic Luster

Mineral	Chemical Composition	Color	Hardness Specific Gravity	Other Features	Comments
Quartz	SiO_2	Colorless, white, gray, pink, green	7 2.67	6-sided crystals; no cleavage; conchoidal fracture	A common rock-forming mineral in all rock groups and hydrothermal rocks. Also occurs in varieties known as chert, flint, agate, and chalcedony.
Siderite	$FeCO_3$	Yellow, brown	4 3.8–4.0	3 cleavages at oblique angles; cleaves into rhombs	Found mostly in concretions and sedimentary rocks associated with coal.
Smectite	$(Al,Mg)_8(Si_4O_{10})_3(OH)_{10} \cdot 12H_2O$	Gray, buff, white	1–1.5 2.5	Earthy masses; particles too small to observe properties	A clay mineral with the unique property of swelling and contracting as it absorbs and releases water.
Sphalerite	ZnS	Yellow, brown, black	3.5–4 4.0–4.1	6 cleavages; cleaves into dodecahedra	The most important ore of zinc. Commonly found with galena in hydrothermal rocks.
Talc	$Mg_3Si_4O_{10}(OH)_2$	White, green	1 2.82	1 cleavage direction; usually in compact masses	Formed by the alteration of magnesium silicates. Mostly in metamorphic rocks. Used in ceramics, cosmetics, and as a filler in paints.
Topaz	$Al_2SiO_4(OH,F)$	Colorless, white, yellow, blue	8 3.5–3.6	High specific gravity; 1 cleavage direction	Found in pegmatites, granites, and hydrothermal rocks. An important gemstone.
Zircon	$SrSiO_4$	Brown, gray	7.5 3.9–4.7	4-sided, elongate crystals	Most common as an accessory in granitic rocks. An ore of zirconium, and used as a gemstone.

Topographic Maps

Nearly everyone has used a map of one kind or another, and is probably aware that a map is a scaled-down version of the area depicted. For a map to be of any use, however, one must understand what is shown on a map and how to read it. A particularly useful type of map for geologists, and people in many other professions, is a *topographic map*, which shows the three-dimensional configuration of Earth's surface on a two-dimensional sheet of paper.

Maps showing relief—differences in elevation in adjacent areas—are actually models of Earth's surface. Such maps are available for some areas, but they are expensive, difficult to carry, and impossible to record data on. Thus, paper sheets showing relief by using lines of equal elevation known as contours are most commonly used. Topographic maps depict (1) relief, which includes hills, mountains, valleys, canyons, and plains; (2) bodies of water, such as rivers, lakes, and swamps; (3) natural features, such as forests, grasslands, and glaciers; and (4) various cultural features, including communities, highways, railroads, land boundaries, canals, and power transmission lines.

Topographic maps known as quadrangles are published by the U.S. Geological Survey (USGS). The area depicted on a topographic map can be identified by referring to the map's name in the upper right and lower right corners, which is usually derived from some prominent geographic feature (Lincoln Creek Quadrangle, Idaho) or community (Mt. Pleasant Quadrangle, Michigan). In addition, most maps have a state outline map along the bottom margin, and shown within the outline is a small black rectangle indicating the part of the state represented by the map.

Contours

Contour lines, or simply contours, are lines of equal elevation used to show topography. Think of contours as the lines formed where imaginary horizontal planes intersect Earth's surface at specific elevations. On maps, contours are brown, and every fifth contour, called an *index contour*, is darker than adjacent ones and labeled with its elevation (Figure D1). Elevations on most USGS topographic maps are in feet, although a few use meters; in either case, the specified elevation is above or below mean sea level. Because contours are defined as lines of equal elevation they cannot divide or cross one another, although they will converge and appear to join in areas with vertical or overhanging cliffs. Notice in Figure D1 that where contours cross a stream they form a V that points upstream toward higher elevations.

The vertical distance between contours is the *contour interval*. If an area has considerable relief, a large contour interval is used, perhaps 80 or 100 feet, whereas a small interval such as 5, 10, or 20 feet is used in areas with little relief. The values recorded on index contours are always multiples of the map's contour interval, shown at the bottom of the map. For instance, if a map has a contour interval of 10 feet, index contour values such as 3600, 3650, and 3700 feet might be shown (Figure D1). In addition to contours, specific elevations are shown at some places on maps and may be indicated by a small *x*, next to which is a number. A specific elevation might also be shown adjacent to the designation BM (benchmark), a place where the elevation and location are precisely known.

Contour spacing depends on slope, so in areas with steep slopes, contours are closely spaced because there is a considerable increase in elevation in a short distance. In contrast, if slopes are gentle, contours are widely spaced (Figure D1). Furthermore, if contour spacing is uniform the slope angle remains constant, but if spacing changes slope angle changes. However, one must be careful in comparing slopes on maps with different contour intervals or different scales.

Topographic features such as hills, valleys, plains, and so on can easily be shown by contours. For instance, a hill is shown by a concentric pattern of contours with the highest elevation in the central part of the pattern. All contours must close on themselves, but they may do so beyond the confines of a particular map. A concentric contour pattern also might show a closed depression, but in this case special contours with short bars perpendicular to the contour pointing toward the central part of the depression, are used (Figure D1).

Map Scales

Any map is a scaled-down version of the area shown, so to be of any use a map must have a scale. Highway maps, for example, commonly have a scale such as "1 inch equals 10 miles," by which one can readily determine distances. Two

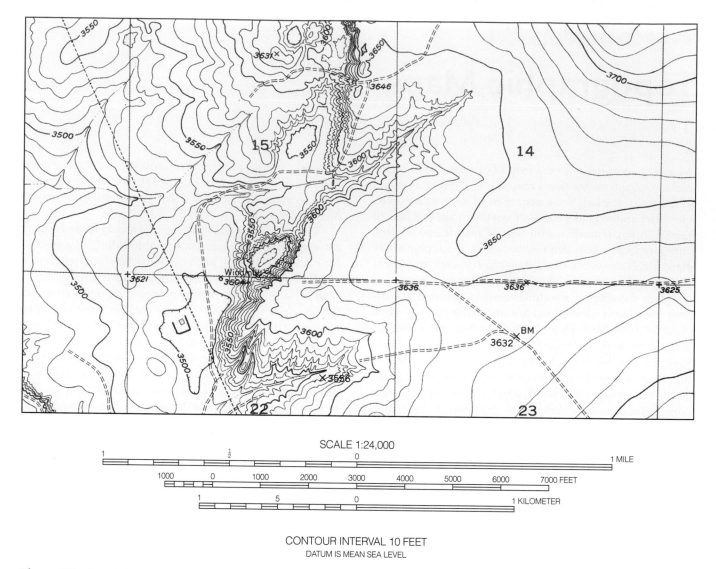

SCALE 1:24,000

CONTOUR INTERVAL 10 FEET
DATUM IS MEAN SEA LEVEL

Figure D1 *Part of the Bottomless Lakes Quadrangle, New Mexico, which has a contour interval of 10 feet; every fifth contour is darker and labeled with its elevation. Notice that contours are widely spaced where slopes are gentle and more closely spaced where they are steeper, as in the central part of the map. Hills are shown by contours that close on themselves, whereas depressions are indicated by contours with hachure marks pointing toward the center of the depression. The dashed blue lines on the map represent intermittent streams; notice that where contours cross a stream's channel they form a V that points upstream.*

types of scales are used on topographic maps. The first and most easily understood is a graphic scale, which is simply a bar subdivided into appropriate units of length. This scale appears at the bottom center of the map and may show miles, feet, kilometers, or meters. Indeed, graphic scales on USGS topographic maps generally show both English and metric distance units.

A ratio or fractional scale, which represents the degree of reduction of the area depicted, appears above the graphic scale. On a map with a ratio scale of 1:24,000, for instance, the area shown on the map is 1/24,000th the size of the actual land area. Another way to express this relationship is to say that any unit of length on the map equals 24,000 of the same units on the ground. Thus, 1 inch on the map equals 24,000 inches on the ground, which is more meaningful if one converts inches to feet, making 1 inch equal to 2000 feet. A few maps have scales of 1:63,360, which converts to 1 inch equals 5280 feet, or 1 inch equals 1 mile.

USGS topographic maps are published in a variety of scales such as 1:50,000, 1:62,500, 1:125,000, and 1:250,000. One should also realize that large scale maps cover less area than small scale maps, and the former show much more detail than the latter. For example, a large scale map (1:24,000) shows more surface features in greater detail than does a small scale map (1:125,000) for the same area.

Map Locations

Location on topographic maps can be determined in two ways. First, the borders of maps correspond to lines of latitude and longitude. Latitude is measured north and south of the equator in degrees, minutes, and seconds, whereas the same units are used to designate longitude east and west of the prime meridian, which passes through Greenwich, England. Maps depicting all areas within the United States are noted in north latitude and west longitude. Latitude and longitude are noted in degrees and minutes at the corners of maps, but usually only minutes and seconds are shown along the margins. Many USGS topographic maps cover 7 1/2 or 15 minutes of latitude and longitude and are thus referred to as 7 1/2 and 15 minute quadrangles.

Beginning in 1812, the General Land Office (now known as the Bureau of Land Management) developed a standardized method for accurately defining the location of property in the United States. This method, known as the General Land Office Grid System has been used for all states except those along the eastern seaboard (except Florida), parts of Ohio, Tennessee, Kentucky, West Virginia, and Texas.

As new land acquired by the United States was surveyed, the surveyors laid out north-south lines they called *principle meridians*, and east-west lines known as *base lines*. These intersecting lines form a set of coordinates for locating specific pieces of property. The basic unit in the General Land Office Grid System is the *township*, an area measuring 6 miles on a side, and thus covering 36 square miles (Figure. D2). Townships are numbered north and south of base lines, and are designated as T.1N., T.1S., and so on. Rows of townships known as *ranges* are numbered east and west of principle meridians; R.2W and R.4E, for example. Note in Figure D2 that each township has a unique designation of township and range numbers.

Townships are subdivided into 36 1-square-mile (640-acre) *sections* numbered from 1 to 36. Because of surveying errors and the adjustments necessary to make a grid system

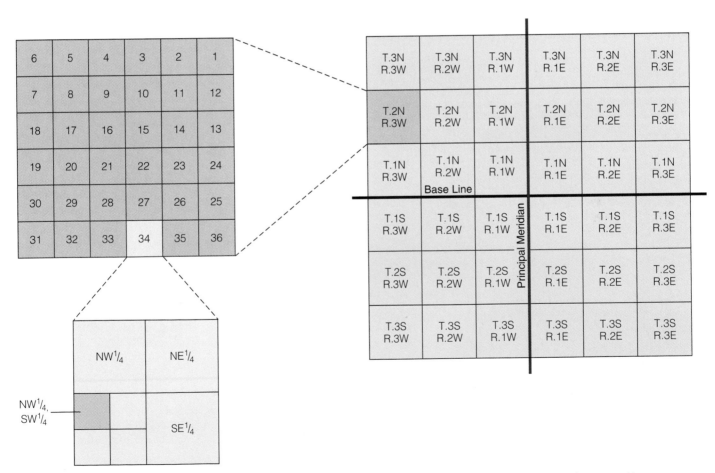

Figure D2 *The General Land Office Grid System. Each 36-square-mile township is designated by township and range numbers. Townships are subdivided into sections, which can be further subdivided into quarter sections and quarter-quarter sections.*

conform to Earth's curved surface, not all sections are exactly 1 mile square. Nevertheless, each section can be further subdivided into half sections and quarter sections designated NE 1/4, NW 1/4, SE 1/4, and SW 1/4, and each quarter section can be further divided into quarter-quarter sections. To show the complete designation for an area, the smallest unit is noted first (quarter-quarter section) followed by quarter section, section number, township, and range. For example, the area shown in pink in Figure D2 is the NW 1/4, SW 1/4, Sec. 34, T.2N., R3W.

Because only a few principle meridians and base lines were established, they do not appear on most topographic maps. Nevertheless, township and range numbers are printed along the margins of 7/12 and 15 minute quadrangles, and a grid consisting of red land boundaries depicts sections. In addition, each section number is shown in red within the map. However, small scale maps show only township and range.

Where to Obtain Topographic Maps

Many people find topographic maps useful. Land use planners, personnel in various local, state, and federal agencies, as well as engineers and real estate developers might use these maps for a variety of reasons. In addition, hikers, backpackers and others interested in exploring undeveloped areas commonly use topographic maps because trails are shown by black dashed lines. Furthermore, map users can readily determine their location by interpreting the topographic features depicted by contours, and anticipate the type of terrain they will encounter during off-road excursions.

Topographic maps for local areas are available at some sporting goods stores, at National Park Visitor Centers, and from some state geological surveys. Free index maps showing the names and locations of all quadrangles for each state are available from the USGS to anyone uncertain of which specific map is needed. Any published topographic map can be purchased from two main sources. For maps of areas east of the Mississippi River write to

Branch of Distribution
U.S. Geological Survey
1200 S. Eads Street
Arlington, Virginia 22202

Maps for areas west of the Mississippi River can be obtained from

Branch of Distribution
U.S. Geological Survey
Box 25286 Federal Center
Denver, Colorado 80225

Answers to Multiple-Choice Review Questions

Chapter 1
1. d; 2. b; 3. c; 4. a; 5. b; 6. e; 7. b;
8. a; 9. c; 10. b; 11. d; 12. b; 13. d.

Chapter 2
1. d; 2. a; 3. b; 4. c; 5. e; 6. c; 7. a;
8. c; 9. d; 10. a; 11. c; 12. e.

Chapter 3
1. b; 2. b; 3. e; 4. c; 5. d; 6. a; 7. a;
8. e; 9. b; 10. b; 11. a; 12. c.

Chapter 4
1. e; 2. c; 3. b; 4. e; 5. a; 6. c; 7. b;
8. c; 9. e; 10. d; 11. b; 12. a.

Chapter 5
1. a; 2. d; 3. c; 4. c; 5. c; 6. a; 7. a;
8. b; 9. e; 10. a; 11. a; 12. c; 13. c;
14. b.

Chapter 6
1. d; 2. e; 3. e; 4. b; 5. e; 6. c; 7. b;
8. c; 9. d; 10. e; 11. c; 12. d; 13. d;
14. a.

Chapter 7
1. b; 2. d; 3. c; 4. a; 5. c; 6. d; 7. e;
8. c; 9. c; 10. e; 11. c; 12. c; 13. d;
14. a.

Chapter 8
1. b; 2. e; 3. a; 4. b; 5. c; 6. d; 7. a;
8. b; 9. c; 10. b; 11. a; 12. d.

Chapter 9
1. b; 2. e; 3. a; 4. d; 5. b; 6. e; 7. b;
8. a; 9. c; 10. a; 11. c; 12. d; 13. c;
14. e.

Chapter 10
1. c; 2. e; 3. d; 4. c; 5. a; 6. b; 7. e;
8. c; 9. a; 10. d; 11. b; 12. e.

Chapter 11
1. b; 2. a; 3. e; 4. e; 5. a; 6. c; 7. d;
8. e; 9. d; 10. e 11. e; 12. d; 13. c.

Chapter 12
1. b; 2. e; 3. b; 4. a; 5. b; 6. d; 7. e;
8. c; 9. d; 10. a; 11. c; 12. d.

Chapter 13
1. d; 2. e; 3. b; 4. d; 5. c; 6. a; 7. c;
8. c; 9. e; 10. c; 11. e; 12. b; 13. c.

Chapter 14
1. a; 2. e; 3. b; 4. c; 5. d; 6. b; 7. e;
8. b; 9. c; 10. c; 11. d; 12. d.

Chapter 15
1. d; 2. a; 3. d; 4. a; 5. c; 6. d; 7. e;
8. a; 9. e; 10. b; 11. b; 12. d.

Chapter 16
1. b; 2. d; 3. c; 4. a; 5. b; 6. e; 7. a;
8. c; 9. a; 10. d; 11. b; 12. a.

Chapter 17
1. b; 2. b; 3. b; 4. b; 5. c; 6. b; 7. a;
8. d; 9. e; 10. b; 11. b; 12. c.

Chapter 18
1. b; 2. d; 3. a; 4. c; 5. a; 6. b; 7. a;
8. d; 9. e; 10. c; 11. b; 12. d.

Chapter 19
1. d; 2. c; 3. d; 4. e; 5. a; 6. d; 7. c;
8. a; 9. c; 10. d; 11. a; 12. e.

Chapter 20
1. a; 2. c; 3. e; 4. b; 5. a; 6. b; 7. e;
8. c; 9. b; 10. a; 11. b; 12. c.

Chapter 21
1. b; 2. d; 3. b; 4. c; 5. b; 6. e; 7. a;
8. e; 9. b; 10. a; 11. d; 12. c; 13. d.

Chapter 22
1. d; 2. d; 3. c; 4. e; 5. d; 6. a; 7. e;
8. c; 9. d; 10. c; 11. b; 12. a; 13. d.

Chapter 23
1. e; 2. d; 3. e; 4. c; 5. d; 6. b; 7. b;
8. a; 9. b; 10. c; 11. b; 12. c.

Chapter 24
1. b; 2. c; 3. e; 4. a; 5. c; 6. c; 7. a;
8. d; 9. c; 10. a; 11. e; 12. e; 13. b;
14. d.

Glossary

A

aa A lava flow with a surface of rough, jagged angular blocks and fragments.

abrasion The process whereby exposed rock is worn and scraped by the impact of solid particles.

Absaroka sequence A widespread sequence of Upper Mississippian to Lower Jurassic sedimentary rocks bounded above and below by unconformities; deposited during a transgressive-regressive cycle of the Absaroka Sea.

absolute dating The process of assigning ages in years before the present to rocks. Various radioactive decay dating techniques yield absolute ages. (*See* relative dating)

abyssal plain Vast flat area on the seafloor adjacent to the continental rise of a passive continental margin.

Acadian orogeny A Devonian orogeny in the northern Appalachian mobile belt resulting from a collision of Baltica with Laurentia.

acanthodian Any of the fish first having a jaw or jawlike mechanism; a class of fishes (class Acanthodii) appearing during the Early Silurian and becoming extinct during the Permian.

active continental margin A continental margin characterized by volcanism and seismicity at the leading edge of a continental plate where oceanic lithosphere is subducted. (*See also* passive continental margin)

allele Alternative form of a gene controlling the same trait.

allopatric speciation Model for the origin of a new species from a small population that became geographically isolated from its parent population.

alluvial fan A cone-shaped alluvial deposit; generally formed where a stream flows from mountains onto an adjacent lowland.

alluvium A general term for all detrital material transported and deposited by streams.

alpha decay A type of radioactive decay involving the emission of a particle consisting of two protons and two neutrons from the nucleus of an atom; decreases the atomic number by two and the atomic mass number by four.

amniote egg An egg in which the embryo develops in a liquid-filled cavity called the amnion. The embryo is also supplied with a yolk sac and a waste sac. The amniote egg is shelled in reptiles, birds, and egg-laying mammals and is retained but modified in all other mammals.

anaerobic A term referring to organisms that are not dependent on oxygen for respiration.

analogous organ Body part, such as wings of insects and birds, that serve the same function, but differ in structure and development. (*See* homologous organ)

Ancestral Rockies Late Paleozoic uplift in the southwestern part of the North American craton.

angiosperm Vascular plants having flowers and seeds; the flowering plants.

angular unconformity An unconformity below which older strata dip at a different angle (usually steeper) than the overlying younger strata. (*See* disconformity *and* nonconformity)

anticline An up-arched fold in which the oldest exposed rocks coincide with the fold axis and all strata dip away from the axis.

Antler orogeny A Late Devonian to Mississippian orogeny that affected the Cordilleran mobile belt; deformation extended from Nevada to Alberta, Canada.

aphanitic texture An igneous texture in igneous rocks in which individual mineral grains are too small to be seen without magnification; results from rapid cooling of magma and generally indicates an extrusive origin.

Appalachian mobile belt A mobile belt along the eastern margin of the North American craton; extends from Newfoundland to Georgia; probably continuous to the southwest with the Ouachita mobile belt.

aquiclude Any layer that prevents the movement of groundwater.

aquifer A permeable layer that allows the movement of groundwater.

archosaur One of a group of animals including dinosaurs, flying reptiles (pterosaurs), crocodiles, and birds.

arête A narrow, serrated ridge separating two glacial valleys or adjacent cirques.

artesian system A confined groundwater system in which high hydrostatic (fluid) pressure builds up causing water to rise above the level of the aquifer.

artificial selection The practice of selective breeding of plants and animals for desirable traits.

artiodactyl Any member of the order Artiodactyla, the even-toed hoofed mammals. Living artiodactyls include swine, sheep, goats, camels, deer, bison, and antelope.

aseismic ridge A ridge or broad, plateaulike feature rising as much as 2 to 3 km above the surrounding seafloor and lacking seismic activity.

assimilation A process in which magma changes composition as it reacts with country rock with which it comes in contact.

asthenosphere The part of the mantle that lies below the lithosphere; behaves plastically and flows.

atom The smallest unit of matter that retains the characteristics of an element.

atomic mass number The total number of protons and neutrons in the nucleus of an atom.

atomic number The number of protons in the nucleus of an atom.

aureole A zone surrounding a pluton in which contact metamorphism has taken place.

australopithecine A term referring to several extinct species of the genus *Australopithecus* that existed in South and East Africa during the Pliocene and Pleistocene epochs.

autotrophic Describes organisms that synthesize their organic nutrients from inorganic raw materials; photosynthesizing bacteria and plants are autotrophs. (*See* heterotrophic)

B

backshore The area of a beach that is usually dry, being covered by water only by storm waves or exceptionally high tides.

Baltica One of six major Paleozoic continents; composed of Russia west of the Ural Mountains, Scandinavia, Poland, and northern Germany.

banded iron formation (BIF)
Sedimentary rocks consisting of alternating thin layers of silica (chert) and iron minerals (mostly the iron oxides hematite and magnetite).

barchan dune A crescent-shaped dune whose tips point downwind.

barrier island A long, narrow island composed of sand oriented parallel to a shoreline, but separated from the mainland by a lagoon.

basal slip A type of glacial movement in which a glacier slides over the underlying surface.

basalt plateau A large plateau built up by numerous lava flows from fissure eruptions.

base level The lowest limit to which a stream can erode.

basin The circular equivalent of a syncline. All strata in a basin dip toward a central point, and the youngest exposed rocks are in the center.

Basin and Range Province An area centered on Nevada but extending into adjacent states and northern Mexico; characterized by Cenozoic block-faulting.

batholith A discordant, irregularly shaped pluton with a surface area of at least 100 km².

baymouth bar A spit that has grown until it cuts off a bay from the open sea.

beach A deposit of unconsolidated sediment extending landward from low tide to a change in topography or where permanent vegetation begins.

beach face The sloping area below the berm that is exposed to wave swash.

bed An individual layer of rock, especially sedimentary rock. (*See also* strata)

bed load The part of a stream's sediment load transported along its bed consisting of sand and gravel.

benthos Any organism that lives on the bottom of seas or lakes; may live upon the bottom or within bottom sediments.

berm The backshore area of a beach consisting of a platform composed of sediment deposited by waves; berms are nearly horizontal or slope gently landward.

beta decay A type of radioactive decay during which a fast-moving electron is emitted from a neutron and thus is converted to a proton; results in an increase of one atomic number, but does not change atomic mass number.

Big Bang A model for the evolution of the universe in which a dense, hot state was followed by expansion, cooling, and a less dense state.

biochemical sedimentary rock A sedimentary rock produced by the chemical activities of organisms, such as some varieties of limestone.

biostratigraphic unit A unit of sedimentary rock defined by its fossil content.

bipedal Walking on two legs as a means of locomotion.

body fossil The actual remains of any prehistoric organism; includes shells, teeth, bones, and, rarely, the soft parts of organisms. (*See* trace fossil)

bonding The process whereby atoms are joined to other atoms.

bony fish A class of fishes (class Osteichthyes) that evolved during the Devonian; the most common fishes; characterized by an internal skeleton of bone; divided into two subgroups, the ray-finned fishes and lobe-finned fishes.

Bowen's reaction series A mechanism accounting for the derivation of intermediate and felsic magmas from a mafic magma. It has a discontinuous branch of ferromagnesian minerals that change from one mineral to another over specific temperature ranges and a continuous branch of plagioclase feldspars whose composition changes as the temperature decreases.

braided stream A stream possessing an intricate network of dividing and rejoining channels. Braiding occurs when sand and gravel bars are deposited within channels.

breaker A wave that steepens as it enters shallow water until the crest plunges forward.

browser An animal that eats tender shoots, twigs, and leaves. *Compare with* grazer.

butte An isolated, steep-sided, pinnacle-like erosional structure; formed by the breaching of a resistant cap rock, which allows rapid erosion of the less resistant underlying rocks.

C

caldera A large, steep-sided, circular or oval volcanic depression usually formed by summit collapse resulting from partial draining of the underlying magma chamber.

Caledonian orogeny A Silurian-Devonian orogeny that occurred along the northwestern margin of Baltica resulting from the collision of Baltica with Laurentia.

Canadian shield The exposed part of the North American craton; mostly in Canada, but also present in Minnesota, Wisconsin, Michigan, and New York.

carbon 14 dating technique An absolute dating method that relies on determining the ratio of C^{14} to C^{12} in a sample; useful back to about 70,000 years ago; can be applied only to organic substances.

carbonate mineral A mineral that contains the negatively charged carbonate ion $(CO_3)^{-2}$ (e.g., calcite $[CaCO_3]$ and dolomite $[CaMg(CO_3)_2]$).

carbonate rock A rock containing mostly carbonate minerals, such as limestone and dolostone.

carnivore-scavenger Any animal that depends on other animals, living or dead, as a source of nutrients.

cartilaginous fish Fishes such as living sharks, rays, and skates, and their extinct relatives that have a skeleton composed of cartilage.

Catskill Delta A Devonian clastic wedge deposited adjacent to the highlands that formed during the Acadian orogeny.

cave A naturally formed subsurface opening that is generally connected to the surface and is large enough for a person to enter.

channel flow Flow confined to a long, troughlike depression, such as a stream channel. (*See also* sheet flow)

chemical sedimentary rock Rock formed of minerals derived from materials dissolved during weathering.

chemical weathering The decomposition of rocks by chemical alteration of parent material.

China One of six major Paleozoic continents; composed of all of southeast Asia, including China, Indochina, part of Thailand, and the Malay Peninsula.

chordate All members of the phylum Chordata; characterized by a notochord, a dorsal, hollow nerve cord, and gill slits at some time during the animal's life cycle.

chromosome Complex, double-stranded, helical molecule of deoxyribonucleic acid (DNA); specific segments of chromosomes are genes.

cinder cone A small steep-sided volcano composed of pyroclastic materials that accumulate around a vent.

circum-Pacific belt A zone of seismic and volcanic activity that nearly encircles the margins of the Pacific Ocean basin.

cirque A steep-walled, bowl-shaped depression at the upper end of a glacial valley.

cladogram A diagram showing possible evolutionary relationships among members of a clade, a group of organisms including their most recent common ancestor.

clastic wedge An extensive accumulation of mostly detrital sediments eroded from and deposited adjacent to an uplifted area; clastic wedges are coarse-grained and thick near the uplift and become finer-grained and thinner away from the uplift, e.g., the Queenston Delta.

cleavage The breaking or splitting of mineral crystals along smooth planes of weakness. Cleavage is determined by the strength of the bonds within minerals.

columnar joint Joints in some igneous rocks consisting of six-sided columns that formed as a result of shrinkage during cooling.

complex movement A combination of different types of mass movements in which one type is not dominant; most complex movements involve sliding and flowing.

composite volcano (stratovolcano) A volcano composed of pyroclastic layers, lava flows typically of intermediate composition, and mudflows. Composite volcanoes are also called stratovolcanoes.

compound A substance resulting from the bonding of two or more different elements, e.g., water (H_2O) and quartz (SiO_2).

compression Stress resulting when rocks are squeezed by external forces directed toward one another.

concordant pluton Refers to plutons whose boundaries parallel the layering in the country rock. (*See also* discordant pluton)

cone of depression A cone-shaped depression in the water table around a well, resulting from pumping water from an aquifer faster than it can be replaced.

contact metamorphism Metamorphism in which a magma body alters the surrounding country rock.

continental-continental plate boundary A convergent plate boundary along which two continental lithospheric plates collide, such as the collision of India with Asia.

continental drift The theory that the continents were once joined into a single landmass that broke apart with the various fragments (continents) moving with respect to one another.

continental glacier A large glacier covering a vast area (at least 50,000 km^2) and unconfined by topography. Also called an ice sheet.

continental margin The area separating the part of a continent above sea level from the deep-seafloor.

continental rise The gently sloping area of the seafloor beyond the base of the continental slope.

continental shelf The area between the shoreline and continental slope where the seafloor slopes gently seaward.

continental slope The relatively steep area between the shelf-slope break (at an average depth of 135 m) and the more gently sloping continental rise or oceanic trench.

convergent evolution The development of similarities in two or more distantly related organisms as a consequence of adapting to a similar life-style, e.g., ichthyosaurs and porpoises. (*See* parallel evolution)

convergent plate boundary The boundary between two plates that are moving toward one another; three types of convergent plate boundaries are recognized. (*See* continental-continental plate boundary, oceanic-continental plate boundary, *and* oceanic-oceanic plate boundary)

Cordilleran orogeny A protracted episode of deformation affecting the western margin of North America from Jurassic to Early Cenozoic time; typically divided into three separate phases called the Nevadan, Sevier, and Laramide orogenies.

Cordilleran mobile belt A mobile belt in western North America bounded on the west by the Pacific Ocean and on the east by the Great Plains; extends north-south from Alaska into central Mexico. (*See* North American Cordillera)

core The interior part of Earth, beginning at a depth of about 2900 km; probably composed mostly of iron and nickel; divided into an outer liquid core and an inner solid core.

Coriolis effect The deflection of winds to the right of their direction of motion (clockwise) in the Northern Hemisphere and to the left of their direction of motion (counterclockwise) in the Southern Hemisphere due to Earth's rotation.

correlation Demonstration of the physical continuity of rock units or biostratigraphic units in different areas, or demonstration of time equivalence as in time-stratigraphic correlation.

country rock Any preexisting rock into which magma is intruded or that is altered by metamorphism.

covalent bond A bond formed by the sharing of electrons between atoms.

crater A circular or oval depression at the summit of a volcano resulting from the extrusion of gases, pyroclastic materials, and lava; connected by a conduit to a magma chamber below Earth's surface.

craton The relatively stable part of a continent; consists of a Precambrian shield and a platform, a buried extension of a shield; the ancient nucleus of a continent.

cratonic sequence A widespread sequence of sedimentary rocks bounded above and below by unconformities; deposited during a transgressive-regressive cycle of an epeiric sea (e.g., the Sauk sequence).

creep A type of mass wasting in which soil or rock moves slowly downslope.

crest The highest part of a wave.

Cretaceous Interior Seaway An interior seaway that existed during the Late Cretaceous; formed when northward-transgressing waters from the Gulf of Mexico joined with southward-transgressing water from the Arctic; effectively divided North America into two large landmasses.

Cro-Magnon A type of *Homo sapiens* that lived mostly in Europe from 35,000 to 10,000 years ago.

cross-bedding Layers in sedimentary rocks deposited at an angle to the surface upon which they accumulated.

crossopterygian A specific type of lobe-finned fish; possessed lungs; ancestral to amphibians.

crust Earth's outermost layer; the upper part of the lithosphere, which is separated from the mantle by the Moho; divided into continental and oceanic crust.

crystal settling The physical separation and concentration of minerals in the lower part of a magma chamber or pluton by crystallization and gravitational settling.

crystalline solid A solid in which the constituent atoms are arranged in a regular, three-dimensional framework.

Curie point The temperature at which iron-bearing minerals in a cooling magma attain their magnetism.

cyclothem A vertical sequence of cyclically repeated sedimentary rocks resulting from alternating periods of marine and nonmarine deposition; commonly contain a coal bed.

cynodont A type of therapsid (advanced mammal-like reptile); the ancestors of mammals.

D

debris flow A mass wasting process involving flowage of a viscous mixture of water, soil, and rocks; much like a mudflow, but at least half the particles are larger than sand.

deflation The removal of loose surface sediment by the wind.

deformation Any change in shape or volume of rocks caused by applied forces. (*See also* strain)

delta An alluvial deposit formed where a stream flows into a lake or the sea.

depositional environment Any area in which sediment is deposited such as a beach or a floodplain.

desert Any area that receives less than 25 cm of rain per year and has a high evaporation rate.

desert pavement A surface mosaic of close-fitting pebbles, cobbles, and boulders found in many dry regions; formed by the removal of sand-sized and smaller particles by wind.

desertification The expansion of deserts into formerly productive lands.

detrital sedimentary rock Rock consisting of the solid particles (detritus) of preexisting rocks, such as sandstone and conglomerate.

differential pressure Pressure that is not applied equally to all sides of a rock body.

differential weathering Weathering of rock at different rates, producing an uneven surface.

dike A tabular or sheetlike discordant pluton.

dip A measure of the maximum angular deviation of an inclined plane from horizontal.

discharge The total volume of water in a stream moving past a particular point in a given period of time.

disconformity An unconformity above and below which the strata are paral-

lel. (*See* angular unconformity *and* nonconformity)

discontinuity A boundary across which seismic wave velocity or direction changes abruptly, such as the mantle-core boundary.

discordant pluton Plutons whose boundaries cut across the layering of country rock. (*See also* concordant pluton)

dissolved load That part of a stream's load consisting of ions in solution.

divergent evolution The diversification of a species into two or more descendant species.

divergent plate boundary The boundary between two plates that are moving apart.

divide A topographically high area that separates adjacent drainage basins.

DNA (deoxyribonucleic acid) The substance of which chromosomes are composed; the genetic material of all organisms except bacteria.

dome A circular equivalent of an anticline. All strata in a dome dip away from a central point, and the oldest exposed rocks are at the dome's center.

drainage basin The surface area drained by a stream and its tributaries.

drainage pattern The regional arrangement of channels in a drainage system.

dripstone Various cave deposits resulting from the deposition of calcite.

drumlin An elongated hill of till formed by the movement of a continental glacier.

dune A mound or ridge of wind-deposited sand.

dynamic metamorphism Metamorphism occurring in fault zones where rocks are subjected to high differential pressures.

E

earthflow A mass wasting process involving downslope flow of water-saturated soil.

earthquake Vibrations caused by the sudden release of energy, usually as a result of the displacement of rocks along faults.

Ediacaran fauna A collective name for all Late Proterozoic faunas containing animal fossils similar to those of the Ediacara fauna of Australia.

elastic rebound theory A theory that explains how energy is suddenly released during earthquakes: when rocks are deformed, they store energy and bend; when the inherent strength of the rocks is exceeded, they rupture

and release energy causing earthquakes.

elastic strain A type of deformation in which the material returns to its original shape when stress is relaxed.

electromagnetic force The force that combines electricity and magnetism into the same force and binds atoms into molecules. It also transmits radiation across the various spectra at wavelengths ranging from gamma rays (shortest) to radio waves (longest) through massless particles called *photons*.

electron A negatively charged particle of very little mass that encircles the nucleus of an atom.

electron capture A type of radioactive decay in which an electron is captured by a proton, and converted to a neutron; results in a loss of one atomic number, but no change in atomic mass number.

electron shell Electrons orbit rapidly around the nuclei of atoms at specific distances known as electron shells.

element A substance composed of all the same atoms; it cannot be changed into another element by ordinary chemical means.

emergent coast A coast where the land has risen with respect to sea level.

end moraine A pile of rubble deposited at the terminus of a glacier. (*See* recessional moraine *and* terminal moraine)

epeiric sea A broad shallow sea that covers part of a continent; six epeiric seas covered parts of North America during the Phanerozoic Eon (e.g., the Sauk Sea).

epicenter The point on Earth's surface vertically above the focus of an earthquake.

erosion The removal of weathered material.

erratic A rock fragment carried some distance from its source by a glacier and usually deposited on materials of a different composition.

esker A long sinuous ridge of stratified drift formed by deposition by running water in tunnels beneath stagnant ice.

eukaryotic cell A type of cell with a membrane-bounded nucleus containing chromosomes; also contains such organelles as plastids and mitochondria that are absent in prokaryotic cells. (*See* prokaryotic cell)

eupantothere Any member of a group of mammals that included the ancestors of both marsupial and placental mammals.

evaporite A sedimentary rock that formed by inorganic chemical precipitation of minerals from solution (e.g., rock salt and rock gypsum).

Exclusive Economic Zone An area extending 371 km seaward from the coast of the United States and its territories in which the United States claims all sovereign rights.

exfoliation dome A large rounded dome of rock resulting from the process of exfoliation.

F

fault A fracture along which movement has occurred parallel to the fracture surface.

fault plane A fracture surface along which blocks of rock on opposite sides have moved relative to one another.

felsic magma A type of magma containing more than 65% silica and considerable sodium, potassium, and aluminum, but little calcium, iron, and magnesium. (*See* mafic magma *and* intermediate magma)

ferromagnesian silicate A silicate mineral containing iron and magnesium or both.

fetch The distance the wind blows over a continuous water surface.

fiord An arm of the sea extending into a U-shaped glacial trough eroded below sea level.

firn Granular snow formed by partial melting and refreezing of snow.

fission track dating The process of dating samples by counting the number of small linear tracks (fission tracks) that result when a mineral crystal is damaged by rapidly moving alpha particles generated by radioactive decay of uranium.

fissure eruption An eruption in which lava or pyroclastic material is emitted from a long, narrow fissure or group of fissures.

floodplain A low-lying, relatively flat area adjacent to a stream, which is covered with water when the stream overflows its banks.

fluid activity An agent of metamorphism in which water and carbon dioxide promote metamorphism by increasing the rate of chemical reactions.

focus The place within Earth where an earthquake originates and energy is released.

foliated texture A texture of metamorphic rocks in which the platy and elongate minerals are arranged in a parallel fashion.

footwall block The block that lies beneath a fault plane.

foreshore The area of a beach covered by water during high tide but exposed during low tide.

fossil Remains or traces of prehistoric organisms preserved in rocks of the crust. (*See* body fossil *and* trace fossil)

fracture A break in a rock resulting from intense applied pressure.

Franklin mobile belt The most northerly mobile belt in North America; extends from northwestern Greenland westward across the Canadian Arctic islands.

frost action The disaggregation of rocks by repeated freezing and thawing of water in cracks and crevices.

G

gene The basic unit of inheritance; a specific segment of a chromosome. (*See* allele)

geologic time scale A chart with the designations for the earliest interval of geologic time at the bottom, followed upward by designations for progressively more recent time intervals.

geology The science concerned with the study of Earth; includes studies of Earth materials (minerals and rocks), surface and internal processes, and Earth history.

geothermal energy Energy that comes from the steam and hot water trapped within the crust.

geothermal gradient The temperature increase with depth; averages 25°C/km near the surface, but varies from area to area.

geyser A hot spring that intermittently ejects hot water and steam.

glacial budget The balance between expansion and contraction of a glacier in response to accumulation and wastage.

glacial drift A collective term for all sediment deposited by glaciers or associated processes; includes till deposited by ice, and outwash deposited by meltwater streams.

glacial ice Water in the solid state within a glacier. Forms as snow partially melts and refreezes and compacts so that it is transformed first into firn and then into glacial ice.

glacial polish A smooth glistening rock surface formed by the movement of a sediment-laden glacier over it.

glacier A mass of ice on land that moves by plastic flow and basal slip.

***Glossopteris* flora** A Late Paleozoic association of plants found only on the Southern Hemisphere continents and India.

Gondwana One of six major Paleozoic continents; composed of the present-day continents of South America, Africa, Antarctica, Australia, and India, and parts of other continents such as southern Europe, Arabia, and Florida.

graded bedding A type of sedimentary bedding in which an individual bed is characterized by a decrease in grain size from bottom to top.

graded stream A stream possessing an equilibrium profile in which a delicate balance exists between gradient, discharge, flow velocity, channel characteristics, and sediment load so that neither significant erosion nor deposition occurs within its channel.

gradient The slope over which a stream flows; expressed in m/km or ft/mi.

gravity The attraction of one body toward another.

gravity anomaly A departure from the expected force of gravity; anomalies might be positive or negative.

grazer Any animal that crops low-growing vegetation, especially grasses. *Compare with* browser.

greenstone belt A linear or podlike association of volcanic and sedimentary rocks particularly common in Archean terranes; typically synclinal and consists of lower and middle volcanic units and an upper sedimentary rock unit.

ground moraine The layer of sediment liberated from melting ice as a glacier's terminus retreats.

groundwater Underground water stored in the pore spaces of rock, sediment, or soil.

guide fossil Any fossil that can be used to determine the relative age of rocks and to correlate rocks of the same relative ages in different areas.

guyot A flat-topped seamount of volcanic origin rising more than 1 km above the seafloor.

gymnosperm The flowerless, seed-bearing land plants.

H

half-life The time required for one-half of the original number of atoms of a radioactive element to decay to a stable daughter product (e.g., the half-life of potassium 40 is 1.3 billion years).

hanging valley A tributary glacial valley whose floor is at a higher level than that of the main glacial valley.

hanging wall block The block that overlies a fault plane.

headland The seaward-projecting part of a shoreline that is eroded on both sides due to wave refraction.

heat An agent of metamorphism; heat comes from increasing depth, magma, and applied pressure.

herbivore An animal that is dependent on plants as a source of nutrients.

Hercynian-Alleghenian orogeny Pennsylvanian to Permian orogenic event during which the Appalachian mobile belt of eastern North America and the Hercynian mobile belt of southern Europe were deformed.

heterotrophic Describes organisms such as animals that depend on preformed organic molecules from the environment as a source of nutrients. (*See* autotrophic)

hominid Abbreviated form of Hominidae; the family to which humans belong. Such bipedal primates as *Australopithecus* and *Homo* are hominids.

Homo The genus to which humans belong; includes *Homo erectus* and *Homo sapiens.*

homologous organ Body parts in different organisms that have a similar structure, similar relationships to other organs, and similar development, but do not necessarily serve the same function (e.g., the wing of a bird and the forelimbs of whales and dogs). (*See* analogous organ)

horn A steep-walled, pyramidal peak formed by the headward erosion of cirques.

hot spot Localized zone of melting below the lithosphere.

hot spring A spring in which the water temperature is warmer than the temperature of the human body (37°C).

hydraulic action The power of moving water.

hydrologic cycle The continuous recycling of water from the oceans, through the atmosphere, to the continents, and back to the oceans.

hydrolysis The chemical reaction between the hydrogen (H^+) ions and hydroxyl (OH^-) ions of water and a mineral's ions.

hydrothermal A term referring to hot water as in hot springs or geysers.

hypothesis A provisional explanation for observations. Subject to continual testing and modification. If well supported by evidence, hypotheses are then generally called theories.

I

Iapetus Ocean A Paleozoic ocean basin that separated North America from Europe; the Iapetus Ocean began closing when North America and Europe started moving toward one another; it was eliminated when these continents collided during the Late Paleozoic.

igneous rock Any rock formed by cooling and crystallization of magma, or by the accumulation and consolidation of pyroclastic materials such as ash.

incised meander A deep, meandering canyon cut into solid bedrock by a stream.

index mineral A mineral that forms only within specific temperature and pressure ranges during metamorphism.

infiltration capacity The maximum rate at which a soil or sediment can absorb water.

inselberg An isolated steep-sided erosional remnant rising above a surrounding desert plain.

intensity The subjective measure of the kind of damage done by an earthquake as well as people's reaction to it.

intermediate magma A magma having a silica content between 53% and 65% and an overall composition intermediate between felsic and mafic magmas.

ion An electrically charged atom produced by adding or removing electrons from the outermost electron shell.

ionic bond A bond that results from the attraction of positively and negatively charged ions.

irons A group of meteorites composed primarily of iron and nickel and accounting for about 6% of all meteorites.

isostasy *See* principle of isostasy.

isostatic rebound The phenomenon in which unloading of Earth's crust causes it to rise upward until equilibrium is again attained.

J

joint A fracture along which no movement has occurred, or where movement has been perpendicular to the fracture surface.

Jovian planet Any of the four planets (Jupiter, Saturn, Uranus, and Neptune) that resemble Jupiter. All are large and have low mean densities, indicating that they are composed mostly of lightweight gases, such as hydrogen and helium, and frozen compounds, such as ammonia and methane.

K

kame Conical hill of stratified drift originally deposited in a depression on a glacier's surface.

karst topography Topography consisting of numerous caves, sinkholes, and solution valleys developed by groundwater solution of rocks. Also characterized by springs, disappearing streams, and underground drainage.

Kaskaskia sequence A widespread sequence of Devonian and Mississippian sedimentary rocks bounded above and below by unconformities; deposited during a transgressive-regressive cycle of the Kaskaskia Sea.

Kazakhstania One of six major Paleozoic continents; a triangular-shaped continent centered on Kazakhstan (part of Asia).

L

labyrinthodont Any of the Devonian to Triassic amphibians, characterized by labyrinthine wrinkling and folding of their teeth.

laccolith A concordant pluton with a mushroomlike geometry.

lahar A mudflow composed of volcanic materials such as ash.

Laramide orogeny The Late Cretaceous to Early Cenozoic phase of the Cordilleran orogeny; responsible for many of the structural features of the present-day Rocky Mountains.

lateral moraine The sediment deposited as a long ridge of till along the margin of a valley glacier.

laterite A red soil, rich in iron or aluminum or both, that forms in the tropics by intense chemical weathering.

Laurasia A Late Paleozoic, Northern Hemisphere continent composed of the present-day continents of North America, Greenland, Europe, and Asia.

Laurentia The name given to a Proterozoic continent that was composed mostly of North America and Greenland, parts of northwestern Scotland, and perhaps parts of the Baltic shield of Scandinavia.

lava Magma at Earth's surface.

lava dome A bulbous, steep-sided structure formed by viscous magma moving upward through a volcanic conduit.

lava flow A stream of magma flowing over Earth's surface.

lava tube A tunnel beneath the solidified surface of a lava flow through which lava continues to flow. Also, the hollow space left when the lava within a tube drains away.

lithification The process of converting sediment into sedimentary rock.

lithosphere Earth's outer, rigid part consisting of the upper mantle, oceanic crust, and continental crust.

lithostatic pressure Pressure exerted on rock by the weight of overlying rocks.

lithostratigraphic unit A unit of sedimentary rock, such as a formation, defined by its physical characteristics rather than its biologic content or time of origin.

loess Windblown silt and clay deposits; derived from three main sources—deserts, Pleistocene glacial outwash deposits, and floodplains of streams in semiarid regions.

longitudinal dune A long ridge of sand generally parallel to the direction of the prevailing wind.

longshore current A current between the breaker zone and the beach that flows parallel to the shoreline and is produced by wave refraction.

Love wave (L-wave) A surface wave in which individual particles of material only move back and forth in a horizontal plane perpendicular to the direction of wave travel.

low-velocity zone The zone within the mantle between the depths of 100 and 250 km where the velocity of both P- and S-waves decreases markedly; it corresponds closely to the asthenosphere.

M

mafic magma A silica-poor magma containing between 45% and 52% silica and proportionately more calcium, iron, and magnesium than an intermediate or felsic magma.

magma Molten rock material generated within Earth.

magma chamber A reservoir of molten rock within Earth's otherwise solid lithosphere.

magma mixing The process of mixing magmas of different composition thereby producing a modified version of the parent magmas.

magnetic anomaly Any change, such as a change in average strength, of Earth's magnetic field.

magnetic reversal The phenomenon in which the north and south magnetic poles are completely reversed.

magnitude The total amount of energy released by an earthquake at its source.

mantle The thick layer between Earth's crust and core.

mantle plume A stationary column of magma that originates deep within the mantle and slowly rises to Earth's surface to form volcanoes or basalt plateaus.

marine regression The withdrawal of the sea from a continent or coastal area resulting in the emergence of land as sea level falls or the land rises with respect to sea level.

marine terrace A wave-cut platform now elevated above sea level.

marine transgression The invasion of coastal areas or much of a continent by the sea resulting from a rise in sea level or subsidence of the land.

marsupial mammal Any of the pouched mammals such as opossums, kangaroos, and wombats. At present, marsupials are common only in Australia.

mass wasting The downslope movement of material under the influence of gravity.

meandering stream A stream possessing a single, sinuous channel with broadly looping curves.

mechanical weathering Disaggregation of rocks by physical processes; yields smaller pieces retaining the same composition as the parent material.

medial moraine A moraine formed where two lateral moraines merge.

Mediterranean belt A zone of seismic and volcanic activity extending westerly from Indonesia through the Himalayas, across Iran and Turkey, and through the Mediterranean region of Europe; about 20% of all active volcanoes and 15% of all earthquakes occur in this belt.

mesa A broad, flat-topped erosional remnant bounded on all sides by steep slopes; forms when resistant cap rock is breached, allowing rapid erosion of the less resistant underlying sedimentary rock.

metamorphic facies A group of metamorphic rocks characterized by particular mineral assemblages formed under the same broad temperature-pressure conditions.

metamorphic rock Any rock type altered by high temperature and pressure and the chemical activities of fluids is said to have been metamorphosed (e.g., slate, gneiss, marble).

metamorphic zone The region between lines of equal metamorphic intensity known as isograds.

meteorite A mass of matter of extraterrestrial origin that has fallen to Earth.

Midcontinent rift A Late Proterozoic intracontinental rift within Laurentia; contains thick accumulations of volcanic rocks and detrital sedimentary rocks.

Milankovitch theory A theory that explains cyclic variations in climate and the onset of ice ages as a result of irregularities in Earth's rotation and orbit.

mineral A naturally occurring, inorganic, crystalline solid having characteristic physical properties and a narrowly defined chemical composition.

mobile belt Elongated area of deformation as indicated by folds and faults; generally located adjacent to a craton (e.g., the Appalachian mobile belt).

modern synthesis A synthesis of the ideas of geneticists, paleontologists, population biologists, and others to yield a neo-Darwinian view of evolution; includes chromosome theory of inheritance, mutation as a source of variation, and gradualism.

Modified Mercalli Intensity Scale A scale having intensity values ranging from I to XII that is used to measure earthquake damage.

Mohorovičić discontinuity (Moho) The boundary between the crust and the mantle. Also called the Moho.

monocline A simple bend or flexure in otherwise horizontal or uniformly dipping rock layers.

monomer A comparatively simple organic molecule, such as an amino acid, that is capable of linking with other monomers to form polymers. (*See* polymer)

monotreme The egg-laying mammals; only two types of monotremes now exist, the platypus and spiny anteater of the Australian region.

mud crack A sedimentary structure found in clay-rich sediment that has dried out. When such sediment dries, it shrinks and forms intersecting fractures.

mudflow A flow consisting of mostly clay- and silt-sized particles and more than 30% water; most common in semiarid and arid environments.

multicellular organism Any organism consisting of many cells as opposed to a single cell; possesses cells specialized to perform specific functions such as reproduction and respiration.

mutation Any change in the hereditary information in genes of organisms; some of the inheritable variation in populations upon which natural selection acts arises from mutations in sex cells.

N

native element A mineral composed of a single element.

natural levee A ridge of sandy alluvium deposited along the margins of a stream channel during floods.

natural selection A mechanism proposed by Charles Darwin and Alfred Russel Wallace to account for evolution; as a result of natural selection, organisms best adapted to their environment are more likely to survive and reproduce.

Neanderthal A type of human that inhabited Europe and the Near East from 150,000 to 32,000 years ago; considered by some to be a variety or subspecies (*Homo sapiens neanderthalensis*) of *Homo sapiens* and by some as a separate species (*Homo neanderthalensis*).

nearshore sediment budget The balance between additions and losses of sediment in the nearshore zone.

nekton Actively swimming organisms (e.g., fishes, whales, and squids). (*See* plankton)

neutron An electrically neutral particle found in the nucleus of an atom.

Nevadan orogeny Late Jurassic to Cretaceous phase of the Cordilleran orogeny; most strongly affected the western part of the Cordilleran mobile belt.

nonconformity An unconformity in which stratified sedimentary rocks above an erosion surface overlie igneous or metamorphic rocks. (*See* angular unconformity *and* disconformity)

nonferromagnesian silicate A silicate mineral that does not contain iron or magnesium.

nonfoliated texture A metamorphic texture in which there is no discernible preferred orientation of mineral grains.

nonvascular Plants lacking specialized tissues for conducting fluids.

normal fault A dip-slip fault on which the hanging wall block has moved downward relative to the footwall block. (*See* reverse fault)

North American Cordillera A complex mountainous region in western North America extending from Alaska into central Mexico. (*See* Cordilleran mobile belt)

nucleus The central part of an atom consisting of one or more protons and neutrons.

nuée ardente A mobile dense cloud of hot pyroclastic materials and gases ejected from a volcano.

O

oblique-slip fault A fault having both dip-slip and strike-slip movement.

oceanic-continental plate boundary A type of convergent plate boundary along which oceanic lithosphere and continental lithosphere collide; characterized by subduction of an oceanic plate beneath a continental plate and by volcanism and seismicity.

oceanic-oceanic plate boundary A type of convergent plate boundary along which two oceanic lithospheric plates collide and one is subducted beneath the other.

oceanic ridge A submarine mountain system found in all of the oceans; it is composed of volcanic rock (mostly basalt) and displays features produced by tension.

oceanic trench A long, narrow feature restricted to active continental margins and along which subduction occurs.

ooze Deep-sea sediment composed mostly of shells of marine animals and plants.

ophiolite A sequence of igneous rocks thought to represent a fragment of oceanic lithosphere; composed of peridotite overlain successively by gabbro, sheeted basalt dikes, and pillow basalts.

organic evolution See theory of evolution.

organic reef A wave-resistant limestone structure with a structural framework of animal skeletons (e.g., stromatoporoid reef or coral reef).

Ornithischia An order of dinosaurs with a birdlike pelvis. (*See* Saurischia)

orogen A linear part of Earth's crust that was deformed during an orogeny. (*See* orogeny)

orogeny The process of forming mountains, especially by folding and thrust faulting; an episode of mountain building.

ostracoderm The "bony-skinned" fish; first appeared during the Late Cambrian and thus are the oldest known vertebrates; characterized by a lack of jaws and teeth and presence of bony armor.

Ouachita mobile belt A mobile belt located along the southern margin of the North American craton; probably continuous with the Appalachian mobile belt.

Ouachita orogeny An orogeny that deformed the Ouachita mobile belt during the Pennsylvanian Period.

outgassing The process whereby gases derived from Earth's interior are released into the atmosphere by volcanic activity.

outwash plain The sediment deposited by meltwater discharging from a continental glacier's terminus.

oxbow lake A cutoff meander filled with water.

oxidation The reaction of oxygen with other atoms to form oxides or, if water is present, hydroxides.

P

pahoehoe A type of lava flow with a smooth, ropy surface.

paleomagnetism Remnant magnetism in rocks studied to determine the intensity and direction of Earth's past magnetic field.

Pangaea The name Alfred Wegener proposed for a supercontinent consisting of all Earth's landmasses that existed at the end of the Paleozoic Era.

Panthalassa Ocean The Late Paleozoic worldwide ocean that surrounded the supercontinent Pangaea.

parabolic dune A crescent-shaped dune in which the tips point upwind.

parallel evolution The development of similarities in two or more closely related but separate lines of descent as a consequence of similar adaptations. (*See* convergent evolution)

parent material The material that is chemically and mechanically weathered to yield sediment and soil.

passive continental margin The trailing edge of a continental plate consisting of a broad continental shelf and a continental slope and rise. A vast, flat abyssal plain is commonly present adjacent to the rise.

pedalfer A soil that develops in humid regions and has an organic-rich A horizon and aluminum-rich clays and iron oxides in horizon B.

pediment An erosion surface of low relief gently sloping away from a mountain base.

pedocal A soil characteristic of arid and semiarid regions with a thin A horizon and a calcium carbonate-rich B horizon.

pegmatite A coarse-grained igneous rock usually of granitic composition.

pelagic clay Generally brown or reddish deep-sea sediment composed of clay-sized particles derived from the continents and oceanic islands.

pelycosaur Pennsylvanian to Permian "finback reptiles"; possessed some mammalian characteristics.

perissodactyl Any member of the order Perissodactyla, the odd-toed hoofed mammals; living perissodactyls include horses, rhinoceroses, and tapirs. (*See* artiodactyl)

permafrost Ground that remains permanently frozen.

permeability A material's capacity for transmitting fluids.

phaneritic texture A course-grained texture in igneous rocks in which the mineral grains are easily visible without magnification; generally indicates an intrusive origin.

photochemical dissociation A process whereby water molecules in the upper atmosphere are disrupted by ultraviolet radiation, yielding oxygen (O_2) and hydrogen (H_2).

photosynthesis The metabolic process of synthesizing organic molecules from water and carbon dioxide, using the radiant energy of sunlight captured by chlorophyll-containing cells.

phyletic gradualism An evolutionary concept holding that a species evolves gradually and continuously through

time to give rise to new species. (*See* punctuated equilibrium)

pillow lava Bulbous masses of basalt resembling pillows formed when lava is rapidly chilled under water.

placental mammal Any of the mammals that have a placenta to nourish the embryo; fusion of the amnion of the amniote egg with the walls of the uterus forms the placenta; most mammals, living and fossil, are placentals.

placoderm The "plate-skinned" fishes; Late Silurian through Permian; characterized by jaws and bony armor especially in the head-shoulder region.

plankton Animals and plants that float passively (e.g., phytoplankton and zooplankton). (*See* nekton)

plastic flow The flow that occurs in response to pressure and causes permanent deformation.

plastic strain The result of stress in which a material cannot recover its original shape and retains the configuration produced by the stress such as by folding of rocks.

plate An individual piece of lithosphere that moves over the asthenosphere.

plate tectonic theory The theory that large segments of the outer part of Earth (lithospheric plates) move relative to one another.

platform The broad area extending from a shield but covered by younger rocks. A platform and shield form a craton.

playa A dry lake bed found in deserts.

pluton An intrusive igneous body that forms when magma cools and crystallizes within Earth's crust.

plutonic (intrusive igneous) rock Igneous rock that crystallizes from magma intruded into or formed in place within the crust.

point bar The sediment body deposited on the gently sloping side of a meander loop.

polymer A comparatively complex organic molecule, such as a nucleic acid and protein, formed by monomers linking together. (*See* monomer)

porosity The percentage of a material's total volume that is pore space.

porphyritic texture An igneous texture with mineral grains of markedly different sizes.

pressure release A mechanical weathering process in which rocks that formed under pressure expand upon being exposed at the surface.

pressure ridge A buckled area on the surface of a lava flow that forms

because of pressure on the partly solid crust of a moving flow.

primary producer Those organisms in a food chain, such as green plants and bacteria, upon which all other members of the food chain depend directly or indirectly; those organisms not dependent on an external source of nutrients. (*See* autotrophic)

primate Any of the mammals belonging to the order Primates. Includes prosimians (lemurs and tarsiers), monkeys, apes, and humans.

principle of cross-cutting relationships A principle used to determine the relative ages of events; holds that an igneous intrusion or fault must be younger than the rocks it intrudes or cuts.

principle of fossil succession A principle that holds that fossils, and especially assemblages of fossils, succeed one another through time in a regular and determinable order.

principle of inclusions A principle that holds that inclusions, or fragments, in a rock unit are older than the rock unit itself (e.g., granite fragments in a sandstone are older than the sandstone rock unit).

principle of isostasy The theoretical concept of Earth's crust "floating" on a dense underlying layer. (*See* isostatic rebound)

principle of lateral continuity A principle that holds that sediment layers extend outward in all directions until they terminate.

principle of original horizontality A principle that holds that sediment layers are deposited horizontally or very nearly so.

principle of superposition A principle that holds that younger rocks are deposited on top of older layers.

principle of uniformitarianism A principle that holds that we can interpret past events by understanding present-day processes; based on the assumption that natural laws have not changed through time.

prokaryotic cell A type of cell having no nucleus and lacking such organelles as plastids and mitochondria; cells of bacteria and cyanobacteria (blue-green algae) are prokaryotic. (*See* eukaryotic cell)

protorothyrid A loosely grouped category of all early reptiles. Protorothyrids were small animals about the same size and shape as today's lizards.

proton A positively charged particle found in the nucleus of an atom.

punctuated equilibrium An evolutionary concept that holds that a new species evolves rapidly, perhaps in a few thousands of years, then remains much the same during its several millions of years of existence.

P-wave A compressional, or push-pull wave; the fastest seismic wave and one that can travel through solids, liquids, and gases; also known as a primary wave.

P-wave shadow zone The area between 103° and 143° from an earthquake focus where little P-wave energy is recorded by seismographs.

pyroclastic (fragmental) texture A fragmental texture characteristic of igneous rocks composed of pyroclastic materials.

pyroclastic materials Fragmental substances, such as ash, explosively ejected from a volcano.

pyroclastic sheet deposit Vast sheetlike deposit of felsic pyroclastic materials erupted from fissures.

Q

quadrupedal Referring to locomotion on all four legs. (*See* bipedal)

Queenston Delta The clastic wedge resulting from the erosion of highlands formed during the Taconic orogeny; deposited on the west side of the Taconic Highlands.

quick clay A clay that spontaneously liquefies and flows like water when disturbed.

R

radioactive decay The spontaneous decay of an atom to an atom of a different element by emission of a particle from its nucleus (alpha and beta decay) or by electron capture.

rainshadow desert A desert found on the lee side of a mountain range; forms because moist marine air moving inland forms clouds and produces precipitation on the windward side of the mountain range so that the air descending on the leeward side is much warmer and drier.

rapid mass movement A type of mass movement involving a visible movement of material.

Rayleigh wave (R-wave) A surface wave in which the individual particles of material move in an elliptical path within a vertical plane oriented in the direction of wave movement.

recessional moraine A type of end moraine formed when a glacier's terminus retreats, then stabilizes and till is deposited. (*See also* end moraine *and* terminal moraine)

red beds Sedimentary rocks, mostly sandstone and shale, with red coloration due to the presence of ferric oxides.

reef A moundlike, wave-resistant structure composed of the skeletons of organisms.

reflection The return to the surface of some of a seismic wave's energy when it encounters a boundary separating materials of different density or elasticity.

refraction The change in direction and velocity of a seismic wave when it travels from one material into another of different density and elasticity.

regional metamorphism Metamorphism that occurs over a large area resulting from tremendous temperatures, pressures, and the chemical activity of fluids within Earth's crust.

regolith The layer of unconsolidated rock and mineral fragments that covers most of Earth's surface.

relative dating The process of determining the age of an event relative to other events; involves placing geologic events in their correct chronologic order, but involves no consideration of when the events occurred in terms of numbers of years ago. (*See* absolute dating)

reserve That part of the resource base that can be extracted economically.

resource A concentration of naturally occurring solid, liquid, or gaseous material in or on Earth's crust in such form and amount that economic extraction of a commodity from the concentration is currently or potentially feasible.

reverse fault A dip-slip fault in which the hanging wall block has moved upward relative to the footwall block. (*See* normal fault)

Richter Magnitude Scale An open-ended scale that measures the amount of energy released during an earthquake.

rip current A narrow surface current that flows out to sea through the breaker zone.

ripple mark Wavelike (undulating) structure produced in granular sediment such as sand; formed by wind,

unidirectional water currents, or wave currents.

rock An aggregate of one or more minerals, as in limestone and granite, or a consolidated aggregate of rock fragments, as in conglomerate. Although exceptions to this definition, coal and natural glass are also considered rocks.

rock cycle A group of processes through which Earth materials may pass as they are transformed from one rock type to another.

rockfall A common type of extremely rapid mass movement in which rocks of any size fall through the air.

rock-forming mineral A mineral common in rocks, which is important in their identification and classification.

rock slide A type of rapid mass movement in which rocks move downslope along a more or less planar surface.

runoff The surface flow of streams.

S

salt crystal growth A mechanical weathering process in which rocks are disaggregated by the growth of salt crystals in crevices and pores.

saltwater incursion The displacement of freshwater by saltwater as a result of excessive pumping of groundwater in coastal areas.

sandstone-carbonate-shale assemblage An association of sedimentary rocks characteristic of passive continental margins.

Sauk sequence A widespread sequence of sedimentary rocks bounded above and below by unconformities; deposited during a latest Proterozoic to Early Ordovician transgressive-regressive cycle of the Sauk Sea.

Saurischia An order of dinosaurs with a lizardlike pelvis. (*See* Ornithischia)

scientific method A logical, orderly approach that involves data gathering, formulating and testing hypotheses, and proposing theories.

seafloor spreading The theory that the seafloor moves away from spreading ridges and is eventually subducted and consumed at subduction zones.

seamount A structure of volcanic origin rising more than 1 km above the seafloor.

sediment Loose aggregate of solids derived from preexisting rocks, or solids precipitated from solution by inorganic chemical processes, or extracted from solution by organisms.

sediment-deposit feeder Any animal that ingests sediment and extracts the nutrients from it.

sedimentary facies Any aspect of a sedimentary rock unit that makes it recognizably different from adjacent sedimentary rocks of the same, or approximately same age (e.g., a sandstone facies).

sedimentary rock Any rock composed of sediment such as sandstone and limestone.

sedimentary structure Any structure in sedimentary rock formed at or shortly after the time of deposition (e.g., cross-bedding, mud cracks, and animal burrows).

seedless vascular plant A type of land plant with vascular tissues for transport of fluids and nutrients throughout the plant; reproduces by spores rather than seeds (e.g., ferns and horsetail rushes).

seismograph An instrument that detects, records, and measures the various waves produced by an earthquake.

seismology The study of earthquakes.

sequence stratigraphy The study of rock relationships within a time-stratigraphic framework of related facies bounded by widespread unconformities.

Sevier orogeny The Cretaceous phase of the Cordilleran orogeny that affected the continental shelf and slope areas of the Cordilleran mobile belt.

shear strength The resisting forces helping to maintain slope stability.

shear stress The result of forces acting parallel to one another but in opposite directions; results in deformation by displacement of adjacent layers along closely spaced planes.

sheet flow More or less continuous sheet of water moving over a surface. (*See also* channel flow).

shield A vast area of exposed Precambrian rocks on a continent that has been relatively stable for a long period of time (e.g., the Canadian shield). A shield and its adjacent platform constitute a craton.

shield volcano A dome-shaped volcano with a low, rounded profile built up mostly of overlapping basalt lava flows.

shoreline The line of intersection between the sea or a lake and the land.

Siberia One of six major Paleozoic continents; composed of Russia east of the Ural Mountains, and Asia north of Kazakhstan and south of Mongolia.

silica A compound of silicon and oxygen atoms.

silica tetrahedron The basic building block of all silicate minerals. It consists of one silicon atom and four oxygen atoms.

silicate A mineral containing silica.

sill A tabular or sheetlike concordant pluton.

sinkhole A depression in the ground that forms in karst regions by the solution of underlying carbonate rocks or the collapse of a cave roof.

slide A type of mass wasting involving movement of material along one or more surfaces of failure.

slow mass movement Mass movement that advances at an imperceptible rate and is usually only detectable by the effects of its movement.

slump A type of mass wasting that takes place along a curved surface of failure and results in backward rotation of the slump mass.

soil Regolith consisting of weathered material, water, air, and organic matter that can support plants.

soil degradation Any process leading to a loss of soil productivity; may involve erosion, chemical pollution, or compaction.

soil horizon A distinct soil layer that differs from other soil layers in texture, structure, composition, and color.

solifluction A type of mass wasting involving the slow downslope movement of water-saturated surface materials; especially the flow at high latitudes or high elevations where the flow is underlain by frozen soil.

solar nebular theory A theory for the evolution of the solar system from a rotating cloud of gas.

solution A reaction in which the ions of a substance separate from one another in a liquid, and the solid substance dissolves.

Sonoma orogeny A Permian-Triassic orogeny caused by the collision of an island arc with the southwestern margin of North America.

spatter cone A small, steep-sided cone that forms when gases escaping from a lava flow hurl globs of molten lava into the air that fall back to the surface and adhere to one another.

species A population of similar individuals that in nature can reproduce and produce fertile offspring.

spheroidal weathering A type of chemical weathering in which corners and sharp edges of rocks weather more rapidly than flat surfaces, thus yielding spherical shapes.

spit A continuation of a beach forming a point that projects into a body of water, commonly a bay.

spring A place where groundwater flows or seeps out of the ground.

stock An irregularly-shaped discordant pluton with a surface area less than 100 km^2. Many stocks are simply the exposed parts of much larger plutons.

stones A group of meteorites composed of iron and magnesium silicate minerals and comprising about 93% of all meteorites.

stony-irons A group of meteorites composed of nearly equal amounts of iron and nickel and silicate minerals and comprising about 1% of all meteorites.

stoping A process in which rising magma detaches and engulfs pieces of the surrounding country rock.

strain Deformation caused by stress.

strata (singular, stratum) Refers to layers in sedimentary rocks.

stratified drift Glacial drift displaying both sorting and stratification.

stream terrace An erosional remnant of a floodplain that formed when a stream was flowing at a higher level.

stress The force per unit area applied to a material such as rock.

strike The direction of a line formed by the intersection of a horizontal plane with an inclined plane, such as a rock layer.

strike-slip fault A fault involving horizontal movement so that blocks on opposite sides of a fault plane slide sideways past one another.

stromatolite A structure in sedimentary rocks, especially limestones, produced by entrapment of sediment grains on sticky mats of photosynthesizing bacteria; a biogenic sedimentary structure.

strong nuclear force The force that binds protons and neutrons together in the nucleus of an atom.

subduction zone A long, narrow zone at a convergent plate boundary where an oceanic plate descends relative to another plate.

submarine canyon A steep-sided canyon cut into the continental shelf and slope.

submarine fan A cone-shaped sedimentary deposit that accumulates on the continental slope and rise.

submergent coast A coast along which sea level rises with respect to the land or the land subsides.

Sundance Sea A wide seaway that existed in western North America during the Middle Jurassic Period.

superposed stream A stream that once flowed on a higher surface and eroded downward into resistant rocks, while still maintaining its course.

suspended load The smallest particles carried by a stream, such as silt and clay, which are kept suspended by fluid turbulence.

suspension feeder An animal that consumes microscopic plants, animals, or dissolved nutrients from water.

S-wave A shear wave that moves material perpendicular to the direction of travel, thereby producing shear stresses in the material it moves through; also known as a secondary wave, an S-wave only travels through solids.

S-wave shadow zone Those areas more than 103° from an earthquake focus where no S-waves are recorded.

syncline A down-arched fold in which the youngest exposed rocks coincide with the fold axis and all strata dip inward toward the axis.

system A combination of related parts that interact in an organized fashion. Earth systems include the atmosphere, hydrosphere, biosphere, and the solid Earth.

T

Taconic orogeny An Ordovician orogeny that resulted in deformation of the Appalachian mobile belt.

talus Weathered material that accumulates at the bases of slopes.

Tejas epeiric sea A Cenozoic epeiric sea that was largely restricted to the Atlantic and Gulf Coastal plains and parts of coastal California, but did extend into the continental interior in the Mississippi Valley.

tension A type of stress in which forces act in opposite directions but along the same line, thus tending to stretch an object.

terminal moraine A type of end moraine; the moraine most distant from a glacier's source, thus marking the greatest extent of the glacier. (*See also* end moraine *and* recessional moraine)

terrane Small lithospheric block accreted to continents that differs markedly from surrounding rocks.

terrestrial planet Any of the four innermost planets (Mercury, Venus, Earth, and Mars). They are all small and have high mean densities, indicating that they are composed of rock and metallic elements.

theory An explanation for some natural phenomenon that has a large body of supporting evidence; to be considered scientific, a theory must be testable (e.g., plate tectonic theory).

theory of evolution The theory that all organisms are related and that they descended with modification from organisms that lived during the past.

therapsid Permian to Triassic reptiles that possessed mammalian characteristics and thus are called mammal-like reptiles; one group of therapsids, the cynodonts, gave rise to mammals.

thermal convection cell A type of circulation of material in the asthenosphere during which hot material rises, moves laterally, cools and sinks, and is reheated and reenters the cycle.

thermal expansion and contraction A type of mechanical weathering in which the volume of rocks changes in response to heating and cooling.

thrust fault A type of reverse fault in which the fault plane dips less than 45°.

tide The regular fluctuation in the sea's surface in response to the gravitational attraction of the Moon and Sun.

till All sediment deposited directly by glacial ice.

time-distance graph A graph showing the average travel times for P- and S-waves for any specific distance from an earthquake's focus.

time-stratigraphic unit A unit of strata that was deposited during a specific interval of geologic time (e.g., the Devonian System, a time-stratigraphic unit, was deposited during that part of geologic time designated as the Devonian Period).

time unit Any of the units such as eon, era, period, epoch, and age used to refer to specific intervals of geologic time.

Tippecanoe sequence A widespread sequence of sedimentary rocks bound-ed above and below by unconformities; deposited during a Middle Ordovician to Early Devonian transgressive-regressive cycle of the Tippecanoe Sea.

tombolo A type of spit that extends out into the sea and connects an island to the mainland.

trace fossil Any indication of prehistoric organic activity such as tracks, trails, burrows, borings, or nests. (*See* body fossil)

Transcontinental Arch An area consisting of several large islands extending from New Mexico to Minnesota that was above sea level during the Cambrian transgression of the Sauk Sea.

transform fault A type of fault along which one type of motion is transformed into another. Commonly displace oceanic ridges, but movement on opposite sides of the fault between displaced ridge segments is the opposite of the apparent displacement. On land recognized as strike-slip faults, such as the San Andreas fault.

transform plate boundary Plate boundary along which plates slide past one another, and crust is neither produced nor destroyed.

transport The mechanism by which weathered material is moved from one place to another, commonly by running water, wind, or glaciers.

transverse dune A long ridge of sand perpendicular to the prevailing wind direction.

tree-ring dating The process of determining the age of a tree or wood in structures by counting the number of annual growth rings.

trough The lowest point between wave crests.

tsunami A destructive sea wave that is usually produced by an earthquake but can also be caused by submarine landslides or volcanic eruptions.

turbidity current A sediment-water mixture denser than normal seawater that flows downslope to the deep-seafloor.

U

unaltered remains Fossil remains that retain their original composition and structure.

unconformity An erosion surface that separates younger strata from older rocks. (*See* angular unconformity, disconformity, *and* nonconformity)

ungulate An informal term referring to the hoofed mammals, especially the orders Artiodactyla and Perissodactyla.

U-shaped glacial trough A valley with steep or vertical walls and a broad, rather flat floor. Formed by the movement of a glacier through a stream valley.

V

valley A linear depression bounded by higher areas such as ridges, hills, or mountains. Most are eroded by running water, although mass wasting is important in their origin and evolution.

valley glacier A glacier confined to a mountain valley or to an interconnected system of mountain valleys.

valley train A long, narrow deposit of stratified drift confined within a glacial valley.

vascular Land plants possessing specialized tissues for transporting fluids.

velocity A measure of the downstream distance water travels per unit of time. Velocity varies considerably among streams and even within the same stream.

ventifact A stone whose surface has been polished, pitted, grooved, or faceted by wind abrasion.

vertebrate Any animal having a segmented vertebral column; members of the subphylum Vertebrata; includes fishes, amphibians, reptiles, mammals, and birds.

vesicle A small hole or cavity formed by gas trapped in cooling lava.

vestigial structure A body part that serves little or no function (e.g., the dewclaws of dogs); a vestige of a structure or organ that was well developed and functional in some ancestor.

viscosity A fluid's resistance to flow.

volcanic explosivity index (VEI) A semiquantitative scale for the size of volcanic eruptions based on evaluation of such criteria as volume of material explosively ejected and the height of the eruption cloud.

volcanic island arc A curved chain of volcanic islands parallel to a deep-sea trench where oceanic lithosphere is subducted causing volcanism and the origin of volcanic islands.

volcanic neck An erosional remnant of the material that solidified in a volcanic pipe.

volcanic pipe The conduit connecting the crater of a volcano with an underlying magma chamber.

volcanic (extrusive igneous) rock An igneous rock that forms when magma is extruded onto Earth's surface where it cools and crystallizes, or when pyroclastic materials become consolidated.

volcanic tremor Ground motion lasting from minutes to hours resulting from magma moving below the surface.

volcanism The process whereby magma and its associated gases rise through the crust and are extruded onto the surface or into the atmosphere.

volcano A mountain formed around a vent as a result of the eruption of lava and pyroclastic materials.

W

water table The surface separating the zone of aeration from the underlying zone of saturation.

water well A well made by digging or drilling into the zone of saturation.

wave An undulation on the surface of a body of water, resulting in the water surface rising and falling.

wave base A depth of about one-half wavelength, where the diameter of the orbits of water in waves is essentially zero; the depth below which water is unaffected by surface waves.

wave-cut platform A beveled surface that slopes gently seaward; formed by the retreat of a sea cliff.

wave refraction The bending of a wave so that it more nearly parallels the shoreline.

weak nuclear force The force responsible for the breakdown of an atom's nucleus, producing radioactive decay.

weathering The physical breakdown and chemical alteration of rocks and minerals at or near Earth's surface.

Z

zone of accumulation Part of a glacier where additions exceed losses and the glacier's surface is perennially covered by snow.

zone of aeration The zone above the water table that contains both water and air within the pore spaces of the rock or soil.

zone of saturation The zone below the zone of aeration in which all the pore spaces are filled with groundwater.

zone of wastage The part of a glacier where losses from melting, sublimation, and calving of icebergs exceed the rate of accumulations.

Credits

Chapter 1

Figure 1.5: THE FAR SIDE. Copyright © 1991 FARWORKS, INC. Distributed by Universal Press Syndicate. Reprinted with permission. All rights reserved. **Figure 1.13:** Modified from Figure 12, Dietrich, R. V., 1979, *Geology and Michigan: Forty-nine Questions and Answers.*

Chapter 5

Perspective 5.1, Figure 1(a): From *Dust Bowl: The Southern Plains in the 1930s,* by Donald Worster. Copyright © 1979 by Oxford University Press, Inc. Reprinted by permission.

Chapter 6

Figure 6.3a: From C. Gillen, *Metamorphic Geology,* Figure 4.4, p. 73. Copyright © 1982. Reprinted with the kind permission of Kluwer Academic Publishers and C. Gillen. **Figure 6.19:** From H. L. James, *G. S. A. Bulletin,* vol. 66, plate 1, page 1454, with permission of the publisher, the Geological Society of America, Boulder, Colorado, USA. Copyright © 1955 Geological Society of America. **Figure 6.20:** From AGI Data Sheet 35.4, *AGI Data Sheets,* 3rd edition (1989) with the kind permission of the American Geological Institute. **Figure 6.22:** From "Effects of Late Jurassic–Early Tertiary Subduction in California," San Joaquin Geological Society Short Course, 1977, 66, Figure 5–9.

Chapter 7

Figure 7.7: Data from National Oceanic and Atmospheric Administration. **Figure 7.11:** From *Nuclear Explosions and Earthquakes: The Parted Veil,* by Bruce A. Bolt. Copyright © 1976 by W. H. Freeman and Company. Used with permission. **Figure 7.12:** From *Nuclear Explosions and Earthquakes: The Parted Veil,* by Bruce A. Bolt. Copyright © 1976 by W. H. Freeman and Company. Used with permission. **Figure 7.14:** Based on data from C. F. Richter, *Elementary Seismology,* 1958. W. H. Freeman and Company. **Figure 7.17:** From *Earthquakes,* by Bruce A. Bolt. Copyright © 1988 by W. H. Freeman and Company. Used with permission. **Perspective 7.2, Figure 2:** From *Earthquakes,* by Bruce A. Bolt. Copyright © 1988 by W. H. Freeman and Company. Used with permission. **Figure 7.25:** From Figure 6, page 17, *Geotimes Vol. 10, No. 9* (1966) with the kind permission of the American Geological Institute. **Figure 7.28:** From G. C. Brown and A. E. Musset, *The Inaccessible Earth* (London: Chapman & Hall, 1981), Figure 7.11. **Figure 7.31:** From G. C. Brown and A. E. Musset, The Inaccessible Earth (London: Chapman & Hall, 1981), Figure 7.11. **Figure 7.32:** From "Journey to the Center of the Earth," by T. A. Heppenheimer, *Discover,* v. 8, no. 11, Nov. 1987. Illustration by Andrew Christie, copyright © 1987. Reprinted with permission of *Discover Magazine.*

Chapter 8

Figure 8.1: From Phyllis Young Forsyth, *Atlantis: The Making of Myth* (Montreal: McGill–Queen's University Press): 13, Figure 2. **Figure 8.11:** From Alyn and Alison Duxbury, *An Introduction to the World's Oceans.* Copyright © 1984 Addison–Wesley Publishing Company, Inc.

Chapter 9

Figure 9.4: From *General Geology,* 5/e by R. J. Foster. Copyright © 1988. Reprinted by permission of Prentice–Hall, Inc., Upper Saddle River, NJ. **Figure 9.7:** Modified from E. H. Colert, *Wandering Lands and Animals* (1973): 72, Figure 31. **Figure 9.10:** From A. Cox and R. R. Doell, "Review of Paleomagnetism," *G. S. A. Bulletin,* vol. 71, figure 33, page 758, with permission of the publisher, the Geological Society of America, Boulder, Colorado. USA. Copyright © 1955 Geological Society of America. **Figure 9.12:** Reprinted with permission from A. Cox, "Geomagnetic Reversals," *Science 163,* January 17, 1969. Copyright © 1969 American Association for the Advancement of Science. **Figure 9.14:** From Larson, R. L. et al. (1985). *The Bedrock Geology of the World,* W. H. Freeman and Co., New York, NY.

Chapter 10

Figure 10.25: From Peter Molnar, "The Geological History and the Structure of the Himalayas." *American Scientist 74:* 148–149, Figure 4, Journal of Sigma Xi, The Scientific Research Group. **Figure 10.26:** From Zvi Ben-Avraham, "The Movement of Continents." *American Scientist 69:* 291–299, Figure 9, p. 298, Journal of Sigma Xi, The Scientific Research Group.

Chapter 11

Perspective 11.2, Figure 1: From G. A. Kiersch, *"Vaiont Reservoir Disaster,"* Civil Engineering 34 (1964). **Perspective 11.2, Figure 3:** From G. A. Kiersch, *"Vaiont Reservoir Disaster,"* Civil Engineering 34 (1964).

Chapter 13

Figure 13.1: From *Trapped* by Robert K. Murray and Roger W. Brucker. Copyright © 1979 by Murray and Brucker, copyright renewed. **Figure 13.2:** Modified from *U.S. News and World Report* (18 March 1991): 72–73. **Figure 13.18:** From J. B. Weeks et al., *U. S. Geological Survey Professional Paper 1400-A,* 1988. **Perspective 13.1, Figure 1:** From USGS Internet site: http://co.water.usgs.gov/trace/pubs/fs-063-00/fig1. **Perspective 13.1, Figure 2:** From USGS Internet site: http://co.water.usgs.gov/trace/pubs/fs-063-00/fig3.

Chapter 17

Figure 17.1: Modified from A. R. Palmer, "The Decade of North American Geology, 1983 Geologic Time Scale." *Geology* (Geological Society of America, 1983), p. 504. **Figure 17.19:** Based on data from S. M. Richardson and H. Y. McSween, Jr., *Geochemistry—Pathways and Processes,* Prentice–Hall. **Figure 17.25:** From E. K. Ralph, H. N. Michael, and M. C. Han, "Radiocarbon Dates and Reality," *MASCA Newsletter* 9 (1973), page 5, figure 8. Used by permission. **Figure 17.26:** From *An Introduction to Tree-Ring Dating,* by Stokes and Smiley, 1968, p. 6. Reprinted by permission of University of Chicago Press.

Chapter 18

Figure 18.10: Reprinted with permission from Starr and Taggart, *Biology: The Unity and Diversity of Life,* 1989, p. 556. Copyright © Wadsworth Publishing Company. **Figure 18.15b:** Reprinted with permission from O. C. Marsh, "Recent Polydactyle Horses," *American Journal of Science,* V. XLIII, no. 256, 1892, p. 343, Fig 6. **Figure 18.18:** Reprinted with permission from R. L. Carroll, *Patterns and Processes of Vertebrate Evolution,* p. 156 (Fig 7.4). Copyright © 1988 Cambridge University Press.

Chapter 20

Table 20.1: Reprinted with permission from H. L. James, "Stratigraphic Commission: Note 40—Subdivision of Precambrian: An Interim Scheme to be Used by USGS," *AAPG,* v. 56, no. 6, 1972, p. 1130 (Table 1), and from Harrison and Peterman, "North American Commission on Stratigraphic Nomenclature: Report 9—Adoption of Geochronometric Units for Division of Precambrian Time," *AAPG,* v. 66, no. 6, 1982, p. 802 (Fig 1), American Association of Petroleum Geologists. **Figure 20.3:** Reprinted with permission from A.M. Goodwin, "The Most Ancient Continental Margins," in *The Geology of Continental Margins,* Burk and Drake (eds), 1974, p. 768 (Fig 1), © Springer–Verlag. **Figure 20.6:** Reprinted from K. C. Condie, *Plate Tectonics and Crustal Evolution,* 2nd ed., © 1982, p. 87 (Fig. 5.10). **Figure 20.7:** Reprinted with permission from Dickinson and Luth, "A Model for Plate Tectonic Evolution of Mantle Layers," *Science,* v. 174, no. 4007, 1971, p. 402 (Fig 1), © 1992 by the AAAS. **Table 20.3:** Reprinted with permission from Kontinen, *Precambrian Research,* © 1987, Elsevier Science Publishers. **Figure 20.8:** Reprinted from K. C. Condie, *Plate Tectonics and Crustal Evolution,* 4th edition, p. 65 (Fig. 2.26), copyright © 1997 Butterworth–Heinemann. **Figure 20.9:** Reprinted with permission from J. L. Anderson "Proterozoic Anorogenic Granite Plutonism of North America," Proterozoic Geology: Selected Papers from an International Symposium, p. 135 (Fig 1). **Figure 20.12a:** Reprinted with permission from G. M. Young, "Tectono-Sedimentary History of Early Proterozoic Rocks of the Northern Great Lakes Region," in *Early Proterozoic Rocks Geology of the Northern Great Lakes Region,* Medaris (ed.), GSA Memoir 160, 1983, p. 16 (Fig. 1), Geological Society of America. **Figure 20.14:** Reprinted from L. A. Frakes, *Climates Throughout Geologic Time,* p. 39. Copyright © 1979 Elsevier Science Publishers. **Figure 20.18:** From S. L. Miller, "The Formation of Organic Compounds on the Primitive Earth," in *Modern Ideas of Spontaneous Generation,* Nigrelli (ed.), Annuals of the New York Academy of Sciences, v. 69, Art. 2, Aug, 30, 1957, p. 261 (Fig 1). **Figure 20.20:** From R. L. Anstey and T. L. Chase, *Environments Through Time.* Copyright © 1974 Macmillan Publishing Company. **Table 20.4:** Reprinted with permission from Kontinen, *Precambrian Research,* © 1987, Elsevier Science Publishers. **Figure 20.22:** From Lynn Margulis, "Symbiosis and Evolution," copyright © 1971 by Scientific American, Inc. All rights reserved. **Figure 20.25:** Reprinted with permission from Kontinen, *Precambrian Research,* © 1987, Elsevier Science Publishers. **Perspective 20.1, Figure 1:** From *Symbiosis in Cell Evolution: Life and its Environment on the Early Earth,* by Lynn Margulis. Copyright © 1981 by W. H. Freeman and Company. Reprinted by permission.

Chapter 21

Figure 21.2: Based on data from The Geological Society Publishing House and American Scientist. **Figure 21.3:** Based on data from The Geological Society Publishing House and American Scientist. **Figure 21.4:** Based on data from The Geological Society Publishing House and American Scientist. **Figure 21.11a:** From K. J. Mesolella, J. D. Robinson, L. M. McCormick, and A. R. Ormiston, "Cyclic Deposition of Silurian Carbonates and Evaporites in Michigan Basin," *AAPG Bulletin,* Vol. 58, No. 1, AAPG © 1958, Fig 6, p. 40. Reprinted by permission of the American Association of Petroleum Geologists whose permission is required for future use. **Figure 21.30:** Based on data from the U.S. Geological Survey.

Chapter 22

Figure 22.3: Reproduced with permission from H. Tappan, "Proceedings of the North American Paleontological Convention, Part H," E. Yochelson (ed.), p. 1064 (Fig. 2), 1970. **Figure 22.12:** Reproduced with permission from "Mass Extinction in the Marine Fossil Record," by D. M. Raup and J. J. Sepkoski, *Science,* Vol 215, p. 1502 (Fig 2). Copyright © 1982 American Association for the Advancement of Science. **Figure 22.13:** Reproduced with permission from "The Evolution of Early Land Plants," by P. Genseld and S. Andrews, *American Scientist,* Vol 75, no. 5, p. 480 (Fig 2). Copyright © 1987 American Scientist.

Chapter 23

Figure 23.2: Reproduced with permission from R. S. Dietz and J. C. Holden, *Journal of Geophysical Research,* Vol 75, no. 26, pp. 4939–4956. Copyright © 1970 by the American Geophysical Union. **Figure 23.3:** Reproduced with permission from K. Burke, *Geology,* Vol 3, no. 614, p. 614. Copyright © 1975 by the Geological Society of America. **Figure 23.10:** Reprinted with permission from R. A. Schweickert and D. S. Cowan, "Tectonic Evolution of the Western Sierra Nevada, California," in *GSA Bulletin* 86, 1975, p. 1334 (Fig. 3), Geological Society of America. **Figure 23.29:** Reproduced with permission from J. S. Hobson, "The Mammal-Like Reptiles: A Study of Transitional Fossils," *The American Biology Teacher,* v. 49, no. 1, 1987, pp. 18–22. National Association of Biology Teachers, Reston, Virginia. **Figure 23.30:** Reproduced with permission from J. S. Hobson, "The Mammal-Like Reptiles: A Study of Transitional Fossils," *The American Biology Teacher,* v. 49, no. 1, 1987, pp. 18–22. National Association of Biology Teachers, Reston, Virginia. **Figure 23.31:** Adapted with permission from J. Benes, *Prehistoric Animals and Plants,* 1979, pp. 174–175, illustration by Zdenek Burian, published by Hippocrene Books, Inc.

Chapter 24

Figure 24.4: Reproduced with permission from J. B. Murphy, G. L. Oppliger, G. H. Brimhall, Jr., and A. Hynes, "Mantle Plumes and Mountains, *American Scientist,* Vol 87, no. 2, p. 152 (March–April 1999). **Figure 24.5:** From D. L. Blackstone, Jr., *Traveler's Guide to the Geology of Wyoming:* Geological Survey of Wyoming Bulletin 67, 1988, pp. 43–44. **Figure 24.9:** Reprinted with permission from W. R. Dickinson, "Cenozoic Plate Tectonic Setting of the Cordilleran Region in the Western United States," *Cenozoic Paleogeography of the Western United States,* Pacific Coast Symposium 3, 1979, p. 2 (Fig 1). **Figure 24.11:** Reprinted with permission from S. W. Lowman, "Sedimentary Facies in the Gulf Coast," *AAPG* v. 33, no 12 1949, p. 1972 (Fig 23). Reprinted by permission of the American Association of Petroleum Geologists whose permission is required for future use. **Figure 24.13:** Columbia University Press, 562 West 113th Street, New York, New York 10025. **Figure 24.14:** Reprinted with permission from Grow and Sheridan, "U.S. Continental Margin," *The Geology of North America,* v. 1–2, p. 2–4, USGS. **Figure 24.22:** From Alfred S. Romer, *The Vertebrate Body,* 3rd ed., Fig 224, copyright © 1962 by Saunders College Publishing and renewed 1980 by Alfred S. Romer, reproduced by permission of the publisher. **Figure 24.23:** Reprinted with permission from *Mammalian Evolution in Major Features of Vertebrate Evolution,* Short Course on Paleontology, no. 7, Dr. R. Prothero and R. M. Schoch (eds.). **Figure 24.26:** Reprinted with permission from *Mammalian Evolution in Major Features of Vertebrate Evolution,* Short Course on Paleontology, no. 7, Dr. R. Prothero and R. M. Schoch (eds.).

Photo Credits

Chapter 1

Chapter Opener: NASA. **Figure 1.3:** © Jean Miele/The Stock Market. **Figure 1.3:** Collection of the New York Public Library, Astor, Lenox, and Tilden Foundations. **Figure 1.6:** Photo composite by Elliott Hill. **Figure 1.14a–f:** Courtesy of Sue Monroe. **Perspective 1.1, Figure 1:** NASA. **Perspective 1.1, Figure 2:** Black Star Publishing Co., Inc., © 1990 David Turnley.

Chapter 2

Chapter Opener: Los Angeles County Museum Specimen, © Harold and Erica Van Pelt. **Figure 2.1:** National Museum of Natural History (NMNH) Specimen #G7101. Photo by D. Penland, courtesy of Smithsonian Institution. **Figure 2.2a:** Jerry Jacka. **Figure 2.2b:** Courtesy of Sue Monroe. **Figure 2.3:** James S. Monroe. **Figure 2.13a–b:** Courtesy of Sue Monroe. **Figure 2.14a–d:** Courtesy of Sue Monroe. **Figure 2.15a–b:** Courtesy of Sue Monroe. **Figure 2.16a–c:** Courtesy of Sue Monroe. **Figure 2.19a–c:** Courtesy of Sue Monroe. **Figure 2.20:** © Gregory G. Dimijian 1992, Photo Researchers, Inc. **Figure 2.20a/inset:** Courtesy of Sue Monroe. **Figure 2.21b/inset:** Courtesy of Sue Monroe. **Figure 2.22a:** Courtesy of Cleveland Clifts Iron Company. **Figure 2.22b:** Courtesy of Sue Monroe. **Perspective 2.1, Figure 1a:** National Museum of Natural History (NMNH) Specimen #R121297. Photo of D. Penland, courtesy of Smithsonian Institution. **Perspective 2.1, Figure 1b:** © Ken Lucas/Visuals Unlimited. **Perspective 2.1, Figure 1(c):** © E. R. Degginger, Photo Researchers, Inc. **Perspective 2.1, Figure 2a:** James S. Monroe. **Perspective 2.1, Figure 2b:** Courtesy of Sue Monroe. **Perspective 2.2, Figure 1a:** National Museum of Natural History, (NMNH) Specimen # R12804, Photo by Chip Clark, courtesy of Smithsonian Institution. **Perspective 2.2, Figure 1b–g:** Courtesy of Sue Monroe.

Chapter 3

Chapter Opener: © Richard Thom/Visuals Unlimited. **Figure 3.1a:** Courtesy of Steve Stahl. **Figure 3.1b:** Courtesy of Sue Monroe. **Figure 3.1c:** © Alissa Crandall/CORBIS. **Figure 3.2:** P. Mouginis-Mark. **Figure 3.6b:** James S. Monroe. **Figure 3.8b/d-f:** Courtesy of Sue Monroe. **Figure 3.9a–b:** Courtesy of Sue Monroe. **Figure 3.9c:** James S. Monroe. **Figure 3.11:** Courtesy of Sue Monroe. **Figure 3.12a–b:** Courtesy of Sue Monroe. **Figure 3.13a–b:** Courtesy of Sue Monroe. **Figure 3.14a–b:** Courtesy of Sue Monroe. **Figure 3.15a:** Courtesy of Steve Stahl. **Figure 3.15b:** Courtesy of Sue Monroe. **Figure 3.17a:** Courtesy of David J. Matty. **Figure 3.17b–d:** Courtesy of Sue Monroe. **Figure 3.19:** © Martin R. Miller/Visuals Unlimited. **Figure 3.20a:** Courtesy of Richard L. Chambers. **Figure 3.20b:** Courtesy of Sue Monroe. **Perspective 3.1, Figure 1:** Courtesy of Wendell E. Wilson. **Perspective 3.1, Figure 2:** W. T. Schaller/USGS. **Perspective 3.2, Figure 1:** © Michael Nicholson/CORBIS. **Perspective 3.2, Figure 2a:** Courtesy of Sue Monroe. **Perspective 3.2, Figure 2b:** Courtesy of Frank Hannah. **Perspective 3.2, Figure 3:** James S. Monroe.

Chapter 4

Chapter Opener: J. D. Griggs/USGS. **Figure 4.1b:** © 1995 Stephen Cottrell. **Figure 4.2a–b:** Courtesy of Sue Monroe. **Figure 4.3a:** Courtesy of Hawaii Volcanoes National Park/USGS. **Figure 4.3b:** K. V. Cashman/USGS. **Figure 4.4a:** J. D. Griggs/USGS. **Figure 4.4b:** J. D. Stokes/USGS. **Figure 4.5a:** T. J. Takahashi/USGS. **Figure 4.5b:** G. Brad Lewis/Omjalla Images. **Figure 4.6:** Courtesy of Sue Monroe. **Figure 4.7a:** Reproduced by permission of Marie Tharp, 1 Washington Avenue, South Nyack, NY 10960. **Figure 4.7b:** James S. Monroe. **Figure 4.8:** Courtesy of Sue Monroe. **Figure 4.9:** Courtesy of Sue Monroe. **Figure 4.10e:** Courtesy of Sue Monroe. **Figure 4.11b:** James S. Monroe. **Figure 4.11c:** James S. Monroe. **Figure 4.12a–b:** James S. Monroe. **Figure 4.12c:** Solarfilma/GeoScience Features. **Figure 4.13b:** R. Solkoski, Consulting Geologists, Vancouver, WA. **Figure 4.13c:** D. R. Crandel/USGS. **Figure 4.14a:** Reuters/CORBIS. **Figure 4.14b:** U.S. Department of Interior, USGS, David A. Johnston, Cascades Volcano Observatory, Vancouver, WA. **Figure 4.15:** Courtesy of Sue Monroe. **Figure 4.16a:** Bettmann/CORBIS. **Figure 4.16b:** Neg./Transparency no. 256108 (photo by E. O. Hovey), Courtesy Department of Library Services, American Museum of Natural History. **Figure 4.17a:** Courtesy of Ward's Natural Science Est., Inc. **Figure 4.18:** W. E. Scott/USGS. **Perspective 4.1, Figure 1:** © Richard Cummins/CORBIS. **Perspective 4.1, Figure 2a, Figure 2b:** James S. Monroe. **Perspective 4.2, Figure 1b:** Courtesy of Keith Ronnholm. **Perspective 4.2, Figure 1c, Figure 2:** © Stone/Richard During.

Chapter 5

Chapter Opener: © Bill Ross/CORBIS. **Figure 5.1:** Courtesy of Sue Monroe. **Figure 5.2a–b:** Courtesy of Sue Monroe. **Figure 5.3:** Courtesy of Sue Monroe. **Figure 5.4b:** James S. Monroe. **Figure 5.5a:** Courtesy of Sue Monroe. **Figure 5.5b:** © Mark Gibson/Visuals Unlimited. **Figure 5.6:** Courtesy of W. D. Lowry. **Figure 5.7:** James S. Monroe. **Figure 5.9:** James S. Monroe. **Figure 5.11d:** James S. Monroe. **Figure 5.14:** © Walt Anderson/Visuals Unlimited. **Figure 5.15a:** James S. Monroe. **Figure 5.15b:** H. H. Waldron/USGS. **Figure 5.16:** © Science VU/Visuals Unlimited. **Figure 5.17a:** Courtesy of R. V. Dietrich. **Figure 5.17b:** Courtesy of Sue Monroe. **Figure 5.20a–d:** Courtesy of Sue Monroe. **Figure 5.21a–b:** Courtesy of Sue Monroe. **Figure 5.21c:** Courtesy of Rex Elliott. **Figure 5.22a–d:** Courtesy of Sue Monroe. **Figure 5.25a–b:** James S. Monroe. **Figure 5.27b:** James S. Monroe. **Figure 5.27d:** James S. Monroe.

Figure 5.28a: Courtesy of R. V. Dietrich. Figure 5.28b: Alan L. Mayo, GeoPhoto Publishing Company. Figure 5.29a-b: Courtesy of Sue Monroe. Figure 5.30: James S. Monroe. Figure 5.31: Alan L. Mayo, GeoPhoto Publishing Company. Perspective 5.1, Figure 1b: The Granger Collection, Ltd., New York. Perspective 5.1, Figure 2: Bettmann/CORBIS. Perspective 5.2, Figure 1a: Courtesy of Sue Monroe. Perspective 5.2, Figure 1b: Courtesy Mike Johnson. Perspective 5.2, Figure 1c: Courtesy of Sue Monroe. Perspective 5.2, Figure 2a-b: Courtesy of Sue Monroe. Perspective 5.2, Figure 3a: Martin Land, Science Photo Library/Photo Researchers, Inc. Perspective 5.2, Figure 3b: Courtesy of Sue Monroe.

Chapter 6
Chapter Opener: Gary Hobart/GEOImagery. Perspective 6.1, Figure 1: Collection of the J. Paul Getty Museum, Malibu, CA. Figure 6.1: Collection of the J. Paul Getty Museum, Malibu, CA. Figure 6.3b: Courtesy of David J. Matty & Jane M. Matty. Figure 6.4: Courtesy of Eric Johnson. Figure 6.6: Courtesy of David J. Matty. Figure 6.7: James S. Monroe. Figure 6.8: Courtesy of Eric Johnson. Figure 6.10b: Reed Wicander. Figure 6.11a: Courtesy of Sue Monroe. Figure 6.11b: Courtesy of R. V. Dietrich. Figure 6.11c: James S. Monroe. Figure 6.12: Courtesy of Sue Monroe. Figure 6.13a-b: Courtesy of Sue Monroe. Figure 6.14: Reed Wicander. Figure 6.15: Ed Bartram, courtesy of R. V. Dietrich. Figure 6.16: Reed Wicander. Figure 6.17a-b: Courtesy of Sue Monroe. Figure 6.18a-b: Courtesy of Sue Monroe. Figure 6.23: Peter Hulme/© Ecoscene/CORBIS.

Chapter 7
Chapter Opener: REUTERS/Savita Kirloskar. © Reuters New Media Inc./CORBIS. Figure 7.1b: AP/Wide World Photos/Murad Sezer. Figure 7.2: Bettmann/CORBIS. Figure 7.3b: James S. Monroe. Figure 7.4: Reproduced by the Trustees of the Science Museum, London. Figure 7.5a: James S. Monroe, graphic work by Precision Graphics. Figure 7.9a: Courtesy of Dennis Fox. Figure 7.9b: © Bunyo Ishikawa/Sygma/CORBIS. Figure 7.10: Courtesy of South Caroliniana Library, University of South Carolina, Columbia. Perspective 7.1, Figure 1a-b: James S. Monroe. Perspective 7.1, Figure 2: Steinbrugge Collection, Earthquake Engineering Research Center, University of California, Berkeley. Perspective 7.1, Figure 3a: Don Bloomer/Time. Perspective 7.1, Figure 3b: © Ted Soqui/Sygma/CORBIS. Perspective 7.1, Figure 3c: © Lee Stone/Sygma/CORBIS. Perspective 7.2, Figure 3: M. Celebi/USGS. Figure 7.19: National Geophysical Data Center. Figure 7.20: Courtesy of ChinastockPhoto. Figure 7.21: Martin E. Klimek, *Marin Independent Journal.* Figure 7.22: Courtesy of Bishop Museum. Figure 7.23a: James S. Monroe. Figure 7.23b: Reed Wicander.

Chapter 8
Chapter Opener: © Woods Hole Oceanographic Institute. Figure 8.2: Painting by Lloyd K. Townsend, copyright © National Geographic Society. Figure 8.5a: Ocean Drilling Program, Texas A & M University. Figure 8.5b: Visuals Unlimited/WHOI/R. Catarach Figure 8.16a: Institute of Oceanographic Sciences/Nerc/Science Library/Photo Researchers, Inc. Figure 8.16b: Dr. James Andrews, University of Hawaii. Figure 8.17: © 1990 Carl Roessler. Figure 8.19a: Courtesy of Sue Monroe. Figure 8.19b: © Douglas Faulkner, Science Source/Photo Researchers, Inc. Perspective 8.1, Figure 1a: Peter Ryan/Scripps/Science Photo Library/Photo Reseachers, Inc. Perspective 8.1, Figure 1b: Visuals Unlimited/WHOI/D. Foster. Perspective 8.2, Figure 2: Courtesy of T. Scott, Florida Geological Survey.

Chapter 9
Chapter Opener: NASA. Figure 9.1a-b: Courtesy of Patricia G. Gensel, University of North Carolina. Figure 9.2: Bildarchiv Preussischer Kulterbesitz. Figure 9.6a: Courtesy of Scott Katz. Figure 9.8: Brian A. Roberts, Courtesy of Reed Wicander. Figure 9.13: Courtesy of ALCOA. Figure 9.17: Reproduced by permission of Marie Tharp, 1 Washington Avenue, South Nyack, NY 10960. Figure 9.19b: © Robert Caputo/Aurora. Figure 9.20b: © Stone/Japan satellite. Figure 9.21b: NASA. Figure 9.22b: NASA. Figure 9.25 inset: © Stone/James Balogs. Figure 9.30: Courtesy of R. V. Dietrich.

Chapter 10
Chapter Opener: James S. Monroe. Figure 10.1a: James S. Monroe, Graphic work by Precision Graphics. Figure 10.2a: Courtesy of Steve Stahl. Figure 10.2b: James S. Monroe. Figure 10.5: Courtesy of John S. Shelton. Figure 10.6a: Courtesy of Sue Monroe. Figure 10.6b: James S. Monroe. Figure 10.8b: Courtesy of Kevin O'Brien. Figure 10.9: Reed Wicander. Figure 10.11: Martin F. Schmidt, Jr. Figure 10.14c: Courtesy of John S. Shelton. Figure 10.16a: © Galen Rowell/Peter Arnold, Inc. Figure 10.16b: Courtesy of C. G. Tillman. Figure 10.16c: James S. Monroe. Figure 10.17b: Courtesy of David J. Matty. Figure 10.17c: James S. Monroe. Figure 10.19a: Courtesy of Sue Monroe. Figure 10.19b: © Martin Miller/Visuals Unlimited. Figure 10.21b: Courtesy of Sue Monroe.

Perspective 10.2, Figure 1c-d: James S. Monroe. Perspective 10.2, Figure 1e: James S. Monroe.

Chapter 11
Chapter Opener: AP/Wide World Photos. Figure 11.3c: Reed Wicander. Figure 11.4: James S. Monroe. Figure 11.5d: Courtesy of R. V. Dietrich. Figure 11.6: Boris Yaro, *Los Angeles Times.* Figure 11.8: © Rod Rolle/Liaison Agency, Inc. Figure 11.10: W. R. Hansen/USGS. Figure 11.11a: Courtesy of Sue Monroe. Figure 11.11b: James S. Monroe. Figure 11.13: Courtesy of John S. Shelton. Perspective 11.1, F1: T. Spencer/Colorific. Figure 11.15a: Courtesy of Eleanora I. Robbins, USGS. Figure 11.15b: Stephen R. Lower, GeoPhoto Publishing Company. Figure 11.16b: B. Bradley and the University of Colorado's Geology Department, National Geophysical Data Center, NOAA, Boulder, CO. Figure 11.17: James S. Monroe. Figure 11.18: B. Pipkin, University of Southern California. Figure 11.19b: Reed Wicander. Figure 11.20: Canadian Air Force. Figure 11.21b: Photograph from "Alaska Earthquake Collection," no. 43ct/ USGS. Figure 11.22b: B. Bradley and the University of Colorado's Geology Department. National Geophysical Data Center, NOAA, Boulder, CO. Figure 11.23: O. J. Ferrains, Jr./USGS. Figure 11.24b-c: B. Bradley and the University of Colorado's Geology Department. National Geophysical Data Center, NOAA, Boulder, CO. Figure 11.24d: Courtesy of David J. Matty. Figure 11.26: George Plafker, USGS. Figure 11.28b: Reed Wicander. Figure 11.30b: © John D. Cunningham/Visuals Unlimited. Figure 11.31b: © Dell R. Foutz/Visuals Unlimited. Figure 11.32b: Reed Wicander. Perspective 11.2, Figure 2: Bettmann/CORBIS.

Chapter 12
Chapter Opener: Courtesy of Sue Monroe. Figure 12.1a: © Schenectady Museum; Hall of Electrical History/CORBIS. Figure 12.1b: Bettman/CORBIS. Figure 12.2a-b: Security Pacific Collection/Los Angeles Public Library. Figure 12.3: © Lowell Georgia/Science Source/Photo Researchers, Inc. Figure 12.9a: Courtesy of Sue Monroe. Figure 12.9b: Courtesy of R. V. Dietrich. Figure 12.10a: James S. Monroe. Figure 12.10b: Courtesy of Sue Monroe. Figure 12.12a-b: Courtesy of Sue Monroe. Figure 12.13: Courtesy of John S. Shelton. Figure 12.14b-c: James S. Monroe. Figure 12.18b: James S. Monroe. Figure 12.20: Courtesy of John S. Shelton. Figure 12.21: © 1997 Jamie Sabau. Figure 12.22: Courtesy of Sue Monroe. Figure 12.23a: James S. Monroe. Figure 12.25b: Courtesy of Kentucky Department of Parks. Figure 12.28a-b: Courtesy of Sue Monroe. Figure 12.31d: J. R. Stacey/USGS. Figure 12.32: Courtesy of John S. Shelton. Perspective 12.2, Figure 1a: Courtesy of Michael Lawton. Perspective 12.2, Figure 1b: Courtesy of Sue Monroe.

Chapter 13
Chapter Opener: © John Elk III/Stock, Boston/PictureQuest. Figure 13.1c: Brown Brothers. Figure 13.5: Courtesy of Nassau Department of Public Works. Figure 13.7a: Courtesy of Sue Monroe. Figure 13.7b: John A. Karachewski / geoscapseshotography.com. Figure 13.8a-b: Courtesy of Sue Monroe. Figure 13.10: J. R. Stacey, USGS. Figure 13.13a: Frank Kujawa, University of Central Florida, GeoPhoto Publishing Company. Figure 13.13b: James S. Monroe. Figure 13.15a: Reed Wicander. Figure 13.15b: Courtesy of John S. Shelton. Figure 13.17: © Danial W. Gotshall/Visuals Unlimited. Figure 13.21: USGS. Figure 13.22: © Stone/Sarah Stone. Figure 13.23: Courtesy of R. V. Dietrich. Figure 13.24: City of Long Beach Department of Oil Properties. Figure 13.26a: Reed Wicander. Figure 13.26b-d: James S. Monroe. Figure 13.27: British Tourist Authority. Figure 13.28a-b: James S. Monroe. Figure 13.30a-b: Reed Wicander. Perspective 13.1, Figure 1: USGS. Perspective 13.1, Figure 2: USGS.

Chapter 14
Chapter Opener: © Ric Ergenbright/CORBIS. Figure 14.1: Oeffentliche Kunstammlung Basel, Martin Bühler. Figure 14.4: Engineering Mechanics, Virginia Polytechnic Institute and State University. Figure 14.5: © Frank Awbrey/Visuals Unlimited. Figure 14.8b: National Park Service photograph by Ruth and Louis Kirk. Figure 14.9: James S. Monroe. Figure 14.10a-c: James S. Monroe. Figure 14.12a: James S. Monroe. Figure 14.12b: © Brian Vikander/CORBIS. Figure 14.13: © Phil Schermeister/CORBIS. Figure 14.14a: Courtesy of Sue Monroe. Figure 14.14b: Courtesy of John S. Shelton. Figure 14.14c: Courtesy Frank Hanna. Figure 14.15: Switzerland Tourism. Figure 14.16: Alan Kesselhein/Mary Pat Ziter, © JLM Visuals. Figure 14.17: Courtesy of R. V. Dietrich. Figure 14.19a-b: Courtesy of Sue Monroe. Figure 14.20a: Engineering Mechanics, Virginia Polytechnic Institute. Figure 14.20b: © Peter Kresan. Figure 14.21: Courtesy Carl Guell Slide Collection, Department of Geography, University of Wisconsin, Oshkosh. Figure 14.23a: Courtesy of John S. Shelton. Figure 14.23b: Courtesy of B. M. C. Pape. Figure 14.23c: © Tom Bean/CORBIS. Figure 14.24: Courtesy of Canadian Geological Survey. Perspective 14.1, Figure 1a: James S. Monroe. Perspective 14.1, Figure 1b: © Ron Watts/CORBIS. Perspective 14.1, Figure 1c: Courtesy of P. Weiss, USGS.

Perspective 14.1, Figure 2b-c: Courtesy of James S. Monroe. **Perspective 14.1, Figure 3:** Carl H. Key, Northern Rocky Mountain Science Center, Glacier Field Station/USGS.

Chapter 15

Chapter Opener: Courtesy of Edward Mann. **Figure 15.1:** © George Gerster/The National Audubon Society/Photo Researchers, Inc. **Figure 15.3:** © Martin G. Miller/Visuals Unlimited. **Figure 15.4:** © O. Alamany & E. Vicens/CORBIS. **Figure 15.5b:** © Martin G. Miller/Visuals Unlimited. **Figure 15.6:** Courtesy of Marion A. Whitney. **Figure 15.7:** © Martin G. Miller/Visuals Unlimited. **Figure 15.8c:** Courtesy of David J. Matty. **Figure 15.9:** Digital Imagery © copyright 1999 PhotoDisc, Inc. **Figure 15.11:** Reed Wicander. **Figure 15.12b:** Courtesy of John S. Shelton. **Figure 15.13b:** © 1994 CNES, Provided by SPOT Image Corporation. **Figure 15.14b:** © W. J. Weber/Visuals Unlimited. **Figure 15.15:** © Stone/Willard Clay. **Figure 15.16b:** Reed Wicander. **Figure 15.17b:** © Nigel J. Dennis/Photo Researchers, Inc. **Figure 15.18:** © Steve McCutcheon/Visuals Unlimited. **Figure 15.23:** © Charlie Ott, The National Audobon Society Collection/Photo Researchers, Inc. **Figure 15.24:** James S. Monroe. **Figure 15.25a-b:** Courtesy of John S. Shelton. **Figure 15.26:** © Martin G. Miller/Visuals Unlimited. **Figure 15.27:** Alan L. and Linda D. Mayo, GeoPhoto Publishing Company. **Figure 15.28b:** Reed Wicander. **Figure 15.29a:** © Adriel Heisey. **Figure 15.29b:** © Tom Bean/CORBIS. **Perspective 15.1, Figure 2a-b:** Reed Wicander.

Chapter 16

Chapter Opener: Courtesy of James S. Monroe. **Figure 16.1a:** Rosenberg Library, Galveston, Texas. **Figure 16.1b:** Rosenberg Library, Galveston, Texas. **Figure 16.2a:** NASA. **Figure 16.2b:** © Tom Till. **Figure 16.3a:** Courtesy of Karl Kuhn. **Figure 16.3b:** Courtesy of Karl Kuhn. **Figure 16.5b-c:** James S. Monroe. **Figure 16.6b-c:** Tom Servais, *Surfer* Magazine. **Figure 16.7:** Courtesy of Sue Monroe, graphic work by Precision Graphics. **Figure 16.8:** Courtesy of John S. Shelton. **Figure 16.9a:** James S. Monroe. **Figure 16.9b:** Courtesy of Michael Slear. **Figure 16.10b:** James S. Monroe. **Figure 16.11b:** Courtesy of John S. Shelton. **Figure 16.12b-c:** James S. Monroe. **Figure 16.13b-c:** James S. Monroe. **Figure 16.14b-c:** Courtesy of Sue Monroe. **Figure 16.15a:** USGS, EROS Data Center, Sioux Falls, SD. **Figure 16.15b:** © Richard Stockton 1998. Duke Photo Department. **Figure 16.18b:** Courtesy of John S. Shelton. **Figure 16.19:** James S. Monroe. **Figure 16.20b:** Courtesy of Associate Professor Nick Harvey, University of Adelaide, South Australia. **Figure 16.20c:** James S. Monroe. **Figure 16.22a-b:** U.S. Army Corps of Engineers. **Figure 16.23a:** Courtesy of Sue Monroe. **Figure 16.23b:** Courtesy of Dr. Stanley R. Riggs, Department of Geology, Graham Bldg. East Carolina University, Greenville, NC 27858 Tel. 252/328-6015, Fax 252/328-4391. **Figure 16.24:** GEOPIC, Earth Satellite Corporation. **Figure 16.25:** Courtesy of Sue Monroe. **Perspective 16.1, Figure 2:** © La médiathèque EDF/Gerard Halary.

Chapter 17

Chapter Opener: Reed Wicander. **Figure 17.2:** Reed Wicander. **Figure 17.3a:** James S. Monroe. **Figure 17.3b:** Reed Wicander. **Figure 17.4c:** Reed Wicander. **Figure 17.5c:** James S. Monroe. **Figure 17.8b:** James S. Monroe. **Figure 17.9b:** James S. Monroe. **Figure 17.10b:** James S. Monroe. **Figure 17.14a-c:** Reed Wicander. **Figure 17.23:** Courtesy of Charles W. Naeser/USGS. **Figure 17.28:** James S. Monroe.

Chapter 18

Chapter Opener: top, © Francois Gohier/Photo Researchers, Inc.; **bottom,** The Field Museum of Natural History Chicago, and the artist, Charles R. Knight. Neg. #CKL21T. **Figure 18.2:** © Historical Pictures/Stock Montage. **Figure 18.15a:** Reed Wicander. **Figure 18.16:** © M. W. Tweedie/Photo Researchers, Inc. **Figure 18.17a-b/d-f:** Courtesy of Sue Monroe. **Figure 18.17c:** Courtesy of Wayne E. Moore.

Chapter 19

Chapter Opener: NASA. **Figure 19.2:** Courtesy of Dana Berry. **Figure 19.3b-c:** Fotosmith. **Figure 19.3d:** Meteor Crater, Northern Arizona. **Figure 19.4a:** D. J. Roddy, USGS. **Figure 19.4b:** Courtesy Paul Dimare. **Figure 19.5a:** NASA. **Figure 19.5c:** Courtesy of Victor Rover. **Figure 19.6a:** Finley Holiday Films. **Figure 19.6b:** NASA. **Figure 19.6c:** AP/Wide World Photos. **Figure 19.7b:** JPL/NASA. **Figure 19.7c:** NASA/JPL. **Figure 19.8:** Courtesy Dana Berry. **Figure 19.10:** Courtesy of Finley Holiday Film Corp./JPL/NASA. **Figure 19.11:** NASA. **Figure 19.12a-b:** NASA. **Figure 19.14a:** UCO/Lick Observatory image. **Figure 19.15:** From 22.17, *M. A. Foundations of Astronomy,* published by Brooks/Cole, Graphics by Precision Graphics. **Perspective 19.1, Figure 1a:** Damon Simonelli and Joseph Vererka, Cornell University/NASA. **Perspective 19.1, Figure 1b:** NASA.

Chapter 20

Chapter Opener: Courtesy of D. D. Trent. **Figure 20.1:** Herb Orth, *Life* Magazine. © Time Warner Inc.: painting by Chesley Bonestell, © The Estate of Chesley Bonestell. **Figure 20.4a:** Courtesy R. V. Dietrich. **Figure 20.4b:** Alan L. Mayo, GeoPhoto Publishing Company. **Figure 20.5b:** James S. Monroe. **Figure 20.10:** Courtesy of R. V. Dietrich. **Figure 20.11b-c:** Courtesy of Asko Kontinen, Geological Survey of Finland. **Figure 20.12a, c:** James S. Monroe. **Figure 20.13b:** James S. Monroe, graphic work by Precision Graphics. **Figure 20.15a-b:** James S. Monroe. **Figure 20.19a:** Courtesy of Sidney W. Fox, Coastal Research and Development Institute, University of South Alabama. **Figure 20.19b:** Courtesy of Sidney W. Fox, Coastal Research and Development Institute, University of South Alabama. **Figure 20.20a:** Courtesy of Phillip E. Playford, Geological Survey of Western Australia. **Figure 20.21:** Courtesy of J. William Schopf, University of California, Los Angeles. **Figure 20.23:** Courtesy of Robert Horodyski. **Figure 20.24a:** Courtesy Andrew H. Knoll, Harvard University. **Figure 20.24b:** Courtesy Andrew H. Knoll, Harvard University. **Figure 20.24c:** Courtesy of Bonnie Bloeser. **Figure 20.26a-c:** Courtesy of Neville Pledge, South Australian Museum. **Figure 20.27:** Courtesy of Sun Weiguo, Nanjing Institute of Geology and Palaeontology, Academia Sinica, Nanjing, People's Republic of China.

Chapter 21

Chapter Opener: © Bill Ross/CORBIS. **Figure 21.8b:** Wisconsin Department of Natural Resources, Bureau of Parks and Recreation. **Figure 21.9a:** Courtesy of L. J. Lipke, Amoco Production Company. **Figure 21.11b-d:** Courtesy of Sue Monroe. **Figure 21.15b:** Reed Wicander. **Figure 21.17:** James S. Monroe. **Figure 21.19b:** Courtesy of Wayne E. Moore. **Figure 21.19d:** © Patricia Caulfield/Photo Researchers, Inc. **Figure 21.20c:** © Tom Bean, 1993/Tom & Susan Bean, Inc. **Figure 21.23:** Rubin's Studio of Photography, Courtesy The Petroleum Museum, Midland, TX 79701. **Figure 21.24:** © David Muench/CORBIS. **Perspective 21.1, Figure 1:** James S. Monroe.

Chapter 22

Chapter Opener: Smithsonian Institution, Transparency No. 86-13471A. **Figure 22.1a-c:** Courtesy of Simon Conway Morris and Stefan Bengtson, University of Cambridge, England. **Figure 22.4:** Carnegie Museum of Natural History. **Figure 22.6a-c:** Douglas H. Erwin/National Museum of Natural History/Smithsonian Institution. **Figure 22.6d:** © Chip Clark, 1995/ National Museum of Natural History/Smithsonian Institution. **Figure 22.7:** Field Museum, Chicago, #Geo80820c. **Figure 22.8:** The Field Museum, Chicago, #Geo80821c. **Figure 22.9:** The Field Museum, Chicago, #Geo80819c. **Figure 22.10:** © American Museum of Natural History, Trans. # K10257. **Figure 22.11:** Neg./Transp. no. K10269, Courtesy Department of Library Services, American Museum of Natural History. **Figure 22.14:** Courtesy of Dr. Lars Ramsköld. **Figure 22.29a:** Courtesy of Dianne Edwards, University College, Wales. **Figure 22.32:** Reed Wicander. **Perspective 22.1, Figure 1:** Courtesy of Sue Monroe.

Chapter 23

Chapter Opener: Reed Wicander. **Figure 23.1:** Corbis/Bettmann. **Figure 23.8d:** Courtesy of John Faivre. **Figure 23.14:** Courtesy of John S. Shelton. **Figure 23.15:** © Stephen J. Kraseman/Photo Researchers, Inc. **Figure 23.16:** Reed Wicander. **Figure 23.17b:** Reed Wicander. **Figure 23.19a-b:** Scanning electron photomicrograph courtesy of Merton E. Hill. **Figure 23.19c-d:** Scanning electron photomicrograph courtesy of John Barron, USGS. **Figure 23.19e-f:** Scanning electron photomicrograph courtesy of John H. Wrenn, Louisiana State University. **Figure 23.20a:** Courtesy of Sue Monroe. **Figure 23.20b:** Tom McHugh/ Photo Researchers. **Figure 23.21b:** Courtesy of Leo J. Hickey, Yale University. **Figure 23.24b:** Photo of a restoration at the California Academy of Sciences, San Francisco, courtesy of Sue Monroe. **Figure 23.24c:** © Archive Photos. **Figure 23.25b:** Courtesy of Sue Monroe. **Figure 23.28:** © Tom McHugh/Photo Researchers, Inc. **Figure 23.32:** Courtesy D. J. Nichols, USGS. **Perspective 23.1 Figure 1a:** Museum of the Rockies. **Perspective 23.1 Figure 1b:** Courtesy of Douglas Henderson.

Chapter 24

Chapter Opener: James S. Monroe. **Figure 24.1:** James S. Monroe. **Figure 24.2a:** Courtesy of Sue Monroe. **Figure 24.2b:** © Roy Anderson. **Figure 24.6a/c:** Courtesy of Sue Monroe. **Figure 24.6b:** Courtesy of Ward's Natural Science Est., Inc. **Figure 24.7:** Courtesy of Sue Monroe. **Figure 24.8a:** Reed Wicander. **Figure 24.8b-c:** Courtesy of Sue Monroe. **Figure 24.10:** Courtesy of Sue Monroe. **Figure 24.17a-b:** Courtesy of B. A. Masters. **Figure 24.18:** Courtesy of Sue Monroe. **Figure 24.20:** The Field Museum, Chicago, Neg. # CK46T. **Figure 24.21:** Courtesy of the Smithsonian Institution. Reconstruction painting of *Amebeledon* and *Teleoceras* mammals by Jay H. Matternes, © Copyright 1982. **Figure 24.27:** Field Museum/ Photo Researchers, Inc. **Figure 24.31:** The Field Museum, Chicago, Neg. # A76851c. **Figure 24.32:** Painting by Ronald Bowen; photo courtesy of Robert Harding Picture Library. **Perspective 24.1, Figure 2:** The Field Museum, Chicago, Neg. # CK1T. **Figure 24.30:** Darwen and Valley Hennings.

Index

Italic pages indicate illustrations. Page numbers followed by *n* refer to a note at the bottom of the page.

A

Aa flow, 87–88, *88*
Aberfan, Wales, 306, *306*, 326
Abrasion, 342, *343*, 407, 431, 432–433, *433*
Absaroka Sea, 610, 612, 613, 615
Absaroka sequence, 611–615, 664, 667
Absolute dating methods, 75, 487, 500–507, 562
Abyssal plains, 219, 220, *221*, 225
Acadian orogeny, 594, 608, 617–618
Acanthodians, 643, 644, 645, *645*
Accretionary wedge, 286, 289
Acritarchs, 582, *583*
Active continental margins, 219, 220, *220*, 289, *290*
Adirondack Mountains, 283, 564
African plate, 83–84, 106, 108, 256, 703
Aftershocks, 173, 181, 190, 197, 199
Agassiz, Louis, 711–712
Agate, 27, 42, *43*
Agriculture, 369, 374, 376, 381–382, 384, 400, 429, 430, 444
Alcatraz terrane, *292*, 293, *293*
Aleutian Islands, 249, 251
Alice Springs orogeny, 448
Alleghenian orogeny, 596, 619, 667
Alleles, 519, 522
Allopatric speciation, 523, *524*
Alluvial fans, 348–349, *351*, 445, *446*, 448
Alluvium, 342, 347
Alpha decay, 501, *502*
Alpine-Himalayan orogenic belt, 286, *287*, 703
Alps, 254, 269, 286, 289, 291, 302, 400, 703
Aluminum, 30, 128
Alvin, 215, 216, 222
Amethyst, 27, 28, 42, *43*
Ammonites, *142*, 680, 693, 717
Ammonoids, 639, *640*
Amnion, 649
Amniote eggs, 649, *649*, 681, 717
Amphibians, 3, 632, 646, 647–648, *648*, 649, *649*, 651
Amphibole minerals, 40, *40*, 41, *41*, 42, 45, 49, 60, 119, 152
Amphibolite, 159, 161, 163
Amu Dar'ya River, 6, *6*, 7
Anaerobic bacteria, 580
Analogous organs, 529, *531*
Ancestral Rockies, 611, *614*
Andesite, 65, 66–67, *67*
Andes Mountains
    convergent plate boundaries and, 18, 249, 252, *253*, 254, 259, 269, 703
    glaciers and, 400
    orogenic activity, 289, *290*
    rainshadow effect, 443, 596
Angiosperms, 679, 680–681, *681*
Angular unconformity, 492, 493, *495*, 497
Ankylosaurs, 683, 686, *688*
Anthracite, 135, 143, 159, 163, 622, 677

Anthropoids, 723, 726
Anticlines, 273–274, *274*, 275, *275*, 276, 277, 278, 708
Anticontinents, 205
Antimountains, 205
Antler orogeny, 594, 620, *620*, 670
Apatosaurus, 683
Aphanitic texture, 63, *64*, 65
Appalachian mobile belt, 593, *593*, 607, 611, 616–619, *618*, 620
Appalachian Mountains
    Cenozoic Era, 703, 710, 711, *711*
    continental crust and, 291
    convergent plate boundaries and, 254, 269, 286, 289
    creep and, 319
    deformation and, 283
    Mesozoic Era, 362, 667, 711
    plate tectonic theory and, 19, 238
    Precambrian, 564
Aquiclude, 370, 374–375, *376*
Aquifer, 370
Arabian plate, 172, 256, 260, 703
Aragonite, 34, 47
Aral Sea, 6–7, *6*, *7*
Archaeocyathids, 601, 636, 637, *637*
Archaeopteryx, 533–534, 689, 690, *690*
Archaic mammals, 717, *718*
Archean Eon
    Archean rocks of, 565–566
    atmosphere, 572, 577, 578
    continental nuclei and, 564
    cratons and, 566–568, *568*, 569
    geologic time and, 562, *562*, 563
    greenstone belts, 570
    organisms, 578, 580
    resources, 583, 584–585
Arches, 113, 592, 605, 708
Arches National Park, 113, *114*, 278, *279*
Archosaurs, 681–682
Arctic Ocean, 213, *213*, 221, 669
Ardipithecus ramidus, 726
Arêtes, *409*, 410, *414*, 415
Arid regions, 442, *442*, 596
Aristotle, 173, 517
Arkoses, 131, 448
Armadillos, *514*, 515
Arsenic, 388–389, *388*, *389*
Artesian-pressure surface, 375, *376*
Artesian systems, 374–376, *375*, *376*
Artiodactyls, 718, 719, *721*, 724
Asbestos, 151, 152, *152*
Aseismic ridges, *221*, 225, 257
Ash, 68, 89, 92, *92*, 101, 316, 545, 702, 716
Ash fall, 89, 507, 509
Ash flow, 92, 569
Assemblage range zones, 500, *501*, 509
Assimilation, 62, 77
Asteroids, 541, 545, *547*, 549
Asthenosphere, 15, *15*, *16*, 17, 18, *18*, 204, 248, 257, 258

Atlantic Ocean. *See also* Mid-Atlantic Ridge
    abyssal plains and, 220
    Atlantis and, 211
    Columbus and, 214
    continental drift and, 237, 239
    Mesozoic Era, 3, 664, *667*, 668, 693
    as ocean, 213, *213*
    Pangaea and, 665
    supercontinent cycle and, 249
    volcanic island arcs and, 251
    volcanoes and, 107
Atlantic-Pacific Ridge, 221
Atlantis, 211–212, *211*
Atmosphere
    Archean Eon, 572, 577, 578
    chemical weathering, 117
    circulation of, 441, *441*
    earthquakes and, 173
    Earth system and, 5, *5*, 8, 17, *18*, 21
    gravity and, 540, 574
    life's origin and, 578, 581
    origin of, 212
    of planets, 545, 552, 553
    plate tectonic theory and, 236
    Precambrian, 564, 573–574, 577–578
    running water and, 335
    volcanoes and, 85
    weathering and, 115
Atolls, 227, *228*, 229
Atomic clocks, 485
Atomic mass number, 31, 501
Atomic number, 31, 501
Atoms, 30–33, *31*, 34, 35, 39, 500, 541
Augite, *41*, 45, *46*, 49
Aureoles, 155, *155*
Australopithecines, 726–727, *728*
Austropithecus afarensis, 726, 727, 729
Austropithecus africanus, 726, 727–728
Austropithecus anamensis, 726, 727
Austropithecus boisei, 726, 727
Austropithecus robustus, 726, 727
Autotrophs, 580, 635
Au Train Formation, 602
Avalanches, 302, 309
Ayers, Henry, 448

B

Baby Arch, 113–114, *114*
Back-arc basin, 251, 286, *288*, 289
Back-arc marginal basins, 566
Backshore, 467, *468*
Badlands, 113, 704
Bajada, 445, *446*
Balanced rocks, 113, *114*
Baltica, 593, 594, 616, 617, *618*, 619, *619*, 620–621
Baltic Sea, 27
Banded iron formations, 145, 572, 576, 578, 585, 586, 678
Barchan dunes, 435, 437, *437*, 438
Barchanoid dunes, 438, *438*

Ectothermic, 683–684, 685
Ediacaran faunas, 583, *586*, 633
Elasticity, 182, 200
Elastic rebound theory, 174–175, *174*
Elastic strain, 270
Electromagnetic force, 541
Electron capture, 502, *502*
Electrons, 30–31, *31*, 32, 33, 500, 541
Electron shells, 30, 32
Elements, 30, 31, A-0–A-1
Emergent coasts, 479–480, *480*
End moraines, 416–417, *418, 419*
Endothermic, 650, 683–684, 685, 686, 692
Energy resources, 9, 11, 12, 49, 369, 393,
   460–461, 622
Environmental issues, 11, 12–13, 22, 125, 369,
   384–386
Environmental Protection Agency (EPA), 152
Eocene Epoch, 702, 703, 705, 712, 717, 719,
   719, 722, 723
Epeiric seas, 592, 594, 600, 609–610, 664, 667
Epicenter location, 175, *177*, 178, 183–185,
   186, 189
Equus, 719, 721, 722
Erosion
   arches and, 113–114
   barrier islands and, 471, 476
   beaches and, 467
   Cenozoic Era, 701
   concretions and, 140
   correlation and, 496
   definition of, 115
   disconformity and, 492
   folded rock layers and, 273, *276*
   glaciers and, 335, 401, 403, 404, 407–408,
      409, 410, 411, 412, 415
   granitic rocks and, 55
   groundwater and, 342, 367, 369, 376–381
   isostatic rebound and, 295, *296*
   joints and, 278
   Laramide orogeny and, 706
   mass wasting and, 323
   mountains and, 267, 283, *283*, 286, 295
   Precambrian rocks and, 564
   running water and, 55, 115, 124, 335,
      341–342, 407, 410, 432, 711
   Sauk sequence and, 600
   sedimentary rocks and, *112*, 113
   sheet erosion, 339
   shorelines and, 458, 474–476, *475*, 478–479
   slumps and, 311
   soil formation and, 124
   stream terraces and, 360
   topography and, 302
   turbulent flow and, 338
   undercutting and, 304, *305*
   vegetation and, 358
   water erosion, 125
   waves and, 463
   wind and, 124, 126, 432–435, 437–441, 445
Erratics, 415, *417*, 711, 713
Eskers, 419–420, *421, 422*, 716
Eukaryotic cells, 580, 581–582, *582*, 584–585,
   584, 585
Eupantotheres, 692–693
European plate, 84, 106
Eusthenopteron, 647, *647*

Evaporites
   Cenozoic Era, 716
   as chemical sedimentary rocks, 133
   formation of, 244
   Mesozoic Era, 665, 667, *667*, 668
   Paleozoic Era, 592, 593, 594, 603, *606, 607*,
      612–613, 615, 622
   as resource, 143, 622
Evolution
   biological evidence, 529, 531–532
   of coral reefs, 214
   Earth systems and, 4
   evidence in support of, 528–535
   extinctions and, 517, 527–528
   fossils and, 516, 517, 528, 532–535, 632
   genetics and, 519–520
   geologic time and, 488
   Mesozoic Era, 664
   modern view of, 521–528
   organic, 4, 21–22
   Paleozoic Era, 632, 635, 642–655
   plate tectonic theory and, 236, 262–263
   primates, 722–723, 726–730
   solar nebula theory and, 545
   speciation and, 522–524
   theory development, 517–519
   types of, 524–525
   variation and, 522
   vestigial structures, 531, 532
Evolutionary novelties, 525, 527
Exclusive Economic Zone (EEZ), 228, *230*
Exfoliation domes, 116, *117*
Extinctions, 517, 527–528, 545, 632, 730–731.
   See also Mass extinctions
Extrusive igneous rocks, 20, *20*, 21, 57, 63, 66,
   85, *108*, 155

F

Falls, 310–311, 316
Farallon Plate, 293, 705–706, *705*, 708
Fault-block basins, 667, 672
Faults. *See also* San Andreas fault
   definition of, 278
   deformation, 278–283, *280*
   dynamic metamorphism and, 157
   earthquakes and, 173, 175, 182, 197
   joints distinguished from, 278
   mountains and, 267
   plate tectonic theory and, 17
   porosity and, 369
   slumps and, 312
   thrust faults, 216, 279, *281*, 286, 289
   transform faults, 219, 224, *224*, 255, *255, 256*
Feldspars, 41, *41*, 42, 44, *45*, 47–50, 60, 70,
   115, 117, 128
Felsic magma, 57, 58–59, 61, 62, 63, 86, 564
Felsic rocks, 67–68, *67*
Ferromagnesian silicates, 40–41, *41*, 42, 46, 49,
   60, 119, 121
Fetch, 464
Finches, 515, *516*
Fiords, 408, *413*, 479
Fire, 180, 181, *181*, 188, 193, 305
Firn, 401
Firn limit, 404, 405, *405*
Fish, 632, 643, 643–647, *644, 646*
Fission track dating, 506, *506*

Fissure eruptions, 102–103, *103*
Fissures, 189, 310, 323
Fixity of species, 515, 517, 518
Floodplains, 333, 342, 345, *345*, 347, *347, 348*,
   360
Floods
   beneficial floods, 335
   control of, 349–350, 352
   deposits and, 342
   deserts and, 444
   glaciers and, 713
   mass wasting and, *300*, 301, *307*, 324
   property damage, 347, 352
   running water and, 333
   shorelines and, 455, 458
Floodways, 350, *354*
Flows, 310, 314, 316–317, 319, 322
Fluid activity, 150, 153–154, 155, 165, 197,
   198–199
Focus, of earthquake, 175–176, *177*, 184
Folded rock layers, 272–277, *275*, 278, 286, 592
Foliated metamorphic rocks, 20, *20*, 158–162,
   159
Food webs, 635, *636*
Footwall block, 278, 279, *280*
Foreset beds, 347–348, *349*
Foreshore, 467
Fossilization, 533, *534*, 535
Fossils
   Burgess Shale, 631, 637, *638*
   Cambrian explosion and, 632–633
   cladistics and, 527
   concretions and, 141
   continental drift and, 17, 238, 240–242, *241*
   correlation and, 498
   Darwin and, 515
   definition of, 22
   depositional environment and, 138, 517, 532
   Devonian Period, 608–609
   evolution and, 516, 517, 528, 532–535, 632
   extinctions and, 527
   photosynthesis and, 578
   Precambrian, 562
   properties of, 27
   Proterozoic Eon, 583, *586*
   relative dating and, 490–492, 517, 532
   seafloor spreading and, 246
   sea level and, 218
   sedimentary rocks and, 128, 138, *142*
   sequence of, 533, *535*
   terranes and, 290
   types of preservation, 533, *534*
Fossil succession, 489, 491, *492*, 498, 507
Fractures
   caves and, 378
   deformation and, 268, 269, 271, *271*
   joints and, 277
   of minerals, 45
   oceanic ridges and, 224, *224*, 250
   porosity and, 369, 370
   seafloor and, 223–224, *224*
   slope stability and, *308*, 324–325
   transform plate boundaries and, 254–255
Franciscan Complex, 165, *166*, 290–291, 293,
   294, 671
Franklin mobile belt, 593, *593*
Frost action, 116, 314

Frost wedging, 116, *117*, 310, 412, 444
Fumaroles, 100, 390

G

Gabbro, *65, 66, 66*, 107, *108*, 206, 216, 221, 291
Galena, *35*, 41, 44, *44, 45*, 46
Galveston Island, 455, *456*, 478
Garnets, 27, 28, *35*, 43, 151, *160*, 165
Gemstones, 27–28, 30, 70, 151
Genes, 518, 519, 520
Genetics, 519–520, *521*
Geodes, 140–141, 141, *141*
Geographic poles, 242, *243*
Geologic structures, 272, 284, 286
Geologic time
    absolute-dating methods and, 75, 487, 500–507
    climate and, 487, 488
    correlation and, 495–496, 498, 500
    geologic time scale, 22, *22, 486*, 487, 507, 509
    Grand Canyon and, 591
    Precambrian, 562, 703
    relative-dating methods and, 486–487, 489–494
    stratigraphy and, 509–510
    uniformitarianism and, 22–23, 488
Geologic time scale, 22, *22, 486*, 487, 507, 509
Geology, 5, 8, 9, 10–12, 16, 23
Geothermal energy, 338, 339, 369, 393, 407
Geothermal gradient, 60, *60*, 205–206, 390
Geyserite, 393, *394*
Geysers, 390, 393, *393*
Giant's Causeway, 89, 90–91, *91*
Glacial budget, 404–405, *405*, 415
Glacial drift, 415
Glacial grooves, 407, *409*
Glacial ice, 401
Glacial-marine sediments, 225, *226*
Glacial polish, 407, *409*
Glacial stages, 713–714
Glacial striations, 239, *240*, 407, *409*, 415, 594
Glacial surges, 406–407
Glaciation, 22, 420–424, 570, 572, 594, 713–714
Glacier National Park, *139*, 403, 410–411, *410, 411, 412*
Glaciers
    continental drift and, 239–240, *240*, 242
    definition of, 400
    deposition and, 218, 267, 404, 411, 416–420
    drainage patterns and, 354
    erosion and, 335, 401, 403, 404, 407–408, *409*, 410, 411, 412, 415
    formation of, 401, *401*
    glacial budget, 404–405, *405*, 416
    Gondwana and, 239, *240*, 594, 638
    granitic rocks and, 55
    ice ages and, 420–424
    isostatic rebound and, 295
    loess and, 440
    mountains and, 267, 286
    movement of, 401–402, *402*, 405–407, *406*
    Paleozoic Era, 239–240, 599
    Pleistocene Epoch, 218, 399, 400, 404, 421, 564, 701, 703, 711–714, *714*

quick clays and, 317
sea level and, 476, 479, 592, 594, 599, 641, 713
sediment and, 129, 239, 713
sorting and, 136
transport and, 225, 267, 407, 411, 416
types of, 402–404
water contained in, 335, 400, 401
Global Seismic Hazard Assessment Map, 197, *197*
Global warming, 12–13, 22, 228, 411, 431, 476, 488, 592, 642
Glossopteris flora, 237, *237, 239*, 240–241, 655
Gneiss, *20*, 42, 159, 160–161, *161*, 565, *565*
Gold
    Archean Eon, 583, 584
    bonding and, 32, 33
    Cenozoic Era, 716
    gold rush, 663–664
    granitic rocks and, 76
    Mesozoic Era, 678
    as mineral resource, 50, 663, *663*
    as native element, 35
    ore deposits of, 128
    Paleozoic Era, 622
    placer deposits, 143
    plate tectonics and, 259
    as precious metal, 36, *36, 37*
    specific gravity, 46
Gondwana
    dinosaurs, 693
    glaciers and, 239, *240*, 594, 638
    Laurasia collision, 612, 619, 620
    Pangaea and, 665
    Piedmont microplate, 621
    plants, 655
    plate tectonic theory and, 237, 238, 239, *239, 240*, 241, *241*, 242
    Proterozoic Eon, 569
Grabens, 286, *287*
Graded bedding, 137, *138*, 218, 566
Graded streams, 356–358
Gradient, of running water, 340, *340*, 354, 356
Grand Canyon, 9–10, 142, *143*, 496, *499*, 564, 591, *709*
Grand Teton, 267, 403, 415
Granite
    Archean rocks and, 565
    classification of, *65*, 67–68, *67*
    as igneous rock, *20*, 46, 55
    magma and, 62
    migmatites and, 161–162
    quartz and, 42, 48, *48*, 55, 115
Granitic rocks, 55–56, 67–68, 76, *76*, 77, 206, 212, 286
Granitization, 76–77
Graphite, 28, 33, *34*, 35, 45, 46, *47*, 165, 167
Gravel, 128, 143, 716
Gravimeter, 294
Gravity
    atmosphere and, 540, 574
    evolution and, 648, 651
    as force of universe, 541
    gravity anomalies, 294, *295*
    groundwater and, 371
    mass wasting and, 302

original horizontality and, 489
planets and, 545, 546, 550
running water and, 336, 360
specific gravity, 45–46
tides and, 458, 459, 462, *462*
Gravity anomalies, 294, *295*
Grazers, 718, 719, *720*
Great Lakes region, 571, *574*, 603
Great Plains, 434, 441, 709, *710*
Greek kouros, 149, *150*
Green River Formation, 144, 715
Greenstone, 159, 162
Greenstone belts, 565–566, *566*, 568, 569, 570
Grenville orogeny, 569, 570, *571, 572*
Groins, 467, *469*, 479
Ground moraine, 417, *418*
Ground shaking, 189, 192–193, *192*
Groundwater
    acidic nature of, 118, 370, 377, 378
    arsenic in, 388–389, *388, 389*
    artesian systems, 374–376
    contamination of, 381, 384–386, *387*
    deposition and, 376–381
    deserts and, 445
    erosion and, 342, 367, 369, 376–381
    geologic maps and, 285
    glacial drift and, 415
    glaciers and, 401
    humans' effect on, 381–387, 390
    hydrothermal activity and, 390, 393
    laminar flow and, 338
    movement of, 369, 370, 371–372, *371*
    precipitation and, 336, 369, 370, 375
    quality of, 386, 390
    saltwater incursions and, 477
    springs, 372, *373*, 374
    streams and, 340
    water contained in, 335, 369
    water table, 370–371, *371*
Guide fossils, 498, *500*
Gulf Coastal Plain, 709–711, *711*, 715
Gulf of Mexico, 665, 667, *667*, 669, 711
Gulf Stream, 460
Gullies, 124, *125*, 358
Guyots, 225, *225*
Gymnosperms, 652–653, *653*, 655, 680–681
Gypsum, 41, *44*, 45, 47, 49, 50, 154

H

Hadean, 562, 564
Half-life, 502
Halides, 39, 41, 49
Halite, 32, *33*, 35, *35*, 41, 44, *44, 45*, 46, 47, 49, 117
Hanging valleys, 408, *409, 413*
Hanging wall block, 278, 279, *280, 282*
Harney Peak Granite, 55, *56*, 71
Headlands, 475, 476, *476*
Headward erosion, 358, *359*, 415, 440
Hearne craton, 569, *571*
Heat, 150, 153, 155
Heat flow, 220
Herbivores, 634, *635*, 683, 684, 717, 718, 724
Hercynian-Alleghenian orogeny, 619
Hercynian mobile belt, 620
Hercynian orogeny, 596, 619

Hess, Harry, 244, 245, 246
Heterotrophic bacteria, 580
Hiatus, 492, *493*
Highway excavations, 304, *307*
Himalayas, 249, 254, 269, 289, 291, *291*, 400, 703, 704
Hominids, 726–730, *728*
Homo, 726, 727
Homo erectus, 728, 729
Homo habilis, 727–728
Homologous organs, 529, *531*
Homonoids, 726
Homo sapiens, 728
Hook (spit), 469
Hoplitomeryx, 724, *724*
Hornblende, *41*, 45, *46*, 49, *160*
Hornfels, 159, 163
Horns, *409*, 410, 415
Horsts, 286, *287*
Hot spots, 257, 283, 549
Hot springs, 390, *391*, *394*
Humans and human activity. *See also*
    Environmental issues; Global warming
    deserts and, 429
    Earth system and, 5, 23
    evolution and, 726–730, *728*
    geology and, 9–12
    groundwater and, 381–387, 390
    mass extinctions, 731
    rock slides and, 314
    running water and, 335, 356
    slope oversteepening and, 311
    soil formation and, 124
Humus, 122, 123, 124, 370
Hurricanes, 455, 458, 476
Hutton, James, 488, 489, 492, 500
Hydraulic action, 342
Hydroelectric power, 335, *336*, 338–339, *338*, *339*, 460–461, *461*
Hydrologic cycle, *336*, *337*, 369, 401
Hydrosphere, 5, *5*, 8, 17, *18*, 21, 85, 115, 173, 236, 335, 458
Hydrostatic pressure, 374, 381
Hydrothermal activity, 165, 259, 390, *393*
Hydrothermal alteration, 157
Hydrothermal vents, 220, 222–223, *223*, 229, 259, 579, 584
Hypotheses, 16

I

Iapetus Ocean, 594, 616, 617, *618*, 619, 639
Ice ages, 399, 420–424, 711–714
Icebergs, 225, 401
Ice caps, 404, 406
Ice fall, 406, *406*
Ice-scoured plains, 415, *417*
Ice streams, 407
Ichthyosaurs, 686–687, *689*
Idaho batholith, 76, 673
Igneous rocks. *See also* Intrusive igneous rocks
    absolute dating and, 503, 507, *508*
    batholiths, 76–78
    Cenozoic Era, 702, 709
    characteristics of, 63–69
    classification of, 64–69, *65*
    continental crust and, 57, 151, 291
    convergent plate boundaries and, 254

crust and, 128
extrusive igneous rocks, 20, *20*, 21, 57, 63, 66, 85, *108*, 155
formation of, 19, 20, *20*, 21, *21*, 57
geysers, 390, *393*
gold and, 664
Laramid orogeny, 706
magma and, 20, *20*, 21, 56, 57–63
magnetic reversals and, 243
minerals and, 46, 48
Moon and, 556–557
paleogeography and, 244
permeability of, 370
porosity and, 369
Precambrian, 592
Proterozoic Eon, 569, 570, *572*
relative dating and, 490
silver and, 37
thunder eggs, 141
volcanoes and, 89, 107–108
Illinoian glaciation, 713
Illinois Basin, 610, 611
Incised meanders, 360, *361*
Inclined folds, 274, *276*
Inclusions, 62, *62*, 77, 489, 490, *491*, 493
Index fossils, 498
Index minerals, 157–158, *158*, 163
Indian Ocean, 213, *213*, 221, 230
Indian plate, 289
Indian Ridge, 107, 108, 221
Infiltration capacity, 338–339
Inheritance of acquired characteristics, 518, 520, 521
Inselbergs, 447
Intensity, 185–186, *187*
Interglacial stages, 712, 713
Intermediate magma, 57, 58, 59, 61, 62, 63
Intrusive igneous rocks
    Cenozoic Era, 706
    characteristics of, 72–76
    contact metamorphism and, 155, *156*, 157
    continental accretion and, 290
    dikes as, 68, 72–73, *72*, *73*, 74, 107, *108*
    formation of, 20, *20*, 21, 57, 60, *61*, 62
    mantle and, 204
    natural resources and, 165
    relative dating and, 489, 493, 507
    texture of, 63
    Triassic Period, 676
Invertebrate marine life, 634–642, *635*, 650, 679–680, *679*, 694, 716–717
Ionic bonding, 32, *33*
Ions, 32, *38*, 39
Irish elk, 722, 724–725, *725*
Iron meteorite, 545, *546*
Iron ores, 41, 50, *50*, 128, 143, 167, 259, 583, 586, 622, 678
Isostatic rebound, 295, *296*, 480
Isotelus, 498, *500*
Isotopes, 31, *32*, 501
Izmit, Turkey, earthquake, 171, *171*, 172, 175–176

J

Japanese Islands, 251, *253*, 286
Jaw evolution, 643, *645*, 691
Jeffreys, Harold, 203

John Day Fossil Beds National Monument, *89*, 701, *701*, 702, *702*
JOIDES *Resolution*, 214, *215*
Joint Oceanographic Institutions for Deep Earth Sampling (JOIDES), 214
Joints, 277–278, *279*, 310, 312, 314, 324, 353. *See also* Columnar joints; Sheet joints
Jovian planets, 543, 544, 549–550, 552–554, *552*, *553*
Juan de Fuca plate, 255, *256*, 708
Juan de Fuca Ridge, 223, 229
Jupiter, *538*, 541, 542, 543, 545, 549–550, 552, *552*, *553*
Jurassic Period
    Absaroka sequence and, 611
    birds, 689, 690
    continental drift and, 238
    dinosaurs, 678, 679
    fossils of, 533
    granitic rock and, 55
    Gulf Coast Plain and, 710
    Laurasia and, 693
    Nevada orogeny and, 673
    North America and, 667
    North American Cordillera and, 704
    paleogeography, *666*, *670*
    Pangaea and, 665
    plants, 655, 680
    pterosaurs, *688*
    sediment, 676
    stratotypes of, 664
    subduction zones, 671

K

Kames, 419, *421*, *422*
Kansan glaciation, 713
Karst topography, 377–378, *377*, *379*
Kaskaskia sequence, 607–610
Kata Tijuṭa (The Olgas), 448, *448*, *449*
Kazakhstania, 594
Kettles, 419, 420, *421*
Key beds, 495, *498*
Kilauea Volcano, 58, 82, 85, 87, 92, 93, 95, 105, 107, 108
Kimberlite pipes, 204, 206, 678
Krakatau, 85, 86, 98, 194

L

Labyrinthodonts, 648, *648*, 649, *649*
Laccoliths, 72, 73, 75
Lahars, 96
Lakes, 356, 458, 464, 470
Lake Superior, 41, 128, 145, 602, *602*
Lamarck, Jean-Baptiste de, 517–518, 519, 521
Laminar flow, 336, *337*, 338
Landfills, 385, *387*, 390
Landslides, 188, 195, *196*, 286, 302, *302*, 303, 323, 464. *See also* Mass wasting
Laramide orogeny, 674, *674*, 704, 705–706, *705*, *706*
Lassen Peak, 58, 99, 100, *100*
Lassen Volcanic National Park, 85, 86, *86*, 94, 95, *391*
Lateral continuity, 489, *490*, 495
Lateral erosion, 358, *359*, 360, 447
Lateral moraines, 416–417, *420*
Laterite, 123, *124*, 128